Nervous System: A Molecular Biological Approach

Nervous System: A Molecular Biological Approach

Editor: Jenna Stephenson

hayle medical

New York

Hayle Medical,
750 Third Avenue, 9th Floor,
New York, NY 10017, USA

Visit us on the World Wide Web at:
www.haylemedical.com

ISBN 978-1-64647-579-7 (Hardback)

Cataloging-in-Publication Data

Nervous system : a molecular biological approach / edited by Jenna Stephenson.
p. cm.
Includes bibliographical references and index.
ISBN 978-1-64647-579-7
1. Molecular neurobiology. 2. Molecular biology. 3. Neurobiology.
4. Nervous system--Diseases--Molecular aspects. 5. Neurology.
I. Stephenson, Jenna.
QP356.2 .N47 2023
591.188--dc23

Contents

Preface

Over the recent decade, advancements and applications have progressed exponentially. This has led to the increased interest in this field and projects are being conducted to enhance knowledge. The main objective of this book is to present some of the critical challenges and provide insights into possible solutions. This book will answer the varied questions that arise in the field and also provide an increased scope for furthering studies.

The nervous system is the primary mechanism for regulating and controlling the actions as well as communicating the sensory information within the body. It consists of the spinal cord, brain and a complex network of nerves. All mental activities such as memory, thought and learning are centered on the nervous system. This system relays messages between the body and the brain. The nervous system is divided into two parts which include the central nervous system (CNS) and the peripheral nervous system (PNS). The spinal cord and brain make up the CNS whereas the autonomic and somatic nervous systems make up the PNS. The majority of multicellular organisms have nervous systems but their complexity varies substantially. This book explores all the important aspects of nervous system from a molecular biological perspective. It is appropriate for students seeking detailed information in this area of study as well as for experts. Different approaches, evaluations, and advanced studies on this topic have been included in this book.

I hope that this book, with its visionary approach, will be a valuable addition and will promote interest among readers. Each of the authors has provided their extraordinary competence in their specific fields by providing different perspectives as they come from diverse nations and regions. I thank them for their contributions.

Editor

1

Circulating cell-free microRNA as biomarkers for screening, diagnosis and monitoring of neurodegenerative diseases and other neurologic pathologies

Kira S. Sheinerman and Samuil R. Umansky*

DiamiR, LLC, Princeton, NJ, USA

Correspondence:
Samuil R. Umansky, DiamiR, LLC, 3 Orchid Court, Princeton, NJ 08540, USA
e-mail: sumansky@diamirbio.com

Many neurodegenerative diseases, such as Alzheimer's disease, Parkinson disease, vascular and frontotemporal dementias, as well as other chronic neurological pathologies, are characterized by slow development with a long asymptomatic period followed by a stage with mild clinical symptoms. As a consequence, these serious pathologies are diagnosed late in the course of a disease, when massive death of neurons has already occurred and effective therapeutic intervention is problematic. Thus, the development of screening tests capable of detecting neurodegenerative diseases during early, preferably asymptomatic, stages is a high unmet need. Since such tests are to be used for screening of large populations, they should be non-invasive and relatively inexpensive. Further, while subjects identified by screening tests can be further tested with more invasive and expensive methods, e.g., analysis of cerebrospinal fluid or imaging techniques, to be of practical utility screening tests should have high sensitivity and specificity. In this review, we discuss advantages and disadvantages of various approaches to developing screening tests based on analysis of circulating cell-free microRNA (miRNA). Applications of circulating miRNA-based tests for diagnosis of acute and chronic brain pathologies, for research of normal brain aging, and for disease and treatment monitoring are also discussed.

Keywords: miRNA, neurodegeneration, Alzheimer's disease, neurologic pathology, screening, biomarker, plasma, CSF

INTRODUCTION

There are several basic types of clinical tests: (i) genetic tests that help predict predisposition to a particular disease; (ii) screening tests, which are applied to a large population for early detection of a disease, preferably prior to its clinical manifestation; (iii) diagnostic tests, which are applied when a person has clinical symptoms of a disease or when the pathology has been detected by a screening test; (iv) predictive tests aimed to predict the disease outcome and drug sensitivity; and (v) disease and treatment monitoring tests. Screening tests are most important for the early detection and successful treatment of a disease, especially if its progression leads to significant changes in the nature of underlying pathological processes. Cancers and neurodegenerative diseases are most common examples of such diseases. Cancer progression leads to tumor invasion into surrounding tissues, metastasis and clonal evolution, which dramatically complicates treatment (Richards, 2009; Caldas, 2012). Similarly, progression of neurodegenerative diseases, leads to a switch from metabolic abnormalities to neurite and synapse destruction and finally to irreversible neuronal death (Bredesen, 2009; Snyder et al., 2012). Further, disease progression leads to involvement of new brain areas and cell types in the pathology (von Bernhardi and Inestrosa, 2008). The new diagnostic methods based on imaging techniques and analysis of proteins and other components in the cerebrospinal fluid (CSF) have been developed recently (Apostolova et al., 2010; Fagan et al., 2011; Mori et al., 2012); these methods, however, are not suitable for the first line screening due to their invasiveness and relatively high cost.

EARLY DIAGNOSIS OF NEUROLOGIC DISEASES

There are hundreds of brain disorders and many ways to classify them; however, from the viewpoint of diagnostics it is helpful to consider the diseases based on the need for different types of diagnostics, and on the underlying pathologic processes that need to be accurately and specifically detected by successful diagnostic tests.

First of all, all neurologic diseases fall into two large groups: acute and chronic pathologies. Obviously, there is a need in diagnostic and monitoring, but not screening tests for acute disorders, such as stroke, brain traumatic injury, and infections (of course, screening tests for predisposition to acute diseases, such as stroke, would be useful). On the other hand, numerous studies have demonstrated that, due to high compensatory potential of the brain, symptoms of chronic neurodegenerative diseases, such as Alzheimer's (AD), Parkinson (PD), Huntington (HD) diseases, vascular and frontotemporal (FTD) dementias occur 10–20 years after the beginning of the pathology. The massive neuronal death, characteristic of late stages of neurodegenerative diseases, makes any therapeutic treatment late in the course of a disease extremely difficult, as illustrated by recent failures of anti-AD therapies in late stage clinical trials (Sperling et al., 2011; Pillai and Cummings, 2013). This phenomenon stresses the importance of the development of screening tests capable of detecting such diseases in early asymptomatic stage.

Further, it is instructive to group neurologic diseases according to the nature of the underlying physiological and pathological processes. For example, although molecular mechanisms of various neurodegenerative diseases are different, many processes, e.g., neurite retraction, dysfunction, and destruction of synapses, followed by neuronal death, are characteristic of neurodegeneration in general (von Bernhardi and Inestrosa, 2008); massive death of neurons and glial cells accompany initial stages of stroke and brain traumatic damage (McIntosh et al., 1998; Zheng et al., 2003); dysfunction of dopaminergic neurons is characteristic of addiction to many drugs (Kreek et al., 2012), etc.

Finally, physical location of pathological processes occurring during different stages of disease development is important for the focused search for specific biomarkers. The neurodegenerative diseases provide a good example of the diversity of physical locations involved: although many underlying processes are common for AD, PD, and FTD, initial pathological events characteristic for different diseases are localized in different brain areas: hippocampus for AD, midbrain for PD, and frontal lobe for FTD. At later stages, each disease spreads to new areas of the brain, which significantly worsens clinical symptoms and narrows treatment options; this expansion can in principle be used for disease monitoring.

A large number of studies focus on the development of minimally invasive molecular tests for early detection of AD, PD, and other brain pathologies. These include reports of some encouraging early data on the development of blood-based assays for AD diagnosis based on analysis of a large number of proteins or antibodies in human blood (Ray et al., 2007; Nagele et al., 2011; Reddy et al., 2011). In this review we describe new approaches to developing screening, diagnostic and monitoring tests for neurodegenerative diseases, and other brain pathologies based on analysis of cell-free microRNA in bodily fluids.

ROLES AND PROPERTIES OF miRNA

miRNA is a class of non-coding RNA, whose final product is an approximately 22 nucleotide-long functional RNA molecule. miRNA repress translation and regulate degradation of their target mRNA by binding to complementary regions of messenger transcripts (Griffiths-Jones et al., 2006; Bartel, 2009). There are several programs for *in silico* analysis of complementarity between miRNA and mRNA; the lists of possible targets for a miRNA frequently include hundreds of genes. Hence, based on sequence analysis alone, a given miRNA can potentially be involved in numerous different pathologies (for example, see miR-Ontology Database: http://ferrolab.dmi.unict.it/miro/). All these predictions must be validated *in vivo*; in most cases this has not been accomplished yet, even though experimental data on miRNA roles in epigenetic regulation of numerous cellular processes are rapidly accumulating (Siegel et al., 2011; McNeill and Van Vactor, 2012). The following miRNA properties make them attractive for using in development of various diagnostic tests.

miRNA are small molecules with higher, e.g., compared to proteins, chances to cross blood–brain, placental, and other barriers and to appear in bodily fluids. Cell-free miRNA have been detected in plasma, serum, urine, saliva, and milk, where they are protected by membranes in exosomes and other microparticles, by various proteins, lipids and, possibly, other molecules (reviews: Sun et al., 2012; Zandberga et al., 2013). Although the organ, tissue, or cell origin of circulating miRNA is difficult to determine *in vivo*, it has been shown that many of those circulating miRNA do not originate in blood cells or cells present in or contacting with respective bodily fluids (Weber et al., 2010; Duttagupta et al., 2011). Numerous data on circulating miRNA have recently been assembled (Russo et al., 2012) and made available via miRandola database: http://atlas.dmi.unict.it/mirandola/index.html.

More than 1,500 human miRNA have been discovered to date, and although there are no miRNA exclusively present in a single tissue or organ, many of them are highly enriched in particular organs, tissues, and cell types (Barad et al., 2004; Liu et al., 2004; Baskerville and Bartel, 2005; Beuvink et al., 2007; Landgraf et al., 2007; Liang et al., 2007; Wang et al., 2007; Castellano and Stebbing, 2013). Many of miRNA are enriched in the brain and, importantly for the development of diagnostic tests, various miRNA are enriched in different brain areas as well as in different cell types (neurons and glial cells; Sempere et al., 2004; Smirnova et al., 2005; Deo et al., 2006; Bak et al., 2008; Trivedi and Ramakrishna, 2009; Weng et al., 2011; He et al., 2012). Moreover, certain miRNA are present or even enriched in particular intracellular compartments, such as synapses, dendrites, and axons (Schratt et al., 2006; Lugli et al., 2008; Schratt, 2009; Edbauer et al., 2010; Natera-Naranjo et al., 2010; Strickland et al., 2011; Wu et al., 2011; Pichardo-Casas et al., 2012). A number of the listed above detailed studies of miRNA expression in various organs, tissue, and cell types have been performed prior to identification of many miRNA, and thus there is an obvious need for additional studies of expression profiles of these newly discovered miRNA. In particular, *in situ* hybridization studies would be very useful for mapping miRNA expression in different brain areas, cell types and intracellular compartments.

Intracellular concentration of miRNA changes in various physiological and pathological processes due to modifications in their transcription, maturation, and stability (Iorio and Croce, 2012). Such changes of miRNA levels in different brain areas are characteristic of many neurodegenerative diseases and other brain pathologies (reviews: Saugstad, 2010; Fiore et al., 2011).

A number of recent papers review the available data on changes in miRNA expression in different brain areas involved in AD development (Fiore et al., 2011). A comprehensive review by Tan et al. (2013) presents an analysis of the published evidence for the involvement of miRNA in the four processes playing critical roles in AD pathogenesis: accumulation of amyloid-β, tau toxicity, inflammation, and neuronal death. While some of the results appear compelling, many more studies are necessary to further elucidate the precise roles of individual miRNA in AD pathogenesis and their involvement in different stages of the disease. Some of the published data appear contradictory, possibly due to differences in methods employed for miRNA measurement and normalization in different studies. For example, both activation and inhibition of the brain-enriched miR-9 expression in hippocampus of AD patients have been reported (review: Jin et al., 2013).

Investigations of the miRNA involvement in PD have focused on analysis of miRNA expression in the midbrain and of miRNA role in functioning of dopaminergic neurons and the α-synuclein synthesis. Downregulation of miR-133b in midbrain of PD patients (Kim et al., 2007) as well as in mouse models of PD has been reported in several studies (reviews: Harraz et al., 2011; Filatova et al., 2012; Mouradian, 2012). miR-7 and miR-153 have been found to downregulate synthesis of α-synuclein (Junn et al., 2009; Doxakis, 2010); suppression of the expression of these miRNA in midbrain of PD patients is expected but has not been demonstrated yet.

Deregulation of 15 miRNA in the brain of the mouse model of prion-induced neurodegeneration has been demonstrated (Saba et al., 2008). Further, changes in miRNA expression caused by or at least accompanying schizophrenia, autism, cognitive dysfunction, drug addiction, neuroblastoma, and other neurologic disorders have been reported (review: Jin et al., 2013).

For the use of miRNA in diagnostics, it is also important that miRNA secretion varies depending on cellular physiology (Wang et al., 2009; Pigati et al., 2010; Palma et al., 2012). In addition to miRNA release into extracellular space and subsequent appearance in the bodily fluids due to cell death, miRNA appear in circulation due to blebbing of apoptotic bodies, budding and shedding of microvesicles, active secretion in the form of exosomes and of miRNA complexes with proteins (AGO2, NPM1, and others) and high density lipoproteins (HDL; reviews: Sun et al., 2012; Zandberga et al., 2013). All these forms of cell-free miRNA are highly stable in the bloodstream and other bodily fluids. The secretion of miRNA is selective and can be significantly changed by various pathological processes. For example, changes in the spectrum of miRNA secreted in exosomes from prion-infected neuronal cells, as compared to uninfected cells, have been demonstrated (Bellingham et al., 2012).

IDENTIFICATION OF POTENTIAL miRNA BIOMARKERS IN BODILY FLUIDS

The development of screening and diagnostic tests for various diseases based on analysis of miRNA in bodily fluids – mainly in plasma or serum, but also in urine, saliva, CSF, and milk – is a very active area of research. Below are three approaches commonly used in such studies, including studies in the area of neurodegenerative diseases:

(1) Measurement of hundreds miRNA in a bodily fluid from patients with a pathology of interest and from control subjects using miRNA array or next generation sequencing (NGS; Qin et al., 2013). An obvious advantage of this approach is that huge numbers of various miRNA can be analyzed. Detection of miRNA sequence variations, which can be informative, e.g., for tumor diagnostics, is another advantage of NGS (Williams et al., 2013). Many potential biomarkers have been found by these techniques. However, the miRNA array-based and sequencing techniques are not sufficiently sensitive to detect many miRNA whose concentration in bodily fluids is relatively low. Usually only 30–40% of plasma or serum miRNA detectable by an individual RT-PCR is measurable by various miRNA arrays. This is also true for the RT-PCR-based array, largely due to significantly lower amounts of plasma analyzed by array compared to individual RT-PCR. In addition, reproducibility is very important for screening tests addressing early disease stages when changes in biomarker concentrations usually are not high. Unfortunately, variability of miRNA array tests is significantly higher than that of the individual miRNA RT-PCR (Leidner et al., 2013). As a consequence, potential biomarkers selected by array analysis have to be confirmed by RT-PCR, and indeed in many cases the array data is not confirmed. Further, even when miRNA array-selected potential biomarkers are validated by RT-PCR, they cannot be automatically considered useful for screening or diagnostic purposes. Most of the miRNA detectable in bodily fluids by arrays are ubiquitous miRNA expressed in all or many tissues, and many of them derive from blood cells (Pritchard et al., 2012; Leidner et al., 2013); the detection of changes in their concentrations in patients with one pathology does not mean that the same miRNA cannot be involved in other diseases of different organs. Many miRNA are associated with a particular pathology type, such as cancer, inflammation, hypoxia, etc., and changes in their concentration in bodily fluids can be associated with diseases of different organs. For example, changes of miR-155 concentrations were found in the bloodstream of patients with breast, esophageal, lung, pancreatic cancers and lymphomas (Blair and Yan, 2012; Xie et al., 2013). Level of miR-21 increases in plasma/serum of patients with osteosarcoma, bladder, esophageal, gastric, lung, breast, colorectal cancers, neck squamous cell carcinoma, and other tumors (Farazi et al., 2011; Blair and Yan, 2012; Xie et al., 2013). It follows that the potential biomarkers found by miRNA arrays should be also tested in other pathologies, not only in healthy control subjects.

(2) The second approach is based on analysis of disease-specific miRNA identified by comparison of miRNA isolated from pathologic and normal tissue, organ, or cell type. Here, subsequent to identification of disease-specific miRNA, e.g., by array followed by RT-PCR, their presence in bodily fluids is analyzed. There are obvious advantages to this strategy. First, a limited number of circulating miRNA should be tested, which makes the use of individual RT-PCR appropriate, thus increasing sensitivity and reproducibility of the analysis. Second, an observation of a correlation between changes in miRNA concentrations in a bodily fluid and in an organ involved in pathology directly suggests that the miRNA is a suitable biomarker for screening tests. However, such a correlation does not always exist and sometimes miRNA concentrations in a bodily fluid and in a target organ change in opposite directions (Boeri et al., 2011; Cuk et al., 2013). This phenomenon can be explained by several factors: (i) if pathology is caused by, or associated with, the change in concentration of a ubiquitous miRNA, the effect of the pathology on the concentration of this miRNA in circulation could be very limited, since only a small fraction of the miRNA in circulation comes from the affected organ or tissue; (ii) changes in miRNA concentration due to pathology development can be accompanied by much more prominent opposite changes in miRNA secretion/excretion, which neutralizes or even overcomes the effect of changed miRNA expression. Of course, these limitations do not preclude successful use of the described approach for the discovery of circulating miRNA biomarkers. In the area of neurodegenerative and other neurologic pathologies this approach can be especially productive for the miRNA biomarker discovery in CSF (see below) because: (i) CSF is not separated from brain by the blood–brain barrier;

and (ii) the amount of miRNA secreted or excreted from other organs to CSF is very limited (Cogswell et al., 2008).

(3) The third approach was proposed recently for searching for biomarkers of mild cognitive impairment (MCI), AD, and other neurodegenerative diseases (Sheinerman et al., 2012; Sheinerman and Umansky, 2013). It is based on analysis of circulating brain-enriched miRNA in bodily fluids, e.g., in plasma. Restricting the analysis to brain-enriched miRNA significantly increases the chances that any observed changes in concentrations of these miRNA in plasma are caused by brain-related processes. In addition, it is proposed to include miRNA enriched or at least present in neurites and synapses, i.e., neuronal compartments, which are involved in neurodegenerative diseases from early stages of pathology development, years before massive death of neurons. It could be also useful to search for miRNA enriched in the brain area, which suffers first during development of a particular disease, e.g., in hippocampus for AD and in midbrain for PD potential biomarkers. As a number of miRNA to be investigated is limited, RT-PCR can be used here, which is critically important since plasma concentration of many brain-enriched neurite/synapse miRNA is too low to be reliably detectable by miRNA arrays. The advantages and disadvantages of the three approaches described above are summarized in **Table 1**.

NORMALIZATION OF miRNA CONCENTRATION IN BODILY FLUIDS

The concentration of miRNA detected in bodily fluids depends on many biological and technical factors. Biological factors include miRNA levels in various tissues, intensity of secretion and excretion into extracellular space, forms of circulating miRNA (exosomes and other vesicles, complexes with proteins and lipids) affecting their ability to cross various barriers, e.g., blood–brain, placental, and kidney barriers, and miRNA stability and half-life in the bloodstream. Technical factors include variability during bodily fluid collection and storage, methods used for miRNA extraction, and the presence in bodily fluids of various factors affecting miRNA purification and RT-PCR. As a consequence, the importance of miRNA normalization is currently broadly recognized (Meyer et al., 2010). At the same time no single normalization method is commonly accepted. Several methods of circulating miRNA normalization have been described in literature:

(1) Normalization per spiked miRNA, which is absent in the investigated species, e.g., *C. elegans* or plant miRNA absent in mammalian cells (Sarkar et al., 2009; Kroh et al., 2010; Sanders et al., 2012). Such miRNA is spiked into a bodily fluid after addition of a lysing solution that inhibits nuclease activity and then is used as a normalizer by the ΔCt approach compensating for variability caused by RNA extraction and the possible presence of RT-PCR inhibitors.

(2) Normalization per ubiquitous and least variable circulating miRNA (Peltier and Latham, 2008; Latham, 2010; Lardizábal et al., 2012). This method is used for comparing circulating miRNA concentrations in controls and in one or a relatively small number of pathologies, which are not associated with changes of a normalizer miRNA. For example, miR-16 is widely used as such normalizer (Kroh et al., 2010). One needs to be mindful, however, that miR-16 is involved in regulation of apoptosis and its expression is changed in many pathologic processes, which leads to changes of its concentration in plasma and other bodily fluids (Schaefer et al., 2010; Katsuura et al., 2012; Wang et al., 2012b). Similar considerations apply to other miRNA normalizers of this type.

(3) Normalization per other (not miRNA) small RNA, mainly small nuclear or small nucleolar RNA. This approach was often used in earlier studies of circulating miRNA, and is still used although more rarely (Sanders et al., 2012). In addition to problems similar to the ones described in the previous paragraph (changes in the RNA concentration associated with some pathologies),the utility of this approach is limited by the fact that these RNA are larger than miRNA, bound to different proteins, and their stability is different from that of miRNA (Lardizábal et al., 2012).

Table 1 | Current approaches to analysis of miRNA in plasma. The same approaches can be used for miRNA analysis in different body fluids (urine, saliva, and milk).

miRNA array/NGS	Disease-specific analysis	Organ/cell-specific analysis
Steps		
1. miRNA isolated from plasma	1. miRNA isolated from pathologic and normal tissue	1. miRNA isolated from plasma
2. Array analysis	2. Array analysis	2. RT-PCR analysis of organ, cell-enriched miRNA
3. RT-PCR verification	3. RT-PCR verification	3. Validation
4. Validation	4. Selected miRNA analyzed in plasma	
	5. Validation	
Advantages		
• 100s of miRNA tested	• Disease-related miRNA tested	• Organ/cell-specific miRNA tested
	• High sensitivity	• High sensitivity
Disadvantages		
• Low sensitivity	• Same miRNA can be involved in pathologies of various organs	• miRNA biomarkers not enriched in target organs or cells can be missed
• High variability	• miRNA levels in tissue and plasma do not always correlate	
	• Selected miRNA can be undetectable in plasma	

(4) If a large number of miRNA is analyzed, usually by miRNA arrays, normalization per average of all miRNA can be used (Mestdagh et al., 2009; D'haene et al., 2012; Blondal et al., 2013).

(5) Finally, in recent years application of "miRNA pair" approach is becoming increasingly popular (Boeri et al., 2011; Hennessey et al., 2012; Matthaei et al., 2012; Sheinerman et al., 2012). In this case, ratios (ΔCt) of each miRNA to each other miRNA measured in a given experiment is calculated, and pairs, which provide the highest accuracy in differentiating patients with a disease of interest from controls, are selected. A variant of the approach is based on the use of miRNA enriched in the organ of interest (Sheinerman et al., 2012; Sheinerman and Umansky, 2013). The rationale for the approach is as follows. First, any pathology is usually associated with upregulation of some miRNA and downregulation of other miRNA, thus considering miRNA pairs may increase test sensitivity and specificity. Second, use of the pair of, rather than one, miRNA enriched in the same organ decreases potential overlap with pathologies of other organs. Third, one can expect that changes unrelated to or non-specific for a pathology of interest, such as changes in blood supply, blood–brain permeability (for neurological pathologies) and others, will be better compensated for by using the pair of miRNA enriched in the same organ. This approach is particularly promising for neurodegenerative and other neurologic diseases. For example, combination of neurite/synapse miRNA with a neuronal body miRNA may be informative for evaluating disease progression from synapse destruction to neuronal death. The changes in relative concentrations of miRNA enriched in different brain areas or different cell types (e.g., neurons and glial cells) may be an indicator of disease progression, and so on.

In summary, data normalization is an extremely important step in developing tests based on analysis of circulating miRNA, and as such it represents an active area of research.

SEARCH FOR miRNA BIOMARKERS FOR NEUROLOGIC DISEASES

There are numerous recent reviews on miRNA role in brain development, normal brain aging, and various neurological disorders (Krichevsky et al., 2003; Gascon and Gao, 2012; Mellios and Sur, 2012; Salta and De Strooper, 2012; Smith-Vikos and Slack, 2012; Wang et al., 2012a; Xu et al., 2012; Dimmeler and Nicotera, 2013; Moreau et al., 2013). Here we concentrate on circulating cell-free miRNA as potential biomarkers for diagnosing neurologic diseases. Despite well-recognized need in early detection of neurodegenerative diseases and other neurological disorders and great promise of circulating miRNA as potential biomarkers, there are relatively few publications in the area. There could be several explanations for this phenomenon, including an observation that concentration of many circulating cell-free brain-enriched miRNA in plasma, serum, and other bodily fluids is relatively low and thus they are not easily detectable by commonly used miRNA arrays. As mechanisms of miRNA appearance in CSF and bloodstream are different, below we separately describe the existing data for these two bodily fluids.

CELL-FREE miRNA IN CSF AND CNS DISORDERS

The presence of miRNA in CSF was first demonstrated by Cogswell et al. (2008). Cogswell et al. (2008) also reported changes in levels of many miRNA in CSF of AD patients when compared to controls. The authors came to a conclusion, however, that the major source of miRNA detected in CSF are immune cells present in CSF since "there was no obvious relationship between the miRNAs altered in CSF and the absolute levels in sites of AD mediated destruction or the directional changes in those regions." Many miRNA, including those that are highly enriched in brain, were not detectable in CSF by methods used. In another study devoted to the same matter the increase in levels of pro-inflammatory miR-146a and miR-155 in CSF of AD patients compared to age-matched controls as well as upregulation of neuron-enriched miR-9 and miR-125b was reported (Alexandrov et al., 2012). Interestingly, Lukiw et al. (2012) also demonstrated upregulation of miR-145b and miR-155 in human neuronal–glial primary cocultures 12 and 36 h after stress treatment and proposed that extracellular miRNA can be involved in spreading of AD inflammatory signaling.

Several studies describe changes in concentration of CSF miRNA in patients with different brain tumors. miR-15b and miR-21 were differentially expressed in CSF from patients with gliomas, compared to controls with various neurologic pathologies, including patients with primary CNS lymphoma and carcinomatous brain metastasis (Baraniskin et al., 2012). The changes in expression and plasma/serum levels of both miR-15b and miR-21 are commonly associated with different types of cancers and are not specific for gliomas. However, it seems unlikely that their concentrations in CSF can be strongly affected by non-CNS tumors. This conjecture needs to be proven for these miRNA to be used as glioma biomarkers. Teplyuk et al. (2012) found a significant increase of miR-10b and miR-21 levels in CSF of patients with glioblastoma and brain metastasis of breast and lung cancers, compared with various non-neoplastic conditions, such as memory problem, dementia, PD, encephalitis, and others. These two miRNA were also significantly upregulated in glioblastoma, compared to normal brain. The use of CSF levels of seven miRNA – miR-10b, miR-21, miR-125b, miR-141, miR-200a, miR-200b, and miR-200c – permitted the authors to achieve high accuracy in separation between all classes of samples. Again, only miR-125b from this list is a brain-enriched miRNA; all other miRNA are associated with carcinogenesis itself, so their usefulness as potential biomarkers is based on their presence in CSF, which is isolated from tumors located in other organs outside of CNS.

Microparticles containing miRNA, including brain-enriched ones, have been found in CSF and the spectrum of miRNA detectable in CSF is changed after brain injury (Patz et al., 2013). miR-451 was detected in CSF microparticles only after brain injury, which can be explained by more effective secretion of this miRNA from abnormal cells (Pigati et al., 2010).

CELL-FREE miRNA IN PLASMA AND SERUM AND CNS DISORDERS

Due to massive damage and death of neurons and glial cells the concentration of circulating brain-enriched miRNA is dramatically increased in the bloodstream after a stroke. For example, in the rat model, the plasma level of miR-124 is up to 150 times higher than in plasma of sham-operated animals (Laterza et al.,

2009). Of course, one cannot expect such large changes of miRNA biomarker levels in plasma/serum from patients with chronic, slowly developing neurodegenerative diseases, such as AD. Perhaps, this consideration explains why, in spite of the huge need in a minimally invasive test for early detection of AD, there are only two publications devoted to the use of circulating miRNA for diagnosis of this pathology.

Geekiyanage et al. (2012) compared concentrations of five miRNA, namely miR-137, miR-181c, miR-9, miR-29a, and miR-29b in serum of MCI and AD patients with their levels in serum of age-matched controls. The choice of miRNA was based on the previous study (Geekiyanage and Chan, 2011) that demonstrated their involvement in AD pathogenesis and downregulation of these miRNA in the brain cortex of sporadic AD patients. Using RT-PCR and normalization per spiked cel-miR-39 and internal miR-22, miR-191, and miR-126 the authors demonstrated that the levels of miR-137, miR-181c, miR-9, miR-29a, and miR-29b are significantly lower in serum of MCI and AD patients, compared to controls. Since each group of patients and controls included only seven subjects, sensitivity and specificity of MCI and AD detection could not be calculated.

Sheinerman et al. (2012) investigated the use of circulating miRNA for early detection of MCI. Since neuronal death, a late event in the development of AD and other neurodegenerative diseases, is preceded by metabolic changes, neurite retraction, synaptic dysfunction, and synapse loss, the authors suggested that these processes could cause excessive secretion and excretion of miRNA from brain areas involved in the pathology. Thus, brain-enriched miRNA, including neurite- and synapse-enriched miRNA were measured in plasma of MCI and AD patients and age-matched control subjects. Described above "miRNA pair" approach was used for data normalization. In the preliminary experiments 32 miRNA were analyzed, then 13 most promising miRNA were selected for the feasibility study. Finally, two sets of biomarker miRNA pairs were identified: the "miR-132 family" (miR-128/miR-491-5p, miR-132/miR-491-5p, and mir-874/miR-491-5p) and the "miR-134 family" (miR-134/miR-370, miR-323-3p/miR-370, and miR-382/miR-370). Each biomarker pair included as numerator neurite/synapse-enriched miRNA. These potential biomarkers differentiated MCI from age-matched controls with sensitivity and specificity of 79–100% (miR-132 family) and 79–95% (miR-134 family). In a small longitudinal study (19 subjects), the identified miRNA biomarker pairs successfully detected MCI in majority of patients at asymptomatic stage 1–5 years prior to clinical diagnosis. MCI is a heterogeneous syndrome. On average, 10–15% of MCI patients annually convert to dementia. About 80% of dementias are caused by AD and the rest of dementia patients are diagnosed with other neurodegenerative diseases. It is perhaps not surprising then that both sets of miRNA pairs also differentiated AD from age-matched control. The biomarkers do not distinguish AD from MCI, however, indicating that these biomarker miRNA pairs detect processes characteristic of various neurodegenerative pathologies (synapse destruction?) but not AD-specific events. Interestingly, these biomarker pairs also appear useful for detecting age-related brain changes. The reported results await confirmation in larger clinical studies.

One study (Gaughwin et al., 2011) described a search for a plasma miRNA, which could be used for early diagnosis of HD. The authors tested many miRNA regulated by mutant HTT protein and found that one of these miRNA, miR-34b, is significantly elevated in plasma from carriers of mutant HTT prior to symptom onset. However, it is unclear if miR-34b can be a specific biomarker for HD, since this miRNA is expressed in many tissues, being especially enriched in pulmonary system, Fallopian tubes, and testicles, and is also deregulated in various pathologies.

Several groups (Haghikia et al., 2012; Siegel et al., 2012; Gandhi et al., 2013; Ridolfi et al., 2013) reported promising results regarding potential use of circulating miRNA in CSF or bloodstream for diagnosis of multiple sclerosis (MS). Since MS is an autoimmune neurodegenerative disease, one can expect that concentration of both brain-enriched miRNA and of inflammation and immune system-associated miRNA could be changed in CSF and plasma/serum. Currently, the majority of the discovered potential biomarkers belong to the latter group: inflammation and immune system associated miRNA.

A limited number of studies report analyses of circulating miRNA in plasma or serum of patients with various neurological and psychological disorders, such as depression (Li et al., 2013), bipolar disorder (Rong et al., 2011), and schizophrenia (Shi et al., 2012), suggesting that the study of circulating miRNA in plasma/serum for diagnosis of neurological disorders represents a promising direction for further more detailed investigations.

SUMMARY AND PERSPECTIVES

In spite of the rapidly growing number of publications on diagnostic applications of circulating cell-free miRNA, their use for screening of CNS diseases is in early stages of development.

One factor impeding the progress in the field is the difficulty of comparing the data reported by different groups due to the use of different methods for searching for potential circulating miRNA biomarkers, different techniques for miRNA measurement and data normalization. Broad acceptance of best practices and uniform methods for analysis of circulating miRNA and the development of statistical apparatus for comparing the results obtained by different techniques will be important to overcome the problem.

Further, when miRNA arrays are used for initial miRNA screening, many tissue-enriched miRNA are not detectable and, as a consequence, ubiquitous miRNA, including miRNA associated with common pathologic processes such as carcinogenesis or inflammation, are often selected as potential biomarkers. These miRNA can successfully differentiate patients with a particular disease from healthy control subjects, but not necessarily from patients with similar pathologies of other organs, since it is highly likely that the plasma or serum levels of these miRNA will also be affected in such patients. From this perspective, searching for miRNA biomarkers in CSF has advantages for detecting CNS disorders, similar to advantages of using stool miRNA for detecting colon diseases (Ahmed et al., 2013): in each of these cases miRNA come from the organ of interest. The invasiveness of CSF collection, however, precludes using it for primary screening, and thus emphasis needs to be placed on the development of screening tests based on analysis of plasma or serum.

An analysis of data in literature suggests a number of optimization steps to increase the chances of success for such efforts. First, the use of brain-enriched miRNA for CNS disorders should increase biomarker specificity. This thesis is supported by data obtained for other organs. For example, increase in plasma/serum concentration of liver-enriched miR-122 is very common in patients with pathologies of this organ (Zhang et al., 2010), and increase in plasma/serum concentration of heart-enriched miR-1, miR-133a, miR-133b, or miR-499-5p is characteristic of acute myocardial infarction and some other cardiac pathologies (Tijsen et al., 2012). Second, combination of several circulating miRNA biomarkers can increase test specificity, since even expression of ubiquitous miRNA varies from organ to organ and one can expect that spectrums of concentration of several miRNA biomarkers will be different for various pathologies of different organs. For example, Li et al. (2011) made an interesting observation of the age-dependent increase of miR-34a concentration in brain, peripheral blood mononuclear cells, and plasma. This miRNA, however, cannot be used as a sole biomarker of aging, since its concentration in plasma is changed in patients with tumors of various locations and with other pathologies. At the same time, in combination with mir-132 and miR-134 biomarker families described above, this miRNA could be a helpful additional biomarker. Finally, it can be useful to combine tissue-enriched miRNA with miRNA associated with a common pathology type, e.g., carcinogenesis.

The data reviewed herein suggests strongly that the analysis of cell-free miRNA in bodily fluids is a highly promising approach for developing minimally invasive screening tests for CNS disorders. A different question is whether circulating miRNA represent a promising class of molecules for the prognosis of neurological disease outcomes and for disease and treatment monitoring. Naturally, diagnostic biomarkers can be used for disease monitoring if same pathologic processes underlie the development of the disease during different disease stages. The use of diagnostic biomarkers for disease monitoring is more problematic, however, if disease progression is caused by involvement of new pathologic processes. AD is a good example of such a disease: (i) the pathology is initiated by not fully understood metabolic abnormalities, (ii) these are followed by morphological changes in particular brain areas, such as formation of amyloid plaques and tau protein tangles, neurite retraction, dysfunction, and destruction of synapses, (iii) finally neurons are dying, and pathology expands to new brain areas and cell types. Thus, it is unlikely that different stages of AD can be differentiated by the same biomarkers. The situation is also complicated by the fact that some processes, e.g., synapse destruction, are common for different pathologies, including normal brain aging and other neurodegenerative diseases. The detection of such common processes is clearly useful for monitoring of normal brain aging and diagnosis of MCI, which is a syndrome characteristic of early stage of various neurodegenerative diseases. However, such a test will not predict MCI outcome. This goal could be accomplished by other tests, such as CSF protein analysis and/or imaging techniques, or by different miRNA biomarkers specific for various AD stages. For example, AD expansion to new brain areas and cell types can be monitored by circulating miRNA enriched in those brain areas and cell types. The switch from synapse destruction to neuronal death can potentially be detected by changes of ratios in plasma of cell-free synapse-enriched and neuronal body miRNA, and so on. Of course, all such proposals need be tested, preferably in longitudinal studies.

We hope the advances in the analyses of circulating miRNA reviewed here will lead to more efforts toward using miRNA as biomarkers of neurodegeneration, of other neurologic pathologies, and of normal brain aging, and will facilitate the development of screening, predictive, and monitoring tests for these diseases.

REFERENCES

Ahmed, F. E., Ahmed, N. C., Vos, P. W., Bonnerup, C., Atkins, J. N., Casey, M., et al. (2013). Diagnostic microRNA markers to screen for sporadic human colon cancer in stool: I. Proof of principle. *Cancer Genomics Proteomics* 10, 93–113.

Alexandrov, P. N., Dua, P., Hill, J. M., Bhattacharjee, S., Zhao, Y., and Lukiw, W. J. (2012). microRNA (miRNA) speciation in Alzheimer's disease (AD) cerebrospinal fluid (CSF) and extracellular fluid (ECF). *Int. J. Biochem. Mol. Biol.* 3, 365–373.

Apostolova, L. G., Hwang, K. S., Andrawis, J. P., Green, A. E., Babakchanian, S., Morra, J. H., et al. (2010). 3D PIB and CSF biomarker associations with hippocampal atrophy in ADNI subjects. *Neurobiol. Aging* 31, 1284–1303. doi: 10.1016/j.neurobiolaging.2010.05.003

Bak, M., Silahtaroglu, A., Møller, M., Christensen, M., Rath, M. F., Skryabin, B., et al. (2008). MicroRNA expression in the adult mouse central nervous system. *RNA* 14, 432–444. doi: 10.1261/rna.783108

Barad, O., Meiri, E., Avniel, A., Aharonov, R., Barzilai, A., Bentwich, I., et al. (2004). MicroRNA expression detected by oligonucleotide microarrays: system establishment and expression profiling in human tissues. *Genome Res.* 14, 2486–2494.

Baraniskin, A., Kuhnhenn, J., Schlegel, U., Maghnouj, A., Zöllner, H., Schmiegel, W., et al. (2012). Identification of microRNAs in the cerebrospinal fluid as biomarker for the diagnosis of glioma. *Neuro Oncol.* 14, 29–33. doi: 10.1093/neuonc/nor169

Bartel, D. P. (2009). MicroRNAs: target recognition and regulatory functions. *Cell* 136, 215–233. doi: 10.1016/j.cell.2009.01.002

Baskerville, S., and Bartel, D. P. (2005). Microarray profiling of microRNAs reveals frequent coexpression with neighboring miRNAs and host genes. *RNA* 11, 241–247.

Bellingham, S. A., Coleman, B. M., and Hill, A. F. (2012). Small RNA deep sequencing reveals a distinct miRNA signature released in exosomes from prion-infected neuronal cells. *Nucleic Acids Res.* 40, 10937–10949. doi: 10.1093/nar/gks832

Beuvink, I., Kolb, F. A., Budach, W., Garnier, A., Lange, J., Natt, F., et al. (2007). A novel microarray approach reveals new tissue-specific signatures of known and predicted mammalian microRNAs. *Nucleic Acids Res.* 35, e52. doi: 10.1093/nar/gkl1118

Blair, L. P., and Yan, Q. (2012). Epigenetic mechanisms in commonly occurring cancers. *DNA Cell Biol.* 31(Suppl. 1) S49–S61. doi: 10.1089/dna.2012.1654

Blondal, T., Jensby Nielsen, S., Baker, A., Andreasen, D., Mouritzen, P., Wrang Teilum, M., et al. (2013). Assessing sample and miRNA profile quality in serum and plasma or other biofluids. *Methods* 59, S1–S6. doi: 10.1016/j.ymeth.2012.09.015

Boeri, M., Verri, C., Conte, D., Roz, L., Modena, P., Facchinetti, F., et al. (2011). MicroRNA signatures in tissues and plasma predict development and prognosis of computed tomography detected lung cancer. *Proc. Natl. Acad. Sci. U.S.A.* 108, 3713–3718. doi: 10.1073/pnas.1100048108

Bredesen, D. E. (2009). Neurodegeneration in Alzheimer's disease: caspases and synaptic element interdependence. *Mol. Neurodegener.* 4, 27. doi: 10.1186/1750-1326-4-27

Caldas, C. (2012). Cancer sequencing unravels clonal evolution. *Nat. Biotechnol.* 30, 408–410. doi: 10.1038/nbt.2213

Castellano, L., and Stebbing, J. (2013). Deep sequencing of small RNAs identifies canonical and non-canonical miRNA and endogenous siRNAs in mammalian somatic tissues. *Nucleic Acids Res.* 41, 3339–3351. doi: 10.1093/nar/gks1474

Cogswell, J. P., Ward, J., Taylor, I. A., Waters, M., Shi, Y., Cannon, B., et al. (2008). Identification of miRNA changes in Alzheimer's disease brain and CSF yields putative biomarkers and insights into disease pathways. *J. Alzheimers Dis.* 14, 27–41. doi:10.1016/j.jalz.2008.05.420

Cuk, K., Zucknick, M., Heil, J., Madhavan, D., Schott, S., Turchinovich, A., et al. (2013). Circulating microRNAs in plasma as early detection markers for breast cancer. *Int. J. Cancer* 132, 1602–1612. doi: 10.1002/ijc.27799

Deo, M., Yu, J. Y., Chung, K. H., Tippens, M., and Turner, D. L. (2006). Detection of mammalian microRNA expression by in situ hybridization with RNA oligonucleotides. *Dev. Dyn.* 235, 2538–2548.

D'haene, B., Mestdagh, P., Hellemans, J., and Vandesompele, J. (2012). miRNA expression profiling: from reference genes to global mean normalization. *Methods Mol. Biol.* 822, 261–272. doi: 10.1007/978-1-61779-427-8_18

Dimmeler, S., and Nicotera, P. (2013). MicroRNAs in age-related diseases. *EMBO Mol. Med.* 5, 180–190. doi: 10.1002/emmm.201201986

Doxakis, E. (2010). Post-transcriptional regulation of alpha-synuclein expression by mir-7 and mir-153. *J. Biol. Chem.* 285, 12726–12734. doi: 10.1074/jbc.M109.086827

Duttagupta, R., Jiang, R., Gollub, J., Getts, R. C., and Jones, K. W. (2011). Impact of cellular miRNAs on circulating miRNA biomarker signatures. *PLoS ONE* 6:e20769. doi: 10.1371/journal.pone.0020769

Edbauer, D., Neilson, J. R., Foster, K. A., Wang, C. F., Seeburg, D. P., Batterton, M. N., et al. (2010). Regulation of synaptic structure and function by FMRP-associated microRNAs miR-125b and miR-132. *Neuron* 65, 373–384. doi: 10.1016/j.neuron.2010.01.005

Fagan, A. M., Shaw, L. M., Xiong, C., Vanderstichele, H., Mintun, M. A., Trojanowski, J. Q., et al. (2011). Comparison of analytical platforms for cerebrospinal fluid measures of β-amyloid 1–42, total tau, and P-tau(181) for identifying Alzheimer disease amyloid plaque pathology. *Arch. Neurol.* 68, 1137–1144. doi: 10.1001/archneurol.2011.105

Farazi, T. A., Spitzer, J. I., Morozov, P., and Tuschl, T. (2011). miRNAs in human cancer. *J. Pathol.* 223, 102–115. doi: 10.1002/path.2806

Filatova, E. V., Alieva, A. Kh., Shadrina, M. I., and Slominsky, P. A. (2012). MicroRNAs: possible role in pathogenesis of Parkinson's disease. *Biochemistry (Mosc.)* 77, 813–819. doi: 10.1134/S0006297912080020

Fiore, R., Khudayberdiev, S., Saba, R., and Schratt, G. (2011). MicroRNA function in the nervous system. *Prog. Mol. Biol. Transl. Sci.* 102, 47–100. doi: 10.1016/B978-0-12-415795-8.00004–0

Gandhi, R., Healy, B., Gholipour, T., Egorova, S., Musallam, A., Shuja, M., et al. (2013). Circulating microRNAs as biomarkers for disease staging in multiple sclerosis. *Ann. Neurol.* 73, 729–740. doi: 10.1002/ana.23880

Gascon, E., and Gao, F. B. (2012). Cause or effect: misregulation of microRNA pathways in neurodegeneration. *Front. Neurosci.* 6:48. doi: 10.3389/fnins.2012.00048

Gaughwin, P. M., Ciesla, M., Lahiri, N., Tabrizi, S. J., Brundin, P., and Björkqvist, M. (2011). Hsa-miR-34b is a plasma-stable microRNA that is elevated in pre-manifest Huntington's disease. *Hum. Mol. Genet.* 20, 2225–2237. doi: 10.1093/hmg/ddr111

Geekiyanage, H., and Chan, C. (2011). MicroRNA-137/181c regulates serine palmitoyltransferase and in turn amyloid β, novel targets in sporadic Alzheimer's disease. *J. Neurosci.* 31, 14820–14830. doi: 10.1523/JNEUROSCI.3883-11.2011

Geekiyanage, H., Jicha, G. A., Nelson, P. T., and Chan, C. (2012). Blood serum miRNA: non-invasive biomarkers for Alzheimer's disease. *Exp. Neurol.* 235, 491–496. doi: 10.1016/j.expneurol.2011.11.026

Griffiths-Jones, S., Grocock, R. J., van Dongen, S., Bateman, A., and Enright, A. J. (2006). miRBase: microRNA sequences, targets and gene nomenclature. *Nucleic Acids Res.* 34, D140–D144.

Haghikia, A., Haghikia, A., Hellwig, K., Baraniskin, A., Holzmann, A., Décard, B. F., et al. (2012). Regulated microRNAs in the CSF of patients with multiple sclerosis: a case-control study. *Neurology* 79, 2166–2170. doi: 10.1212/WNL.0b013e3182759621

Harraz, M. M., Dawson, T. M., and Dawson, V. L. (2011). MicroRNAs in Parkinson's disease. *J. Chem. Neuroanat.* 42, 127–130. doi: 10.1016/j.jchemneu.2011.01.005

He, M., Liu, Y., Wang, X., Zhang, M. Q., Hannon, G. J., and Huang, Z. J. (2012). Cell-type-based analysis of microRNA profiles in the mouse brain. *Neuron* 73, 35–48. doi: 10.1016/j.neuron.2011.11.010

Hennessey, P. T., Sanford, T., Choudhary, A., Mydlarz, W. W., Brown, D., Adai, A. T., et al. (2012). Serum microRNA biomarkers for detection of non-small cell lung cancer. *PLoS ONE* 7:e32307. doi: 10.1371/journal.pone.0032307

Iorio, M. V., and Croce, C. M. (2012). Causes and consequences of microRNA dysregulation. *Cancer J.* 18, 215–222. doi: 10.1097/PPO.0b013e318250c001

Jin, X. F., Wu, N., Wang, L., and Li, J. (2013). Circulating MicroRNAs: a novel class of potential biomarkers for diagnosing and prognosing central nervous system diseases. *Cell. Mol. Neurobiol.* 33, 601–613. doi: 10.1007/s10571-013-9940–9949

Junn, E., Lee, K. W., Jeong, B. S., Chan, T. W., Im, J. Y., and Mouradian, M. M. (2009). Repression of alpha-synuclein expression and toxicity by microRNA-7. *Proc. Natl. Acad. Sci. U.S.A.* 106, 13052–13057. doi: 10.1073/pnas.0906277106

Katsuura, S., Kuwano, Y., Yamagishi, N., Kurokawa, K., Kajita, K., Akaike, Y., et al. (2012). MicroRNAs miR-144/144* and miR-16 in peripheral blood are potential biomarkers for naturalistic stress in healthy Japanese medical students. *Neurosci. Lett.* 516, 79–84. doi: 10.1016/j.neulet.2012.03.062

Kim, J., Inoue, K., Ishii, J., Vanti, W. B., Voronov, S. V., Murchison, E., et al. (2007). A microRNA feedback circuit in midbrain dopamine neurons. *Science* 317, 1220–1224. doi: 10.1126/science.1140481

Kreek, M. J., Levran, O., Reed, B., Schlussman, S. D., Zhou, Y., and Butelman, E. R. (2012). Opiate addiction and cocaine addiction: underlying molecular neurobiology and genetics. *J. Clin. Invest.* 122, 3387–3393. doi: 10.1172/JCI60390

Krichevsky, A. M., King, K. S., Donahue, C. P., Khrapko, K., and Kosik, K. S. (2003). A microRNA array reveals extensive regulation of microRNAs during brain development. *RNA* 9, 1274–1281. doi: 10.1261/rna.5980303

Kroh, E. M., Parkin, R. K., Mitchell, P. S., and Tewari, M. (2010). Analysis of circulating microRNA biomarkers in plasma and serum using quantitative reverse transcription-PCR (qRT-PCR). *Methods* 50, 298–301. doi: 10.1016/j.ymeth.2010.01.032

Landgraf, P., Rusu, M., Sheridan, R., Sewer, A., Iovino, N., Aravin, A., et al. (2007). A mammalian microRNA expression atlas based on small RNA library sequencing. *Cell* 129, 1401–1414. doi: 10.1016/j.cell.2007.04.040

Lardizábal, M. N., Nocito, A. L., Daniele, S. M., Ornella, L. A., Palatnik, J. F., and Veggi, L. M. (2012). Reference genes for real-time PCR quantification of microRNAs and messenger RNAs in rat models of hepatotoxicity. *PLoS ONE* 7:e36323. doi:10.1371/journal.pone.0036323

Laterza, O. F., Lim, L., Garrett-Engele, P. W., Vlasakova, K., Muniappa, N., Tanaka, W. K., et al. (2009). Plasma MicroRNAs as sensitive and specific biomarkers of tissue injury. *Clin. Chem.* 55, 1977–1983. doi: 10.1373/clinchem.2009.131797

Latham, G. J. (2010). Normalization of microRNA quantitative RT-PCR data in reduced scale experimental designs. *Methods Mol. Biol.* 667, 19–31. doi: 10.1007/978-1-60761-811-9_2

Leidner, R. S., Li, L., and Thompson, C. L. (2013). Dampening enthusiasm for circulating microRNA in breast cancer. *PLoS ONE* 8:57841. doi: 10.1371/journal.pone.0057841

Li, X., Khanna, A., Li, N., and Wang, E. (2011). Circulatory miR34a as an RNAbased, noninvasive biomarker for brain aging. *Aging (Albany NY)* 3, 985–1002.

Li, Y. J., Xu, M., Gao, Z. H., Wang, Y. Q., Yue, Z., Zhang, Y. X., et al. (2013). Alterations of serum levels of BDNF-related miRNAs in patients with depression. *PLoS ONE* 8:e63648. doi: 10.1371/journal.pone.0063648

Liang, Y., Ridzon, D., Wong, L., and Chen, C. (2007). Characterization of microRNA expression profiles in normal human tissues. *BMC Genomics* 8:166. doi: 10.1186/1471-2164-8-166

Liu, C. G., Calin, G. A., Meloon, B., Gamliel, N., Sevignani, C., Ferracin, M., et al. (2004). An oligonucleotide microchip for genome-wide microRNA profiling in human and mouse tissues. *Proc. Natl. Acad. Sci. U.S.A.* 101, 9740–9744. doi: 10.1073/pnas.0403293101

Lugli, G., Torvik, V. I., Larson, J., and Smalheiser, N. R. (2008). Expression of microRNAs and their precursors in synaptic fractions of adult mouse forebrain. *J. Neurochem.* 106, 650–661. doi: 10.1111/j.1471-4159.2008.05413.x

Lukiw, W. J., Alexandrov, P. N., Zhao, Y., Hill, J. M., and Bhattacharjee, S. (2012). Spreading of Alzheimer's disease inflammatory signaling through soluble micro-RNA. *Neuroreport* 23, 621–626. doi: 10.1097/WNR.0b013e32835542b0

Matthaei, H., Wylie, D., Lloyd, M. B., Dal Molin, M., Kemppainen, J., Mayo, S. C., et al. (2012). miRNA biomarkers in cyst fluid augment the diagnosis and management of pancreatic cysts. *Clin. Cancer Res.* 18, 4713–4724. doi: 10.1158/1078-0432.CCR-12–0035

McIntosh, T. K., Saatman, K. E., Raghupathi, R., Graham, D. I., Smith, D. H., Lee, V. M., et al. (1998). The Dorothy Russell Memorial Lecture. The molecular and cellular sequelae of experimental traumatic brain injury: pathogenetic mechanisms. *Neuropathol. Appl. Neurobiol.* 24, 251–267. doi: 10.1046/j.1365-2990.1998.00121.x

McNeill, E., and Van Vactor, D. (2012). MicroRNAs shape the neuronal landscape. *Neuron* 75, 363–379. doi: 10.1016/j.neuron.2012.07.005

Mellios, N., and Sur, M. (2012). The emerging role of microRNAs in schizophrenia and autism spectrum disorders. *Front. Psychiatry* 3:39. doi: 10.3389/fpsyt.2012.00039

Mestdagh, P., Van Vlierberghe, P., De Weer, A., Muth, D., Westermann, F., Speleman, F., et al. (2009). A novel and universal method for microRNA RT-qPCR data normalization. *Genome Biol.* 10, R64. doi: 10.1186/gb-2009-10-6-r64

Meyer, S. U., Pfaffl, M. W., and Ulbrich, S. E. (2010). Normalization strategies for microRNA profiling experiments: a 'normal' way to a hidden layer of complexity? *Biotechnol. Lett.* 32, 1777–1788. doi: 10.1007/s10529-010-0380-z

Moreau, M. P., Bruse, S. E., Jornsten, R., Liu, Y., and Brzustowicz, L. M. (2013). Chronological changes in microRNA expression in the developing human brain. *PLoS ONE* 8:e60480. doi: 10.1371/journal.pone.0060480

Mori, T., Maeda, J., Shimada, H., Higuchi, M., Shinotoh, H., Ueno, S., et al. (2012). Molecular imaging of dementia. *Psychogeriatrics* 12, 106–114. doi: 10.1111/j.1479-8301.2012.00409.x

Mouradian, M. M. (2012). MicroRNAs in Parkinson's disease. *Neurobiol. Dis.* 46, 279–284. doi: 10.1016/j.nbd.2011.12.046

Nagele, E., Han, M., Demarshall, C., Belinka, B., and Nagele, R. (2011). Diagnosis of Alzheimer's disease based on disease-specific autoantibody profiles in human sera. *PLoS ONE* 6:e23112. doi: 10.1371/journal.pone.0023112

Natera-Naranjo, O., Aschrafi, A., Gioio, A. E., and Kaplan, B. B. (2010). Identification and quantitative analyses of microRNAs located in the distal axons of sympathetic neurons. *RNA* 16, 1516–1529. doi: 10.1261/rna.1833310

Palma, J., Yaddanapudi, S. C., Pigati, L., Havens, M. A., Jeong, S., Weiner, G. A., et al. (2012). MicroRNAs are exported from malignant cells in customized particles. *Nucleic Acids Res.* 40, 9125–9138. doi: 10.1093/nar/gks656

Patz, S., Trattnig, C., Grünbacher, G., Ebner, B., Gülly, C., Novak, A., et al. (2013). More than cell dust: microparticles isolated from cerebrospinal fluid of brain injured patients are messengers carrying mRNAs, miRNAs and proteins. *J. Neurotrauma* 30, 1232–1242. doi: 10.1089/neu.2012.2596

Peltier, H. J., and Latham, G. J., (2008). Normalization of microRNA expression levels in quantitative RT-PCR assays: identification of suitable reference RNA targets in normal and cancerous human solid tissues. *RNA* 14, 844–852. doi: 10.1261/rna.939908

Pichardo-Casas, I., Goff, L. A., Swerdel, M. R., Athie, A., Davila, J., Ramos-Brossier, M., et al. (2012). Expression profiling of synaptic microRNAs from the adult rat brain identifies regional differences and seizure-induced dynamic modulation. *Brain Res.* 1436, 20–33. doi: 10.1016/j.brainres.2011.12.001

Pigati, L., Yaddanapudi, S. C., Iyengar, R., Kim, D. J., Hearn, S. A., Danforth, D., et al. (2010). Selective release of microRNA species from normal and malignant mammary epithelial cells. *PLoS ONE* 5:e13515. doi: 10.1371/journal.pone.0013515

Pillai, J. A., and Cummings, J. L. (2013). Clinical trials in predementia stages of Alzheimer disease. *Med. Clin. North Am.* 97, 439–457. doi: 10.1016/j.mcna.2013.01.002

Pritchard, C. C., Kroh, E., Wood, B., Arroyo, J. D., Dougherty, K. J., Miyaji, M. M., et al. (2012). Blood cell origin of circulating microRNAs: a cautionary note for cancer biomarker studies. *Cancer Prev. Res. (Phila).* 5, 492–497. doi: 10.1158/1940-6207.CAPR-11–0370

Qin, L. X., Tuschl, T., and Singer, S. (2013). An empirical evaluation of normalization methods for microRNA arrays in a liposarcoma study. *Cancer Inform.* 12, 83–101. doi: 10.4137/CIN.S11384

Ray, S., Britschgi, M., Herbert, C., Takeda-Uchimura, Y., Boxer, A., Blennow, K., et al. (2007). Classification and prediction of clinical Alzheimer's diagnosis based on plasma signaling proteins. *Nat. Med.* 13, 1359–1362. doi: 10.1038/nm1653

Reddy, M. M., Wilson, R., Wilson, J., Connell, S., Gocke, A., Hynan, L., et al. (2011). Identification of candidate IgG biomarkers for Alzheimer's disease via combinatorial library screening. *Cell* 144, 132–142. doi: 10.1016/j.cell.2010.11.054

Richards, M. A. (2009). The National Awareness and Early Diagnosis Initiative in England: assembling the evidence. *Br. J. Cancer* 101(Suppl. 2), S1–S4. doi: 10.1038/sj.bjc.6605382

Ridolfi, E., Fenoglio, C., Cantoni, C., Calvi, A., De Riz, M., Pietroboni, A., et al. (2013). Expression and genetic analysis of microRNAs involved in multiple sclerosis. *Int. J. Mol. Sci.* 14, 4375–4384. doi: 10.3390/ijms14034375

Rong, H., Liu, T. B., Yang, K. J., Yang, H. C., Wu, D. H., Liao, C. P., et al. (2011). MicroRNA-134 plasma levels before and after treatment for bipolar mania. *J. Psychiatr. Res.* 45, 92–95. doi: 10.1016/j.jpsychires

Russo, F., Di Bella, S., Nigita, G., Macca, V., Laganà, A., Giugno, R., et al. (2012). miRandola: extracellular circulating microRNAs database. *PLoS ONE* 7:e47786. doi: 10.1371/journal.pone.0047786

Saba, R., Goodman, C. D., Huzarewich, R. L., Robertson, C., and Booth, S. A. (2008). A miRNA signature of prion induced neurodegeneration. *PLoS ONE* 3:e3652. doi: 10.1371/journal.pone.0003652

Salta, E., and De Strooper, B. (2012). Non-coding RNAs with essential roles in neurodegenerative disorders. *Lancet Neurol.* 11, 189–200. doi: 10.1016/S1474-4422(11)70286–1

Sanders, I., Holdenrieder, S., Walgenbach-Brünagel, G., von Ruecker, A., Kristiansen, G., Müller, S. C., et al. (2012). Evaluation of reference genes for the analysis of serum miRNA in patients with prostate cancer, bladder cancer and renal cell carcinoma. *Int. J. Urol.* 19, 1017–1025. doi: 10.1111/j.1442-2042.2012.03082.x

Sarkar, D., Parkin, R., Wyman, S., Bendoraite, A., Sather, C., Delrow, J., et al. (2009). Quality assessment and data analysis for microRNA expression arrays. *Nucleic Acids Res.* 37, e17. doi: 10.1093/nar/gkn932

Saugstad, J. A. (2010). MicroRNAs as effectors of brain function with roles in ischemia and injury, neuroprotection, and neurodegeneration. *J. Cereb. Blood Flow Metab.* 30, 1564–1576. doi: 10.1038/jcbfm.2010.101

Schaefer, A., Jung, M., Miller, K., Lein, M., Kristiansen, G., Erbersdobler, A., et al. (2010). Suitable reference genes for relative quantification of miRNA expression in prostate cancer. *Exp. Mol. Med.* 42, 749–758. doi: 10.3858/emm.2010.42.11.076

Schratt, G. (2009). microRNAs at the synapse. *Nat. Rev. Neurosci.* 10, 842–849. doi: 10.1038/nrn2763

Schratt, G. M., Tuebing, F., Nigh, E. A., Kane, C. G., Sabatini, M. E., Kiebler, M., et al. (2006). A brain-specific microRNA regulates dendritic spine development. *Nature* 439, 283–289. doi:10.1038/nature04367

Sempere, L. F., Freemantle, S., Pitha-Rowe, I., Moss, E., Dmitrovsky, E., and Ambros, V. (2004). Expression profiling of mammalian microRNAs uncovers a subset of brain-expressed microRNAs with possible roles in murine and human neuronal differentiation. *Genome Biol.* 5, R13.

Sheinerman, K. S., Tsivinsky, V. G., Crawford, F., Mullan, M. J., Abdullah, L., and Umansky, S. R. (2012). Plasma microRNA biomarkers for detection of mild cognitive impairment. *Aging (Albany NY)* 4, 590–605.

Sheinerman, K. S., and Umansky, S. R. (2013). Early detection of neurodegenerative diseases: circulating brain-enriched microRNA. *Cell Cycle* 12, 1–2. doi: 10.4161/cc.23067

Shi, W., Du, J., Qi, Y., Liang, G., Wang, T., Li, S., et al. (2012). Aberrant expression of serum miRNAs in schizophrenia. *J. Psychiatr. Res.* 46, 198–204. doi: 10.1016/j.jpsychires.2011

Siegel, S. R., Mackenzie, J., Chaplin, G., Jablonski, N. G., and Griffiths, L. (2012). Circulating microRNAs involved in multiple sclerosis. *Mol. Biol. Rep.* 39, 6219–6225. doi: 10.1007/s11033-011-1441–1447

Siegel, G., Saba, R., and Schratt, G. (2011). microRNAs in neurons: manifold regulatory roles at the synapse. *Curr. Opin. Genet. Dev.* 21, 491–497. doi: 10.1016/j.gde.2011.04.008

Smirnova, L., Gräfe, A., Seiler, A., Schumacher, S., Nitsch, R., and Wulczyn, F. G. (2005). Regulation of miRNA expression during neural cell specification. *Eur. J. Neurosci.* 21, 1469–1477. doi: 10.1111/j.1460-9568.2005.03978.x

Smith-Vikos, T., and Slack, F. J. (2012). MicroRNAs and their roles in aging. *J. Cell Sci.* 125(Pt 1), 7–17. doi: 10.1242/jcs.099200

Snyder, E. M., Olin, J., and David, F. S. (2012). Maximizing the value of diagnostics in Alzheimer's disease drug development. *Nat. Rev. Drug Discov.* 11, 183–184. doi: 10.1038/nrd3535

Sperling, R. A., Jack, C. R. Jr., and Aisen, P. S. (2011). Testing the right target and right drug at the right stage. *Sci. Transl. Med.* 3, 111cm33. doi: 10.1126/scitranslmed.3002609

Strickland, I. T., Richards, L., Holmes, F. E., Wynick, D., Uney, J. B., and Wong, L. F. (2011). Axotomy-induced miR-21 promotes axon growth in adult dorsal root ganglion neurons. *PLoS ONE* 6:e23423. doi: 10.1371/journal.pone.0023423

Sun, Y., Zhang, K., Fan, G., and Li, J. (2012). Identification of circulating microRNAs as biomarkers in cancers: what have we got? *Clin. Chem. Lab. Med.* 50, 2121–2126.

Tan, L., Yu, J. T., Hu, N., and Tan, L. (2013). Non-coding RNAs in Alzheimer's disease. *Mol. Neurobiol.* 47, 382–393. doi: 10.1007/s12035-012-8359–8355

Teplyuk, N. M., Mollenhauer, B., Gabriely, G., Giese, A., Kim, E., Smolsky, M., et al. (2012). MicroRNAs in cerebrospinal fluid identify glioblastoma and metastatic brain cancers and reflect disease activity. *Neuro Oncol.* 14, 689–700. doi: 10.1093/neuonc/nos074

Tijsen, A. J., Pinto, Y. M., and Creemers, E. E. (2012). Circulating microRNAs as diagnostic biomarkers for cardiovascular diseases. *Circ. Res.* 110, 483–495. doi: 10.1161/CIRCRESAHA.111.247452

Trivedi, S., and Ramakrishna, G. (2009). miRNA and neurons. *Int. J. Neurosci.* 119, 1995–2016.

von Bernhardi, R., and Inestrosa, N. C., eds. (2008). *Neurodegenerative Diseases: From Molecular Concepts to Therapeutic Targets.* Hauppauge: Nova Publishers.

Wang, W., Kwon, E. J., and Tsai, L. H. (2012a). MicroRNAs in learning, memory, and neurological diseases. *Learn. Mem.* 19, 359–368. doi: 10.1101/lm.026492.112

Wang, H., Zhang, P., Chen, W., Feng, D., Jia, Y., and Xie, L. X. (2012b). Evidence for serum miR-15a and miR-16 levels as biomarkers that distinguish sepsis from systemic inflammatory response syndrome in human subjects. *Clin. Chem. Lab. Med.* 50, 1423–1428. doi: 10.1515/cclm-2011-0826

Wang, Y., Weng, T., Gou, D., Chen, Z., Chintagari, N. R., and Liu, L. (2007). Identification of rat lung-specific microRNAs by micoRNA microarray: valuable discoveries for the facilitation of lung research. *BMC Genomics* 8:29. doi: 10.1186/1471-2164-8-29

Wang, K., Zhang, S., Weber, J., Baxter, D., and Galas, D. J. (2009). Mammalian cells in culture actively export specific microRNAs. *Nat. Precedings.* doi: 10101/npre.2009.3718.1

Weber, J. A., Baxter, D. H., Zhang, S., Huang, D. Y., Huang, K. H., Lee, M. J., et al. (2010). The microRNA spectrum in 12 body fluids. *Clin. Chem.* 56, 1733–1741. doi: 10.1373/clinchem.2010.147405

Weng, H., Shen, C., Hirokawa, G., Ji, X., Takahashi, R., Shimada, K., et al. (2011). Plasma miR-124 as a biomarker for cerebral infarction. *Biomed. Res.* 32, 135–141.

Williams, Z., Ben-Dov, I. Z., Elias, R., Mihailovic, A., Brown, M., Rosenwaks, Z., et al. (2013). Comprehensive profiling of circulating microRNA via small RNA sequencing of cDNA libraries reveals biomarker potential and limitations. *Proc. Natl. Acad. Sci. U.S.A.* 110, 4255–4260. doi: 10.1073/pnas.1214046110

Wu, D., Raafat, M., Pak, E., Hammond, S., and Murashov, A. K. (2011). MicroRNA machinery responds to peripheral nerve lesion in an injury-regulated pattern. *Neuroscience* 190, 386–397. doi: 10.1016/j.neuroscience.2011.06.017

Xie, B., Ding, Q., Han, H., and Wu, D. (2013). miRCancer: a microRNA-cancer association database constructed by text mining on literature. *Bioinformatics* 29, 638–644. doi: 10.1093/bioinformatics/btt014

Xu, B., Hsu, P. K., Karayiorgou, M., and Gogos, J. A. (2012). MicroRNA dysregulation in neuropsychiatric disorders and cognitive dysfunction. *Neurobiol. Dis.* 46, 291–301. doi: 10.1016/j.nbd.2012.02.016

Zandberga, E., Kozirovskis, V., Ābols, A., Andrējeva, D., Purkalne, G., and Linē, A. (2013). Cell-free microRNAs as diagnostic, prognostic, and predictive biomarkers for lung cancer. *Genes Chromosomes Cancer* 52, 356–369. doi: 10.1002/gcc.22032

Zhang, Y., Jia, Y., Zheng, R., Guo, Y., Wang, Y., Guo, H., et al. (2010). Plasma microRNA-122 as a biomarker for viral-, alcohol-, and chemical-related hepatic diseases. *Clin. Chem.* 56, 1830–1838. doi: 10.1373/clinchem.2010.147850

Zheng, Z., Lee, J. E., and Yenari, M. A. (2003). Stroke: molecular mechanisms and potential targets for treatment. *Curr. Mol. Med.* 3, 361–372. doi: 10.2174/1566524033479717

2

Pluripotent stem cells as a model to study non-coding RNAs function in human neurogenesis

Alexandra Benchoua[1]* *and Marc Peschanski*[2]

[1] Centre d'Etude des Cellules Souches, Institut des cellules Souches pour le Traitement et l'Étude des Maladies monogéniques, Association Française contre les Myopathies, Evry, France

[2] INSERM/UEVE UMR 86, Institut des cellules Souches pour le Traitement et l'Étude des Maladies monogéniques, Association Française contre les Myopathies, Evry, France

Correspondence:
Alexandra Benchoua, Centre d'Etude des Cellules Souches, Institut des cellules Souches pour le Traitement et l'Étude des Maladies monogéniques, Association Française contre les Myopathies, 5 rue Henri Desbrueres-Genopole campus 1 91030 Evry Cedex, France
e-mail: abenchoua@istem.fr

As fine regulators of gene expression, non-coding RNAs, and more particularly micro-RNAs (miRNAs), have emerged as key players in the development of the nervous system. *In vivo* experiments manipulating miRNAs expression as neurogenesis proceeds are very challenging in the mammalian embryo and totally impossible in the human. Human pluripotent stem cells (hPSCs), from embryonic origin (hESCs) or induced from adult somatic cells (iPSCs), represent an opportunity to study the role of miRNAs in the earliest steps of human neurogenesis in both physiological and pathological contexts. Robust protocols are now available to convert pluripotent stem cells into several sub-types of fully functional neurons, recapitulating key developmental milestones along differentiation. This provides a convenient cellular system for dissecting the role of miRNAs in phenotypic transitions critical to brain development and plasticity that may be impaired in neurological diseases with onset during development. The aim of this review is to illustrate how hPSCs can be used to recapitulate early steps of human neurogenesis and summarize recent reports of their contribution to the study of the role of miRNA in regulating development of the nervous system.

Keywords: pluripotent stem cells, micro-RNA, neurogenesis, neuro-developmental diseases, psychiatry

INTRODUCTION

Human neurogenesis is the result of a tightly controlled sequence of events that associates environmental signals and intra-cellular molecular mechanisms. Impaired neurogenesis is at the origin of the so-called neuro-developmental disorders, such as autism spectrum disorders (ASDs) or Down's syndrome, and is thought to be involved in the etiology of psychiatric disorders such as schizophrenia or bipolar disorders (Moreno-De-Luca et al., 2013). Micro-RNAs (miRNAs) are abundant, short-lived, double strand non-coding RNAs (nc-RNAs) of 22 nucleotides that act as post-transcriptional repressors targeting multiple mRNAs (Yates et al., 2013). They account for an additional level of intricacy to gene regulation and, therefore, represent attractive candidates to interpret subtle developmental regulations. Genetic manipulation of the miRNAs machinery in rodent models severely impaired several aspect of neurogenesis (Sun et al., 2013). However, to evaluate whether this may be relevant to human neurogenesis, both physiologically and in a pathological context, these experiments need to be replicated and supplemented in human models of neurogenesis.

Over the last decade, human pluripotent stem cells (hPSCs) have emerged as powerful tools with the potential to further illuminate key mechanisms underlying neuronal development (Chamberlain et al., 2008; Crook and Kobayashi, 2008). These cells allow investigators to more specifically assess aspects of human neurogenesis that were previously hardly attainable due to technical obstacles at accessing human embryonic and fetal tissues. Pluripotent stem cell (PSC) can be obtained from the embryo inner cell mass or reprogramed from any adult somatic cells (Gokhale and Andrews, 2012). Their self-renewal property offers the opportunity to amplify them until reaching the cell mass necessary to perform large throughput studies including miRNA whole genome profiling. As PSCs, they can virtually give rise *in vitro* to any cell type of the human body including neurons. Of interest, the differentiation paradigms of hPSC into different sub-types of neurons recapitulates the key milestones of human neurogenesis including: (i) early neural commitment and neuro-epithelial cells differentiation, (ii) regionalization of the early neuro-epithelial cells into more specialized neural progenitors, (iii) terminal differentiation of the specialized progenitors into specific sub-types of neurons and maturation of these neurons until the formation of functional synapses and complex networks. The involvement of miRNAs in each milestone can therefore be assessed functionally using genetic manipulation to achieve loss or gain of function into the cells.

While human embryonic stem cells (hESCs) remain the gold standard to study the physiological aspects of early human development, PSCs reprogramed from adult somatic cells (iPSC) offer the opportunity to address important issues regarding miRNA participation to disease etiology in a patient-related genetic background. Here, we summarize the different steps of human neurogenesis that can be recapitulated from hPSC and the current protocols that have been implemented to obtain them. We will next review the current knowledge regarding miRNA-dependent regulation in PSC-derived models of neurogenesis. Finally, we will discuss the current success of iPSC derivation from patient with neuro-developmental or early onset psychiatric disorders and how these new cellular models could help increasing our understanding about the involvement of miRNAs in human diseases.

PLURIPOTENT STEM CELLS AS A CELLULAR SYSTEM TO RECAPITULATE KEY STEPS OF EARLY HUMAN NEUROGENESIS

ENGAGEMENT OF PLURIPOTENT STEM CELLS INTO THE NEURAL LINEAGE

The earlier event of human neurogenesis that can be modeled *in vitro* using PSC is the conversion of PSCs into the neural lineage to form the first population of neural progenitors found in the neural plate, the neuro-epithelial cells. hPSC may be coaxed along the neural lineage and differentiate into a population of bipolar neuro-epithelial cells that express the main markers of the neural tube, Sox1 and Sox2, and organize in rosette-like multicellular structures using different methods. One efficient and convenient protocol was revealed in parallel by Chambers et al. (2009) and our group (Boissart et al., 2012). PSC are first cultivated as a monolayer then the medium changed to a neural induction medium containing a combination of inhibitors of both the bone morphogenetic proteins (BMPs) and transforming growth factor-beta (TGF-β), Smad-dependent, pathways. After 10–15 days, the neural conversion of PSC into Pax6/Sox1 positive neuro-epithelial cells is fully achieved. When used to inhibit TGF-β-mediated pluripotency networks, the small molecule SB431542 promotes exit of cells from the pluripotent compartment and suppresses mesendodermal fates by inhibiting endogenous activin and nodal signals. Neural conversion of the resulting ectodermal cells was achieved with addition of the BMP inhibitor Noggin (Chambers et al., 2009; Boissart et al., 2012). The resulting neuro-epithelial cells are competent to form neural "rosettes" that morphologically mimic the neural tube cells and could further be differentiated into different sub-types of functional neurons. Clear advantage of this system is its dramatic efficiency and relative simplicity. The efficiency of the phenotypic transition between pluripotency and neural commitment can be easily monitored in real-time following the morphological changes characteristic of the formation of the neuro-epithelium, the so-called neural "rosettes." Quantification can be achieved measuring the number of cells expressing the canonical neural markers Sox1 and Pax6 but down-regulating the pluripotency markers Oct-4 and Nanog. This allows the differential large-scale profiling of miRNA expression since both pluripotent and neuro-epithelial cells can be obtain to near-purity. Finally, the monolayer culture mode is perfectly adapted to transfection methods and functional validation can easily be conducted (Boissart et al., 2012). Neural commitment can also be achieved using embryoid bodies (EBs) where PSCs are differentiated *in vitro* by spontaneously self-assembling in suspension into 3D cell aggregates. This technique of differentiation promotes the formation of the three embryonic germ layers in parallel. It can be more challenging to manipulate these 3D structures than cells differentiated as monolayer; however, the contribution of a given miRNA to influence the balance between the different embryonic fates can be addressed (Xu et al., 2009).

REGIONAL PATTERNING OF NEURO-EPITHELIAL CELLS

In addition to the acquisition of an early neural fate, neuro-epithelial cells will progressively adopt a specific regional identity along the neural tube axes in response to exogenous factors. The resulting cells have a more restricted potential and produce only specific sub-types of neurons according to their position along the rostro-caudal and dorso-ventral (DV) axis. Fundamental to the existence of divergent structures in the brain is the early region-specific molecular programing. In mammal embryos, the anterio-posterior (AP) axis is specified as neural commitment proceeds. The closing neural tube quickly divides into three primary vesicles: the anterior forebrain, the midbrain, and the posterior hindbrain. The forebrain will further sub-divide into two structures, the rostral telencephalon and the diencephalon (Pombero and Martinez, 2009), whereas the caudal hindbrain will form the rhombencephalon and the spinal cord. Secondary patterning sequences will further specify DV domains inside each structure (Lupo et al., 2006). The organization of these secondary vesicles prefigures the future brain structures. The telencephalon will give rise to the cortex in its dorsal part and to basal ganglia in its ventral part. The thalamus and hypothalamus will emerge from the ventral diencephalon, the substantia nigra from the ventral mesencephalon, the cerebellum from the rhombencephalon, spinal motor neurons will form from the ventral part of the spinal cord whereas sensorial neurons will differentiate from the dorsal part (de Graaf-Peters and Hadders-Algra, 2006).

In vitro regionalization of PSC-derived neuro-epithelial cells has been achieved successfully for some representative neuronal populations by translating knowledge from embryogenesis. During embryogenesis, regional patterning is under the control of extra-cellular signals that provide a group of neural progenitors with the unique competency to produce specific neuronal sub-types. The activity of these signals is spatiotemporally integrated by neural progenitors to determine the specific combinations of transcription factors activated in distinct AP and DV compartments of the central nervous system (CNS; Vieira et al., 2010). Master factors include Wnt, sonic hedgehog (SHH), fibroblast growth factors (FGFs), BMPs, and retinoic acid (RA) acting in gradient of concentrations. Anterior fates have been obtained from PSC using default protocols, in the absence of any morphogen (Li et al., 2009; Zeng et al., 2010). However, the efficiency can be greatly improved by using inhibitors of the Wnt pathway. The resulting population of neural progenitors expresses high levels of the anterior marker FoxG1 and can be further patterned dorsally or ventrally using the interplay between Wnt and SHH pathways. Further differentiation of primitive neuro-epithelial cells in the absence of SHH spontaneously produce neural progenitors that express the dorsal markers Pax6 and Emx1 and will ultimately give rise to glutamatergic projection neurons of the different cortical layers (Shi et al., 2012). In contrast, gradual activation of SHH-dependent pathway allows the successful production of progenitors from the ventral ganglionic eminence, expressing the markers Nkx2.1 or Gsh-2, and the corresponding cortical or striatal GABAergic interneurons (Carri et al., 2013; Maroof et al., 2013).

Interplays between RA, FGF-8, and Wnt pathways promote more caudal fates. Midbrain dopaminergic (mDA) neurons of the substantia nigra and the ventral tegmental area emerge from progenitors located in the floor plate of the ventral part of the embryonic midbrain (Ono et al., 2007). Accordingly, most protocols that aim at deriving DA neurons progenitors from hPSC rely on the exposition of early neuro-epithelial to SHH and to FGF-8, a weak but sufficient caudalizing factor (Perrier et al., 2004; Andersson et al., 2006; Friling et al., 2009; Rhee et al., 2011). Recently, the small molecule CHIR99021, a glycogen synthase kinase-3 inhibitor that mimics Wnt pathway activation, has been identified as a more potent caudalizing agent than FGF-8, and has been used to produce high yields of mDA neurons in combination with a modified form of SHH (Kirkeby et al., 2011; Kriks et al., 2011). Spinal cord motor neurons originate from the motoneuron progenitor domain located in the ventral developing spinal cord (Soula et al., 2001). In order to obtain efficient spinal motor neurons *in vitro*, primitive neuro-epithelial cells need to be caudalized as the neural induction proceeds, using high concentrations of RA then ventralized using SHH (Chipman et al., 2012; Takazawa et al., 2012).

Most of these protocols of directed differentiation are efficient enough to yield largely enriched population of a given progenitor sub-type (**Figure 1**). Differential miRNA profiling experiments comparing different progenitor populations differentiated from the same neuro-epithelial cells can help identifying miRNAs specifically involved in progenitor specification. In addition, the involvement of miRNA in progenitor response to patterning molecules can easily be assessed in a dose-dependent manner.

IN VITRO MODELING OF THE BALANCE BETWEEN SELF-RENEWAL AND NEURONAL DIFFERENTIATION

During embryogenesis, progenitors committed to the neural lineage and regionalized undergo several rounds of symmetric divisions (self-renewal) before giving rise to terminally differentiated

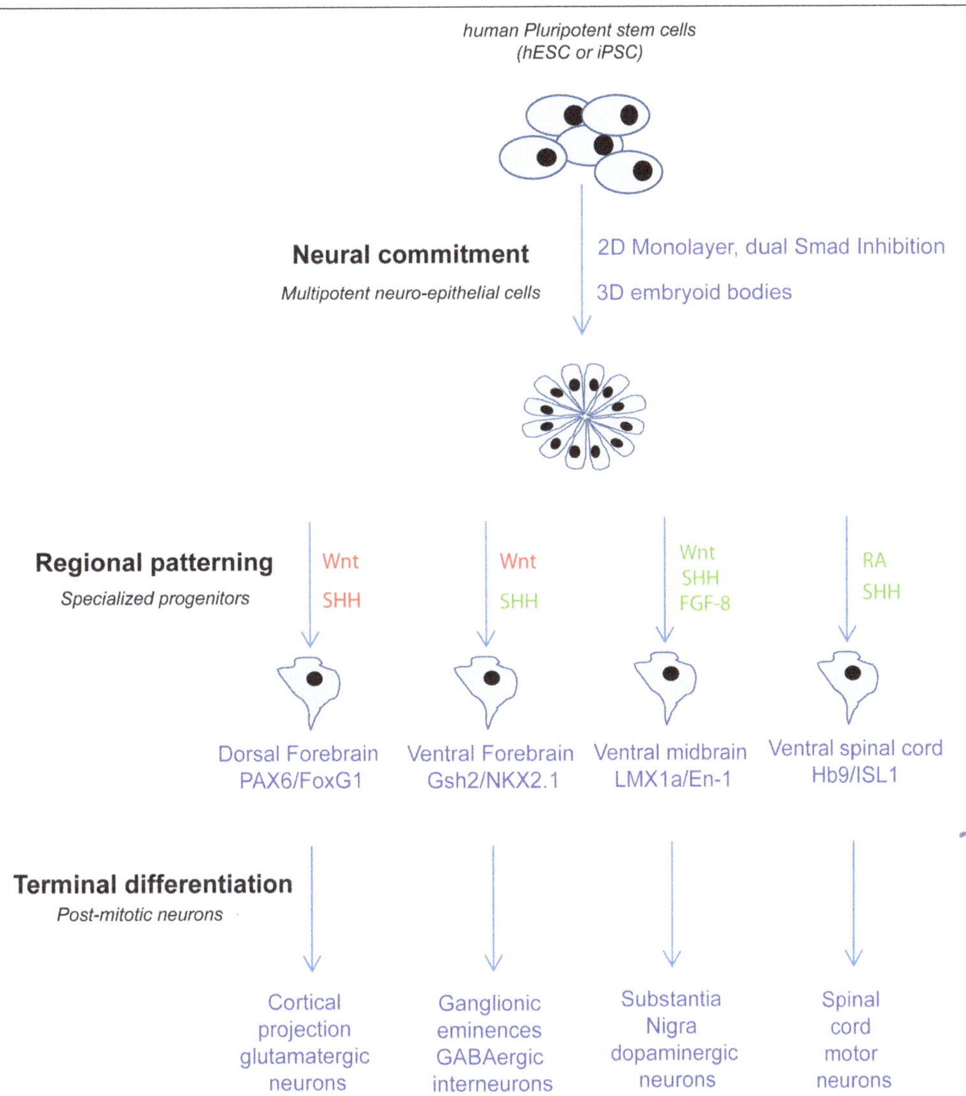

FIGURE 1 | Schematic depicting the different strategies used to produce different neuronal sub-types from PSC. Related pathway activations are in green and inhibitions are in red. SHH, sonic hedgehog; FGF-8, fibroblast growth factors 8; RA, retinoic acid.

post-mitotic neurons. This active phase of self-renewal is necessary to constitute then maintain a large pool of progenitors. In addition, for many brain structures, waves of terminal differentiation occur at different time points allowing the genesis of different sub-types of neurons from the same starting progenitors (McConnell, 1995). This is particularly illustrated by the formation of the mammal neocortex layers. Glutamatergic projection neurons of the neocortex organize as six layers in the human brain. All layers are formed from the same pool of progenitors located in the cortical sub-ventricular zone. However, early born neurons will form the deeper layers (V–VI) whereas later born neurons will constitute the upper layers (II–IV; Rakic, 1974; Frantz and McConnell, 1996). Studying the regulation of this balance between self-renewal and terminal differentiation is therefore particularly meaningful to improve our ability to grow PSC-derived neural progenitors *in vitro* in order to control the production of relevant neuronal sub-types but also because subtle disturbance of this balance may be involved in the genesis of many neuro-developmental diseases.

Pluripotent stem cell-derived self-renewing neural progenitors can be obtained by placing the early neural rosettes, which contain neuro-epithelial cells, in culture media supplemented with mitogenic factors. When cultivated in presence of Notch ligands and SHH, early neuro-epithelial cells retain both the morphological organization (epithelial structures) and molecular signature of naïve neuro-epithelial cells (Elkabetz et al., 2008). These rosette-derived neural stem cells (R-NSC) can be maintained upon several rounds of symmetric division and produce different sub-types of neurons after a short exposure to relevant patterning molecules. Successful amplification of neuro-epithelial was also achieved using FGF-2 alone or in combination with epidermal growth factor (Delaloy et al., 2010; Falk et al., 2012) with cells responding at least in part to patterning molecules. It remains to establish whether the molecules used to amplify neural progenitors by the mean of self-renewal are relevant to the physiological situation. However, upon mitogens withdrawal, these neural progenitors quickly exit the proliferative compartment to engage the final program of differentiation as post-mitotic neurons. It is therefore possible to screen for miRNAs involved in the maintenance of the proliferative state or, in contrast, in the decision of terminal differentiation.

SYNAPTOGENESIS AND FUNCTIONAL NETWORK

To be considered as fully functional, neurons have to form electrically active synapses and organize as complex neuronal networks. Formation of electrically functional synapses has been recorded in most of PSC-derived neuronal sub-types. Mixed population of forebrain neurons (which include GABAergic and glutamatergic neurons) can self-organize as a network forming functional synapses (Kim et al., 2011b). When the differentiation is directed to form cortical pyramidal neurons, formation of glutamatergic synapses can be monitored by measuring the assembly of cellular contact where presynaptic proteins are co-localized with PSD-95, a protein of the post-synaptic densities specific of glutamatergic synapses. Electrophysiology experiments showed neuronal responses to glutamate challenges. The recording of spontaneous action potentials indicated that these pyramidal neurons organized as complex autonomous networks (Shi et al., 2012). Telencephalic GABAergic interneurons, with a striatal or a cortical identity, also form spontaneously vesicular GABA transporter (VGAT)/gephyrin-expressing GABAergic synapses that are electrically active (Carri et al., 2013; Maroof et al., 2013). PSC-derived mesencephalic neurons efficiently release dopamine and exhibit spontaneous, network-mediated, electrical activity (Kriks et al., 2011). Finally, several studies reported that PSC-derived spinal cord motor neurons are able to form functional neuromuscular junctions when co-cultivated with myotubes (Umbach et al., 2012). In all these culture systems, synapse formation and activity can easily be monitored both morphologically, following clustering and co-expression of specific molecular markers, and functionally, measuring calcium entry in response to pharmacological stimulations or by the mean of electrophysiological recordings. This opens the path to functional studies of miRNA involvement in the very early steps of synaptogenesis as well as in modulating synaptic activity.

PLURIPOTENT STEM CELLS TO STUDY miRNA FUNCTION IN PHYSIOLOGICAL NEUROGENESIS

So far, hPSC and their neural progeny have contributed to increase our knowledge regarding the involvement of miRNAs in two neurogenesis steps that are hardly accessible *in vivo*: the very early stage of engagement into the neural lineage and the governance of self-renewal/migration/neuronal differentiation of neural progenitors (**Figure 2**).

miRNA REGULATING THE COMMITMENT INTO THE NEURAL

At least two miRNAs have been demonstrated to promote the conversion of PSCs into neuro-epithelial cells, miR-125 and miR-145. Xu et al. (2009) used the EBs model to identify miRNAs regulating hESC differentiation. Using a whole genome approach based on Taqman qPCR, they showed that miR-145 expression quickly increased as hESC enter the differentiation process. Forced expression of miR-145 favored the differentiation along the ectodermal lineage including the neural fate. In contrast, when miR-145 activity was blocked using locked nucleic acid (LNA)-antimiR oligos, endodermal differentiation occurred. The authors identified Oct-4 and KLF-4 as relevant targets of miR-145. miR-125 was also reported as a key regulator of hESC neural conversion. We induced the neural conversion of hESCs by treating them with the two Smad inhibitors: Noggin and SB431542 (Boissart et al., 2012). Using this model, the kinetics of activation of three brain-expressed miRNAs, miR-9, miR-124, and miR-125, was analyzed over time. Only miR-125 was found to be activated in a time window compatible with a role in the neural commitment decision. Functional studies confirmed that miR-125 activity was necessary to fully achieve an efficient engagement of hESC into the neural lineage by both promoting hESC differentiation and blocking alternative, non-neural fate choices. Silencing by miR-125 of Smad-4, the key co-factor of activin- and BMP-dependent Smad pathways, was central to its role in the promotion on neural commitment.

Next to miRNAs actively promoting neural conversion are miRNAs that block this critical decision in order to maintain the cells in a pluripotent state or foster alternative fates. miRNAs of the miR-302/miR-367 family are particularly enriched in PSC. miR-302/miR-367 target several endogenous inhibitors of BMP and

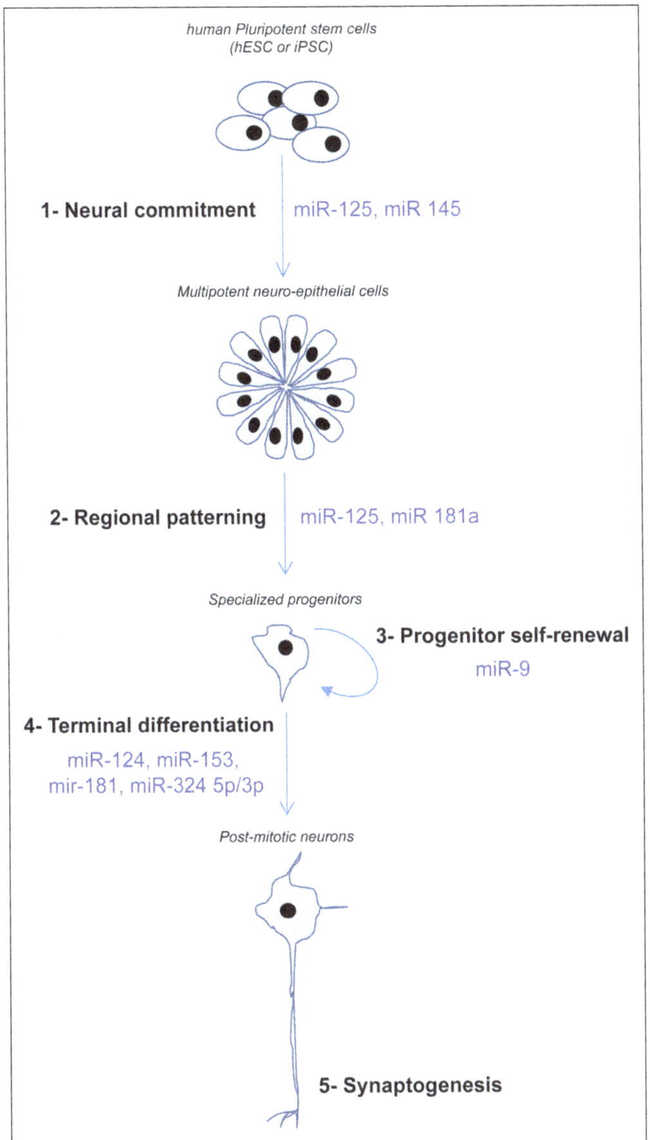

FIGURE 2 | Schematic depicting the different steps of neurogenesis recapitulated *in vitro* using hPSC and the miRNA demonstrated as regulators of these steps.

activin pathways including Lefty, TOB2, DAZAP2, and SLAIN1. Consequently, Smad activity is elevated in cells expressing these miRNAs which render neural conversion impossible (Lipchina et al., 2011). More recently, miR-302 was shown to directly target NR2F2, a transcription factor involved in the very early triggering of the neural genetic program, suggesting that miR-302 may more specifically avoid a spontaneous commitment of PSCs into the neural lineage (Rosa and Brivanlou, 2011). Next to miR-302 family members, the miR-371/miR-372/miR-373 cluster is also considered as a potent inhibitor of the neural lineage commitment (Kim et al., 2011a). Expression of miR-371 is induced by the pluripotency-associated transcription factor KLF-4. PSC lines exhibiting high levels of endogenous miR-371 showed an altered neurogenic potential. Efficient neural conversion was restored upon miR-371 inhibition using LNA-antagomiR oligos demonstrating a causal role of miR-371 in repressing the neural fate. Interestingly, miR-371 activity did not compromise differentiation into other lineages but seemed rather specific of the neural lineage as it controlled the sensitivity of the PSC response to BMPs signal.

Taken together, the identification of miRNAs regulating the efficiency of PSCs neural conversion has highlighted the importance of the fine tuning of Smad-dependent pathways. miR-125, miR-302, and miR-371 both target proteins involved directly in signaling mediated by receptors of the TGF-beta family and modulate finely the strength of the signal transduction. miRNAs promoting the neural conversion target directly the Smad proteins whereas miRNAs favoring alternative fates contribute to secure the activation of these pathways by targeting their endogenous inhibitors. The hPSC model of neural differentiation has therefore contributed to illustrate how subtle the decision of lineage commitment is regulated.

miRNA REGULATING PROLIFERATION AND DIFFERENTIATION OF NEURAL PROGENITORS

A crucial role of miRNAs in regulating the pool of neural progenitors has been well established in rodent models (Volvert et al., 2012). The brain-enriched miR-9 is mainly expressed in neurogenic areas during development suggesting its involvement in molecular mechanisms regulating self-renewal and proliferation of neural progenitors (Deo et al., 2006). Accordingly, miR-9 has been found highly expressed in self-renewing progenitors stably established from the multipotent and immature hESC-derived neuro-epithelial cells (Delaloy et al., 2010; Boissart et al., 2012). In hESC-derived progenitors amplified as neurospheres using FGF-2, miR-9 plays a crucial role in the maintenance of the capacity of proliferation and migration of these progenitors (Delaloy et al., 2010). Interestingly, the authors also assessed the role of miR-9 in the transition between immature multipotent neuro-epithelial cells and the fate restricted mature neural progenitors. They showed that miR-9 activity was essential to this maturation step by directly targeting the cytosolic protein Stathmin.

In contrast to the proliferation-promoting action of miR-9, several miRNAs have been identified as regulators of the decision of terminal neuronal differentiation. The neuronal-specific miR-124 was found enriched in culture of hPSC-derived post-mitotic neurons but not in the proliferative neural progenitors from which they were differentiated (Delaloy et al., 2010; Stappert et al., 2013). Functional studies showed that its forced expression increased the rate of neuronal differentiation whereas blocking its activity resulted in an impaired neuronal production (Stappert et al., 2013).

Next to the study of miRNAs already described as brain-specific, a differential, whole genome, miRNAs profiling was performed comparing self-renewing hESC-derived multipotent neuro-epithelial stem cells (lt-NES) to their neuronal progeny (Stappert et al., 2013). This extensive profiling pointed to additional miRNAs enriched in differentiated neurons, miR-125b, miR-153, miR-181a/181a*, and the cluster miR-324-5p/3p. Ectopic expression of miR-153, miR-181a/181a*, and miR-324-5p/3p shifted lt-NES cells from self-renewal to neuronal differentiation. lt-NES represent ventral hindbrain precursors that mainly give rise of GABAergic interneurons. However, they can also produce small amount of tyrosine hydroxylase (TH) neurons, an enzyme converting L-tyrosine into L-DOPA, considered a marker of catecholaminergic neurons. The authors asked whether, next to promoting neuronal differentiation, these miRNA can also influence the neuronal fate. miR-125b and miR-181a both increased the number of TH positive neurons. In contrast, miR-181a* inhibited the formation of TH neurons and promoted the production of GABAergic neuronal cells.

These pioneer studies illustrate how hPSC can greatly improve our knowledge about miRNAs involvement in physiological neurogenesis by helping elucidating the functional impact of miRNA activities and identifying their targets.

DECIPHERING THE IMPACT OF miRNAs IN NEURO-DEVELOPMENTAL DISEASES USING PATIENT-DERIVED iPSCs

THE iPSC BREAKTHROUGH

In 2007, Yamanaka and colleagues made the breakthrough discovery of a simple method to reprogram human somatic adult cells into fully PSCs, a type of cells now widely known as iPSCs (Takahashi et al., 2007). Since iPSC can be derived from virtually all nucleated cell types of the body, it suddenly removed the barriers raised until this date by the use of PSC derived from embryos that limited investigations to cells with unknown clinical status (the so-called wild-type cells) or to lines carrying mutations of the few diseases eligible for a pre-implantation genetic diagnosis (PGD). PSC induced from somatic cells offer the unlimited possibility to model neurogenesis from any patient including those for which the cause of the disease is not fully identified – e.g., multifactorial disorders – but with a well-documented clinical characterization (Bellin et al., 2012).

Although promising, the field of iPSC is still considered "at work" and questions remain regarding the accuracy of modeling disease with a strong epigenetic origin using genetically reprogramed cells. Indeed, concerns have been raised about the differences in genes expression, including miRNA profiles, between iPSC and hESC, suggesting that, next to the disease context, iPSC behavior may also be influenced by the technique and efficiency of reprograming as well as by the cell type from which the iPSC line was produced (Chin et al., 2009; Marchetto et al., 2009; Vitale et al., 2012). Consequently, iPSC may model neuro-developmental diseases with strong epigenetic components differently than hESC (Urbach et al., 2010). However, some evidences indicate that certain epigenetic marks are conserved with reprograming, including parental imprinting (Chamberlain et al., 2010; Yang et al., 2010). To date, several iPSC lines have been derived from patients with various neuro-developmental disorders including Rett's syndrome (RS; Kim et al., 2011c), fragile X syndrome (FXS; Urbach et al., 2010), Down's syndrome (Briggs et al., 2013; Weick et al., 2013), Timothy's syndrome (Pasca et al., 2011), Angelman's syndrome (Chamberlain et al., 2010), Prader–Willi's syndrome (Yang et al., 2010), and Schizophrenia (Brennand et al., 2011). Some of those have been shown to recapitulate *in vitro* important features of the diseases which make them attractive tools to further study mechanisms leading to pathological phenotypes including the influence of miRNAs already described as dysregulated in brains of patients or in animal models (**Table 1**).

MONOGENIC SYNDROMES OF AUTISM SPECTRUM DISORDERS AND MENTAL RETARDATION

Micro-RNAs, as fine regulators of protein translation, influence directly the level of gene expression. A central role of synaptic gene dosage in the emergence of ASDs and mental retardation (MR) is now well established (Toro et al., 2010), suggesting that miRNAs studies in iPSC models may bring some light into the dark areas of these early onset neuro-developmental disorders. More particularly, understanding the link between genes responsible for monogenic forms of ASD/MR and the miRNA machinery may help understand why patients with the same genetic mutation in coding sequences can develop differentially severe symptoms. So far, iPSC have been successfully derived from individuals with two monogenic forms of ASD/MR: the RS and the FXS. RS is

Table 1 | Summary of iPSC lines in which the role of miRNAs dysregulated in animal models or human brains can be further investigated.

Disease	Origin	References of iPSC lines	Phenotype of iPSC-derived neurons	miRNAs of interest
Fragile X syndrome	Loss of function of FMRP (FMR1 gene)	Urbach et al. (2010), Sheridan et al. (2011)	Hyper-excitability of glutamatergic synapses	DICER and AGO-1 complexes
Rett's syndrome	Loss of function of MeCP2 transcriptional repressor	Marchetto et al. (2010), Kim et al. (2011c), Cheung et al. (2012)	Decreased soma size, neurite atrophy, decreased efficiency of glutamatergic synapses	miR-132, miR-184, miR-483-5p, miR-212
Schizophrenia	Multifactorial	Urbach et al. (2010); Brennand et al. (2011), Paulsen Bda et al. (2012), Robicsek et al. (2013)	Diminished neuronal connectivity	miR-17-5p, miR-34a, miR-107, miR-122, miR-132, miR-134, miR-137
Down's syndrome	Additional copy of chromosome 21	Briggs et al. (2013), Weick et al. (2013)	Reduced synaptic activity, increased sensitivity to oxidative stress	miR-99a, miR-125b, miR-155, miR-802, Ret 7c

an inherited neuro-developmental disorder with an X-linked gene inheritance mainly affecting women. Most forms of RS are due to a loss of function mutation in the transcriptional repressor MeCP2 (methyl CpG binding protein 2; Ravn et al., 2011). RS iPSC replicate some prototypical features found in animal models including decreased neuronal soma size, neuritic atrophy and decreased efficiency of glutamatergic synapses (Marchetto et al., 2010; Kim et al., 2011c; Cheung et al., 2012). Disruption of MeCP2 gene in mice leads to the dysregulation of a set of miRNA potentially of influence in neurogenesis including miR-132, miR-184, miR-483-5p, and miR-212 (Nomura et al., 2008; Im et al., 2010; Urdinguio et al., 2010; Han et al., 2013). The role of each miRNA in the development of typical RS neuronal features can be addressed using iPSC-derived neurons. Similarly, several iPSC lines have been derived from individuals with the FXS. In FXS, abnormal expansion of a CGG triplet in the 5′UTR of the FMR1 gene leads to the defective translation of the FMR1 gene and to the loss of the resulting protein fragile X mental retardation protein (FMRP; Wijetunge et al., 2013). FXS iPSC lines-derived neurons recapitulate the typical hyper-excitability of glutamatergic synapses and developmental defects described in animal models (Urbach et al., 2010; Sheridan et al., 2011). FMRP has been shown to directly control miRNA biogenesis through direct interaction with DICER and AGO-1 complexes (Jin et al., 2004). Of interest, functional investigations of miRNAs regulated by MeCP2 or FMRP proteins in iPSC-derived neurogenesis models may help address whether the dysregulation of these miRNAs have a real impact on the strength of the pathologic phenotypes and whether this can be reversed.

MULTIFACTORIAL SYNDROMES

One of the main interests of recapitulating neurogenesis with patient-derived iPSC is the opportunity to address the role of miRNAs in a human genetic background permissive to the development of multifactorial diseases such as schizophrenia or Down's syndrome. Although psychiatric disorders, such as schizophrenia, affect several brain regions and produce a complex array of clinical symptoms, basic phenotypes likely exist at the level of single neurons and simple networks. Being highly heritable, it is hypothesized that these disorders are amenable to cell-based studies *in vitro* (Brennand et al., 2012). Accordingly, the human-induced PSC (hiPSC) technology makes it possible to study schizophrenia and other psychiatric disorders using live human neurons with a genetic predisposition without knowledge of the genes interacting to produce the disease state. Genome-wide profiling has listed a number of changes in miRNAs expression levels in the brain of patients with a diagnosis of schizophrenia (Kim et al., 2010; Moreau et al., 2011; Santarelli et al., 2011). These include miR-17-5p, miR-34a, miR-107, miR-122, the brain-specific miR-132, the synaptic miR-134, miR-185, miR-382, and miR-652. Interestingly, a single-nucleotide polymorphism in miR-137, a miRNA previously reported as a regulator of neuronal maturation, was consistently found to be one of the common alleles associated with a high risk of developing schizophrenia (Whalley et al., 2012). To date, several groups have obtained iPSC from individuals diagnosed with schizophrenia (Urbach et al., 2010; Brennand et al., 2011; Paulsen Bda et al., 2012; Robicsek et al., 2013) and have described impaired neurogenesis in these lines opening the path to further investigations regarding miRNA active participation to the initiation and the progression of the disease.

Another multifactorial disease for which hiPSC would be a valuable study tool is Down's syndrome, also known as trisomy 21. In the human, Down's syndrome sums MR, craniofacial morphological abnormalities and heart failure due to an additional copy of the long arm of chromosome 21 (Hsa21). Hsa21 contains approximately 552 genes, 166 of which are orthologous to genes localized in syntenic regions of three mouse chromosomes: Mmu16 (110 orthologous genes), Mmu17 (19 orthologous genes), and Mmu10 (37 orthologous genes; Rueda et al., 2012). Based on these homologies, several mouse models that are trisomic for different sets of Hsa21 genes have been developed but failed to properly recapitulate the complete neurological symptoms of Down's syndrome, suggesting that Hsa21 might contain additional elements that are human-specific. Hsa21 has been predicted to contain at least five nc-RNAs, miR-99a, miR-125b, miR-155, miR-802, and Ret-7c (Kuhn et al., 2008). The recently published trisomic iPSC lines (Briggs et al., 2013; Weick et al., 2013) will probably help addressing the question of the importance of gene dosage, including miRNA, in the development of the neurological features of Down's syndrome.

CONCLUSION

By their ability to recapitulate human neurogenesis *in vitro* and their flexibility regarding genetic manipulation, hPSC have revealed valuable tools to help deciphering the role of miRNAs in the earlier events of human neurogenesis.

So far, hESC remain the "gold standard" to faithfully investigate the functional consequences of miRNA activity on different steps of human neurogenesis since they still represent the closest cellular model to the physiological situation. However, the iPSC have revolutionized the field of PSCs used as models of neurogenesis in two ways. Firstly, their somatic origin and the relative simplicity of the reprograming process has dramatically expanded the use of PSCs in general by removing the ethical constraint linked to embryo destruction. Secondly, iPSC derived from individuals with a known clinical profile represent real "patients in a dish" and offer for the first time the opportunity to address the role of miRNAs in the etiology of complex neuro-developmental disorders using patient-derived neurons. Many questions regarding the iPSC system remain to be answered and more particularly whether the reprograming step can actually compromise the proper modeling of a disease by erasing epigenetic signatures including the genuine miRNA expression profile.

While still in its infancy, the hPSC field has already demonstrated its usefulness to elucidate the function of miRNA in critical aspect of human neurogenesis. It should reveal its full potential in the coming years to become a standard that will complement studies performed in animal models.

ACKNOWLEDGMENTS

Authors want to thank members of the I-STEM institute for their constant support and expertise.

REFERENCES

Andersson, E., Tryggvason, U., Deng, Q., Friling, S., Alekseenko, Z., Robert, B., et al. (2006). Identification of intrinsic determinants of midbrain dopamine neurons. *Cell* 124, 393–405. doi: 10.1016/j.cell.2005.10.037

Bellin, M., Marchetto, M. C., Gage, F. H., and Mummery, C. L. (2012). Induced pluripotent stem cells: the new patient? *Nat. Rev. Mol. Cell Biol.* 13, 713–726. doi: 10.1038/nrm3448

Boissart, C., Nissan, X., Giraud-Triboult, K., Peschanski, M., and Benchoua, A. (2012). miR-125 potentiates early neural specification of human embryonic stem cells. *Development* 139, 1247–1257. doi: 10.1242/dev.073627

Brennand, K. J., Simone, A., Jou, J., Gelboin-Burkhart, C., Tran, N., Sangar, S., et al. (2011). Modelling schizophrenia using human induced pluripotent stem cells. *Nature* 473, 221–225. doi: 10.1038/nature09915

Brennand, K. J., Simone, A., Tran, N., and Gage, F. H. (2012). Modeling psychiatric disorders at the cellular and network levels. *Mol. Psychiatry* 17, 1239–1253. doi: 10.1038/mp.2012.20

Briggs, J. A., Sun, J., Shepherd, J., Ovchinnikov, D. A., Chung, T. L., Nayler, S. P., et al. (2013). Integration-free induced pluripotent stem cells model genetic and neural developmental features of Down syndrome etiology. *Stem Cells* 31, 467–478. doi: 10.1002/stem.1297

Carri, A. D., Onorati, M., Lelos, M. J., Castiglioni, V., Faedo, A., Menon, R., et al. (2013). Developmentally coordinated extrinsic signals drive human pluripotent stem cell differentiation toward authentic DARPP-32+ medium-sized spiny neurons. *Development* 140, 301–312. doi: 10.1242/dev.084608

Chamberlain, S. J., Chen, P. F., Ng, K. Y., Bourgois-Rocha, F., Lemtiri-Chlieh, F., Levine, E. S., et al. (2010). Induced pluripotent stem cell models of the genomic imprinting disorders Angelman and Prader–Willi syndromes. *Proc. Natl. Acad. Sci. U.S.A.* 107, 17668–17673. doi: 10.1073/pnas.1004487107

Chamberlain, S. J., Li, X. J., and Lalande, M. (2008). Induced pluripotent stem (iPS) cells as *in vitro* models of human neurogenetic disorders. *Neurogenetics* 9, 227–235. doi: 10.1007/s10048-008-0147-z

Chambers, S. M., Fasano, C. A., Papapetrou, E. P., Tomishima, M., Sadelain, M., and Studer, L. (2009). Highly efficient neural conversion of human ES and iPS cells by dual inhibition of SMAD signaling. *Nat. Biotechnol.* 27, 275–280. doi: 10.1038/nbt.1529

Cheung, A. Y., Horvath, L. M., Carrel, L., and Ellis, J. (2012). X-chromosome inactivation in Rett syndrome human induced pluripotent stem cells. *Front. Psychiatry* 3:24. doi: 10.3389/fpsyt.2012.00024

Chin, M. H., Mason, M. J., Xie, W., Volinia, S., Singer, M., Peterson, C., et al. (2009). Induced pluripotent stem cells and embryonic stem cells are distinguished by gene expression signatures. *Cell Stem Cell* 5, 111–123. doi: 10.1016/j.stem.2009.06.008

Chipman, P. H., Toma, J. S., and Rafuse, V. F. (2012). Generation of motor neurons from pluripotent stem cells. *Prog. Brain Res.* 201, 313–331. doi: 10.1016/B978-0-444-59544-7.00015-9

Crook, J. M., and Kobayashi, N. R. (2008). Human stem cells for modeling neurological disorders: accelerating the drug discovery pipeline. *J. Cell. Biochem.* 105, 1361–1366. doi: 10.1002/jcb.21967

de Graaf-Peters, V. B., and Hadders-Algra, M. (2006). Ontogeny of the human central nervous system: what is happening when? *Early Hum. Dev.* 82, 257–266. doi: 10.1016/j.earlhumdev.2005.10.013

Delaloy, C., Liu, L., Lee, J. A., Su, H., Shen, F., Yang, G. Y., et al. (2010). MicroRNA-9 coordinates proliferation and migration of human embryonic stem cell-derived neural progenitors. *Cell Stem Cell* 6, 323–335. doi: 10.1016/j.stem.2010.02.015

Deo, M., Yu, J. Y., Chung, K. H., Tippens, M., and Turner, D. L. (2006). Detection of mammalian microRNA expression by in situ hybridization with RNA oligonucleotides. *Dev. Dyn.* 235, 2538–2548. doi: 10.1002/dvdy.20847

Elkabetz, Y., Panagiotakos, G., Al Shamy, G., Socci, N. D., Tabar, V., and Studer, L. (2008). Human ES cell-derived neural rosettes reveal a functionally distinct early neural stem cell stage. *Genes Dev.* 22, 152–165. doi: 10.1101/gad.1616208

Falk, A., Koch, P., Kesavan, J., Takashima, Y., Ladewig, J., Alexander, M., et al. (2012). Capture of neuroepithelial-like stem cells from pluripotent stem cells provides a versatile system for *in vitro* production of human neurons. *PLoS ONE* 7:e29597. doi: 10.1371/journal.pone.0029597

Frantz, G. D., and McConnell, S. K. (1996). Restriction of late cerebral cortical progenitors to an upper-layer fate. *Neuron* 17, 55–61. doi: 10.1016/S0896-6273(00)80280-9

Friling, S., Andersson, E., Thompson, L. H., Jonsson, M. E., Hebsgaard, J. B., Nanou, E., et al. (2009). Efficient production of mesencephalic dopamine neurons by Lmx1a expression in embryonic stem cells. *Proc. Natl. Acad. Sci. U.S.A.* 106, 7613–7618. doi: 10.1073/pnas.0902396106

Gokhale, P. J., and Andrews, P. W. (2012). The development of pluripotent stem cells. *Curr. Opin. Genet. Dev.* 22, 403–408. doi: 10.1016/j.gde.2012.07.006

Han, K., Gennarino, V. A., Lee, Y., Pang, K., Hashimoto-Torii, K., Choufani, S., et al. (2013). Human-specific regulation of MeCP2 levels in fetal brains by microRNA miR-483-5p. *Genes Dev.* 27, 485–490. doi: 10.1101/gad.207456.112

Im, H. I., Hollander, J. A., Bali, P., and Kenny, P. J. (2010). MeCP2 controls BDNF expression and cocaine intake through homeostatic interactions with microRNA-212. *Nat. Neurosci.* 13, 1120–1127. doi: 10.1038/nn.2615

Jin, P., Zarnescu, D. C., Ceman, S., Nakamoto, M., Mowrey, J., Jongens, T. A., et al. (2004). Biochemical and genetic interaction between the fragile X mental retardation protein and the microRNA pathway. *Nat. Neurosci.* 7, 113–117. doi: 10.1038/nn1174

Kim, A. H., Reimers, M., Maher, B., Williamson, V., Mcmichael, O., Mcclay, J. L., et al. (2010). MicroRNA expression profiling in the prefrontal cortex of individuals affected with schizophrenia and bipolar disorders. *Schizophr. Res.* 124, 183–191. doi: 10.1016/j.schres.2010.07.002

Kim, H., Lee, G., Ganat, Y., Papapetrou, E. P., Lipchina, I., Socci, N. D., et al. (2011a). miR-371-3 expression predicts neural differentiation propensity in human pluripotent stem cells. *Cell Stem Cell* 8, 695–706. doi: 10.1016/j.stem.2011.04.002

Kim, J. E., O'Sullivan, M. L., Sanchez, C. A., Hwang, M., Israel, M. A., Brennand, K., et al. (2011b). Investigating synapse formation and function using human pluripotent stem cell-derived neurons. *Proc. Natl. Acad. Sci. U.S.A.* 108, 3005–3010. doi: 10.1073/pnas.1007753108

Kim, K. Y., Hysolli, E., and Park, I. H. (2011c). Neuronal maturation defect in induced pluripotent stem cells from patients with Rett syndrome. *Proc. Natl. Acad. Sci. U.S.A.* 108, 14169–14174. doi: 10.1073/pnas.1018979108

Kirkeby, A., Grealish, S., Wolf, D. A., Nelander, J., Wood, J., Lundblad, M., et al. (2011). Generation of regionally specified neural progenitors and functional neurons from human embryonic stem cells under defined conditions. *Cell Rep.* 1, 703–714. doi: 10.1016/j.celrep.2012.04.009

Kriks, S., Shim, J. W., Piao, J., Ganat, Y. M., Wakeman, D. R., Xie, Z., et al. (2011). Dopamine neurons derived from human ES cells efficiently engraft in animal models of Parkinson's disease. *Nature* 480, 547–551. doi: 10.1038/nature10648

Kuhn, D. E., Nuovo, G. J., Martin, M. M., Malana, G. E., Pleister, A. P., Jiang, J., et al. (2008). Human chromosome 21-derived miRNAs are overexpressed in Down syndrome brains and hearts. *Biochem. Biophys. Res. Commun.* 370, 473–477. doi: 10.1016/j.bbrc.2008.03.120

Li, X. J., Zhang, X., Johnson, M. A., Wang, Z. B., Lavaute, T., and Zhang, S. C. (2009). Coordination of sonic hedgehog and Wnt signaling determines ventral and dorsal telencephalic neuron types from human embryonic stem cells. *Development* 136, 4055–4063. doi: 10.1242/dev.036624

Lipchina, I., Elkabetz, Y., Hafner, M., Sheridan, R., Mihailovic, A., Tuschl, T., et al. (2011). Genome-wide identification of microRNA targets in human ES cells reveals a role for miR-302 in modulating BMP response. *Genes Dev.* 25, 2173–2186. doi: 10.1101/gad.17221311

Lupo, G., Harris, W. A., and Lewis, K. E. (2006). Mechanisms of ventral patterning in the vertebrate nervous system. *Nat. Rev. Neurosci.* 7, 103–114. doi: 10.1038/nrn1843

Marchetto, M. C., Carromeu, C., Acab, A., Yu, D., Yeo, G. W., Mu, Y., et al. (2010). A model for neural development and treatment of Rett syndrome using human induced pluripotent stem cells. *Cell* 143, 527–539. doi: 10.1016/j.cell.2010.10.016

Marchetto, M. C., Yeo, G. W., Kainohana, O., Marsala, M., Gage, F. H., and Muotri, A. R. (2009). Transcriptional signature and memory retention of human-induced pluripotent stem cells. *PLoS ONE* 4:e7076. doi: 10.1371/journal.pone.0007076

Maroof, A. M., Keros, S., Tyson, J. A., Ying, S. W., Ganat, Y. M., Merkle, F. T., et al. (2013). Directed differentiation and functional maturation of cortical interneurons from human embryonic stem cells. *Cell Stem Cell* 12, 559–572. doi: 10.1016/j.stem.2013.04.008

McConnell, S. K. (1995). Strategies for the generation of neuronal diversity in the developing central nervous system. *J. Neurosci.* 15, 6987–6998.

Moreau, M. P., Bruse, S. E., David-Rus, R., Buyske, S., and Brzustowicz,

L. M. (2011). Altered microRNA expression profiles in postmortem brain samples from individuals with schizophrenia and bipolar disorder. *Biol. Psychiatry* 69, 188–193. doi: 10.1016/j.biopsych.2010.09.039

Moreno-De-Luca, A., Myers, S. M., Challman, T. D., Moreno-De-Luca, D., Evans, D. W., and Ledbetter, D. H. (2013). Developmental brain dysfunction: revival and expansion of old concepts based on new genetic evidence. *Lancet Neurol.* 12, 406–414. doi: 10.1016/S1474-4422(13)70011-5

Nomura, T., Kimura, M., Horii, T., Morita, S., Soejima, H., Kudo, S., et al. (2008). MeCP2-dependent repression of an imprinted miR-184 released by depolarization. *Hum. Mol. Genet.* 17, 1192–1199. doi: 10.1093/hmg/ddn011

Ono, Y., Nakatani, T., Sakamoto, Y., Mizuhara, E., Minaki, Y., Kumai, M., et al. (2007). Differences in neurogenic potential in floor plate cells along an anteroposterior location: midbrain dopaminergic neurons originate from mesencephalic floor plate cells. *Development* 134, 3213–3225. doi: 10.1242/dev.02879

Pasca, S. P., Portmann, T., Voineagu, I., Yazawa, M., Shcheglovitov, A., Pasca, A. M., et al. (2011). Using iPSC-derived neurons to uncover cellular phenotypes associated with Timothy syndrome. *Nat. Med.* 17, 1657–1662. doi: 10.1038/nm.2576

Paulsen Bda, S., De Moraes Maciel, R., Galina, A., Souza Da Silveira, M., Dos Santos Souza, C., Drummond, H., et al. (2012). Altered oxygen metabolism associated to neurogenesis of induced pluripotent stem cells derived from a schizophrenic patient. *Cell Transplant.* 21, 1547–1559. doi: 10.3727/096368911X600957

Perrier, A. L., Tabar, V., Barberi, T., Rubio, M. E., Bruses, J., Topf, N., et al. (2004). Derivation of midbrain dopamine neurons from human embryonic stem cells. *Proc. Natl. Acad. Sci. U.S.A.* 101, 12543–12548. doi: 10.1073/pnas.0404700101

Pombero, A., and Martinez, S. (2009). Telencephalic morphogenesis during the process of neurulation: an experimental study using quail-chick chimeras. *J. Comp. Neurol.* 512, 784–797. doi: 10.1002/cne.21933

Rakic, P. (1974). Neurons in rhesus monkey visual cortex: systematic relation between time of origin and eventual disposition. *Science* 183, 425–427. doi: 10.1126/science.183.4123.425

Ravn, K., Roende, G., Duno, M., Fuglsang, K., Eiklid, K. L., Tumer, Z., et al. (2011). Two new Rett syndrome families and review of the literature: expanding the knowledge of MECP2 frameshift mutations. *Orphanet J. Rare Dis.* 6, 58. doi: 10.1186/1750-1172-6-58

Rhee, Y. H., Ko, J. Y., Chang, M. Y., Yi, S. H., Kim, D., Kim, C. H., et al. (2011). Protein-based human iPS cells efficiently generate functional dopamine neurons and can treat a rat model of Parkinson disease. *J. Clin. Invest.* 121, 2326–2335. doi: 10.1172/JCI45794

Robicsek, O., Karry, R., Petit, I., Salman-Kesner, N., Muller, F. J., Klein, E., et al. (2013). Abnormal neuronal differentiation and mitochondrial dysfunction in hair follicle-derived induced pluripotent stem cells of schizophrenia patients. *Mol. Psychiatry.* doi: 10.1038/mp.2013.67 [Epub ahead of print].

Rosa, A., and Brivanlou, A. H. (2011). A regulatory circuitry comprised of miR-302 and the transcription factors OCT4 and NR2F2 regulates human embryonic stem cell differentiation. *EMBO J.* 30, 237–248. doi: 10.1038/emboj.2010.319

Rueda, N., Florez, J., and Martinez-Cue, C. (2012). Mouse models of Down syndrome as a tool to unravel the causes of mental disabilities. *Neural Plast.* 2012, 584071. doi: 10.1155/2012/584071

Santarelli, D. M., Beveridge, N. J., Tooney, P. A., and Cairns, M. J. (2011). Upregulation of dicer and microRNA expression in the dorsolateral prefrontal cortex Brodmann area 46 in schizophrenia. *Biol. Psychiatry* 69, 180–187. doi: 10.1016/j.biopsych.2010.09.030

Sheridan, S. D., Theriault, K. M., Reis, S. A., Zhou, F., Madison, J. M., Daheron, L., et al. (2011). Epigenetic characterization of the FMR1 gene and aberrant neurodevelopment in human induced pluripotent stem cell models of fragile X syndrome. *PLoS ONE* 6:e26203. doi: 10.1371/journal.pone.0026203

Shi, Y., Kirwan, P., Smith, J., Robinson, H. P., and Livesey, F. J. (2012). Human cerebral cortex development from pluripotent stem cells to functional excitatory synapses. *Nat. Neurosci.* 15, 477–486, S471. doi: 10.1038/nn.3041

Soula, C., Danesin, C., Kan, P., Grob, M., Poncet, C., and Cochard, P. (2001). Distinct sites of origin of oligodendrocytes and somatic motoneurons in the chick spinal cord: oligodendrocytes arise from Nkx2.2-expressing progenitors by a Shh-dependent mechanism. *Development* 128, 1369–1379.

Stappert, L., Borghese, L., Roese-Koerner, B., Weinhold, S., Koch, P., Terstegge, S., et al. (2013). MicroRNA-based promotion of human neuronal differentiation and subtype specification. *PLoS ONE* 8:e59011. doi: 10.1371/journal.pone.0059011

Sun, A. X., Crabtree, G. R., and Yoo, A. S. (2013). MicroRNAs: regulators of neuronal fate. *Curr. Opin. Cell Biol.* 25, 215–221. doi: 10.1016/j.ceb.2012.12.007

Takahashi, K., Tanabe, K., Ohnuki, M., Narita, M., Ichisaka, T., Tomoda, K., et al. (2007). Induction of pluripotent stem cells from adult human fibroblasts by defined factors. *Cell* 131, 861–872. doi: 10.1016/j.cell.2007.11.019

Takazawa, T., Croft, G. F., Amoroso, M. W., Studer, L., Wichterle, H., and Macdermott, A. B. (2012). Maturation of spinal motor neurons derived from human embryonic stem cells. *PLoS ONE* 7:e40154. doi: 10.1371/journal.pone.0040154

Toro, R., Konyukh, M., Delorme, R., Leblond, C., Chaste, P., Fauchereau, F., et al. (2010). Key role for gene dosage and synaptic homeostasis in autism spectrum disorders. *Trends Genet.* 26, 363–372. doi: 10.1016/j.tig.2010.05.007

Umbach, J. A., Adams, K. L., Gundersen, C. B., and Novitch, B. G. (2012). Functional neuromuscular junctions formed by embryonic stem cell-derived motor neurons. *PLoS ONE* 7:e36049. doi: 10.1371/journal.pone.0036049

Urbach, A., Bar-Nur, O., Daley, G. Q., and Benvenisty, N. (2010). Differential modeling of fragile X syndrome by human embryonic stem cells and induced pluripotent stem cells. *Cell Stem Cell* 6, 407–411. doi: 10.1016/j.stem.2010.04.005

Urdinguio, R. G., Fernandez, A. F., Lopez-Nieva, P., Rossi, S., Huertas, D., Kulis, M., et al. (2010). Disrupted microRNA expression caused by Mecp2 loss in a mouse model of Rett syndrome. *Epigenetics* 5, 656–663. doi: 10.4161/epi.5.7.13055

Vieira, C., Pombero, A., Garcia-Lopez, R., Gimeno, L., Echevarria, D., and Martinez, S. (2010). Molecular mechanisms controlling brain development: an overview of neuroepithelial secondary organizers. *Int. J. Dev. Biol.* 54, 7–20. doi: 10.1387/ijdb.092853cv

Vitale, A. M., Matigian, N. A., Ravishankar, S., Bellette, B., Wood, S. A., Wolvetang, E. J., et al. (2012). Variability in the generation of induced pluripotent stem cells: importance for disease modeling. *Stem Cells Transl. Med.* 1, 641–650. doi: 10.5966/sctm.2012-0043

Volvert, M. L., Rogister, F., Moonen, G., Malgrange, B., and Nguyen, L. (2012). MicroRNAs tune cerebral cortical neurogenesis. *Cell Death Differ.* 19, 1573–1581. doi: 10.1038/cdd.2012.96

Weick, J. P., Held, D. L., Bonadurer, G. F., III, Doers, M. E., Liu, Y., Maguire, C., et al. (2013). Deficits in human trisomy 21 iPSCs and neurons. *Proc. Natl. Acad. Sci. U.S.A.* 110, 9962–9967. doi: 10.1073/pnas.1216575110

Whalley, H. C., Papmeyer, M., Romaniuk, L., Sprooten, E., Johnstone, E. C., Hall, J., et al. (2012). Impact of a microRNA MIR137 susceptibility variant on brain function in people at high genetic risk of schizophrenia or bipolar disorder. *Neuropsychopharmacology* 37, 2720–2729. doi: 10.1038/npp.2012.137

Wijetunge, L. S., Chattarji, S., Wyllie, D. J., and Kind, P. C. (2013). Fragile X syndrome: from targets to treatments. *Neuropharmacology* 68, 83–96. doi: 10.1016/j.neuropharm.2012.11.028

Xu, N., Papagiannakopoulos, T., Pan, G., Thomson, J. A., and Kosik, K. S. (2009). MicroRNA-145 regulates OCT4, SOX2, and KLF4 and represses pluripotency in human embryonic stem cells. *Cell* 137, 647–658. doi: 10.1016/j.cell.2009.02.038

Yang, J., Cai, J., Zhang, Y., Wang, X., Li, W., Xu, J., et al. (2010). Induced pluripotent stem cells can be used to model the genomic imprinting disorder Prader–Willi syndrome. *J. Biol. Chem.* 285, 40303–40311. doi: 10.1074/jbc.M110.183392

Yates, L. A., Norbury, C. J., and Gilbert, R. J. (2013). The long and short of microRNA. *Cell* 153, 516–519. doi: 10.1016/j.cell.2013.04.003

Zeng, H., Guo, M., Martins-Taylor, K., Wang, X., Zhang, Z., Park, J. W., et al. (2010). Specification of region-specific neurons including forebrain glutamatergic neurons from human induced pluripotent stem cells. *PLoS ONE* 5:e11853. doi: 10.1371/journal.pone.0011853

3

microRNA regulation and dysregulation in epilepsy

Danyella B. Dogini, Simoni H. Avansini, Andre S. Vieira and Iscia Lopes-Cendes*

Department of Medical Genetics, School of Medical Sciences, University of Campinas, Campinas, São Paulo, Brazil

Correspondence:
Iscia Lopes-Cendes, Department of Medical Genetics, School of Medical Sciences, University of Campinas, Tessália Vieira de Camargo, 126, Campinas, São Paulo 13083-887, Brazil
e-mail:icendes@unicamp.br

Epilepsy, one of the most frequent neurological disorders, represents a group of diseases that have in common the clinical occurrence of seizures. The pathogenesis of different types of epilepsy involves many important biological pathways; some of which have been shown to be regulated by microRNAs (miRNAs). In this paper, we will critically review relevant studies regarding the role of miRNAs in epilepsy. Overall, the most common type of epilepsy in the adult population is temporal lobe epilepsy (TLE), and the form associated with mesial temporal sclerosis (MTS), called mesial TLE, is particularly relevant due to the high frequency of resistance to clinical treatment. There are several target studies, as well few genome-wide miRNA expression profiling studies reporting abnormal miRNA expression in tissue with MTS, both in patients and in animal models. Overall, these studies show a fine correlation between miRNA regulation/dysregulation and inflammation, seizure-induced neuronal death and other relevant biological pathways. Furthermore, expression of many miRNAs is dynamically regulated during neurogenesis and its dysregulation may play a role in the process of cerebral corticogenesis leading to malformations of cortical development (MCD), which represent one of the major causes of drug-resistant epilepsy. In addition, there are reports of miRNAs involved in cell proliferation, fate specification, and neuronal maturation and these processes are tightly linked to the pathogenesis of MCD. Large-scale analyzes of miRNA expression in animal models with induced status epilepticus have demonstrated changes in a selected group of miRNAs thought to be involved in the regulation of cell death, synaptic reorganization, neuroinflammation, and neural excitability. In addition, knocking-down specific miRNAs in these animals have demonstrated that this may consist in a promising therapeutic intervention.

Keywords: microRNAs, epilepsy, temporal lobe, cortical malformations, animal models

MicroRNAs IN HUMAN MESIAL TEMPORAL LOBE EPILEPSY

Epileptic seizures are the clinical manifestations that reflect a temporary dysfunction of a set of neurons in the brain (Engel, 2001). Epilepsy has a high prevalence in the population, about 1.5–2% and it is considered a public health problem since it has important social and economic impact (Annegers et al., 1996; Borges et al., 2004). Because of its high prevalence and severity, temporal lobe epilepsy (TLE) is one of the most studied types of epilepsy. In TLE complete seizure control with drug treatment is achieved in less than 50% of patients (Sander, 1993; Mattson, 1994). The most common form of TLE is mesial TLE (MTLE), which has the symptoms generated by the involvement of the medial temporal lobe structures (Engel, 2001). Resistance to drug treatment is a crucial problem for patients with MTLE and surgery to remove the affected brain area is, in many cases, a successful therapeutic strategy (Engel, 2001). Surgical specimens in MTLE most frequently show mesial temporal sclerosis (MTS), which is a pathological condition with specific features, including selective neural loss and gliosis in the CA1 hippocampal region (Wieser, 2004). Other changes may include dispersion of the granule cells in the *dentate gyrus*, neurogenesis of granule cell and synaptic reorganization of the mossy fibers (Thom, 2004). Focal lesions and malformations of cortical development (MCD; cortical dysplasia) may represent other findings in patients with drug refractory MTLE (Blumcke et al., 2002; Thom, 2004).

It has been demonstrated that different microRNAs (miRNAs) may have different expression pattern in different brain regions, and these differences in distribution may be related to the preferential concentration of synaptically localized mRNA targeted by these miRNAs (Pichardo-Casas et al., 2012). Furthermore, these differences in concentration could be modulated by epileptogenic activity (Pichardo-Casas et al., 2012). McKiernan et al. (2012a) detected a significant expression of about 200 miRNAs in healthy human hippocampus. However, when working with tissue obtained from patients with MTLE and using TaqMan® low-density arrays (TLDAs) they found a large-scale reduction of miRNA expression, with 51% of miRNAs tested expressed at lower levels than in controls and about 24% not detectable in epileptic tissue. In addition, these authors showed that a possible mechanism involved in failure of mature miRNA expression was a significant decreased expression of DICER, an enzyme required for the generation of mature miRNAs (McKiernan et al., 2012a).

MicroRNA may also have a significant role in inflammation pathways which have been shown to be involved in MTLE (Vezzani et al., 2013). MiR-146a is significantly up-regulated in tissue obtained from patients with MTLE (Aronica et al., 2010; Omran et al., 2012). MiR-146a has been implicated in regulation of astrocyte-mediated inflammatory response (Iyer et al., 2012). In addition, *in vitro* experiments showed a significant up-regulation of miR-146a in astrocytes when exposed to interleukin-1 beta (IL-1b) stimulation, which is known to be up-regulated in the acute phase of some animal models of MTLE (Aronica and Crino, 2011). Another miRNA that has been associated with inflammatory pathways in MTLE is miR-155 (Ashhab et al., 2013). It has been demonstrated an increase in the expression of miR-155 in hippocampal tissue from children with MTLE, as well as in an immature rat epilepsy model. Moreover, the observed increase in miR-155 expression correlates with an increase in TNF-α in the nervous tissue (Ashhab et al., 2013).

It is well known that neuronal death related to seizures involves direct glutamate-driven excitotoxic necrosis. MiR-34a, which belongs to a conserved miRNA family, appears to have a direct pro-apoptotic effect in cells and regulates p53 (Hermeking, 2010). In addition, up-regulation or overexpression of this miR-34a promotes apoptosis in a variety of non-neuronal cell (Chang et al., 2007). Therefore, it has been suggested recently, that miR-34a could represent a key player in the mechanism underlying neuronal death induced by seizures (Hu et al., 2012; Sano et al., 2012).

MicroRNAs may also be involved in enzyme-related epileptic pathology. It is known that adenosine is an endogenous regulator of hippocampal activity and that it has a potent anti-ictogenic and neuroprotective properties (Bjorklund et al., 2008), as well as it is crucial for astrocyte physiology (Boison, 2009). Synaptic levels of adenosine in adult brain are largely regulated by an astrocyte-based adenosine-cycle (Boison, 2009). Adenosine is rapidly phosphorylated by adenosine kinase (ADK), which is almost exclusively expressed in astrocytes (Studer et al., 2006). According to the ADK hypothesis of epileptogenesis (Boison, 2009), any type of brain injury can produce astrogliosis, which leads to the up-regulation of ADK, creating focal adenosine deficiency as a direct cause of seizures. Using lentiviral vectors in human mesenchymal stem cells coexpressing miRNA against ADK transduction, Ren and Boison (2010) found about 80% of ADK down-regulation. These results suggest that miRNAs are important regulators of seizure-induced neuronal death and that these molecules might be used as novel therapeutic targets in the treatment of epilepsy. Some other miRNAs, such as miR-124, miR-134, miR-132, miR-196b (You et al., 2012; Peng et al., 2013) have also been reported to be involved in epilepsy (**Table 1**).

MicroRNAs AND MALFORMATIONS OF CORTICAL DEVELOPMENT

Malformations of cortical development are a frequent cause of medically intractable epilepsy. It has been estimated that 25–40% of drug-resistant epilepsies are caused by MCD (Guerrini et al., 2003). The development of the human cerebral cortex is a dynamic and complex process. These processes are orchestrated by interactions between extracellular and intracellular signaling cues and any disruption of these cellular processes can result in cortical malformations (Sisodiya, 2004; Guillemot et al., 2006; Guerrini et al., 2008; McLoughlin et al., 2012).

Molecular biology and genetic studies have greatly expanded knowledge on cortical neurogenesis so that several disorders of cortical development have been recognized and, for some of them, specific causative genetic defects have been identified (Aronica et al., 2012). Furthermore, recent data support a major role for miRNAs in fine-tuning of signaling pathways that control the concomitant phases of corticogenesis. Supporting this notion, we have previously shown that groups of miRNAs are differentially regulated during normal mouse brain development (Dogini et al., 2008). Small alterations of their expression have been associated with a variety of neurological disorders (Volvert et al., 2012). Nevertheless, few studies have investigated the possible role of miRNAs in the pathogenesis and/or epileptogenesis of MCDs. Therefore, we aim in the next few paragraphs to summarize current knowledge about miRNAs and cerebral corticogenesis (**Figure 1**) and how its dysregulation may play a role in the process leading to MCDs and ultimately to epileptogenesis as seen in some of these lesions (**Table 1**).

MicroRNAs IN NEURONAL AND GLIAL PROLIFERATION AND DIFFERENTIATION

The first step of cortical development is cellular proliferation and differentiation, which takes place between the 5th week and 20th week of gestation (Sidman and Rakic, 1973; Guerrini and Barba, 2010). Microcephaly, tuberous sclerosis, and focal cortical dysplasia (FCD) have been considered to be malformations of these phases. MiR-9, miR-124, miR-137, miR-184, and let-7b were shown to control cell proliferation in the cortex (Krichevsky et al., 2006; Makeyev et al., 2007; Silber et al., 2008; Liu et al., 2010a; Zhao et al., 2010). In addition, loss of miR-9 expression, a brain-specific miRNA, suppresses the proliferation and promotes the migration of human embryonic neural progenitors, cultured *in vitro*, by targeting stathmin, which increases microtubule instability in migrating neuroblasts (Delaloy et al., 2010). In the mouse embryonic brain, miR-9 suppressed TLX expression, resulting in a reduction of neural stem cell proliferation and an acceleration of neural differentiation (Zhao et al., 2009).

The cellular complexity of the cerebral cortex emerges through specification of cortical progenitors into distinct subtypes of neurons and glia that reach cortical layers (Kriegstein and Alvarez-Buylla, 2009). Changes in gene expression underlie the transition from progenitors to neurons (Guillemot et al., 2006). Conditional removal of Dicer in the cortex affects this process. Kawase-Koga et al. (2009) reported that the cerebral cortex of deficient Dicer-mice showed a significant reduction in cortical thickness, caused by a reduction in neural stem cells and neural progenitors with increased apoptosis and impaired neuronal differentiation. In the same way, it has been observed an inability to generate both neurons and glial cells in the embryonic cerebral cortex of a Dicer-null mouse, and that this enzyme plays a role in maintaining the phenotype of neural stem cells during neuronal differentiation (Andersson et al., 2010). Other miRNAs have also been reported as critical for neural differentiation. These include miR-137, miR34a, miR-153, miR-324, and miR-181a (Smrt et al., 2010; Agostini et al., 2011; Stappert et al., 2013).

Focal cortical dysplasia is characterized by a spectrum of abnormalities in the development of the laminar structure of the human

Table 1 | MicroRNAs potentially involved in epilepsy.

MicroRNA	Human studies/ experimental models	Potential role in epilepsy	Reference
miR-124	Human; immature rat	Potential role in mesial temporal lobe epilepsy; control cell proliferation	Makeyev et al. (2007); Peng et al. (2013)
miR-132	Human; mouse kainic acid	Associated to neuronal activation and synaptic plasticity	Vo et al. (2005)
miR-134	Human (*in vitro* experiments); mouse kainic acid	Suppresses evoked seizures; regulates cell migration	Jimenez-Mateos et al. (2012)
miR-137	Human; rat	Regulates cell proliferation; critical for neural differentiation	Krichevsky et al. (2006); Smrt et al. (2010).
miR-146	Human; mouse; rat	Regulation of astrocyte-mediated inflammatory response; neural inflammation	Lukiw et al. (2008); Nakasa et al. (2008), Pauley et al. (2008); Sonkoly et al. (2008), Aronica and Crino (2011); Iyer et al. (2012), Cheng et al. (2013)
miR-153, miR-324, miR-181a	Human; rat	Critical role in neural differentiation	Smrt et al. (2010); Agostini et al. (2011), Stappert et al. (2013)
miR-184	Human; mouse kainic acid	Regulates cell proliferation; neuroprotective effect	Krichevsky et al. (2006); Makeyev et al. (2007), Silber et al. (2008); Liu et al. (2010a), Zhao et al. (2010); McKiernan et al. (2012b)
miR-196b	Human	Associated with the occurrence of seizures	You et al. (2012)
miR-21	Rat pilocarpine	Possible associated with increased neuronal loss following *status epilepticus*	Risbud and Porter (2013)
miR-34a	Human; rat pilocarpine; mouse kainic acid	Involved in seizure-induced neuronal death; critical for neural differentiation	Agostini et al. (2011); Hu et al. (2012), Sano et al. (2012)
miR-9	Human (*in vitro* experiments)	Regulates cell proliferation; promotes cell migration; accelerates neural differentiation	Krichevsky et al. (2006); Delaloy et al. (2010)
let-7b	Human; rat kainic acid	Regulates cell proliferation	Krichevsky et al. (2006); Makeyev et al. (2007), Silber et al. (2008); Liu et al. (2010b), Zhao et al. (2010)

cerebral cortex. Microscopically, FCD is usually associated with cell abnormalities, giant/dysmorphic neurons and balloon cells (Palmini et al., 2004; Guerrini et al., 2008; Sisodiya et al., 2009; Blumcke et al., 2011). As FCDs are the most frequent epileptogenic malformation, susceptible to surgical treatment, it is of great importance to understand the mechanisms underlying epileptogenesis in FCDs (Aronica et al., 2012; Hauptman and Mathern, 2012; Sakakibara et al., 2012). In this context, Iyer et al. (2012) evaluated function of miR-146a in response to pro-inflammatory stimuli and found, by using *in situ* hybridization, increased expression of miR-146a in reactive astrocytes which are abundantly present within the dysplastic cortex in FCD IIb. This observation suggests a role for miR-146a in an astrocyte-mediated mechanism predisposing to seizure in FCDs.

MicroRNAs IN NEURONAL MIGRATION

In humans, neuronal migration occurs from 6th–7th weeks till approximately 20th–24th weeks of gestation (Sidman and Rakic, 1973; Guerrini and Barba, 2010). Abnormalities disrupting neuronal migration result in highly epileptogenic lesions, causing severe neurological impairment, such as those found in periventricular nodular heterotopia, subcortical heterotopias, and lissencephaly (Guerrini and Parrini, 2010). Doublecortin (Dcx) regulates tangential and radial neuron migration and has been implicated in the pathogenesis of lissencephaly and subcortical heterotopias (Reiner et al., 2006). Gaughwin et al. (2011) demonstrate that miR-134 regulates cell migration *in vitro* and down-regulates Dcx protein *in vivo*, thereby attenuating neuronal migration.

Experiments using neural stem cells of embryonic mouse brains suggest that miR-137 triggered premature differentiation and outward migration through regulation of a lysine-specific histone demethylase (LSD1; Sun et al., 2011). Moreover, the transfection of exogenous miR-125b increased migration of neural stem/progenitor cells compared to a control group (Cui et al., 2012).

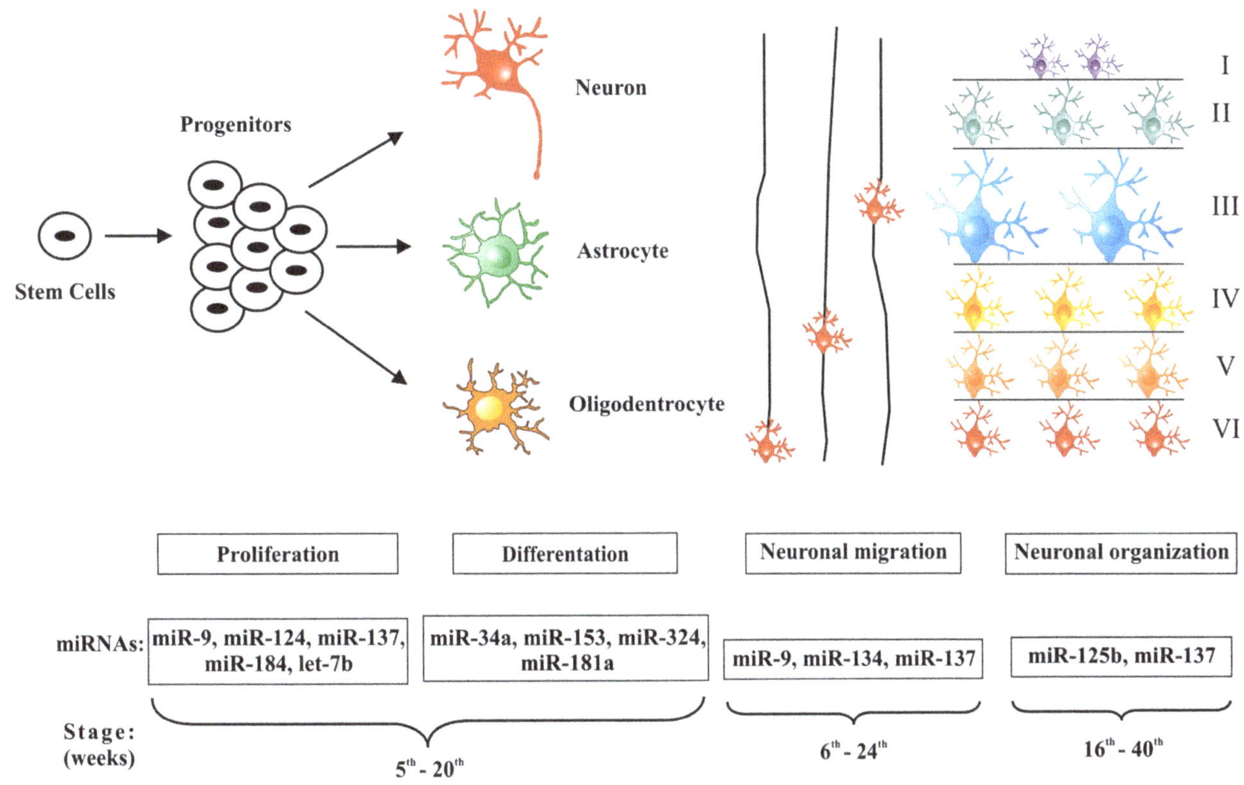

FIGURE 1 | MicroRNAs involved in the regulation of cerebral cortex development. The figure demonstrates microRNAs that have been associated with the three main phases of cortical development. In the first stage, stem cells generate progenitors that are not yet committed to differentiation and can produce neurons, astrocytes, and oligodendrocytes; the concomitant steps of proliferation and differentiation (5th–20th weeks of gestation) are regulated by a set of microRNAs: miR-9, miR-124, miR-137, miR-184, let-7b and miR-34a, miR-153, miR-324, miR-181a. Successive waves of neurons migrate (6th–24th weeks of gestation) from the ventricular regions, along radial glial cells, toward the more external areas of the cortex, these processes are regulated by miR-9, miR-134, and miR-137. Finally, the organization of cortical layers (16th–40th, weeks of gestation) is regulated at this stage through miR-137 and miR-125b.

A mice model constructed with Dicer depletion, by the Nestin-Cre system revealed a critical role for Dicer in cortical migration (McLoughlin et al., 2012). There was a sevenfold increase in Dcx expression that may have contributed to the premature maturation of neurons in inappropriate regions, which in turn may led to complete cortical disorganization (McLoughlin et al., 2012). Shibata et al. (2011) observed, after reduction of miR-9 expression, that cortical layers were reduced and that the tangential migration of interneurons from basal forebrain was impaired.

MicroRNAs IN NEURONAL ORGANIZATION

The third stage in cortical development is cortical organization. When migration is complete, the cortex is a six-layered structure, with each layer containing different types of neurons (Guerrini and Barba, 2010). Polymicrogyria and schizencephaly have been considered to be malformations of this post-migrational cortical organization stage. Two miRNAs have been shown to regulate key processes at this stage. MiR-137 which regulates neuronal maturation by inhibiting dendrite formation through binding Mind bomb 1 (Mibl; Smrt et al., 2010), and miR-125b which seems to have a similar role, since overexpression of miR-125b leads to longer and thinner dendritic spines (Edbauer et al., 2010).

MicroRNAs AND ANIMAL MODELS OF EPILEPSY

Induced animal models are one of the most used tools to study the pathophysiology of different types of epilepsy and they have been most frequently used in MTLE. In these models, animals present behavioral, electroencephalographic, and neuropathological features in the limbic structures similar to those observed in patients with MTLE (Avanzini et al., 1993; Lothman et al., 1995; Engel, 1996).

One of the first miRNAs shown to be differentially expressed in the hippocampus in an induced animal model was miR-132 (Nudelman et al., 2010). These authors observed an increase in the expression of miR-132 in the hippocampus 8 h after the administration of the convulsant drug pilocarpine in mice. In neurons, miR-132 expression is induced by electrical activity and the action of neurotrophins, consequently its proposed role would be the regulation of synaptic plasticity-related genes (Vo et al., 2005; Wayman et al., 2008). Another miRNA that was initially explored in epilepsy experimental models was miR-146a (Aronica et al., 2010). This miRNA can be induced by pro-inflammatory cytokines, such as IL-1b, and it is up-regulated in various human disorders associated with inflammatory response (Lukiw et al., 2008; Nakasa et al., 2008; Pauley et al., 2008; Sonkoly et al., 2008;

Cheng et al., 2013). In a rat model of MTLE induced by repetitive electrical stimulation of the perforant pathway, it was observed that miR-146a was up-regulated in the CA3 hippocampus subfield 1 week (latent phase) and 3 months (chronic phase) after the episode of *status epilepticus*. In these experiments, the observation by *in situ* hybridization of miR-146 expression in hippocampus reactive astrocytes further indicated a possible role for this miRNA in neural inflammation. However, the exact genes regulated by miR-146 in the hippocampus remains to be determined.

Subsequently, with an increasing interest in the possible role of regulatory RNAs in epilepsy, large-scale analyzes of miRNA expression profile by either hybridization or TaqMan® arrays were undertaken in the hippocampus of animals with induced epilepsy (Liu et al., 2010b; Jimenez-Mateos et al., 2011; Song et al., 2011; Hu et al., 2012; McKiernan et al., 2012b; Pichardo-Casas et al., 2012; Peng et al., 2013; Risbud and Porter, 2013). Analyzes were performed on the lithium-pilocarpine model (Song et al., 2011; Hu et al., 2012), systemic pilocarpine (Risbud and Porter, 2013), systemic kainic acid (Liu et al., 2010b; McKiernan et al., 2012b; Pichardo-Casas et al., 2012), intra-amygdala kainic acid (Jimenez-Mateos et al., 2011), with time points ranging from a few hours (McKiernan et al., 2012b) to months after *status epilepticus* (Song et al., 2011; Hu et al., 2012). All studies found a significant number of miRNAs differentially regulated in the epileptic state when compared to control animals, indicating a tight regulation of miRNAs associated with the events observed in induced epilepsy models. Some miRNAs were found to be differentially expressed, such as miR-34a (Hu et al., 2012; Sano et al., 2012) or miR-132 (Nudelman et al., 2010; Jimenez-Mateos et al., 2011). However, a coherent interpretation of the results produced by the above mentioned experiments is hindered by the still incomplete knowledge of miRNAs regulated genes in the hippocampus and by the heterogeneity of findings obtained by different studies.

The apparent lack of reproducibility in the miRNA expression profile experiments may be explained by the diversity in animal models, time points, and even hippocampal structures analyzed. Moreover, miRNA expression was profiled employing microarrays (Song et al., 2011; Hu et al., 2012; Pichardo-Casas et al., 2012; Risbud and Porter, 2013) or TLDAs (Liu et al., 2010b; Eacker et al., 2011; Jimenez-Mateos et al., 2011; McKiernan et al., 2012b). As a consequence, differences on the sensibility and specificity of both techniques may be responsible for part of the diversity observed in the published literature. In addition, a critical point to be considered is that some studies analyzed whole hippocampus homogenates (Liu et al., 2010b; Song et al., 2011; Hu et al., 2012; Pichardo-Casas et al., 2012; Peng et al., 2013; Risbud and Porter, 2013) and others were restricted to the CA3 subfield (Jimenez-Mateos et al., 2011; McKiernan et al., 2012b). It is known that the different hippocampus subfields are molecularly diverse (Lein, 2004; Greene et al., 2009). Therefore, analyzes of whole hippocampus homogenates certainly dilutes subfield-specific changes that may take place in these epilepsy models. Strategies such as laser capture microdissection of different hippocampus subfields could circumvent the exposed shortcomings of whole homogenate strategies, improving the ability of an experiment to detect more subtle and spatially restricted changes in miRNA regulation. Furthermore, since different hippocampus subfields have different functional characteristics, sensibility to neurodegeneration and contributions to the establishment of an epileptic state (Becker et al., 2003; Majores et al., 2004), a separate analyzes of miRNA profile in each structure certainly would facilitate data interpretation. Another point to be considered is that the translation of these animal models miRNA expression findings to human MTLE could be hindered by the fact that many patients do not present an initial precipitating event (Van Paesschen et al., 1997). Moreover the occurrence of an episode of *status epilepticus* is uncommon in human MTLE. Such a diversity of models and analyzes strategies present in the literature poses an advantage, since the differentially regulated miRNAs common to all studies may indicate the presence of a common mechanism underlying the epileptogenic process. However, care should be taken when employing rodent data in the effort of understanding human MTLE miRNA associated mechanisms due to the existence of many primate-specific miRNAs (Bentwich et al., 2005). Therefore some mechanisms may only be found with the direct analysis of tissue from patients that undergo epilepsy surgery.

As already noted, many of the functional implications of the identified differentially expressed miRNAs in the hippocampus of animals with induced epilepsy are still unknown. Antagomirs are stable, locked nucleic acids, engineered RNA oligonucleotides that can recognize, based on sequence complementarity, specific miRNAs, inducing its degradation (Krutzfeldt et al., 2005, 2007). These engineered molecules consist in valuable tools for probing miRNAs function *in vivo*, and indeed, functional studies were undertaken in some epilepsy animal models. The induction of low intensity seizures renders animals resistant to subsequent induction of an epileptic state, a phenomena termed epileptic tolerance (for a review see Jimenez-Mateos and Henshall, 2009). It was observed that miR-132 was down-regulated in mice CA3 subfield after seizure preconditioning (Jimenez-Mateos et al., 2011). In the same study, the authors observed that the reduction in expression of miR-132, by the intracerebroventricular administration of an antagomir directed to this miRNA, reduced neuronal loss in the hippocampus after the induction of *status epilepticus* in mice. In the hippocampus miR-132 regulates mRNAs such as acetylcholinesterase or the GTPase activator p250GAP (Hanin and Soreq, 2011; Shaltiel et al., 2013). Furthermore, miR-132 has been previously associated with synaptic plasticity (Vo et al., 2005; Wayman et al., 2008). However, the miR-132 gene targets responsible for the facilitation of neuronal death remain to be determined. Yet another study exploring the role of miRNAs in epileptic tolerance, found an increase in the expression of miR-184 after preconditioning by systemic administration of a low dose of kainic acid (McKiernan et al., 2012b). Subsequently, these authors demonstrated that reduction of miR-184 by intracerebroventricular administration of an antagomir directed to this miRNA reduced the neuroprotective effect of preconditioning on hippocampal neurons, restoring the levels of neuronal death observed when *status epilepticus* was induced without preconditioning. The mRNAs that may interact *in vivo* with miR-184 in the hippocampus are not determined and the mechanism responsible for this miRNA-mediated neuroprotection in the hippocampal CA3 subfield is also unknown. Finally, miR-34a was shown to be up-regulated in different epilepsy animal models and its involvement in neuronal death in the hippocampus was probed with the use of antagomirs (Hu et al., 2012; Sano et al., 2012). The down-regulation of miR-34a by intracerebroventricular injection of antagomirs reduced neuronal death observed in the hippocampus in a lithium-pilocarpine epilepsy model (Hu et al., 2012), but it had no effect on an intra-amygdala kainic acid injection model in mice (Sano et al., 2012). The difference in the experiments outcome may be related to the different models, species and time points analyzed. It is believed that miR-34a may regulate expression of apoptosis-related genes in the hippocampus; however, further experiments are needed to confirm these observations.

Among the functional studies involving miRNAs, the one that explored the role of miR-134 in experimental epilepsy is noteworthy. In an intra-amygdala kainic acid injection epilepsy model in mice, it was observed an increase in the expression level of miR-134 following *status epilepticus.* Furthermore, this miRNA was shown to be expressed by pyramidal neurons in CA3, by interneurons in the hilus and by neocortical as well as amygdala neurons (Jimenez-Mateos et al., 2012). In the same study, the reduction of miR-134 expression by intracerebroventricular injection of antagomirs induced a decrease in CA3 pyramidal neurons spine density and, remarkably, it significantly reduced the severity of the induced seizures following intra-amygdala kainic acid injection. The authors also demonstrated that the induced down-regulation of this single miRNA enhanced resistance to evoked seizures resulting in reduction in all events associated with experimental induction of epilepsy, namely neuronal loss, gliosis, sprouting, and subsequent spontaneous recurrent seizures.

In conclusion, miRNAs are emerging as key regulators of sets of genes involved in the events that take place during epileptogenesis and chronic epilepsy states. Additionally, functional studies employing antagomirs indicate that these regulatory RNAs as promising targets for new possible strategies in the treatment of epilepsy.

ACKNOWLEDGMENTS

We are grateful to Mrs. Mercedes de Fátima Santos for her technical assistance with the art work. This work was supported by FAPESP (Fundação de Amparo à Pesquisa do Estado de São Paulo, BRAZIL), grant # CEPID 2013/07559-3.

REFERENCES

Agostini, M., Tucci, P., Killick, R., Candi, E., Sayan, B. S., Rivetti di Val Cervo, P., et al. (2011). Neuronal differentiation by TAp73 is mediated by microRNA-34a regulation of synaptic protein targets. *Proc. Natl. Acad. Sci. U.S.A.* 108, 21093–21098. doi: 10.1073/pnas.1112061109

Andersson, T., Rahman, S., Sansom, S. N., Alsio, J. M., Kaneda, M., Smith, J., et al. (2010). Reversible block of mouse neural stem cell differentiation in the absence of dicer and microRNAs. *PLoS ONE* 5:e13453. doi: 10.1371/journal.pone.0013453

Annegers, J. F., Rocca, W. A., and Hauser, W. A. (1996). Causes of epilepsy: contributions of the Rochester epidemiology project. *Mayo Clin. Proc.* 71, 570–575. doi: 10.1016/S0025-6196(11)64114-1

Aronica, E., Becker, A. J., and Spreafico, R. (2012). Malformations of cortical development. *Brain Pathol.* 22, 380–401. doi: 10.1111/j.1750-3639.2012.00581.x

Aronica, E., and Crino, P. B. (2011). Inflammation in epilepsy: clinical observations. *Epilepsia* 52(Suppl. 3), 26–32. doi: 10.1111/j.1528-1167.2011.03033.x

Aronica, E., Fluiter, K., Iyer, A., Zurolo, E., Vreijling, J., van Vliet, E. A., et al. (2010). Expression pattern of miR-146a, an inflammation-associated microRNA, in experimental and human temporal lobe epilepsy. *Eur. J. Neurosci.* 31, 1100–1107. doi: 10.1111/j.1460-9568.2010.07122.x

Ashhab, M. U., Omran, A., Kong, H., Gan, N., He, F., Peng, J., et al. (2013). Expressions of tumor necrosis factor alpha and microRNA-155 in immature rat model of status epilepticus and children with mesial temporal lobe epilepsy. *J. Mol. Neurosci.* doi: 10.1007/s12031-013-0013-9 [Epub ahead of print].

Avanzini, G., Vergnes, M., Spreafico, R., and Marescaux, C. (1993). Calcium-dependent regulation of genetically determined spike and waves by the reticular thalamic nucleus of rats. *Epilepsia* 34, 1–7. doi: 10.1111/j.1528-1157.1993.tb02369.x

Becker, A. J., Chen, J., Zien, A., Sochivko, D., Normann, S., Schramm, J., et al. (2003). Correlated stage- and subfield-associated hippocampal gene expression patterns in experimental and human temporal lobe epilepsy. *Eur. J. Neurosci.* 18, 2792–2802. doi: 10.1111/j.1460-9568.2003.02993.x

Bentwich, I., Avniel, A., Karov, Y., Aharonov, R., Gilad, S., Barad, O., et al. (2005). Identification of hundreds of conserved and nonconserved human microRNAs. *Nat. Genet.* 37, 766–770. doi: 10.1038/ng1590

Bjorklund, O., Shang, M., Tonazzini, I., Dare, E., and Fredholm, B. B. (2008). Adenosine A1 and A3 receptors protect astrocytes from hypoxic damage. *Eur. J. Pharmacol.* 596, 6–13. doi: 10.1016/j.ejphar.2008.08.002

Blumcke, I., Thom, M., Aronica, E., Armstrong, D. D., Vinters, H. V., Palmini, A., et al. (2011). The clinicopathologic spectrum of focal cortical dysplasias: a consensus classification proposed by an ad hoc Task Force of the ILAE Diagnostic Methods Commission. *Epilepsia* 52, 158–174. doi: 10.1111/j.1528-1167.2010.02777.x

Blumcke, I., Thom, M., and Wiestler, O. D. (2002). Ammon's horn sclerosis: a maldevelopmental disorder associated with temporal lobe epilepsy. *Brain Pathol.* 12, 199–211. doi: 10.1111/j.1750-3639.2002.tb00436.x

Boison, D. (2009). Engineered adenosine-releasing cells for epilepsy therapy: human mesenchymal stem cells and human embryonic stem cells. *Neurotherapeutics* 6, 278–283. doi: 10.1016/j.nurt.2008.12.001

Borges, M. A., Min, L. L., Guerreiro, C. A., Yacubian, E. M., Cordeiro, J. A., Tognola, W. A., et al. (2004). Urban prevalence of epilepsy: populational study in Sao Jose do Rio Preto, a medium-sized city in Brazil. *Arq. Neuropsiquiatr.* 62, 199–204. doi: 10.1590/S0004-282X2004000200002

Chang, T. C., Wentzel, E. A., Kent, O. A., Ramachandran, K., Mullendore, M., Lee, K. H., et al. (2007). Transactivation of miR-34a by p53 broadly influences gene expression and promotes apoptosis. *Mol. Cell* 26, 745–752. doi: 10.1016/j.molcel.2007.05.010

Cheng, H. S., Sivachandran, N., Lau, A., Boudreau, E., Zhao, J. L., Baltimore, D., et al. (2013). MicroRNA-146 represses endothelial activation by inhibiting pro-inflammatory pathways. *EMBO Mol. Med.* 5, 949–966. doi: 10.1002/emmm.201202318

Cui, Y., Xiao, Z., Han, J., Sun, J., Ding, W., Zhao, Y., et al. (2012). MiR-125b orchestrates cell proliferation, differentiation and migration in neural stem/progenitor cells by targeting Nestin. *BMC Neurosci.* 13:116. doi: 10.1186/1471-2202-13-116

Delaloy, C., Liu, L., Lee, J. A., Su, H., Shen, F., Yang, G. Y., et al. (2010). MicroRNA-9 coordinates proliferation and migration of human embryonic stem cell-derived neural progenitors. *Cell Stem Cell* 6, 323–335. doi: 10.1016/j.stem.2010.02.015

Dogini, D. B., Ribeiro, P. A., Rocha, C., Pereira, T. C., and Lopes-Cendes, I. (2008). MicroRNA expression profile in murine central nervous system development. *J. Mol. Neurosci.* 35, 331–337. doi: 10.1007/s12031-008-9068-4

Eacker, S. M., Keuss, M. J., Berezikov, E., Dawson, V. L., and Dawson, T. M. (2011). Neuronal activity regulates hippocampal miRNA expression. *PloS ONE* 6:e25068. doi: 10.1371/journal.pone.0025068

Edbauer, D., Neilson, J. R., Foster, K. A., Wang, C. F., Seeburg, D. P., Batterton, M. N., et al. (2010). Regulation of synaptic structure and function by FMRP-associated microRNAs miR-125b and miR-132. *Neuron* 65, 373–384. doi: 10.1016/j.neuron.2010.01.005

Engel, J. Jr. (1996). Introduction to temporal lobe epilepsy. *Epilepsy Res.* 26, 141–150. doi: 10.1016/S0920-1211(96)00043-5

Engel, J. Jr. (2001). Mesial temporal lobe epilepsy: what have we learned? *Neuroscientist* 7, 340–352. doi: 10.1177/107385840100700410

Gaughwin, P., Ciesla, M., Yang, H., Lim, B., and Brundin, P. (2011). Stage-specific modulation of cortical neuronal development by Mmu-miR-134. *Cereb. Cortex* 21, 1857–1869. doi: 10.1093/cercor/bhq262

Greene, J. G., Borges, K., and Dingledine, R. (2009). Quantitative transcriptional neuroanatomy of the rat hippocampus: evidence for wide-ranging, pathway-specific heterogeneity among three principal cell layers. *Hippocampus* 19, 253–264. doi: 10.1002/hipo.20502

Guerrini, R., and Barba, C. (2010). Malformations of cortical development and aberrant cortical networks: epileptogenesis and functional organization. *J. Clin. Neurophysiol.* 27, 372–379. doi: 10.1097/WNP.0b013e3181fe0585

Guerrini, R., Dobyns, W. B., and Barkovich, A. J. (2008). Abnormal development of the human cerebral cortex: genetics, functional consequences and treatment options. *Trends Neurosci.* 31, 154–162. doi: 10.1016/j.tins.2007.12.004

Guerrini, R., and Parrini, E. (2010). Neuronal migration disorders. *Neurobiol. Dis.* 38, 154–166. doi: 10.1016/j.nbd.2009.02.008

Guerrini, R., Sicca, F., and Parmeggiani, L. (2003). Epilepsy and malformations of the cerebral cortex. *Epileptic Disord.* 5(Suppl. 2), S9–S26.

Guillemot, F., Molnar, Z., Tarabykin, V., and Stoykova, A. (2006). Molecular mechanisms of cortical differentiation. *Eur. J. Neurosci.* 23, 857–868. doi: 10.1111/j.1460-9568.2006.04626.x

Hanin, G., and Soreq, H. (2011). Cholinesterase-targeting microRNAs identified in silico affect specific biological processes. *Front. Mol. Neurosci.* 4:28. doi: 10.3389/fnmol.2011.00028

Hauptman, J. S., and Mathern, G. W. (2012). Surgical treatment of epilepsy associated with cortical dysplasia: 2012 update. *Epilepsia* 53(Suppl. 4), 98–104. doi: 10.1111/j.1528-1167.2012.03619.x

Hermeking, H. (2010). The miR-34 family in cancer and apoptosis. *Cell Death Differ.* 17, 193–199. doi: 10.1038/cdd.2009.56

Hu, K., Xie, Y. Y., Zhang, C., Ouyang, D. S., Long, H. Y., Sun, D. N., et al. (2012). MicroRNA expression profile of the hippocampus in a rat model of temporal lobe epilepsy and miR-34a-targeted neuroprotection against hippocampal neurone cell apoptosis post-status epilepticus. *BMC Neurosci.* 13:115. doi: 10.1186/1471-2202-13-115

Iyer, A., Zurolo, E., Prabowo, A., Fluiter, K., Spliet, W. G., van Rijen, P. C., et al. (2012). MicroRNA-146a: a key regulator of astrocyte-mediated inflammatory response. *PloS ONE* 7:e44789. doi: 10.1371/journal.pone.0044789

Jimenez-Mateos, E. M., Bray, I., Sanz-Rodriguez, A., Engel, T., McKiernan, R. C., Mouri, G., et al. (2011). miRNA Expression profile after status epilepticus and hippocampal neuroprotection by targeting miR-132. *Am. J. Pathol.* 179, 2519–2532. doi: 10.1016/j.ajpath.2011.07.036

Jimenez-Mateos, E. M., Engel, T., Merino-Serrais, P., McKiernan, R. C., Tanaka, K., Mouri, G., et al. (2012). Silencing microRNA-134 produces neuroprotective and prolonged seizure-suppressive effects. *Nat. Med.* 18, 1087–1094. doi: 10.1038/nm.2834

Jimenez-Mateos, E. M., and Henshall, D. C. (2009). Seizure preconditioning and epileptic tolerance: models and mechanisms. *Int. J. Physiol. Pathophysiol. Pharmacol.* 1, 180–191.

Kawase-Koga, Y., Otaegi, G., and Sun, T. (2009). Different timings of Dicer deletion affect neurogenesis and gliogenesis in the developing mouse central nervous system. *Dev. Dyn.* 238, 2800–2812. doi: 10.1002/dvdy.22109

Krichevsky, A. M., Sonntag, K. C., Isacson, O., and Kosik, K. S. (2006). Specific microRNAs modulate embryonic stem cell-derived neurogenesis. *Stem Cells* 24, 857–864. doi: 10.1634/stemcells.2005-0441

Kriegstein, A., and Alvarez-Buylla, A. (2009). The glial nature of embryonic and adult neural stem cells. *Annu. Rev. Neurosci.* 32, 149–184. doi: 10.1146/annurev.neuro.051508.135600

Krutzfeldt, J., Kuwajima, S., Braich, R., Rajeev, K. G., Pena, J., Tuschl, T., et al. (2007). Specificity, duplex degradation and subcellular localization of antagomirs. *Nucleic Acids Res.* 35, 2885–2892. doi: 10.1093/nar/gkm024

Krutzfeldt, J., Rajewsky, N., Braich, R., Rajeev, K. G., Tuschl, T., Manoharan, M., et al. (2005). Silencing of microRNAs in vivo with 'antagomirs'. *Nature* 438, 685–689. doi: 10.1038/nature04303

Lein, E. S. (2004). Defining a molecular atlas of the hippocampus using DNA microarrays and high-throughput in situ hybridization. *J. Neurosci.* 24, 3879–3889. doi: 10.1523/JNEUROSCI.4710-03.2004

Liu, C., Teng, Z. Q., Santistevan, N. J., Szulwach, K. E., Guo, W., Jin, P., et al. (2010a). Epigenetic regulation of miR-184 by MBD1 governs neural stem cell proliferation and differentiation. *Cell Stem Cell* 6, 433–444. doi: 10.1016/j.stem.2010.02.017

Liu, D. Z., Tian, Y., Ander, B. P., Xu, H., Stamova, B. S., Zhan, X., et al. (2010b). Brain and blood microRNA expression profiling of ischemic stroke, intracerebral hemorrhage, and kainate seizures. *J. Cereb. Blood Flow Metab.* 30, 92–101. doi: 10.1038/jcbfm.2009.186

Lothman, E. W., Rempe, D. A., and Mangan, P. S. (1995). Changes in excitatory neurotransmission in the CA1 region and dentate gyrus in a chronic model of temporal lobe epilepsy. *J. Neurophysiol.* 74, 841–848.

Lukiw, W. J., Zhao, Y., and Cui, J. G. (2008). An NF-kappaB-sensitive micro RNA-146a-mediated inflammatory circuit in Alzheimer disease and in stressed human brain cells. *J. Biol. Chem.* 283, 31315–31322. doi: 10.1074/jbc.M805371200

Majores, M., Eils, J., Wiestler, O. D., and Becker, A. J. (2004). Molecular profiling of temporal lobe epilepsy: comparison of data from human tissue samples and animal models. *Epilepsy Res.* 60, 173–178. doi: 10.1016/j.eplepsyres.2004.07.002

Makeyev, E. V., Zhang, J., Carrasco, M. A., and Maniatis, T. (2007). The microRNA miR-124 promotes neuronal differentiation by triggering brain-specific alternative pre-mRNA splicing. *Mol. Cell* 27, 435–448. doi: 10.1016/j.molcel.2007.07.015

Mattson, R. H. (1994). Current challenges in the treatment of epilepsy. *Neurology* 44(Suppl. 5), S4–S9.

McKiernan, R. C., Jimenez-Mateos, E. M., Bray, I., Engel, T., Brennan, G. P., Sano, T., et al. (2012a). Reduced mature microRNA levels in association with dicer loss in human temporal lobe epilepsy with hippocampal sclerosis. *PloS ONE* 7:e35921. doi: 10.1371/journal.pone.0035921

McKiernan, R. C., Jimenez-Mateos, E. M., Sano, T., Bray, I., Stallings, R. L., Simon, R. P., et al. (2012b). Expression profiling the microRNA response to epileptic preconditioning identifies miR-184 as a modulator of seizure-induced neuronal death. *Exp. Neurol.* 237, 346–354. doi: 10.1016/j.expneurol.2012.06.029

McLoughlin, H. S., Fineberg, S. K., Ghosh, L. L., Tecedor, L., and Davidson, B. L. (2012). Dicer is required for proliferation, viability, migration and differentiation in corticoneurogenesis. *Neuroscience* 223, 285–295. doi: 10.1016/j.neuroscience.2012.08.009

Nakasa, T., Miyaki, S., Okubo, A., Hashimoto, M., Nishida, K., Ochi, M., et al. (2008). Expression of microRNA-146 in rheumatoid arthritis synovial tissue. *Arthritis Rheum.* 58, 1284–1292. doi: 10.1002/art.23429

Nudelman, A. S., DiRocco, D. P., Lambert, T. J., Garelick, M. G., Le, J., Nathanson, N. M., et al. (2010). Neuronal activity rapidly induces transcription of the CREB-regulated microRNA-132, in vivo. *Hippocampus* 20, 492–498. doi: 10.1002/hipo.20646

Omran, A., Peng, J., Zhang, C., Xiang, Q. L., Xue, J., Gan, N., et al. (2012). Interleukin-1beta and microRNA-146a in an immature rat model and children with mesial temporal lobe epilepsy. *Epilepsia* 53, 1215–1224. doi: 10.1111/j.1528-1167.2012.03540.x

Palmini, A., Najm, I., Avanzini, G., Babb, T., Guerrini, R., Foldvary-Schaefer, N., et al. (2004). Terminology and classification of the cortical dysplasias. *Neurology* 62, S2–S8.

Pauley, K. M., Satoh, M., Chan, A. L., Bubb, M. R., Reeves, W. H., and Chan, E. K. (2008). Upregulated miR-146a expression in peripheral blood mononuclear cells from rheumatoid arthritis patients. *Arthritis Res. Ther.* 10, R101. doi: 10.1186/ar2493

Peng, J., Omran, A., Ashhab, M. U., Kong, H., Gan, N., He, F., et al. (2013). Expression Patterns of miR-124, miR-134, miR-132, and miR-21 in an immature rat model and children with mesial temporal lobe epilepsy. *J. Mol. Neurosci.* 50, 291–297. doi: 10.1007/s12031-013-9953-3

Pichardo-Casas, I., Goff, L. A., Swerdel, M. R., Athie, A., Davila, J., Ramos-Brossier, M., et al. (2012). Expression profiling of synaptic microRNAs from the adult rat brain identifies regional differences and seizure-induced dynamic modulation. *Brain Res.* 1436, 20–33. doi: 10.1016/j.brainres.2011.12.001

Reiner, O., Coquelle, F. M., Peter, B., Levy, T., Kaplan, A., Sapir, T., et al. (2006). The evolving doublecortin (DCX) superfamily. *BMC Genomics* 7:188. doi: 10.1186/1471-2164-7-188

Ren, G., and Boison, D. (2010). Engineering human mesenchymal stem cells to release adenosine using miRNA technology. *Methods Mol. Biol.* 650, 225–240. doi: 10.1007/978-1-60761-769-3_17

Risbud, R. M., and Porter, B. E. (2013). Changes in microRNA expression in the whole hippocampus and hippocampal synaptoneurosome fraction following pilocarpine induced status epilepticus. *PloS ONE* 8:e53464. doi: 10.1371/journal.pone.0053464

Sakakibara, T., Sukigara, S., Saito, T., Otsuki, T., Takahashi, A., Kaneko, Y., et al. (2012). Delayed maturation and differentiation of neurons in focal cortical dysplasia with the transmantle sign: analysis of layer-specific marker expression. *J. Neuropathol. Exp. Neurol.* 71, 741–749. doi: 10.1097/NEN.0b013e318262e41a

Sander, J. W. (1993). Some aspects of prognosis in the epilepsies: a review. *Epilepsia* 34, 1007–1016. doi: 10.1111/j.1528-1157.1993.tb02126.x

Sano, T., Reynolds, J. P., Jimenez-Mateos, E. M., Matsushima, S., Taki, W., and Henshall, D. C. (2012). MicroRNA-34a upregulation during seizure-induced neuronal death. *Cell Death Dis.* 3:e287. doi: 10.1038/cddis.2012.23

Shaltiel, G., Hanan, M., Wolf, Y., Barbash, S., Kovalev, E., Shoham, S., et al. (2013). Hippocampal microRNA-132 mediates stress-inducible cognitive deficits through its acetylcholinesterase target. *Brain Struct. Funct.* 218, 59–72. doi: 10.1007/s00429-011-0376-z

Shibata, M., Nakao, H., Kiyonari, H., Abe, T., and Aizawa, S. (2011). MicroRNA-9 regulates neurogenesis in mouse telencephalon by targeting multiple transcription factors. *J. Neurosci.* 31, 3407–3422. doi: 10.1523/JNEUROSCI.5085-10.2011

Sidman, R. L., and Rakic, P. (1973). Neuronal migration, with special reference to developing human brain: a review. *Brain Res.* 62, 1–35. doi: 10.1016/0006-8993(73)90617-3

Silber, J., Lim, D. A., Petritsch, C., Persson, A. I., Maunakea, A. K., Yu, M., et al. (2008). miR-124 and miR-137 inhibit proliferation of glioblastoma multiforme cells and induce differentiation of brain tumor stem cells. *BMC Med.* 6:14. doi: 10.1186/1741-7015-6-14

Sisodiya, S. M. (2004). Malformations of cortical development: burdens and insights from important causes of human epilepsy. *Lancet Neurol.* 3, 29–38. doi: 10.1016/S1474-4422(03)00620-3

Sisodiya, S. M., Fauser, S., Cross, J. H., and Thom, M. (2009). Focal cortical dysplasia type II: biological features and clinical perspectives. *Lancet Neurol.* 8, 830–843. doi: 10.1016/S1474-4422(09)70201-7

Smrt, R. D., Szulwach, K. E., Pfeiffer, R. L., Li, X., Guo, W., Pathania, M., et al. (2010). MicroRNA miR-137 regulates neuronal maturation by targeting ubiquitin ligase mind bomb-1. *Stem Cells* 28, 1060–1070. doi: 10.1002/stem.431

Song, Y. J., Tian, X. B., Zhang, S., Zhang, Y. X., Li, X., Li, D., et al. (2011). Temporal lobe epilepsy induces differential expression of hippocampal miRNAs including let-7e and miR-23a/b. *Brain Res.* 1387, 134–140. doi: 10.1016/j.brainres.2011.02.073

Sonkoly, E., Stahle, M., and Pivarcsi, A. (2008). MicroRNAs and immunity: novel players in the regulation of normal immune function and inflammation. *Semin. Cancer Biol.* 18, 131–140. doi: 10.1016/j.semcancer.2008.01.005

Stappert, L., Borghese, L., Roese-Koerner, B., Weinhold, S., Koch, P., Terstegge, S., et al. (2013). MicroRNA-based promotion of human neuronal differentiation and subtype specification. *PloS ONE* 8:e59011. doi: 10.1371/journal.pone.0059011

Studer, F. E., Fedele, D. E., Marowsky, A., Schwerdel, C., Wernli, K., Vogt, K., et al. (2006). Shift of adenosine kinase expression from neurons to astrocytes during postnatal development suggests dual functionality of the enzyme. *Neuroscience* 142, 125–137. doi: 10.1016/j.neuroscience.2006.06.016

Sun, G., Ye, P., Murai, K., Lang, M. F., Li, S., Zhang, H., et al. (2011). miR-137 forms a regulatory loop with nuclear receptor TLX and LSD1 in neural stem cells. *Nat. Commun.* 2, 529. doi: 10.1038/ncomms1532

Thom, M. (2004). Recent advances in the neuropathology of focal lesions in epilepsy. *Expert Rev. Neurother.* 4, 973–984. doi: 10.1586/14737175.4.6.973

Van Paesschen, W., Duncan, J. S., Stevens, J. M., and Connelly, A. (1997). Etiology and early prognosis of newly diagnosed partial seizures in adults: a quantitative hippocampal MRI study. *Neurology* 49, 753–757. doi: 10.1212/WNL.49.3.753

Vezzani, A., Friedman, A., and Dingledine, R. J. (2013). The role of inflammation in epileptogenesis. *Neuropharmacology* 69, 16–24. doi: 10.1016/j.neuropharm.2012.04.004

Vo, N., Klein, M. E., Varlamova, O., Keller, D. M., Yamamoto, T., Goodman, R. H., et al. (2005). A cAMP-response element binding protein-induced microRNA regulates neuronal morphogenesis. *Proc. Natl. Acad. Sci. U.S.A.* 102, 16426–16431. doi: 10.1073/pnas.0508448102

Volvert, M. L., Rogister, F., Moonen, G., Malgrange, B., and Nguyen, L. (2012). MicroRNAs tune cerebral cortical neurogenesis. *Cell Death Differ.* 19, 1573–1581. doi: 10.1038/cdd.2012.96

Wayman, G. A., Davare, M., Ando, H., Fortin, D., Varlamova, O., Cheng, H. Y., et al. (2008). An activity-regulated microRNA controls dendritic plasticity by down-regulating p250GAP. *Proc. Natl. Acad. Sci. U.S.A.* 105, 9093–9098. doi: 10.1073/pnas.0803072105

Wieser, H. G. (2004). Epilepsy ICoNo. ILAE Commission Report. Mesial temporal lobe epilepsy with hippocampal sclerosis. *Epilepsia* 45, 695–714. doi: 10.1111/j.0013-9580.2004.09004.x

You, G., Yan, W., Zhang, W., Wang, Y., Bao, Z., Li, S., et al. (2012). Significance of miR-196b in tumor-related epilepsy of patients with gliomas. *PloS ONE* 7:e46218. doi: 10.1371/journal.pone.0046218

Zhao, C., Sun, G., Li, S., Lang, M. F., Yang, S., Li, W., et al. (2010). MicroRNA let-7b regulates neural stem cell proliferation and differentiation by targeting nuclear receptor TLX signaling. *Proc. Natl. Acad. Sci. U.S.A.* 107, 1876–1881. doi: 10.1073/pnas.0908750107

Zhao, C., Sun, G., Li, S., and Shi, Y. (2009). A feedback regulatory loop involving microRNA-9 and nuclear receptor TLX in neural stem cell fate determination. *Nat. Struct. Mol. Biol.* 16, 365–371. doi: 10.1038/nsmb.1576

4

Plasmid-based target protectors allow specific blockade of miRNA silencing activity in mammalian developmental systems

Jennifer L. Knauss[1,2], Shan Bian[1] and Tao Sun[1,2]*

[1] Department of Cell and Developmental Biology, Weill Medical College of Cornell University, New York, NY, USA
[2] Weill Cornell Graduate School of Medical Sciences, New York, NY, USA

***Correspondence:**
Tao Sun, Department of Cell and Developmental Biology, Weill Medical College of Cornell University, 1300 York Avenue, Box 60, New York, NY 10065, USA
e-mail: tas2009@med.cornell.edu

Over the past decade, microRNAs (miRNAs) have emerged as essential posttranscriptional regulators of gene expression. Though a great deal has been discovered about miRNA genomics, biogenesis, mechanisms, and functions, the challenge of attributing phenotypes of altered miRNA expression to specific targets still remains. Here, we apply the existing target protector concept of blocking miRNA action at a single binding site in the 3′untranslated region (3′UTR) of its target to a plasmid-based approach. We optimize and demonstrate target protector efficacy *in vitro*, where it blocks repression of a luciferase construct and an endogenous protein. Using the developing mouse cortex as a model, we validate that target protectors are effective *in vivo*, where protectors for the miR-19a binding sites in the *Pten* 3′UTR alter proliferation and specification of neural progenitors, phenocopying Pten ectopic expression phenotypes. Our study introduces a new tool for analyzing specific miRNA:target interactions across mammalian developmental systems, facilitating further miRNA functional discoveries.

Keywords: miRNA, miR-19, Pten, mRNA protector, neural development

INTRODUCTION

MicroRNAs (miRNAs) are an extensive class of small non-coding RNAs that are critically important throughout development (Carrington and Ambros, 2003; Bartel, 2004; Alvarez-Garcia and Miska, 2005; Kloosterman and Plasterk, 2006). They regulate gene expression posttranscriptionally through incorporation into a RNA-induced silencing complex (RISC), which uses partial sequence complementarity to bind target messenger RNAs (mRNAs), usually in the 3′untranslated region (3′UTR; Lewis et al., 2003). RISC binding can lead to blocking translation or enhancing the decay of target mRNAs, the final result being downregulation of the target protein (Bartel, 2004). Hundreds of miRNAs have been identified in the genomes of mammals including rodents and humans, and each miRNA can regulate many mRNAs in a cell-type specific manner, leading to a complex network of interactions (Landgraf et al., 2007). It is estimated that between 25 and 60% of human transcripts are regulated by miRNAs (Lewis et al., 2005; Lim et al., 2005; Friedman et al., 2009).

Conditional knockout and knockdown techniques have been used to examine the roles of miRNAs; however, there can be difficulty in interpreting molecular mechanisms underlying the observed phenotypes due to the large number of potential targets of each miRNA. Groups working in model systems such as zebrafish and *Xenopus* have bypassed this obstacle through a morpholino target protector approach, using sequence complementarity to block a miRNA from binding to a specific site (Choi et al., 2007; Bonev et al., 2011; Stanton and Giraldez, 2011). However, morpholinos are not a tractable tool in mammals and their short length of activity limits their application in developmental systems.

We here have developed a plasmid-based target protector system to tease apart the physiological roles of miRNAs in mammalian systems. Previous work in our lab has shown that in the developing cortex, *miR-19a* targets *Pten* mRNA (Bian et al., 2013). In the developing mouse cortex, Pten functions to repress progenitor expansion; therefore its repression by *miR-19a* results in increased proliferation (Groszer et al., 2001; Zheng et al., 2008). Thus, the *miR-19a*:*Pten* relationship provides an ideal readout for testing *Pten* derepression through target protectors. Here, we have designed and optimized target protectors for the miR-19a binding sites in the *Pten* 3′UTR. We demonstrate that these target protectors can be electroporated *in utero* to allow functional investigation of a specific miRNA:mRNA interaction during cortical development *in vivo*. Our results provide a useful tool for investigation of long term, specific miRNA–target interactions both *in vitro* and *in vivo* using a plasmid-based target protector system.

MATERIALS AND METHODS

TARGET PROTECTOR DESIGN

Protectors were designed as perfectly complementary sequence covering the miR-19a binding sites in the *Pten* 3′UTR. The miRNA seed binding sequence was centered in the target protector, with complementary sequence on each side. Outside of the complementary sequence, restriction sites can be added as necessary for a cloning strategy.

For the second miR-19a binding site, target protectors with three lengths of complementarity to the *Pten* 3′UTR were designed: 22, 40, and 60 nucleotides (nt; **Figure 2A**). All of the target protectors were designed to be the same total length as the 60 nt protector and included junk sequences to increase their length as necessary, keeping the target protector in the middle of the construct. We ordered the target protectors as complementary oligonucleotides. After annealing, protectors were subcloned and inserted into the pCAGIG vector for electroporation and pCDNA3.1 for the luciferase assay.

miR-19a EXPRESSION CONSTRUCT

The precursor hairpin sequence of miR-19a and ~100 nt of genomic sequence flanking each side of the hairpin sequence was amplified by PCR from the genomic locus of the mouse miR-17-92 cluster. Sequences of primers are as following: miR-19a: F: 5′-CAGCTCGAGCAATCCAAGTCA-3′, R: 5′-GCAGGCTCTACATCGACAC-3′. To generate the miR-19a expression construct, the miRNA fragment was inserted into pcDNA3.1 for transfection in cell lines, and pCAGIG for electroporation.

LUCIFERASE ASSAY

pGL4.13 firefly luciferase (Promega) vector was used for making constructs containing amplified 3′UTRs of targets. pGL4.73 renilla luciferase (Promega) was used as a transfection control. Plasmid DNA was quantified by UV spectrophotometry and used for transfection in a 6:2:1 ratio (protector:miRNA:target luciferase constructs) in Neuro2a (N2a) cells using Lipofectamine 2000 (Invitrogen) according to the manufacturer's protocol. Luciferase was activated using the Dual-Luciferase Reporter Assay kit (Promega) using the manufacturer's protocol and read on a Victor3 1420 multilabel counter (Perkin Elmer). Results were shown as firefly luciferase activity normalized to renilla as controls.

To make the 3′UTR construct for the luciferase assay, a cDNA fragment encoding the mouse Pten 3′UTR was amplified and subcloned into the pGL4.13 luciferase vector. The first miR-19a binding site was mutated using QuikChange II Site-Directed Mutagenesis Kit (Agilent Technologies) according to manufacturer's instructions. All the primers for cloning of targets in the 3′UTR and their mutation are listed as the following: Pten-3′UTR: F: 5′-CATCTAGAATACATCCACAGGGTTTTGACA-3′, R: 5′-TTGAAGCCCTAATCCCAACTCT-3′; Pten-3′UTR-miR-19a-mut1: 5′-CCGGGTTCACGTCCTACCCCATTACAATTGT GGCAACAGATAAGTTT-3′.

NORTHERN BLOT ANALYSIS

Total RNA was isolated from N2a cells transfected with either the 60 nt target protector or the pcDNA3.1 empty vector using Trizol reagent (Invitrogen) according to manufacturer's instructions. RNA samples and 0.1–2 kb RNA ladder (Invitrogen) were denatured at 70°C for 15 min and cooled on ice. Ethidium bromide was added to the RNA ladder for visualization. The DNA control sample was denatured at 95°C and cooled on ice. Samples were loaded onto a 1% formaldehyde agarose gel and separated at room temperature. After running, the ladder band locations were marked on the gel.

Samples were transferred onto a nitrocellulose membrane using a semi-dry transfer method overnight. After transfer, the ladder band locations were marked on the membrane. After cross-linking for 4 h at 80°C, the membrane was hybridized at 50°C overnight using a denatured DNA probe for the 60 nt target protector. The probe was body-labeled with digoxygenin (DIG)-labeled nucleotides using the DIG DNA Labeling Kit (Roche), following manufacturer instructions. After washing, the RNA was detected using the CDP-star chemiluminescent substrate (Roche).

WESTERN BLOT ANALYSIS

Expression levels of Pten were analyzed by the Western blot analysis. Protein extracts were harvested by lysing N2a cells transfected with combinations miR-19a, 60 nt target protector, and empty vector with RIPA lysis buffer (150 mM NaCl, 1 mM $Na_4P_2O_7$, 1 mM NaF, 1 mM EDTA, 1 mM PMSF, 2 mM Na_3VO_4, 1% NP-40, 50 mM Tris, pH 7.5) with complete^{TM} EDTA-free protease inhibitor mixture (Roche Diagnostics, Indianapolis, IN, USA). The protein samples were boiled in SDS sample buffer for 10 min before loading onto 10% Tris–glycine gels as 10 μg for each lane and transferred onto PVDF membrane (Pall Corporation, Pensacola, FL, USA). For immunoblotting, membranes were blocked with 5% (w/v) non-fat milk powder in 0.05% TBST [50 mM Tris–Cl, pH 7.5, 150 mM NaCl, with 0.05% (v/v) Tween-20] and incubated at 4°C overnight with the following primary antibodies which were diluted in 0.05% TBST with 5% non-fat milk: Pten and actin. After washing with TBST, membranes were incubated with specific HRP-conjugated secondary antibodies for 1 h at room temperature followed with extended washes with TBST. Immunoblot reactions were visualized using chemiluminescent substrate (Pierce, Rockford, IL, USA) on Kodak BioMax light films (Kodak, Rochester, NY, USA). The intensities of the bands were densitometrically quantified with the image software ImageJ.

The following antibodies were used: anti-Pten (rabbit, 1:1,000; Cell Signaling) and anti-Actin (rabbit, 1:400; Sigma).

IN UTERO ELECTROPORATION

In utero electroporation was performed as described by Saito (2006). Briefly, electroporation was conducted at E13.5 and the brain tissues were harvested 24 h later at E14.5. Plasmid DNA was prepared using the EndoFree plasmid maxi kit (Qiagen) according to manufacturer's instructions, and diluted to 2.5 μg/μl. DNA solution was injected into the lateral ventricle of the cerebral cortex, and electroporated with five 50-ms pulses at 35 V using an ECM830 electrosquareporator (BTX).

Wild type CD-1 mice were used for all experiments. For staging of embryos, midday of the day of vaginal-plug formation was considered E0.5; the first 24 h after birth were defined as P0. Animal use was overseen by the Animal Facility at Weill Cornell Medical College.

TISSUE PREPARATION AND IMMUNOHISTOCHEMISTRY

Mouse brains were fixed in 4% paraformaldehyde (PFA) in phosphate-buffered saline (PBS) over night, incubated in 25–30% sucrose in PBS, embedded in OCT and stored at −80°C until use. Brains were sectioned (10–14 μm) using a cryostat. For antigen recovery, sections were incubated in heated (95–100°C) antigen recovery solution (1 mM EDTA, 5 mM Tris, pH 8.0) for 15–20 min, and cooled down for 20–30 min. Before applying antibodies, sections were blocked in 10% normal goat serum (NGS) in PBS with 0.1% Tween-20 (PBT) for 1 h. Sections were incubated with primary antibodies at 4°C overnight and visualized using goat anti-rabbit IgG-Alexa-Fluor-488 and/or goat anti-mouse IgG-Alexa-Fluor-546 (1:350, Molecular Probes) for 1.5 h at room temperature. Images were captured using a Leica digital camera under a Zeiss confocal microscope.

Primary antibodies against the following antigens were used: bromodeoxyuridine (BrdU; 1:50, DSHB), Ki67 (1:500, Abcam), Pax6 (1:200, Covance, rabbit), Tbr2 (1:500, Abcam), GFP (1:1000, Abcam, chicken), and GFP (1:1000, Rockland, rabbit).

CELL COUNTING

Coronal sections were collected in the medial cortical region (at levels between the anterior commissure and the anterior hippocampus). At least four sections from each brain and three brains from different litters were chosen for antibody labeling. Positive cells were quantified in fixed areas of 318 μm × 318 μm and normalized to the averaged empty vector control value.

STATISTICS

For the luciferase and Western blot assays, three independent experiments were performed. For electroporated mouse sections, at least three brains from each group were analyzed. Statistical comparison was made by an analysis of variance (unpaired Student's *t*-test).

RESULTS

TARGET PROTECTOR DESIGN AND *IN VITRO* TESTING

In a typical cell, miRNAs target multiple mRNAs through partial sequence complementarity (**Figure 1A**). The concept behind target protectors is that a construct with perfect complementarity to a specific miRNA-binding site will outcompete the miRNA for binding at that site (**Figure 1B**). Thus, a single miRNA target is derepressed while the others remain regulated, allowing analysis of the effects of a single miRNA:mRNA relationship. To apply this concept in mammalian development, we designed plasmid-based mRNA target protectors using miR-19a:Pten regulation, since our previous work and others have demonstrated the targeting effect of miR-19a on Pten (Mu et al., 2009; Olive et al., 2009; Mavrakis et al., 2010; Bian et al., 2013).

To optimize the minimum length of sequence complementarity necessary for maximum protector efficacy, we designed target protectors with three lengths of complementarity for the second miR-19a binding site in the *Pten* 3′UTR (**Figure 2A**). We chose the shortest length of 22 nt because this is the approximate size of most miRNAs (**Figure 2A**). We also designed protectors of 40 and 60 nt to increase complementarity, binding strength, and specificity (**Figure 2A**). Our basic rules for protector design are: (1) each protector is centered over the predicted miRNA seed binding site, (2) the protectors have perfect complementarity along the length of the 3′UTR, and (3) the protector must not overlap any other known functional sites in the 3′UTR. For the 22 and 40 nt target protectors, junk DNA sequences were inserted outside of the complementary region so that the overall length of all three constructs was constant with the 60 nt protector.

We hypothesized that an effective target protector would block miR-19a activity in the 3′UTR of *Pten* in a luciferase assay, resulting

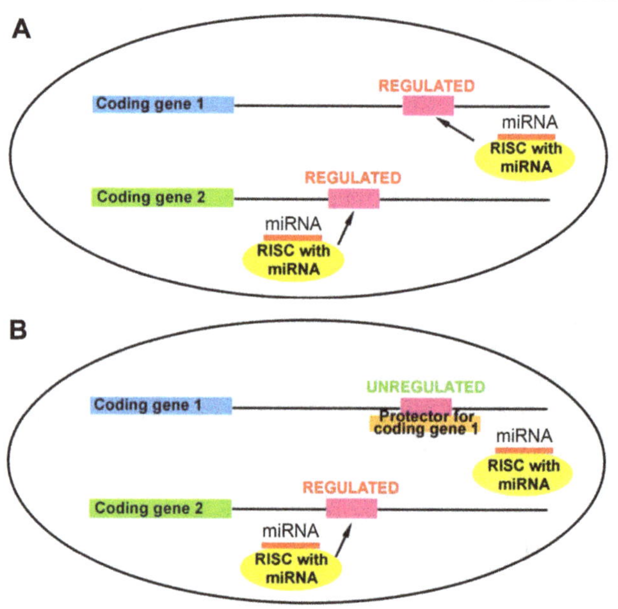

FIGURE 1 | Target protectors block miRNA action at a specific mRNA binding site. (A) miRNAs regulate multiple targets through sequence-specific binding in the 3′UTR. **(B)** Target protectors block miRNA action at a single target, while other targets remain regulated.

in a recovery of luciferase activity. To test the target protectors, we used a luciferase vector containing only the second miR-19a binding site of the *Pten* 3′UTR (the first miR-19a binding site was mutated in the *Pten* 3′UTR), and cotransfected it with miR-19a and each target protector in N2a cells. In the absence of any target protector, luciferase activity of the *Pten* 3′UTR was significantly decreased (**Figure 2B**). The 22 and 40 nt protectors did not show any significant recovery of luciferase activity, while the 60 nt protector showed a significant, almost complete recovery of activity (**Figure 2B**). Neither miR-19a nor any of the target protectors had an effect on the luciferase activity of the *Pten* 3′UTR containing a mutation in both miR-19a binding sites (**Figure 2B**). Our results indicate that the 60 nt target protector is the most effective at blocking miR-19a activity at the *Pten* 3′UTR

60 nt TARGET PROTECTOR BLOCKS miR-19a ACTIVITY *IN VITRO*

To ensure that the 60 nt target protector was transcribed and expressed as expected, we performed a northern blot assay using RNA extracted from N2a cells transfected with either the target protector or an empty vector. As a positive control, we also included digested vector DNA containing the protector. Based on the insert size and the predicted transcription start site and polyadenylation signals of the vector, the expect RNA size to be about 305 nt (**Figure 3A**). The protector is detected at the expected size in target protector transfected RNA and is not detected in the empty vector-transfected RNA, indicating that the protector is transcribed and expressed (**Figure 3B**).

We previously established that miR-19a regulates *Pten* posttranscriptionally by preventing its translation (Bian et al., 2013). Thus, we hypothesized that transfection of the protector should result in

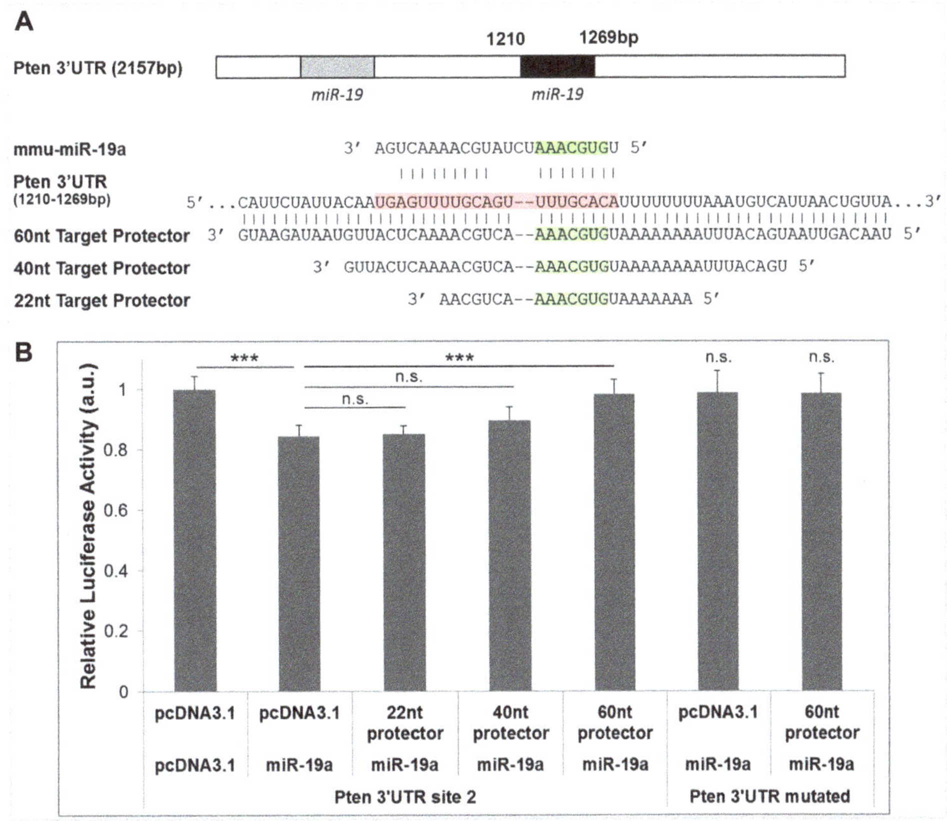

FIGURE 2 | Target protectors for *Pten* block miR-19a-induced repression. **(A)** Binding sites of miR-19a in the *Pten* 3′UTR and complementary target protector sequences for the second miR-19a binding site. The seed binding sequence of miR-19a is highlighted in green, and the entire length of miR-19a along the *Pten* 3′UTR is highlighted in red. **(B)** Luciferase assays of target protector effects on miR-19a repression of the second miR-19a binding site in the *Pten* 3′UTR. miR-19a reduced luciferase activity in the absence of target protector. The 60 nt target protector but not the 22 or 40 nt target protector recovered luciferase activity of the *Pten* 3′UTR. Neither miR-19a nor the 60 nt target protector had an effect on the luciferase activity of the full length *Pten* 3′UTR when the miR-19a binding sites were mutated. Data are presented as mean ± SEM; $n = 3$ luciferase assays; p values in relation to control (***$p < 0.01$). n.s., not significant.

increased endogenous Pten protein in N2a cells. Indeed, a Western blot assay showed that transfection of only exogenous miR-19a results in a decrease of endogenous Pten, while cotransfection of exogenous miR-19a and the target protector, or the protector alone significantly rescues the endogenous Pten protein levels (**Figures 3C,D**). Our results suggest that the plasmid-based target protectors are transcribed and work to block posttranscriptional regulation by miRNAs.

60 nt TARGET PROTECTOR BLOCKS miR-19a ACTIVITY *IN VIVO*

Having established that the plasmid-based target protector is effective *in vitro*, we sought to apply it *in vivo*, where such a tool can provide insight to the function of specific miRNA–mRNA interactions during development. We previously established that miR-19a targeting of *Pten* promotes progenitor cell expansion in the developing mouse cortex (Bian et al., 2013). We expected that blocking miR-19a activity with the target protector will result in decreased proliferation of progenitors. To observe the maximum effect of blocking miR-19a binding to *Pten*, we also designed a 60 nt target protector for the first binding site in the *Pten* 3′UTR and used *in utero* electroporation to introduce protectors for both miR-19a binding sites in the *Pten* 3′UTR into the embryonic day 13.5 (E13.5) cortex, analyzed at E14.5. The numbers of BrdU$^+$ and Ki67$^+$ cells were significantly decreased after *Pten* target protectors were electroporated compared to empty vector electroporation, suggesting a functional result of the blockade of miR-19a silencing effect on Pten expression (**Figure 4**).

To test further whether decreased proliferation of neural progenitors in target protector electroporation is caused by up-regulation of Pten, which is usually suppressed by miR-19a, we electroporated full length *Pten* containing the 3′UTR with miR-19a binding sites (*Pten-FL-3′ UTR*). We observed a significant decrease in both BrdU$^+$ and Ki67$^+$ cells upon electroporation of the *Pten* overexpression construct compared to empty vector electroporation (**Figures 4B,D**). When *Pten-FL-3′ UTR* was co-electroporated with exogenous miR-19a, the numbers of BrdU$^+$ and Ki67$^+$ cells were recovered to wild type levels, indicating that the change is dependent on miR-19a activity on the *Pten* 3′UTR (**Figure 4**). Our results demonstrate that plasmid-based target protectors can be electroporated into developing cortices to block miRNA action at a specific target.

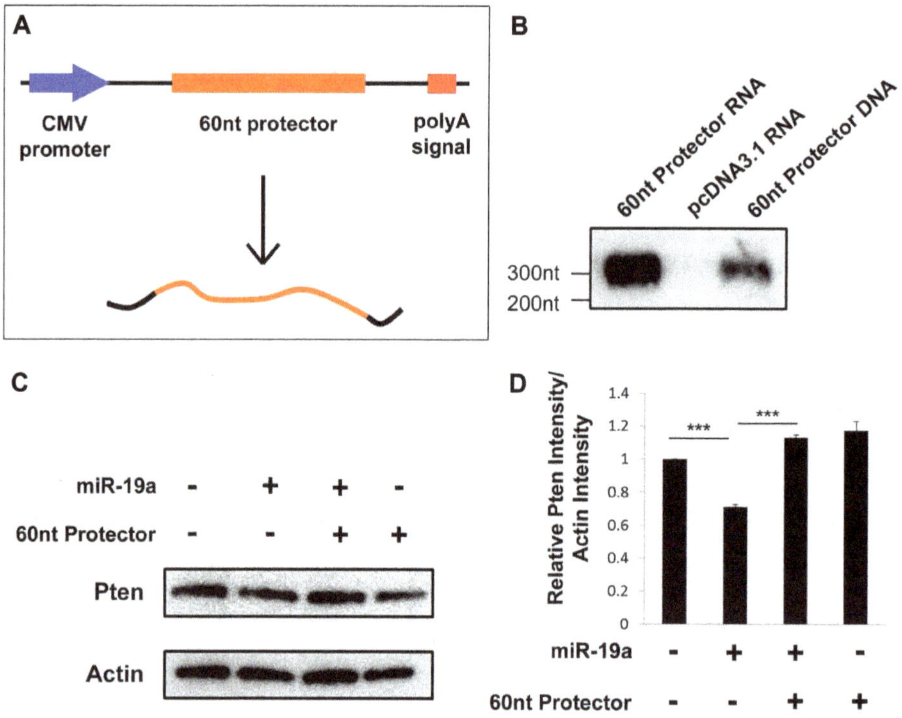

FIGURE 3 | Target protectors are transcribed and effective *in vitro*. (A) Schematic vector map of the 60 nt target protector in pcDNA3.1. **(B)** The 60 nt target protector and control target protector DNA were detected by Northern blot assays at the expected size. No target protector was detected in RNA extracted from empty vector-transfected cells. **(C,D)** The endogenous protein levels of Pten were decreased upon transfection of miR-19a but recovered upon cotransfection of miR-19a and the 60 nt target protector. Transfection of the 60 nt target protector alone also showed an increase in Pten. Data are presented as mean ± SD; $n = 3$ separate protein extractions; p values in relation to control (***$p < 0.001$).

TARGET PROTECTORS REVEAL THE ROLE OF miR-19a REPRESSION OF Pten IN THE DEVELOPING CORTEX

During corticogenesis, the transition of proliferative radial glia cells (RGCs) and intermediate progenitor (IP) into postmitotic neurons is tightly regulated in order to generate an appropriate amount of neurons while maintaining a progenitor pool. Our previous work has shown that the miRNA family miR-17-92, containing miR-19a, promotes RGC proliferation (Bian et al., 2013). To validate that miR-19a targeting of *Pten* is responsible for RGC proliferation, we introduced *Pten* target protectors into E13.5 cortices, which should result in a decrease of RGCs. We found that the number of Pax6$^+$ RGCs is significantly decreased, while the number of Tbr2$^+$ IPs is not changed (**Figure 5**).

To confirm further that *Pten* is responsible for the decrease of RGCs, we electroporated *Pten-FL-3′ UTR*, which we expected to show a similar phenotype as the *Pten* target protectors, in E13.5 cortices collected at E14.5. Indeed, the number Pax6$^+$ RGCs was decreased and the number of Tbr2$^+$ IPs was not changed (**Figures 5B,D**). When *Pten-FL-3′ UTR* and miR-19a were co-electroporated, the number of Pax6$^+$ RGCs was recovered, while the number of Tbr2$^+$ IPs remained unchanged (**Figure 5**). These results show that miR-19a targeting of *Pten* is critical for proliferation of RGCs in the developing cortex but does not affect the IP cell population. Using our *Pten* target protectors, we further demonstrate that the specific effects of miR-19a on RGC expansion occur through silencing *Pten*.

DISCUSSION

miRNAs have been proven to play critical roles in development of invertebrates and vertebrates (Carrington and Ambros, 2003; Alvarez-Garcia and Miska, 2005; Kloosterman and Plasterk, 2006). Since one miRNA has multiple targets, it has been a daunting task to demonstrate which genes are major targets in specific cells or tissues during development. In this study, we have designed a plasmid-based tool for analyzing specific miRNA:target relationships, shown that it works effectively *in vitro* and *in vivo*, and applied it to a miRNA:mRNA pair in the developing mouse cortex.

In the past, many studies have examined the roles of individual miRNAs by removing them globally or conditionally. In *Drosophila* and mice, genetic mutants have been generated, while in other systems such as zebrafish or *Xenopus*, morpholinos or other antisense oligonucleotides have been used to knockdown miRNAs (Sokol, 2005; Karres et al., 2007; Ventura et al., 2008; Woltering and Durston, 2008; Conte et al., 2010; Rasmussen et al., 2010). miRNA sponges with multiple miRNA-binding sites have also been used to soak up mature miRNAs (Ebert et al., 2007; Loya et al., 2009; Ebert and Sharp, 2010). While these methods provide a genetic approach to explore the overall function of a miRNA, they lack the power to attribute any phenotype to a particular mRNA target. Later, morpholino-based target protectors were applied in zebrafish and *Xenopus* to examine the importance of miRNA repression of a specific target (Choi et al., 2007; Bonev et al., 2011; Stanton and Giraldez, 2011). While this technique has been useful

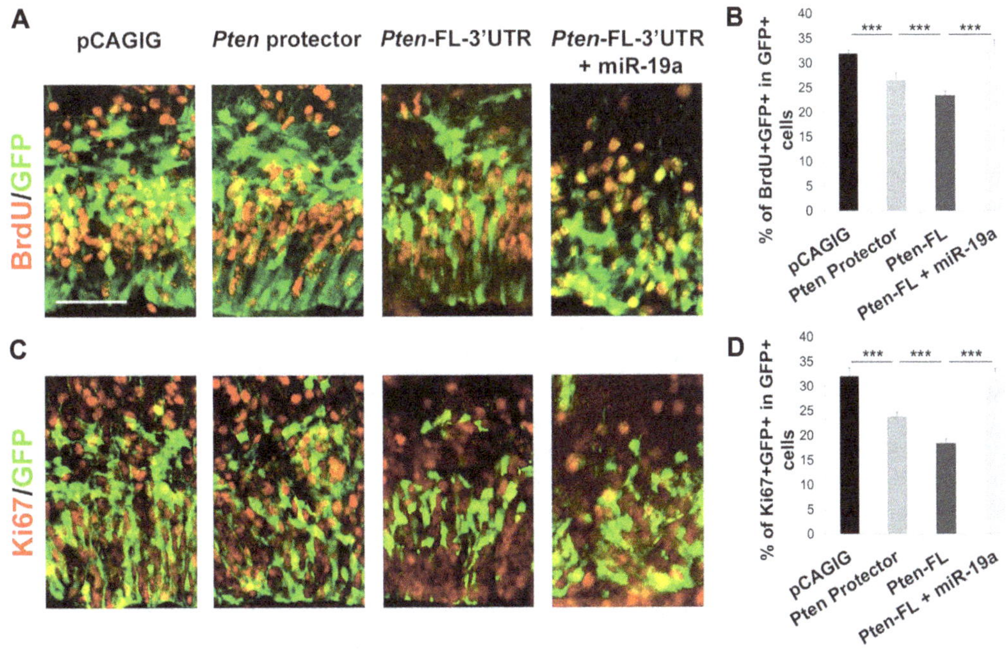

FIGURE 4 | Target protectors block miR-19a repression of *Pten* in the developing mouse cortex. (A–D) Electroporation of the 60 nt target protector at E13.5 for analysis at E14.5 significantly decreased the number of BrdU-incorporating or Ki67$^+$ proliferative cells colabeled with GFP in the cortex. Ectopic expression of full length *Pten* containing the 3′UTR (*Pten-FL-3′ UTR*) showed a similar effect, while co-electroporation of *Pten-FL-3′ UTR* and miR-19a ablated this effect. Data are presented as mean ± SD; $n \geq 3$ for all electroporations; p values in relation to control (***$p < 0.001$). Scale bar: 50 μm.

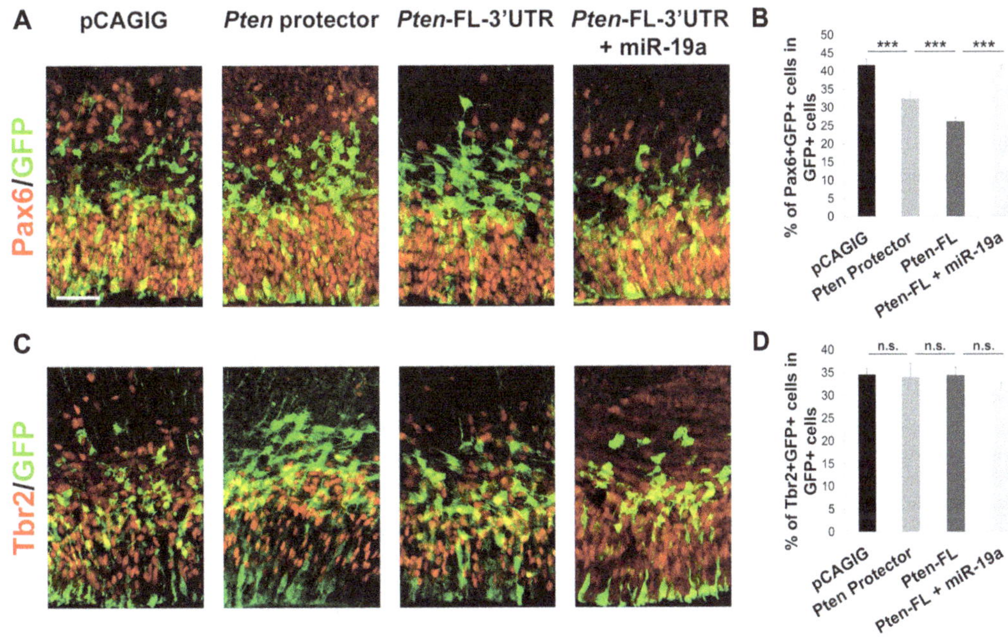

FIGURE 5 | miR-19a targeting of *Pten* is responsible for RGC expansion in the developing mouse cortex. (A,B) miR-19a targeting of *Pten* is important for RGC expansion. Introduction of the 60 nt target protector at E13.5 for analysis at E14.5 significantly decreased the number of RGCs colabeled with GFP and Pax6. Ectopic expression of *Pten-FL-3′ UTR* showed a similar effect, while co-electroporation of *Pten-FL-3′ UTR* and miR-19a ablated this effect. **(C,D)** Ectopic expression of the 60 nt target protector, *Pten-FL-3′ UTR*, or *Pten-FL-3′ UTR* and miR-19a had no effect on the number of IPs colabeled with GFP and Tbr2. Data are presented as mean ± SEM; $n \geq 3$ for all electroporations; p values in relation to control (***$p < 0.001$). n.s., not significant. Scale bar: 50 μm.

in some model organisms, it is not applicable to many mammalian systems.

We have shown that plasmid-based target protectors are effective in mammalian systems. They are a simple and clean method for blocking miRNA action at a single target, eliminating uncertainty surrounding which mRNA targets are responsible for a phenotype. Because they are plasmid-based, these target protectors can be applied across a wide variety of model systems and tissues. They also have long lasting expression and can be used to study gene functions at any stage of development.

Here, we used *in utero* electroporation to introduce target protectors into the developing cortex; however, this is not the only application of plasmid-based target protectors. These protectors could also be used in viral transfection, or they could be incorporated to create transgenic lines. For any method of delivery, a conditional expression vector can be used to analyze a miRNA:target relationship in a specific cell type and at a specific stage. These target protectors also have disease treatment potential, as they could be used to increase expression of a dysregulated gene by blocking a miRNA.

While our plasmid-based target protectors are shown to be very effective for dissecting the miR-19a:*Pten* relationship, we have applied them to other miRNA:target pairs with mixed results. First, it is important to reiterate that these protectors are not applicable to every miRNA-binding site: if the protector overlaps another miRNA-binding site, protein binding site, or interferes with other RNA processing signals, then the results will be difficult to interpret. Another factor that may influence protector function is RNA secondary structure, either of the target RNA or the protector itself. To alleviate this problem, we recommend designing and testing target protectors of multiple lengths; though we found the 60 nt target protector to be the most effective, this may not be true in every case.

Another reason that a single target protector might not show a phenotype is that miRNAs often target multiple members of the same pathway. Thus, derepression of a single pathway member may be masked by continued regulation of the rest of the pathway. In this study, we successfully introduced target protectors for two miR-19a binding sites and we recommend this combinatorial approach for derepression of multiple pathway members. However, this method is limited by the maximum amount of DNA that can be introduced at a time. A large amount of each target protector is required to see an effect, which would be diluted by including multiple target protectors.

Plasmid-based target protectors open myriad opportunities in the miRNA field for dissecting specific miRNA:target interactions in mammalian model systems. These target protectors are applicable across tissues and developmental systems, and can be introduced in many ways. Here, we have presented basic concepts of target protector design and shown their application both *in vitro* and *in vivo*. We also revealed an important role of the miR-19a:*Pten* interaction in the developing cortex, demonstrating the great potential of this essential new tool.

ACKNOWLEDGMENTS

We thank Andrew Pollock for helpful discussions of protector designs. This work was supported by the NIH T32 Training Grant in Developmental Biology at Weill Cornell Medical College (Jennifer L. Knauss), the Hirschl/Weill-Caulier Trust (Tao Sun), an NPRP grant (09-1011-3-260) from the Qatar National Research Fund and an R01-MH083680 grant from the NIH/NIMH (Tao Sun).

REFERENCES

Alvarez-Garcia, I., and Miska, E. A. (2005). MicroRNA functions in animal development and human disease. *Development* 132, 4653–4662. doi: 10.1242/dev.02073

Bartel, D. P. (2004). MicroRNAs: genomics, biogenesis, mechanism, and function. *Cell* 116, 281–297. doi: 10.1016/S0092-8674(04)00045-5

Bian, S., Hong, J., Li, Q., Schebelle, L., Pollock, A., Knauss, J. L., et al. (2013). MicroRNA cluster miR-17-92 regulates neural stem cell expansion and transition to intermediate progenitors in the developing mouse neocortex. *Cell Rep.* 3, 1398–1406. doi: 10.1016/j.celrep.2013.03.037

Bonev, B., Pisco, A., and Papalopulu, N. (2011). MicroRNA-9 reveals regional diversity of neural progenitors along the anterior–posterior axis. *Dev. Cell* 20, 19–32. doi: 10.1016/j.devcel.2010.11.018

Carrington, J. C., and Ambros, V. (2003). Role of microRNAs in plant and animal development. *Science* 301, 336–338. doi: 10.1126/science.1085242

Choi, W.-Y., Giraldez, A. J., and Schier, A. F. (2007). Target protectors reveal dampening and balancing of nodal agonist and antagonist by miR-430. *Science* 318, 271–274. doi: 10.1126/science.1147535

Conte, I., Carrella, S., Avellino, R., Karali, M., Marco-Ferreres, R., Bovolenta, P., et al. (2010). miR-204 is required for lens and retinal development via Meis2 targeting. *Proc. Natl. Acad. Sci. U.S.A.* 107, 15491–15496. doi: 10.1073/pnas.0914785107

Ebert, M. S., Neilson, J. R., and Sharp, P. A. (2007). MicroRNA sponges: competitive inhibitors of small RNAs in mammalian cells. *Nat. Methods* 4, 721–726. doi: 10.1038/nmeth1079

Ebert, M. S., and Sharp, P. A. (2010). MicroRNA sponges: progress and possibilities. *RNA* 16, 2043–2050. doi: 10.1261/rna.2414110

Friedman, R. C., Farh, K. K., Burge, C. B., and Bartel, D. P. (2009). Most mammalian mRNAs are conserved targets of microRNAs. *Genome Res.* 19, 92–105. doi: 10.1101/gr.082701.108

Groszer, M., Erickson, R., Scripture-Adams, D. D., Lesche, R., Trumpp, A., Zack, J. A., et al. (2001). Negative regulation of neural stem/progenitor cell proliferation by the pten tumor suppressor gene *in vivo*. *Science* 294, 2186–2189. doi: 10.1126/science.1065518

Karres, J. S., Hilgers, V., Carrera, I., Treisman, J., and Cohen, S. M. (2007). The conserved microRNA miR-8 tunes atrophin levels to prevent neurodegeneration in *Drosophila*. *Cell* 131, 136–145. doi: 10.1016/j.cell.2007.09.020

Kloosterman, W. P., and Plasterk, R. H. (2006). The diverse functions of microRNAs in animal development and disease. *Dev. Cell* 11, 441–450. doi: 10.1016/j.devcel.2006.09.009

Landgraf, P., Rusu, M., Sheridan, R., Sewer, A., Iovino, N., Aravin, A., et al. (2007). A mammalian microRNA expression atlas based on small RNA library sequencing. *Cell* 129, 1401–1414. doi: 10.1016/j.cell.2007.04.040

Lewis, B. P., Burge, C. B., and Bartel, D. P. (2005). Conserved seed pairing, often flanked by adenosines, indicates that thousands of human genes are MicroRNA targets. *Cell* 120, 15–20. doi: 10.1016/j.cell.2004.12.035

Lewis, B. P., Shih, I., Jones-Rhoades, M. W., Bartel, D. P., and Burge, C. B. (2003). Prediction of mammalian MicroRNA targets. *Cell* 115, 787–798. doi: 10.1016/S0092-8674(03)01018-3

Lim, L. P., Lau, N. C., Garrett-Engele, P., Grimson, A., Schelter, J. M., Castle, J., et al. (2005). Microarray analysis shows that some microRNAs downregulate large numbers of target mRNAs. *Nature* 433, 769–773. doi: 10.1038/nature03315

Loya, C. M., Lu, C. S., Vanvactor, D., and Fulga, T. A. (2009). Transgenic microRNA inhibition with spatiotemporal specificity in intact organisms. *Nat. Methods* 6, 807–903. doi: 10.1038/nmeth.1402

Mavrakis, K. J., Wolfe, A. L., Oricchio, E., Palomero, T., De Keersmaecker, K., McJunkin, K., et al. (2010). Genome-wide RNA-mediated interference screen identifies miR-19 targets in Notch-induced T-cell acute lymphoblastic leukaemia. *Nat. Cell Biol.* 12, 372–379. doi: 10.1038/ncb2037

Mu, P., Han, Y. C., Betel, D., Yao, E., Squatrito, M., Ogrodowski, P., et al. (2009). Genetic dissection of the miR-17–92 cluster of microRNAs in Myc-induced B-cell lymphomas. *Genes Dev.* 23, 2806–2811. doi: 10.1101/gad.1872909

Olive, V., Bennett, M. J., Walker, J. C., Ma, C., Jiang, I., Cordon-Cardo, C., et al. (2009). miR-19 is a key oncogenic component of mir-17-92. *Genes Dev.* 23, 2839–2849. doi: 10.1101/gad.1861409

Rasmussen, K. D., Simmini, S., Abreu-Goodger, C., Bartonicek, N., Di Giacomo, M., Bilbao-Cortes, D., et al. (2010). The miR-144/451 locus is required for erythroid homeostasis. *J. Exp. Med.* 207, 1351–1358. doi: 10.1084/jem.20100458

Saito, T. (2006). *In vivo* electroporation in the embryonic mouse central nervous system. *Nat. Protoc.* 1, 1552–1558. doi: 10.1038/nprot.2006.276

Sokol, N. S. (2005). Mesodermally expressed *Drosophila* microRNA-1 is regulated by Twist and is required in muscles during larval growth. *Genes Dev.* 19, 2343–2354. doi: 10.1101/gad.1356105

Stanton, A. A., and Giraldez, A. J. (2011). Use of target protector morpholinos to analyze the physiological roles of specific miRNA-mRNA pairs *in vivo*. *Nat. Protoc.* 6, 2035–2049. doi: 10.1038/nprot.2011.423

Ventura, A., Young, A. G., Winslow, M. M., Lintault, L., Meissner, A., Erkeland, S. J., et al. (2008). Targeted deletion reveals essential and overlapping functions of the miR-17–92 family of miRNA clusters. *Cell* 132, 875–886. doi: 10.1016/j.cell.2008.02.019

Woltering, J. M., and Durston, A. J. (2008). MiR-10 represses HoxB1a and HoxB3a in zebrafish. *PLoS ONE* 3:e1396. doi: 10.1371/journal.pone.0001396

Zheng, H., Ying, H., Yan, H., Kimmelman, A. C., Hiller, D. J., Chen, A. J., et al. (2008). p53 and Pten control neural and glioma stem/progenitor cell renewal and differentiation. *Nature* 455, 1129–1133. doi: 10.1038/nature07443

5

microRNAs in the axon and presynaptic nerve terminal

Barry B. Kaplan[1]*, Amar N. Kar[1], Anthony E. Gioio[1] and Armaz Aschrafi[2]

[1] Laboratory of Molecular Biology, National Institute of Mental Health, National Institutes of Health, Bethesda, MD, USA
[2] Department of Cognitive Neuroscience, Donders Institute for Brain, Cognition, and Behavior, Radboud University Nijmegen Medical Center, Nijmegen, Netherlands

***Correspondence:**
Barry B. Kaplan, Laboratory of Molecular Biology, National Institute of Mental Health, National Institutes of Health, 9000 Rockville Pike, Building 10, Room 4N213, Bethesda, MD 20892-1381, USA
e-mail: barry.kaplan@nih.gov

The distal structural/functional domains of the neuron, to include the axon and presynaptic nerve terminal, contain a large, heterogeneous population of mRNAs and an active protein synthetic system. These local components of the genetic expression machinery play a critical role in the development, function, and long-term viability of the neuron. In addition to the local mRNA populations these presynaptic domains contain a significant number of non-coding RNAs that regulate gene expression post-transcriptionally. Here, we review a small, but rapidly evolving literature on the composition and function of microRNAs that regulate gene expression locally in the axon and nerve terminal. In this capacity, these small regulatory RNAs have a profound effect on axonal protein synthesis, local energy metabolism, and the modulation of axonal outgrowth and branching.

Keywords: axonal mRNAs, mitochondrial mRNAs, intra-axonal protein synthesis, translational regulation, energy metabolism, ROS generation, axonal growth

INTRODUCTION

The distal structural/functional domains of the neuron (i.e., axon, presynaptic nerve terminal, and dendrite) are well known to contain a highly diverse population of mRNAs and an active protein synthetic system. In large, asymmetric neurons, these basic components of the genetic machinery play an especially important role in the development and maintenance of its cellular polarity, as well as in its synaptic plasticity, regeneration, and repair (for review, see Jung et al., 2012).

Initial estimates of the diversity of these mRNA populations, as derived from invertebrate model systems, suggested the presence of 200–400 different mRNAs (Perrone-Capano et al., 1987; Moccia et al., 2003). However, the study of mammalian neurons and the resolution afforded by advanced gene profiling methodology has revealed a remarkable complexity in the axonal transcriptome (Willis et al., 2005; Taylor et al., 2009; Zivraj et al., 2010; Gumy et al., 2011). These messengers encode a complex set of proteins that can be organized into several functional categories to include: cytoskeletal and scaffolding proteins, translation factors and ribosomal proteins, molecular motors and chaperones, and metabolic enzymes.

In addition to these diverse mRNA populations, axons and nerve terminals also contain numerous microRNAs (miRNAs). These highly conserved, small, non-coding RNAs play a key role in the post-transcriptional regulation of gene expression. In general the bio-genesis and function of miRNAs in the nervous system has been well reviewed (Schratt, 2009; Siegel et al., 2011; Olde Loohuis et al., 2012). In this brief communication, attention will be focused on those miRNAs that function to post-transcriptionally regulate gene expression locally in the distal structural domains of the neuron.

IDENTIFICATION OF miRNAs IN THE AXON

In a recent study, Natera-Naranjo et al. (2010) employed primary sympathetic neurons cultured in compartmentalized Campenot chambers to obtain a pure axonal RNA fraction to identify the component miRNAs by microarray analysis. Surprisingly, this study revealed considerable complexity in the miRNA population present in this cellular compartment. The relative abundance of several of these miRNAs was found to be highly enriched in the axon as compared to the parental cell bodies, a finding that raised the possibility that there could be a selective transport of these molecules into the axon (see below).

Of course, estimation of the number of miRNAs present in the axon, as derived from microarray profiling, is totally dependent upon the numerical threshold used to determine a positive signal (i.e., the signal to noise ratio). To address this issue, the authors employed the convergence of the microarray data with the results obtained from quantitative RT-PCR analyses. High correlations between the microarray data and the results of quantitative RT-PCR were obtained for miRNAs with microarray signal intensities greater than one standard deviation above median background values. Hence, using this as a "cut-off" value it was determined that there was approximately 137 different miRNAs in the axons of rat superior cervical ganglion neurons (Natera-Naranjo et al., 2010). It bears mention here that the cell culture system used in these studies precludes one from distinguishing between the axon and nerve terminal, and hence reference to these two cellular compartments will be combined throughout the review.

Bioinformatic search for putative mRNA targets of the axonally abundant miRNAs revealed an enrichment of transcripts that encode proteins that function in neuronal signaling, mRNA and protein transport, as well as, mRNA transcription and translation. Interestingly, this analysis also identified a subset of axonal miRNAs that potentially target multiple mRNAs involved in related cellular functions. These observations suggest that these small, non-coding RNAs in the axon could function to integrate/orchestrate multiple functions in the axon.

Many miRNAs are expressed as clusters on a single polycistronic transcript. Interestingly, the relative abundance of some miRNAs derived from polycistronic transcripts differ markedly in the axon as compared to their parental cell soma (Natera-Naranjo et al., 2010). For example, the miRNA-17-92 cluster is comprised of six individual miRNAs. The relative abundance of two of the mature miRNAs were two- to fourfold greater in the axon than in the corresponding cell bodies (Natera-Naranjo et al., 2010; Zhang et al., 2013). After over-expression of the miRNA-17-92 cluster in embryonic cortical neurons cultured in microfluidity chambers, the relative abundance of miR-19b and miR-20a, two components of the polycistronic transcript, were approximately 5- to 20-fold greater in the axon compared to the cell soma (Zhang et al., 2013). The mechanism underlying the differential or selective transport of these miRNAs is currently unknown, but is eminently worthy of future investigation.

ACTIVITY OF THE AXONAL PROTEIN SYNTHETIC SYSTEM IS MODULATED BY miRNAs

One of the most abundant miRNAs present in the axons of sympathetic neurons is miR-16. Results of a bioinformatics search for mRNAs whose 3′untranslated regions (UTRs) contained miR-16 binding sites revealed two mRNAs that encoded eukaryotic translation initiation factors, eIF2B2 and eIF4G2 (Kar et al., 2013). The cognate mRNAs for both these factors are present in the axon and the local expression of eIF2B2 and eIF4G2 proteins can be modulated by miR-16. The regulation of the expression of these factors was shown to be effected through the binding of miR-16 to the 3′UTR with subsequent degradation of the mRNAs. The transfection of the precursor miRNA directly into the axon greatly reduced the levels of these factors and markedly inhibited the activity of the local protein synthetic system, as judged by metabolic labeling studies. Similar effects on local protein synthesis and axon growth were observed after small interfering RNA-mediated knockdown of axonal eIF2B2 and eIF4G2 mRNA. Taken together, these findings demonstrated that the expression of miRNAs in the distal axon could modulate local protein synthesis by regulating the expression of key components of the translation system.

miRNAs REGULATE ENERGY METABOLISM IN THE AXON

One surprising feature of all axons studied to date is the presence of a large number of nuclear-encoded mitochondrial mRNAs. It has been estimated that in large invertebrate axons and presynaptic nerve terminal nearly one-quarter of the newly synthesized protein is destined for mitochondria, and that the membrane potential and activity of this organelle is highly dependent on the local synthesis of these rapidly turning over proteins (Hillefors et al., 2007; Kaplan et al., 2009).

It is noteworthy that several of these nuclear-encoded mitochondrial mRNAs code for proteins that play a key role in oxidative phosphorylation, a finding that suggests that their local synthesis might contribute to the regulation of ATP production. At least two of these mitochondrial mRNAs, cytochrome c-oxidase (CoxIV) and ATP synthase (ATP5G1) contain a binding site for miR-338 in their 3′UTRs. Both of these binding sites are situated in a hairpin-loop structure that could facilitate miRNA accessibility (**Figure 1**). Axonal transfection studies conducted with chimeric reporter gene constructs containing these putative binding sites established that they were *bona fide* targets of miR-338 (Aschrafi et al., 2008, 2012). Consistent with these findings, the over-expression of miR-338 in the axon greatly reduced the levels of endogenous CoxIV and ATP5G1 mRNA and protein in the axon, and resulted in marked

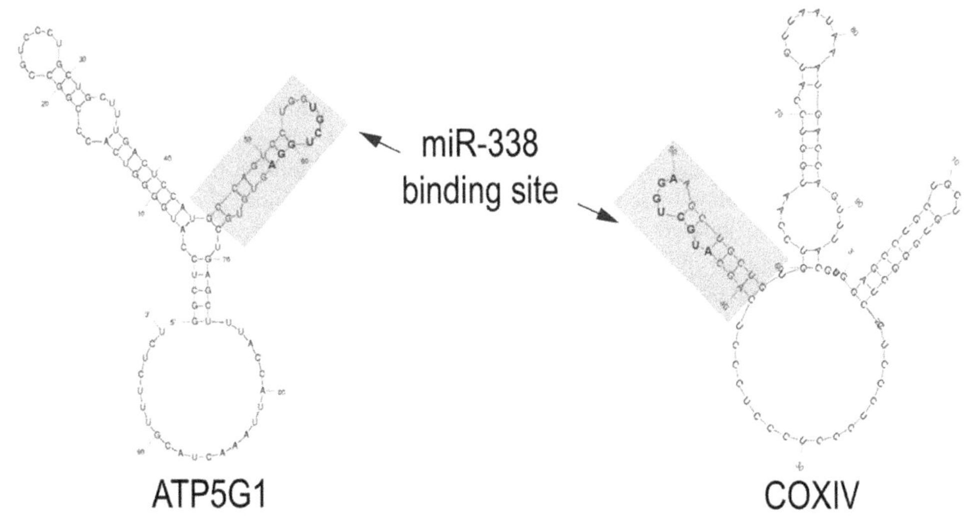

FIGURE 1 | miR-338 binding sites in nuclear-encoded mitochondrial mRNAs present in the axon. The secondary structure of the 3′UTRs of CoxIV and ATP5G1 mRNAs were predicted by Mfold analyses. The miR-338 binding site present in a 38-nucleotide stem-loop structure is indicated in gray. Nucleotides comprising the microRNA targeting sequence are bolded. Modified from Aschrafi et al. (2012).

reduction in local ATP levels and elevation in the production of reactive oxygen species (ROS). In contrast, inhibition of endogenous miR-338 function had the opposite effects on axonal ATP levels and ROS generation.

To delineate the impact of the modulation of local levels of miR-338 on the metabolic rate and function of axons of noradrenergic sympathetic neurons, mitochondrial oxygen consumption, as estimated by reduction of the redox dye Alamar Blue, as well as norepinephrine uptake into the axon was assessed after introducing precursor miR-338 or anti-miR-338 directly into the axon. Under the cell culture conditions employed in these experiments, modulation of the local levels of miR-338 had a profound effect on the metabolic rate and function of the axon (Aschrafi et al., 2008).

LOCAL EXPRESSION OF miRNAs EFFECT AXONAL GROWTH AND BRANCHING

In light of the fact that miRNAs can influence the activity of the intra-axonal protein synthetic system, as well as local energy metabolism, it is not surprising that the expression of these small regulatory RNAs could have profound effects on the growth and branching of the axon. For example, elevation of the levels of miR-338 or miR-16 in the axons of cultured primary sympathetic neurons inhibits their rate of elongation (Aschrafi et al., 2008, 2012; Kar et al., 2013). The attenuation in axonal outgrowth could be attributed, at least in part, to the dysregulation of mitochondrial function with consequent elevation in the production of ROS in the axon. In this regard, partial restoration of axonal growth could be effected by the application of anti-oxidants to the culture media (Natera-Naranjo et al., 2012).

In a recent study, Zhang et al. (2013) reported the expression of the components of the miR-17-92 cluster in the distal axons of primary embryonic cortical neurons. Over-expression of this cluster substantially increased axonal outgrowth, whereas the inhibition of endogenous miR-19a, a key component of the cluster, suppressed axonal growth. The local effects of this miRNA were attributed to the modulation of phosphatase and tension homolog (PTEN) protein levels by miR-19a.

In an elegant series of experiments, Dajas-Bailador et al. (2012) demonstrated that inhibition of local miR-9 in primary embryonic cortical neurons facilitated axonal outgrowth and inhibited branching of the axon. Ostensibly, these effects were mediated through the regulation of one of its targets, microtubule-associated protein 1b (MAP1B). Interestingly, although miR-9 was detected in dendrites, the over-expression of this miRNA had no effect on their length, a finding which could reflect the preferential localization of Map1b mRNA to the developing axon.

To assess the role of miR-9 *in vivo*, Dajas-Bailador et al. (2012) introduced a specific miR-9 inhibitor into the cerebral cortex of 14.5 day mouse embryos. The inhibition of endogenous miR-9 activity resulted in a severe disruption of neuronal migration and differentiation.

Taken together, these findings indicate that miRNAs play an important role in the function and development of the axon through the post-transcriptional modulation of the expression of key constituents of the local mRNA population.

DISCUSSION

Recent research has revealed the presence of a diverse population of miRNAs in the axon and presynaptic nerve terminal. Nonetheless, only a few of these miRNAs and their target genes have been characterized, and hence the function of the vast majority of these non-coding RNAs remains unknown. Some of these miRNAs are expressed specifically in brain and the relative abundance of others differs markedly in the various cellular compartments of the neuron. These observations suggest a unique regulatory role for these RNAs in the development, maintenance and function of the distal regions of this highly polarized cell.

One remarkable feature of miRNAs is that they can coordinately regulate the expression of multiple mRNAs that encode proteins with related cellular functions. In this regard, miRNAs might be envisaged as "master regulators" of the expression of batteries of genes which comprise functional networks and/or pathways (for review, see Olde Loohuis et al., 2012). One case in point, is miR-338 which targets several mRNAs that code for proteins that are key components of the enzymatic complexes that comprise the oxidative phosphorylation chain. It is important to note here, that the effects of miR-338 on CoxIV and ATP5G1 expression are additive and hence, this miRNA could serve to "fine-tune" the local production of ATP in response to neuronal activity.

On a more global level, non-coding RNAs, such as miR-16, have been shown to regulate the activity of the axonal protein synthetic system, itself. This effect was mediated through the modulation of the expression of two translation initiation factors (**Figure 2**). Thus, through this mechanism, miR-16 could affect the local expression of large numbers of gene products. One might speculate that some of these locally synthesized proteins are likely to play an important role in the activity-dependent modulation of the function of the axon and nerve terminal, as well as in synaptic plasticity.

In view of the effects of miRNA on protein synthesis and local energy metabolism, it is not surprising that these non-coding RNAs could have a profound effect on the growth and development of the axon. These findings support the hypothesis that miRNAs play a central role in gene regulatory networks involved in axon development, as well the plasticity of the presynaptic nerve terminal. These observations also raise the possibility that dysregulation of miRNA function might play a role in the pathophysiology of neurological and psychiatric disorders (for review, see Qureshi and Mehler, 2012; Rege et al., 2013).

Despite the evidence for the existence of a host of miRNAs in the axon, there is a dearth of knowledge regarding the mechanism(s) underlying the transport of these RNAs to the distal regions of the neuron. Several models have been proposed for the selective shuttling of miRNAs to the synapse (for example, see Kosik, 2006). These working hypotheses include: co-delivery of the miRNAs with their cognate mRNA targets via RNA granules; mRNA-independent delivery of mature/functional miRNAs; and the delivery of precursor miRNAs followed by their sequential local processing to the mature, functional form of the non-coding RNA. This last model is supported by several reports of the presence of miRNA processing machinery in mammalian axons (Hengst et al., 2006; Aschrafi et al., 2008; Zhang et al., 2013). In addition, the

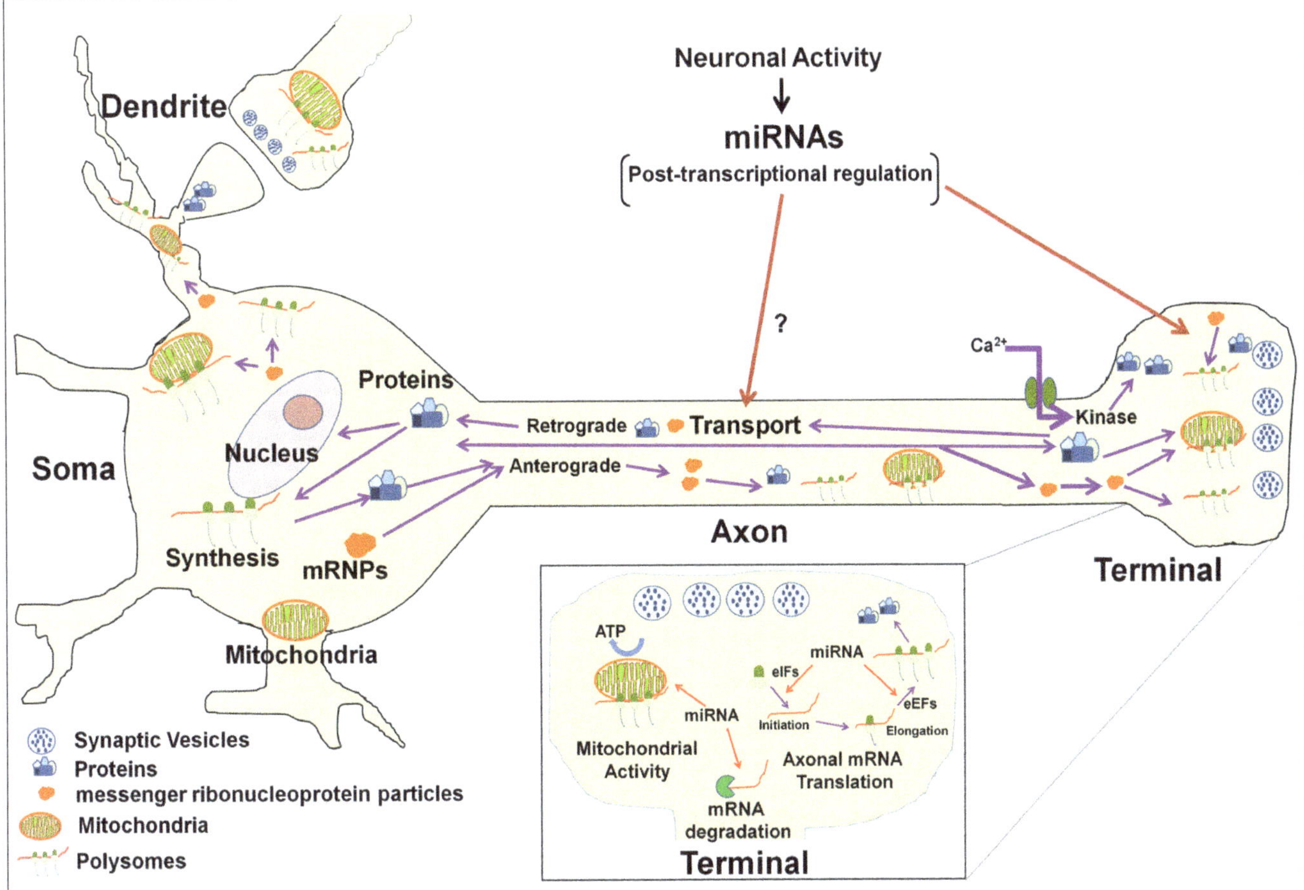

FIGURE 2 | Model for miRNA-mediated regulation of neuronal function. In the neuron, protein synthesis occurs in multiple compartments that include the cell body, dendrite, axon, and presynaptic nerve terminal. A subset of mRNAs transcribed in the nucleus are packaged into stable messenger ribonucleoproteins complexes (mRNPs) and are selectively and rapidly transported to the distal structural/functional domains of the neuron. The selective translation of these localized mRNAs plays key roles in neuronal development, axon growth and maintenance, and synaptic plasticity. Neuronal miRNAs function at multiple levels within the neuronal gene expression system to modulate neuronal activity and function. It has been shown that miRNAs regulate the local post-transcriptional gene expression of specific target mRNAs that encode factors affecting mitochondrial activity (see inset) as well as axonal growth and branching. In addition, miRNAs can also modulate translation of multiple mRNAs in the axon and nerve terminal by regulating local expression of eukaryotic translation factors (see inset). It is also conceivable that miRNAs might regulate the local synthesis and retrograde transport of transcription factors in response to growth factors or neural injury and hence influence gene transcription in the parental soma. Last, miRNA control of the local synthesis of cytoskeletal and/or motor proteins might facilitate the regulation of their own anterograde transport to their ultimate sites of function.

introduction of precursor miRNAs directly into the axon results in a marked increase in the levels of the mature, functional forms of the molecule within hours (Aschrafi et al., 2008). These observations clearly indicate that axons have the capability to process precursor miRNAs to mature forms of the molecule. One interesting question for future research is whether synaptic activity could regulate axonal miRNA trafficking or alternatively regulate the retrograde transport of protein(s) that could influence miRNA transcription in their parental neurons (**Figure 2**). In this regard, it is well known that transcription factors can be locally synthesized in the axon and retrogradely transported to the neuronal cell nucleus, especially in response to neurotrophic signaling and/ or neuronal injury (Cox et al., 2008; Gumy et al., 2010; Ben-Yaakov et al., 2012). However, to date, no studies have investigated the role that miRNAs might play in the local regulation of axon regeneration.

Future characterization of the axonal miRNA regulatory landscape will generate further insights into the molecular basis of axonal activity, maintenance, and neuronal function. Moreover, recent studies have delineated the functional significance of miRNAs in the regulation of local translation in developing axons, dysregulation of which might ultimately underlie the etiology of neurodevelopmental disorders and mental illness.

ACKNOWLEDGMENTS

The authors work was supported by the Division of Intramural Research Programs of the National Institute of Mental Health (ZO1MH00276). We thank Noreen Gervasi, Mi Hellifors, Margaret Macgibeny, Orlangie Natera-Naranjo, and Sanah Vohra for invaluable contributions to the authors' research endeavors.

REFERENCES

Aschrafi, A., Schwechter, A. D., Mameza, M. G., Natera-Naranjo, O., Gioio, A. E., and Kaplan, B. B. (2008). MicroRNA-338 regulates local cytochrome c oxidase IV mRNA levels and oxidative phosphorylation in the axons of sympathetic neurons. *J. Neurosci.* 28, 12581–12590. doi: 10.1523/JNEUROSCI.3338-08.2008

Aschrafi, A., Kar, A. N., Natera-Naranjo, O., Macgibeny, M. A., Gioio, A. E., and Kaplan, B. B. (2012). MicroRNA-338 regulates the axonal expression of multiple nuclear-encoded mitochondrial mRNAs encoding subunits of the oxidative phosphorylation machinery. *Cell. Mol. Life. Sci.* 6, 4017–4027. doi: 10.1007/s00018-012-1064-8

Ben-Yaakov, K., Dagan, S. Y., Segal-Ruder, Y., Shalem, O., Vuppalanchi, D., Willis, D. E., et al. (2012). Axonal transcription factors signal retrogradely in lesioned peripheral nerve. *EMBO J.* 31, 1350–1363. doi: 10.1038/emboj.2011.494

Cox, L. J., Hengst, U., Gurskaya, N. G., Lukyanov, K. A., and Jaffrey, S. R. (2008). Intra-axonal translation and retrograde trafficking of CREB promotes neuronal survival. *Nat. Cell Biol.* 10, 149–159. doi: 10.1038/ncb1677

Dajas-Bailador, F., Bonev, B., Garcez, P., Stanley, P., Guillemot, F., and Papalopulu, N. (2012). microRNA-9 regulates axon extension and branching by targeting Map1b in mouse cortical neurons. *Nat. Neurosci.* 15, 697–699. doi: 10.1038/nn.3082

Gumy, L. F., Tan, C. L., and Fawcett, J. W. (2010). The role of local protein synthesis and degradation in axon regeneration. *Exp. Neurol.* 223, 28–37. doi: 10.1016/j.expneurol.2009.06.004

Gumy, L. F., Yeo, G. S., Tung, Y. C., Zivraj, K. H., Willis, D., Coppola, G., et al. (2011). Transcriptome analysis of embryonic and adult sensory axons reveals changes in mRNA repertoire localization. *RNA* 17, 85–98. doi: 10.1261/rna.2386111

Hengst, U., Cox, L. J., Macosko, E. Z., and Jaffrey, S. R. (2006). Functional and selective RNA interference in developing axons and growth cones. *J. Neurosci.* 26, 5727–5732. doi: 10.1523/JNEUROSCI.5229-05.2006

Hillefors, M., Gioio, A. E., Mameza, M. G., and Kaplan, B. B. (2007). Axon viability and mitochondrial function are dependent on local protein synthesis in sympathetic neurons. *Cell. Mol. Neurobiol.* 27, 701–716. doi: 10.1007/s10571-007-9148-y

Jung, H., Yoon, B. C., and Holt, C. E. (2012). Axonal mRNA localization and local protein synthesis in nervous system assembly, maintenance and repair. *Nat. Rev. Neurosci.* 13, 308–324. doi: 10.1038/nrn3274

Kaplan, B. B., Gioio, A. E., Hillefors, M., and Aschrafi, A. (2009). Axonal protein synthesis and the regulation of local mitochondrial function. *Results Probl. Cell Differ.* 48, 225–242. doi: 10.1007/400_2009_1

Kar, A. N., Macgibeny, M. A., Gervasi, N. M., Gioio, A. E., and Kaplan, B. B. (2013). Intra-axonal synthesis of eukaryotic translation initiation factors regulates local protein synthesis and axon growth in rat sympathetic neurons. *J. Neurosci.* 33, 7165–7174. doi: 10.1523/JNEUROSCI.2040-12.2013

Kosik, K. S. (2006). The neuronal microRNA system. *Nat. Rev. Neurosci.* 7, 911–920. doi: 10.1038/nrn2037

Moccia, R., Chen, D., Lyles, V., Kapuya, E. E., Kalachikov, S., Spahn, C. M., et al. (2003). An unbiased cDNA library prepared from isolated Aplysia sensory neuron processes is enriched for cytoskeletal and translational mRNAs. *J. Neurosci.* 23, 9409–9417.

Natera-Naranjo, O., Aschrafi, A., Gioio, A. E., and Kaplan, B. B. (2010). Identification and quantitative analyses of microRNAs located in the distal axons of sympathetic neurons. *RNA* 16, 1516–1529. doi: 10.1261/rna.1833310

Natera-Naranjo, O., Kar, A. N., Aschrafi, A., Gervasi, N. M., MacGibeny, M. A., Gioio, A. E., et al. (2012). Local translation of ATP synthase subunit 9 mRNA alters ATP levels and the production of ROS in the axon. *Mol. Cell. Neurosci.* 49, 263–270. doi: 10.1016/j.mcn.2011.12.006

Olde Loohuis, N. F., Kos, A., Martens, G. J., Van Bokhoven, H., Nadif Kasri, N., and Aschrafi, A. (2012). MicroRNA networks direct neuronal development and plasticity. *Cell. Mol. Life Sci.* 69, 89–102. doi: 10.1007/s00018-011-0788-1

Perrone-Capano, C. P., Giuditta, A., Castigli, E., and Kaplan, B. B. (1987). Occurrence and sequence complexity of polyadenylated RNA in squid axoplasm. *J. Neurochem.* 49, 698–704. doi: 10.1111/j.1471-4159.1987.tb00950.x

Qureshi, I. A., and Mehler, M. F. (2012). Emerging roles of non-coding RNAs in brain evolution, development, plasticity, and disease. *Nat. Rev. Neurosci.* 13, 528–541. doi: 10.1038/nrn3234

Rege, S. D., Geetha, T., Pondugula, S. R., Zizza, C. A., Wernette, C. M., and Ramesh Babu, J. (2013). Non-coding RNAs in neurodegenerative diseases. *ISRN Neurol.* 2013, 375852. doi: 10.1155/2013/375852

Schratt, G. (2009). microRNAs at the synapse. *Nat. Rev. Neurosci.* 10, 842–849. doi: 10.1038/nrn2763

Siegel, G., Saba, R., and Schratt, G. (2011). microRNAs in neurons: manifold regulatory roles at the synapse. *Curr. Opin. Genet. Dev.* 21, 491–497. doi: 10.1016/j.gde.2011.04.008

Taylor, A. M., Berchtold, N. C., Perreau, V. M., Tu, C. H., Jeon, N. L., and Cotman, C. W. (2009). Axonal mRNA in uninjured and regenerating cortical mammalian axons. *J. Neurosci.* 29, 4697–4707. doi: 10.1523/JNEUROSCI.6130-08.2009

Willis, D., Li, K. W., Zheng, J. Q., Chang, J. H., Smit, A., Kelly, J., et al. (2005). Differential transport and local translation of cytoskeletal, injury-response, and neurodegeneration protein mRNAs in axons. *J. Neurosci.* 25, 778–791. doi: 10.1523/JNEUROSCI.4235-04.2005

Zhang, Y., Ueno, Y., Liu, X. S., Buller, B., Wang, X., Chopp, M., et al. (2013). The microRNA-17-92 cluster enhances axonal outgrowth in embryonic cortical neurons. *J. Neurosci.* 33, 6885–6894. doi: 10.1523/JNEUROSCI.5180-12.2013

Zivraj, K. H., Tung, Y. C., Piper, M., Gumy, L., Fawcett, J. W., Yeo, G. S., et al. (2010). Subcellular profiling reveals distinct and developmentally regulated repertoire of growth cone mRNAs. *J. Neurosci.* 30, 15564–15478. doi: 10.1523/JNEUROSCI.1800-10.2010

6

Neuronal dark matter: the emerging role of microRNAs in neurodegeneration

Emily F. Goodall*, Paul R. Heath, Oliver Bandmann, Janine Kirby and Pamela J. Shaw

Department of Neuroscience, Sheffield Institute for Translational Neuroscience, University of Sheffield, Sheffield, UK

***Correspondence:**
Emily F. Goodall, Sheffield Institute for Translational Neuroscience, University of Sheffield, 385A Glossop Road, Sheffield S10 2HQ, UK
e-mail: e.goodall@sheffield.ac.uk

MicroRNAs (miRNAs) are small, abundant RNA molecules that constitute part of the cell's non-coding RNA "dark matter." In recent years, the discovery of miRNAs has revolutionised the traditional view of gene expression and our understanding of miRNA biogenesis and function has expanded. Altered expression of miRNAs is increasingly recognized as a feature of many disease states, including neurodegeneration. Here, we review the emerging role for miRNA dysfunction in Alzheimer's disease, Parkinson's disease, amyotrophic lateral sclerosis (ALS) and Huntington's disease pathogenesis. We emphasize the complex nature of gene regulatory networks and the need for systematic studies, with larger sample cohorts than have so far been reported, to reveal the most important miRNA regulators in disease. Finally, miRNA diversity and their potential to target multiple pathways, offers novel clinical applications for miRNAs as biomarkers and therapeutic agents in neurodegenerative diseases.

Keywords: microRNA, neurodegeneration, Alzheimer's disease, Parkinson's disease, amyotrophic lateral sclerosis, Huntington's disease

INTRODUCTION

Arguably one of the most important discoveries in molecular biology in recent years has been the finding and characterization of regulatory RNAs. The majority of the human genome is transcribed, however, less than 1.5% encodes protein. A vast amount appears to be biologically active, non-coding RNAs which are often referred to as the "dark matter" of the cell (Mattick, 2005). Progressively more advanced RNA sequencing techniques have uncovered many classes of small regulatory RNA, however, there is general recognition of three main types: microRNAs (miRNAs), short interfering RNAs (siRNAs) and piwi-interacting RNAs (piRNAs). The full range of these RNA species has been reviewed elsewhere (Carthew and Sontheimer, 2009; Kapranov and St Laurent, 2012). The aim of this article is to focus on miRNAs and their emerging role in neurodegeneration.

Traditionally conditions such as Parkinson's disease, Alzheimer's disease and amyotrophic lateral sclerosis (ALS) have been considered as distinct entities, however, there is increasing evidence of clinical, pathological and genetic overlap. Neurodegenerative diseases can therefore be considered a spectrum of aetiologies culminating in a final common final pathway of neuronal cell death. The pathogenic mechanisms underlying neurodegeneration are complex, but the universal risk factor is aging and there are common themes across the disorders, including protein aggregation, neuroinflammation and mitochondrial dysfunction. There are also common challenges across these conditions including the lack of early diagnostic testing and a large proportion of patients having sporadic forms of the disease (excepting Huntington's disease). Unraveling the similarities and differences between these conditions, and understanding cell type specific vulnerability, will be key to developing new therapeutic interventions.

BASIC BIOLOGY OF miRNA

MiRNAs are a novel class of small (18–25 nucleotides), non-coding RNA molecules predicted to post-transcriptionally regulate at least half the human transcriptome (Friedman et al., 2009). The discovery, and subsequent characterization, of miRNAs has revealed an intriguing additional level of gene regulation that is fundamental in a diverse range of pathways including development, differentiation and pathological processes. Each miRNA is estimated to regulate around 200 targets, and mRNA transcripts may be regulated by multiple miRNAs (Lewis et al., 2003; Krek et al., 2005; Lim et al., 2005). The miRNA biogenesis pathway is highly conserved, as are many miRNA sequences and their target binding sites, highlighting their importance across evolution (Berezikov et al., 2005; Friedman et al., 2009).

MiRNA genes are encoded either in intergenic regions under control of their own promoter, within the introns of protein coding genes or are exonic, overlapping with coding regions and transcribed by the host promoter (Rodriguez et al., 2004). The majority of miRNAs in humans are transcribed independently and putative promoters for the most of these have been identified (Zhou et al., 2007; Ozsolak et al., 2008). Over 40% of human miRNAs are found in clusters that are co-transcribed as polycistronic transcriptional units (Lee et al., 2002; Griffiths-Jones et al., 2008). Many miRNAs are highly temporally and spatially regulated, either via transcription factors or epigenetic mechanisms including DNA methylation and histone modification (Chuang and Jones, 2007). Overall, the mechanisms that control miRNA expression are similar to those of protein-coding genes with a trend toward regulation by their target mRNAs and double-negative feedback loops (Carthew and Sontheimer, 2009).

miRNA BIOGENESIS

CANONICAL PATHWAY

The bulk of miRNAs are generated via the typical, canonical pathway of miRNA biogenesis (**Figure 1**). MiRNA genes are transcribed by RNA polymerase II (pol II) to generate long primary transcripts (pri-miRNAs), which can be several kilobases long. The pri-miRNAs are capped, spliced and polyadenylated. They may encode a single miRNA, clusters of distinct miRNAs, or a protein and can therefore also act as mRNA precursors (Carthew and Sontheimer, 2009). The next step also takes place in the nucleus and is orchestrated by the microprocessor complex. The principal components of this complex are the RNase III enzyme known as Drosha and its binding partner DiGeorge syndrome critical region gene 8 (DGCR8), a double-stranded RNA-binding protein (Denli et al., 2004). Drosha digests pri-miRNAs to release hairpin structures called precursor miRNAs (pre-miRNAs), which are 60–70 nucleotides in length. Exportin-5 interacts directly with the pre-miRNAs to mediate their export into the cytoplasm, where a second RNase III enzyme named Dicer, cleaves the pre-miRNA to generate a double-stranded miRNA duplex of ~22 nucleotides. Following Dicer processing the miRNA duplex is rapidly unwound as it associates with Argonaute (Ago) proteins, one strand is retained to become the mature miRNA and is loaded into RNA-induced silencing complexes (RISCs) to participate in mRNA regulation. The complementary strand, which is found at lower concentrations within the cell and is sometimes called the * sequence, was believed to be non-functional and rapidly degraded. However, recent studies have demonstrated that several miRNA* sequences associate with different Ago protein complexes to also become active (Czech and Hannon, 2011).

NON-CANONICAL PATHWAYS

The advent of deep-sequencing technologies has led to the discovery of many miRNAs that are generated via alternative mechanisms, by-passing the usual Drosha/Dicer two-step processing (for in depth review see Miyoshi et al., 2010). In mammals four Drosha independent pathways have been identified, namely the mirtron pathway, small nucleolar RNA-derived, tRNA-derived and short hairpin RNA-derived pathways (Babiarz et al., 2008; Ender et al., 2008; Saraiya and Wang, 2008). The most common of these replaces the microprocessor step with a splicing event to produce short hairpin introns known as mirtrons that can be transported by Exportin-5 and cleaved by Dicer (Ruby et al., 2007). Mirtrons are relatively uncommon compared to canonical miRNAs, but have been identified throughout the animal kingdom and there is evidence to suggest a particular importance of mirtrons in the primate nervous system (Berezikov et al., 2007). In addition, there are two Dicer independent miRNA processing pathways. These are very rare with a single miRNA (miR-451) known to be produced via direct pre-miRNA loading onto Ago2 and miRNA-like small RNA sequences generated from tRNAs, with RNaseZ cleavage of pre-miRNAs in place of Dicer (Lee et al., 2009; Cheloufi et al., 2010; Haussecker et al., 2010).

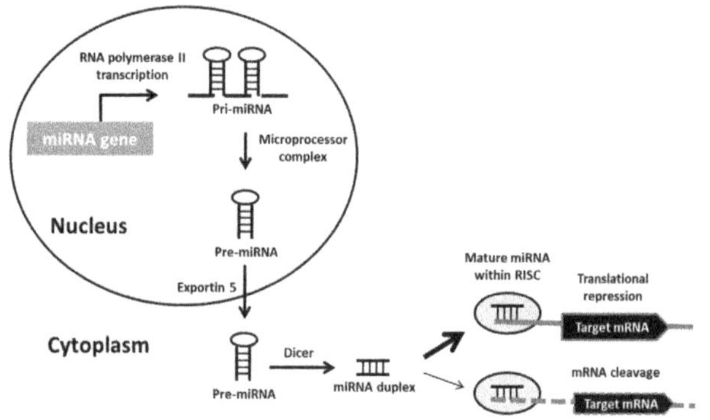

FIGURE 1 | Canonical miRNA biogenesis pathway. Primary miRNA (pri-miRNA) transcripts are transcribed by RNA polymerase II. Pri-miRNAs are processed by the microprocessor complex into precursor miRNA (pre-miRNA) hairpins. These are transported into the cytoplasm, where they are further processed by Dicer into miRNA duplexes. Following strand separation, the mature miRNAs are loaded into RNA-induced silencing complexes (RISCs) to guide the repression of protein synthesis or mRNA degradation.

miRNA MECHANISM OF ACTION

RISC is a generic term for a family of heterogeneous complexes containing Ago proteins that are involved with gene silencing (Pratt and MacRae, 2009). Once incorporated into the RISC, mature miRNAs act as a guide to direct target recognition via base-pairing interactions with mRNA transcripts, which are often located in the 3'UTR region (Bartel, 2009). The majority of animal miRNAs do not match their target sequences exactly, however, nucleotides 2–6 of the miRNA are known as the "seed region" and are critical for target recognition (Lewis et al., 2003, 2005). The extent of complementarity between a miRNA and its target mRNA sequence influences the downstream regulatory mechanism, with perfect matches leading to degradation, while mismatches result in translational repression. In humans the Ago2 protein catalyses target mRNA cleavage and subsequent degeneration of miRNA, although translational repression is the most prevalent mode of action for miRNAs in animals (Liu et al., 2004). The exact mechanism for repression remains unclear. There is evidence to support disruption of translation initiation, promotion of target mRNA deadenylation, sequestration of miRNAs and their targets to processing (P) bodies and stress granules or RISC-mediated protein degradation after translation (Tang et al., 2008). Translational repression by miRNAs is therefore complex and usually produces a fine tuning effect, with a typical miRNA-target interaction producing <2-fold reduction in protein level (Ebert and Sharp, 2012). An additional level of regulation has recently been hypothesized whereby mRNA transcripts compete for common miRNAs by sharing miRNA binding sites. These competing endogenous RNAs (ceRNAs) could be pseudogenes that have the ability to co-regulate gene expression in intricate ceRNA networks (Salmena et al., 2011). Further experimental evidence is, however, required to validate this theory. MiRNAs are recognized as negative regulators of gene expression but there are reports of target activation by miRNAs under certain conditions such as cellular stress (Bhattacharyya et al., 2006; Vasudevan et al., 2007; Orom et al., 2008).

In contrast to miRNA biogenesis, which has been extensively studied and well-defined, the regulation of miRNA degradation and turnover is less clear. MiRNAs are generally considered to be highly stable molecules with a long half-life (Krol et al., 2010b). However, recent studies indicate that miRNA turnover can vary widely among miRNAs and cell types, with rapid miRNA decay a common feature of neuronal cells (Krol et al., 2010a). There is also evidence of miRNA recycling by the cell which may help to explain their capacity to regulate large numbers of transcripts (Baccarini et al., 2011). Mature miRNAs are protected by binding to Ago proteins and the presence of mRNA target sequences are believed to be an important factor in preventing their release from RISC complexes and subsequent degradation (Diederichs and Haber, 2007). Hence, in the absence of complementary mRNA targets, miRNAs could be specifically released to make RISC available for loading new miRNAs. Two families of exonuclease enzymes have so far been identified as mediators of miRNA decay, namely small RNA degrading nuclease (SDN) genes in plants and exoribonuclease 2 (XRN2) in animals (Ramachandran and Chen, 2008; Chatterjee and Grosshans, 2009).

miRNAs IN NEURODEGENERATIVE DISORDERS

MiRNAs are found in high abundance within the nervous system where they are key regulators of functions such as neurite outgrowth, dendritic spine morphology, neuronal differentiation and synaptic plasticity. The dysfunction of miRNAs in neurodegenerative disorders is increasing recognized, see **Table 1** for a summary of the miRNAs discussed within this review.

ALZHEIMER'S DISEASE

Alzheimer's disease is a complex neurodegenerative disorder and the most common form of dementia in the elderly (Avramopoulos, 2009; Schonrock and Gotz, 2012). The clinical signs of disease are a slow, progressive loss of cognitive function and memory loss, due to destruction of synapses and neurons, which ultimately leads to dementia and death. Alzheimer's disease is progressive with different brain regions and cells affected in a sequential process of increasing deposition of amyloid-β (Aβ) plaques and neurofibrillary tangles of hyperphosphorylated tau as described by Braak staging (Braak and Braak, 1995). Aβ is a mainly 40–42 amino acid fragment derived from the membrane spanning amyloid precursor protein (APP) by proteolytic cleavage by the β-site APP cleaving enzyme (BACE1) and presenilin dependent γ-secretase (Delay et al., 2012).

Less than 1% of Alzheimer's disease cases are familial, with autosomal dominant mutations described in only three genes that lead to early onset disease; *APP*, presenilin 1 *(PSEN1)* and presenilin 2 *(PSEN2)*, both of the latter encoding components of the γ-secretase pathway (Schonrock and Gotz, 2012). No other candidate genes have been identified for familial Alzheimer's disease, although over 500 polymorphisms have been proposed to be risk alleles (Bertram et al., 2007, 2010; Tanzi, 2012). Possession of the ε4 allele of the Apolipoprotein E (*ApoE*) genotype is known to have a modifying influence on the genotype and is associated with a predisposition for the disease. The vast majority of Alzheimer's disease is sporadic, with no obvious genetic component, suggesting that other mechanisms are responsible. Recent studies have demonstrated that alterations in the network of miRNAs contribute to the disease process.

Several studies have used profiling strategies to show miRNA dysregulation in Alzheimer's disease patient brain tissues [see Schonrock and Gotz (2012) for a detailed review of these]. However, little overlap in the specific miRNA changes identified has been observed, which might result from differences in experimental technique, but it is likely that much of this variation derives from differences in the tissue examined and diagnostic features. Comparative miRNA expression in gray and white matter of normal individuals and early stage Alzheimer's disease revealed that most of the disease associated miRNA changes were found in the gray matter. This work highlights that cellular composition of the regions has a marked effect upon the miRNA expression profile, for instance the white matter profile is markedly influenced by the oligodendrocyte content of the tissue (Wang et al., 2011). The use of tissue homogenates, with diverse cell type compositions, and various regions of tissue at different Braak stages, makes comparing the results of individual studies challenging. Therefore, systematic studies investigating the expression of these miRNAs in the different regions of the brain in relation to Braak staging are needed to clarify their significance in relation to the pathogenesis of Alzheimer's disease. However, there are miRNAs that have consistently been identified as dysregulated including miR-107, miR-29, miR-9, miR-181, miR-34, miR-106, and miR-146 (Schonrock and Gotz, 2012). Many of these have been linked to altered regulation of key genes known to be involved with Alzheimer's disease.

Down regulation of miR-107 at an early stage of Alzheimer's disease has been observed in temporal cortex and correlated with the up regulation of BACE1 in two studies, which could impact upon Aβ production (Wang et al., 2008b; Nelson and Wang, 2010) This finding was confirmed as being specific to miR-107 (and not a family member such as miR-103) and demonstrated that as miR-107 declines with advancing pathology, BACE1 increases along with neuritic plaque density (Wang et al., 2008b). Interestingly, miR-107 and miR-124a, two miRNAs experimentally proven to target BACE1 also regulate other aspects of APP metabolism, thus demonstrating the capacity for single miRNAs to influence several components of the same pathway and the potential to produce additive effects. MiR-107 directly targets a disintegrin and metalloproteinase 10 (ADAM10), another secretase enzyme which processes APP, and miR-124a is involved in the regulation of APP mRNA alternative splicing via direct targeting of polypyrimidine tract binding protein 1 (PTBP1) (Smith et al., 2011; Augustin et al., 2012).

The miR-29 family of miRNAs have target sites on BACE1 mRNA and loss of this cluster is negatively correlated with BACE1 expression in a subset of sporadic Alzheimer's disease cases (Hebert et al., 2008; Zong et al., 2011). The correlation was Alzheimer's disease specific and was verified in HEK293 and SH-SY5Y cell culture models, where an increase in Aβ production was also observed as a result. Whilst not specific for brain regions particularly associated with Alzheimer's disease, as demonstrated by analysis of material taken from the cerebellum (a brain area not typically affected by the disease), it is an important additional relationship between an miRNA and

Table 1 | Dysregulated miRNAs discussed within this review article.

miRNA	Neurodegenerative disease	Reported findings	Proposed biological effects	References
Let-7	PD	Reduced in two *C.elegans* models of PD (alpha-synuclein transgenic and pdr-1 strains), antagonised by pathogenic LRRK2 in *Drosophila* model	Increased dopaminergic neuronal cell death via increased expression of E2F1 and DP transcription factors	Asikainen et al., 2010; Gehrke et al., 2010
miR-106a*	PD	Increased in human PD substantia nigra	Chaperone mediated autophagy pathway	Kim et al., 2007; Alvarez-Erviti et al., 2013
miR-106a/b	AD	Down regulation in human AD temporal cortex and cerebral cortex of APPswe/PSΔE9 AD mouse model	Regulation of APP and ABCA1, impact on Aβ production	Hebert et al., 2008, 2009; Patel et al., 2008; Kim et al., 2012
miR-107	AD	Down regulation in human AD temporal cortex	Regulation of BACE1 and ADAM10, impact on Aβ production	Wang et al., 2008b; Nelson and Wang, 2010; Augustin et al., 2012
miR-124a	AD	Down regulated in human AD temporal cortex	Targets BACE1 and PTBP1, impact on Aβ production	Smith et al., 2011
	ALS	Reduced in spinal cord from SOD1 mouse model of ALS	Regulates glutamate transport protein EAAT2/GLT1	Morel et al., 2013
	HD	Down regulated in human HD frontal cortex and striatum	Down regulated by REST	Marti et al., 2010
miR-125b	HD	Down regulated in Hdh^{Q111}/Hdh^{Q111} cell model of HD	Targets HTT, regulates p53 and is predicted to target TBP	Sinha et al., 2010, 2011; Ghose et al., 2011
miR-132	AD	Down regulation in human AD hippocampus, cerebellum and medial frontal gyrus	Regulates several AD associated genes including *SIRT1, AChE, PTEN, FOXO3a* and *p300*	Cogswell et al., 2008; Shaked et al., 2009; Wong et al., 2013
	HD	Down regulated in human HD cortex	Down regulated by REST	Johnson et al., 2008
miR-133b	PD	Down regulated in PD patient midbrain	Negative feedback circuit with Pitx3, function of midbrain dopaminergic neurons	Kim et al., 2007
miR-146a	AD	Up regulated in AD hippocampus and temporal cortex	Targets inflammatory pathways including NFkB and TSPAN12 to potentially impact upon Aβ production	Lukiw et al., 2008; Sethi and Lukiw, 2009; Li et al., 2011
	HD	Down regulated in Hdh^{Q111}/Hdh^{Q111} cell model of HD	Targets HTT, TBP and is regulated by the p53 pathway	Sinha et al., 2010, 2011; Ghose et al., 2011
miR-146a*	ALS	Up regulated in human spinal cord homogenates	Targets NFL	Campos-Melo et al., 2013
miR-150	HD	Down regulated in Hdh^{Q111}/Hdh^{Q111} cell model of HD	Targets HTT regulates p53 and is predicted to target TBP	Sinha et al., 2010, 2011
miR-153	AD	Down regulated in cerebral cortex of APPswe/PSΔE9 AD mouse model	Regulation of APP and APLP2, impact on Aβ production	Liang et al., 2012; Long et al., 2012
	PD	Interaction with SNCA in cell culture models of PD	Targets SNCA	Doxakis, 2010
miR-181c	AD	Down regulated in human temporal cortex and patient serum. Down regulated by addition of Aβ to primary hippocampal neurons	Targets SIRT1, impact on tau metabolism	Hebert et al., 2008; Schonrock et al., 2010, 2012; Geekiyanage et al., 2012

(Continued)

Table 1 | Continued

miRNA	Neurodegenerative disease	Reported findings	Proposed biological effects	References
miR-184*	PD	Antagonised by pathogenic LRRK2 in *Drosophila* model of PD	Increased dopaminergic neuronal cell death via increased expression of E2F1 and DP transcription factors	Gehrke et al., 2010
miR-200a/c	HD	Up regulated cerebral cortex of N171-82Q HD mouse model	Predicted to target genes involved in neuronal function	Jin et al., 2012
miR-205	PD	Down regulated in PD cerebral cortex tissue	Targets LRRK2	Cho et al., 2013
miR-206	ALS	Up regulation in muscle from ALS patients and ALS mouse model	Nerve-muscle communication and promotes reinnervation following nerve damage	Williams et al., 2009
miR-21*	PD	Increased in human PD substantia nigra	Chaperone mediated autophagy pathway	Alvarez-Erviti et al., 2013
miR-224	PD	Increased in human PD substantia nigra and amygdala	Chaperone mediated autophagy pathway, predicted to target lamp-2a mRNA	Kim et al., 2007; Alvarez-Erviti et al., 2013
miR-26b	PD	Increased in human PD substantia nigra	Chaperone mediated autophagy pathway, predicted to target hsc70 mRNA	Kim et al., 2007; Alvarez-Erviti et al., 2013
miR-29a/b	AD	Down regulation in human AD temporal cortex, cerebellum and patient serum	Regulation of BACE1 and impact on Aβ production, regulation of microglia in the aged brain	Hebert et al., 2008; Geekiyanage et al., 2012; Fenn et al., 2013
miR-29c	AD	Up regulation cerebral cortex of APPswe/PSΔE9 AD mouse model	Regulation of BACE1 and impact on Aβ production	Zong et al., 2011
miR-301b	PD	Increased in human PD substantia nigra	Chaperone mediated autophagy pathway	Alvarez-Erviti et al., 2013
miR-338-3p	ALS	Up regulated in human ALS frontal cortex and ALS leukocytes	Predicted to target neurotransmitter signalling pathways	Shioya et al., 2010; De Felice et al., 2012
miR-34a	AD	Up regulated in human AD hippocampus, cerebellum, medial frontal gyrus and white blood cells. Up regulated in cerebral cortex of APPswe/PSΔE9 AD mouse model	Targets BCL2 and potentially increases apoptotic cell death, targets SIRT1, impact on tau metabolism	Schipper et al., 2007; Cogswell et al., 2008; Wang et al., 2009
miR-34b	HD	Up regulated in mutant HTT transfected NT2 cell model of HD and elevated in human HD patient plasma	Interaction with p53 pathway	Gaughwin et al., 2011
miR-34b/c	PD	Down regulation in PD amygdala, frontal cortex and substantia nigra	Altered mitochondrial function and oxidative stress, also linked to brain ageing	Minones-Moyano et al., 2011; Liu et al., 2012
miR-373*	PD	Increased in human PD substantia nigra and amygdala	Chaperone mediated autophagy pathway, predicted to target lamp-2a mRNA	Kim et al., 2007; Alvarez-Erviti et al., 2013
miR-433	PD	Polymorphism in *FGF20* binding site linked to increased risk of PD	Increase FGF20 expression and downstream up regulation of SNCA	Wang et al., 2008a

(Continued)

Table 1 | Continued

miRNA	Neurodegenerative disease	Reported findings	Proposed biological effects	References
miR-64/65	PD	Reduced in two *C.elegans* model of PD (alpha-synuclein transgenic and cat-1 strains)	Unknown, target candidates include transcription factor mdl-1 and the development gene ptc-1	Asikainen et al., 2010
miR-7	PD	Interaction with SNCA and neuroprotective role in cell culture models of PD	Targets SNCA and supresses SNCA mediated toxicity	Junn et al., 2009; Doxakis, 2010
miR-9	AD	Up and down regulation reported in human AD brain tissue. Down regulated in patient serum and by the addition of Aβ to primary hippocampal neurons	Targets include NFH and SIRT1, plague and tangle formation	Detailed review see Geekiyanage et al., 2012; Schonrock and Gotz, 2012
	ALS	Down regulation in Dicer knock out mature motor neurons and up regulation in SMA mouse model	Targets NFH	Haramati et al., 2010
	HD	Down regulated in human HD cortex	Down regulated by REST and targets REST in a double negative feedback loop	Packer et al., 2008
miR-9*	HD	Down regulated in human HD cortex	Down regulated by REST and targets CoREST in a double negative feedback loop	Packer et al., 2008

AD, Alzheimer's disease; PD; Parkinson's disease; ALS, Amyotrophic lateral sclerosis; HD, Huntington's disease.

mRNA expression (Hebert et al., 2008). In addition to regulating BACE1, miR-29a/b are increased in the aging brain and linked to modulation of microglial activation (Fenn et al., 2013). The miR-29 cluster has been sequenced in a cohort of sporadic and familial patients and variants were found within the cluster that significantly associated with Alzheimer's disease (Bettens et al., 2009). However, this finding requires further validation in additional cohorts and the functional effects of these variants remains unclear.

APP is also a target for miRNA regulation, miR-106a and miR-106b directly bind to APP mRNA and are down regulated in the anterior temporal cortex of Alzheimer's disease patients (Hebert et al., 2008, 2009). Interestingly, miR-106 has also been found to regulate ATP-binding cassette transporter A1 (ABCA1), a lipid transporter implicated in ApoE lipidation and the production of Aβ, suggesting that this miRNA could influence the Aβ generation via more than one route (Kim et al., 2012).

Recent studies have examined possible associations of miR-153 with Alzheimer's disease after functional studies confirmed an interaction with APP and amyloid beta precursor-like protein 2 (APLP2) mRNA transcripts (Liang et al., 2012; Long et al., 2012). Levels of miR-153 were significantly decreased at early and late stages of disease in the APPswe/PSΔE9 double mutant mouse model. Furthermore, the interaction has been demonstrated *in vitro* using HeLa and primary human fetal brain cells where delivery of miR-153 down regulated endogenous expression of APP and APLP2 (Long et al., 2012). miR-153 levels were significantly decreased in the cohort of advanced Alzheimer's disease post-mortem brain specimens with neocortical neurofibrillary tangle pathology (Braak III–VI) as compared with specimens lacking neocortical neurofibrillary tangle pathology (control and Braak stage I/II specimens). Importantly, an inverse co-regulation of miR-153 and APP in human frontal cortex was observed at the protein level (Long et al., 2012). Thus, evidence indicates that miR-153 contributes to post-transcriptional regulation of APP/APLP2 and may therefore have a role in Alzheimer's disease, although further validation of this potential interaction is required.

MiR-9 is a highly conserved, brain enriched miRNA and the most frequently identified misregulated miRNA in Alzheimer's disease to date, although there are inconsistencies regarding up or down regulation as both have been reported (Schonrock and Gotz, 2012). Addition of Aβ to primary neuron cultures results in a rapid decrease of miR-9 *in vitro* and suggests that deregulation may be related to plaque formation (Schonrock et al., 2010). The targets for miR-9 include neurofilament heavy chain (NFH), a protein found in neurofibrillary tangles, and sirtuin (SIRT1), a de-acetylase that interacts with tau and is linked to accumulation of hyperphosphorylated forms of tau in the disease (Haramati et al., 2010; Saunders et al., 2010; Liu et al., 2011; Schonrock et al., 2012). Three other miRNAs have been found to supress SIRT1, namely miR-181c, miR-34, and miR-132, all of which show consistent altered expression in Alzheimer's disease brain (Schonrock and Gotz, 2012; Wong et al., 2013). Furthermore, miR-132 has several direct targets of relevance to Alzheimer's disease pathogenesis including Tensin Homolog (*PTEN*), Forkhead Box O3a (*FOXO3a*), and E1A binding protein p300 (*P300*), which all have a role in neural apoptosis, and the acetylcholinesterase enzyme (AChE), inhibition of which is a standard treatment in Alzheimer's disease and links into the cholinergic anti-inflammatory pathway (Shaked et al., 2009; Wong et al., 2013).

An additional miRNA linked to both inflammation and Alzheimer's disease is miR-146a. This key regulator of innate immunity is up regulated in brain regions affected by Alzheimer's pathology, including the hippocampus and temporal cortex, yet remains unchanged in unaffected regions (Lukiw et al., 2008; Sethi and Lukiw, 2009). Experimentally proven targets of miR-146a include complement factor H (*CFH*), interleukin-1 receptor-associated kinase-1 (*IRAK1*) and TNF receptor-associated factor 6 (*TRAF6*), all associated with innate immunity and inflammatory pathways which are dysregulated in Alzheimer's disease (Wang et al., 2012). Interestingly, miR-146a also targets transmembrane spanning tetraspanin 12 (*TSPAN12*), a key regulator of ADAM10 and therefore has the potential to impact upon Aβ metabolism (Li et al., 2011). These findings further demonstrate the capacity of miRNAs to influence several pathways and mediate cross-talk between pathogenic mechanisms.

PARKINSON'S DISEASE

Parkinson's disease is characterized clinically by bradykinesia, tremor and rigidity. This is caused by the progressive loss of dopaminergic neurons in the substantia nigra pars compacta. The majority of cases are idiopathic, however, around 20% of patients have a positive family history. The most important and widely accepted monogenically inherited Parkinson's disease genes are α-synuclein (*SNCA*) and leucine-rich repeat kinase 2 (*LRRK2*) for late-onset disease and Parkin *(PARK2)*, oncogene DJ1 (*DJ1*) and PTEN Induced Putative Kinase 1 (*PINK1*) for early onset (Coppede, 2012). The neuropathology of Parkinson's disease is characterized by cellular inclusions known as Lewy bodies in neurons, the main components of which are α-synuclein, neurofilament and ubiquitin.

Recent studies suggest that miRNAs may be involved in the development of Parkinson's disease. Deletion of Dicer in dopaminergic neurons in transgenic mice led to reduced locomotion and symptoms reminiscent of human Parkinson's disease (Kim et al., 2007). Expression profiling of miRNAs from patient midbrain samples revealed a significant decrease in miR-133b. MiR-133b targets Pixt3, a transcription factor enriched in dopaminergic neurons, which is deficient in the aphakia mouse model of Parkinson's disease (Hwang et al., 2003). A negative feedback model has been proposed to explain the relationship, in which, Pitx3 specifically induces transcription of miR-133b and Pitx3 activity is directly down regulated by miR-133b (Kim and Kim, 2007). However, the impact of miR-133b *in vivo* remains unclear, miR-133b null mice display normal midbrain dopaminergic neuronal development and function with a lack of disease phenotype (Heyer et al., 2012).

MiRNA profiling to evaluate dysregulation of miRNAs in various regions of human Parkinson's disease brain tissue has also reported a widespread reduction in the miR-34b/c cluster, which could be detected early in the disease course. Depletion of these miRNAs in dopaminergic neuronal cells led to a reduction of cell viability accompanied by mitochondrial dysfunction (Minones-Moyano et al., 2011). Interestingly, miR-34 has been linked with aging in *Drosophila*, as identified by comparing brain miRNA profiles at three time points, 3, 30, and 60 days. Loss of this age-modulated miRNA in transgenic flies resulted in a late-onset brain degeneration and a striking decline in survival (Liu et al., 2012).

In a recent study Asikainen et al. (2010) used global analysis of miRNAs in three *C.elegans* models of Parkinson's disease. Reduced expression of miR-64 and miR-65 was observed in *SNCA* transgenic and vesicular catecholamine transporter mutant strains, while members of the let-7 family were dysregulated in the *SNCA* and Parkin mutant strains (Asikainen et al., 2010). Let-7 miRNAs are highly conserved and abundant in the central nervous system (CNS) (Lagos-Quintana et al., 2002). Unfortunately there is no literature to describe the function of the miR-64/65 cluster and these results are yet to be validated in rodent models or human tissue.

One of the most important factors in Parkinson's disease pathology is α-synuclein protein accumulation. Mutations and multiplications of the *SNCA* gene are found in familial forms of the disease and polymorphisms in the gene are linked to greater susceptibility in sporadic cases (Hardy et al., 2009). Examination of the *SNCA* gene has revealed an unusually highly conserved and long 3′UTR sequence which is important in the post-translational control of the gene and strongly suggests a role for miRNA regulation (Sotiriou et al., 2009). Two miRNAs have been identified to date as directly targeting *SNCA*, namely miR-7 and miR-153. These brain enriched miRNAs have been found to bind directly to *SNCA* mRNA and down regulate expression, with an additive effect (Doxakis, 2010). In addition, miR-7 suppresses *SNCA* mediated cytotoxicity in neuronal cell models (Junn et al., 2009). Other miRNAs found to be significantly increased in Parkinson's disease brain tissue include six (miR-21*, miR-224, miR-373*, miR-26b, miR-106a*, and miR-301b) that target components of the chaperone-mediated autophagy pathway (Alvarez-Erviti et al., 2013). Defects in this pathway have the potential to disrupt α-synuclein protein degradation and have been proposed as a mechanism for Lewy body pathology (Winslow and Rubinsztein, 2011).

Mutations in the *LRRK2* gene are the most common cause of Parkinson's disease identified to date, but the pathogenic mechanism remains unclear. The LRRK2 protein has been found to directly associate with components of the miRNA processing pathway, including Ago proteins (Dachsel et al., 2007; Gehrke et al., 2010). Pathogenic LRRK2 in *Drosophila* antagonises at least two miRNAs, let-7 and miR-184*, leading to greater dopaminergic neuronal cell death via increased expression of E2F1 and DP transcription factors (Gehrke et al., 2010). The pathogenic effects of *LRRK2* mutations were age-dependent. However, this mechanism has yet to be investigated in vertebrate systems and awaits confirmation in human patient tissue models such as *LRRK2* mutant fibroblasts or induced pluripotent stem cell-derived neurons. Interrogation of the *LRRK2* gene sequence has revealed a highly conserved binding site for miR-205 in the 3'UTR. In human and mouse brain tissue the level of miR-205 inversely correlated with LRRK2 protein. Further investigation in Parkinson's disease cases revealed a significant decrease of miR-205 in the frontal cortex compared to controls and *in vitro* luciferase assays confirmed a direct interaction of this miRNA with LRRK2 mRNA (Cho et al., 2013). This novel regulatory mechanism for LRRK2 suggests miR-205 may serve as a therapeutic target for Parkinson's disease.

Another gene associated with increased risk of Parkinson's disease in some populations is fibroblast growth factor 20 (*FGF20*) (Itoh and Ohta, 2013). One polymorphism (rs12720208) is predicted to disrupt the binding site for miR-433 in the 3'UTR of the gene, leading to increased expression of FGF20 and a downstream up regulation of *SNCA* (Wang et al., 2008a). An additional miR-433 putative binding site polymorphism has also been identified in the *SNCA* 3′UTR, however, no difference in allele distribution between patients and controls has been found, and a regulatory effect for miR-433 on *SNCA* expression could not be confirmed (Schmitt et al., 2012).

AMYOTROPHIC LATERAL SCLEROSIS

ALS is characterized by the progressive loss of upper and lower motor neurons from the motor cortex, brain stem and spinal cord. For the patient, this results in severe muscle atrophy leading to paralysis and death usually within 2–5 years of symptom onset (McDermott and Shaw, 2008). A family history of ALS is found in 5% of patients, with the remaining 95% of cases sporadic in nature. Clinically, familial and sporadic ALS are very similar, with the exception of an earlier than the typical mid-life onset in some familial cases. Several genes have now been identified as causative in ALS of which the most frequent are *C9ORF72*, superoxide dismutase 1 (*SOD1*), transactive response DNA-binding protein (*TARDBP*) and fused in sarcoma (*FUS*) (Goodall et al., 2012). The proteins encoded by the latter 3 genes, SOD1, TDP-43, and FUS, have been found within the ubiquitinated inclusions that are pathological hallmarks of ALS (Al-Chalabi et al., 2012).

To determine if miRNAs are essential to motor neuron survival, Haramati et al. (2010) used Dicer knockdown to generate transgenic mice lacking the ability to produce mature miRNAs in a subset of their post mitotic motor neurons. The transgenic animals showed progressive locomotor defects and denervation muscle atrophy caused by motor neuron loss. Further work revealed a specific increase in NFH expression, which was at least in part attributed to the loss of miR-9. This is a miRNA highly expressed in the brain and found to be up regulated in mouse models of the juvenile motor neuron disorder known as spinal muscular atrophy (SMA) (Haramati et al., 2010). In addition, miRNAs that directly target neurofilament light chain (NFL) have been found to be altered in ALS. Up regulation of miR-146a* and down regulation of miRNAs 524-5p and 582-3p were reported in SALS spinal cord compared to controls. However, the study used whole spinal cord tissue homogenates so the contribution of differing cell type composition between cases and controls may have influenced the miRNA expression profile differences (Campos-Melo et al., 2013).

The ALS associated proteins TDP-43 and FUS have been found to directly bind key components of the miRNA processing pathway, implicating miRNA dysregulation in disease pathogenesis. Drosha forms two distinct protein complexes, one with DGCR8 which is responsible for the bulk of miRNA processing in the cell (the microprocessor) and a larger complex of at least 17 polypeptides, including TDP-43 and FUS, with limited pri-miRNA processing activity (Gregory et al., 2004). In addition, TDP-43 can directly bind Dicer, Ago2, subsets of pri-miRNAs in the nucleus and pre-miRNAs in the cytoplasm (Kawahara and Mieda-Sato, 2012). Depletion of TDP-43 and FUS protein *in vitro* affects the generation of specific subsets of miRNAs, some of which are implicated in neuromuscular development, neuronal function and survival (Buratti et al., 2010; Kawahara and Mieda-Sato, 2012; Morlando et al., 2012). The mislocalisation of TDP-43 and FUS to cytoplasmic inclusions in ALS is therefore likely to reduce their availability to bind miRNA processing components and affect the production of at least a subset of miRNAs, the consequences of which for neuronal cells have yet to be investigated.

Changes in miRNAs have also been seen in peripheral ALS tissues. Williams et al. (2009) profiled the miRNAs present in the muscle from mutant SOD1 mouse models of ALS. A dramatic increase in the miR-206 was observed in transgenic mice at the time of symptom onset and was found to be a direct result of denervation (Williams et al., 2009). miR-206 is a skeletal muscle enriched miRNA that has fundamental roles in muscle development and plasticity (McCarthy, 2008). A similar increase in miR-206 has also been observed in human ALS patient muscle tissue (Russell et al., 2012). The loss of miR-206 from transgenic SOD1 mice accelerated the rate of disease progression, most likely because miR-206 is a key player in nerve-muscle communication and therefore essential for reinnervation following nerve damage (Williams et al., 2009).

The role of miRNAs as mediators of intercellular communication via exosomes has also been observed in the CNS. Exosomes are small membrane bound vesicles secreted by a variety of cell types including astrocytes and neurons (Raposo and Stoorvogel, 2013). There is evidence that neuronal miRNAs packaged in exosomes can be internalized by astrocytes where they influence protein expression (Morel et al., 2013). Interestingly, this mechanism of regulation has been observed for the main CNS glutamate transporter EAAT2/GLT1. Defects in glutamate transport are well-documented in ALS and a specific decrease in EAAT2/GLT1 levels has been observed in ALS patient samples and the SOD1 mouse model, though the cause of this defect remains elusive (Robberecht and Philips, 2013). Recent work has shown that miR-124a from neuronal exosomes is internalized by astrocytes to result in specific increased expression of EAAT2/GLT1 protein levels via an indirect mechanism. Levels of miR-124a in the spinal cord of mutant SOD1 mouse models is decreased at the end stage of disease and *in vivo* injection of artificial miR-124a oligos into the spinal cord of these mice led to a 30% increase in EAAT2/GLT1 expression. These exciting findings open up the potential for miRNA mediated therapy in ALS to combat the excitotoxicity seen in the disease (Morel et al., 2013).

HUNTINGTON'S DISEASE

Huntington's disease is an autosomal dominant inherited disorder caused by an elongated CAG repeat expansion in the huntingtin (*HTT*) gene. The classical motor symptom of chorea is not present in all patients, whilst other motor features such as impaired balance or abnormal fine finger movements are more likely to interfere with the patient's quality of life. Huntington's disease patients frequently develop neuropsychiatric complications such as progressive cognitive decline, personality change and depression. Pathologically, there is severe degeneration of the corpus striatum and atrophy of several brain regions, including the caudate nucleus, putamen and globus pallidus, but also the cortex itself (Zuccato et al., 2010). Medium spiny neurons of the striatum are particularly vulnerable to the *HTT* mutation, which is believed to predominantly cause a toxic gain of function. Although *HTT* is ubiquitously expressed, the aggregates of mutant HTT protein, which are a pathological hallmark of the disease, are restricted to neuronal cells (Imarisio et al., 2008).

There are widespread gene expression changes in Huntington's disease and evidence suggests these can be attributed partly to miRNA dysregulation (Seredenina and Luthi-Carter, 2012). The HTT protein directly interacts with Ago2 and is found to localize to P bodies. Depletion of wild type HTT compromises miRNA mediated gene silencing and the mutant protein disrupts neuronal P body integrity (Savas et al., 2008). There is also evidence to suggest other key components of miRNA biogenesis are dysregulated in mouse models of the disease, including Dicer, Drosha and Exportin-5, at different stages of the disease course (Lee et al., 2011). However, these findings are yet to be further validated.

An alternative mechanism of aberrant transcriptional regulation in Huntington's disease is increased nuclear localization of RE1-Silencing Transcription Factor (REST). REST is a transcriptional repressor that acts to silence neuronal gene expression in non-neuronal cells. In healthy neurons REST is sequestered in the cytoplasm, but in Huntington's disease there is increased nuclear translocation of REST in neurons leading to increased gene repression, which has a negative effect on survival (Zuccato et al., 2007). In addition to targeting mRNA, REST has been shown to regulate miRNAs, including a neuronal miRNA family containing miR-124a, miR-132, miR-9, and miR-9* (Conaco et al., 2006; Johnson et al., 2008; Marti et al., 2010). MiR-124a and miR-132 are highly expressed in the CNS and are crucial regulators of neural identity and function (Conaco et al., 2006; Wanet et al., 2012). Further investigation into miR-9/miR-9* has revealed that they directly target two components of the REST complex to form a double negative feedback network (Packer et al., 2008). The majority of REST-regulated miRNAs identified to date have displayed reduced expression in Huntington's patient brain tissue and models of the disease (Johnson et al., 2008; Packer et al., 2008).

Studies to profile miRNA expression in human tissue, mouse models of disease and cellular systems have revealed numerous expression changes in miRNAs not under REST control, suggesting that miRNA dysregulation is extensive in Huntington's disease (Marti et al., 2010; Sinha et al., 2010; Ghose et al., 2011; Jin et al., 2012). More specifically, the miR-200 family is altered in the cortex of mutant HTT mouse models at early stages of disease, which may compromise a network of genes involved in neuronal plasticity and survival (Jin et al., 2012). In cellular models of Huntington's disease, miR-146a, miR-125b, and miR-150 are down regulated while miR-34b was elevated by the presence of mutant HTT protein (Sinha et al., 2010; Gaughwin et al., 2011). Further investigation revealed complex interplay between these miRNAs and several transcriptions factors, including p53, RelA, and NFkB, (Gaughwin et al., 2011; Ghose et al., 2011). Interestingly, miR-146a, miR-150, and miR-125b also targeted HTT and were predicted to interact with tata binding protein (TBP), a protein known to be recruited into mutant HTT aggregates and were shown to modulate aggregate formation (Sinha et al., 2010, 2011). The relevance of this observation in relation to the pathogenesis of Huntington's disease remains unknown and represents an interesting subject for further investigation (Sinha et al., 2011).

CLINICAL APPLICATIONS OF miRNA

BIOMARKERS

There is an urgent need for effective biomarkers in neurodegenerative disease. For the majority of these conditions, diagnosis relies upon clinical assessment and monitoring the progression of symptoms, which causes substantial delay. Once a neurodegenerative disease has manifested, significant neuronal loss and CNS damage will already be present, therefore early diagnosis is essential to maximize the effectiveness of disease modifying therapies. In addition, neurodegeneration is clinically heterogeneous, with multiple subtypes associated with different survival times, rates of progression and symptoms. Robust biomarkers would be valuable not only for the initial diagnosis, but the classification of various subtypes of disease, monitoring responses to therapeutic agents and tracking disease progression (Shi et al., 2009).

Recent studies have demonstrated the existence of miRNAs in the body fluids including blood, cerebrospinal fluid (CSF) and saliva, at detectable levels where they are exceptionally stable and potential candidates for biomarker discovery (Chen et al., 2008). These extracellular miRNAs are proposed to originate from passive leakage from damaged tissue as the result of cell lysis or apoptosis, active transport from cells via microvesicles such as exosomes or bound within RISC protein complexes (Etheridge et al., 2011).

Blood is an attractive source of biomarkers as it interacts with every tissue in the body and sample collection is already part of standard clinical practice. There has therefore been a recent focus on circulating miRNAs as biomarkers, both extracellular and those expressed in white blood cells, with a number of studies investigating these in neurodegenerative disease patients.

Several studies have interrogated blood-based miRNAs in Parkinson's disease (Margis and Rieder, 2011; Martins et al., 2011; Khoo et al., 2012; Cardo et al., 2013; Soreq et al., 2013). The first determined miRNA expression profiles of peripheral blood mononuclear cells from 19 patients and 13 controls using Exiqon miRCURY LNA assays and identified a panel of 18 significantly dysregulated miRNAs. These were all under-expressed and could differentiate patients from healthy controls (Martins et al., 2011). In order to place these miRNAs in a wider biological context, the authors performed pathway analysis of the predicted target genes of these miRNAs and revealed an over-representation in pathways previously linked to Parkinson's disease, including semaphorin signaling in neurons and transcriptional repression signaling (Martins et al., 2011). A second study investigated 85 miRNAs in whole blood samples using real-time PCR assays from 8 patients to reveal a set of three miRNAs, miR-1, miR-22*, and miR-29a, with reduced expression when compared to 8 control subjects. A second set of miRNAs, miR-16-2*, miR-26a-2*, and miR-30a, was identified as increased in response to levodopa treatment, suggesting a role for anti-parkinsonian drugs in altering miRNA expression (Margis and Rieder, 2011). Two studies have interrogated Parkinson's disease patient plasma samples, Cardo et al. profiled 384 miRNAs from 31 patients at onset of symptoms and 25 controls were compared using TaqMan real-time PCR assays. The study revealed only one significantly up regulated miRNA, namely miR-331-5p, in Parkinson's disease cases (Cardo et al., 2013). Khoo et al. used Agilent microarrays followed by TaqMan QPCR validation to identify a panel of plasma biomarkers for Parkinson's disease consisting of miR-1826, miR-450b-3p, miR-626, and miR-505, which provided 91% sensitivity and 100% specificity (Khoo et al., 2012). Lastly, circulating miRNAs have been profiled in Parkinson's disease patients before and after deep brain stimulation treatment. This study compared leukocyte miRNA expression profiles using SOLiD sequencing in 7 patients before treatment and 6 healthy controls to reveal 16 dysregulated miRNAs, including miR-16, miR-20a, and miR-320. Interestingly, following deep brain stimulation 5 of the 11 leukocyte miRNAs that were significantly altered matched those changed by disease but in the opposite direction (Soreq et al., 2013).

Overall, there is a lack of overlap between these studies and little concordance with the findings from miRNA profiling in CNS tissue in Parkinson's disease, which highlights the difficulties of analysing different sample types and comparing different methodologies.

In ALS, a study of 8 patients and 12 healthy controls revealed 8 miRNAs with significantly altered expression in leukocytes (De Felice et al., 2012). One of these, miR-338-3p is predicted to target genes involved in neurotransmitter signaling pathways and had previously been described as up regulated ALS patient brain tissue, a finding that failed to validate in an enlarged study population and awaits further experimentation (Shioya et al., 2010; De Felice et al., 2012). MiRNA profiling of peripheral monocytes in the SOD1 mouse model of ALS and in ALS patients showed a pro-inflammatory phenotype with high expression of miR-27a, miR-155, miR-146a, and 532-3p in sporadic ALS patients and not in healthy control or multiple sclerosis subjects (Butovsky et al., 2012). A similar profile was also observed in 4 familial ALS patients with *SOD1* mutations, which may represent a common abnormality in the immune system of different forms of ALS (Butovsky et al., 2012).

Huntington's disease is an inherited disorder and can therefore be diagnosed using genetic testing, however, biomarkers are still required for the pre-symptomatic period as this coincides with an opportunity for therapeutic interventions and biomarkers are needed to track disease progression. Circulating levels of miR-34b have been observed at the pre-clinical stage in a small study of Huntington's disease plasma samples when compared to healthy controls. Moreover, miR-34b is induced by the expression of mutant *HTT* gene in neuronally differentiated cell lines (Gaughwin et al., 2011). The study proposes the use of miR-34b as a biomarker for the onset of Huntington's disease, however, the cohort size was small and this findings has yet to be replicated.

In Alzheimer's disease the levels of disease associated miRNAs miR-29a/b, miR-181c, and miR-9 have been reported as down regulated in patient serum samples compared to healthy controls (Geekiyanage et al., 2012). However, the study was conducted in a small study cohort of 7 per group and further validation is required. CSF miRNA signatures have been investigated in Alzheimer's disease patients. Cogswell et al. (2008) recovered miRNAs from CSF samples from 10 Braak stage V Alzheimer's disease patients and 10 Braak stage I patients. Sixty miRNAs were significantly differentially regulated between the different Braak stages, including Let-7 family members, a finding which has since been replicated (Cogswell et al., 2008; Lehmann et al., 2012). Interestingly, extracellular let-7 was shown to activate the RNA-sensing Toll-like receptor (TLR) 7 to mediate neurodegeneration, demonstrating a role for miRNAs as signaling molecules, a function that is independent of their conventional role in gene regulation (Lehmann et al., 2012). In peripheral blood mononuclear cells of Alzheimer's disease patients compared to controls several miRNAs have been identified as differentially expressed including miR-34a and miR-29b, both of which have been found to be dysregulated in brain tissue (Schipper et al., 2007; Villa et al., 2013). Levels of miR-29a were inversely related to SP1, a transcription factor associated with Alzheimer's disease, and is the first reported incidence of a miRNA and its target acting in cooperation as potential biomarkers (Villa et al., 2013). A direct interaction between them, however, remains untested.

Investigation of miRNA-based biomarkers in neurodegenerative disease is in its infancy and has thus far been confounded by small sample sizes, lack of replication and a wide range of methodologies for extraction and quantification of miRNAs. To fully investigate miRNA potential as biomarkers, improved study design, including longitudinal experiments at various disease stages, and standardization of sample preparation and detection methods is required. However, these early studies have highlighted the potential of using circulating biomarkers to measure the effects of disease modifying treatments, an attractive prospect for future neurodegenerative disease clinical trials.

THERAPY

A further clinical application for miRNA is the development of miRNA-based therapy and there are currently several clinical trials testing the therapeutic efficacy of miRNA modulation in other disease areas, such as cancer and chronic hepatitis C viral infection, with more expected with the next few years (Elmen et al., 2008; Nana-Sinkam and Croce, 2013). The therapeutic application of miRNAs can be summarized by two broad strategies, RNA interference (RNAi) using miRNA mimics and miRNA inhibition via miRNA antagonists (including antimiRNA oligonucleotides and sponges).

The use of RNAi techniques to target disease-associated genes, such as *BACE1*, *APP*, *HTT*, *SOD1*, and *SNCA*, holds great promise for neurodegeneration, with a number of studies demonstrating beneficial effects in animal models (Gonzalez-Alegre, 2007; Ling et al., 2011). RNAi faces the same challenges as traditional drug development, including pharmacokinetics, target specificity, efficacy and toxicity (Nana-Sinkam and Croce, 2013). MiRNA strategies are likely to be less toxic, given that they mimic naturally occurring RNAi mechanisms, and there is evidence of reduced immune activation compared to other short hairpin RNAs (shRNAs) when used to treat neurodegenerative disease. McBride et al. (2008) screened several shRNAs that targeted HTT in mouse models of Huntington's disease and found unexpected neurotoxicity caused by microglial activation and astrogliosis. Toxicity was notably reduced when shRNAs were placed into artificial miRNA expression systems (McBride et al., 2008).

One of the advantages of miRNAs as therapeutic agents is their ability to influence multiple target genes and pathways. However, this can also be disadvantageous due to potential off-target effects such as secondary and tertiary consequences of modulating complex miRNA networks. Each miRNA can target several hundred mRNAs, thus understanding the effects of unwanted interactions between the miRNA and endogenous RNAs are important. Another consideration is that artificially introduced miRNAs could overwhelm the biogenesis machinery and impair the effectiveness of endogenous miRNAs (Khan et al., 2009). These saturation-based effects may be of particular importance for neurodegeneration given the evidence of impaired microprocessor function in these conditions. One potential strategy to minimize saturation is to employ non-canonical miRNAs, such as mirtrons, which bypass the microprocessor complex. Proof of principle has been demonstrated by work in Parkinson's disease, where RNAi sequences to *LRRK2* and *SNCA* were incorporated into the miR-1224 mirtron backbone. The artificial mirtron mimics could directly silence human *LRRK2* and *SNCA* in a cell-type specific manner, by using human synapsin promoter in the neuronal SH-SY5Y cell line (Sibley et al., 2012). Artificial mirtrons are therefore an attractive approach for the future, however, their efficacy *in vivo* is yet to be tested.

MiRNA antagonists have also been investigated in models of neurodegeneration. In the ALS SOD1 mouse model,

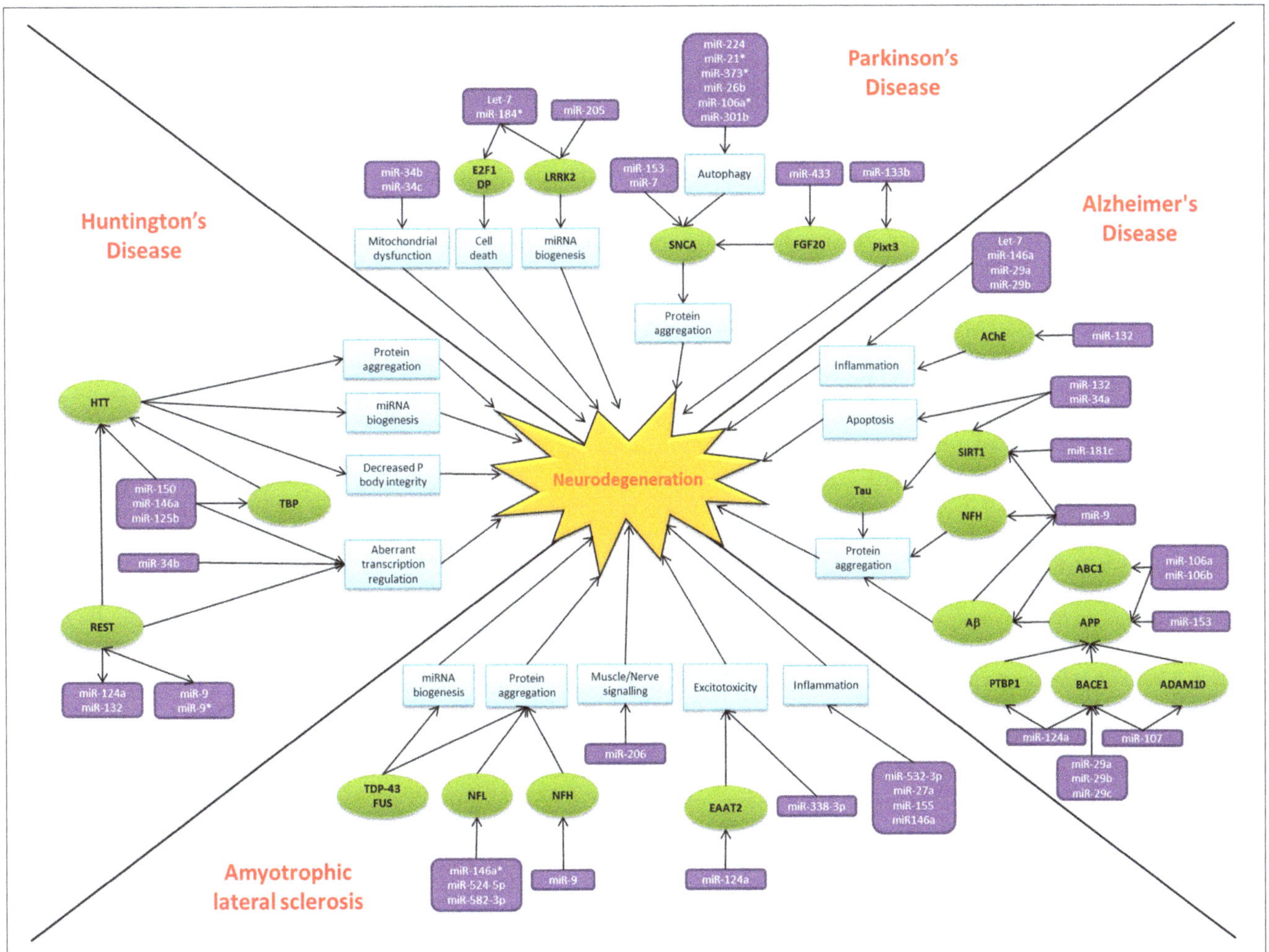

FIGURE 2 | MiRNAs implicated in neurodegeneration. A diagram to summarize the miRNA dysfunction networks in Alzheimer's disease, Parkinson's disease, amyotrophic lateral sclerosis and Huntington's disease. Blue boxes indicate pathogenic processes, green circles are affected genes/proteins and the miRNAs are in purple boxes. Arrows indicate the direction of the interactions, culminating in the common pathway of neurodegeneration.

oligonucleotide-based miRNA inhibitors (anti-miRs) to miR-155 have been used to prolong survival and disease duration by 38%. miR-155 was previously identified as up regulated in SOD1 mouse and human ALS patients spinal cord tissues, in addition to patient peripheral blood cells, and is linked to altered inflammation in the disease (Butovsky et al., 2012; Koval et al., 2013). Another experimental strategy to inhibit miRNA function is miRNA sponges, which are based upon competing endogenous RNAs (Ebert and Sharp, 2010). Sponge RNAs contain complementary binding sites to a miRNA of interest and specifically hamper the activity of miRNAs with a common seed sequence. These have yet to be tested for therapeutic applications and have thus far remained in experimental settings.

Effective treatment of neurodegenerative disorders will most likely require manipulation of multiple targets and biochemical pathways. The capacity of miRNAs to modify multiple targets is an attractive feature for developing therapeutic strategies in the future.

CONCLUDING REMARKS

Over the past two decades there has been an explosion of research focused on small non-coding RNAs, the so called "dark matter" of the cell. MiRNAs have emerged as key players in regulating gene expression and their dysregulation is common to many disease states, including neurodegeneration. The alteration of miRNA-mediated regulatory activity potentially upsets the delicate balance required for neuronal cell survival, thereby contributing to pathogenesis and disease progression (**Figure 2**). In common with proposed disease mechanisms and pathological features, overlap in the dysregulated miRNAs between neurodegenerative conditions are beginning to emerge. Examples of miRNAs with perhaps a more general role in neurodegeneration include miR-9, miR-132, miR-124a, and miR-34. MiR-9 has been found to be dysregulated in Alzheimer's disease, Huntington's disease and animal models of SMA. Reported targets important in terms of neurodegeneration include NFH, SIRT1, BACE1 and REST. Moreover, miR-9 is regulated by Aβ and REST in complex feedback regulatory mechanisms (Packer et al., 2008; Haramati et al., 2010; Schonrock and Gotz, 2012). MiR-132 has been linked to AKT survival signaling, anti-inflammatory pathways and acetylcholine metabolism. There are reports of down regulation causing neuronal death in cell culture models and reduced expression in Alzheimer's disease and Huntington's disease patient tissue (Cogswell et al., 2008; Johnson et al., 2008; Shaked et al., 2009; Wong et al., 2013). MiR-124a is another miRNA which targets BACE1 and has an additional role in excitotoxicity via regulating the glutamate transporter EAAT2 and is itself affected by the Huntington's disease associated transcription factor REST (Marti et al., 2010; Smith et al., 2011; Morel et al., 2013). Lastly, the miR-34 family target SIRT1 to affect tau metabolism, are decreased in Parkinson's disease patients and show increased expression in the presence of mutant HTT (Wang et al., 2009; Schonrock and Gotz, 2012). Overall it is too early to gain an understanding of the scope for miRNAs across the spectrum of neurodegenerative disease. Nevertheless, it is interesting to note that these four miRNAs have been linked to ageing, a key risk factor for neurodegenerative disease, and neuroinflammation, a common pathogenic mechanism (Soreq and Wolf, 2011; Nissan et al., 2012). However, these miRNAs are also reported as brain enriched or neuron specific and may therefore be affected by publication bias as they are the most frequently investigated in the field to date.

The studies highlighted in this review generally have small sample sizes, and results may reflect individual variability within the cohort rather than true disease specific changes in miRNA expression. In addition, many of the studies have focused on miRNA targets related to already known disease genes, such as *LRRK2* in Parkinson's disease and *BACE1* in Alzheimer's disease. A key feature of miRNAs is their short length, making them ideal candidates for non-biased expression profiling techniques, such as next generation sequencing. Such approaches would also address the concern of publication bias that may have affected the field to date. The challenge of unraveling complex gene regulatory networks calls for large, systematic studies of miRNAs in the CNS and continuous development of robust experimental approaches for studying miRNA function. These will need to take into account the issue of disparate CNS/brain regions with divergent cell type composition. Techniques such as laser capture microdissection and induced pluripotent stem cells, in combination with the increased availability of more sophisticated sequencing technologies, means that we can anticipate larger, non-biased, cell type enriched or specific studies of miRNAs for neurodegeneration in the near future.

While the clinical application of miRNAs as biomarkers and therapies in neurodegeneration is perhaps premature, the rate of discovery is promising. In an era of personalized medicine, the use of miRNA expression signatures to subclassify neurodegenerative disease, provide markers for therapeutic effectiveness and prognosis prediction, is an attractive prospect. Despite the anticipated off-target effects which cannot be fully predicted, saturation of the miRNA biogenesis pathway and possible immune activation, miRNA-based therapy has shown promise in animal models of neurodegeneration. Considerably more groundwork is needed in terms of functional studies to characterize miRNA targets and identify the most appropriate candidates before their potential in clinic can be realized.

ACKNOWLEDGMENTS

Emily F. Goodall is supported by the Motor Neurone Disease Association. Paul R. Heath receives funding from the Medical Research Council, Biotechnology and Biological Sciences Research Council and Alzheimer's Research Trust. Oliver Bandmann is supported by Parkinson's UK funding. Janine Kirby and Pamela J. Shaw are supported by an EU Framework 7 grant (Euromotor No259867) and SOPHIA, a project funded by EU Joint Programme—Neurodegenerative Disease Research and the Medical Research Council. Pamela J. Shaw is also supported as an NIHR Senior Investigator.

REFERENCES

Al-Chalabi, A., Jones, A., Troakes, C., King, A., Al-Sarraj, S., and van den Berg, L. H. (2012). The genetics and neuropathology of amyotrophic lateral sclerosis. *Acta Neuropathol.* 124, 339–352. doi: 10.1007/s00401-012-1022-4

Alvarez-Erviti, L., Seow, Y., Schapira, A. H., Rodriguez-Oroz, M. C., Obeso, J. A., and Cooper, J. M. (2013). Influence of microRNA deregulation on chaperone-mediated autophagy and alpha-synuclein pathology in Parkinson's disease. *Cell Death Dis.* 4, e545. doi: 10.1038/cddis.2013.73

Asikainen, S., Rudgalvyte, M., Heikkinen, L., Louhiranta, K., Lakso, M., Wong, G., et al. (2010). Global microRNA expression profiling of Caenorhabditis elegans Parkinson's disease models. *J. Mol. Neurosci.* 41, 210–218. doi: 10.1007/s12031-009-9325-1

Augustin, R., Endres, K., Reinhardt, S., Kuhn, P. H., Lichtenthaler, S. F., Hansen, J., et al. (2012). Computational identification and experimental validation of microRNAs binding to the Alzheimer-related gene ADAM10. *BMC Med. Genet.* 13:35. doi: 10.1186/1471-2350-13-35

Avramopoulos, D. (2009). Genetics of Alzheimer's disease: recent advances. *Genome Med.* 1, 34. doi: 10.1186/gm34

Babiarz, J. E., Ruby, J. G., Wang, Y., Bartel, D. P., and Blelloch, R. (2008). Mouse ES cells express endogenous shRNAs, siRNAs, and other Microprocessor-independent, Dicer-dependent small RNAs. *Genes Dev.* 22, 2773–2785. doi: 10.1101/gad.1705308

Baccarini, A., Chauhan, H., Gardner, T. J., Jayaprakash, A. D., Sachidanandam, R., and Brown, B. D. (2011). Kinetic analysis reveals the fate of a microRNA following target regulation in mammalian cells. *Curr. Biol.* 21, 369–376. doi: 10.1016/j.cub.2011.01.067

Bartel, D. P. (2009). MicroRNAs: target recognition and regulatory functions. *Cell* 136, 215–233. doi: 10.1016/j.cell.2009.01.002

Berezikov, E., Chung, W. J., Willis, J., Cuppen, E., and Lai, E. C. (2007). Mammalian mirtron genes. *Mol. Cell* 28, 328–336. doi: 10.1016/j.molcel.2007.09.028

Berezikov, E., Guryev, V., van de Belt, J., Wienholds, E., Plasterk, R. H., and Cuppen, E. (2005). Phylogenetic shadowing and computational identification of human microRNA genes. *Cell* 120, 21–24. doi: 10.1016/j.cell.2004.12.031

Bertram, L., Lill, C. M., and Tanzi, R. E. (2010). The genetics of Alzheimer disease: back to the future. *Neuron* 68, 270–281. doi: 10.1016/j.neuron.2010.10.013

Bertram, L., McQueen, M. B., Mullin, K., Blacker, D., and Tanzi, R. E. (2007). Systematic meta-analyses of Alzheimer disease genetic association studies: the AlzGene database. *Nat. Genet.* 39, 17–23. doi: 10.1038/ng1934

Bettens, K., Brouwers, N., Engelborghs, S., Van Miegroet, H., De Deyn, P. P., Theuns, J., et al. (2009). APP and BACE1 miRNA genetic variability has no major role in risk for Alzheimer disease. *Hum. Mutat.* 30, 1207–1213. doi: 10.1002/humu.21027

Bhattacharyya, S. N., Habermacher, R., Martine, U., Closs, E. I., and Filipowicz, W. (2006). Relief of microRNA-mediated translational repression in human cells subjected to stress. *Cell* 125, 1111–1124. doi: 10.1016/j.cell.2006.04.031

Braak, H., and Braak, E. (1995). Staging of Alzheimer's disease-related neurofibrillary changes. *Neurobiol. Aging* 16, 271–278. discussion: 278–284. doi: 10.1016/0197-4580(95)00021-6

Buratti, E., De Conti, L., Stuani, C., Romano, M., Baralle, M., and Baralle, F. (2010). Nuclear factor TDP-43 can affect selected microRNA levels. *FEBS J.* 277, 2268–2281. doi: 10.1111/j.1742-4658.2010.07643.x

Butovsky, O., Siddiqui, S., Gabriely, G., Lanser, A. J., Dake, B., Murugaiyan, G., et al. (2012). Modulating inflammatory monocytes with a unique microRNA gene signature ameliorates murine ALS. *J. Clin. Invest.* 122, 3063–3087. doi: 10.1172/JCI62636

Campos-Melo, D., Droppelmann, C. A., He, Z., Volkening, K., and Strong, M. J. (2013). Altered microRNA expression profile in amyotrophic lateral sclerosis: a role in the regulation of NFL mRNA levels. *Mol. Brain* 6, 26. doi: 10.1186/1756-6606-6-26

Cardo, L. F., Coto, E., de Mena, L., Ribacoba, R., Moris, G., Menendez, M., et al. (2013). Profile of microRNAs in the plasma of Parkinson's disease patients and healthy controls. *J. Neurol.* 260, 1420–1422. doi: 10.1007/s00415-013-6900-8

Carthew, R. W., and Sontheimer, E. J. (2009). Origins and Mechanisms of miRNAs and siRNAs. *Cell* 136, 642–655. doi: 10.1016/j.cell.2009.01.035.

Chatterjee, S., and Grosshans, H. (2009). Active turnover modulates mature microRNA activity in Caenorhabditis elegans. *Nature* 461, 546–549. doi: 10.1038/nature08349

Cheloufi, S., Dos Santos, C. O., Chong, M. M., and Hannon, G. J. (2010). A dicer-independent miRNA biogenesis pathway that requires Ago catalysis. *Nature* 465, 584–589. doi: 10.1038/nature09092

Chen, X., Ba, Y., Ma, L., Cai, X., Yin, Y., Wang, K., et al. (2008). Characterization of microRNAs in serum: a novel class of biomarkers for diagnosis of cancer and other diseases. *Cell Res.* 18, 997–1006. doi: 10.1038/cr.2008.282

Cho, H. J., Liu, G., Jin, S. M., Parisiadou, L., Xie, C., Yu, J., et al. (2013). MicroRNA-205 regulates the expression of Parkinson's disease-related leucine-rich repeat kinase 2 protein. *Hum. Mol. Genet.* 22, 608–620. doi: 10.1093/hmg/dds470

Chuang, J. C., and Jones, P. A. (2007). Epigenetics and microRNAs. *Pediatr. Res.* 61, 24R–29R. doi: 10.1203/pdr.0b013e3180457684

Cogswell, J. P., Ward, J., Taylor, I. A., Waters, M., Shi, Y., Cannon, B., et al. (2008). Identification of miRNA changes in Alzheimer's disease brain and CSF yields putative biomarkers and insights into disease pathways. *J. Alzheimers Dis.* 14, 27–41.

Conaco, C., Otto, S., Han, J. J., and Mandel, G. (2006). Reciprocal actions of REST and a microRNA promote neuronal identity. *Proc. Natl. Acad. Sci. U.S.A.* 103, 2422–2427. doi: 10.1073/pnas.0511041103

Coppede, F. (2012). Genetics and epigenetics of Parkinson's disease. *ScientificWorldJournal* 2012, 489830. doi: 10.1100/2012/489830

Czech, B., and Hannon, G. J. (2011). Small RNA sorting: matchmaking for Argonautes. *Nat. Rev. Genet.* 12, 19–31. doi: 10.1038/nrg2916

Dachsel, J. C., Taylor, J. P., Mok, S. S., Ross, O. A., Hinkle, K. M., Bailey, R. M., et al. (2007). Identification of potential protein interactors of Lrrk2. *Parkinsonism Relat. Disord.* 13, 382–385. doi: 10.1016/j.parkreldis.2007.01.008

De Felice, B., Guida, M., Coppola, C., De Mieri, G., and Cotrufo, R. (2012). A miRNA signature in leukocytes from sporadic amyotrophic lateral sclerosis. *Gene* 508, 35–40. doi: 10.1016/j.gene.2012.07.058

Delay, C., Mandemakers, W., and Hebert, S. S. (2012). MicroRNAs in Alzheimer's disease. *Neurobiol. Dis.* 46, 285–290. doi: 10.1016/j.nbd.2012.01.003

Denli, A. M., Tops, B. B., Plasterk, R. H., Ketting, R. F., and Hannon, G. J. (2004). Processing of primary microRNAs by the Microprocessor complex. *Nature* 432, 231–235. doi: 10.1038/nature03049

Diederichs, S., and Haber, D. A. (2007). Dual role for argonautes in microRNA processing and posttranscriptional regulation of microRNA expression. *Cell* 131, 1097–1108. doi: 10.1016/j.cell.2007.10.032

Doxakis, E. (2010). Post-transcriptional regulation of alpha-synuclein expression by mir-7 and mir-153. *J. Biol. Chem.* 285, 12726–12734. doi: 10.1074/jbc.M109.086827

Ebert, M. S., and Sharp, P. A. (2010). MicroRNA sponges: progress and possibilities. *RNA* 16, 2043–2050. doi: 10.1261/rna.2414110

Ebert, M. S., and Sharp, P. A. (2012). Roles for microRNAs in conferring robustness to biological processes. *Cell* 149, 515–524. doi: 10.1016/j.cell.2012.04.005

Elmen, J., Lindow, M., Schutz, S., Lawrence, M., Petri, A., Obad, S., et al. (2008). LNA-mediated microRNA silencing in non-human primates. *Nature* 452, 896–899. doi: 10.1038/nature06783

Ender, C., Krek, A., Friedlander, M. R., Beitzinger, M., Weinmann, L., Chen, W., et al. (2008). A human snoRNA with microRNA-like functions. *Mol. Cell* 32, 519–528. doi: 10.1016/j.molcel.2008.10.017

Etheridge, A., Lee, I., Hood, L., Galas, D., and Wang, K. (2011). Extracellular microRNA: a new source of biomarkers. *Mutat. Res.* 717, 85–90. doi: 10.1016/j.mrfmmm.2011.03.004

Fenn, A. M., Smith, K. M., Lovett-Racke, A. E., Guerau-de-Arellano, M., Whitacre, C. C., and Godbout, J. P. (2013). Increased micro-RNA 29b in the aged brain correlates with the reduction of insulin-like growth factor-1 and fractalkine ligand. *Neurobiol. Aging* 34, 2748–2758. doi: 10.1016/j.neurobiolaging.2013.06.007

Friedman, R. C., Farh, K. K., Burge, C. B., and Bartel, D. P. (2009). Most mammalian mRNAs are conserved targets of microRNAs. *Genome Res.* 19, 92–105. doi: 10.1101/gr.082701.108

Gaughwin, P. M., Ciesla, M., Lahiri, N., Tabrizi, S. J., Brundin, P., and Bjorkqvist, M. (2011).

Hsa-miR-34b is a plasma-stable microRNA that is elevated in pre-manifest Huntington's disease. *Hum. Mol. Genet.* 20, 2225–2237. doi: 10.1093/hmg/ddr111

Geekiyanage, H., Jicha, G. A., Nelson, P. T., and Chan, C. (2012). Blood serum miRNA: non-invasive biomarkers for Alzheimer's disease. *Exp. Neurol.* 235, 491–496. doi: 10.1016/j.expneurol.2011.11.026

Gehrke, S., Imai, Y., Sokol, N., and Lu, B. (2010). Pathogenic LRRK2 negatively regulates microRNA-mediated translational repression. *Nature* 466, 637–641. doi: 10.1038/nature09191

Ghose, J., Sinha, M., Das, E., Jana, N. R., and Bhattacharyya, N. P. (2011). Regulation of miR-146a by RelA/NFkB and p53 in STHdh(Q111)/Hdh(Q111) cells, a cell model of Huntington's disease. *PLoS ONE* 6:e23837. doi: 10.1371/journal.pone.0023837

Gonzalez-Alegre, P. (2007). Therapeutic RNA interference for neurodegenerative diseases: From promise to progress. *Pharmacol. Ther.* 114, 34–55. doi: 10.1016/j.pharmthera.2007.01.003

Goodall, E. F., Bury, J. J., Cooper-Knock, J., Shaw, P. J., and Kirby, J. (2012). "Genetics of familial amyotrophic lateral sclerosis, amyotrophic lateral sclerosis," in *Amyotrophic Lateral Sclerosis*, ed M. H. Maurer (Rijeka: InTech), 517–536. doi: 10.5772/32498

Gregory, R. I., Yan, K. P., Amuthan, G., Chendrimada, T., Doratotaj, B., Cooch, N., et al. (2004). The Microprocessor complex mediates the genesis of microRNAs. *Nature* 432, 235–240. doi: 10.1038/nature03120

Griffiths-Jones, S., Saini, H. K., van Dongen, S., and Enright, A. J. (2008). miRBase: tools for microRNA genomics. *Nucleic Acids Res.* 36, D154–D158. doi: 10.1093/nar/gkm952

Haramati, S., Chapnik, E., Sztainberg, Y., Eilam, R., Zwang, R., Gershoni, N., et al. (2010). miRNA malfunction causes spinal motor neuron disease. *Proc. Natl. Acad. Sci. U.S.A.* 107, 13111–13116. doi: 10.1073/pnas.1006151107

Hardy, J., Lewis, P., Revesz, T., Lees, A., and Paisan-Ruiz, C. (2009). The genetics of Parkinson's syndromes: a critical review. *Curr. Opin. Genet. Dev.* 19, 254–265. doi: 10.1016/j.gde.2009.03.008

Haussecker, D., Huang, Y., Lau, A., Parameswaran, P., Fire, A. Z., and Kay, M. A. (2010). Human tRNA-derived small RNAs in the global regulation of RNA silencing. *RNA* 16, 673–695. doi: 10.1261/rna.2000810

Hebert, S. S., Horre, K., Nicolai, L., Bergmans, B., Papadopoulou, A. S., Delacourte, A., et al. (2009). MicroRNA regulation of Alzheimer's Amyloid precursor protein expression. *Neurobiol. Dis.* 33, 422–428. doi: 10.1016/j.nbd.2008.11.009

Hebert, S. S., Horre, K., Nicolai, L., Papadopoulou, A. S., Mandemakers, W., Silahtaroglu, A. N., et al. (2008). Loss of microRNA cluster miR-29a/b-1 in sporadic Alzheimer's disease correlates with increased BACE1/beta-secretase expression. *Proc. Natl. Acad. Sci. U.S.A.* 105, 6415–6420. doi: 10.1073/pnas.0710263105

Heyer, M. P., Pani, A. K., Smeyne, R. J., Kenny, P. J., and Feng, G. (2012). Normal midbrain dopaminergic neuron development and function in miR-133b mutant mice. *J. Neurosci.* 32, 10887–10894. doi: 10.1523/JNEUROSCI.1732-12.2012

Hwang, D. Y., Ardayfio, P., Kang, U. J., Semina, E. V., and Kim, K. S. (2003). Selective loss of dopaminergic neurons in the substantia nigra of Pitx3-deficient aphakia mice. *Brain Res. Mol. Brain Res.* 114, 123–131. doi: 10.1016/S0169-328X(03)00162-1

Imarisio, S., Carmichael, J., Korolchuk, V., Chen, C. W., Saiki, S., Rose, C., et al. (2008). Huntington's disease: from pathology and genetics to potential therapies. *Biochem. J.* 412, 191–209. doi: 10.1042/BJ20071619

Itoh, N., and Ohta, H. (2013). Roles of FGF20 in dopaminergic neurons and Parkinson's disease. *Front. Mol. Neurosci.* 6:15. doi: 10.3389/fnmol.2013.00015

Jin, J., Cheng, Y., Zhang, Y., Wood, W., Peng, Q., Hutchison, E., et al. (2012). Interrogation of brain miRNA and mRNA expression profiles reveals a molecular regulatory network that is perturbed by mutant huntingtin. *J. Neurochem.* 123, 477–490. doi: 10.1111/j.1471-4159.2012.07925.x

Johnson, R., Zuccato, C., Belyaev, N. D., Guest, D. J., Cattaneo, E., and Buckley, N. J. (2008). A microRNA-based gene dysregulation pathway in Huntington's disease. *Neurobiol. Dis.* 29, 438–445. doi: 10.1016/j.nbd.2007.11.001

Junn, E., Lee, K. W., Jeong, B. S., Chan, T. W., Im, J. Y., and Mouradian, M. M. (2009). Repression of alpha-synuclein expression and toxicity by microRNA-7. *Proc. Natl. Acad. Sci. U.S.A.* 106, 13052–13057. doi: 10.1073/pnas.0906277106

Kapranov, P., and St Laurent, G. (2012). Dark matter RNA: existence, function, and controversy. *Front. Genet.* 3:60. doi: 10.3389/fgene.2012.00060

Kawahara, Y., and Mieda-Sato, A. (2012). TDP-43 promotes microRNA biogenesis as a component of the Drosha and Dicer complexes. *Proc. Natl. Acad. Sci. U.S.A.* 109, 3347–3352. doi: 10.1073/pnas.1112427109

Khan, A. A., Betel, D., Miller, M. L., Sander, C., Leslie, C. S., and Marks, D. S. (2009). Transfection of small RNAs globally perturbs gene regulation by endogenous microRNAs. *Nat. Biotechnol.* 27, 549–555. doi: 10.1038/nbt.1543

Khoo, S. K., Petillo, D., Kang, U. J., Resau, J. H., Berryhill, B., Linder, J., et al. (2012). Plasma-based circulating MicroRNA biomarkers for parkinson's disease. *J. Parkinsons Dis.* 2, 321–331. doi: 10.3233/JPD-012144

Kim, J., Inoue, K., Ishii, J., Vanti, W. B., Voronov, S. V., Murchison, E., et al. (2007). A MicroRNA feedback circuit in midbrain dopamine neurons. *Science* 317, 1220–1224. doi: 10.1126/science.1140481

Kim, J., Yoon, H., Ramirez, C. M., Lee, S. M., Hoe, H. S., and Fernandez-Hernando, C. (2012). MiR-106b impairs cholesterol efflux and increases Abeta levels by repressing ABCA1 expression. *Exp. Neurol.* 235, 476–483. doi: 10.1016/j.expneurol.2011.11.010

Kim, Y. K., and Kim, V. N. (2007). Processing of intronic microRNAs. *EMBO J.* 26, 775–783. doi: 10.1038/sj.emboj.7601512

Koval, E. D., Shaner, C., Zhang, P., du Maine, X., Fischer, K., Tay, J., et al. (2013). Method for widespread microRNA-155 inhibition prolongs survival in ALS-model mice. *Hum. Mol. Genet.* 22, 4127–4135. doi: 10.1093/hmg/ddt261

Krek, A., Grun, D., Poy, M. N., Wolf, R., Rosenberg, L., Epstein, E. J., et al. (2005). Combinatorial microRNA target predictions. *Nat. Genet.* 37, 495–500. doi: 10.1038/ng1536

Krol, J., Busskamp, V., Markiewicz, I., Stadler, M. B., Ribi, S., Richter, J., et al. (2010a). Characterizing light-regulated retinal microRNAs reveals rapid turnover as a common property of neuronal microRNAs. *Cell* 141, 618–631. doi: 10.1016/j.cell.2010.03.039

Krol, J., Loedige, I., and Filipowicz, W. (2010b). The widespread regulation of microRNA biogenesis, function and decay. *Nat. Rev. Genet.* 11, 597–610. doi: 10.1038/nrg2843

Lagos-Quintana, M., Rauhut, R., Yalcin, A., Meyer, J., Lendeckel, W., and Tuschl, T. (2002). Identification of tissue-specific microRNAs from mouse. *Curr. Biol.* 12, 735–739. doi: 10.1016/S0960-9822(02)00809-6

Lee, S. T., Chu, K., Im, W. S., Yoon, H. J., Im, J. Y., Park, J. E., et al. (2011). Altered microRNA regulation in Huntington's disease models. *Exp. Neurol.* 227, 172–179. doi: 10.1016/j.expneurol.2010.10.012

Lee, Y., Jeon, K., Lee, J. T., Kim, S., and Kim, V. N. (2002). MicroRNA maturation: stepwise processing and subcellular localization. *EMBO J.* 21, 4663–4670. doi: 10.1093/emboj/cdf476

Lee, Y. S., Shibata, Y., Malhotra, A., and Dutta, A. (2009). A novel class of small RNAs: tRNA-derived RNA fragments (tRFs). *Genes Dev.* 23, 2639–2649. doi: 10.1101/gad.1837609

Lehmann, S. M., Kruger, C., Park, B., Derkow, K., Rosenberger, K., Baumgart, J., et al. (2012). An unconventional role for miRNA: let-7 activates Toll-like receptor 7 and causes neurodegeneration. *Nat. Neurosci.* 15, 827–835. doi: 10.1038/nn.3113

Lewis, B. P., Burge, C. B., and Bartel, D. P. (2005). Conserved seed pairing, often flanked by adenosines, indicates that thousands of human genes are microRNA targets. *Cell* 120, 15–20. doi: 10.1016/j.cell.2004.12.035

Lewis, B. P., Shih, I. H., Jones-Rhoades, M. W., Bartel, D. P., and Burge, C. B. (2003). Prediction of mammalian microRNA targets. *Cell* 115, 787–798. doi: 10.1016/S0092-8674(03)01018-3

Li, Y. Y., Cui, J. G., Dua, P., Pogue, A. I., Bhattacharjee, S., and Lukiw, W. J. (2011). Differential expression of miRNA-146a-regulated inflammatory genes in human primary neural, astroglial and microglial cells. *Neurosci. Lett.* 499, 109–113. doi: 10.1016/j.neulet.2011.05.044

Liang, C., Zhu, H., Xu, Y., Huang, L., Ma, C., Deng, W., et al. (2012). MicroRNA-153 negatively regulates the expression of amyloid precursor protein and amyloid precursor-like protein 2. *Brain Res.* 1455, 103–113. doi: 10.1016/j.brainres.2011.10.051

Lim, L. P., Lau, N. C., Garrett-Engele, P., Grimson, A., Schelter, J. M., Castle, J., et al. (2005). Microarray analysis shows that some microRNAs downregulate large numbers of target mRNAs. *Nature* 433, 769–773. doi: 10.1038/nature03315

Ling, S., Zhou, J., Rudd, J. A., Hu, Z., and Fang, M. (2011). The recent updates of therapeutic approaches against abeta for the treatment of Alzheimer's disease. *Anat. Rec. (Hoboken)* 294, 1307–1318. doi: 10.1002/ar.21425

Liu, J., Carmell, M. A., Rivas, F. V., Marsden, C. G., Thomson, J. M., Song, J. J., et al. (2004). Argonaute2 is the catalytic engine of mammalian RNAi. *Science* 305, 1437–1441. doi: 10.1126/science.1102513

Liu, N., Landreh, M., Cao, K., Abe, M., Hendriks, G. J., Kennerdell, J. R., et al. (2012). The microRNA miR-34 modulates ageing and neurodegeneration in Drosophila. *Nature* 482, 519–523. doi: 10.1038/nature10810

Liu, Q., Xie, F., Alvarado-Diaz, A., Smith, M. A., Moreira, P. I., Zhu, X., et al. (2011). Neurofilamentopathy in neurodegenerative diseases. *Open Neurol. J.* 5, 58–62. doi: 10.2174/1874205X01105010058

Long, J. M., Ray, B., and Lahiri, D. K. (2012). MicroRNA-153 physiologically inhibits expression of amyloid-beta precursor protein in cultured human fetal brain cells and is dysregulated in a subset of Alzheimer disease patients. *J. Biol. Chem.* 287, 31298–31310. doi: 10.1074/jbc.M112.366336

Lukiw, W. J., Zhao, Y., and Cui, J. G. (2008). An NF-kappaB-sensitive micro RNA-146a-mediated inflammatory circuit in Alzheimer disease and in stressed human brain cells. *J. Biol. Chem.* 283, 31315–31322. doi: 10.1074/jbc.M805371200

Margis, R., and Rieder, C. R. (2011). Identification of blood microRNAs associated to Parkinson's disease. *J. Biotechnol.* 152, 96–101. doi: 10.1016/j.jbiotec.2011.01.023

Marti, E., Pantano, L., Banez-Coronel, M., Llorens, F., Minones-Moyano, E., Porta, S., et al. (2010). A myriad of miRNA variants in control and Huntington's disease brain regions detected by massively parallel sequencing. *Nucleic Acids Res.* 38, 7219–7235. doi: 10.1093/nar/gkq575

Martins, M., Rosa, A., Guedes, L. C., Fonseca, B. V., Gotovac, K., Violante, S., et al. (2011). Convergence of miRNA expression profiling, alpha-synuclein interacton and GWAS in Parkinson's disease. *PLoS ONE* 6:e25443. doi: 10.1371/journal.pone.0025443

Mattick, J. S. (2005). The functional genomics of noncoding RNA. *Science* 309, 1527–1528. doi: 10.1126/science.1117806

McBride, J. L., Boudreau, R. L., Harper, S. Q., Staber, P. D., Monteys, A. M., Martins, I., et al. (2008). Artificial miRNAs mitigate shRNA-mediated toxicity in the brain: implications for the therapeutic development of RNAi. *Proc. Natl. Acad. Sci. U.S.A.* 105, 5868–5873. doi: 10.1073/pnas.0801775105

McCarthy, J. J. (2008). MicroRNA-206: the skeletal muscle-specific myomiR. *Biochim. Biophys. Acta* 1779, 682–691. doi: 10.1016/j.bbagrm.2008.03.001

McDermott, C. J., and Shaw, P. J. (2008). Diagnosis and management of motor neurone disease. *BMJ* 336, 658–662. doi: 10.1136/bmj.39493.511759.BE

Minones-Moyano, E., Porta, S., Escaramis, G., Rabionet, R., Iraola, S., Kagerbauer, B., et al. (2011). MicroRNA profiling of Parkinson's disease brains identifies early downregulation of miR-34b/c which modulate mitochondrial function. *Hum. Mol. Genet.* 20, 3067–3078. doi: 10.1093/hmg/ddr210

Miyoshi, K., Miyoshi, T., and Siomi, H. (2010). Many ways to generate microRNA-like small RNAs: non-canonical pathways for microRNA production. *Mol. Genet. Genomics.* 284, 95–103. doi: 10.1007/s00438-010-0556-1

Morel, L., Regan, M., Higashimori, H., Ng, S. K., Esau, C., Vidensky, S., et al. (2013). Neuronal exosomal miRNA-dependent translational regulation of astroglial glutamate transporter GLT1. *J. Biol. Chem.* 288, 7105–7116. doi: 10.1074/jbc.M112.410944

Morlando, M., Dini Modigliani, S., Torrelli, G., Rosa, A., Di Carlo, V., Caffarelli, E., et al. (2012). FUS stimulates microRNA biogenesis by facilitating co-transcriptional Drosha recruitment. *EMBO J.* 31, 4502–4510. doi: 10.1038/emboj.2012.319

Nana-Sinkam, S. P., and Croce, C. M. (2013). Clinical applications for microRNAs in cancer. *Clin. Pharmacol. Ther.* 93, 98–104. doi: 10.1038/clpt.2012.192

Nelson, P. T., and Wang, W. X. (2010). MiR-107 is reduced in Alzheimer's disease brain neocortex: validation study. *J. Alzheimers Dis.* 21, 75–79. doi: 10.3233/JAD-2010-091603

Nissan, X., Blondel, S., Navarro, C., Maury, Y., Denis, C., Girard, M., et al. (2012). Unique preservation of neural cells in Hutchinson- Gilford progeria syndrome is due to the expression of the neural-specific miR-9 microRNA. *Cell Rep.* 2, 1–9. doi: 10.1016/j.celrep.2012.05.015

Orom, U. A., Nielsen, F. C., and Lund, A. H. (2008). MicroRNA-10a binds the 5'UTR of ribosomal protein mRNAs and enhances their translation. *Mol. Cell* 30, 460–471. doi: 10.1016/j.molcel.2008.05.001

Ozsolak, F., Poling, L. L., Wang, Z., Liu, H., Liu, X. S., Roeder, R. G., et al. (2008). Chromatin structure analyses identify miRNA promoters. *Genes Dev.* 22, 3172–3183. doi: 10.1101/gad.1706508

Packer, A. N., Xing, Y., Harper, S. Q., Jones, L., and Davidson, B. L. (2008). The bifunctional microRNA miR-9/miR-9* regulates REST and CoREST and is downregulated in Huntington's disease. *J. Neurosci.* 28, 14341–14346. doi: 10.1523/JNEUROSCI.2390-08.2008

Patel, N., Hoang, D., Miller, N., Ansaloni, S., Huang, Q., Rogers, J. T., et al. (2008). MicroRNAs can regulate human APP levels. *Mol. Neurodegener.* 3:10. doi: 10.1186/1750-1326-3-10

Pratt, A. J., and MacRae, I. J. (2009). The RNA-induced silencing complex: a versatile gene-silencing machine. *J. Biol. Chem.* 284, 17897–17901. doi: 10.1074/jbc.R900012200

Ramachandran, V., and Chen, X. (2008). Degradation of microRNAs by a family of exoribonucleases in Arabidopsis. *Science* 321, 1490–1492. doi: 10.1126/science.1163728

Raposo, G., and Stoorvogel, W. (2013). Extracellular vesicles: exosomes, microvesicles, and friends. *J. Cell Biol.* 200, 373–383. doi: 10.1083/jcb.201211138

Robberecht, W., and Philips, T. (2013). The changing scene of amyotrophic lateral sclerosis. *Nat. Rev. Neurosci.* 14, 248–264. doi: 10.1038/nrn3430

Rodriguez, A., Griffiths-Jones, S., Ashurst, J. L., and Bradley, A. (2004). Identification of mammalian microRNA host genes and transcription units. *Genome Res.* 14, 1902–1910. doi: 10.1101/gr.2722704

Ruby, J. G., Jan, C. H., and Bartel, D. P. (2007). Intronic microRNA precursors that bypass Drosha processing. *Nature* 448, 83–86. doi: 10.1038/nature05983

Russell, A. P., Wada, S., Vergani, L., Hock, M. B., Lamon, S., Leger, B., et al. (2012). Disruption of skeletal muscle mitochondrial network genes and miRNAs in amyotrophic lateral sclerosis. *Neurobiol. Dis.* 49C, 107–117. doi: 10.1016/j.nbd.2012.08.015

Salmena, L., Poliseno, L., Tay, Y., Kats, L., and Pandolfi, P. P. (2011). A ceRNA hypothesis: the rosetta stone of a hidden RNA language. *Cell* 146, 353–358. doi: 10.1016/j.cell.2011.07.014

Saraiya, A. A., and Wang, C. C. (2008). snoRNA, a novel precursor of microRNA in Giardia lamblia. *PLoS Pathog.* 4:e1000224. doi: 10.1371/journal.ppat.1000224

Saunders, L. R., Sharma, A. D., Tawney, J., Nakagawa, M., Okita, K., Yamanaka, S., et al. (2010). miRNAs regulate SIRT1 expression during mouse embryonic stem cell differentiation and in adult mouse tissues. *Aging (Albany NY)* 2, 415–431.

Savas, J. N., Makusky, A., Ottosen, S., Baillat, D., Then, F., Krainc, D., et al. (2008). Huntington's disease protein contributes to RNA-mediated gene silencing through association with Argonaute and P bodies. *Proc. Natl. Acad. Sci. U.S.A.* 105, 10820–10825. doi: 10.1073/pnas.0800658105

Schipper, H. M., Maes, O. C., Chertkow, H. M., and Wang, E. (2007). MicroRNA expression in Alzheimer blood mononuclear cells. *Gene Regul. Syst. Biol.* 1, 263–274.

Schmitt, I., Wullner, U., van Rooyen, J. P., Khazneh, H., Becker, J., Volk, A., et al. (2012). Variants in the 3'UTR of SNCA do not affect miRNA-433 binding and alpha-synuclein expression. *Eur. J. Hum. Genet.* 20, 1265–1269. doi: 10.1038/ejhg.2012.84

Schonrock, N., and Gotz, J. (2012). Decoding the non-coding RNAs in Alzheimer's disease. *Cell. Mol. Life Sci.* 69, 3543–3559. doi: 10.1007/s00018-012-1125-z

Schonrock, N., Humphreys, D. T., Preiss, T., and Gotz, J. (2012). Target gene repression mediated by miRNAs miR-181c and miR-9 both of which are down-regulated by amyloid-beta. *J. Mol. Neurosci.* 46, 324–335. doi: 10.1007/s12031-011-9587-2

Schonrock, N., Ke, Y. D., Humphreys, D., Staufenbiel, M., Ittner, L. M., Preiss, T., et al. (2010). Neuronal microRNA deregulation in response to Alzheimer's disease amyloid-beta. *PLoS ONE* 5:e11070. doi: 10.1371/journal.pone.0011070

Seredenina, T., and Luthi-Carter, R. (2012). What have we learned from gene expression profiles in Huntington's disease. *Neurobiol. Dis.* 45, 83–98. doi: 10.1016/j.nbd.2011.07.001

Sethi, P., and Lukiw, W. J. (2009). Micro-RNA abundance and stability in human brain: specific alterations in Alzheimer's disease temporal lobe neocortex.

Neurosci. Lett. 459, 100–104. doi: 10.1016/j.neulet.2009.04.052

Shaked, I., Meerson, A., Wolf, Y., Avni, R., Greenberg, D., Gilboa-Geffen, A., et al. (2009). MicroRNA-132 potentiates cholinergic anti-inflammatory signaling by targeting acetylcholinesterase. *Immunity* 31, 965–973. doi: 10.1016/j.immuni.2009.09.019

Shi, M., Caudle, W. M., and Zhang, J. (2009). Biomarker discovery in neurodegenerative diseases: a proteomic approach. *Neurobiol. Dis.* 35, 157–164. doi: 10.1016/j.nbd.2008.09.004

Shioya, M., Obayashi, S., Tabunoki, H., Arima, K., Saito, Y., Ishida, T., et al. (2010). Aberrant microRNA expression in the brains of neurodegenerative diseases: miR-29a decreased in Alzheimer disease brains targets neurone navigator 3. *Neuropathol. Appl. Neurobiol.* 36, 320–330. doi: 10.1111/j.1365-2990.2010.01076.x

Sibley, C. R., Seow, Y., Curtis, H., Weinberg, M. S., and Wood, M. J. (2012). Silencing of Parkinson's disease-associated genes with artificial mirtron mimics of miR-1224. *Nucleic Acids Res.* 40, 9863–9875. doi: 10.1093/nar/gks712

Sinha, M., Ghose, J., and Bhattarcharyya, N. P. (2011). Micro RNA -214, -150, -146a and-125b target Huntingtin gene. *RNA Biol.* 8, 1005–1021. doi: 10.4161/rna.8.6.16035

Sinha, M., Ghose, J., Das, E., and Bhattarcharyya, N. P. (2010). Altered microRNAs in STHdh(Q111)/Hdh(Q111) cells: miR-146a targets TBP. *Biochem. Biophys. Res. Commun.* 396, 742–747. doi: 10.1016/j.bbrc.2010.05.007

Smith, P., Al Hashimi, A., Girard, J., Delay, C., and Hebert, S. S. (2011). In vivo regulation of amyloid precursor protein neuronal splicing by microRNAs. *J. Neurochem.* 116, 240–247. doi: 10.1111/j.1471-4159.2010.07097.x

Soreq, H., and Wolf, Y. (2011). NeurimmiRs: microRNAs in the neuroimmune interface. *Trends Mol. Med.* 17, 548–555. doi: 10.1016/j.molmed.2011.06.009

Soreq, L., Salomonis, N., Bronstein, M., Greenberg, D. S., Israel, Z., Bergman, H., et al. (2013). Small RNA sequencing-microarray analyses in Parkinson leukocytes reveal deep brain stimulation-induced splicing changes that classify brain region transcriptomes. *Front. Mol. Neurosci.* 6:10. doi: 10.3389/fnmol.2013.00010

Sotiriou, S., Gibney, G., Baxevanis, A. D., and Nussbaum, R. L. (2009). A single nucleotide polymorphism in the 3'UTR of the SNCA gene encoding alpha-synuclein is a new potential susceptibility locus for Parkinson disease. *Neurosci. Lett.* 461, 196–201. doi: 10.1016/j.neulet.2009.06.034

Tang, G., Tang, X., Mendu, V., Jia, X., Chen, Q. J., and He, L. (2008). The art of microRNA: various strategies leading to gene silencing via an ancient pathway. *Biochim. Biophys. Acta* 1779, 655–662. doi: 10.1016/j.bbagrm.2008.06.006

Tanzi, R. E. (2012). The genetics of Alzheimer disease. *Cold Spring Harb. Perspect. Med.* 2, 6296–6306. doi: 10.1101/cshperspect.a006296

Vasudevan, S., Tong, Y., and Steitz, J. A. (2007). Switching from repression to activation: microRNAs can up-regulate translation. *Science* 318, 1931–1934. doi: 10.1126/science.1149460

Villa, C., Ridolfi, E., Fenoglio, C., Ghezzi, L., Vimercati, R., Clerici, F., et al. (2013). Expression of the transcription factor Sp1 and its regulatory hsa-miR-29b in peripheral blood mononuclear cells from patients with Alzheimer's disease. *J. Alzheimers Dis.* 35, 487–494. doi: 10.3233/JAD-122263

Wanet, A., Tacheny, A., Arnould, T., and Renard, P. (2012). miR-212/132 expression and functions: within and beyond the neuronal compartment. *Nucleic Acids Res.* 40, 4742–4753. doi: 10.1093/nar/gks151

Wang, G., van der Walt, J. M., Mayhew, G., Li, Y. J., Zuchner, S., Scott, W. K., et al. (2008a). Variation in the miRNA-433 binding site of FGF20 confers risk for Parkinson disease by overexpression of alpha-synuclein. *Am. J. Hum. Genet.* 82, 283–289. doi: 10.1016/j.ajhg.2007.09.021

Wang, W. X., Rajeev, B. W., Stromberg, A. J., Ren, N., Tang, G., Huang, Q., et al. (2008b). The expression of microRNA miR-107 decreases early in Alzheimer's disease and may accelerate disease progression through regulation of beta-site amyloid precursor protein-cleaving enzyme 1. *J. Neurosci.* 28, 1213–1223. doi: 10.1523/JNEUROSCI.5065-07.2008

Wang, L. L., Huang, Y., Wang, G., and Chen, S. D. (2012). The potential role of microRNA-146 in Alzheimer's disease: biomarker or therapeutic target. *Med. Hypotheses* 78, 398–401. doi: 10.1016/j.mehy.2011.11.019

Wang, W. X., Huang, Q., Hu, Y., Stromberg, A. J., and Nelson, P. T. (2011). Patterns of microRNA expression in normal and early Alzheimer's disease human temporal cortex: white matter versus gray matter. *Acta Neuropathol.* 121, 193–205. doi: 10.1007/s00401-010-0756-0

Wang, X., Liu, P., Zhu, H., Xu, Y., Ma, C., Dai, X., et al. (2009). miR-34a, a microRNA up-regulated in a double transgenic mouse model of Alzheimer's disease, inhibits bcl2 translation. *Brain Res. Bull.* 80, 268–273. doi: 10.1016/j.brainresbull.2009.08.006

Williams, A. H., Valdez, G., Moresi, V., Qi, X., McAnally, J., Elliott, J. L., et al. (2009). MicroRNA-206 delays ALS progression and promotes regeneration of neuromuscular synapses in mice. *Science* 326, 1549–1554. doi: 10.1126/science.1181046

Winslow, A. R., and Rubinsztein, D. C. (2011). The Parkinson disease protein alpha-synuclein inhibits autophagy. *Autophagy* 7, 429–431. doi: 10.4161/auto.7.4.14393

Wong, H. K., Veremeyko, T., Patel, N., Lemere, C. A., Walsh, D. M., Esau, C., et al. (2013). De-repression of FOXO3a death axis by microRNA-132 and -212 causes neuronal apoptosis in Alzheimer's disease. *Hum. Mol. Genet.* 22, 3077–3092. doi: 10.1093/hmg/ddt164

Zhou, X., Ruan, J., Wang, G., and Zhang, W. (2007). Characterization and identification of microRNA core promoters in four model species. *PLoS Comput. Biol.* 3:e37. doi: 10.1371/journal.pcbi.0030037

Zong, Y., Wang, H., Dong, W., Quan, X., Zhu, H., Xu, Y., et al. (2011). miR-29c regulates BACE1 protein expression. *Brain Res.* 1395, 108–115. doi: 10.1016/j.brainres.2011.04.035

Zuccato, C., Belyaev, N., Conforti, P., Ooi, L., Tartari, M., Papadimou, E., et al. (2007). Widespread disruption of repressor element-1 silencing transcription factor/neuron-restrictive silencer factor occupancy at its target genes in Huntington's disease. *J. Neurosci.* 27, 6972–6983. doi: 10.1523/JNEUROSCI.4278-06.2007

Zuccato, C., Valenza, M., and Cattaneo, E. (2010). Molecular mechanisms and potential therapeutical targets in Huntington's disease. *Physiol. Rev.* 90, 905–981. doi: 10.1152/physrev.00041.2009

microRNAs and cell fate in cortical and retinal development

Federico Cremisi*

Scuola Normale Superiore, Pisa, Italy

Correspondence:
Federico Cremisi, Scuola Normale Superiore, Piazza dei Cavalieri 7, 56126 Pisa, Italy
e-mail: f.cremisi@sns.it

MicroRNAs (miRNAs) are involved in crucial steps of neurogenesis, neural differentiation, and neuronal plasticity. Here we review experimental evidence suggesting that miRNAs may regulate the histogenesis of the cerebral cortex and neural retina. Both cortical and retinal early progenitor cells are multipotent, that is, they can generate different types of cortical or retinal cells, respectively, in one lineage. In both cortical and retinal development, the precise timing of activation of cell fate transcription factors results in a stereotyped schedule of generation of the different types of neurons. Emerging evidence indicates that miRNAs may play an important role in regulating such temporal programing of neuronal differentiation. Neuronal subtypes of the cortex and retina exhibit distinct miRNA signatures, implying that miRNA codes may be used to specify different types of neurons. Interfering with global miRNA activity changes the ratio of the different types of neurons produced. In fact, there are examples of cell fate genes that are regulated at the translational level, both in retinogenesis and in corticogenesis. A model depicting how miRNAs might orchestrate both the type and the birth of different neurons is presented and discussed.

Glossary

- Lineage: the temporally ordered cell progeny of an individual progenitor cell.
- Specification: the (reversible) process by which a cell becomes capable of, and biased toward, a particular fate.
- Commitment: the process by which cell fate is fully determined and can no longer be affected by external cues.
- Competence: a cell condition linked to temporal identity. It can be defined as the ability of a progenitor cell to respond to a signal and become a particular type of neuron.
- Potency: the entire complement of cells that a progenitor can ultimately produce.
- Multipotency: the ability to give rise to more than one cell type.
- Progenitor: a dividing cell that, in contrast to a stem cell, cannot proliferate indefinitely.
- Antago-miR: modified antisense oligonucleotide that blocks the activity of a miRNA.
- Heterochronic neuron: type of neurons that is generated at inappropriate times of development.
- Neuron birth date: the time of the last mitosis of a neuronal cell.

Keywords: cortex, retina, cell-fate, heterochronic, timing, cell birth date, development

GENERAL IMPLICATIONS OF miRNAs IN NEURAL DEVELOPMENT

MicroRNAs (miRNAs) are a large family of non-coding RNAs of approximately 21 nucleotides in length, which inhibit gene expression at the translational level and are involved in the control of many developmental and cellular processes in eukaryotic organisms, including vertebrate neural development (Krol et al., 2010). miRNAs have been found to regulate many aspects of neural development, including the early steps in neurogenesis, the specification and differentiation of neural progenitor cells, brain patterning, and the plasticity of mature neurons (Coolen and Bally-Cuif, 2009; Fineberg et al., 2009; Bian and Sun, 2011).

Examples of miRNAs involved in the specification of distinct types of mature neurons have also been described. miR-7a is expressed in a gradient opposing Pax6 along the ventricular walls and restricts its translation in the dorsal aspect. *In vivo* inhibition of miR-7a in Pax6-negative regions of the lateral wall induced Pax6 protein expression and increased dopaminergic neurons in the olfactory bulb (De Chevigny et al., 2012). miR-132 plays a key role in the differentiation of dopamine neurons by directly regulating the expression of Nurr1, which is one of the most important transcription factors in determining dopamine neuron development and differentiation (Yang et al., 2012). The over-expression of miR-181a and miR-125b increases the expression of dopaminergic markers and the ratio of tyrosine hydroxylase (TH) positive neurons generated by neural stem cells derived from human embryonic stem cells, whereas the inhibition of these miRNAs impairs the generation of the dopaminergic subtype (Stappert et al., 2013). miR-9, which is reiteratively used in patterning, neurogenesis, and differentiation (Coolen and Bally-Cuif, 2009), also has a role in establishing distinct types of motor neurons. miR-9 is transiently expressed in a motor neuron subtype together with its target gene FoxP1, which determines distinct motor neuron subtypes. Consequently, miR-9 overexpression or knockdown switches columnar identities in developing chick spinal cords (Otaegi et al., 2011).

Recent observations suggest that combinatorial miRNA expression may contribute to specifying neuron identity. The expression of a large fraction of known miRNAs with distinct expression profiles in glutamatergic and subtypes of GABAergic neurons has recently been demonstrated (He et al., 2012). In the mouse retina, a comprehensive survey of miRNA expression was achieved by *in situ* hybridization, revealing the expression of specific sets of miRNAs in distinct neuronal subtypes (Karali et al., 2010). Here we discuss the role that miRNAs may play in the generation of distinct types of neurons at different times in the development of layered structures. We will focus on the histogenesis of the neural retina and the cerebral cortex, where the role of miRNAs has been most widely investigated.

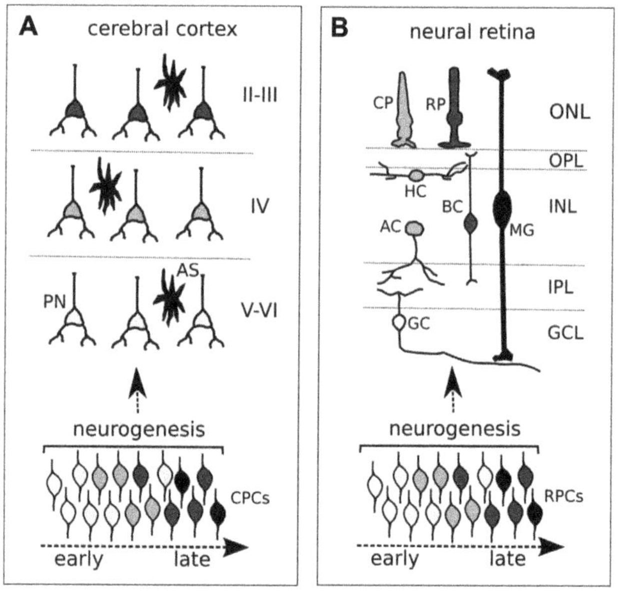

FIGURE 1 | Neurogenic timing in the developing cortex (A) and retina (B). **(A,B)** Different degrees of gray depict distinct neuronal identities in cortex **(A)** and retina **(B)**. Both cortical and retinal progenitor cells (CPCs and RPCs, respectively) change competence over time (different degrees of gray from early to late). Although an overlap in neuronal cell birth periods is shown, the time of exit from the cell cycle (neurogenesis, or cell birth date) influences the acquisition of distinct cell identities of post-mitotic neurons. **(A)** CPCs comprise both ventricular (primary) and subventricular (secondary) progenitor cells. PN, projecting neuron; AS, astrocyte. Roman numerals indicate cortical layers. **(B)** Different retinal neurons and glia. CP, cone photoreceptor; RP, rod photoreceptor; HC, horizontal cell; BC, bipolar cell; MG, Müller glia; AC, amacrine cell; GC, ganglion cell; ONL, outer nuclear layer; OPL, outer plexiform layer; INL, inner nuclear layer; IPL, inner plexiform layer; GCL, ganglion cell layer.

CORTICOGENESIS AND RETINOGENESIS SHOW SIMILAR MECHANISMS FOR ESTABLISHING DISTINCT CELL FATES

One main characteristic of the both retina and the cortex is that the identity of a certain type of mature neuron correlates with the time of its last division (cell birth date). Cortical projection neurons are derived from progenitor cells of the dorsal forebrain. After an initial phase of expansion, which is realized by symmetric divisions, progenitor cells of the ventricular zone (radial glia) start asymmetric divisions that generate new radial glia and either post-mitotic neurons (direct neurogenesis) or secondary (intermediate) progenitors. The net result is that the pool of progenitors does not deplete over the time and a single progenitor can generate a lineage made of different types of neurons with different birth dates. In the cortex, neurons with early birth dates are produced by primary (early) progenitor cells of the ventricular zone (radial glia) and populate the deep layers VI–V. Neurons with late birth dates, which fill the superficial layers II–III, are primarily generated by Tbr2-positive secondary progenitor cells of the subventricular zone (Leone et al., 2008; Sessa et al., 2008, 2010; **Figure 1A**). By the time a young neuron has progressed through its final mitotic division, the cell has acquired the information needed to migrate to the layer typical of its birth date, independent of the environment. Cellular studies by transplantation experiments suggest a progressive restriction in the developmental potential of cortical cells. Early progenitors, which normally produce deep-layer neurons, are multipotent: these cells can directly produce upper-layer neurons when transplanted into an older brain environment (McConnell and Kaznowski, 1991). Conversely, the progenitors of layer IV–II neurons have lost the ability to form layer VI neurons if transplanted into younger brains (Frantz and McConnell, 1996; Desai and McConnell, 2000).

In the retina, landmark studies of lineage-tracing have shown that early progenitor cells are multipotent and, likewise, early cortical progenitors can generate lineages containing different types of neurons (Turner and Cepko, 1987; Holt et al., 1988; Wetts and Fraser, 1988). The six types of neurons and the Müller glia making up the vertebrate retina are generated in a stereotyped sequence, with a correlation between cell birth date and cell fate, though with some overlap in the production of retinal cell types at any given time. Retinal ganglion cells (RGCs) are generated first, followed by the production of cone photoreceptors, horizontal cells, and amacrine neurons. Rod photoreceptors, bipolar neurons, and Müller glia are generated last (**Figure 1B**). Retinal progenitors generate these different cell types by proceeding through intrinsically defined competence states, with a certain degree of influence of environmental cues.

A growing list of transcription factors has emerged as key intrinsic regulators of cortical and retinal cell fate. Cortical progenitors sequentially activate a number of transcription factor genes that have the potential to determine the fates of their daughter cells. Early progenitor cells produce deep-layer neurons that express Fezf2 and Ctip2, which specify subcortically projecting neurons. Late progenitors generate upper-layer neurons expressing Satb2, which is required for the formation of axonal projections that connect the two cerebral hemispheres. Fezf2/Ctip2 and Satb2 pathways appear to be mutually repressive, thus ensuring that individual neurons adopt either a subcortical or callosal projection neuron identity (Leone et al., 2008). The molecular nature of this cross-repression is under scrutiny (Srinivasan et al., 2012). Interestingly, the Satb2 protein, in contrast to mRNA, was not detected in late progenitors, but was detected in post-mitotic cells of the cortical plate, suggesting that a Satb2 translation block might occur in the progenitor cell (Britanova et al., 2005).

Retinal cell fate specification is mainly regulated by combinations of bHLH and homeobox genes. In mice, Atoh7 (bHL) and Pou4f2 (homeobox) cooperate to regulate RGC genesis. The expression of Prox1 (homeobox) is essential for horizontal cell generation, while a number of factors, including Neurod1 and Neurod4 (bHL), Pax6 and Six3 (homeobox), regulate the production of amacrine cells. Crx (homeobox) is crucial for specifying photoreceptors, and Vsx2 (also named Chx10, homeobox) is required for bipolar cell genesis (Ohsawa and Kageyama, 2008). Notably, the *Xenopus* homologs of Crx and Vsx2 (Xotx5b and Xvsx, respectively) coordinate the production of photoreceptors and bipolar cells via a translational control mechanism (Decembrini et al., 2006). The sequential expression of the two Sry-related HMG box proteins Sox11 and Sox4, during retinogenesis, leads to the fine adjustment of retinal differentiation. Overexpression of Sox11 and Sox4 in retinal progenitors increases the number of cone cells and dramatically decreases the number of rod cells and Müller glia, by acting through epigenetic mechanisms (Usui et al., 2013).

Although key transcription factors of cell fate are known, how they are activated in distinct cells at specific developmental times is not clear. Consequently, the mechanisms responsible for shifts in competence over time in the lineage of a progenitor cell remain largely elusive. One important feature shared by the cortex and retina is that the potency of progenitor cells diminishes and their competence changes as they "age" during embryonic development. We do not know the precise sort of "clock" that measures a progenitor's age, though one possible way would be through the length of its cell cycle. In fact, during neural development the proliferation rate decreases over time as the progenitor cell cycle length increases (Caviness et al., 1995; Alexiades and Cepko, 1996; Decembrini et al., 2006).

The proliferation rate of neural progenitor cells is regulated by the activation of a number of growth factor pathways. The activation of Wnt and fibroblast growth factor (FGF) pathways during cortical development supports the expression of cyclinD1 and shortens the cell cycle of progenitors, thus promoting proliferation, expansion of apical progenitors, and reduced generation of basal progenitors (Salomoni and Calegari, 2010). Wnts and FGFs, together with bone morphogenic proteins (BMPs), play a crucial role also in cortical patterning (Rubenstein, 2011) but they have not been shown to directly affect the establishment of distinct neuronal fates. The Shh pathway supports cell cycle progression, both in the retina (Wang et al., 2005; Locker et al., 2006) and in the mouse cerebral cortex (Komada et al., 2008). Interestingly, blocking the Shh pathway affects the histogenesis of both the *Xenopus* retina (Decembrini et al., 2009) and mouse cortex (Komada et al., 2008). In the *Xenopus* retina, this is caused by release from translational inhibition of Xotx5b and Xvsx, which are necessary for specifying the bipolar fate. Notably, shortening the cell cycle by E2F overexpression exerts opposite effects, thus supporting the idea that Shh acts on cell fate through the cell cycle machinery (Decembrini et al., 2006). Whether (and how) cell cycle progression relates to the clock controlling the competence of differentiation, and how this clock in turn regulates activation of the transcription factors that specify the distinct neuron types remain open issues.

miRNAs AND CORTICAL HISTOGENESIS

Most of our knowledge on the role of miRNAs in cortical and retinal histogenesis comes from analyzing the phenotypes observed after global loss of miRNA regulation, which is induced by disrupting the pre-miRNA processing enzyme Dicer. Conditional knock-out (CKO) of Dicer in the cortex was achieved after breeding Dicer:lox/lox mice with distinct forebrain Cre-driver mouse strains, including Nestin:Cre, Emx1:Cre, or FoxG1:Cre (De Pietri Tonelli et al., 2008; Kawase-Koga et al., 2009; Nowakowski et al., 2011; **Table 1**). A general effect common to different mouse strains driving early inactivation of Dicer in the cortex is the induction of cell death, because miRNAs target several players of the DNA-damage response signal-transduction network (Bailey et al., 2010). However, Dicer CKO also has profound effects on cortical layering.

FoxG1:Cre;Dicer:lox/lox embryos deactivate Dicer from E8, and the effects on the expression of mature miRNAs are detectable by E11.5 in most forebrain cells. In these mice, neuroepithelial stem cell identity is not affected, but expression of the markers of radial glia Nestin, Sox9, and ErbB2 is abnormally low. Early telencephalic progenitors generate correct proportions of neurons after Dicer deletion, but many of those neurons migrate abnormally, possibly due to a defect in radial glia-guided migration. Moreover, the population of secondary (basal) progenitors, which are generated by the radial glia, is disorganized and expanded (Nowakowski et al., 2011). The depletion of miR-92b may play a crucial role in generating this phenotype. In fact, this miRNA is predicted to target the 3′ untranslated region (UTR) of the transcription factor Tbr2, which regulates the generation of intermediate progenitors. Acute miR-92b gain of function causes rapid reductions in the ratio of Tbr2-expressing cells, whereas acute miR-92b loss of function has opposite effects (Nowakowski et al., 2013).

Dicer CKO in dorsal forebrain cells has been achieved with Cre expression from around E10 to E10.5 in Emx1:Cre;Dicer:lox/lox and Nestin:Cre;Dicer:lox/lox mice. The Nestin:Cre strain drove a milder and later inactivation of Dicer as compared to the Emx1:Cre strain. Emx1:Cre;Dicer:lox/lox showed overproduction of early-born neurons and a reduced number of Brn1-expressing upper-layer neurons as compared with controls, and the remaining ones were intermingled with Tbr1-expressing deep-layer neurons. Nestin:Cre;Dicer:lox/lox mice had no defects in the production of early-born neurons, but exhibited affected generation and migration of late-born neuron (De Pietri Tonelli et al., 2008; Kawase-Koga et al., 2009).

Dicer CKO in post-mitotic neurons of CamKII:Cre;Dicer:loxP/loxP mice caused reduced dendritic branch elaboration, but generated normal cortical layering (Davis et al., 2008), indicating that a late inactivation of Dicer cannot affect layer identity.

Altogether, these results show that mature miRNAs are required at different times in corticogenesis to fine-tune cell fate and, depending on the time of Dicer inactivation, different cell types and layers are affected. Unfortunately, these studies did not address the question of whether the translation of key transcription factors of cortical cell fate was affected.

Table 1 | Dicer-CKO phenotypes in the cortex and retina.

Cre transgene	Cre expression	Main phenotype	Reference
Cortex			
Foxg1-Cre	From E8 in most forebrain cells	Altered balance of apical and basal progenitors	Nowakowski et al. (2011)
Emx1-Cre	From E10 to E10.5 in most cells of the dorsal telencephalon	Overproduction of early-born neurons and reduced number of Brn1-expressing upper-layer neurons	Kawase-Koga et al. (2009)
Nestin-Cre	From E10 to E10.5 in forebrain stem cells and progenitors	Affected late-born neuron generation and migration	De Pietri Tonelli et al. (2008)
CamKII-Cre	From E15.5 in post-mitotic neurons of the cortex and hippocampus	Normal layering; reduced dendritic branch elaboration	Davis et al. (2008)
Retina			
Chx10-Cre	Mosaic pattern, from E14.5 in progenitors of all retinal layers	Decreased ERG responses, retinal disorganization, progressive retinal degeneration from P16	Damiani et al. (2008)
αPax6-Cre	Peripheral retina from day E10.5; differentiated amacrine cells, by E14.5	Overproduction of ganglion cells, failure to generate late cell types such as M′uller glia and rod photoreceptors	Georgi and Reh (2010)
Dkk3-Cre	From E10.5 in progenitors of all neuroretinal cell types	Microphthalmia, massive apoptosis	Iida et al. (2011)
Rx-Cre	Ubiquitously in the developing neuroretina and optic stalk; later in the optic chiasm	Microphthalmia, massive apoptosis, defects in retinal ganglion cell axon pathfinding	Pinter and Hindges (2010)

miRNAs AND RETINAL HISTOGENESIS

Four different Dicer-CKO mouse models have recently allowed investigating the effects of global miRNA down-regulation in mouse retinal development (**Table 1**). Cre-mediated Dicer excision in retinal progenitors resulted in phenotypes of variable severity, likely dependent on the time and the extent of Dicer deletion. Accordingly, when Dicer excision began earlier in retinal development, or when Cre was more uniformly expressed throughout the developing retina, more severe phenotypes were consistently observed. Likewise, when driving Dicer CKO in the developing cortex, a general effect of cell death was observed at different extents and times in all the retina CKOs (Damiani et al., 2008; Pinter and Hindges, 2010; Iida et al., 2011; Nowakowski et al., 2013).

Chx10-Cre expression exhibits a mosaic pattern and begins before embryonic day 14.5 in progenitors of all retinal layers. Dicer CKO driven by the Chx10-Cre transgene led to decreased electroretinogram (ERG) responses, morphological anomalies, and formation of photoreceptor rosettes at post-natal day 16. This phenotype progressed to more general cellular disorganization and widespread degeneration of retinal cell types as the animals aged (Damiani et al., 2008).

αPax6-Cre is expressed in peripheral regions of the developing retina, beginning on embryonic day 10.5. Dicer CKO driven by αPax6-Cre, which inactivated Dicer in a less mature population of retinal progenitors than Chx10-Cre, generated a more severe phenotype, consisting in the abnormal differentiation of retinal cell types. The production of early generated cell types (RGC and horizontal cells) was increased. Interestingly, ganglion cells (GCs) were generated beyond their normal competence window and, probably as a consequence, the Dicer-deleted areas of the retina showed a decrease in later generated cell types (amacrine cells and rod photoreceptors). These results indicate that miRNAs are required for shifts in the competence of retinal progenitors over time (Georgi and Reh, 2010).

Dkk3-Cre is ubiquitously expressed in all retinal progenitors beginning on embryonic day 10.5. Dicer CKO by this transgene produced massive death of retinal progenitor cells (RPCs), resulting in microphthalmia and the absence of layers. *In vitro* reaggregation culture of Dicer-CKO retinal cells revealed that cell death and the suppression of proliferation by Dicer inactivation occurred in a cell-autonomous manner (Iida et al., 2011). Such results are consistent with the phenotype observed after early inactivation of Dicer by morpholino microinjection in *Xenopus* (Decembrini et al., 2008).

Rx-Cre is ubiquitously expressed in the developing neuroretina. Dicer CKO by Rx-driven Cre activation caused cell death and a reduction in overall eye size. However, a RGC layer formed and no defects were observed in the formation of the optic disc, which is the exit point for RGC axons from the retina. Interestingly, mutants showed a marked increase in ipsilateral projections, with RGC axons extending outside the optic chiasm or showing aberrant projections, indicating a miRNA role in ensuring correct axon guidance decisions. Notably, these phenotypes were not the result of a mis-patterning of the eye (or the chiasm), suggesting that miRNAs have direct functions in the intracellular processes needed for axon growth and pathfinding (Pinter and Hindges, 2010).

Recent observations suggest that distinct miRNAs might be responsible for the cell death observed after Dicer CKO. In *Xenopus*, the inhibition of miR-24a, which is predicted to target the pro-apoptotic factors caspase-9 and protease-activating factor 1 (apaf1), resulted in increased apoptosis of retinal progenitors and microphthalmia (Walker and Harland, 2009). In mice, the knock-out of miR-124 caused apoptosis of newly differentiated cone photoreceptors (Sanuki et al., 2011). Individual miRNAs controlling retinal cell identity are emerging. miR-204 has an active role in establishing dorsoventral (D/V) polarity of the optic cup of medaka fish. When miR-204 activity was blocked by antagomiR, the expression domain of ventral markers was reduced or absent, whereas the expression domains of the dorsal markers were expanded ventrally. A reciprocal molecular phenotype was observed after miR-204 overexpression. These phenotypes were associated with concomitant up- or down-regulation of olMeis2, which is a target of miR-204 and mediates its effects on D/V eye polarity (Conte et al., 2010).

DISTINCT miRNAs AND mRNAs REGULATE THE TIMING OF RETINOGENESIS

A defined temporal sequence of gene expression that could explain the chronological order of cell birth in different neuronal lineages was first described in *Drosophila* (Isshiki et al., 2001). Further studies have confirmed the generality of this strategy, with different sequences of transcription factors being used in different structures of the *Drosophila* nervous system to generate neuronal diversity, according to a well-defined time schedule (Bayraktar and Doe, 2013; Li et al., 2013; Suzuki et al., 2013). Homologs of key transcription factors defining the temporal identity of *Drosophila* neuroblasts have now been detected in the developing mammalian retina. One of them, IKAROS family zinc finger 1 (Ikzf1/Ikaros), is a mouse ortholog of hunchback (hb), which is necessary and sufficient to specify early-born neurons in *Drosophila.* Ikaros is both necessary and sufficient to confer early temporal competence to mouse RPCs. In fact, mis-expression of Ikaros is sufficient to generate early-born neurons at inappropriate times: after viral Ikaros transduction in late RPCs, heterochronic amacrine and horizontal cells were generated *in vivo* and GCs in cell culture. In addition, Ikaros mis-expression caused a reduction in late-born neurons (bipolar cells) and prevented Müller glia formation (**Figure 2A**). Consistent with this, Ikaros-deficient retinas exhibited a permanent reduction in most early-born cell types. Cones were not affected by the gain or loss of Ikaros, suggesting that different regulatory mechanisms control the timing of their production (Elliott et al., 2008). These findings indicate that Ikaros is required for progression to a late temporal state. Surprisingly, the timing of Ikaros activation is due to regulated translational repression, because Ikaros mRNA is expressed throughout retinal development, whereas the protein is present only in early RPCs (**Figure 2A**). Although not currently proven, key mediators of this repression might be miRNAs, as suggested by the similarity of the phenotypes observed after Ikaros mis-expression and Dicer CKO by αPax6-Cre transgene (see above).

A central role of Ikaros in determining the temporal fate of neurons in mouse was recently indicated also by a study of cortical development. Ikaros is expressed in progenitor cells of the mouse cerebral cortex at high level during the early stages of neurogenesis and thereafter its expression decreases over time. Sustained Ikaros expression prolonged the period of the generation of deep-layer neurons and delayed the production of late-born neurons. However, there is no direct evidence that Ikaros expression during corticogenesis is regulated at the post-transcriptional level as in the developing retina. In fact, Ikaros mRNA level is high at early stages and decreases by over 80% from embryonic E10.5 to E15.5. A possible role of miRNA in mediating the decrease of Ikaros mRNA level during cortical development was discussed (Alsiö and Tarchini, 2013).

Distinct miRNAs that can rescue Pax6-Cre driven Dicer CKO have recently been found. These miRNAs, let7, miR-9, and miR-125, are expressed in early retinal progenitors and serve as key regulators of the early to late developmental transition in retinal progenitors. When down-regulated, they cause an increase in GCs, whereas their up-regulation accelerates retinogenesis, increasing the ratio of late photoreceptor cells (rods) at the expense of early neurons (ganglion and horizontal cells). Let7, miR-9, and miR-125 target Protogenin (Prtg) and Lin-28b, two proteins that are crucial for maintaining an early competence state of RPCs. In fact, overexpression of Prtg and Lin-28b from E16 caused an extra number of heterochronic GCs that were generated at late times in retinogenesis (**Figure 2A**). Ikaros and Lin-28/Prtg seem to constitute two parallel pathways for the control of developmental timing, because let7, miR-9, and miR-125 do not appear to directly regulate the expression of Ikaros. However, there are conserved binding sites for miR-125 in the 3′UTR of two members of the Ikaros family, Ikzf3 and Ikzf5. These two genes show small increases of expression in the Dicer-CKO retina and the possibility that they play a role in retinal development has to be considered (La Torre et al., 2013).

Finally, key transcription factors of late retinal cell identity that are regulated at the translational level have been described in *Xenopus.* Xotx5b is the *Xenopus* homolog of the mammalian homeobox gene Crx and specifies photoreceptor identity. Xotx2 and Xvsx1 are the *Xenopus* counterparts of the mammalian Otx2 and Vsx2 homeobox genes, respectively, and support the differentiation of bipolar cells in *Xenopus* (Viczian et al., 2003; D'Autilia et al., 2006; Decembrini et al., 2006). Xotx5b, Xvsx1, and Xotx2 are transcribed since the early stages of retinogenesis in multipotent progenitor cells, but their translation is inhibited until later stages, when the generation of photoreceptor and bipolar cells begins. This translational inhibition is due to signals in the 3′UTR and is controlled by progression of the cell cycle (Decembrini et al., 2006). We have identified a set of four miRNAs that inhibit the translation of Xvsx1 and Xotx2 by binding to their 3′UTR. The four miRNAs (miR-129, miR-155, miR-214, and miR-222) are down-regulated as retinal development proceeds. Interestingly, their expression is decreased in early progenitors by the inhibition of the Shh pathway, which has the effect of lengthening the cell cycle, and is increased in progenitors forced into the S-phase. These treatments, respectively, accelerate and block the translation of Xvsx1 and Xotx2. We have proposed that cell cycle length, which is known to increase as retinogenesis progresses (Alexiades and Cepko, 1996), provides an intrinsic timer that regulates cell birth through miRNA activity (Decembrini et al., 2009; Pitto and Cremisi, 2010; **Figure 2B**). Shh

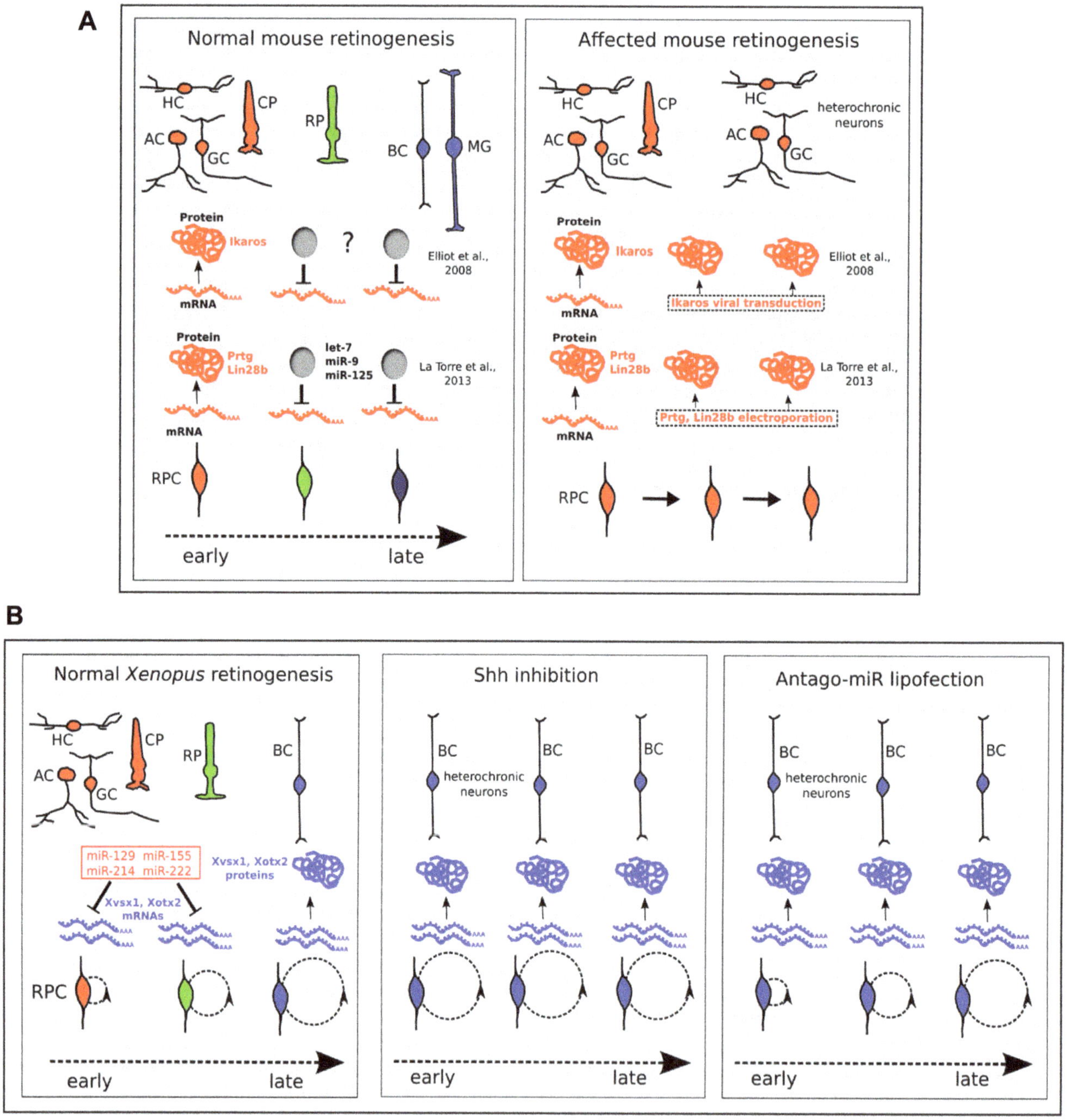

FIGURE 2 | The temporal identity of retinal progenitor cells (RPCs) is defined through the translational regulation of key proteins. (A) In mice, Ikaros, Prtg, and Lin-28b are transcribed throughout retinogenesis, but are translated only in early RPCs. While the molecular nature of the inhibitor of Ikaros translation ("?" label) is unknown, Prtg and Lin-28b are targeted by let-7, miR-9, and miR-125. When the protein expression of Ikaros, or Prtg and Lin-28b, is forced throughout retinogenesis, heterochronic neurons of the early-born type (HC, horizontal cells; AC, amacrine cells; GC, ganglion cells) are generated at late times in retinogenesis (Elliott et al., 2008; La Torre et al., 2013). CP, cone photoreceptor; RP, rod photoreceptor; BC, bipolar cell; MG, Müller glia. **(B)** In *Xenopus*, bipolar fate is driven by the homeobox Xvsx1 and Xotx2 genes, which are transcribed in RPCs from early developmental stages (15 and 25, respectively), but are translated only from late stages 37 and 38–39, respectively (Decembrini et al., 2006). A set of four cell cycle-regulated miRNAs (miR-129, miR-155, miR-214, and miR-222, in red) bind the 3′UTR of Xvsx1 and Xotx2, inhibiting their translation in early RPCs. In normal *Xenopus* retinogenesis, the duration of the cell cycle (indicated by dashed circles) inversely correlates with the expression of the four miRNAs. Lengthening the cell cycle by treatment with the Shh signaling inhibitor cyclopamine (Shh inhibition) down-regulates this set of miRNAs, leads to earlier translation of Xvsx1 and Xotx2 and causes the generation of heterochronic bipolar cells. Antago-miR lipofection in early RPCs inhibits the activity of the four miRNAs. Compared to cyclopamine treatment, the lipofection exerts similar effects on the translation of Xvsx1 and Xotx2, and on the generation of bipolar cells, but does not affect progression of the cell cycle (Decembrini et al., 2009). This favors the hypothesis that cell cycle progression may affect neuronal fate through the set of four miRNAs. In these experiments, the effect of miRNAs on Müller glia was not examined. CP, cone photoreceptor; RP, rod photoreceptor; HC, horizontal cell; BC, bipolar cell; AC, amacrine cell; GC, ganglion cell.

is a possible mediator of this process, as it regulates the cell cycle length in the retina (Locker et al., 2006).

CONCLUSION

The generation of distinct types of neurons in the cerebral cortex and neural retina relies on the ordered activation of cell fate genes over time. Studies in *Xenopus* and mouse retinal development described key proteins of neuronal identity whose expression is regulated at the translational level. Distinct miRNAs target these proteins and are crucial for early or late competence of progenitor cells in retinogenesis. Although no specific miRNA has been found to control the translation of key factors of cell fate in the cortex, the involvement of miRNAs in the control of the competence of cortical progenitor cells (CPCs) is strongly suggested by the results of Dicer down-regulation in CKO mice. In both the retina and cortex, expression of miRNAs is necessary for the transition from early to late development. However, in *Xenopus* retinogenesis there is evidence that distinct miRNAs must also be down-regulated to generate the latest neuron types.

An intriguing hypothesis is that the multipotency of early progenitor cells results from the transcription of mRNAs that serve to specify different neuronal identities, but are repressed by miRNAs. The release from the translational inhibition of distinct types of such mRNAs might determine what type of neuron is generated, and when. In *Xenopus*, release from the translational inhibition of Xvsx1 and Xotx2 is due to cell cycle lengthening, which causes the down-regulation of the four miRNAs targeting Xvsx1 and Xotx2. A similar mechanism, which makes use of cell-cycle-dependent miRNAs, might provide an intrinsic timer to regulate the cell birth of different types of neurons (**Figure 2B**). Shh, which regulates the cell cycle length in both the cortex and retina, might play a key role in this regard, and its function in temporally regulated aspects of retinogenesis and corticogenesis warrants further study.

ACKNOWLEDGMENTS

I am grateful to Giuseppina Barsacchi for her helpful discussions and suggestions. This work has been supported by grant no. 2011.0251 of the Cassa di Risparmio di Trento e Rovereto.

REFERENCES

Alexiades, M. R., and Cepko, C. (1996). Quantitative analysis of proliferation and cell cycle length during development of the rat retina. *Dev. Dyn.* 205, 293–307. doi: 10.1002/(SICI)1097-0177(199603)205:3<293::AID-AJA9>3.0.CO;2-D

Alsiö, J., and Tarchini, B. (2013). Ikaros promotes early-born neuronal fates in the cerebral cortex. *Proc. Natl. Acad. Sci. U.S.A.* 110, E716–E725. doi: 10.1073/pnas.1215707110

Bailey, S. G., Sanchez-Elsner, T., Stephanou, A., Cragg, M. S., and Townsend, P. A. (2010). Regulating the genome surveillance system: miRNAs and the p53 super family. *Apoptosis* 15, 541–552. doi: 10.1007/s10495-010-0456-1

Bayraktar, O. A., and Doe, C. Q. (2013). Combinatorial temporal patterning in progenitors expands neural diversity. *Nature* 498, 449–455. doi: 10.1038/nature12266

Bian, S., and Sun, T. (2011). Functions of noncoding RNAs in neural development and neurological diseases. *Mol. Neurobiol.* 44, 359–373. doi: 10.1007/s12035-011-8211-3

Britanova, O., Akopov, S., Lukyanov, S., Gruss, P., and Tarabykin, V. (2005). Novel transcription factor Satb2 interacts with matrix attachment region DNA elements in a tissue-specific manner and demonstrates cell-type-dependent expression in the developing mouse CNS. *Eur. J. Neurosci.* 21, 658–668. doi: 10.1111/j.1460-9568.2005.03897.x

Caviness, V. S., Takahashi, T., and Nowakowski, R. S. (1995). Numbers, time and neocortical neuronogenesis: a general developmental and evolutionary model. *Trends Neurosci.* 18, 379–383. doi: 10.1016/0166-2236(95)93933-O

Conte, I., Carrella, S., Avellino, R., Karali, M., Marco-Ferreres, R., Bovolenta, P., et al. (2010). miR-204 is required for lens and retinal development via Meis2 targeting. *Proc. Natl. Acad. Sci. U.S.A.* 107, 15491–15496. doi: 10.1073/pnas.0914785107

Coolen, M., and Bally-Cuif, L. (2009). MicroRNAs in brain development and physiology. *Curr. Opin. Neurobiol.* 19, 461–470. doi: 10.1016/j.conb.2009.09.006

Damiani, D., Alexander, J. J., O'Rourke, J. R., McManus, M., Jadhav, A. P., Cepko, C. L., et al. (2008). Dicer inactivation leads to progressive functional and structural degeneration of the mouse retina. *J. Neurosci.* 28, 4878–4887. doi: 10.1523/JNEUROSCI.0828-08.2008

D'Autilia, S., Decembrini, S., Casarosa, S., He, R.-Q., Barsacchi, G., Cremisi, F., et al. (2006). Cloning and developmental expression of the *Xenopus* homeobox gene Xvsx1. *Dev. Genes Evol.* 216, 829–834. doi: 10.1002/(SICI)1097-0177(199603)205:3<293::AID-AJA9>3.0.CO;2-D

Davis, T. H., Cuellar, T. L., Koch, S. M., Barker, A. J., Harfe, B. D., McManus, M. T., et al. (2008). Conditional loss of Dicer disrupts cellular and tissue morphogenesis in the cortex and hippocampus. *J. Neurosci.* 28, 4322–4330. doi: 10.1523/JNEUROSCI.4815-07.2008

Decembrini, S., Andreazzoli, M., Barsacchi, G., and Cremisi, F. (2008). Dicer inactivation causes heterochronic retinogenesis in *Xenopus laevis. Int. J. Dev. Biol.* 52, 1099–1103. doi: 10.1387/ijdb.082646sd

Decembrini, S., Andreazzoli, M., Vignali, R., Barsacchi, G., and Cremisi, F. (2006). Timing the generation of distinct retinal cells by homeobox proteins. *PLoS Biol.* 4:e272. doi: 10.1371/journal.pbio.0040272

Decembrini, S., Bressan, D., Vignali, R., Pitto, L., Mariotti, S., Rainaldi, G., et al. (2009). MicroRNAs couple cell fate and developmental timing in retina. *Proc. Natl. Acad. Sci. U.S.A.* 106, 21179–21184. doi: 10.1073/pnas.0909167106

De Chevigny, A., Coré, N., Follert, P., Gaudin, M., Barbry, P., Béclin, C., et al. (2012). miR-7a regulation of Pax6 controls spatial origin of forebrain dopaminergic neurons. *Nat. Neurosci.* 15, 1120–1126. doi: 10.1038/nn.3142

De Pietri Tonelli, D., Pulvers, J. N., Haffner, C., Murchison, E. P., Hannon, G. J., and Huttner, W. B. (2008). miRNAs are essential for survival and differentiation of newborn neurons but not for expansion of neural progenitors during early neurogenesis in the mouse embryonic neocortex. *Development* 135, 3911–3921. doi: 10.1242/dev.025080

Desai, A. R., and McConnell, S. K. (2000). Progressive restriction in fate potential by neural progenitors during cerebral cortical development. *Development* 127, 2863–2872. doi: 10.1073/pnas.1215707110

Elliott, J., Jolicoeur, C., Ramamurthy, V., and Cayouette, M. (2008). Ikaros confers early temporal competence to mouse retinal progenitor cells. *Neuron* 60, 26–39. doi: 10.1016/j.neuron.2008.08.008

Fineberg, S. K., Kosik, K. S., and Davidson, B. L. (2009). MicroRNAs potentiate neural development. *Neuron* 64, 303–309. doi: 10.1016/j.neuron.2009.10.020

Frantz, G. D., and McConnell, S. K. (1996). Restriction of late cerebral cortical progenitors to an upper-layer fate. *Neuron* 17, 55–61. doi: 10.1016/S0896-6273(00)80280-9

Georgi, S. A, and Reh, T. A. (2010). Dicer is required for the transition from early to late progenitor state in the developing mouse retina. *J. Neurosci.* 30, 4048–4061. doi: 10.1523/JNEUROSCI.4982-09.2010

He, M., Liu, Y., Wang, X., and Zhang, M. (2012). Cell-type based analysis of microRNA profiles in the mouse brain. *Neuron* 73, 35–48. doi: 10.1016/j.neuron.2011.11.010

Holt, C. E., Bertsch, T. W., Ellis, H. M., and Harris, W. A. (1988). Cellular determination in the *Xenopus* retina is independent of lineage and birth date. *Neuron* 1, 15–26. doi: 10.1016/0896-6273(88)90205-X

Iida, A., Shinoe, T., Baba, Y., Mano, H., and Watanabe, S. (2011). Dicer plays essential roles for retinal development by regulation of survival and differentiation. *Invest. Ophthalmol. Vis. Sci.* 52, 3008–3017. doi: 10.1167/iovs.10-6428

Isshiki, T., Pearson, B., Holbrook, S., and Doe, C. Q. (2001). *Drosophila* neuroblasts sequentially express transcription factors which specify the temporal identity of their neuronal progeny. *Cell* 106, 511–521. doi: 10.1016/S0092-8674(01)00465-2

Karali, M., Peluso, I., Gennarino, V. A, Bilio, M., Verde, R.,

Lago, G., et al. (2010). miRNeye: a microRNA expression atlas of the mouse eye. *BMC Genomics* 11:715. doi: 10.1186/1471-2164-11-715

Kawase-Koga, Y., Otaegi, G., and Sun, T. (2009). Different timings of Dicer deletion affect neurogenesis and gliogenesis in the developing mouse central nervous system. *Dev. Dyn.* 238, 2800–2812. doi: 10.1002/dvdy.22109

Komada, M., Saitsu, H., Kinboshi, M., Miura, T., Shiota, K., and Ishibashi, M. (2008). Hedgehog signaling is involved in development of the neocortex. *Development* 135, 2717–2727. doi: 10.1242/dev.015891

Krol, J., Loedige, I., and Filipowicz, W. (2010). The widespread regulation of microRNA biogenesis, function and decay. *Nat. Rev. Genet.* 11, 597–610.

La Torre, A., La, Georgi, S., and Reh, T. A. (2013). Conserved microRNA pathway regulates developmental timing of retinal neurogenesis. *Proc. Natl. Acad. Sci. U.S.A.* 110, E2362–E2370. doi: 10.1073/pnas.1301837110

Leone, D., Srinivasan, K., and Chen, B. (2008). The determination of projection neuron identity in the developing cerebral cortex. *Curr. Opin. Neurobiol.* 18, 28–35. doi: 10.1016/j.conb.2008.05.006

Li, X., Erclik, T., Bertet, C., Chen, Z., Voutev, R., Venkatesh, S., et al. (2013). Temporal patterning of *Drosophila* medulla neuroblasts controls neural fates. *Nature* 498, 456–462. doi: 10.1038/nature12319

Locker, M., Agathocleous, M., Amato, M. A., Parain, K., Harris, W. A., and Perron, M. (2006). Hedgehog signaling and the retina: insights into the mechanisms controlling the proliferative properties of neural precursors. *Genes Dev.* 20, 3036–3048. doi: 10.1101/gad.391106

McConnell, S. K., and Kaznowski, C. E. (1991). Cell cycle dependence of laminar determination in developing neocortex. *Science* 254, 282–285. doi: 10.1126/science.1925583

Nowakowski, T. J., Fotaki, V., Pollock, A., Sun, T., Pratt, T., and Price, D. J. (2013). MicroRNA-92b regulates the development of intermediate cortical progenitors in embryonic mouse brain. *Proc. Natl. Acad. Sci. U.S.A.* 110, 7056–7061. doi: 10.1073/pnas.1219385110

Nowakowski, T. J., Mysiak, K. S., Pratt, T., and Price, D. J. (2011). Functional Dicer is necessary for appropriate specification of radial glia during early development of mouse telencephalon. *PLoS ONE* 6: e23013. doi: 10.1371/journal.pone.0023013

Ohsawa, R., and Kageyama, R. (2008). Regulation of retinal cell fate specification by multiple transcription factors. *Brain Res.* 1192, 90–98. doi: 10.1016/j.brainres.2007.04.014

Otaegi, G., Pollock, A., Hong, J., and Sun, T. (2011). MicroRNA miR-9 modifies motor neuron columns by a tuning regulation of FoxP1 levels in developing spinal cords. *J. Neurosci.* 31, 809–818. doi: 10.1523/JNEUROSCI.4330-10.2011

Pinter, R., and Hindges, R. (2010). Perturbations of microRNA function in mouse Dicer mutants produce retinal defects and lead to aberrant axon pathfinding at the optic chiasm. *PLoS ONE* 5:e10021. doi: 10.1371/journal.pone.0010021

Pitto, L., and Cremisi, F. (2010). Timing neurogenesis by cell cycle? *Cell Cycle* 9, 434–435. doi: 10.4161/cc.9.3.10762

Rubenstein, J. L. R. (2011). Annual research review: development of the cerebral cortex: implications for neurodevelopmental disorders. *J. Child Psychol. Psychiatry* 52, 339–355. doi: 10.1111/j.1469-7610.2010.02307.x

Salomoni, P., and Calegari, F. (2010). Cell cycle control of mammalian neural stem cells: putting a speed limit on G1. *Trends Cell Biol.* 20, 233–243. doi: 10.1016/j.tcb.2010.01.006

Sanuki, R., Onishi, A., Koike, C., Muramatsu, R., Watanabe, S., Muranishi, Y., et al. (2011). miR-124a is required for hippocampal axogenesis and retinal cone survival through Lhx2 suppression. *Nat. Neurosci.* 14, 1125–1134. doi: 10.1038/nn.2897

Sessa, A., Mao, C.-A., Colasante, G., Nini, A., Klein, W. H., and Broccoli, V. (2010). Tbr2-positive intermediate (basal) neuronal progenitors safeguard cerebral cortex expansion by controlling amplification of pallial glutamatergic neurons and attraction of subpallial GABAergic interneurons. *Genes Dev.* 24, 1816–1826. doi: 10.1101/gad.575410

Sessa, A., Mao, C.-A., Hadjantonakis, A.-K., Klein, W. H., and Broccoli, V. (2008). Tbr2 directs conversion of radial glia into basal precursors and guides neuronal amplification by indirect neurogenesis in the developing neocortex. *Neuron* 60, 56–69. doi: 10.1016/j.neuron.2008.09.028

Srinivasan, K., Leone, D. P., Bateson, R. K., Dobreva, G., Kohwi, Y., Kohwi-Shigematsu, T., et al. (2012). A network of genetic repression and derepression specifies projection fates in the developing neocortex. *Proc. Natl. Acad. Sci. U.S.A.* 109, 19071–19078. doi: 10.1073/pnas.1216793109

Stappert, L., Borghese, L., Roese-Koerner, B., Weinhold, S., Koch, P., Terstegge, S., et al. (2013). MicroRNA-based promotion of human neuronal differentiation and subtype specification. *PLoS ONE* 8:e59011. doi: 10.1371/journal.pone.0059011

Suzuki, T., Kaido, M., Takayama, R., and Sato, M. (2013). A temporal mechanism that produces neuronal diversity in the *Drosophila* visual center. *Dev. Biol.* 380, 12–24. doi: 10.1016/j.ydbio.2013.05.002

Turner, D. L., and Cepko, C. L. (1987). A common progenitor for neurons and glia persists in rat retina late in development. *Nature* 328, 131–136. doi: 10.1038/328131a0

Usui, A., Mochizuki, Y., Iida, A., Miyauchi, E., Satoh, S., Sock, E., et al. (2013). The early retinal progenitor-expressed gene Sox11 regulates the timing of the differentiation of retinal cells. *Development* 140, 740–750. doi: 10.1242/dev.090274

Viczian, A. S., Vignali, R., Zuber, M. E., Barsacchi, G., and Harris, W. A. (2003). XOtx5b and XOtx2 regulate photoreceptor and bipolar fates in the *Xenopus* retina. *Development* 130, 1281–1294. doi: 10.1242/dev.00343

Walker, J. C., and Harland, R. M. (2009). microRNA-24a is required to repress apoptosis in the developing neural retina. *Genes Dev.* 23, 1046–1051. doi: 10.1101/gad.1777709

Wang, Y., Dakubo, G. D., Thurig, S., Mazerolle, C. J., and Wallace, V. A. (2005). Retinal ganglion cell-derived sonic hedgehog locally controls proliferation and the timing of RGC development in the embryonic mouse retina. *Development* 132, 5103–5113. doi: 10.1242/dev.02096

Wetts, R., and Fraser, S. E. (1988). Multipotent precursors can give rise to all major cell types of the frog retina. *Science* 239, 1142–1145. doi: 10.1126/science.2449732

Yang, D., Li, T., Wang, Y., Tang, Y., Cui, H., Tang, Y., et al. (2012). miR-132 regulates the differentiation of dopamine neurons by directly targeting Nurr1 expression. *J. Cell Sci.* 125(Pt 7), 1673–1682. doi: 10.1242/jcs.086421

8

microRNA function in left–right neuronal asymmetry: perspectives from *C. elegans*

Amel Alqadah[1,2†], Yi-Wen Hsieh[1†] and Chiou-Fen Chuang[1]*

[1] *Division of Developmental Biology, Cincinnati Children's Hospital Research Foundation, Cincinnati, OH, USA*
[2] *Molecular and Developmental Biology Graduate Program, University of Cincinnati, Cincinnati, OH, USA*

***Correspondence:**
Chiou-Fen Chuang, Division of Developmental Biology, Cincinnati Children's Hospital Research Foundation, 240 Albert Sabin Way, Cincinnati, OH 45229, USA
e-mail: chiou-fen.chuang@cchmc.org

†Amel Alqadah and Yi-Wen Hsieh have contributed equally to this work.

Left–right asymmetry in anatomical structures and functions of the nervous system is present throughout the animal kingdom. For example, language centers are localized in the left side of the human brain, while spatial recognition functions are found in the right hemisphere in the majority of the population. Disruption of asymmetry in the nervous system is correlated with neurological disorders. Although anatomical and functional asymmetries are observed in mammalian nervous systems, it has been a challenge to identify the molecular basis of these asymmetries. *C. elegans* has emerged as a prime model organism to investigate molecular asymmetries in the nervous system, as it has been shown to display functional asymmetries clearly correlated to asymmetric distribution and regulation of biologically relevant molecules. Small non-coding RNAs have been recently implicated in various aspects of neural development. Here, we review cases in which microRNAs are crucial for establishing left–right asymmetries in the *C. elegans* nervous system. These studies may provide insight into how molecular and functional asymmetries are established in the human brain.

Keywords: microRNA, neuronal asymmetry, sensory neurons, calcium signaling, *C. elegans*

INTRODUCTION

microRNAs (miRNAs) are endogenous 20–24 nt small non-coding RNAs that regulate gene expression through binding to complementary sequences in target messenger RNAs (mRNAs), leading to translational repression and/or cleavage of target mRNAs (Ambros, 2004; He and Hannon, 2004; Bartel, 2009; Chekulaeva and Filipowicz, 2009; Ghildiyal and Zamore, 2009). While most miRNAs downregulate gene expression, there are examples of miRNA-mediated upregulation of target gene expression during cell cycle arrest, suggesting that miRNA function is complex and context dependent (Vasudevan et al., 2007; Orom et al., 2008). miRNAs have been implicated in many aspects of development and disease including cell cycle, cell differentiation, apoptosis, life span, developmental timing, stress responses, neural development and regeneration, cancers, and neurodegenerative disorders (Boehm and Slack, 2005; Bushati and Cohen, 2007; Chang et al., 2009; Ambros, 2011; Sayed and Abdellatif, 2011; Zhang et al., 2011; Boulias and Horvitz, 2012; Cochella and Hobert, 2012a; Saito and Saito, 2012; Zou et al., 2012, 2013).

FUNCTIONS OF miRNAs IN NEURAL DEVELOPMENT

The importance of miRNAs in various aspects of neuronal development has been demonstrated in several animal models. In zebrafish, maternal–zygotic *dicer* mutants, that disrupt the processing of precursor miRNAs into mature miRNAs, display deleterious effects on the development of the brain, and injection of mature miR-430 rescues the early brain patterning defects (Giraldez et al., 2005). In mice, the neuron-specific miRNA miR-124 induces neuronal differentiation by directly targeting a global repressor of alternative pre-mRNA splicing and triggering a downstream switch to neuron-specific alternative splicing (Makeyev et al., 2007). In the *Xenopus* retina, a number of cell cycle related miRNAs target Xotx2 and Xvsx1 in early retinal progenitor cells to inhibit bipolar cell differentiation (Decembrini et al., 2009). In *Drosophila*, miR-9a targets senseless to inhibit neuronal fate in non-sensory organ precursors (Li et al., 2006). In *C. elegans*, the miRNA *lin-4* targets the LIN-14 transcription factor to inhibit netrin-mediated axon attraction (Chang et al., 2004a; Zou et al., 2012), and the miRNA *let-7* contributes to a developmental decline in neuronal regeneration (Zou et al., 2013). In addition, miRNAs *lsy-6*, *mir-273*, and *mir-71* function in asymmetric differentiation of two pairs of *C. elegans* sensory neurons, which will be discussed later (Johnston and Hobert, 2003; Chang et al., 2004b; Hsieh et al., 2012). Thus, miRNAs are important factors that control neuronal development across the animal kingdom.

LEFT–RIGHT ASYMMETRY OF THE NERVOUS SYSTEM

Although the nervous systems of animals are largely symmetric across the left–right axis, there have been several observations of anatomical and functional brain lateralization throughout the animal kingdom. For example, zebrafish display asymmetry in the epithalamus (Snelson and Gamse, 2009; Taylor et al., 2010); mice have been shown to have paw preferences, indicating the presence of a dominant hemisphere in motor control (Signore et al., 1991; Biddle et al., 1993); and the majority of humans have language centers such as Wernicke's and Broca's area located in the left hemisphere of the brain (Sun and Walsh, 2006). This lateralization of the nervous system is thought to be beneficial, as it allows for an increase of functional capacity (Rogers et al., 2012).

Disruption of asymmetry in the brain is seen in a number of neurodevelopmental diseases, including dyslexia, schizophrenia, autism, Alzheimer's disease, and attention deficit/hyperactivity disorder (ADHD; Herbert et al., 2005; Derflinger et al., 2011; Oertel-Knochel and Linden, 2011; Renteria, 2012). In children diagnosed with ADHD, the prefrontal cortex was shown to have a loss in rightward asymmetric distribution of prefrontal cortex volume as compared to typically developing children (Shaw et al., 2009). In dyslexic patients, the planum temporale shows alteration in asymmetry, as the right temporale appears physically larger than the left counterpart (Galaburda et al., 1985).

There have been several reports on functional and anatomical nervous system asymmetries; however there have been comparatively fewer studies on identifying the molecular mechanisms that establish lateralization. Here we narrow our focus to the function and regulation of miRNAs in the development of neuronal asymmetry. The most interesting evidence of miRNAs in vertebrate neuronal asymmetry comes from the investigation of miRNA function in neocortex development. The simplest form of asymmetry is the division of a cell to give rise to two asymmetric fates. In the case of the mouse neocortex, a progenitor cell divides asymmetrically to give rise to a progenitor cell and a neuron. Asymmetric localization of the TRIM-NHL protein TRIM32 is observed in the daughter cell that becomes a neuron, while the cell lacking TRIM32 remains a progenitor (Schwamborn et al., 2009). TRIM32 increases the activity of specific miRNAs through binding of the RNAse argonaute-1 (Hammell et al., 2009; Schwamborn et al., 2009). It was further shown that the miRNA Let-7a, one of the TRIM32 targets, is required and sufficient for neuronal differentiation (Schwamborn et al., 2009).

A subset of zebrafish olfactory bulb output neurons called the mitral cells send axons asymmetrically to the right habenula, which is an asymmetric part of the brain where the higher olfactory processing center is located (Miyasaka et al., 2009; Taylor et al., 2010). Several molecules are asymmetrically expressed in adult-born olfactory neurons in the ventricular–subventricular zone (V–SVZ) of the zebrafish brain. The transcription factors Myt1 and Neurogenin1 are predominantly expressed in the left V–SVZ, while DeltaA and hairy/enhancer of split-related protein (a Notch effector) are mainly expressed on the right side (Kishimoto et al., 2013). In addition, members of the miR-200 family are involved in the proper differentiation of olfactory neurons of both mice and zebrafish (Choi et al., 2008). It would be interesting to see whether the miR-200 family or other miRNAs influences the asymmetric and lateral projection of olfactory axons to the higher olfactory centers, and whether Myt1, Neurogenin1, and Delta/Notch are potential targets of miR-200.

A study found that 27 genes are differentially expressed in the embryonic human cerebral cortex, and the Lim domain transcription factor LMO4 is more abundant in the right perisylvian cortex than the left, and may be involved in asymmetric development of the cortex (Sun et al., 2005). The hypothesis that differential expression of these genes between left and right sides of the cortex may be regulated by miRNAs is plausible and worth further investigation.

Although there are limited reports of the involvement of miRNAs in the development of neuronal asymmetry in vertebrates, *C. elegans* has proved to be a powerful model organism to study lateralization of the nervous system due to its genetic amenability, as well as evidence of functional asymmetry having clear molecular correlates. In this review, we highlight two cases of the roles that miRNAs play in establishing left/right asymmetry in the *C. elegans* nervous system. Both involve the specification of two types of chemosensory neurons: the pair of amphid neurons, single ciliated endings (ASE) taste neurons, in which *lsy-6* and *mir-273* miRNAs are involved, and the pair of amphid wing "C" (AWC) olfactory neurons, where *mir-71* is crucial for establishment of asymmetry.

miRNAs IN TASTE NEURON ASYMMETRY

Like other animals, the *C. elegans* nervous system appears generally symmetric. However, the pair of taste neurons, called ASE left (ASEL) and ASE right (ASER) displays molecular and functional asymmetries. The ASE neurons are located in the nerve ring of the nematode, which is the brain equivalent in the worm. Although the neurons are derived from different cell lineages, they are anatomically symmetric in terms of cell position, morphologies, and axonal projections (White et al., 1986). The ASEL neuron, however, differentially expresses the putative chemoreceptor *gcy-7*, while the ASER neuron expresses another putative chemoreceptor *gcy-5* (Yu et al., 1997). The pair of ASE neurons also senses different chemicals, as the ASEL neuron functions to detect sodium, while ASER senses chloride (Pierce-Shimomura et al., 2001).

Over the past few years, there have been many studies investigating the molecular mechanism on how ASE asymmetry is established. Intriguingly, the first step of breaking symmetry of the taste neurons occurs several divisions before the ASEL and ASER neurons are born, during the early embryonic stage (Poole and Hobert, 2006). In the ASEL lineage, a pair of redundant T-box transcription factors TBX-37/38 are transiently expressed six cell divisions before the birth of ASEL (Good et al., 2004; **Figure 1**). These transcription factors work to "prime" a miRNA called *lsy-6*, which promotes the ASEL cell fate. This is achieved by binding of TBX-37/38 to a downstream primer element of *lsy-6* and results in physical opening up of the *lsy-6* chromatin (Cochella and Hobert, 2012b). The priming event is then "remembered" several cell divisions later in the ASEL mother cell. The open chromatin status of *lsy-6* allows for the CHE-1 zinc finger transcription factor to bind to an upstream booster element of the *lsy-6* locus (Cochella and Hobert, 2012b). This induces "boosting" of *lsy-6* expression levels in the ASEL neuron. The "prime and boost" model is essential for establishing ASE asymmetry.

In the ASER lineage, a Delta/Notch signal in the ASER precursor cell causes the T-box transcription factors TBX-37/38 to be repressed (Good et al., 2004; Priess, 2005). Therefore, the *lsy-6* "priming" event does not occur in the ASER lineage, allowing the *lsy-6* chromatin to remain in a compact form (Cochella and Hobert, 2012b). This in turn leads to the inability of the CHE-1 transcription factor to physically bind the upstream element of *lsy-6*, and no boosting of the miRNA expression levels occurs in ASER. Overall, this causes asymmetric distribution of the *lsy-6* miRNA in the ASEL neuron.

The *lsy-6* miRNA functions in a double negative feedback loop to control the asymmetry of ASEL/R neurons (Johnston et al.,

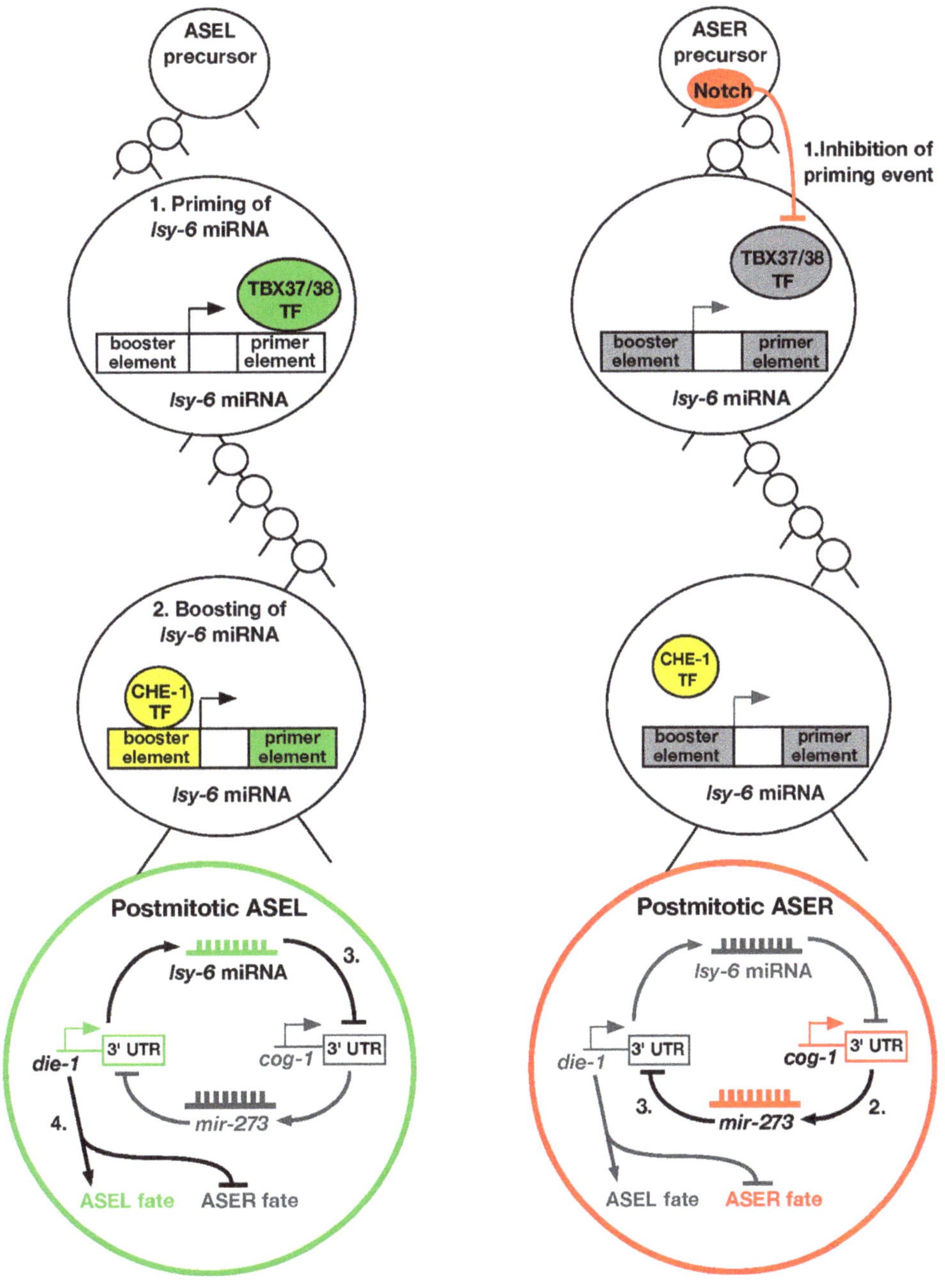

FIGURE 1 | Priming and boosting model of ASE asymmetry. ASEL: (1) The first step of promoting the ASEL cell fate occurs six cell divisions before the postmitotic ASEL is born. A transient pair of T-box transcription factors TBX-37/38 acts on the downstream primer element of *lsy-6* miRNA to open up the chromatin. (2) The priming event allows boosting of *lsy-6* miRNA levels by the CHE-1 transcription factor in the ASEL mother cell, leading to high levels of the miRNA in the postmitotic ASEL neuron. (3) The *lsy-6* miRNA directly inhibits the ASER promoting transcription factor COG-1, and allows the ASEL promoting transcription factor DIE-1 to be expressed. (4) DIE-1 then activates ASEL effector genes and represses the ASER cell fate. ASER: (1) Notch is expressed in the ASER precursor and inhibits the *lsy-6* priming event. Therefore, the *lsy-6* miRNA expression level cannot be boosted as the chromatin remains compact. (2) Consequently, in the postmitotic ASER neuron, the COG-1 transcription factor is expressed, which activates another miRNA, *mir-273*. (3) *mir-273* directly inhibits the ASEL promoting transcription factor DIE-1, and therefore promotes the ASER fate.

2005). In the ASEL neuron, *lsy-6* directly represses an ASER promoting transcription factor COG-1, through physical binding of complementary bases in the *cog-1* 3′UTR (Johnston and Hobert, 2003). Repression of COG-1 allows for yet another transcription factor DIE-1 to be expressed, which is the output of the feedback loop. DIE-1 then activates ASEL effector genes and suppresses ASER effector genes (Johnston et al., 2005; **Figure 1**).

In the ASER neuron, *lsy-6* expression is relatively low, and cannot effectively repress the COG-1 transcription factor. This allows activation of another miRNA, *mir-273*, which displays complementarity to the 3′UTR of the previously described *die-1* output (Chang et al., 2004b). *mir-273* therefore represses DIE-1, resulting in the de-repression of the ASER fate (**Figure 1**). Mutating any of the factors involved results in a loss of asymmetry, with both cells adopting either the ASEL or ASER cell fate.

miRNAs IN OLFACTORY NEURON ASYMMETRY

Like the ASE neurons, the *C. elegans* AWC olfactory neurons are bilaterally symmetrical at the morphological level. However, AWC left (AWCL) and AWC right (AWCR) neurons express different odorant receptors and sense different odors (**Figure 2**; Troemel et al., 1999). The AWC^{ON} neuron expresses the odorant receptor gene *str-2* and specifically senses the odor butanone; while the contralateral AWC^{OFF} neuron expresses the odorant receptor gene *srsx-3* and specifically senses the odor pentanedione (**Figure 2**; Troemel et al., 1999; Wes and Bargmann, 2001; Bauer Huang et al., 2007). In wild-type animals, only one of the AWC neurons expresses *str-2*. AWC asymmetry is stochastic and coordinated, so that 50% of the worms in a population express *str-2* in AWCL, while the other 50% express *str-2* in AWCR (Troemel et al., 1999; Taylor et al., 2010). The default state of the AWC neurons is AWC^{OFF}, which is specified by a calcium-regulated and microtubule-dependent MAP kinase pathway including UNC-43/CaMKII, TIR-1/Sarm1, NSY-1/MAPKKK, and SEK-1/MAPKK (**Figure 2**; Sagasti et al., 2001; Tanaka-Hino et al., 2002; Chuang and Bargmann, 2005; Chang et al., 2011). NSY-4 claudin-like protein and NSY-5 gap junction protein act in parallel to inhibit the calcium signaling pathway in the induced AWC^{ON} cell (Vanhoven et al., 2006; Chuang et al., 2007). In addition, intercellular calcium signaling between AWCs and non-AWC neurons via a NSY-5 gap junction-dependent neural network coordinates precise AWC asymmetry (Schumacher et al., 2012).

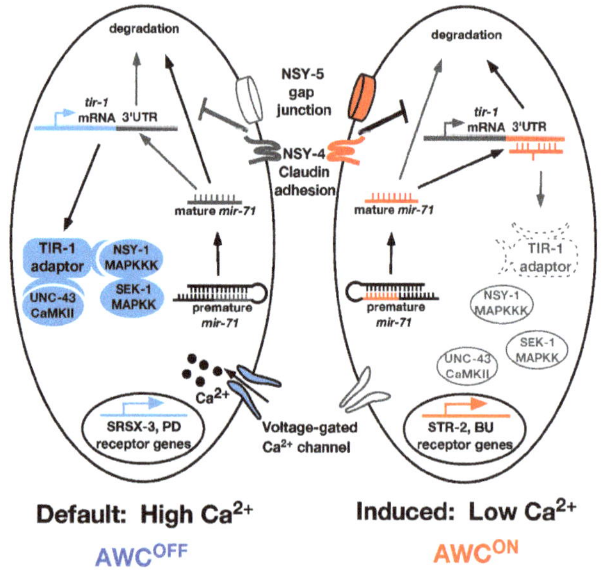

FIGURE 2 | Model of *mir-71* function in AWC asymmetry. In the AWC^{OFF} cell, high calcium level activates the calcium-regulated UNC-43/TIR-1/MAPK cascade, leading to *srsx-3* expression. In the AWC^{ON} cell, *nsy-4* and *nsy-5* promote the stability of mature *mir-71* through an unknown mechanism. The mature *mir-71* miRNA targets the 3′UTR of *tir-1* mRNA for degradation, leading to the inhibition of the calcium signaling pathway and the subsequent induction of *str-2* expression. Orange color represents factors that promote AWC^{ON}. Blue color represents factors that promote AWC^{OFF}. Gray/white color represents inactive or less active factors. Dotted line represents factors being downregulated. PD, pentanedione; BU, butanone.

One of the important questions on AWC asymmetry is how the calcium signaling pathway is downregulated by *nsy-4* and *nsy-5* in the AWC^{ON} cell. A recent study showed that the miRNA *mir-71* acts downstream of *nsy-4* and *nsy-5* to promote AWC^{ON} in a cell autonomous manner through inhibiting the expression of the calcium signaling adaptor protein gene *tir-1* (**Figure 2**; Hsieh et al., 2012). The TIR-1/Sarm1 adaptor protein assembles a calcium-signaling complex to cell-autonomously specify the default AWC^{OFF} identity (Chuang and Bargmann, 2005). Thus downregulation of *tir-1* expression by *mir-71* is an efficient mechanism to inhibit calcium signaling in the cell becoming AWC^{ON}.

mir-71 is regulated at both transcriptional and post-transcriptional levels in AWC (Hsieh et al., 2012). At the transcriptional level, the expression level of *mir-71* is higher in the AWC^{ON} cell than in the AWC^{OFF} cell. This transcriptional bias of *mir-71* is not dependent on *nsy-4* or *nsy-5*; thus, the mechanisms that regulate differential expression of *mir-71* in the two AWC cells are yet to be elucidated. At the post-transcriptional level, the stability of mature *mir-71* is dependent on *nsy-4* and *nsy-5*. The *C. elegans* 5′ → 3′ exoribonuclease XRN-2 has been shown to be involved in degradation of mature miRNAs (Chatterjee and Grosshans, 2009). It is possible that *nsy-4* and *nsy-5* may antagonize the *xrn-2*-mediated miRNA turnover pathway to increase the level of mature *mir-71*. However, RNA interference (RNAi) knockdown of *xrn-2* did not cause a defect in AWC asymmetry, suggesting that the stability of mature *mir-71* may be independent of *xrn-2*. In support of this idea, not all miRNAs accumulate in *xrn-2* RNAi worms (Chatterjee and Grosshans, 2009), suggesting the existence of alternative miRNA turnover pathways that may be inhibited by *nsy-4* and *nsy-5*.

PERSPECTIVES

Understanding the molecular mechanisms involved in establishing left–right asymmetry in the *C. elegans* nervous system can lay the groundwork for identifying the processes used in higher organisms, as the methods used may be evolutionarily conserved. Because of the highly conserved nature of miRNAs, insights into how they are involved to control asymmetric fates will help facilitate our understanding in vertebrate neuronal asymmetry. The involvement of miRNAs in asymmetry may also be reflective of the principles these small non-coding RNAs use in directing other neurodevelopmental processes.

ACKNOWLEDGMENTS

We thank Grethel Millington and Chieh Chang for comments on the manuscript. Amel Alqadah is supported by a Choose Ohio First Scholarship, Yi-Wen Hsieh by a NIH Organogenesis Training Grant, and Chiou-Fen Chuang by a Whitehall Foundation Research Award, an Alfred P. Sloan Research Fellowship, and a NIH R01GM098026 grant.

REFERENCES

Ambros, V. (2004). The functions of animal microRNAs. *Nature* 431, 350–355. doi: 10.1038/nature02871

Ambros, V. (2011). MicroRNAs and developmental timing. *Curr. Opin. Genet. Dev.* 21, 511–517. doi: 10.1016/j.gde.2011.04.003

Bartel, D. P. (2009). MicroRNAs: target recognition and regulatory functions. *Cell* 136, 215–233. doi: 10.1016/j.cell.2009.01.002

Bauer Huang, S. L., Saheki, Y., VanHoven, M. K., Torayama, I., Ishihara, T., Katsura, I., et al. (2007). Left–right olfactory asymmetry results from antagonistic functions of voltage-activated calcium channels and the Raw repeat protein OLRN-1 in *C. elegans*. *Neural Dev.* 2, 24. doi: 10.1186/1749-8104-2-24

Biddle, F. G., Coffaro, C. M., Ziehr, J. E., and Eales, B. A. (1993). Genetic variation in paw preference (handedness) in the mouse. *Genome* 36, 935–943. doi: 10.1139/g93-123

Boehm, M., and Slack, F. (2005). A developmental timing microRNA and its target regulate life span in *C. elegans*. *Science* 310, 1954–1957. doi: 10.1126/science.1115596

Boulias, K., and Horvitz, H. R. (2012). The *C. elegans* microRNA mir-71 acts in neurons to promote germline-mediated longevity through regulation of DAF-16/FOXO. *Cell Metab.* 15, 439–450. doi: 10.1016/j.cmet.2012.02.014

Bushati, N., and Cohen, S. M. (2007). microRNA functions. *Annu. Rev. Cell Dev. Biol.* 23, 175–205. doi: 10.1146/annurev.cellbio.23.090506.123406

Chang, C., Hsieh, Y. W., Lesch, B. J., Bargmann, C. I., and Chuang, C. F. (2011). Microtubule-based localization of a synaptic calcium-signaling complex is required for left–right neuronal asymmetry in *C. elegans*. *Development* 138, 3509–3518. doi: 10.1242/dev.069740

Chang, C., Yu, T. W., Bargmann, C. I., and Tessier-Lavigne, M. (2004a). Inhibition of netrin-mediated axon attraction by a receptor protein tyrosine phosphatase. *Science* 305, 103–106. doi: 10.1126/science.1096983

Chang, S., Johnston, R. J. Jr., Frokjaer-Jensen, C., Lockery, S., and Hobert, O. (2004b). MicroRNAs act sequentially and asymmetrically to control chemosensory laterality in the nematode. *Nature* 430, 785–789. doi: 10.1038/nature02752

Chang, S., Wen, S., Chen, D., and Jin, P. (2009). Small regulatory RNAs in neurodevelopmental disorders. *Hum. Mol. Genet.* 18, R18–R26. doi: 10.1093/hmg/ddp072

Chatterjee, S., and Grosshans, H. (2009). Active turnover modulates mature microRNA activity in *Caenorhabditis elegans*. *Nature* 461, 546–549. doi: 10.1038/nature08349

Chekulaeva, M., and Filipowicz, W. (2009). Mechanisms of miRNA-mediated post-transcriptional regulation in animal cells. *Curr. Opin. Cell Biol.* 21, 452–460. doi: 10.1016/j.ceb.2009.04.009

Choi, P. S., Zakhary, L., Choi, W. Y., Caron, S., Alvarez-Saavedra, E., Miska, E. A., et al. (2008). Members of the miRNA-200 family regulate olfactory neurogenesis. *Neuron* 57, 41–55. doi: 10.1016/j.neuron.2007.11.018

Chuang, C. F., and Bargmann, C. I. (2005). A toll-interleukin 1 repeat protein at the synapse specifies asymmetric odorant receptor expression via ASK1 MAPKKK signaling. *Genes Dev.* 19, 270–281. doi: 10.1101/gad.1276505

Chuang, C. F., Vanhoven, M. K., Fetter, R. D., Verselis, V. K., and Bargmann, C. I. (2007). An innexin-dependent cell network establishes left–right neuronal asymmetry in *C. elegans*. *Cell* 129, 787–799. doi: 10.1016/j.cell.2007.02.052

Cochella, L., and Hobert, O. (2012a). Diverse functions of microRNAs in nervous system development. *Curr. Top. Dev. Biol.* 99, 115–143. doi: 10.1016/B978-0-12-387038-4.00005-7

Cochella, L., and Hobert, O. (2012b). Embryonic priming of a miRNA locus predetermines postmitotic neuronal left/right asymmetry in *C. elegans*. *Cell* 151, 1229–1242. doi: 10.1016/j.cell.2012.10.049

Decembrini, S., Bressan, D., Vignali, R., Pitto, L., Mariotti, S., Rainaldi, G., et al. (2009). MicroRNAs couple cell fate and developmental timing in retina. *Proc. Natl. Acad. Sci. U.S.A.* 106, 21179–21184. doi: 10.1073/pnas.0909167106

Derflinger, S., Sorg, C., Gaser, C., Myers, N., Arsic, M., Kurz, A., et al. (2011). Grey-matter atrophy in Alzheimer's disease is asymmetric but not lateralized. *J. Alzheimers Dis.* 25, 347–357.

Galaburda, A. M., Sherman, G. F., Rosen, G. D., Aboitiz, F., and Geschwind, N. (1985). Developmental dyslexia: four consecutive patients with cortical anomalies. *Ann. Neurol.* 18, 222–233. doi: 10.1002/ana.410180210

Ghildiyal, M., and Zamore, P. D. (2009). Small silencing RNAs: an expanding universe. *Nat. Rev. Genet.* 10, 94–108. doi: 10.1038/nrg2504

Giraldez, A. J., Cinalli, R. M., Glasner, M. E., Enright, A. J., Thomson, J. M., Baskerville, S., et al. (2005). MicroRNAs regulate brain morphogenesis in zebrafish. *Science* 308, 833–838. doi: 10.1126/science.1109020

Good, K., Ciosk, R., Nance, J., Neves, A., Hill, R. J., and Priess, J. R. (2004). The T-box transcription factors TBX-37 and TBX-38 link GLP-1/Notch signaling to mesoderm induction in *C. elegans* embryos. *Development* 131, 1967–1978. doi: 10.1242/dev.01088

Hammell, C. M., Lubin, I., Boag, P. R., Blackwell, T. K., and Ambros, V. (2009). nhl-2 modulates microRNA activity in *Caenorhabditis elegans*. *Cell* 136, 926–938. doi: 10.1016/j.cell.2009.01.053

He, L., and Hannon, G. J. (2004). MicroRNAs: small RNAs with a big role in gene regulation. *Nat. Rev. Genet.* 5, 522–531. doi: 10.1038/nrg1379

Herbert, M. R., Ziegler, D. A., Deutsch, C. K., O'Brien, L. M., Kennedy, D. N., Filipek, P. A., et al. (2005). Brain asymmetries in autism and developmental language disorder: a nested whole-brain analysis. *Brain* 128, 213–226. doi: 10.1093/brain/awh330

Hsieh, Y. W., Chang, C., and Chuang, C. F. (2012). The microRNA mir-71 inhibits calcium signaling by targeting the TIR-1/Sarm1 adaptor protein to control stochastic L/R neuronal asymmetry in *C. elegans*. *PLoS Genet.* 8:e1002864. doi: 10.1371/journal.pgen.1002864

Johnston, R. J. Jr., Chang, S., Etchberger, J. F., Ortiz, C. O., and Hobert, O. (2005). MicroRNAs acting in a double-negative feedback loop to control a neuronal cell fate decision. *Proc. Natl. Acad. Sci. U.S.A.* 102, 12449–12454. doi: 10.1073/pnas.0505530102

Johnston, R. J., and Hobert, O. (2003). A microRNA controlling left/right neuronal asymmetry in *Caenorhabditis elegans*. *Nature* 426, 845–849. doi: 10.1038/nature02255

Kishimoto, N., Asakawa, K., Madelaine, R., Blader, P., Kawakami, K., and Sawamoto, K. (2013). Interhemispheric asymmetry of olfactory input-dependent neuronal specification in the adult brain. *Nat. Neurosci.* 16, 884–888. doi: 10.1038/nn.3409

Li, Y., Wang, F., Lee, J. A., and Gao, F. B. (2006). MicroRNA-9a ensures the precise specification of sensory organ precursors in *Drosophila*. *Genes Dev.* 20, 2793–2805. doi: 10.1101/gad.1466306

Makeyev, E. V., Zhang, J., Carrasco, M. A., and Maniatis, T. (2007). The MicroRNA miR-124 promotes neuronal differentiation by triggering brain-specific alternative pre-mRNA splicing. *Mol. Cell* 27, 435–448. doi: 10.1016/j.molcel.2007.07.015

Miyasaka, N., Morimoto, K., Tsubokawa, T., Higashijima, S., Okamoto, H., and Yoshihara, Y. (2009). From the olfactory bulb to higher brain centers: genetic visualization of secondary olfactory pathways in zebrafish. *J. Neurosci.* 29, 4756–4767. doi: 10.1523/JNEUROSCI.0118-09.2009

Oertel-Knochel, V., and Linden, D. E. (2011). Cerebral asymmetry in schizophrenia. *Neuroscientist* 17, 456–467. doi: 10.1177/1073858410386493

Orom, U. A., Nielsen, F. C., and Lund, A. H. (2008). MicroRNA-10a binds the 5′UTR of ribosomal protein mRNAs and enhances their translation. *Mol. Cell* 30, 460–471. doi: 10.1016/j.molcel.2008.05.001

Pierce-Shimomura, J. T., Faumont, S., Gaston, M. R., Pearson, B. J., and Lockery, S. R. (2001). The homeobox gene lim-6 is required for distinct chemosensory representations in *C. elegans*. *Nature* 410, 694–698. doi: 10.1038/35070575

Poole, R. J., and Hobert, O. (2006). Early embryonic programming of neuronal left/right asymmetry in *C. elegans*. *Curr. Biol.* 16, 2279–2292. doi: 10.1016/j.cub.2006.09.041

Priess, J. R. (2005). "Notch signaling in the *C. elegans* embryo," in *WormBook*, ed. The *C. elegans* Research Community (Pasadena, CA: WormBook), 1–16. Available at: http://www.wormbook.org

Renteria, M. E. (2012). Cerebral asymmetry: a quantitative, multifactorial, and plastic brain phenotype. *Twin Res. Hum. Genet.* 15, 401–413. doi: 10.1017/thg.2012.13

Rogers, L. J., Vallortigara, G., and Andrew, R. J. (2012). *Divided Brains: the Biology and Behaviour of Brain Asymmetries.* Cambridge: Cambridge University Press.

Sagasti, A., Hisamoto, N., Hyodo, J., Tanaka-Hino, M., Matsumoto, K., and Bargmann, C. I. (2001). The CaMKII UNC-43 activates the MAPKKK NSY-1 to execute a lateral signaling decision required for asymmetric olfactory neuron fates. *Cell* 105, 221–232. doi: 10.1016/S0092-8674(01)00313-0

Saito, Y., and Saito, H. (2012). MicroRNAs in cancers and neurodegenerative disorders. *Front. Genet.* 3:194. doi: 10.3389/fgene.2012.00194

Sayed, D., and Abdellatif, M. (2011). MicroRNAs in development and disease. *Physiol. Rev.* 91, 827–887. doi: 10.1152/physrev.00006.2010

Schumacher, J. A., Hsieh, Y. W., Chen, S., Pirri, J. K., Alkema, M. J., Li, W. H., et al. (2012). Intercellular calcium signaling in a gap junction-coupled cell network establishes asymmetric neuronal fates in *C. elegans*. *Development* 139, 4191–4201. doi: 10.1242/dev.083428

Schwamborn, J. C., Berezikov, E., and Knoblich, J. A. (2009). The TRIM-NHL protein TRIM32 activates microRNAs and prevents self-renewal in mouse neural progenitors. *Cell* 136, 913–925. doi: 10.1016/j.cell.2008.12.024

Shaw, P., Lalonde, F., Lepage, C., Rabin, C., Eckstrand, K., Sharp, W., et al. (2009). Development of cortical asymmetry in typically developing children and its disruption in attention-deficit/hyperactivity disorder. *Arch. Gen. Psychiatry* 66, 888–896. doi: 10.1001/archgenpsychiatry.2009.103

Signore, P., Nosten-Bertrand, M., Chaoui, M., Roubertoux, P. L., Marchaland, C., and Perez-Diaz, F. (1991). An assessment of handedness in mice. *Physiol. Behav.* 49, 701–704. doi: 10.1016/0031-9384(91)90305-8

Snelson, C. D., and Gamse, J. T. (2009). Building an asymmetric brain: development of the zebrafish epithalamus. *Semin. Cell Dev. Biol.* 20, 491–497. doi: 10.1016/j.semcdb.2008.11.008

Sun, T., Patoine, C., Abu-Khalil, A., Visvader, J., Sum, E., Cherry, T. J., et al. (2005). Early asymmetry of gene transcription in embryonic human left and right cerebral cortex. *Science* 308, 1794–1798. doi: 10.1126/science.1110324

Sun, T., and Walsh, C. A. (2006). Molecular approaches to brain asymmetry and handedness. *Nat. Rev. Neurosci.* 7, 655–662. doi: 10.1038/nrn1930

Tanaka-Hino, M., Sagasti, A., Hisamoto, N., Kawasaki, M., Nakano, S., Ninomiya-Tsuji, J., et al. (2002). SEK-1 MAPKK mediates Ca^{2+} signaling to determine neuronal asymmetric development in *Caenorhabditis elegans*. *EMBO Rep.* 3, 56–62. doi: 10.1093/embo-reports/kvf001

Taylor, R. W., Hsieh, Y. W., Gamse, J. T., and Chuang, C. F. (2010). Making a difference together: reciprocal interactions in *C. elegans* and zebrafish asymmetric neural development. *Development* 137, 681–691. doi: 10.1242/dev.038695

Troemel, E. R., Sagasti, A., and Bargmann, C. I. (1999). Lateral signaling mediated by axon contact and calcium entry regulates asymmetric odorant receptor expression in *C. elegans*. *Cell* 99, 387–398. doi: 10.1016/S0092-8674(00)81525-1

Vanhoven, M. K., Bauer Huang, S. L., Albin, S. D., and Bargmann, C. I. (2006). The claudin superfamily protein nsy-4 biases lateral signaling to generate left–right asymmetry in *C. elegans* olfactory neurons. *Neuron* 51, 291–302. doi: 10.1016/j.neuron.2006.06.029

Vasudevan, S., Tong, Y., and Steitz, J. A. (2007). Switching from repression to activation: microRNAs can up-regulate translation. *Science* 318, 1931–1934. doi: 10.1126/science.1149460

Wes, P. D., and Bargmann, C. I. (2001). *C. elegans* odour discrimination requires asymmetric diversity in olfactory neurons. *Nature* 410, 698–701. doi: 10.1038/35070581

White, J. G., Southgate, E., Thomson, J. N., and Brenner, S. (1986). The structure of the nervous system of the nematode *Caenorhabditis elegans*. *Philos. Trans. R. Soc. Lond. B Biol. Sci.* 314, 1–340. doi: 10.1098/rstb.1986.0056

Yu, S., Avery, L., Baude, E., and Garbers, D. L. (1997). Guanylyl cyclase expression in specific sensory neurons: a new family of chemosensory receptors. *Proc. Natl. Acad. Sci. U.S.A.* 94, 3384–3387. doi: 10.1073/pnas.94.7.3384

Zhang, X., Zabinsky, R., Teng, Y., Cui, M., and Han, M. (2011). microRNAs play critical roles in the survival and recovery of *Caenorhabditis elegans* from starvation-induced L1 diapause. *Proc. Natl. Acad. Sci. U.S.A.* 108, 17997–18002. doi: 10.1073/pnas.1105982108

Zou, Y., Chiu, H., Domenger, D., Chuang, C. F., and Chang, C. (2012). The lin-4 microRNA targets the LIN-14 transcription factor to inhibit netrin-mediated axon attraction. *Sci. Signal.* 5, ra43. doi: 10.1126/scisignal.2002437

Zou, Y., Chiu, H., Zinovyeva, A., Ambros, V., Chuang, C. F., and Chang, C. (2013). Developmental decline in neuronal regeneration by the progressive change of two intrinsic timers. *Science* 340, 372–376. doi: 10.1126/science.1231321

9

Combined fluorescent *in situ* hybridization for detection of microRNAs and immunofluorescent labeling for cell-type markers

Amrita D. Chaudhuri, Sowmya V. Yelamanchili and Howard S. Fox*

Department of Pharmacology and Experimental Neuroscience, University of Nebraska Medical Center, Omaha, NE, USA

****Correspondence:***
Howard S. Fox, Department of Pharmacology and Experimental Neuroscience, University of Nebraska Medical Center, 985800 Nebraska Medical Center - DRC 3008, Omaha, NE 68198-5800, USA
e-mail: hfox@unmc.edu

Identification of the cell type of origin for normal or aberrant gene expression is critical for many studies, and poses a significant problem for some regulatory RNAs such as microRNAs. MicroRNAs are small non-coding RNAs that regulate cellular function by targeting specific mRNAs and reducing the level of their protein product. Aberrant expression of miRNAs in cell-types where they are not normally expressed occurs in several disease conditions. Therefore, it is important to determine not only the expression level of microRNAs, but also where they are expressed. Here we describe a detailed method for fluorescent *in situ* hybridization (FISH) combined with immunofluorescent labeling for cell-type markers in formalin fixed paraffin embedded (FFPE) sections along with modifications required to adapt the protocol for primary neurons grown in culture. We have combined the specificity and stability of locked nucleic acid (LNA) probes with tyramide signal amplification. To prevent loss of small RNA species, we performed post-fixation with ethylcarbodiimide (EDC). Additionally by omitting protease digestion and using only high temperature with sodium citrate buffer for FFPE sections, we were able to perform immunolabeling for proteins concurrently with *in situ* hybridization without compromising efficacy of either procedure.

Keywords: brain, FFPE, neuron, LNA, TSA

INTRODUCTION

Since their discovery in 1993, microRNAs (miRNAs) have emerged as important regulators of cellular function. MiRNAs are synthesized as long primary transcripts (pri-miRNAs) that are processed sequentially by two RNase III enzymes, Drosha in the nucleus and Dicer in the cytoplasm, along with their associated proteins (DGCR8 and TRBP, respectively), to generate the mature form (Kim et al., 2009). These small (~22-nucleotide long) RNAs do not encode proteins. Rather they get incorporated into a protein complex, called the RNA-induced silencing complex (RISC) that regulates translation of target mRNAs. Most often, miRNAs bind to the 3′UTR of their target mRNAs and reduce the corresponding protein level by either degradation of the target or repression of translation (Filipowicz et al., 2008). One miRNA can target hundreds of mRNAs and about 60% of all protein coding genes are predicted to be regulated by miRNAs (Friedman et al., 2009). Target specificity of miRNAs is therefore often determined by their cell type and developmental stage specific expression. In the brain, for example, distinct groups of miRNAs are expressed in neurons, astrocytes, oligodendrocytes, and microglia. MiRNAs-124, 434, and 376a are specifically enriched in neurons while miRNA-143, 146a, 449a, and 193 are enriched in astrocytes (Jovicic et al., 2013). Overexpression of these neuron-enriched miRNAs can drive differentiation of neural stem cells to neurons, while overexpression of glia-enriched miRNAs prevent differentiation to neurons and promote glial differentiation (Lim et al., 2005). Aberrant expression of miRNAs in cell types where they are not expressed under physiological conditions occurs in disease states and following injury. For example, miRNA-21 and miRNA-142, two miRNAs not normally expressed in neurons, are detectable in neurons following nerve crush injury (Wu et al., 2011) and in infection-induced CNS inflammation (Yelamanchili et al., 2010; Chaudhuri et al., 2013). It is therefore important to study both the level of expression of miRNAs and their localization. Fluorescent *in situ* hybridization (FISH) is an excellent method for this purpose, as it can be combined with simultaneous immunofluorescent labeling for cell-type markers.

The earliest methods for *in situ* hybridization detection of miRNAs used digoxigenin labeled ribonucleotide probes (Chen, 2004; Kosman et al., 2004). However, due to the small size of miRNAs, the sensitivity and specificity of ribonucleotide probes is often not sufficient to distinguish true signal from background noise. A number of modifications have therefore been introduced to improve the signal to noise ratio. The first and probably most important of these modifications is the use of locked nucleic acid (LNA) probes instead of ribonucleotide probes (Koshkin et al., 1998). In LNA the 2′ oxygen and the 5′ carbon moieties of ribose are linked by a methylene bridge, locking it in a C3′-endo conformation that is the major conformation in A-form of nucleic acids and is most suitable for hybridization with RNA (Koshkin et al., 1998; Valoczi et al., 2004). This prevents denaturation of the hybridized probe and results in increased stability of the RNA-LNA hybrid, thereby increasing the melting temperature (Tm) by 2–10°C for each LNA monomer included in the probe (Koshkin et al., 1998; Valoczi et al., 2004). Additionally LNA probes have enhanced mismatch discrimination capability (Valoczi et al., 2004; Chou et al., 2005; You et al., 2006), further improving their specificity, and are comparatively resistant to degradation by nucleases (Wahlestedt et al., 2000).

One strategy used to improve sensitivity of detection is to amplify the specific *in situ* hybridization signal using the catalyzed reporter deposition method (Bobrow et al., 1989; Kerstens et al., 1995). In this method, LNA probes labeled with digoxigenin moieties on the 3′ end, 5′ end, or both, are used for hybridization. The digoxigenin moieties are recognized by horseradish peroxidase (HRP)-tagged anti-digoxigenin antibodies. Cyanine 5 (Cy5), cyanine 3 (Cy3) or fluorescein-conjugated tyramides are used as HRP substrate. The tyramides are converted by HRP to highly reactive tyramide radicals that bind covalently to nearby tyrosine residues. These radicals are extremely short-lived, preventing them from diffusing away from the site of synthesis. This technique allows approximately 500-times amplification of the original signal (Qian and Lloyd, 2003).

Recently, a number of methods have been described for combined detection of miRNA and proteins in tissue samples and cultured cells (Zaidi et al., 2000; Nuovo et al., 2009; de Planell-Saguer et al., 2010; Nuovo, 2010; Sempere et al., 2010; Schneider et al., 2011; Wu et al., 2011; Herzer et al., 2012; Nielsen and Holmstrom, 2013; Sempere and Korc, 2013). In most of these methods, immunolabeling of protein markers is performed after completion of all the FISH steps and many of them use proteinase K to break the formaldehyde cross-links and allow for better penetration of FISH probes. Here, we describe a method where IF labeling can be performed simultaneously with FISH once the probe hybridization is complete. Additionally, in our experience we found that proteinase K treatment may destroy some of the epitopes, adversely affecting the IF signal. Therefore, similar to de Planell-Saguer et al. (2010), we have eliminated the use of protease treatment and find that high heat in sodium citrate buffer, which we previously used for combined radioactive *in situ* hybridization of mRNA and colorimetric immunohistochemistry (IHC) of proteins in formalin fixed paraffin embedded (FFPE) tissue sections (Roberts et al., 2003), also sufficiently allows penetration of FISH probes for miRNA, and conserves and enhances immunoreactivity of proteins. Furthermore, to prevent loss of small RNA species during subsequent washes, we find that post-fixation with ethylcarbodiimide (EDC) as described previously (Pena et al., 2009), is beneficial. With combined use of antigen retrieval, EDC post-fixation, LNA probes, and tyramide signal amplification (TSA), we were able to simultaneously detect miRNA and cell-type markers in neurons and other cells types.

MATERIALS

Equipment:

1. Hybridization oven and chamber (Boekel Scientific, Feasterville, PA, USA, Catalog # 241000), or other suitable incubator/oven.

General materials required:

1. Diethylpyrocarbonate (DEPC) (Applichem, Darmstadt, Germany, Catalog # A0881).
2. TRIS (Amresco LLC, Solon, OH, USA, Catalog # 0497).
3. Sodium Chloride (Fisher Scientific, Fair Lawn, NJ, USA, Catalog # S271).
4. Hydrochloric Acid (12 M) (Fisher Scientific, Catalog # A144-212).
5. Ultrapure 20X saline-sodium citrate (SSC) (Invitrogen, Carlsbad, CA, USA, Catalog # 15557).
6. Cy 5 tyramide signal amplification (TSA) kit (Perkin Elmer, Shelton, CT USA, Catalog # NEL745001KT).
7. LNA probes (Exiqon, Woburn, MA, USA), specific for target miRNAs as well as positive and negative controls. We used snRNA U6 as positive control and a scrambled probe with no complementarity to any known miRNAs as negative control.
8. Anti-Digoxigenin-POD, Fab fragments (Roche Diagnostics, Mannheim, Germany, Catalog # 11207733910).
9. Kimwipes (Fisher Scientific, Catalog # 06-666-1A).
10. Prolong Gold anti-fade reagent with DAPI (Invitrogen, Catalog # P36935).

Materials required for deparaffinization, rehydration, and conditioning of FFPE sections:

1. Xylene (Fisher Scientific, Catalog # X5P-1GAL).
2. Ethanol 200 proof (Decon Labs Inc., King of Prussia, PA, USA, Catalog # 2701).
3. Citric acid (Fisher Scientific, Catalog # BP339-500).
4. Sodium Citrate (Fisher Scientific, Catalog # BP327-500).

Materials required for post-fixation:

1. 1-methylimidazole (Sigma-Aldrich, Saint Louis, MO, USA, Catalog # 336092).
2. N-(3-Dimethylaminopropyl)-N′-ethylcarbodiimide hydrochloride (EDC) (Sigma-Aldrich, Catalog # E1769).
3. Glycine (Sigma-Aldrich, Catalog # G7126).

Materials required for hybridization buffer:

1. Formamide (Fisher Scientific, Catalog # BP227-500).
2. AG 501-X8 Resin (Biorad, Hercules, CA, USA, Catalog # 142-6424).
3. Yeast tRNA (Invitrogen, Catalog # 15401-029).
4. Denhardt's solution 50X concentrate (Sigma-Aldrich, Catalog # D2532).
5. Ultrapure 10% SDS solution (Invitrogen, Catalog # 15553-027).
6. Ultrapure 0.5 M EDTA, pH 8.0 (Invitrogen, Catalog # 15575-020).
7. Dextran sulfate (Sigma-Aldrich, Catalog # 8906-100).

Materials required for blocking buffer:

1. Bovine serum albumin (BSA) (Sigma-Aldrich, Catalog # B4287).

2. Normal goat serum (NGS) (Vector laboratories, Burlingame, CA, USA, Catalog # S-1000).
3. Phosphate buffered saline (PBS) (Gibco, Life Technologies, Grand Island, NY, USA, Catalog # 10010).

BUFFERS AND SOLUTIONS

Ensure that all equipment and working surfaces are RNase free. This can be done by autoclaving equipment and wiping working surfaces with RNaseZAP (Ambion, Life technologies, NY, USA, Catalog # AM9780).

The following buffers or solutions can be prepared ahead of time.

1. DEPC-treated water:
   Add 1 mL of DEPC per 1 L of water (0.1% v/v). Incubate overnight and autoclave for 30 min. Store at room temperature.
2. NaCl (3 M):
   Dissolve 87.8 g of NaCl in 500 mL of DEPC treated water.
3. Tris-HCl (1 M):
   Dissolve 121.1 g of Tris base in 800 mL of DEPC treated water. Adjust pH by adding 12 M HCl. For combined FISH and IF experiments Tris-HCl at both pH 7.4 and 8.0 need to be prepared.
4. 10X Tris Buffered Saline:
   410 ml of DEPC-treated water
   500 ml of 1 M Tris-HCl pH 7.4
   90 g of NaCl.

Note: Adding DEPC to premade Tris buffer is not recommended as DEPC forms a complex with the free amino groups of Tris. DEPC-treated water that has been autoclaved can however, be used to dissolve Tris, as DEPC is hydrolyzed during autoclaving.

5. Sodium citrate (0.1 M):
   29.41 g of sodium citrate in 1000 mL of DEPC-treated water. Store at 4°C.
6. Citric acid (0.1 M):
   9.56 g of citric acid in 500 mL of DEPC-treated water. Store at 4°C.
7. Sodium citrate buffer (0.01 M, pH = 6.4):
   Add 41 mL of 0.1 M sodium citrate to 9 mL of 0.1 M citric acid. Make the volume up to 500 mL with DEPC-treated water. Store at 4°C.
8. Hybridization Buffer:
   Deionize formamide by adding 5 g of ion exchange resin per 100 mL of formamide. Stir for 30 min at room temperature and filter through Whatman filter paper. Deionized formamide can be stored at −20°C.
   For 100 mL of hybridization buffer:
   50 mL of deionized formamide (final concentration 50%)
   1 mL of 1 M Tris-HCl, pH = 8.0 (final concentration 10 mM)
   2.5 mL of 10% SDS (final concentration 0.25%)
   200 μg/mL yeast tRNA
   1 X Denhardt's solution
   600 mM NaCl
   1 mM EDTA
   10% Dextran sulfate
   Make up to 100 mL with DEPC-treated water
   Hybridization buffer can be stored in aliquots at −20°C.
9. Blocking Buffer:
   1% BSA, 3% NGS in 1X PBS. Store at 4°C.
10. Diluted SSC:
    Dilute 20X SSC in DEPC-treated water to make 1X, 2X, and 0.2X SSC solutions.

The following buffers have to be prepared fresh on the day of the experiment.

1. Methylimidazole buffer:
   Add 1.6 ml of 1-methylimidazole to 130 ml of DEPC-treated water. Adjust pH to 8.0 by adding ~450 μl 12 M HCl, then add 16 ml 3 M NaCl and DEPC-treated water to a final volume of 160 mL. Final concentrations are 0.13 M 1–methylimidazole, 300 mM NaCl, pH 8.0.
2. EDC Solution:
   Add 307 mg EDC into 10 ml of the above methylimidazole buffer, and then readjust the pH to 8.0 by further addition of ~100 μl 12 M HCl if required. Final concentration of EDC is 0.16 M.

METHODS

Ethics Statement: Animal experiments to obtain the tissues used for these experiments were performed with approval from UNMC Institutional Animal Use and Care Committee.

For combined FISH and IF on FFPE tissue, 5 μm thick sections are cut from the tissue blocks, floated on DEPC-treated water, and picked up on glass slides and air-dried. To ensure tissue adherence slides are baked at 60°C for 1 h and cooled to room temperature the day of or the day before starting the experiment.

For combined FISH and IF on neuronal cultures, neurons grown on poly-D-lysine coated glass coverslips are fixed in 4% paraformaldehyde (PFA) for 15 min at room temperature followed by two washes with PBS for 5 min each at room temperature. Coverslips are then placed overnight in 70% ethanol at 4°C for permeabilization.

DAY 1

1.1. To deparaffinize and rehydrate FFPE sections incubate at room temperature in:

- Xylene, 3 times, 5 min each
- 100% ethanol, twice, 5 min each
- 70% ethanol (diluted in DEPC-treated water), once, 5 min
- 50% ethanol (diluted in DEPC-treated water), once, 5 min
- DEPC water, twice, 3 min each

1.2. Antigen retrieval:

Incubate slides in 0.01 M citrate buffer pH = 6.4, for 40 min at 90°C. Cool slides by incubating in citrate buffer for 20 min room temperature. Wash with TBS, 3 times, 3 min each.

Note: For combined FISH and IF on neuronal culture coverslips, the deparaffinization and antigen retrieval steps are omitted and post-fixation with EDC (next) is the first step to be performed on day 1.

1.3. EDC Treatment

- Incubate slides/coverslips in freshly prepared methylimidazole solutions, twice for 10 min each at room temperature.
- Incubate for 1 h in freshly prepared EDC solution at room temperature in a humidified chamber.
- Wash once for 5 min with 0.2% (w/v) glycine in TBS
- Wash twice, 5 min each with TBS

1.4. Prehybridization
Prepare hybridization chamber by placing 1X SSC-soaked Kimwipes at the bottom of the chamber. Place the slides or plate containing coverslips on the rack on top of the Kimwipes. Pipette enough hybridization buffer onto the tissue section so that it is completely covered. Incubate in hybridization oven at 37°C (or desired hybridization temperature) for 1 h.

1.5. Hybridization
Add 4 picomoles of 5′ and 3′ double digoxigenin-labeled LNA probe per 100 μL of hybridization buffer, mix, and heat at 65°C for 5 min to ensure denaturation of probes. Replace the hybridization buffer without probe (from pre-hybridization step) with the pre-heated hybridization buffer containing the LNA probe. Cover the tissue sections with pieces of Parafilm to prevent evaporation. Alternatively commercially available plastic slides (e.g., from IHCWorld, Woodstock, MD, USA, Catalog # IW-2601) can be used for this purpose. A thin layer of hybridization buffer will be present in between the Parafilm and tissue section and care should be taken that there are no air bubbles. For neurons grown on coverslips, the plate containing the coverslips can be sealed with Parafilm. Hybridize overnight at 37°C or optimized hybridization temperature. Preliminary experiments to optimize the hybridization temperature in order to maximize specific signal and minimize background noise may be required.

DAY 2

2.1. Stringency washes:

- Three times for 20 min each with 2X SSC at 42°C
- Twice for 20 min each with 0.2X SSC 42°C

Note: The temperature may be increased if needed to reduce background.

2.2. Blocking for IF:
Incubate in blocking buffer for 1 h at room temperature in a humidified chamber. A humidified chamber can be prepared by placing the slides on top of Kimwipes soaked in water or PBS placed at the bottom of a box with a lid.

2.3. Incubation with Primary antibodies:
Incubate with primary antibodies diluted in blocking buffer, overnight at 4°C in a humidified chamber. At this step antibodies against the desired cell-type marker can be mixed together with the anti-digoxigenin-POD. We have used up to two cell-type specific markers along with the anti-digoxigenin-POD antibody (to recognize the double digoxigenin labeled probes). The final concentration of the anti-digoxigenin-POD antibody is 1:100 in blocking buffer. Care should be taken that the antibodies for the cell-type specific markers are raised in different species. Dilutions for the cell-type specific antibodies should be determined empirically.

DAY 3

3.1. Wash the slides/coverslips twice in TBS for 2 min each at room temperature.

3.2. Incubation with secondary antibodies:
From this step onwards care should be taken to minimize exposure of the slides to light. Incubate at room temperature for 1 h with fluorochrome-labeled secondary antibodies corresponding to the species in which the primary cell-type specific antibodies were raised. We use Alexa Fluor-labeled antibodies (Invitrogen, CA, USA) for our experiments, the specific fluorochromes should match one's fluorescence filters on the microscope but not overlap with the Cy5 signal from the FISH.

3.3. Wash slides twice in TBS for 2 min each at room temperature.

3.4. FISH signal amplification with peroxidase substrate:
Dilute Cy5 standard from Cy5-TSA kit 1:100 in the provided diluent buffer. Add enough to cover tissue sections and incubate at room temperature for 10 min. Wash slides with 0.1% Tween 20 in TBS three times and twice with TBS, for 5 min each. Dip slides in DEPC-treated water before mounting in prolong gold anti-fade reagent with DAPI (Invitrogen, CA, USA). Let the slides dry at room temperature, overnight, away from light.

DAY 4

4.1. Seal slides with nail polish before imaging in a fluorescence or confocal microscope. We used a Zeiss Observer.Z1 microscope equipped with a monochromatic Axiocam MRm camera using Axiovision 40 v.4.8.0.0 software.

RESULTS AND DISCUSSION

Several methods have been described for *in situ* hybridization of miRNAs in tissue sections (Chen, 2004; Deo et al., 2006; Nelson et al., 2006; Ryan et al., 2006; Obernosterer et al., 2007; Schaefer et al., 2007; Sempere et al., 2007; Silahtaroglu et al., 2007; Thompson et al., 2007; Williams et al., 2007; Bak et al., 2008; Nuovo, 2008; Wang et al., 2008; Lu and Tsourkas, 2009; Nelson and Wilfred, 2009; Pena et al., 2009; Yamamichi et al., 2009; Havelda, 2010; Jorgensen et al., 2010; Song et al., 2010; Soe et al., 2011; Shi et al., 2012) and a some for concurrent detection of proteins using IHC or IF (Zaidi et al., 2000; Nuovo et al., 2009; de Planell-Saguer et al., 2010; Nuovo, 2010; Sempere et al., 2010; Schneider et al., 2011; Wu et al., 2011; Herzer et al., 2012; Nielsen and Holmstrom, 2013; Sempere and Korc, 2013). Among the published methods for co-detection of miRNA and proteins in FFPE sections, Nuovo et al. described a method for colorimetric *in situ* hybridization for miRNAs using digoxigenin-labeled LNA probe (Nuovo et al., 2009). The tissue sections were digested with pepsin to facilitate penetration of the LNA probes. The recommended probe concentration was 2 picomoles/μL of hybridization buffer (i.e., 2 μM). The hybridized probe was recognized by an alkaline phosphatase (AP)-tagged anti-digoxigenin antibody and color development was performed using the AP substrate NBT/BCIP. This was followed by IHC for protein target using an automated system. The different color signals from *in situ* hybridization and IHC were converted to fluorescent signals using the Nuance system for co-expression analysis (Nuovo, 2010). This method did not allow for amplification of the hybridization signal and was therefore less effective for detection of miRNAs with low abundance in the tissues/cells of interest. Nielsen et al. and de Planell-Saguer et al. combined the advantages of LNA technology with signal amplification using TSA (de Planell-Saguer et al., 2010; Nielsen and Holmstrom, 2013). Additionally, de Planell-Saguer et al. eliminated the protease digestion step and performed antigen retrieval using high heat and citrate buffer (de Planell-Saguer et al., 2010). Sempere et al. used the TSA amplification system for detection of both the miRNA and protein of interest (Sempere et al., 2010; Sempere and Korc, 2013). They performed digestion with proteinase K to improve tissue penetration of LNA probes. After hybridization, protein labeling was performed using an automated system (Sempere et al., 2010).

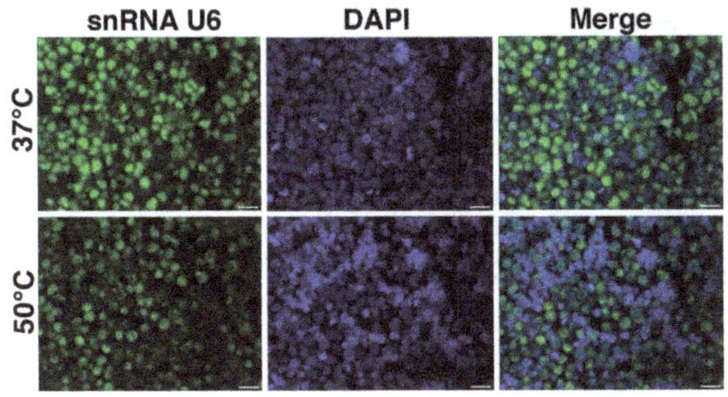

FIGURE 1 | Optimization of hybridization temperature. FISH was performed for snRNA U6 in FFPE sections of BE(2)M17 cells. Two hybridization temperatures were compared. The snRNA U6 signal (green) appeared to be brighter in the sections that were hybridized at 37°C (top panel) compared to those that were hybridized at 50°C (bottom panel). Scale bars are 20 μm.

We developed a method for combined FISH and IF in FFPE sections, that further improves signal to noise ratio by addition of an EDC-crosslinking step to prevent loss of small RNAs. This enabled us to use very low probe concentrations for hybridization (0.04 picomoles/μL or 40 nM) even for low abundant miRNAs, e.g., miRNA-142 in the brain. In our experience, we found that digestion of tissue sections with proteases while improving miRNA FISH signal, also resulted in loss of epitopes and worsened the IF signal. Therefore, similar to de Planell-Saguer et al. (2010) we did not perform proteinase K/pepsin digestion, but performed antigen retrieval with high heat and citrate buffer instead. We have applied this method previously for detecting miRNA localization in specific cell types in archived human brain sections (Yelamanchili et al., 2010) and brain sections from rhesus macaques (Chaudhuri et al., 2013). In this article, we have described in detail this method for combined FISH and IF.

OPTIMIZATION OF HYBRIDIZATION TEMPERATURE

Exiqon provides the Tm for the LNA probes. However, this predicted Tm, may not be the same as the true Tm to immobilized miRNA in cells/sections. Tm is also influenced by components of hybridization buffers (e.g., formamide, salt). Thus,

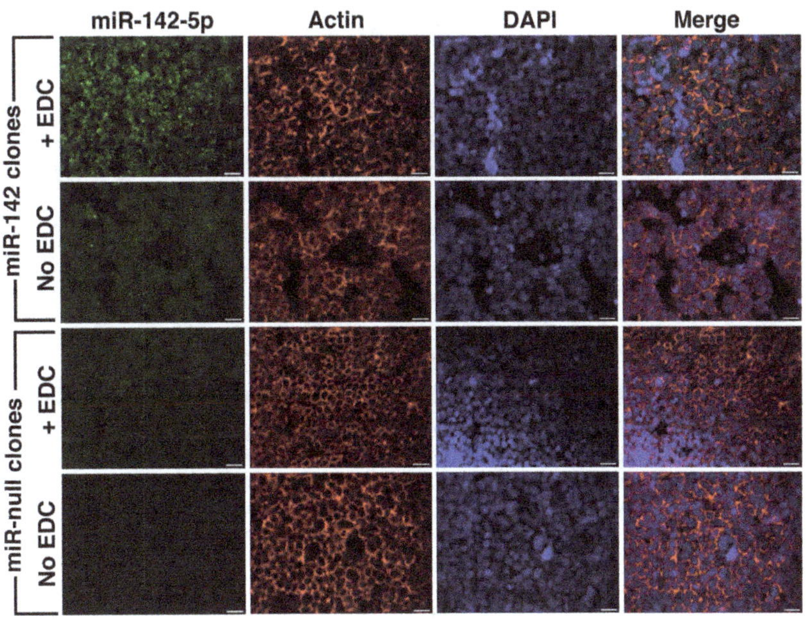

FIGURE 2 | Post-fixation with EDC improves FISH signal. Combined FISH and IF was performed for miRNA-142-5p and actin in FFPE sections of BE(2)M17 stable clones that express miRNA-142 and those that were transfected with a control plasmid miRNA-null. After post-fixation with EDC, adequate miRNA-142-5p signal (green) could be detected in the miRNA-142 clones (top panel). Without EDC post-fixation, miRNA-142-5p signal was very low in the miRNA-142 clones. No miRNA-142-5p signal was detected in the miRNA-null clones. Scale bars are 20 μm.

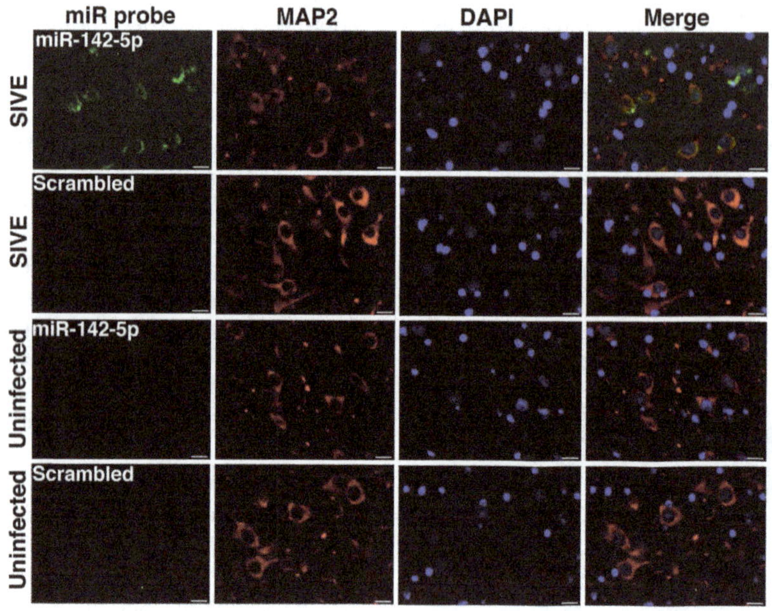

FIGURE 3 | MiRNA-142-5p is expressed in neurons in SIVE. Combined FISH and IF was performed for miRNA-142-5p and MAP2 in FFPE hippocampal sections from rhesus macaques with SIVE and uninfected macaques. MiRNA-142-5p expression (green) was detected within MAP2-labeled neurons (red) only in sections from macaques with SIVE. No miRNA-142-5p signal was detected in uninfected control sections. A scrambled miRNA probe was used as negative control for hybridization. Scale bars are 20 μm.

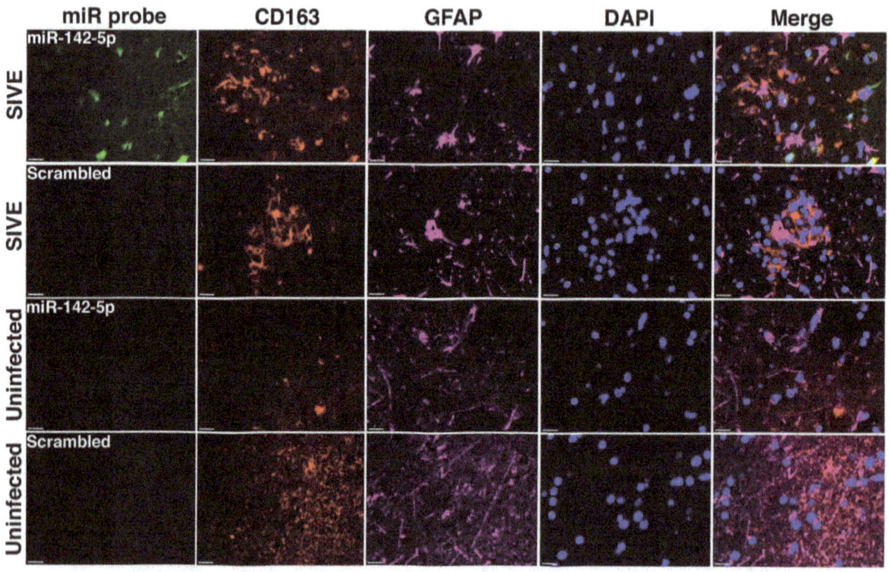

FIGURE 4 | MiRNA-142-5p is expressed in some macrophages/microglia in SIVE. FISH was performed for miRNA-142-5p along with IF labeling for CD163 (microglia/macrophage maker) and GFAP (astrocyte marker). In cortical sections from rhesus macaques with SIVE some CD163 labeled macrophage/microglia (red) expressed miRNA-142-5p (green). No co-localization was observed for miRNA-142-5p and GFAP (magenta). No miRNA-142-5p signal was detected in uninfected control sections. A scrambled miRNA probe was used as negative control for hybridization. Scale bars are 20 μm.

we begin with an empiric determination of hybridization temperature using hybridization at two temperatures, 37 and 50°C and/or additional temperatures. One commonly used temperature is 20°C below the provided Tm. This can be performed on controls, performing only FISH without IF. For example, examining the ubiquitously expressed snRNA U6, a positive control for miRNA FISH, on paraffin embedded BE(2)M17 cells (a neuroblastoma cell line), a much brighter FISH signal was observed at when hybridization was performed at 37°C (**Figure 1**).

POST-FIXATION WITH EDC IMPROVES FISH SIGNAL WITHOUT AFFECTING IF SIGNAL

EDC has been previously used in single miRNA FISH experiments to cross-link and prevent miRNA loss during the FISH procedure (Pena et al., 2009). We performed EDC post-fixation after antigen retrieval followed by combined FISH and IF for miRNA-142-5p and actin. These experiments were performed on FFPE sections of BE(2)M17 stable clones expressing miRNA-142 and clones that were transfected with the control plasmid (miRNA-null). As BE(2)M17 cells do not express any endogenous miRNA-142, these clones provided us with elegant positive and negative controls. As the miRNA-142-5p probe should show specific hybridization only in the miRNA-142 clones, any FISH signal detected in the miRNA-null clones would indicate non-specific hybridization under the conditions used. Similarly, absence of FISH signal in the miRNA-142 clones would indicate a failed hybridization. Using the combine FISH and IF protocol described, we could detect adequate miRNA-142-5p signal in the miRNA-142 clones only when sections were fixed with EDC (**Figure 2**). We could still detect some miRNA-142-5p signal in sections from miRNA-142 clones that were not fixed with EDC, however, the signal intensity was extremely low (**Figure 2**). No miRNA-142-5p signal was detected in the miRNA-null clones in either condition (**Figure 2**). Concurrent IF labeling for actin was performed and signal strength for actin appeared to be similar in all the sections.

DETECTION OF miRNA-142-5p IN FFPE MONKEY BRAIN SECTIONS:

We have applied the method described here to detect miRNA expression in brain sections from rhesus macaques and humans (Yelamanchili et al., 2010; Chaudhuri et al., 2013). **Figures 3, 4** are representative results of such experiments. Sections of different brain regions from rhesus macaques with simian immunodeficiency virus encephalitis (SIVE) or uninfected macaques were hybridized with miRNA-142-5p probe or a scrambled miRNA probe. This was combined with concurrent labeling for microtubule-associated protein 2 (MAP2), neuron-specific marker, in hippocampal sections (**Figure 3**), or with astrocyte specific marker glial fibrillary acidic protein (GFAP) and microglia/macrophage specific marker CD163 (**Figure 4**) in cortical sections. MiRNA-142-5p expression was detected in MAP2-labeled hippocampal neurons in SIVE. In the cortical sections miRNA-142-5p was found to be expressed in some CD163 positive macrophages/microglia (**Figure 4**). MiRNA-142-5p signal was not detected in GFAP labeled astrocytes (**Figure 4**). No signal was detected in uninfected control sections as well as when the sections were hybridized with a scrambled miRNA probe.

CONCLUSION

Here, we have described a method for combined *in situ* detection of miRNAs and IF labeling for cell-type markers. We modified existing methods by adding an EDC post-fixation step that greatly improved FISH signal strength without compromising IF signal. This method can be used to determine cell-types in which miRNAs are expressed in physiological and pathological conditions.

ACKNOWLEDGMENTS

The authors thank Kathleen Emanuel and Brenda Morsey for technical support. This work was supported by NIH grants MH062261 and MH073490.

REFERENCES

Bak, M., Silahtaroglu, A., Moller, M., Christensen, M., Rath, M. F., Skryabin, B., et al. (2008). MicroRNA expression in the adult mouse central nervous system. *RNA* 14, 432–444. doi: 10.1261/rna.783108

Bobrow, M. N., Harris, T. D., Shaughnessy, K. J., and Litt, G. J. (1989). Catalyzed reporter deposition, a novel method of signal amplification application to immunoassays. *J. Immunol. Methods* 125, 279–285. doi: 10.1016/0022-1759(89)90104-X

Chaudhuri, A. D., Yelamanchili, S. V., Marcondes, M. C., and Fox, H. S. (2013). Up-regulation of microRNA-142 in simian immunodeficiency virus encephalitis leads to repression of sirtuin1. *FASEB J.* 9, 3720–3729. doi: 10.1096/fj.13-232678

Chen, X. (2004). A microRNA as a translational repressor of APETALA2 in Arabidopsis flower development. *Science* 303, 2022–2025. doi: 10.1126/science.1088060

Chou, L. S., Meadows, C., Wittwer, C. T., and Lyon, E. (2005). Unlabeled oligonucleotide probes modified with locked nucleic acids for improved mismatch discrimination in genotyping by melting analysis. *Biotechniques* 39, 644, 646, 648 passim. doi: 10.2144/000112050

Deo, M., Yu, J.-Y., Chung, K.-H., Tippens, M., and Turner, D. L. (2006). Detection of mammalian microRNA expression by *in situ* hybridization with RNA oligonucleotides. *Dev. Dyn.* 235, 2538–2548. doi: 10.1002/dvdy.20847

de Planell-Saguer, M., Rodicio, M. C., and Mourelatos, Z. (2010). Rapid *in situ* codetection of noncoding RNAs and proteins in cells and formalin-fixed paraffin-embedded tissue sections without protease treatment. *Nat. Protoc.* 5, 1061–1073. doi: 10.1038/nprot.2010.62

Filipowicz, W., Bhattacharyya, S. N., and Sonenberg, N. (2008). Mechanisms of post-transcriptional regulation by microRNAs: are the answers in sight? *Nat. Rev. Genet.* 9, 102–114. doi: 10.1038/nrg2290

Friedman, R. C., Farh, K. K., Burge, C. B., and Bartel, D. P. (2009). Most mammalian mRNAs are conserved targets of microRNAs. *Genome Res.* 19, 92–105. doi: 10.1101/gr.082701.108

Havelda, Z. (2010). *in situ* detection of miRNAs using LNA probes. *Methods Mol. Biol.* 592, 127–136. doi: 10.1007/978-1-60327-005-2_9

Herzer, S., Silahtaroglu, A., and Meister, B. (2012). Locked nucleic acid-based *in situ* hybridisation reveals miR-7a as a hypothalamus-enriched microRNA with a distinct expression pattern. *J. Neuroendocrinol.* 24, 1492–1504. doi: 10.1111/j.1365-2826.2012.02358.x

Jorgensen, S., Baker, A., Moller, S., and Nielsen, B. S. (2010). Robust one-day *in situ* hybridization protocol for detection of microRNAs in paraffin samples using LNA probes. *Methods* 52, 375–381. doi: 10.1016/j.ymeth.2010.07.002

Jovicic, A., Roshan, R., Moisoi, N., Pradervand, S., Moser, R., Pillai, B., et al. (2013). Comprehensive expression analyses of neural cell-type-specific miRNAs identify new determinants of the specification and maintenance of neuronal phenotypes. *J. Neurosci.* 33, 5127–5137. doi: 10.1523/JNEUROSCI.0600-12.2013

Kerstens, H. M., Poddighe, P. J., and Hanselaar, A. G. (1995). A novel *in situ* hybridization signal amplification method based on the deposition of biotinylated tyramine. *J. Histochem. Cytochem.* 43, 347–352. doi: 10.1177/43.4.7897179

Kim, V. N., Han, J., and Siomi, M. C. (2009). Biogenesis of small RNAs in animals. *Nat. Rev. Mol. Cell Biol.* 10, 126–139. doi: 10.1038/nrm2632

Koshkin, A. A., Singh, S. K., Nielsen, P., Rajwanshi, V. K., Kumar, R., Meldgaard, M., et al. (1998). LNA (Locked Nucleic Acids): synthesis of the adenine, cytosine, guanine, 5-methylcytosine, thymine and uracil bicyclonucleoside monomers, oligomerisation, and unprecedented nucleic acid recognition. *Tetrahedron* 54, 3607–3630. doi: 10.1016/S0040-4020(98)00094-5

Kosman, D., Mizutani, C. M., Lemons, D., Cox, W. G., McGinnis, W., and Bier, E. (2004). Multiplex detection of RNA expression in Drosophila embryos. *Science* 305, 846. doi: 10.1126/science.1099247

Lim, L. P., Lau, N. C., Garrett-Engele, P., Grimson, A., Schelter, J. M., Castle, J., et al. (2005). Microarray analysis shows that some microRNAs downregulate large numbers of target mRNAs. *Nature* 433, 769–773. doi: 10.1038/nature03315

Lu, J., and Tsourkas, A. (2009). Imaging individual microRNAs in single mammalian cells *in situ*. *Nucleic Acids Res.* 37, e100. doi: 10.1093/nar/gkp482

Nelson, P. T., and Wilfred, B. R. (2009). *In situ* hybridization is a necessary experimental complement to microRNA (miRNA) expression profiling in the human brain. *Neurosci. Lett.* 466, 69–72. doi: 10.1016/j.neulet.2009.04.044

Nelson, P. T., Baldwin, D. A., Kloosterman, W. P., Kauppinen, S., Plasterk, R. H., and Mourelatos, Z. (2006). RAKE and LNA-ISH reveal microRNA expression and localization in archival human brain. *RNA* 12, 187–191. doi: 10.1261/rna.2258506

Nielsen, B. S., and Holmstrom, K. (2013). Combined microRNA *in situ* hybridization and immunohistochemical detection of protein markers. *Methods Mol. Biol.* 986, 353–365. doi: 10.1007/978-1-62703-311-4_22

Nuovo, G. J. (2008). *in situ* detection of precursor and mature microRNAs in paraffin embedded, formalin fixed tissues and cell preparations. *Methods* 44, 39–46. doi: 10.1016/j.ymeth.2007.10.008

Nuovo, G. J. (2010). *In situ* detection of microRNAs in paraffin embedded, formalin fixed tissues and the co-localization of their putative targets. *Methods* 52, 307–315. doi: 10.1016/j.ymeth.2010.08.009

Nuovo, G. J., Elton, T. S., Nana-Sinkam, P., Volinia, S., Croce, C. M., and Schmittgen, T. D. (2009). A methodology for the combined *in situ* analyses of the precursor and mature forms of microRNAs and correlation with their putative targets. *Nat. Protoc.* 4, 107–115. doi: 10.1038/nprot.2008.215

Obernosterer, G., Martinez, J., and Alenius, M. (2007). Locked nucleic acid-based *in situ* detection of microRNAs in mouse tissue sections. *Nat. Protoc.* 2, 1508–1514. doi: 10.1038/nprot.2007.153

Pena, J. T., Sohn-Lee, C., Rouhanifard, S. H., Ludwig, J., Hafner, M., Mihailovic, A., et al. (2009). miRNA *in situ* hybridization in formaldehyde and EDC-fixed tissues. *Nat. Methods* 6, 139–141. doi: 10.1038/nmeth.1294

Qian, X., and Lloyd, R. V. (2003). Recent developments in signal amplification methods for *in situ* hybridization. *Diagn. Mol. Pathol.* 12, 1–13. doi: 10.1097/00019606-200303000-00001

Roberts, E. S., Zandonatti, M. A., Watry, D. D., Madden, L. J., Henriksen, S. J., Taffe, M. A., et al. (2003). Induction of pathogenic sets of genes in macrophages and neurons in NeuroAIDS. *Am. J. Pathol.* 162, 2041–2057. doi: 10.1016/S0002-9440(10)64336-2

Ryan, D. G., Oliveira-Fernandes, M., and Lavker, R. M. (2006). MicroRNAs of the mammalian eye display distinct and overlapping tissue specificity. *Mol. Vis.* 12, 1175–1184.

Schaefer, A., O'Carroll, D., Tan, C. L., Hillman, D., Sugimori, M., Llinas, R., et al. (2007). Cerebellar neurodegeneration in the absence of microRNAs. *J. Exp. Med.* 204, 1553–1558. doi: 10.1084/jem.20070823

Schneider, M., Andersen, D. C., Silahtaroglu, A., Lyngbaek, S., Kauppinen, S., Hansen, J. L., et al. (2011). Cell-specific detection of microRNA expression during cardiomyogenesis by combined *in situ* hybridization and immunohistochemistry. *J. Mol. Histol.* 42, 289–299. doi: 10.1007/s10735-011-9332-8

Sempere, L. F., Christensen, M., Silahtaroglu, A., Bak, M., Heath, C. V., Schwartz, G., et al. (2007). Altered MicroRNA expression confined to specific epithelial cell subpopulations in breast cancer. *Cancer Res.* 67, 11612–11620. doi: 10.1158/0008-5472.CAN-07-5019

Sempere, L. F., and Korc, M. (2013). A method for conducting highly sensitive microRNA *in situ* hybridization and immunohistochemical analysis in pancreatic cancer. *Methods Mol. Biol.* 980, 43–59. doi: 10.1007/978-1-62703-287-2_4

Sempere, L. F., Preis, M., Yezefski, T., Ouyang, H., Suriawinata, A. A., Silahtaroglu, A., et al. (2010). Fluorescence-based codetection with protein markers reveals distinct cellular compartments for altered MicroRNA expression in solid tumors. *Clin. Cancer Res.* 16, 4246–4255. doi: 10.1158/1078-0432.CCR-10-1152

Shi, Z., Johnson, J. J., and Stack, M. S. (2012). Fluorescence *in situ* hybridization for microRNA detection in archived oral cancer tissues. *J. Oncol.* 2012, 903581. doi: 10.1155/2012/903581

Silahtaroglu, A. N., Nolting, D., Dyrskjot, L., Berezikov, E., Moller, M., Tommerup, N., et al. (2007). Detection of microRNAs in frozen tissue sections by fluorescence *in situ* hybridization using locked nucleic acid probes and tyramide signal amplification. *Nat. Protoc.* 2, 2520–2528. doi: 10.1038/nprot.2007.313

Soe, M. J., Moller, T., Dufva, M., and Holmstrom, K. (2011). A sensitive alternative for microRNA *in situ* hybridizations using probes of 2'-O-methyl RNA + LNA. *J. Histochem. Cytochem.* 59, 661–672. doi: 10.1369/0022155411409411

Song, R., Ro, S., and Yan, W. (2010). *in situ* hybridization detection of microRNAs. *Methods Mol. Biol.* 629, 287–294. doi: 10.1007/978-1-60761-657-3_18

Thompson, R. C., Deo, M., and Turner, D. L. (2007). Analysis of microRNA expression by *in situ* hybridization with RNA oligonucleotide probes. *Methods* 43, 153–161. doi: 10.1016/j.ymeth.2007.04.008

Valoczi, A., Hornyik, C., Varga, N., Burgyan, J., Kauppinen, S., and Havelda, Z. (2004). Sensitive and specific detection of microRNAs by northern blot analysis using LNA-modified oligonucleotide probes. *Nucleic Acids Res.* 32, e175. doi: 10.1093/nar/gnh171

Wahlestedt, C., Salmi, P., Good, L., Kela, J., Johnsson, T., Hokfelt, T., et al. (2000). Potent and nontoxic antisense oligonucleotides containing locked nucleic acids. *Proc. Natl. Acad. Sci. U.S.A.* 97, 5633–5638. doi: 10.1073/pnas.97.10.5633

Wang, W. X., Rajeev, B. W., Stromberg, A. J., Ren, N., Tang, G., Huang, Q., et al. (2008). The expression of microRNA miR-107 decreases early in Alzheimer's disease and may accelerate disease progression through regulation of beta-site amyloid precursor protein-cleaving enzyme 1. *J. Neurosci.* 28, 1213–1223. doi: 10.1523/JNEUROSCI.5065-07.2008

Williams, A. E., Moschos, S. A., Perry, M. M., Barnes, P. J., and Lindsay, M. A. (2007). Maternally imprinted microRNAs are differentially expressed during mouse and human lung development. *Dev. Dyn.* 236, 572–580. doi: 10.1002/dvdy.21047

Wu, D., Raafat, M., Pak, E., Hammond, S., and Murashov, A. K. (2011). MicroRNA machinery responds to peripheral nerve lesion in an injury-regulated pattern. *Neuroscience* 190, 386–397. doi: 10.1016/j.neuroscience.2011.06.017

Yamamichi, N., Shimomura, R., Inada, K., Sakurai, K., Haraguchi, T., Ozaki, Y., et al. (2009). Locked nucleic acid *in situ* hybridization analysis of miR-21 expression during colorectal cancer development. *Clin. Cancer Res.* 15, 4009–4016. doi: 10.1158/1078-0432.CCR-08-3257

Yelamanchili, S. V., Chaudhuri, A. D., Chen, L. N., Xiong, H., and Fox, H. S. (2010). MicroRNA-21 dysregulates the expression of MEF2C in neurons in monkey and human SIV/HIV neurological disease. *Cell Death Dis.* 1, e77. doi: 10.1038/cddis.2010.56

You, Y., Moreira, B. G., Behlke, M. A., and Owczarzy, R. (2006). Design of LNA probes that improve mismatch discrimination. *Nucleic Acids Res.* 34, e60. doi: 10.1093/nar/gkl175

Zaidi, A. U., Enomoto, H., Milbrandt, J., and Roth, K. A. (2000). Dual fluorescent *in situ* hybridization and immunohistochemical detection with tyramide signal amplification. *J. Histochem. Cytochem.* 48, 1369–1375. doi: 10.1177/002215540004801007

10

The interplay of microRNA and neuronal activity in health and disease

Stephen M. Eacker[1,2], Ted M. Dawson[1,2,3]* and Valina L. Dawson[1,2,3,4]*

[1] Neuroregeneration and Stem Cell Programs, Institute for Cell Engineering, Johns Hopkins University School of Medicine, Baltimore, MD, USA
[2] Department of Neurology, Johns Hopkins University School of Medicine, Baltimore, MD, USA
[3] Solomon H. Snyder Department of Neuroscience, Johns Hopkins University School of Medicine, Baltimore, MD, USA
[4] Department of Physiology, Johns Hopkins University School of Medicine, Baltimore, MD, USA

Correspondence:
Valina L. Dawson and Ted M. Dawson, Neuroregeneration and Stem Cell Programs, Institute for Cell Engineering, Johns Hopkins University School of Medicine, 733 North Broadway, BRB 731, Baltimore, MD 21205, USA
e-mail: vdawson@jhmi.edu; tdawson@jhmi.edu

MicroRNAs (miRNAs) are small 19–23 nucleotide regulatory RNAs that function by modulating mRNA translation and/or turnover in a sequence-specific fashion. In the nervous system, miRNAs regulate the production of numerous proteins involved in synaptic transmission. In turn, neuronal activity can regulate the production and turnover of miRNA through a variety of mechanisms. In this way, miRNAs and neuronal activity are in a reciprocal homeostatic relationship that balances neuronal function. The miRNA function is critical in pathological states related to overexcitation such as epilepsy and stroke, suggesting miRNA's potential as a therapeutic target. We review the current literature relating the interplay of miRNA and neuronal activity and provide future directions for defining miRNA's role in disease.

Keywords: microRNA, synaptic plasticity, stroke, epilepsy, neuroprotective agents

MicroRNAs are endogenously encoded small RNAs that are processed sequentially into mature 19–23 nucleotide (nt) regulatory molecules (Krol et al., 2010b). Once processed to their mature length, miRNAs are bound by the core component of the RNA-induced silencing complex (RISC), the PIWI-domain containing protein Argonaute (AGO). In mammals, there are four members of the Argonaute family (AGO1–4), all of which are capable of silencing mRNAs in a miRNA-dependent fashion. Of these AGO family members, only AGO2 is capable of directly catalyzing endonucleolytic cleavage of RNA targets, and only when there is complete complementarity between the miRNA and its target (Filipowicz et al., 2008). The AGO2-mediated cleavage of target RNAs is thought to be the primary effector of the exogenously supplied small interfering RNA (siRNA) or short hairpin RNA (shRNA). Most endogenous miRNA–mRNA interactions differ in two important ways (Bartel, 2009). First, most miRNAs share only partial complementarity with their mRNA targets, guiding RISC to its targets through interaction between the 5′-most 6–8 nt of the miRNA (the so-called seed sequence) and its mRNA target, usually in the 3′ untranslated region (3′ UTR). Second, though essential for miRNA-mediated silencing, AGO does not directly cleave mRNA targets. Instead, the AGO–miRNA complex recruits a host of additional factors that result in the silencing of mRNA targets.

The mechanism by which miRNAs silence their mRNA targets remains highly controversial (Filipowicz et al., 2008; Djuranovic et al., 2011). There are two primary camps in the mechanism of silencing debate. One camp believes that mRNA silencing by miRNAs occurs completely by mRNA deadenylation and decay. The other camp does not dispute that RISC can catalyze deadenylation and decay of targets, but believes there is an element of translational repression that precedes the destruction of target messages. Not surprisingly, there are excellent experiments that support both of these models that fuel this ongoing debate in the literature. This debate lies beyond the scope of this review, but remains important especially when considering how miRNAs sculpt neuronal function.

There are numerous ways in which miRNAs modulate the function of the nervous system. A variety of studies have shown how miRNA plays an important role in the development and in a wide array of disorders of the nervous system (Sun et al., 2013). The focus of this review is the interplay of miRNA and neuronal activity.

miRNAs REGULATE STRUCTURAL AND SYNAPTIC PLASTICITY

Regulation of neuronal excitability is a major control point for synaptic plasticity, a fundamental component of learning and memory. Alterations in synaptic function are associated with both structural changes at the synapse and changes in synaptic strength (**Figure 1**). Structural changes at the synapse are commonly associated with alterations in the cytoskeletal function, leading either to the establishment or dissolution of synapses (Bosch and Hayashi, 2012; Rochefort and Konnerth, 2012). Altered synapse morphology can also reflect the strength and stability of a synapse owing to the associated size of the post-synaptic density, the site of synaptic transmission reception. In general, larger, mushroom-shaped spines are associated with larger post-synaptic densities, allowing for more stable and stronger synaptic transmission. This is

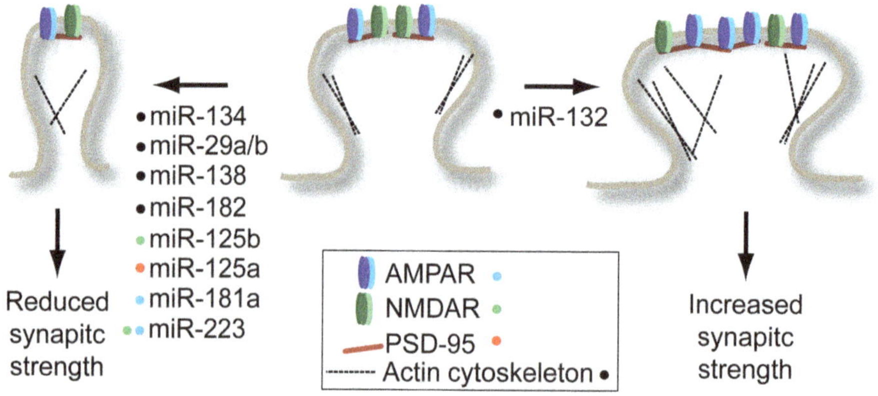

FIGURE 1 | miRNAs regulate structural and synaptic plasticity by targeting synaptic spine proteins. Many miRNAs promote decrease in spine volume and/or glutamate receptor density (left) while only miR-132 has been shown to promote spine maturation (right). Colored dots next to miRNA indicate the target(s) of each miRNA shown to modulate plasticity. See text for details on individual miRNA targets.

in contrast to the more dynamic filopodia-type spines, which are associated with weak or absent synaptic transmission (Yoshihara et al., 2009).

One of the first synaptically enriched miRNAs identified, miR-134, was shown to be a negative regulator of synaptic spine volume (Schratt et al., 2006). This pioneering study showed that miR-134 acts by repressing Limk1 expression, a kinase that regulates spine morphology by regulating ADF/cofilin interactions with the actin cytoskeleton. The reduced spine volume associated with miR-134 overexpression should lead to reduced synaptic strength. Consistent with this hypothesis, mice overexpressing miR-134 show defects in the establishment of long-term potentiation (LTP) in the hippocampus (Gao et al., 2010). However, in this study the authors identified cAMP-response element binding protein (Creb) as a target of miR-134. It is conceivable that both studies found true, but different targets of miR-134: the low level of sequence complementarity required to guide a miRNA to its target means that every miRNA has potentially hundreds of targets. This highlights the somewhat arbitrary process of target identification for miRNAs when performing phenotypic analysis of miRNA overexpression or knockdown. Importantly, both studies found that inhibition of miR-134 increased levels of Limk1 and Creb, respectively, and reversed the synaptic and structural plasticity effects observed when miR-134 is overexpressed.

Structural plasticity-related mRNAs seem to be a prominent target for miRNAs. This observation is supported by a study by Chi et al. (2009) where neural miRNA targets were identified using an unbiased biochemical approach called high-throughput sequencing of RNAs isolated by crosslinking immunoprecipitation (HITS–CLIP). In this approach, the AGO protein was UV-crosslinked to its mRNA targets in a cell suspension derived from the cortex of young mice. The crosslinked mRNA–miRNA–AGO ternary complexes were immunoprecipitated and the RNAs from this purification were subject to high-throughput sequencing. Gene ontology analysis of target mRNAs revealed an enrichment of genes involved in cytoskeleton regulation, particularly overlapping with genes also identified as involved in neuronal differentiation.

The enrichment of miRNA targeting structural plasticity-related genes is found prominently throughout the literature. Actin related protein 2/3 complex, subunit 3 (Arpc3) was identified as a miR-29a/b target in a screen for miRNAs involved in drugs of abuse-related plasticity (Lippi et al., 2011). Overexpression of miR-29a/b increased filopodia-type spines and decreased mushroom-type spines. This spine phenotype is consistent with the phenotype observed when Arpc3 is knocked down. As predicted by the change in morphology of spines, there is a reduction in miniature excitatory post-synaptic current (mEPSC) frequency when miR-29a/b is overexpressed. The activity-induced miR-132 seems to have an opposite impact on structural plasticity. Knockdown of miR-132 activity in newborn neurons in the adult hippocampus or in neurons of the visual cortex leads to a reduction in stable mushroom spines and an increase in filopodia-type spines (Luikart et al., 2011; Mellios et al., 2011). These structural changes were accompanied by decreased frequency, but not amplitude of mEPSCs. Conversely, infusion of miR-132 mimics into the visual cortex following monocular deprivation increased mushroom-type spines and eliminated ocular-dominance associated plasticity (Tognini et al., 2011). The precise level of miR-132 seems to play an important role in plasticity – modest overexpression in the hippocampus improves performance in a Barnes maze test without altering spine density. However, when expressed at supraphysiological levels, miR-132 impairs performance in this memory task, which is accompanied by increased spine density (Hansen et al., 2013). It is worth noting that the target protein(s) that mediate this miR-132-dependent structural plasticity have yet to be identified.

Functional screening of synaptically enriched miRNA identified an interaction between miR-138 and the acyl protein thioesterase 1 (Apt1) mRNA (Siegel et al., 2009). Inhibition of miR-138 was the only synaptically enriched miRNA out of 11 tested to significantly increase the volume of dendritic spines of cultured hippocampal neurons. Surprisingly, this increased volume was accompanied by decreased mEPSC amplitude, which the authors ascribe to a decrease in GluR2 positive clusters found on the dendritic spines. Knockdown of the miR-138 target Apt1 recapitulates the effects on spine density observed in miR-138 overexpression, suggesting that miR-138 acts through Apt1 to regulate spine morphology.

The miRNA regulation of structural plasticity extends beyond the hippocampus and cortex. Auditory fear conditioning downregulates a number of miRNA in the amygdala (Griggs et al., 2013). Bioinformatic analysis suggested that of these downregulated miRNAs, miR-182 could bind to the 3′UTR of a number of key regulators of the actin cytoskeleton in synapses. Infusion of miR-182 mimics into the lateral amygdala led to downregulation of RAC1, cortactin, and to a lesser extent, cofilin. This same infusion protocol impaired long-term amygdala-dependent fear memory suggesting that the miR-182 plays a repressive role in memory formation. However, how precisely miR-182 impacts structural or synaptic plasticity remains to be determined.

In addition to regulating the structural aspects of dendritic spines, a number of miRNAs have been shown to directly regulate components of the post-synaptic density. The synaptically enriched miR-181a can target the GluR2 subunit of the 2-amino-3-(3-hydroxy-5-methyl-isoxazol-4-yl)propanoic acid receptor (AMPAR) through a conserved binding site in its 3′UTR (Saba et al., 2012). Interestingly, miR-181a is induced by dopamine D1/5 receptor agonist SKF-38393 *in vitro* and by amphetamine and cocaine in various anatomical structures *in vivo*. Surface levels of GluR2 are reduced in neurons overexpressing miR-181a; however, only mEPSC frequency and not amplitude is affected. The authors suggest that AMPAR-dependent spine development might be effected, explaining reduced mEPSC frequency. However, the possibility remains that miR-181a may have additional relevant targets in hippocampal neurons.

A number of mRNAs encoding post-synaptic density proteins appear to be shared targets of miRNAs and the fragile-X mental retardation protein (FMRP). The FMRP negatively regulates mRNA translation by directly interacting with target mRNAs. In one study, the authors identified miRNA enriched in FMRP-bound RNA immunoprecipitation experiments (Edbauer et al., 2010). Of these miRNAs enriched on FMRP-bound messages, miR-125b and miR-132 had significant effects on structural and synaptic plasticity when overexpressed. While miR-125b overexpression led to thinner spines and decreased amplitude of mEPSC, miR-132 overexpression led to the formation of short, thicker spines and increased mEPSC amplitude and frequency, roughly consistent with the miR-132 studies described above. Turning their focus to miR-125b, the authors identified the *N*-methyl-D-aspartate (NMDA) receptor 2A (NR2A) as a direct miR-125b target. Overexpression and knockdown of miR-125b had significant impact on the NMDAR currents, consistent with the observed up and downregulation of NR2A.

A second study found that the FMRP-bound PSD-95 transcript is under the regulation of miR-125a (Muddashetty et al., 2011). Interference with miR-125a induces increased PSD-95 and increased spine number. The FMRP-bound transcripts are translationally repressed and in many cases reactivated for translation by activity through the metabotropic glutamate receptors 1 and 5 (mGluR1/5; Darnell and Klann, 2013). Muddashetty et al. (2011) demonstrate that the PSD-95 mRNA is liberated from RISC following stimulation with the mGluR1/5 agonist dihydroxyphenylglycine (DHPG). Interestingly, both the study of Edbauer et al. (2010) and Muddashetty et al. (2011) found that FMRP was required for miRNA-mediated silencing of their targets. Although previous studies in *Drosophila* have shown physical interaction between RISC and FMRP (Ishizuka et al., 2002; Jin et al., 2004), current data suggest that mammalian RISC and FMRP do not physically interact. More likely, FMRP acts to create a permissive environment on the mRNA that allows RISC to bind and silence its targets. One way in which FMRP may accomplish this is by stalling translating ribosomes (Darnell et al., 2011), preventing ribosomes from dislodging RISC from coding sequence target sites. While most effective miRNA target sites are in the 3′UTR of genes, there is evidence of extensive RISC binding in coding regions (Chi et al., 2009; Helwak et al., 2013). Alternatively, FMRP binding in 3′UTRs could occlude the binding of other RNA-binding proteins that could otherwise dislodge RISC from its target.

NEURONAL ACTIVITY ALTERS miRNA BIOGENESIS

The miRNA biogenesis is a multiple-step process that begins with the transcription of a primary miRNA transcript (pri-miRNA; Krol et al., 2010b). The pri-miRNA can be derived from a noncoding transcript containing one or many miRNAs, or can be processed from intronic sequences. A hairpin structure containing the mature miRNA sequence is recognized by the microprocessor complex, which cleaves the hairpin out of the context of the pri-miRNA transcript, yielding a miRNA precursor (pre-miRNA). The 60–80 nt pre-miRNA is processed by the RNase III protein Dicer to produce a double-stranded 19–23 bp miRNA. Then one strand is selectively loaded into AGO, yielding a mature miRNA engaged in RISC. Each of these steps are potential control points for the regulation of the cellular miRNA milieu. In this section, we will review how neuronal activity regulates the steps of miRNA biogenesis (**Figure 2**).

Activity has a strong influence on the transcriptional state of neurons. Creb is one of the primary transcriptional activators that respond to neuronal activity. In a genome-wide screen for Creb binding sites, Vo et al. (2005) identified two consensus binding sites near the miR-212/miR-132 locus. Stimulation of primary neurons with brain-derived neurotrophic factor (BDNF; Vo et al., 2005), KCl depolarization, or bicuculline (Wayman et al., 2008) all induce the production of pri-miR-132 transcript, making it the first recognized miRNA whose expression is regulated by neuronal activation. These observations have been supported by numerous *in vivo* investigations (Nudelman et al., 2010; Eacker et al., 2011; Mellios et al., 2011; Tognini et al., 2011; Wang et al., 2013) among others. Regulation through the cAMP response elements in the putative promoter region of the miR-212/miR-132 cluster were further confirmed by the activity-dependent acquisition of transcription-promoting chromatin marks in the visual cortex after visual stimulation (Tognini et al., 2011). Coupled with the potent effects of miR-132 on structural and synaptic plasticity (see above), it's activity-dependent transcriptional control suggests that miR-132 is a potent regulator of experience-dependent plasticity *in vivo*.

A second, larger cluster of miRNAs, the miR-379~410 cluster, has also shown to be upregulated by neuronal activity (Fiore

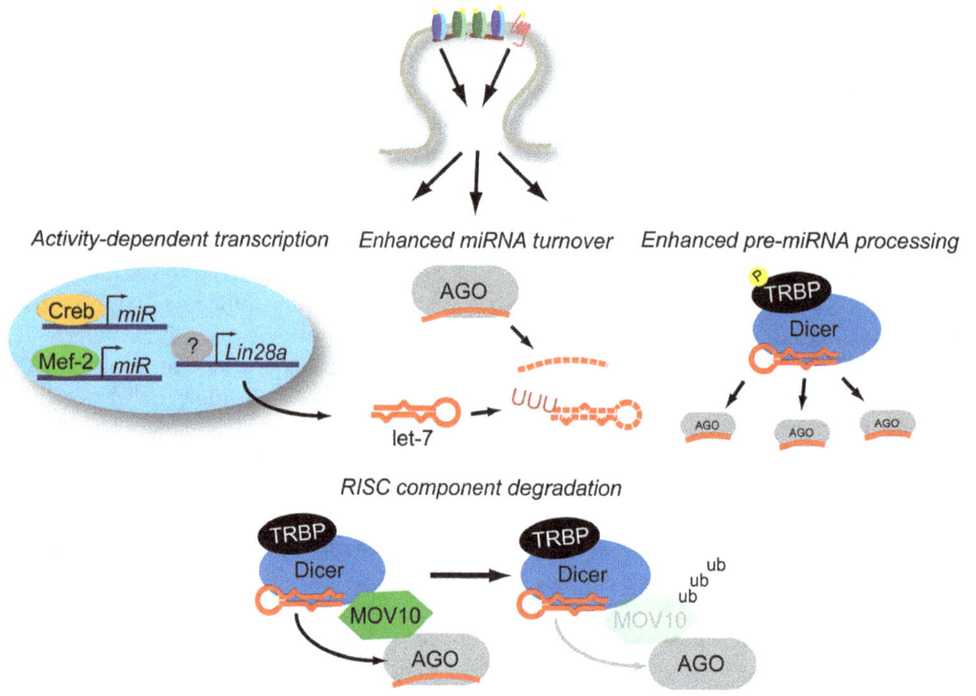

FIGURE 2 | Neuronal activity influences miRNA expression through multiple mechanisms. Activity-dependent transcription induces miRNA production through Mef2 and Creb while Lin28a increases expression through an uncertain combination of transcription-dependent and -independent mechanisms. Increased levels of Lin28a lead to destabilization of pre-let-7 transcripts through poly-uridynylation. Activity also promotes accelerated miRNA turnover through an additional unknown mechanism(s). Activity-dependent phosphorylation of TRBP stabilizes Dicer, leading to increased mature miRNA production. NMDAR-activation promotes the ubiquitinylation of MOV10, a component of RISC, which may lead to decreased levels of mature-miRNA loaded RISC. See text for details.

et al., 2009). This miRNA cluster is regulated both by BDNF and KCl depolarization via the myocyte enhancing factor 2 (Mef2) transcription factor. Among the many miRNAs transcribed from this locus is miR-134, a known regulator of structural plasticity through Limk1. In addition to directly regulating Limk1, this study suggests miR-134 may play a broader role in regulating protein synthesis by regulating the RNA binding protein Pumilio 2 (Pum2).

Increased expression of pri-miRNA transcript does not necessarily result in increased mature miRNA levels. One striking example of how neuronal activity can influence mature miRNA levels independently of transcription is through the induction the Lin28a RNA binding protein by BDNF (Huang et al., 2012). The Lin28 family members can selectively impair the processing of both pri- and pre-let-7 miRNA, though Lin28a primarily targets pre-let-7 through the post-transcriptional addition of poly-uridine to the 3′ end of the transcript (Thornton and Gregory, 2012). In response to bath application of BDNF in the presence of actinomycin D, hippocampal neurons show increased expression of Lin28a leading to a decrease in mature let-7 levels, resulting to increased translation of let-7 target RNAs. Under the same stimulating conditions, Huang et al. (2012) observed an increase in pre-miRNA processing of non-let-7 miRNA. This was the result of increased phosphorylation of the Dicer binding partner transactivation response RNA-binding protein (TRBP). The extracellular signal-regulated kinase (ERK)-dependent phosphorylation of TRBP stabilizes Dicer (Paroo et al., 2009), and in the context of BDNF stimulation leads to increased levels of pre-miRNA processing.

While activity induced by BDNF may lead to increased processing of pre-miRNA, activity through glutamate receptors seems to accelerate mature miRNA turnover (Krol et al., 2010a). Using a variety of transcriptional inhibitors, Krol et al. (2010a) demonstrated that neuronal miRNAs have an unusually high turnover rate compared to other cell types. The turnover rate was further accelerated by addition of glutamate and decelerated by the application of to tetrodotoxin (TTX). These findings are strikingly similar to those made in *Aplysia* where application of serotonin results in the rapid degradation of miR-124 (Rajasethupathy et al., 2009). The mechanism for this rapid, activity-dependent miRNA turnover remains unclear. However, it is worth noting that at timepoints distal to neuronal activation, there are many more miRNAs that are downregulated than upregulated over a number of array-based studies (Eacker et al., 2011; Jimenez-Mateos et al., 2011; Griggs et al., 2013; Risbud and Porter, 2013). This activity-dependent miRNA turnover may allow neurons to rapidly reprogram RISC in a global manner in a way that promotes synaptic plasticity.

There is growing evidence that pre-miRNA processing can occur in dendrites. Biochemical purification of RISC components from synaptosomes showed detectable levels of pre-miRNA (Lugli et al., 2008). Recently, *in situ* hybridization methods that allow for specific detection of pre-miRNA have lent credence to the possibility that dendritically localized pre-miRNA may be an important phenomenon (Bicker et al., 2013). Using a clever biochemical approach, the authors of this study identified DHX36, a DExH-box RNA helicase as an interactor with the loop region of the dendritically localized pre-miR-134 transcript. Knockdown of DHX36 reduced dendritic transport of pre-miR-134 and enhanced the translation of reporters of miR-134 activity. The authors propose that DHX36 stabilizes pre-miR-134 for dendritic processing, perhaps in an activity-dependent manner, though this remains to be determined.

The *Drosophila* protein Armitage (Armi) and its mammalian homolog MOV10 are both implicated in the activity-dependent relief of miRNA repression (Ashraf et al., 2006; Banerjee et al., 2009). Armi and MOV10 are DExH-box RNA helicases that have been shown to be components of RISC (Tomari et al., 2004; Meister et al., 2005). In the case of *Drosophila*, Armi is degraded via the ubiquitin proteasome following activation of the nicotinic acetylcholine receptor. In mammals, MOV10 is also degraded in a proteosomal-dependent manner in response to NMDA receptor activation (Banerjee et al., 2009; Jarome et al., 2011). In neurons, degradation of MOV10/Armi relieves miRNA-mediated repression by an unknown mechanism, allowing for the translation of mRNAs involved in synaptic plasticity (Ashraf et al., 2006; Banerjee et al., 2009). Biochemical data concerning Armi's function suggest that it is required for the loading of AGO with mature miRNAs following Dicer processing (Tomari et al., 2004). However, recent genome-wide studies have identified numerous promiscuous interactions between polyadenylated mRNAs and MOV10 (Castello et al., 2012; Sievers et al., 2012), suggesting additional potential interactions of MOV10 and miRNA-mediated silencing. More research is required to elucidate this potentially important interface between the relief of miRNA-mediated repression and activity-dependent protein synthesis.

miRNAs FUNCTION IN DISEASES OF NEURONAL OVEREXCITATION

Excessive neuronal activity is associated with neuronal cell death in a number of contexts. During stroke, neuronal depolarization leads to excess glutamate release that cannot be compensated for by normal reuptake mechanisms. The resulting excess glutamate results in glutamate receptor hyperactivation and excitotoxic cell death via excess calcium influx. As miRNAs are potent regulators of the cellular stress response (Leung and Sharp, 2010; Mendell and Olson, 2012) and neuronal excitability (see above), they are a logical target for the investigation and enhancement of intrinsic neuroprotective pathways (**Figure 3**). A fair amount of effort has been made toward understanding the global miRNA response to stroke in rodent models. These studies generally rely on comparisons of RNA samples from stroke and sham brains, and subject them to some type of high-throughput array. Not surprisingly, the resulting miRNA expression profiles between different experimental stroke conditions, different laboratories, and different array platforms show almost no overlap. Despite this lack of overlap, there is valuable information that can be gained from a brief review of these studies.

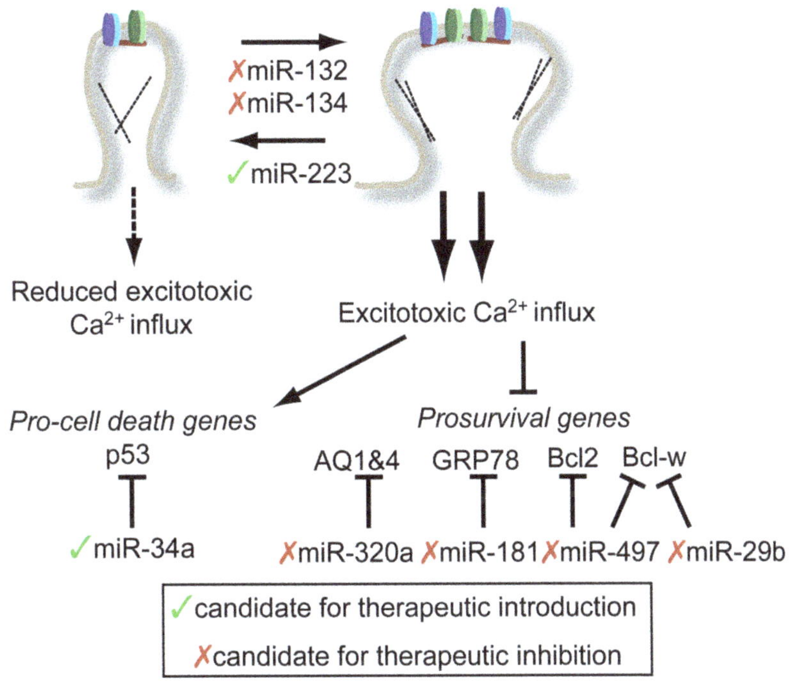

FIGURE 3 | Potential targets for miRNA-mediated therapies in the treatment of stroke and epilepsy. Potential interventions target miRNAs to either reduce excitotoxic calcium influx (top) or reduce the impact of excessive calcium influx (bottom). Potential miRNA targets and be influenced either by introduction therapeutic inhibitors (red x) or mimics (green check). See text for details. Note: Conflicting results of miR-134 inhibition on spine morphology with embryonic neuronal studies described above. See text for details.

One of the best studies that highlights the complexity of the miRNA response to CNS injury was performed on both blood and brain tissue from rats after ischemic stroke, intracerebral hemorrhage, or kainic acid-induced excitotoxicity (Liu et al., 2010). In this study, relatively few miRNAs showed consistent changes in expression in either brain tissue or blood. However, three miRNAs found in the blood (miR-155, miR-298, and miR-362-3p) change expression greater than twofold in response to some of the injuries. This suggests that expression of some miRNAs might be useful biomarkers to identify subtypes of CNS injuries.

In two related studies, the miRNA cortex of mice treated with middle cerebral artery occlusion (MCAO) model of stroke were profiled either by PCR 24 h post-stroke (Yin et al., 2010) or by array over an extended time course (Dharap et al., 2009). There was no significant overlap found between these two studies, perhaps because of methodological differences. Each study did identify some miRNAs that showed reproducible changes in their model. In one case, the authors found that miR-145 showed significant and enduring upregulation following stroke (Dharap et al., 2009). Inhibition of miR-145 with chemically modified antisense oligonucleotides (so-called antagomirs, or anti-miRs) lead to upregulation of superoxide dismutase 2, a predicted miR-145 target gene. Whether miR-145 inhibition had any therapeutic benefit was not determined. In the second study, the authors identified miR-497 as an upregulated miRNA in cortex following MCAO and in neuroblastoma cells subjected to oxygen–glucose deprivation (OGD), a cell culture model of stroke (Yin et al., 2010). Through a series of experiments, the authors show that Bcl-2 and Bcl-w, two anti-apoptotic molecules are targets of mIR-497. Infusion with antagomirs targeting miR-497 prior to MCAO resulted in reduced infarct volume and reduction in the severity of neurological deficits. Similar results were observed for mIR-29b, another miRNA which targets Bcl-w (Shi et al., 2012). The authors observed an increase in miR-29b following MCAO and showed that simple overexpression of miR-29b lead to spontaneous neuronal cell death in a Bcl-w-dependent manner.

Another approach toward finding potentially therapeutic miRNAs for treating stroke is to work backward from a known therapeutic target or pathway and identify miRNA interactors. GRP78 (also known as BIP) is a chaperone that is primarily localized to the ER and plays a key role in the ER-stress response. Ouyang et al. (2012) found that decreased GRP78 levels were accompanied by increased miR-181 in the core of MCAO-induced infarctions. After establishing that miR-181 could directly target GRP78, the author demonstrated that inhibition of mIR-181 through pretreatment with antisense oligo could significantly reduce infarct volume following MCAO. A similar approach identified miR-320a as a regulator of aquaporins 1 and 4 (AQ1 and AQ4) (Sepramaniam et al., 2010). Reduced expression of AQ1 and 4 in actrocytes is associated cerebral edema. By inhibiting miR-320a with antisense oligonucleotides, there was an increase in AQ1 and 4 expressions and a reduction in MCAO infarct volume.

Our lab recently investigated the role of miR-223 in neuroprotection motivated by the disproportionate number of nervous system targets predicted to be miR-223 targets (Harraz et al., 2012). Bioinformatic predictions identified that miR-223 should target glutamate receptor subunits NR2B and GluR2, making it a candidate for protection against excitotoxic insult. Targeted mutation of miR-223 leads increased levels of GluR2 and NR2B in the hippocampus, but not other post-synaptic proteins, and increased mEPSC decay time and amplitude. Consistent with these findings, miR-223 knockdown increased NMDAR-dependent calcium influx while overexpression decreased influx. Overexpression of miR-223 was protective across brain regions, protecting the striatum from NMDA-induced excitotoxicity and hippocampus from bilateral common carotid artery occlusion. Importantly miR-223 mutant mice were highly sensitized to both of these injuries, establishing a definitive role for miR-223 in the endogenous neuroprotective program.

Excitotoxicity also plays an important role in cell death associated with temporal lobe epilepsy (TLE), a chronic intractable form of epilepsy. TLE can be modeled in rodents by injection of pilocarpine, a muscarinic acetylcholine receptor agonist or injection of kainic acid, an agonist of the kainate-type glutamate receptor. Injection of either of these compounds results in establishment of *status epilepticus* (SE) in rodents. A number of studies have profiled miRNA from the hippocampus of mice following establishment of SE. Two studies from different research groups using kainic acid (Sano et al., 2012) and pilocarpine (Hu et al., 2012) model, respectively, found SE-associated upregulation of miR-34a in the hippocampus. MiR-34a expression is associated with p53-dependent pro-apoptotic program (Chang et al., 2007), making its upregulation by SE an attractive target for TLE therapy. Though both groups observed elevated caspase 3 cleavage associated with overexpression of miR-34a, only Hu et al. (2012) observed a protective effect by inhibiting miR-34a. Though methodological differences may explain some of the differences between these results, a robust and reproducible protective effect would be necessary to warrant further investigation of miR-34a in TLE therapeutics.

Two activity-regulated miRNAs, miR-132 and miR-134, are both upregulated in mouse models of TLE (Jimenez-Mateos et al., 2011, 2012). Interestingly, miR-132 induction in this TLE model is suppressed when mice are preconditioned with a low, peripherally administered dose of kainic acid (Jimenez-Mateos et al., 2011). This suggests that impairing miR-132 expression, by whatever means, may be neuroprotective. Indeed this was the case: inhibition of miR-132 by infusion of miR-132 antagomirs before kainic acid injection reduced cell death in the CA3 subfield of the hippocampus. Given the evidence from the synaptic plasticity field regarding miR-132's ability to stimulate stable mushroom-type dendritic spines suggests that anti-miR-132 treatments may reduce cell death by reducing hippocampal neuron's excitability, and therefore susceptibility to excitotoxicity. Similarly, miR-134 inhibition with antagomirs prior to SE induction provided neuroprotection. However, unlike miR-132 inhibition, a single injection of miR-134 antagomir provided long-lasting inhibition of recurrent spontaneous seizures (Jimenez-Mateos et al., 2012). Building on the work of Schratt et al. (2006), this study demonstrated that anti-miR-134 treatment reduced spine density *in vivo*, likely acting via Limk1. The long-lasting nature of the anti-miR-134 protection suggest that miR-134-based therapeutic may hold promise in treating intractable TLE.

CONCLUSIONS AND FUTURE STUDIES

In the relative short time since their discovery, the miRNA field has made a dramatic impact on our understanding of nervous system function. Insights concerning the mechanisms of synaptic plasticity have been of particular interest, potentially coupling activity-dependent synaptic plasticity and protein synthesis. The local translation of proteins has long been considered a mechanism for regulating the connections at individual synapses and miRNA may be a key regulator at this point (Sutton and Schuman, 2006). However attractive this hypothesis is, there have been no clear demonstrations that miRNA can regulate local, activity-dependent protein synthesis. Furthermore, while there is some evidence that miRNA can sequester mRNAs from translating ribosomes, it is not clear that sequestered mRNA can undergo reactivation. In some studies, investigators observe a decrease in miRNA-targeted mRNA levels consistent with the deadenylation and decay model while other see only a change in protein levels, more consistent with a sequestration model. While a difficult problem to interrogate, understanding local control of protein synthesis by miRNA should be a central focus of future investigations.

As discussed above, individual miRNAs can have potentially hundreds of targets owing to the small amount of complementarity required to establish miRNA–mRNA interactions. In virtually all studies, there is a focus on a one miRNA–one target interaction. This is largely for practical purposes – requirements for publication dictate a laser-like focus on a making a bulletproof set of observations. While editors and reviewers require this type of focus, miRNAs do not have such single-minded focus (Baek et al., 2008; Selbach et al., 2008). Having a more comprehensive view of how individual miRNAs regulate the proteome will be necessary, especially if they are to be used in therapeutics.

Virtually all the studies described in this review rely on antisense oligonucleotides or antagomirs to conduct loss-of-function experiments. While these reagents are widely used and proven effective under many circumstances, virtually nothing is known about their distribution and perdurance in the CNS following intracranial injection, making interpretation of how they work problematic. A reasonable complement to antagomir application would the use of traditional targeted mutations, which are becoming increasingly more available to the general scientific research world (Park et al., 2012).

Finally, the most successful candidate miRNAs for translation to the clinic have arisen from rigorous work in the basic sciences. Both miR-132 and miR-134 have a long track record of careful investigation in the basic synaptic plasticity field and have now been shown to have potential in the treatment of epilepsy. This cross-pollination of basic and translational science seems to be the most fruitful way forward for the future of miRNA-based therapeutics.

ACKNOWLEDGMENTS

We apologize to our colleagues for whose results there was not enough space to be reviewed in this manuscript. Funding support: This work was supported by the NIH/NIDA DA000266. Ted M. Dawson is the Leonard and Madlyn Abramson Professor in Neurodegenerative Diseases.

REFERENCES

Ashraf, S. I., McLoon, A. L., Sclarsic, S. M., and Kunes, S. (2006). Synaptic protein synthesis associated with memory is regulated by the RISC pathway in *Drosophila*. *Cell* 124, 191–205. doi: 10.1016/j.cell.2005.12.017

Baek, D., Villén, J., Shin, C., Camargo, F. D., Gygi, S. P., and Bartel, D. P. (2008). The impact of microRNAs on protein output. *Nature* 455, 64–71. doi: 10.1038/nature07242

Banerjee, S., Neveu, P., and Kosik, K. S. (2009). A coordinated local translational control point at the synapse involving relief from silencing and MOV10 degradation. *Neuron* 64, 871–884. doi: 10.1016/j.neuron.2009.11.023

Bartel, D. P. (2009). MicroRNAs: target recognition and regulatory functions. *Cell* 136, 215–233. doi: 10.1016/j.cell.2009.01.002

Bicker, S., Khudayberdiev, S., Weiss, K., Zocher, K., Baumeister, S., and Schratt, G. (2013). The DEAH-box helicase DHX36 mediates dendritic localization of the neuronal precursor-microRNA-134. *Genes Dev.* 27, 991–996. doi: 10.1101/gad.211243.112

Bosch, M., and Hayashi, Y. (2012). Structural plasticity of dendritic spines. *Curr. Opin. Neurobiol.* 22, 383–388. doi: 10.1016/j.conb.2011.09.002

Castello, A., Fischer, B., Eichelbaum, K., Horos, R., Beckmann, B. M., Strein, C., et al. (2012). Insights into RNA biology from an atlas of mammalian mRNA-binding proteins. *Cell* 149, 1393–1406. doi: 10.1016/j.cell.2012.04.031

Chang, T.-C., Wentzel, E. A., Kent, O. A., Ramachandran, K., Mullendore, M., Lee, K. H., et al. (2007). Transactivation of miR-34a by p53 broadly influences gene expression and promotes apoptosis. *Mol. Cell* 26, 745–752. doi: 10.1016/j.molcel.2007.05.010

Chi, S. W., Zang, J. B., Mele, A., and Darnell, R. B. (2009). Argonaute HITS-CLIP decodes microRNA–mRNA interaction maps. *Nature* 460, 479–486. doi: 10.1038/nature08170

Darnell, J. C., and Klann, E. (2013). The translation of translational control by FMRP: therapeutic targets for FXS. *Nat Neurosci.* doi: 10.1038/nn.3379 [Epub ahead of print].

Darnell, J. C., Van Driesche, S. J., Zhang, C., Hung, K. Y. S., Mele, A., Fraser, C. E., et al. (2011). FMRP stalls ribosomal translocation on mRNAs linked to synaptic function and autism. *Cell* 146, 247–261. doi: 10.1016/j.cell.2011.06.013

Dharap, A., Bowen, K., Place, R., Li, L.-C., and Vemuganti, R. (2009). Transient focal ischemia induces extensive temporal changes in rat cerebral microRNAome. *J. Cereb. Blood Flow Metab.* 29, 675–687. doi: 10.1038/jcbfm.2008.157

Djuranovic, S., Nahvi, A., and Green, R. (2011). A parsimonious model for gene regulation by miRNAs. *Science* 331, 550–553. doi: 10.1126/science.1191138

Eacker, S. M., Keuss, M. J., Berezikov, E., Dawson, V. L., and Dawson, T. M. (2011). Neuronal activity regulates hippocampal miRNA expression. *PLoS ONE* 6:e25068. doi: 10.1371/journal.pone.0025068

Edbauer, D., Neilson, J. R., Foster, K. A., Wang, C.-F., Seeburg, D. P., Batterton, M. N., et al. (2010). Regulation of synaptic structure and function by FMRP-associated MicroRNAs miR-125b and miR-132. *Neuron* 65, 373–384. doi: 10.1016/j.neuron.2010.01.005

Filipowicz, W., Bhattacharyya, S. N., and Sonenberg, N. (2008). Mechanisms of post-transcriptional regulation by microRNAs: are the answers in sight? *Nat. Rev. Genet.* 9, 102–114. doi: 10.1038/nrg2290

Fiore, R., Khudayberdiev, S., Christensen, M., Siegel, G., Flavell, S. W., Kim, T. K., et al. (2009). Mef2-mediated transcription of the miR379-410 cluster regulates activity-dependent dendritogenesis by fine-tuning Pumilio2 protein levels. *EMBO J.* 28, 697–710. doi: 10.1038/emboj.2009.10

Gao, J., Wang, W.-Y., Mao, Y.-W., Gräff, J., Guan, J.-S., Pan, L., et al. (2010). A novel pathway regulates memory and plasticity via SIRT1 and miR-134. *Nature* 466, 1105–1109. doi: 10.1038/nature09271

Griggs, E. M., Young, E. J., Rumbaugh, G., and Miller, C. A. (2013). MicroRNA-182 regulates amygdala-dependent memory formation. *J. Neurosci.* 33, 1734–1740. doi: 10.1523/JNEUROSCI.2873-12.2013

Hansen, K. F., Karelina, K., Sakamoto, K., Wayman, G. A., Impey, S., and Obrietan, K. (2013). miRNA-132: a dynamic regulator of cognitive capacity. *Brain Struct. Funct.* 218, 817–831. doi: 10.1007/s00429-012-0431-4

Harraz, M. M., Eacker, S. M., Wang, X., Dawson, T. M., and Dawson, V. L. (2012). MicroRNA-223 is neuroprotective by targeting glutamate receptors. *Proc. Natl. Acad. Sci. U.S.A.* 109, 18962–18967. doi: 10.1073/pnas.1121288109

Helwak, A., Kudla, G., Dudnakova, T., and Tollervey, D. (2013). Mapping the human miRNA interactome by CLASH reveals frequent noncanonical binding. *Cell* 153, 654–665. doi: 10.1016/j.cell.2013.03.043

Hu, K., Xie, Y.-Y., Zhang, C., Ouyang, D.-S., Long, H.-Y., Sun, D.-N., et al. (2012). MicroRNA expression profile of the hippocampus in a rat model of temporal lobe epilepsy and miR-34a-targeted neuroprotection against hippocampal neurone cell apoptosis post-status epilepticus. *BMC Neurosci.* 13:115. doi: 10.1186/1471-2202-13-115

Huang, Y.-W. A., Ruiz, C. R., Eyler, E. C. H., Lin, K., and Meffert, M. K. (2012). Dual regulation of miRNA biogenesis generates target specificity in neurotrophin-induced protein synthesis. *Cell* 148, 933–946. doi: 10.1016/j.cell.2012.01.036

Ishizuka, A., Siomi, M. C., and Siomi, H. (2002). A *Drosophila* fragile X protein interacts with components of RNAi and ribosomal proteins. *Genes Dev.* 16, 2497–2508. doi: 10.1101/gad.1022002

Jarome, T. J., Werner, C. T., Kwapis, J. L., and Helmstetter, F. J. (2011). Activity dependent protein degradation is critical for the formation and stability of fear memory in the amygdala. *PLoS ONE* 6:e24349. doi: 10.1371/journal.pone.0024349

Jimenez-Mateos, E. M., Bray, I., Sanz-Rodriguez, A., Engel, T., McKiernan, R. C., Mouri, G., et al. (2011). miRNA expression profile after status epilepticus and hippocampal neuroprotection by targeting miR-132. *Am. J. Pathol.* 179, 2519–2532. doi: 10.1016/j.ajpath.2011.07.036

Jimenez-Mateos, E. M., Engel, T., Merino-Serrais, P., McKiernan, R. C., Tanaka, K., Mouri, G., et al. (2012). Silencing microRNA-134 produces neuroprotective and prolonged seizure-suppressive effects. *Nat. Med.* 18, 1087–1094. doi: 10.1038/nm.2834

Jin, P., Zarnescu, D. C., Ceman, S., Nakamoto, M., Mowrey, J., Jongens, T. A., et al. (2004). Biochemical and genetic interaction between the fragile X mental retardation protein and the microRNA pathway. *Nature* 7, 113–117.

Krol, J., Busskamp, V., Markiewicz, I., Stadler, M. B., Ribi, S., Richter, J., et al. (2010a). Characterizing light-regulated retinal microRNAs reveals rapid turnover as a common property of neuronal microRNAs. *Cell* 141, 618–631. doi: 10.1016/j.cell.2010.03.039

Krol, J., Loedige, I., and Filipowicz, W. (2010b). The widespread regulation of microRNA biogenesis, function and decay. *Nat. Rev. Genet.* 11, 597–610. doi: 10.1038/nrg2843

Leung, A. K. L., and Sharp, P. A. (2010). MicroRNA functions in stress responses. *Mol. Cell* 40, 205–215. doi: 10.1016/j.molcel.2010.09.027

Lippi, G., Steinert, J. R., Marczylo, E. L., D'Oro, S., Fiore, R., Forsythe, I. D., et al. (2011). Targeting of the Arpc3 actin nucleation factor by miR-29a/b regulates dendritic spine morphology. *J. Cell Biol.* 194, 889–904. doi: 10.1083/jcb.201103006

Liu, D.-Z., Tian, Y., Ander, B. P., Xu, H., Stamova, B. S., Zhan, X., et al. (2010). Brain and blood microRNA expression profiling of ischemic stroke, intracerebral hemorrhage, and kainate seizures. *J. Cereb. Blood Flow Metab.* 30, 92–101. doi: 10.1038/jcbfm.2009.186

Lugli, G., Torvik, V. I., Larson, J., and Smalheiser, N. R. (2008). Expression of microRNAs and their precursors in synaptic fractions of adult mouse forebrain. *J. Neurochem.* 106, 650–661. doi: 10.1111/j.1471-4159.2008.05413.x

Luikart, B. W., Bensen, A. L., Washburn, E. K., Perederiy, J. V., Su, K. G., Li, Y., et al. (2011). miR-132 mediates the integration of newborn neurons into the adult dentate gyrus. *PLoS ONE* 6:e19077. doi: 10.1371/journal.pone.0019077

Meister, G., Landthaler, M., Peters, L., Chen, P. Y., Urlaub, H., Lührmann, R., et al. (2005). Identification of novel argonaute-associated proteins. *Curr. Biol.* 15, 2149–2155. doi: 10.1016/j.cub.2005.10.048

Mellios, N., Sugihara, H., Castro, J., Banerjee, A., Le, C., Kumar, A., et al. (2011). miR-132, an experience-dependent microRNA, is essential for visual cortex plasticity. *Nat. Neurosci.* 14, 1240–1242. doi: 10.1038/nn.2909

Mendell, J. T., and Olson, E. N. (2012). MicroRNAs in stress signaling and human disease. *Cell* 148, 1172–1187. doi: 10.1016/j.cell.2012.02.005

Muddashetty, R. S., Nalavadi, V. C., Gross, C., Yao, X., Xing, L., Laur, O., et al. (2011). Reversible inhibition of PSD-95 mRNA translation by miR-125a, FMRP phosphorylation, and mGluR signaling. *Mol. Cell* 42, 673–688. doi: 10.1016/j.molcel.2011.05.006

Nudelman, A. S., DiRocco, D. P., Lambert, T. J., Garelick, M. G., Le, J., Nathanson, N. M., et al. (2010). Neuronal activity rapidly induces transcription of the CREB-regulated microRNA-132, in vivo. *Hippocampus* 20, 492–498. doi: 10.1002/hipo.20646

Ouyang, Y.-B., Lu, Y., Yue, S., Xu, L.-J., Xiong, X.-X., White, R. E., et al. (2012). miR-181 regulates GRP78 and influences outcome from cerebral ischemia in vitro and in vivo. *Neurobiol. Dis.* 45, 555–563. doi: 10.1016/j.nbd.2011.09.012

Park, C. Y., Jeker, L. T., Carver-Moore, K., Oh, A., Liu, H. J., Cameron, R., et al. (2012). A resource for the conditional ablation of microRNAs in the mouse. *Cell Rep.* 1, 385–391. doi: 10.1016/j.celrep.2012.02.008

Paroo, Z., Ye, X., Chen, S., and Liu, Q. (2009). Phosphorylation of the human microRNA-generating complex mediates MAPK/Erk signaling. *Cell* 139, 112–122. doi: 10.1016/j.cell.2009.06.044

Rajasethupathy, P., Fiumara, F., Sheridan, R., Betel, D., Puthanveettil, S. V., Russo, J. J., et al. (2009). Characterization of small RNAs in aplysia reveals a role for miR-124 in constraining synaptic plasticity through CREB. *Neuron* 63, 803–817. doi: 10.1016/j.neuron.2009.05.029

Risbud, R. M., and Porter, B. E. (2013). Changes in MicroRNA expression in the whole hippocampus and hippocampal synaptoneurosome fraction following pilocarpine induced status epilepticus. *PLoS ONE* 8:e53464. doi: 10.1371/journal.pone.0053464

Rochefort, N. L., and Konnerth, A. (2012). Dendritic spines: from structure to in vivo function. *EMBO Rep.* 13, 699–708. doi: 10.1038/embor.2012.102

Saba, R., Storchel, P. H., Aksoy-Aksel, A., Kepura, F., Lippi, G., Plant, T. D., et al. (2012). Dopamine-regulated MicroRNA MiR-181a controls GluA2 surface expression in hippocampal neurons. *Mol. Cell. Biol.* 32, 619–632. doi: 10.1128/MCB.05896-11

Sano, T., Reynolds, J. P., Jimenez-Mateos, E. M., Matsushima, S., Taki, W., and Henshall, D. C. (2012). MicroRNA-34a upregulation during seizure-induced neuronal death. *Cell Death Dis.* 3, e287. doi: 10.1038/cddis.2012.23

Schratt, G. M., Tuebing, F., Nigh, E. A., Kane, C. G., Sabatini, M. E., Kiebler, M., et al. (2006). A brain-specific microRNA regulates dendritic spine development. *Nature* 439, 283–289. doi: 10.1038/nature04367

Selbach, M., Schwanhausser, B., Thierfelder, N., Fang, Z., Khanin, R., and Rajewsky, N. (2008). Widespread changes in protein synthesis induced by microRNAs. *Nature* 455, 58–63. doi: 10.1038/nature07228

Sepramaniam, S., Armugam, A., Lim, K. Y., Karolina, D. S., Swaminathan, P., Tan, J. R., et al. (2010). MicroRNA 320a functions as a novel endogenous modulator of aquaporins 1 and 4 as well as a potential therapeutic target in cerebral ischemia. *J. Biol. Chem.* 285, 29223–29230. doi: 10.1074/jbc.M110.144576

Shi, G., Liu, Y., Liu, T., Yan, W., Liu, X., Wang, Y., et al. (2012). Upregulated miR-29b promotes neuronal cell death by inhibiting Bcl2L2 after ischemic brain injury. *Exp. Brain Res.* 216, 225–230. doi: 10.1007/s00221-011-2925-3

Siegel, G., Obernosterer, G., Fiore, R., Oehmen, M., Bicker, S., Christensen, M., et al. (2009). A functional screen implicates microRNA-138-dependent regulation of the depalmitoylation enzyme APT1 in dendritic spine morphogenesis. *Nat. Cell Biol.* 11, 705–716. doi: 10.1038/ncb1876

Sievers, C., Schlumpf, T., Sawarkar, R., Comoglio, F., and Paro, R. (2012). Mixture models and wavelet transforms reveal high confidence RNA–protein interaction sites in MOV10 PAR-CLIP data. *Nucleic Acids Res.* 40, e160. doi: 10.1093/nar/gks697

Sun, A. X., Crabtree, G. R., and Yoo, A. S. (2013). MicroRNAs: regulators of neuronal fate. *Curr. Opin. Cell Biol.* 25, 215–221. doi: 10.1016/j.ceb.2012.12.007

Sutton, M. A., and Schuman, E. M. (2006). Dendritic protein synthesis, synaptic plasticity, and memory. *Cell* 127, 49–58. doi: 10.1016/j.cell.2006.09.014

Thornton, J. E., and Gregory, R. I. (2012). How does Lin28 let-7 control development and disease? *Trends Cell Biol.* 22, 474–482. doi: 10.1016/j.tcb.2012.06.001

Tognini, P., Putignano, E., Coatti, A., and Pizzorusso, T. (2011). Experience-dependent expression of miR-132 regulates ocular dominance plasticity. *Nat. Neurosci.* 14, 1237–1239. doi: 10.1038/nn.2920

Tomari, Y., Du, T., Haley, B., Schwarz, D. S., Bennett, R., Cook, H. A., et al. (2004). RISC assembly defects in the *Drosophila* RNAi mutant armitage. *Cell* 116, 831–841. doi: 10.1016/S0092-8674(04)00218-1

Vo, N., Klein, M. E., Varlamova, O., Keller, D. M., Yamamoto, T., Goodman, R. H., et al. (2005). A cAMP-response element binding protein-induced microRNA regulates neuronal morphogenesis. *Proc. Natl. Acad. Sci. U.S.A.* 102, 16426–16431. doi: 10.1073/pnas.0508448102

Wang, R.-Y., Phang, R.-Z., Hsu, P.-H., Wang, W.-H., Huang, H.-T., and Liu, I. Y. (2013). In vivo knockdown of hippocampal miR-132 expression impairs memory acquisition of trace fear conditioning. *Hippocampus* 23, 625–633. doi: 10.1002/hipo.22123

Wayman, G. A., Davare, M., Ando, H., Fortin, D., Varlamova, O., Cheng, H. Y., et al. (2008). An activity-regulated microRNA controls dendritic plasticity by down-regulating p250GAP. *Proc. Natl. Acad. Sci. U.S.A.* 105, 9093–9098. doi: 10.1073/pnas.0803072105

Yin, K.-J., Deng, Z., Huang, H., Hamblin, M., Xie, C., Zhang, J., et al. (2010). miR-497 regulates neuronal death in mouse brain after transient focal cerebral ischemia. *Neurobiol. Dis.* 38, 17–26. doi: 10.1016/j.nbd.2009.12.021

Yoshihara, Y., De Roo, M., and Muller, D. (2009). Dendritic spine formation and stabilization. *Curr. Opin. Neurobiol.* 19, 146–153. doi: 10.1016/j.conb.2009.05.013

11

microRNA function is required for neurite outgrowth of mature neurons in the mouse postnatal cerebral cortex

Janet Hong*[1]*, Haijun Zhang*[1]*, Yoko Kawase-Koga*[1,2] *and Tao Sun*[1]**

[1] *Department of Cell and Developmental Biology, Cornell University Weill Medical College, New York, NY, USA*
[2] *Department of Oral and Maxillofacial Surgery, The University of Tokyo Hospital, Tokyo, Japan*

Correspondence:
Tao Sun, Department of Cell and Developmental Biology, Cornell University Weill Medical College, 1300 York Avenue, Box 60, New York, NY 10065, USA
e-mail: tas2009@med.cornell.edu

The structure of the postnatal mammalian cerebral cortex is an assembly of numerous mature neurons that exhibit proper neurite outgrowth and axonal and dendritic morphology. While many protein coding genes are shown to be involved in neuronal maturation, the role of microRNAs (miRNAs) in this process is also becoming evident. We here report that blocking miRNA biogenesis in differentiated neurons results in microcephaly like phenotypes in the postnatal mouse brain. The smaller brain defect is not caused by defective neurogenesis, altered neuronal migration or significant neuronal cell death. Surprisingly, a dramatic increase in neuronal packing density within the postnatal brain is observed. Loss of miRNA function causes shorter neurite outgrowth and smaller soma size of mature neurons *in vitro*. Our results reveal the impact of miRNAs on normal development of neuronal morphology and brain function. Because neurite outgrowth is critical for neuroregeneration, our studies further highlight the importance of miRNAs in the treatment of neurological diseases.

Keywords: miRNAs, Dicer, neurogenesis, neurite outgrowth, cerebral cortex

INTRODUCTION

In the mammalian cerebral cortex, projection neurons are generated from radial glial cells (RGCs) and intermediate progenitors (IPs) that reside in the ventricular zone (VZ) and subventricular zone (SVZ), respectively (Noctor et al., 2001; Rakic, 2003; Haubensak et al., 2004; Englund et al., 2005; Gotz and Huttner, 2005). Postmitotic neurons (PNs) differentiate and migrate into the cortical plate (CP), in which PNs are organized in an inside–out six layered structure, with earliest born neurons in the deep layers and later born neurons in the upper layers (Guillemot, 2005; Molyneaux et al., 2007). Proper neurite outgrowth and axonal and dendritic morphogenesis are critical for neuronal maturation, synaptic formation, and neuronal function (Frank and Tsai, 2009; Merot et al., 2009). Molecular mechanisms regulating neuronal differentiation and maturation remain an exciting research topic.

The importance of microRNAs (miRNAs)-mediated neurogenesis and neuronal maturation in the central nervous system (CNS) has drawn significant attention (Kosik, 2006; Fineberg et al., 2009; Shi et al., 2010; Bian and Sun, 2011). MiRNAs are approximately 22 nucleotide (nt) endogenous non-coding small RNAs (Lee et al., 1993; Wightman et al., 1993). A mature miRNA recognizes a complementary sequence in the 3′-untranslated region (3′-UTR) of its target messenger RNA (mRNA) and affects mRNA stability and/or silences protein translation (Carthew and Sontheimer, 2009; Kim et al., 2009). Because miRNA precursors are processed into mature miRNAs by the RNAase III enzyme Dicer, the role of miRNAs in neurogenesis has been demonstrated by regional-specific deletion of *Dicer* expression in the CNS using different *Cre* lines (Volvert et al., 2012; Zhang et al., 2012). For example, *Dicer* ablated knockout (Ko) mice in PNs using the *CaMKII-Cre* line display impaired dendritic branching in pyramidal neurons in the CA1 region of the hippocampus (Davis et al., 2008; Hebert et al., 2010). These studies indicate the importance of miRNA functions in morphogenesis of mature neurons in the brain.

In this study, we demonstrate the critical role of miRNAs in neurite outgrowth of mature cortical neurons. Blocking miRNA biogenesis in PNs in the mouse cortex at perinatal stages does not significantly affect neurogenesis, neuronal survival, and layer organization. However, the neuronal packing density is greatly increased in the CP, resulting in a significantly reduced cortical size. Correspondingly, neurite outgrowth and soma size development are significantly reduced in cultured *Dicer* Ko PNs. Our results demonstrate that miRNA functions are required for proper neuronal maturation. Moreover, our studies suggest a potential role of miRNAs in promoting neurite outgrowth in the treatment of neurodegenerative diseases.

MATERIALS AND METHODS

GENERATION OF *Dicer* CONDITIONAL KNOCKOUT MICE

The floxed Dicer transgenic mice (*Dicer*$^{flox/flox}$; C57/BL6 × 129 background; kindly provided by the Greg Hannon's lab at the Cold Spring Harbor Laboratory; Murchison et al., 2005) were bred with *Nex-Cre* mice (C57/BL6 background, provided by Drs M. Schwab and K. Nave at Max-Planck-Institute of Experimental Medicine, Goettingen, Germany; Goebbels et al., 2006) to generate *Nex-Cre-Dicer* Ko (*Nex-Cre*; *Dicer*$^{flox/flox}$) animals.

For staging of embryos, mid-day of the day of vaginal plug formation is considered embryonic day 0.5 (E0.5), and the first 24 h after birth is defined as postnatal day 0 (P0). Animal use was overseen by the Animal Facility at the Weill Cornell Medical College.

GENOTYPING OF *Dicer* CONDITIONAL KNOCKOUT MICE

Mouse tail tip biopsies were used for genotyping by polymerase chain reaction reactions using the following primer pairs: for *Cre*, 5′-TAAAGATATCTCACGTACTGACGGTG-3′ and 5′-TCTCTGACCAGAGTCATCCTTAGC-3′ (product size: 350 bp); for *Dicer*, 5′-ATTGTTACCAGCGCTTAGAATTCC-3′ and 5′-GTACGTCTACAATTGTCTATG-3′ (product sizes: 767 bp from the floxed *Dicer* allele and 560 bp from the wild-type *Dicer* gene).

BREEDING THE *Nex-Cre* LINE WITH FLOXED LacZ REPORTER MICE

To localize the *Cre* activity sites, *Nex-Cre* transgenic mice were crossed with homozygous ROSA26 floxed LacZ reporter mice, obtained from Jackson Laboratories (Bar Harbor, Maine). The ROSA26 mice carry a loxP-flanked transcriptional "STOP" DNA sequence that prevents the transcription of the *LacZ* gene. Only the cells that express the *Cre* recombinase can remove the "STOP" sequence and subsequently activate the transcription of the *LacZ* gene. Cells which express LacZ produce a blue color in the β-galactosidase assay (X-gal staining).

β-GALACTOSIDASE ACTIVITY ASSAY

Mouse brains were dissected in ice-cold 1× phosphate buffered saline (PBS) and placed in 4% paraformaldehyde (PFA) in PBS for 15 min at room temperature. Fixed brains were washed in PBS for 3 × 5 min and sectioned coronally (100 μm) using a Leica vibratome (Leica, VT1000 S). Brain sections were washed three times in a wash solution (0.1 M phosphate buffer and 2 mM $MgCl_2$) and subjected to a 5-bromo-4-chloro-3-indolyl-β-D-galactopyranoside (X-gal) solution (1 mg/ml X-gal and 5 mm potassium ferrocyanide, 5 mm potassium ferricyanide in wash buffer) for 30 min to 1 h at 37°C. The reaction was quenched by washing sections three times in wash solution and incubating them in 4% PFA in PBS for 5 min at room temperature. The sections were washed three times in wash solution and mounted with a coverslip. The images were collected using a Leica digital camera under a dissection scope (Leica, MZ16F).

TISSUE PREPARATION AND IMMUNOHISTOCHEMISTRY

Mouse brains were collected and fixed in 4% PFA in PBS at 4°C overnight, followed by incubating in 30% sucrose in PBS. Brain tissues were embedded in optimal cutting temperature (OCT) and stored at -80°C until use. Brains were sectioned coronally (14 μm) using a Leica cryostat (Leica, CM3050 S).

For antigen recovery, sections were incubated in heated (95–100°C) antigen recovery solution [1 mM ethylenediaminetetraacetic acid (EDTA), 5 mM Tris, pH 8.0] for 15–20 min, and cooled down for 20–30 min. Before applying antibodies, sections were blocked in 10% normal goat serum (NGS) in PBS with 0.1% Tween-20 (PBT) for 1 h. Sections were incubated with primary antibodies at 4°C overnight and visualized using goat anti-rabbit IgG–Alexa-Fluor-488 and/or goat anti-mouse IgG–Alexa-Fluor-546 (1:350, Molecular Probes) for 1.5 h at room temperature. Images were captured using a Leica digital camera under a fluorescent microscope (Leica DMI6000B).

Primary antibodies against the following antigens were used: bromodeoxyuridine (BrdU; 1:50, DSHB), Ki67 (1:500, Abcam), Tbr1 (1:2500, Abcam), Ctip2 (1:1000, Abcam), Cux1 (1:200, Santa Cruz), Satb2 (1:1000, Abcam), β-tubulin III (TuJ1; 1:500, Chemicon), Map2 (1:500, Chemicon), and NeuN (1:300, Chemicon).

NISSL STAINING

Sections (14 μm) were processed through incubation in the following solutions in order: ethanol/chloroform (1:1, overnight), 100% ethanol (30 s), 95% ethanol (30 s), distilled water (30 s, twice), cresyl violet (3–5 min), distilled water (2 min, three times), 50% ethanol (2 min), 95% ethanol (5–30 min), 100% ethanol (5 min, twice), xylene (3 min, twice), and then mounted with a coverslip. The images were collected using a Leica digital camera under a dissection scope (Leica, MZ16F).

IN SITU HYBRIDIZATION

In situ hybridization for miRNA expression was performed according to previously published methods with modifications using locked nucleic acid (LNA) probes (Obernosterer et al., 2007). Briefly, after fixation with 4% PFA, acetylation with acetylation buffer (13.33% triethanolamince, 2.5% acetic anhydride, 20 mM HCl), treatment of proteinase K (10 mg/ml, IBI Scientific) and pre-hybridization [1× saline-sodium citrate (SSC), 50% formamide, 0.1 mg/ml salmon sperm DNA solution, 1× Denhardt, 5 mM EDTA, pH 7.5], brain sections were hybridized with digoxigenin (DIG)-labeled LNA probes at a proper temperature (Tm-22°C) overnight. After washing with pre-cooled wash buffer (1× SSC, 50% formamide, 0.1% Tween-20) and 1× maleic acid buffer containing Tween 20 (MABT), sections were blocked with blocking buffer (1× MABT, 2% blocking solution, 20% heat-inactived sheep serum) and incubated with anti-DIG antibody (1:1,500, Roche) at 4°C overnight. Brain sections were washed with 1× MABT and staining buffer (0.1M NaCl, 50 mM $MgCl_2$, 0.1M Tris–HCl, pH 9.5), stained with BM purple (Roche) at room temperature until ideal intensity. The microRNA LNA probes (Exiqon) were 3′ end labeled with DIG–ddUTP with terminal transferase using the DIG–3′ end labeling kit (Roche).

The images of *in situ* hybridization were collected using a Leica digital camera under a dissection scope (Leica, MZ16F).

BrdU INCORPORATION

To assess proliferation of neural progenitors in the developing cortex, one dose of BrdU (50 μg/g body weight) was administrated by intraperitoneal injection to timed-pregnant female mice half an hour before sacrifice.

TUNEL ASSAY

To identify apoptotic cells in the cortex, we performed a TUNEL (terminal deoxynucleotidyl transferase dUTP nick end labeling) assay using an Apop Tag Fluorescein *in situ* Apoptosis detection kit (Chemicon) on 14-μm frozen sections. This assay was performed according to the manufacturer's instructions.

CELL COUNTING IN THE CORTICAL WALL

Coronal sections were collected in the medial cortical region (at levels between the anterior commissure and the anterior hippocampus). At least four sections from each brain and three brains from different litters were chosen for antibody labeling and

TUNEL assay. For **Figures 2** and **3**, positive cells were quantified in fixed areas of 186 μm × 1200 μm in the cortical wall of P5 and P10 cortices. For **Figure 4**, positive cells were quantified in fixed areas of 186 μm × 186 μm in the cortical wall of P5 and P10 cortices.

PRIMARY NEURONAL CULTURES

Neuronal cultures were performed according to established protocols (Yu et al., 2005) with modifications. Briefly, the dorsal cortex was dissected from the P0 brain, and transferred to pre-cooled Hanks' balanced salt solution (HBSS) medium. Tissue was dissociated with 0.5 mg/ml DNAse I (Sigma D4527) in HBSS for 2 min at 37°C and mechanically triturated with fire-polished Pasteur pipettes into a single cell suspension. Cortical neurons were plated onto poly-L-lysine (PLL) and Laminin treated coverslips at 5 × 10^4 cells/well in 24-well plates. Neuronal cultures were maintained in neuronal medium [Dulbecco's modified Eagle medium (DMEM)/F12, N2, B27, glucose, $NaHCO_3$, HEPES (4-(2-hydroxyethyl)-1-piperazineethanesulfonic acid)] with fibroblast growth factor 2 (FGF-2; 20 ng/ml; Invitrogen) treatment for the first 24 h only. Afterward, cells were cultured in neuronal medium only and medium was changed every 2–3 days.

Primary neurons were fixed after 10 days *in vitro* (DIV 10) with 4% PFA in PBS for 30 min at room temperature. Before applying antibodies, cells were blocked in 10% NGS in PBS with 0.3% Triton X-100 for 1 h. Cells were incubated with primary antibodies at 4°C overnight and visualized using goat anti-rabbit IgG–Alexa-Fluor-488 and/or goat anti-mouse IgG–Alexa-Fluor-546 (1:350, Molecular Probes) for 1.5 h at room temperature. Images were captured using a Leica digital camera under a fluorescent microscope (Leica DMI6000B).

ANALYSIS OF NEURITE GROWTH AND SOMA SIZE

Typically, pictures of 30–50 neurons from three separate coverslips from each experiment were taken using a Leica digital camera under a fluorescent microscope (Leica DMI6000B). Representative cells with strong Map2 and Tuj1 immunoreactivity labeling neurite (axonal and dendritic) processes were analyzed. Neurites that had lengths that were at least twice the diameter of the cell body were measured. Neurite lengths from the soma and soma size area were traced and measured using Image J software and the data were compiled and analyzed using the Excel program (Microsoft).

STATISTIC ANALYSIS

At least three *Nex-Cre-Dicer Ko* (*Ko*) and three control (*Ctrl*) animals were used for all statistical analyses. Data were shown as mean ± SEM. Statistical comparison was made by analysis of variance (unpaired *t*-test or analyses of variance). Additional details regarding the *n* (number of animals) or *N* (number of neurites or cells) are found in the pertinent figure legend.

RESULTS

CORTICAL GROWTH DEFECTS IN *Nex-Cre-Dicer* KNOCKOUT MICE

To examine the role of miRNAs in the maturation of differentiated neurons, we conditionally ablated Dicer expression in PNs in the mouse cerebral cortex utilizing a *Cre*-loxp system. A floxed Dicer mouse line ($Dicer^{flox/flox}$) with two loxP sites flanking exon 22 and exon 23, which encode the RNAase III domains of *Dicer*, were bred with a *Nex-Cre* mouse line to generate *Nex-Cre-Dicer Ko* mice (**Figure 1A**). The *Nex-Cre* line displays activity by E13.5 and is prominently expressed in differentiating neurons of the dorsal telencephalon without affecting proliferating precursor cells of the VZ (Goebbels et al., 2006). Proliferating precursor cells can be detected by labeling cells in the S phase with a 30 min pulse of BrdU, and in the G1, S, G2, and M phase with Ki67. Indeed, quantification of $BrdU^+$ and $Ki67^+$ cells revealed no change in E15.5 *Nex-Cre-Dicer Ko* cortices compared to controls (data not shown). As such, Dicer and consequently miRNA production was conditionally ablated in PNs in the cortex after *Cre* recombination, as demonstrated by X-gal staining in P1 cortices of mice bred between the *Nex-Cre* line and the *Rosa26-LacZ* reporter line (**Figure 1B**).

Inactivation of Dicer in differentiated neurons caused markedly reduced postnatal growth. Moreover, *Nex-Cre-Dicer Ko* mice could not survive past P23, presumably due to starvation and dehydration after weaning. At P1, the brain size of *Nex-Cre-Dicer Ko* mice was comparable to that of controls (data not shown). However, gross brain morphology at P10 revealed a significant size reduction in *Nex-Cre-Dicer Ko* brains compared to controls (**Figure 1C**). Quantification of the body and brain weights of P10 *Dicer Ko* mice showed a significant reduction compared to controls, with a more profound reduction in brain weight (**Figure 1D**). Next, cortical morphology was analyzed in coronal sections of P10 brains by Nissl staining. While overall cortical lamination appeared normal, the thickness of the cortical wall was significantly reduced in *Nex-Cre-Dicer Ko* brains compared to controls (**Figure 1E**).

To verify that the brain phenotypes were caused by miRNA loss, we performed miRNA *in situ* hybridization in control and *Dicer Ko* brains. Three brain-enriched miRNAs, miR-9, Let-7, and miR-128, were utilized. We found that expression levels of all three miRNAs were reduced in P1 cortices and almost diminished in P10 cortices, suggesting a progressive loss of miRNAs due to Dicer deletion (**Figure 1F** and data not shown).

Our results indicate that Dicer deletion in differentiated neurons in the developing brain causes early postnatal death, reduced body and brain weights, and severe reduction of the cortical wall.

DEPLETION OF miRNA FUNCTION IN POSTMITOTIC NEURONS DOES NOT SIGNIFICANTLY AFFECT CORTICAL LAMINATION AND NEURONAL PRODUCTION

Due to the significant reduction in cortical wall thickness, we investigated the effects of Dicer ablation on the generation of early- and late-born neurons. During mouse cortical development, early-born neurons generated at E12.5–E13.5 migrate and form deep cortical layers VI and V, and express Tbr1 and Ctip2, respectively (Arlotta et al., 2005; Kolk et al., 2006; Chen et al., 2008; Han et al., 2011). Late-born neurons are generated at E14.5–E18.5 and migrate to form the upper neuronal layers II–IV above the deep cortical layers, and can be detected by Satb2 and Cux1 expression (Alcamo et al., 2008; Britanova et al., 2008; Cubelos et al., 2010). We first examined early- and late-born neuron production in the P5 cortex. Quantification of

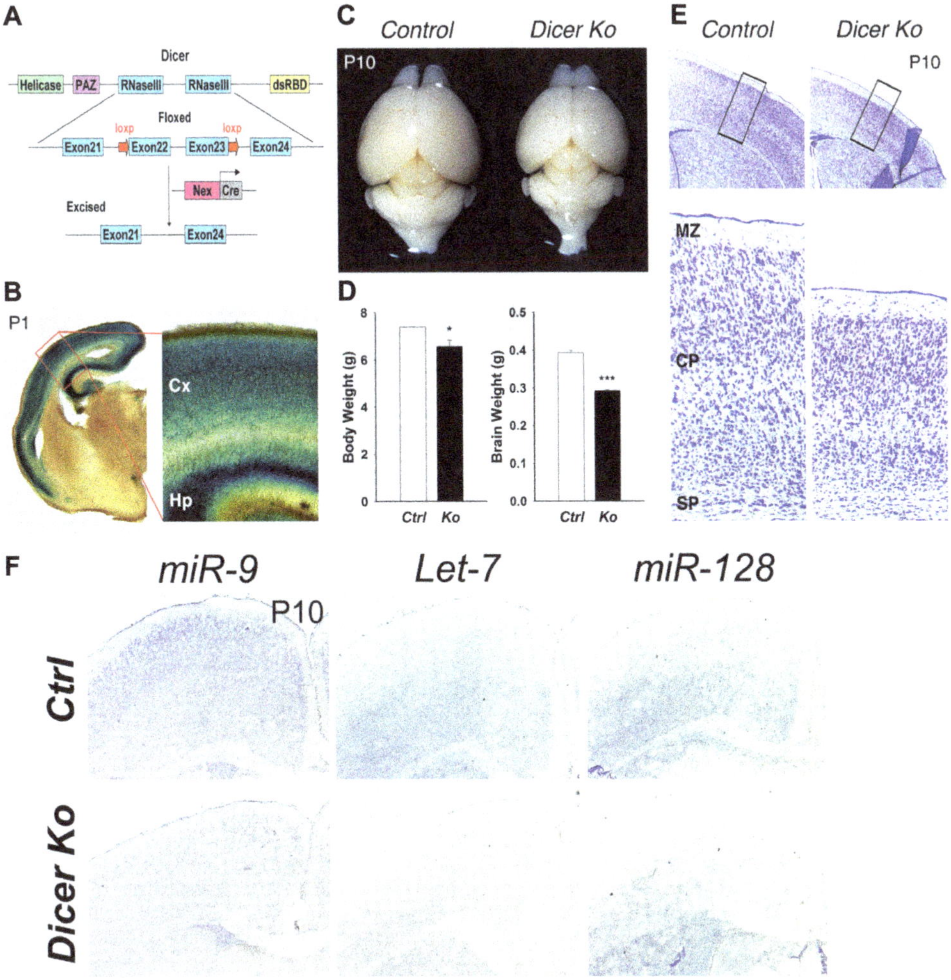

FIGURE 1 | Conditional ablation of Dicer in postmitotic neurons of the cerebral cortex results in a smaller cortex and reduced thickness of the cortical wall. (A) Dicer targeting construct. The N-terminal RNA helicase domain, piwi argonaute and zwille (PAZ) domain, two ribonuclease III domains, and a double-stranded RNA- binding domain (RBD) are labeled. The exon 22 and exon 23 of Dicer are conditionally excised after *Nex-Cre* recombination. **(B)** X-gal staining at the level of the cortex in a *Nex-Cre* and *Rosa26-LacZ*-expressing mouse at P1 illustrating cortical specificity of the *Nex-Cre* line. The red box indicates the region shown at higher magnification. The cortex (Cx) and hippocampus (Hp) are labeled. **(C)** Appearance of representative brains from P10 control and *Nex-Cre-Dicer Ko* mice (litter mates). **(D)** Body and brain weights of control (*Ctrl*) and *Nex-Cre-Dicer Ko* (*Ko*) mice at P10. **(E)** Coronal sections of P10 brains with Nissl staining of control and *Nex-Cre-Dicer Ko* mice. The black boxes indicate the region shown at higher magnification. The subplate (SP), cortical plate (CP), and marginal zone (MZ) are labeled. **(F)** *In situ* hybridization of *miR-9*, *Let-7*, and *miR-128* in control and *Dicer Ko* cortices at P10. Data are presented as mean ± SEM; $n \geq 3$ in all genotypes; p values in relation to control ($*p < 0.05$, $***p < 0.00002$).

early-born neurons with Tbr1$^+$ and Ctip2$^+$ cells revealed no significant difference between control and *Nex-Cre-Dicer Ko* cortices (**Figures 2A,B**). For late-born neurons, Satb2$^+$ cells were unaffected but Cux1$^+$ cells were slightly decreased in *Dicer Ko* cortices compared to controls (**Figures 2A,B**). Next, we analyzed neuronal production in the P5 cortex. Quantification of neuron and cell numbers by NeuN and DAPI immunostaining showed no significant difference in *Nex-Cre-Dicer Ko* cortices compared to controls (**Figures 2C,D**).

Given that Dicer ablation did not reveal a significant defect in cortical lamination and neuronal production despite the reduced cortical thickness, we investigated the possibility of neuronal cell death. Apoptotic cells in the cortex were detected by TUNEL assay. At P5, there was a significant increase in apoptotic cells in *Nex-Cre-Dicer Ko* cortices compared to controls, which was not detected in P1 cortices (**Figures 2E,F** and data not shown). Moreover, TUNEL$^+$ cells in *Dicer Ko* brains were localized in the far-upper cortical layer at the marginal zone boundary, suggesting apoptosis of a subset of late-born neurons (**Figures 2Ea′,b′**).

We further examined cortical lamination, neuronal production, and apoptosis in P10 control and *Nex-Cre-Dicer Ko* brains. Numbers of Tbr1$^+$ and Ctip2$^+$ early-born neurons were increased and decreased in *Dicer Ko* cortices, respectively, compared to controls (**Figures 3A,B**). Conversely, quantification of Satb2$^+$ and Cux1$^+$ late-born neurons revealed no significant difference between control and *Dicer Ko* cortices (**Figures 3A,B**). Subsequently, we analyzed neuron and cell number by NeuN and DAPI immunostaining in the P10 cortex. Interestingly, countings of

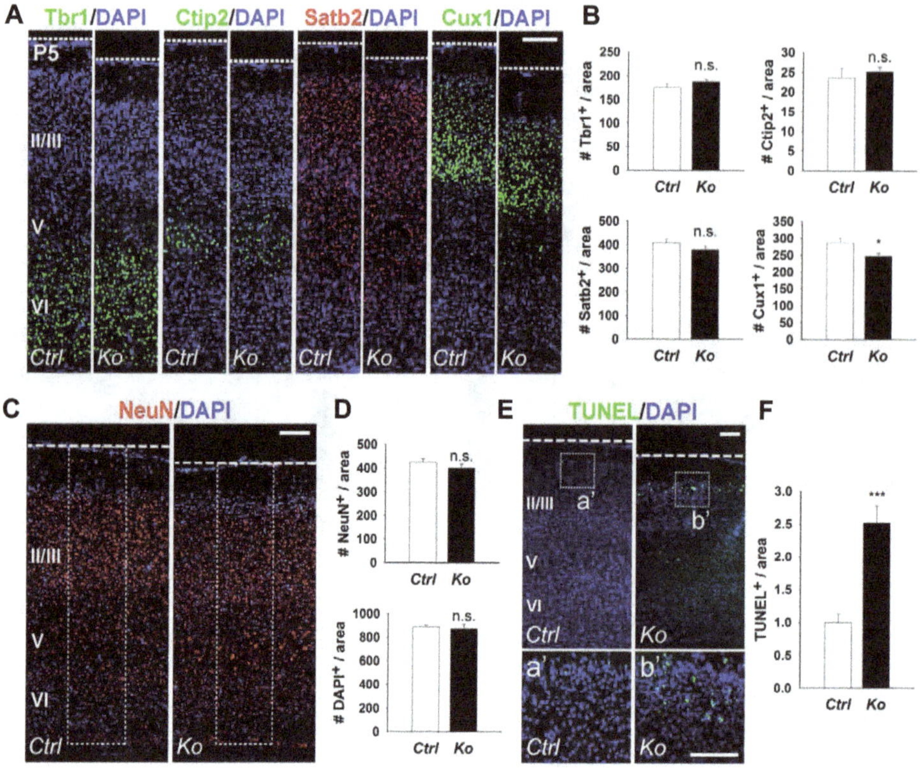

FIGURE 2 | Conditional Dicer loss in mature neurons does not significantly alter cortical lamination and neuronal production, and exhibits a transitory increase in cell death at P5. (A,B) The numbers of early-born neurons labeled with Tbr1 and Ctip2 were unaffected in P5 *Nex-Cre-Dicer Ko* (*Ko*) cortices, compared to controls (*Ctrl*). The numbers of late-born neurons labeled with Satb2 were unaffected and Cux1 were decreased in P5 *Dicer Ko* cortices, compared to controls. **(C,D)** Numbers of $NeuN^+$ and $DAPI^+$ cells were unaffected in the cortical wall of P5 *Dicer Ko* cortices compared to controls. The dashed box indicates the area of quantification. **(E)** TUNEL assay of coronal cryosections of P5 control and *Dicer Ko* cortices. The dashed box indicates the region shown at higher magnification in panel **a′**, **b′**. **(F)** Normalized quantification of $TUNEL^+$ cells per area in the cortical wall of P5 control and *Dicer Ko* brains. Cortical layers (VI), (V), and (II/III) are labeled. Scale bar: 100 μm. Data are presented as mean ± SEM; $n \geq 3$ in all genotypes; *p* values in relation to control ($^*p < 0.04$, $^{***}p < 0.0002$). n.s., not significant.

$NeuN^+$ and $DAPI^+$ cells within fixed columns of the cortical wall revealed no significant alterations in neuronal or cell number in *Nex-Cre-Dicer Ko* cortices, despite its significantly thinner cortex (**Figures 3C,D**). Further TUNEL analysis in the P10 cortex revealed no significant differences in the numbers of apoptotic cells in *Dicer Ko* and control brains (**Figures 3E,F**).

Our results indicate that even though the numbers of early- and late-born neurons, and apoptotic cells show temporal changes in postnatal cortices of *Nex-Cre-Dicer Ko* mice, overall cortical lamination and neuronal production remain undisrupted.

CONDITIONAL *Dicer* ABLATION AFFECTS NEURON AND CELL PACKING DENSITY WITHIN THE CORTEX

Considering that inactivation of Dicer in PNs did not adversely affect cortical lamination and neuronal production and only had a transient effect on cell survival, we investigated the cause of the smaller cortex in *Nex-Cre-Dicer Ko* mice further. We analyzed the density of neurons and cells by quantifying the number of $NeuN^+$ and $DAPI^+$ cells within uniform boxed areas in the upper and lower regions of the CP. At P5, there were no alterations in $NeuN^+$ neuron and $DAPI^+$ cell density in *Dicer Ko* cortices compared to controls (**Figures 4A,B**). However, P10 *Dicer Ko* cortices revealed significantly increased $NeuN^+$ and $DAPI^+$ cell numbers compared to controls, indicating increased density and packing of cells within the cortex during the stage of neuronal maturation (**Figures 4C,D**). These results demonstrate that Dicer ablation in PNs does not cause defective neuronal production but alters the neuronal packing density within the cortex.

LOSS OF *Dicer* CAUSES ABNORMAL NEURONAL MATURATION WITH SHORTER NEURITE OUTGROWTH AND SMALLER CELL BODY SIZE

Given that the packing density of neurons was dramatically increased in the *Nex-Cre-Dicer Ko* cortices, we decided to further analyze neuronal morphology *in vitro*. This was done by harvesting cortical neurons from P0 mouse brains and culturing them under differentiation conditions using previously described methods with modifications (**Figure 5A**; Yu et al., 2005). After 10 days *in vitro* (DIV 10), cultures of differentiated neurons from control and *Dicer Ko* cortices were labeled with antibodies against Map2 and Tuj1 to illustrate neurites. We found that *Dicer Ko* neurons displayed significantly shorter neurites and processes compared to controls (**Figures 5B,C**). We next

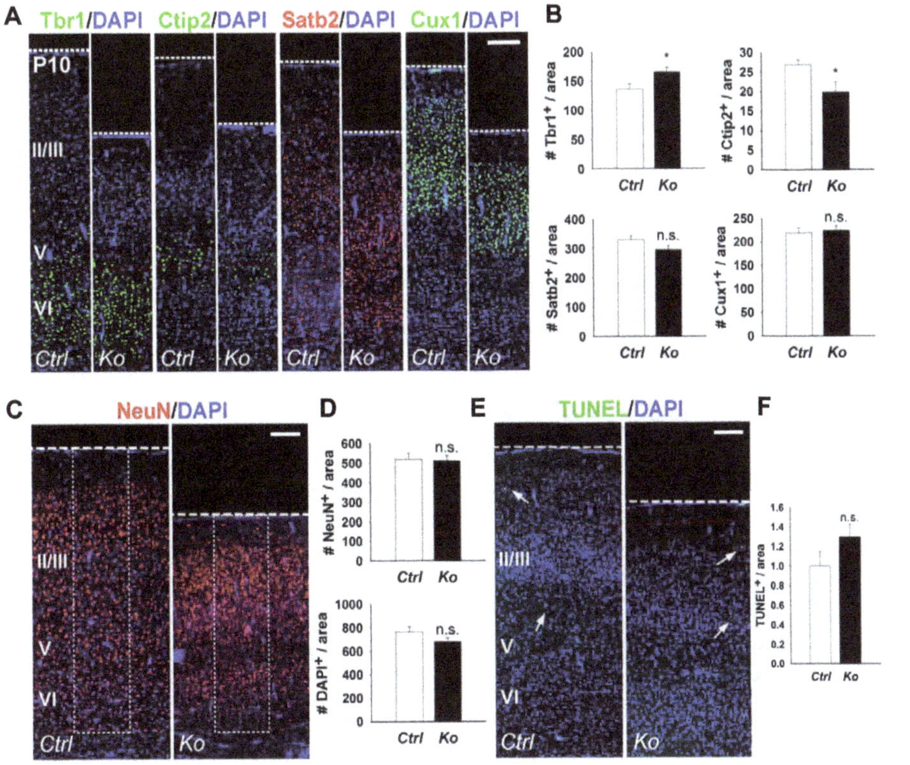

FIGURE 3 | Depletion of miRNA function in postmitotic neurons slightly alters early-born neurons and has no measured effect on neuronal production and cell death at P10. (A,B) The numbers of early-born neurons labeled with Tbr1 were increased and Ctip2 were decreased in P10 *Nex-Cre-Dicer Ko* (*Ko*) cortices, compared to controls (*Ctrl*). The numbers of late-born neurons labeled with Satb2 and Cux1 were unaffected in P10 *Dicer Ko* cortices, compared to controls. **(C,D)** Numbers of NeuN$^+$ and DAPI$^+$ cells were unaffected in the cortical wall of P10 *Dicer Ko* cortices compared to controls. The dashed box indicates the area of quantification. **(E)** TUNEL assay of coronal cryosections of P10 control and *Dicer Ko* cortices. Arrows indicate TUNEL$^+$ cells. **(F)** Normalized quantification of TUNEL$^+$ cells per area in the cortical wall of P10 control and *Dicer Ko* brains. Cortical layers (VI), (V), and (II/III) are labeled. Scale bar: 100 μm. Data are presented as mean ± SEM; $n \geq 3$ in all genotypes; p values in relation to control (*$p < 0.04$). n.s., not significant.

quantified soma size by measuring the cell body area of Map2- and Tuj1-stained neurons. Analysis of soma size revealed that *Nex-Cre-Dicer Ko* neurons displayed significantly smaller cell body area compared to controls (**Figures 6A,B**). Our results indicate that miRNA function is required for proper neurite outgrowth and soma size development of differentiated neurons during maturation.

DISCUSSION

MiRNAs have been found to be crucial for proper development of the CNS. The results reported here underscore the importance of Dicer and miRNAs for neuronal differentiation and maturation. Although removal of Dicer in postmitotic cortical neurons has no immediate impact on neurogenesis, neuronal survival, or layer organization, it has dramatic effects on neurite outgrowth and cortical packing density. Consequently, Dicer-deficient mice exhibited thinner cortical walls and a progressive decline in postnatal growth, resulting in neurodegeneration defects. In conclusion, our results provide evidence that Dicer and miRNAs function is essential for neuronal maturation and that interference with the miRNA pathway results in phenotypes similar to neurodegenerative diseases.

Previous studies have revealed essential roles of miRNAs for neural progenitor proliferation, survival, and differentiation through Dicer ablation during embryonic development of the mouse neocortex (De Pietri Tonelli et al., 2008; Kawase-Koga et al., 2009, 2010; Andersson et al., 2010; Nowakowski et al., 2011). Moreover, limited studies have examined the role of Dicer in specific subpopulations of neurons, such as Purkinje cells, dopaminergic neurons, and excitatory neurons (Kim et al., 2007; Schaefer et al., 2007; Davis et al., 2008). In our mouse model, Dicer is ablated in PNs with the *Nex-Cre* line. Although the *Nex-Cre* line displays activity in the cortex by E13.5 (Goebbels et al., 2006), our model system reveals no significant alterations in brain weight or morphology in *Dicer* deficient mice at P1 (data not shown). This is perhaps caused by a delayed Dicer deletion, which allows a low level of Dicer proteins to continue to process miRNAs and regulate PNs until complete inactivation (Harfe et al., 2005; Kawase-Koga et al., 2009). MiRNAs are expressed in a diverse spectrum and change dynamically during brain development (Lagos-Quintana et al., 2002; Krichevsky et al., 2003; Miska et al., 2004; Sempere et al., 2004; Smirnova et al., 2005). Moreover, conserved complex interactions of multiple genes form a wide regulatory network in the developing cortex (Guillemot, 2005; Molyneaux et al., 2007).

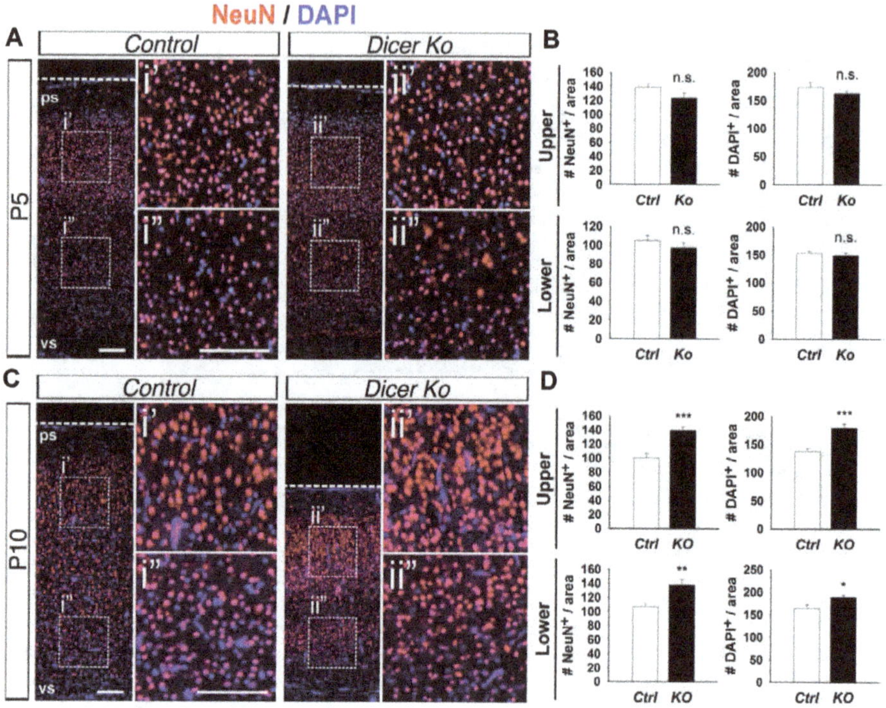

FIGURE 4 | Conditional Dicer ablation increases neuron and cell packing density within the cortex at P10. (A,B) Numbers of NeuN^{+} and DAPI^{+} cells were unaffected in the upper and lower regions of the cortical wall of P5 *Nex-Cre-Dicer Ko* (*Ko*) cortices compared to controls (*Ctrl*). **(C,D)** Numbers of NeuN^{+} and DAPI^{+} cells in the upper and lower regions of the cortical wall were significantly increased in P10 *Dicer Ko* cortices compared to controls. The dashed boxes indicate the region shown at higher magnification in panel i′, i″, ii′, ii″; the boxed area in this region was chosen for subsequent analysis. The ventricular surface (vs) and pial surface (ps) are labeled. Scale bar: 100 μm. Data are presented as mean ± SEM; $n \geq 3$ in all genotypes; p values in relation to control (*$p < 0.02$, **$p < 0.004$, ***$p < 0.0008$). n.s., not significant.

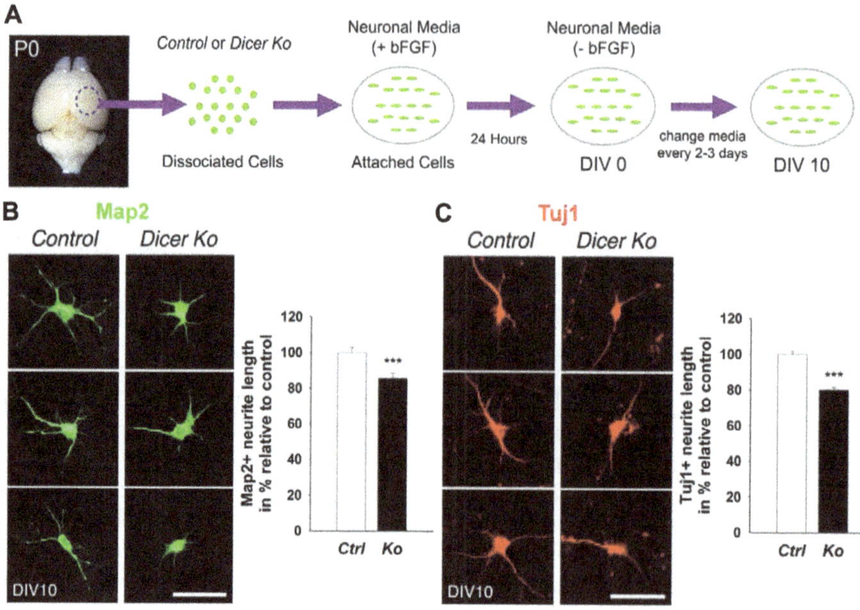

FIGURE 5 | Loss of Dicer in mature neurons delays neurite outgrowth *in vitro*. (A) An illustrative summary of primary neuronal culture derived from control (*Ctrl*) and *Nex-Cre-Dicer Ko* (*Ko*) P0 dorsal cortex. **(B)** Measurements of Map2^{+} processes revealed shorter neurite outgrowth 10 days *in vitro* (DIV 10) in *Dicer Ko* ($N = 113$) neural cultures compared to controls ($N = 158$). **(C)** Measurements of Tuj1^{+} processes displayed shorter neurite outgrowth in DIV 10 *Dicer Ko* ($N = 411$) neural cultures compared to controls ($N = 423$). Scale bar: 50 μm. Data are presented as mean ± SEM; $n \geq 3$ in all genotypes; p values in relation to control (***$p < 0.0008$).

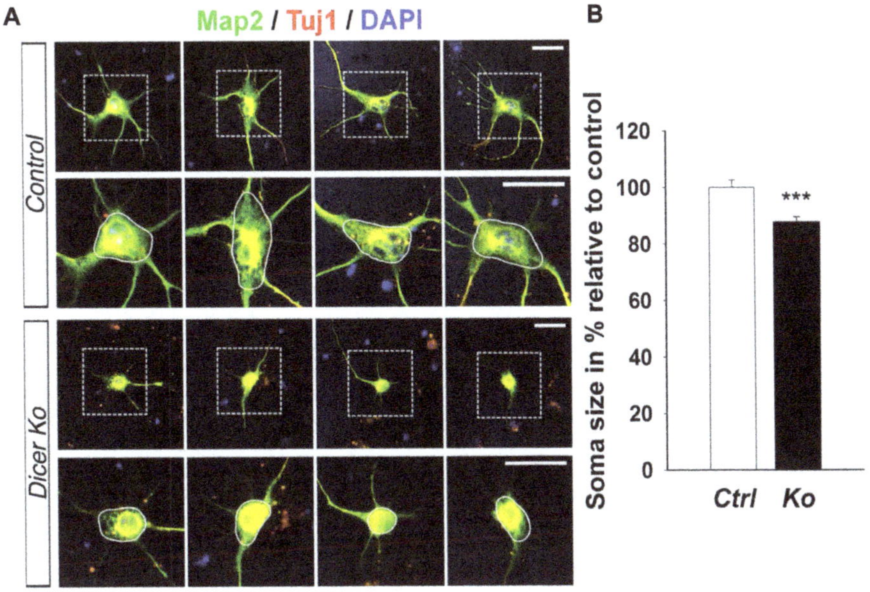

FIGURE 6 | miRNA depletion in maturing cortical neurons causes a reduction of soma size *in vitro*. (A) Immunofluorescence microscopy of control (*Ctrl*) and *Nex-Cre-Dicer Ko* (*Ko*) primary neural cultures at 10 days *in vitro* (DIV 10) showing Map2 (green), Tuj1 (red), and DAPI (Jentarra et al., 2010). The dashed box indicates the region shown at higher magnification. The area encircled by the white line indicates the region of soma size analysis. **(B)** Measurements of soma size area in DIV 10 primary neurons revealed a significant reduction in *Dicer Ko* ($N = 324$) cell body size compared to controls ($N = 259$). Scale bar: 25 μm. Data are presented as mean ± SEM; $n \geq 3$ in all genotypes; *p* values in relation to control (***$p < 0.00009$).

As such, the slight alterations in early- and late-born neuron populations in P5 and P10 *Nex-Cre-Dicer Ko* cortices are perhaps a balanced outcome of a multitude of distinct miRNAs with a variety of regulatory functions and targets.

Given the significant reduction in postnatal cortical growth in *Nex-Cre-Dicer Ko* brains, it is surprising to find preservation of neuronal cell numbers in the cortex. Moreover, despite a temporal increase of apoptotic cells in P5 cortices, *Nex-Cre-Dicer Ko* mice do not exhibit significant cell death. These results are in direct contrast to previous studies of Dicer function in Purkinje neurons and DAT-expressing neurons, which found widespread and continuous neurodegeneration and neuronal cell death (Kim et al., 2007; Schaefer et al., 2007). Moreover, compared to Dicer ablation studies in embryonic neural progenitors, which found dramatic apoptotic and differentiation defects, our studies have shown that loss of Dicer activity in postmitotic cortical neurons has minimal impact on neuronal survival (De Pietri Tonelli et al., 2008; Kawase-Koga et al., 2009). This mild apoptosis defect is similar to observations in *Dicer Ko* mice generated using the *CaMKII-Cre* line (Davis et al., 2008). These results highlight the diverse and variable functions Dicer and miRNAs carry for cell survival of different cell types at different time points during development.

Although blocking miRNA biogenesis in mature neurons reveals no apparent loss of neurons in the cortex, we have found a major increase in neuronal density in the cerebral cortex. This indicates that neuronal cell volume rather than neuron number is altered by depletion of Dicer and miRNAs in postmitotic cortical neurons. Moreover, direct differentiation of PNs from *Dicer* deficient cortices in a cell culture system has shown defects in neurite outgrowth (dendrites and axons) and decreased soma size. Decreased neurite outgrowth and increased packing density may contribute to reduced brain size in our *Nex-Cre-Dicer Ko* mice and in *Dicer Ko* mice generated using the *CaMKII-Cre* line (Davis et al., 2008). Moreover, our findings further support previous work, which have found a causal link between specific miRNAs such as miR-134, miR-34, miR-124, miR-9, and miR-132 with neurite outgrowth and elaboration *in vitro* (Vo et al., 2005; Yu et al., 2008; Agostini et al., 2011; Gaughwin et al., 2011; Clovis et al., 2012; Franke et al., 2012).

In conclusion, our results shed light on the essential role of Dicer-mediated miRNA functions for postmitotic neuronal maturation. Although loss of miRNA function in postmitotic cortical neurons has no definitive impact on neurogenesis, cortical patterning, or cell survival, it causes an atrophic change in neurites (dendrites and axons) and soma size. The aforementioned neurite outgrowth phenotypes are comparable with mouse models of neurodegeneration, which induce generalized atrophy of neuronal soma, dendrites and axons in the brain (Sakai et al., 2006). Increased packing density is also detected in a mouse model of Rett syndrome/X-linked mental retardation (Jentarra et al., 2010). Moreover, abnormally high packing density has been observed in patient brains with Rett syndrome, Williams syndrome, and schizophrenia (Bauman et al., 1995; Selemon et al., 1995; Galaburda et al., 2002). Our model of mature neuron degeneration bears resemblance to cell pathologies associated with schizophrenia and neurodegenerative diseases. As such, understanding the role of specific miRNAs during processes such as neuronal differentiation and maturation may be fundamental to discovering the morphological mechanisms of neurological disorders.

ACKNOWLEDGMENTS

We thank members of the Sun laboratory for their valuable discussions. We appreciate Dr G. Hannon at the Cold Spring Harbor Laboratory and Drs M. Schwab and K. Nave at Max-Planck-Institute of Experimental Medicine for providing mice. This work was supported by the Hirschl/Weill-Caulier Trust (Tao Sun), an NPRP grant (09-1011-3-260) from the Qatar National Research Fund (Tao Sun) and an R01-MH083680 grant from the NIH/NIMH (Tao Sun).

REFERENCES

Agostini, M., Tucci, P., Steinert, J. R., Shalom-Feuerstein, R., Rouleau, M., Aberdam, D., et al. (2011). microRNA-34a regulates neurite outgrowth, spinal morphology, and function. *Proc. Natl. Acad. Sci. U.S.A.* 108, 21099–21104. doi: 10.1073/pnas.1112063108

Alcamo, E. A., Chirivella, L., Dautzenberg, M., Dobreva, G., Farinas, I., Grosschedl, R., et al. (2008). Satb2 regulates callosal projection neuron identity in the developing cerebral cortex. *Neuron* 57, 364–377. doi: 10.1016/j.neuron.2007.12.012

Andersson, T., Rahman, S., Sansom, S. N., Alsio, J. M., Kaneda, M., Smith, J., et al. (2010). Reversible block of mouse neural stem cell differentiation in the absence of dicer and microRNAs. *PLoS ONE* 5:e13453. doi: 10.1371/journal.pone.0013453

Arlotta, P., Molyneaux, B. J., Chen, J., Inoue, J., Kominami, R., and Macklis, J. D. (2005). Neuronal subtype-specific genes that control corticospinal motor neuron development *in vivo*. *Neuron* 45, 207–221. doi: 10.1016/j.neuron.2004.12.036

Bauman, M. L., Kemper, T. L., and Arin, D. M. (1995). Pervasive neuroanatomic abnormalities of the brain in three cases of Rett's syndrome. *Neurology* 45, 1581–1586. doi: 10.1212/WNL.45.8.1581

Bian, S., and Sun, T. (2011). Functions of noncoding RNAs in neural development and neurological diseases. *Mol. Neurobiol.* 44, 359–373. doi: 10.1007/s12035-011-8211-3

Britanova, O., de Juan Romero, C., Cheung, A., Kwan, K. Y., Schwark, M., Gyorgy, A., et al. (2008). Satb2 is a postmitotic determinant for upper-layer neuron specification in the neocortex. *Neuron* 57, 378–392. doi: 10.1016/j.neuron.2007.12.028

Carthew, R. W., and Sontheimer, E. J. (2009). Origins and mechanisms of miRNAs and siRNAs. *Cell* 136, 642–655. doi: 10.1016/j.cell.2009.01.035

Chen, B., Wang, S. S., Hattox, A. M., Rayburn, H., Nelson, S. B., and McConnell, S. K. (2008). The Fezf2-Ctip2 genetic pathway regulates the fate choice of subcortical projection neurons in the developing cerebral cortex. *Proc. Natl. Acad. Sci. U.S.A.* 105, 11382–11387. doi: 10.1073/pnas.0804918105

Clovis, Y. M., Enard, W., Marinaro, F., Huttner, W. B., and De Pietri Tonelli, D. (2012). Convergent repression of Foxp2 3′UTR by miR-9 and miR-132 in embryonic mouse neocortex: implications for radial migration of neurons. *Development* 139, 3332–3342. doi: 10.1242/dev.078063

Cubelos, B., Sebastian-Serrano, A., Beccari, L., Calcagnotto, M. E., Cisneros, E., Kim, S., et al. (2010). Cux1 and Cux2 regulate dendritic branching, spine morphology, and synapses of the upper layer neurons of the cortex. *Neuron* 66, 523–535. doi: 10.1016/j.neuron.2010.04.038

Davis, T. H., Cuellar, T. L., Koch, S. M., Barker, A. J., Harfe, B. D., McManus, M. T., et al. (2008). Conditional loss of Dicer disrupts cellular and tissue morphogenesis in the cortex and hippocampus. *J. Neurosci.* 28, 4322–4330. doi: 10.1523/JNEUROSCI.4815-07.2008

De Pietri Tonelli, D., Pulvers, J. N., Haffner, C., Murchison, E. P., Hannon, G. J., and Huttner, W. B. (2008). miRNAs are essential for survival and differentiation of newborn neurons but not for expansion of neural progenitors during early neurogenesis in the mouse embryonic neocortex. *Development* 135, 3911–3921. doi: 10.1242/dev.025080

Englund, C., Fink, A., Lau, C., Pham, D., Daza, R. A., Bulfone, A., et al. (2005). Pax6, Tbr2, and Tbr1 are expressed sequentially by radial glia, cells, and postmitotic neurons in developing neocortex. *J. Neurosci.* 25, 247–251. doi: 10.1523/JNEUROSCI.2899-04.2005

Fineberg, S. K., Kosik, K. S., and Davidson, B. L. (2009). MicroRNAs potentiate neural development. *Neuron* 64, 303–309. doi: 10.1016/j.neuron.2009.10.020

Frank, C. L., and Tsai, L. H. (2009). Alternative functions of core cell cycle regulators in neuronal migration, neuronal maturation, and synaptic plasticity. *Neuron* 62, 312–326. doi: 10.1016/j.neuron.2009.03.029

Franke, K., Otto, W., Johannes, S., Baumgart, J., Nitsch, R., and Schumacher, S. (2012). miR-124-regulated RhoG reduces neuronal process complexity via ELMO/Dock180/Rac1 and Cdc42 signalling. *EMBO J.* 31, 2908–2921. doi: 10.1038/emboj.2012.130

Galaburda, A. M., Holinger, D. P., Bellugi, U., and Sherman, G. F. (2002). Williams syndrome: neuronal size and neuronal-packing density in primary visual cortex. *Arch. Neurol.* 59, 1461–1467. doi: 10.1001/archneur.59.9.1461

Gaughwin, P., Ciesla, M., Yang, H., Lim, B., and Brundin, P. (2011). Stage-specific modulation of cortical neuronal development by Mmu-miR-134. *Cereb. Cortex* 21, 1857–1869. doi: 10.1093/cercor/bhq262

Goebbels, S., Bormuth, I., Bode, U., Hermanson, O., Schwab, M. H., and Nave, K. A. (2006). Genetic targeting of principal neurons in neocortex and hippocampus of NEX-Cre mice. *Genesis* 44, 611–621. doi: 10.1002/dvg.20256

Gotz, M., and Huttner, W. B. (2005). The cell biology of neurogenesis. *Nat. Rev. Mol. Cell Biol.* 6, 777–788. doi: 10.1038/nrm1739

Guillemot, F. (2005). Cellular and molecular control of neurogenesis in the mammalian telencephalon. *Curr. Opin. Cell Biol.* 17, 639–647. doi: 10.1016/j.ceb.2005.09.006

Han, W., Kwan, K. Y., Shim, S., Lam, M. M., Shin, Y., Xu, X., et al. (2011). TBR1 directly represses Fezf2 to control the laminar origin and development of the corticospinal tract. *Proc. Natl. Acad. Sci. U.S.A.* 108, 3041–3046. doi: 10.1073/pnas.1016723108

Harfe, B. D., McManus, M. T., Mansfield, J. H., Hornstein, E., and Tabin, C. J. (2005). The RNaseIII enzyme Dicer is required for morphogenesis but not patterning of the vertebrate limb. *Proc. Natl. Acad. Sci. U.S.A.* 102, 10898–10903. doi: 10.1073/pnas.0504834102

Haubensak, W., Attardo, A., Denk, W., and Huttner, W. B. (2004). Neurons arise in the basal neuroepithelium of the early mammalian telencephalon: a major site of neurogenesis. *Proc. Natl. Acad. Sci. U.S.A.* 101, 3196–3201. doi: 10.1073/pnas.0308600100

Hebert, S. S., Papadopoulou, A. S., Smith, P., Galas, M. C., Planel, E., Silahtaroglu, A. N., et al. (2010). Genetic ablation of Dicer in adult forebrain neurons results in abnormal tau hyperphosphorylation and neurodegeneration. *Hum. Mol. Genet.* 19, 3959–3969. doi: 10.1093/hmg/ddq311

Jentarra, G. M., Olfers, S. L., Rice, S. G., Srivastava, N., Homanics, G. E., Blue, M., et al. (2010). Abnormalities of cell packing density and dendritic complexity in the MeCP2 A140V mouse model of Rett syndrome/X-linked mental retardation. *BMC Neurosci.* 11:19. doi: 10.1186/1471-2202-11-19

Kawase-Koga, Y., Low, R., Otaegi, G., Pollock, A., Deng, H., Eisenhaber, F., et al. (2010). RNAase-III enzyme Dicer maintains signaling pathways for differentiation and survival in mouse cortical neural stem cells. *J. Cell Sci.* 123, 586–594. doi: 10.1242/jcs.059659

Kawase-Koga, Y., Otaegi, G., and Sun, T. (2009). Different timings of dicer deletion affect neurogenesis and gliogenesis in the developing mouse central nervous system. *Dev. Dyn.* 238, 2800–2812. doi: 10.1002/dvdy.22109

Kim, J., Inoue, K., Ishii, J., Vanti, W. B., Voronov, S. V., Murchison, E., et al. (2007). A microRNA feedback circuit in midbrain dopamine neurons. *Science* 317, 1220–1224. doi: 10.1126/science.1140481

Kim, V. N., Han, J., and Siomi, M. C. (2009). Biogenesis of small RNAs in animals. *Nat. Rev. Mol. Cell Biol.* 10, 126–139. doi: 10.1038/nrm2632

Kolk, S. M., Whitman, M. C., Yun, M. E., Shete, P., and Donoghue, M. J. (2006). A unique subpopulation of Tbr1-expressing deep layer neurons in the developing cerebral cortex. *Mol. Cell. Neurosci.* 32, 200–214. doi: 10.1016/j.mcn.2005.08.022

Kosik, K. S. (2006). The neuronal microRNA system. *Nat. Rev. Neurosci.* 7, 911–920. doi: 10.1038/nrn2037

Krichevsky, A. M., King, K. S., Donahue, C. P., Khrapko, K., and Kosik, K. S. (2003). A microRNA array reveals extensive regulation of microRNAs during brain development. *RNA* 9, 1274–1281. doi: 10.1261/rna.5980303

Lagos-Quintana, M., Rauhut, R., Yalcin, A., Meyer, J., Lendeckel, W., and Tuschl, T. (2002). Identification of tissue-specific microRNAs from mouse. *Curr. Biol.* 12, 735–739. doi: 10.1016/S0960-9822(02)00809-6

Lee, R. C., Feinbaum, R. L., and Ambros, V. (1993). The *C. elegans* heterochronic gene lin-4 encodes small

RNAs with antisense complementarity to lin-14. *Cell* 75, 843–854. doi: 10.1016/0092-8674(93)90529-Y

Merot, Y., Retaux, S., and Heng, J. I. (2009). Molecular mechanisms of projection neuron production and maturation in the developing cerebral cortex. *Semin. Cell Dev. Biol.* 20, 726–734. doi: 10.1016/j.semcdb.2009.04.003

Miska, E. A., Alvarez-Saavedra, E., Townsend, M., Yoshii, A., Sestan, N., Rakic, P., et al. (2004). Microarray analysis of microRNA expression in the developing mammalian brain. *Genome Biol.* 5, R68. doi: 10.1186/gb-2004-5-9-r68

Molyneaux, B. J., Arlotta, P., Menezes, J. R., and Macklis, J. D. (2007). Neuronal subtype specification in the cerebral cortex. *Nat. Rev. Neurosci.* 8, 427–437. doi: 10.1038/nrn2151

Murchison, E. P., Partridge, J. F., Tam, O. H., Cheloufi, S., and Hannon, G. J. (2005). Characterization of Dicer-deficient murine embryonic stem cells. *Proc. Natl. Acad. Sci. U.S.A.* 102, 12135–12140. doi: 10.1073/pnas.0505479102

Noctor, S. C., Flint, A. C., Weissman, T. A., Dammerman, R. S., and Kriegstein, A. R. (2001). Neurons derived from radial glial cells establish radial units in neocortex. *Nature* 409, 714–720. doi: 10.1038/35055553

Nowakowski, T. J., Mysiak, K. S., Pratt, T., and Price, D. J. (2011). Functional dicer is necessary for appropriate specification of radial glia during early development of mouse telencephalon. *PLoS ONE* 6:e23013. doi: 10.1371/journal.pone.0023013

Obernosterer, G., Martinez, J., and Alenius, M. (2007). Locked nucleic acid-based *in situ* detection of microRNAs in mouse tissue sections. *Nat. Protoc.* 2, 1508–1514. doi: 10.1038/nprot.2007.153

Rakic, P. (2003). Developmental and evolutionary adaptations of cortical radial glia. *Cereb. Cortex* 13, 541–549. doi: 10.1093/cercor/13.6.541

Sakai, K., Yamada, M., Sato, T., Tsuji, S., and Takahashi, H. (2006). Neuronal atrophy and synaptic alteration in a mouse model of dentatorubral-pallidoluysian atrophy. *Brain* 129, 2353–2362. doi: 10.1093/brain/awl182

Schaefer, A., O'Carroll, D., Tan, C. L., Hillman, D., Sugimori, M., Llinas, R., et al. (2007). Cerebellar neurodegeneration in the absence of microRNAs. *J. Exp. Med.* 204, 1553–1558. doi: 10.1084/jem.20070823

Selemon, L. D., Rajkowska, G., and Goldman-Rakic, P. S. (1995). Abnormally high neuronal density in the schizophrenic cortex. A morphometric analysis of prefrontal area 9 and occipital area 17. *Arch. Gen. Psychiatry* 52: 805–818; discussion 819–820. doi: 10.1001/archpsyc.1995.03950220015005

Sempere, L. F., Freemantle, S., Pitha-Rowe, I., Moss, E., Dmitrovsky, E., and Ambros, V. (2004). Expression profiling of mammalian microRNAs uncovers a subset of brain-expressed microRNAs with possible roles in murine and human neuronal differentiation. *Genome Biol.* 5, R13. doi: 10.1186/gb-2004-5-3-r13

Shi, Y., Zhao, X., Hsieh, J., Wichterle, H., Impey, S., Banerjee, S., et al. (2010). MicroRNA regulation of neural stem cells and neurogenesis. *J. Neurosci.* 30, 14931–14936. doi: 10.1523/JNEUROSCI.4280-10.2010

Smirnova, L., Grafe, A., Seiler, A., Schumacher, S., Nitsch, R., and Wulczyn, F. G. (2005). Regulation of miRNA expression during neural cell specification. *Eur. J. Neurosci.* 21, 1469–1477. doi: 10.1111/j.1460-9568.2005.03978.x

Vo, N., Klein, M. E., Varlamova, O., Keller, D. M., Yamamoto, T., Goodman, R. H., et al. (2005). A cAMP-response element binding protein-induced microRNA regulates neuronal morphogenesis. *Proc. Natl. Acad. Sci. U.S.A.* 102, 16426–16431. doi: 10.1073/pnas.0508448102

Volvert, M. L., Rogister, F., Moonen, G., Malgrange, B., and Nguyen, L. (2012). MicroRNAs tune cerebral cortical neurogenesis. *Cell Death Differ.* 19, 1573–1581. doi: 10.1038/cdd.2012.96

Wightman, B., Ha, I., and Ruvkun, G. (1993). Posttranscriptional regulation of the heterochronic gene lin-14 by lin-4 mediates temporal pattern formation in *C. elegans*. *Cell* 75, 855–862. doi: 10.1016/0092-8674(93)90530-4

Yu, J. Y., Chung, K. H., Deo, M., Thompson, R. C., and Turner, D. L. (2008). MicroRNA miR-124 regulates neurite outgrowth during neuronal differentiation. *Exp. Cell Res.* 314, 2618–2633. doi: 10.1016/j.yexcr.2008.06.002

Yu, J. Y., Wang, T. W., Vojtek, A. B., Parent, J. M., and Turner, D. L. (2005). Use of short hairpin RNA expression vectors to study mammalian neural development. *Methods Enzymol.* 392, 186–199. doi: 10.1016/S0076-6879(04)92011-3

Zhang, H., Shykind, B., and Sun, T. (2012). Approaches to manipulating microRNAs in neurogenesis. *Front. Neurosci.* 6:196. doi: 10.3389/fnins.2012.00196

12

Identification and function of long non-coding RNA

Carl Ernst [1,2] *** and *Cynthia C. Morton*** [3,4,5]

[1] Douglas Hospital Research Institute, Montreal, QC, Canada
[2] Department of Psychiatry, McGill University, Montreal, QC, Canada
[3] Department of Obstetrics, Gynecology and Reproductive Biology, Brigham and Women's Hospital, Harvard Medical School, Boston, MA, USA
[4] Department of Pathology, Brigham and Women's Hospital, Harvard Medical School, Boston, MA, USA
[5] Medical and Population Genetics Program, The Broad Institute of M.I.T. and Harvard, Cambridge, MA, USA

***Correspondence:**
Carl Ernst, Douglas Hospital Research Institute, 6875 LaSalle Boulevard, Frank Common Building, Room 2101.2, Montreal, QC H4H 1R3, Canada
e-mail: carl.ernst@mcgill.ca

Long non-coding (lnc) RNAs are defined as non-protein coding RNAs distinct from housekeeping RNAs such as tRNAs, rRNAs, and snRNAs, and independent from small RNAs with specific molecular processing machinery such as micro- or piwi-RNAs. Recent studies of lncRNAs across different species have revealed a diverse population of RNA molecules of differing size and function. RNA sequencing studies suggest transcription throughout the genome, so there is a need to understand how sequence relates to functional and structural relationships amongst RNA molecules. Our synthesis of recent studies suggests that neither size, presence of a poly-A tail, splicing, direction of transcription, nor strand specificity are of importance to lncRNA function. Rather, relative genomic position in relation to a target is fundamentally important. In this review, we describe issues of key importance in functional assessment of lncRNA and how this might apply to lncRNAs important in neurodevelopment.

Keywords: non-coding RNA, epigenetics, gene regulation, neurodevelopment

THERE IS A WIDE VARIETY OF NON-CODING RNA IN MANY SPECIES

The co-occurrence of massively parallel sequencing technology applied to RNA and the recognition that non-coding, functional RNA species may not be restricted to X-chromosome inactivation (Jeon et al., 2012; Batista and Chang, 2013) or to protein synthesis machinery, have revealed an RNA universe of remarkable diversity in plant and animal cells. Non-coding (nc) RNAs, those RNA molecules that are not templates for protein synthesis, make up a large portion of the total RNA in the cell suggesting a profound functional importance. Despite their abundance, few ncRNAs have been studied and even fewer have been functionally characterized. These ncRNAs come in many forms: they can be very small or several hundred kilobases long; they may be spliced or unspliced; they can form linear or tertiary structures; they may or may not have a poly-A tail, and some interact with DNA, protein, or other RNA molecules (Novikova et al., 2013). As is described in this review, among various roles, Long non-coding (lnc) RNAs participate in guidance of large protein complexes to DNA leading to influence over locus-specific gene expression, and in the modification of expression or abundance of complementary messenger RNA strands. The wide diversity of function and form of ncRNA, combined with the explosive growth in newly identified ncRNA molecules, has lead to a need to understand better potential relationships of function between ncRNA and to consider a more categorical approach to classification.

Non-protein coding RNA has long been recognized in cells. Transfer RNAs and ribosomal RNAs were identified over 50 years ago; neither encodes a peptide chain, though they are integral components of the machinery for protein synthesis. Identification of these RNAs demonstrated that ncRNAs interact with proteins, perform specific cell functions, and operate autonomously from information transfer. Francis Crick's central dogma (1955) described information transfer amongst DNA, mRNA, and protein, and even at that time was recognized as an oversimplification. Crick himself subsequently built substantial flexibility into the model in 1970 such as the idea that RNA may be prone to "special" and "unknown" transfers of information (Crick, 1970). While he may not have imagined the diversity of RNA (Nakamura et al., 1996), there was a tacit acknowledgment that there was likely more to RNA than was known. Subsequent identification of ncRNAs unrelated to protein synthesis over 25 years ago, specifically, the catalytic ribozymes that formed secondary and tertiary structures thought to be important to early life on earth, re-enforced the diversity of RNA species (Sharp, 1985; Lamond and Gibson, 1990).

Several recent papers have identified new ncRNA species of particular function, and mechanistic insight into some of these different varieties of RNA reveal overlapping features, both in plant and animal cells. This diversity of RNA has been extensively reviewed with respect to small RNA-induced silencing complex (RISC)-related RNAs (e.g., Czech and Hannon, 2011) and lncRNAs (e.g., Rinn and Chang, 2012), with particular emphasis on disease specificity (Qureshi and Mehler, 2012; Sana et al., 2012) and epigenetic function (Lee, 2012). While there are several reviews that categorically describe different studies on RNA (Esteller, 2011; Wan et al., 2011), a more critical analysis of what defines a long ncRNA is lacking and the methods used for this, as well as a synthesis of lncRNA function across cell types. The purpose of the current review is to contextualize lncRNAs more generally, and review their effects in cellular function with respect to mechanism. This information will then be used to frame some of the preliminary studies emerging from studies of neurodevelopment.

CHARACTERIZATION OF lncRNAs

Several recent reviews have delineated ncRNA species into subcategories based on size (less or greater than 200 bases – often used as the definition of long versus short ncRNA), position (e.g., RNA species generated from the 3′UTRs or 5′UTRs), molecular interactions (e.g., Drosha- or Dicer-dependent), and molecular function, a good example of which is competitive antisense (AS) RNA that binds to microRNA and acts as a sponge to inhibit competitively microRNA from binding to a sense mRNA transcript (Cesana et al., 2011). It is unclear whether these categories are empirically determined, or whether they will prove relevant to categorization as future ncRNAs are discovered; indeed, the identification of such a wide diversity of RNA is consistent with what might be expected from an ancient, flexible molecule, capable of forming 3D structures and interacting with DNA, protein, or other RNAs.

What makes a lncRNA a lncRNA rather than some other RNA species? Are they a functionally distinct RNA product or are they a small part of the transcriptome that has been suggested to occur from large portions of the genome, mostly from recent ENCODE data (Carninci et al., 2005; Birney et al., 2007)? Certainly, a recent report (Guttman et al., 2013) suggests that intergenic lncRNAs are indeed non-coding, an issue that has been previously determined using algorithms (Lin et al., 2011) to assess whether different combinations of potential codons are similar to any other previously identified amino acid molecule. Most studies of lncRNA also attempt to determine whether an RNA species is localized to the nucleus, usually using RNA fluorescence *in situ* hybridization (FISH). Because translation occurs in the cytoplasm this might be evidence for the lack of coding potential. This analysis is somewhat arbitrary though, because ncRNA might be identified in the nuclear, chromatin, or cytoplasmic fraction of cells. Compartmentalization of lncRNAs in one of these fractions may be a defining feature of different lncRNA and may help to guide future classification schemes. Functional studies of lncRNA have also led to a proliferation of potential future categories for lncRNA, some of which are listed in **Table 1**, but this categorization creates its own problems in that many lncRNAs have overlapping features. This is a major issue at the moment and one likely to increase in complexity given the number of RNAs that can be detected from so many regions of the genome.

The current classification system will likely evolve as more RNA species are discovered, and classification of each ncRNA might follow a similar trajectory to that of protein coding gene classification. Genes that lead to an mRNA product are not divided up by length, genomic position, whether they are spliced or not for example, and numerous coding genes fit into different classification categories. Instead they are classified by function or conserved domains. Likely it is the novelty of the RNA field, facilitated by the detection of so many transcripts by massively parallel sequencing that is leading to the classification conundrum, but this may diminish as individual RNAs are functionally analyzed.

Several recent reports have carefully documented lncRNAs over a very unique range of function. To understand how lncRNAs are similar or different in both structure and function, we synthesize this information from recent papers to determine if there are any patterns or consistencies across RNA species. We focus on currently defined long RNA (>200 bp) and omit discussion of small RNAs such as microRNA, piwiRNA, or imprinting-related RNA's.

FUNCTIONAL STUDIES OF lncRNAs

The recognition of *HOTAIR* (Rinn et al., 2007) as a lncRNA that regulates gene expression in *cis* and *trans* (it is transcribed on chromosome 12 from the *HoxC* cluster and can regulate the chromosome 2 *HoxD* gene cluster) opened a new chapter for RNA molecules. *HOTAIR* defined a class of molecules distinct from housekeeping RNAs, microRNAs, and others, and which were not involved in fundamental imprinting processes. It hinted at the existence of RNA in the genome with regulatory functions directly related to their particular sequence and

Table 1 | Some examples of categorization of non-coding RNA.

Category	Description	Example	Reference
Intronic	Expressed from the intron of target	DMD lncRNA	Bovolenta et al. (2012)
H3K4me3	Has a methylated H3K4 promoter	lincRNA-p21	Huarte et al. (2010)
Antisense	Expressed from the non-coding strand and acts on the complementary target	BACE1-AS	Faghihi et al. (2008)
Enhancer	Expressed to enhance expression at a locus at some distance from target	p53 eRNAs	Melo et al. (2013)
Promoter	Acting on and expressed from the promoter of target	DBE-T	Cabianca et al. (2012)
Intergenic	Expressed at some distance from coding genes	lincRNA00299	Talkowski et al. (2012)
Trans-acting	Acting at some distance from target	Evf2	Bond et al. (2009)
Cis-acting	Acting on an adjacent target	AIRN	Santoro et al. (2013)
Small	Less than 200 bp in size	microRNA 137	Ripke et al. (2011)
Long	Greater than 200 bp in size	Fendrr	Grote et al. (2013)
5-UTR	Expressed near the 5′UTR of target	5′UTR ELK-1	Rahim et al. (2012)

provided an explanation for the targeting specificity required by ubiquitous binding molecules such as large chromatin modifying complexes. Since *HOTAIR*'s description, the function of many other lncRNAs has been revealed. Shown below is the large diversity of these molecules, from their genomic position in relation to the genes they regulate, their size, processing, and mechanism of action. While this diversity is large, there are also similarities, especially in reference to function. To demonstrate differences and similarities, we have selected all reports from the last 2 years (2011–2013) that have characterized positional, processing, and functional information of specific lncRNAs. **Table 2** lists structural information from lncRNAs that have been characterized functionally and this information is synthesized with the functional characteristics in the concluding remarks.

The lncRNA *COLDAIR* presents a series of themes for lncRNAs with respect to function. *COLDAIR* recruits polycomb repressive complex 2 (PRC2), a complex of proteins that can alter histone chemical groups to decrease gene expression, through an intermediate protein (homolog of Enhancer of zeste, *Drosophila*) and the binding of *COLDAIR* occurs through a CXC domain of this intermediate protein (see **Table 3** for a discussion of RNA:protein interaction domains). *COLDAIR* is expressed at equal ratios over time, despite an increasing repression of the target, suggesting increased affinity for the PRC2 interaction. *COLDAIR* reveals several potential areas of diversity/similarity amongst lncRNAs. What determines expression of the lncRNA itself? Is the lncRNA regulated in conjunction with the target or independently from it? Is the lncRNA action direct on the target or indirect? Is the lncRNA repressive or activating? Does it act on a single target or a cluster of targets at a locus?

The lncRNA *IRT1* differs significantly from the mechanistic action of *COLDAIR*, but also functions in a repressive manner to block expression of the target gene *IME1*. *IRT1* can respond within hours to a cell stressor to aid in the inhibition of gametogenesis, which means the repressive mechanism used by IRT1 may be specific to fast-acting effects. *IRT1* completely covers the 2 kb promoter of the target gene and functions to block transcription factors from binding and promoting transcription and acts in *cis*, similar to *COLDAIR*. Because *IRT1* is transcribed over the promoter of the target gene in the sense direction, it has an identical specificity to the DNA of the *IME1* promoter. The blocking of transcription factors in combination with aiding in the establishment of a repressive chromatin state through histone methyltransferases

Table 2 | Processing and positional diversity of lncRNA (in order described in text).

lncRNA [Ref]	Species	Description of all structural properties reported
COLDAIR (Heo and Sung, 2011)	*A. thaliana*	1100 base RNA expressed from the sense strand relative to target, has no poly-A tail, expressed from an intron of the target gene
IRT1 (van Werven et al., 2012)	*S. cerevisiae*	Expressed from promoter of target, 1.4 kb in length, not spliced and is transcribed from the same strand as the gene it regulates
NeST (Gomez et al., 2013)	*M. musculus*	Encoded on the antisense strand, contains six exons spread over a 45 kb region, transcript is 914 bases
Braveheart (Klattenhoff et al., 2013)	*M. musculus*	590 base RNA with three exons, 33% confined to the nucleus
NeST (Gomez et al., 2013)	*M. musculus*	Encoded on antisense strand, contains six exons spread over a 45 kb region, primary transcript is 914 bases
DBE-T (Cabianca et al., 2012)	*H. sapiens*	9.5 kb is one major product, transcribed from same strand as target genes, transcript contains one of many targets, nuclear and chromatin associated
HOTTIP (Wang et al., 2011)	*H. sapiens*	3,764-nucleotide, spliced and polyadenylated intergenic RNA, ~330 base product, regulates gene cluster
ANRIL (Wan et al., 2013)	*H. sapiens*	~126 kb transcript, spliced, 19 exons with an ~1.1 kb transcript, 13 isoforms transcribed in the antisense orientation of gene cluster, overlaps one target gene
lincMD1 (Cesana et al., 2011)	*M. musculus*	Three exons and two introns in 14 kb of genomic space, spliced product of 521 bases, accumulates as cytoplasmic poly-A+ RNA, transcribed on same strand in same orientation as microRNAs for which it acts as a decoy
TINCR (Kretz et al., 2013)	*H. sapiens*	Three exons, 3.7 kb transcript predominantly cytoplasmically expressed, over 100 different targets dispersed through genome
UCHL1-AS (Carrieri et al., 2012)	*M. musculus*	Four exons spanning 70 kb, overlaps the first 73 bases of *UCHL1*, including the AUG start codon, transcribed in reverse orientation in a head-to-head fashion, second intron of *UCHL1* contains the TSS for *UCHL1-AS*, enriched in the nucleus
1/2sbsRNA1 (Gong and Maquat, 2011)	*H. sapiens*	Present in cytoplasm, poly-A+, two alternative transcripts consist of 688 nucleotides, multiple targets throughout genome

Table 3 | Protein:RNA interacting domains.

Cysteine domains
CXC (redox-like), CXXC (redox), or C-X(6)-X (zinc finger or ring finger) motifs refer to the cysteine residue (C) with any amino acid (X) in between. These Cys residues may be active, meaning they can use their highly active sulfhydryl (SH) group to form a covalent bond with the OH group on the RNA sugar ring. These motifs can also interact with Ser, Thr, or Tyr amino acid residues to form S–S or S–O bonds on other proteins. An example of the cysteine RNA interacting domain are the Enhancer of zeste-related proteins with conserved X(6)-C-X(3)-C-X-C motifs.
WD domains
WD domains refer to peptide domains with rich repeats of tryptophan (W; hydrophobic) and aspartic acid (D; negatively charged) that are present in a large range of proteins. WD domains are non-catalytic and are thought to form a platform for the interaction of different cellular partners.

and deacetylases suggests that IRT1 can physically hinder TFs but also guide repressive chromatin complexes. Here the repressive effects are different than COLDAIR in that repression is due to the deposition of H3K4m2 and H3K36me by factors traveling with the RNA polymerase transcribing IRT1. Little is known about the regulation of IRT1, but it must be under tight control to hinder or allow expression of the target gene within such a timeframe of only hours.

NeST is a lncRNA that functions to increase transcription of the target gene and appears to act in trans although it is physically proximal to its target gene, *Ifng*. Evidence for trans action comes from NeST being genetically unlinked to its target gene and from experimental injection of NeST into cells. NeST action on the target gene is similar to IRT1 and COLDAIR in that it acts through a histone complex, but in this case it physically interacts with WDR5, which has a WD repeat domain of ~40 amino acids (see **Table 3**). WDR5 is a core subunit of complexes that catalyze the methylation of histone H3 at lysine 4, a mark of active gene expression, so NeST interacts directly with the histone modifying complex, unlike COLDAIR. It is likely that NeST functions to physically bring the histone modifying complex in close proximity to the target gene; it is 59 kb downstream from its target in mouse and 166 kb in human.

Bvht is a *cis*-acting lncRNA expressed only in mouse, meaning it may be a lncRNA that has recently gained a function. In a continuing theme for lncRNAs, it interacts directly with SUZ12 (a component of the PRC2 complex) and functions upstream of a key gene in lineage commitment. This differs from the action of *COLDAIR* that requires a binding partner for interaction with PRC2, whereas *bvht* directly interacts with one of the subunits. Notably, *SUZ12* has a zinc finger motif, which may explain the protein/RNA binding (see **Table 3**).

DBE-T is a human lncRNA expressed only in a diseased condition only that acts in *cis* and affects genes in a large chromosomal region, in contrast to COLDAIR, IRT1, NeST, or bvht, which appear to regulate a single target, although these targets often trigger expression of many other genes. DBE-T is transcribed from the first repeat of the D4Z4 repeat domain that is important for recruitment of PRC2. Repression of the region, controlled by PRC2 binding and spreading, commences at the repeat region, thus the basal state in adult cells is the repression of genes at this chromosomal locus. Loss of PRC2 at the repeat region corresponds with the binding of ASHL1, a histone lysine N-methyltransferase that is part of the TrixG group, which recruits DBE-T to chromatin. Thus, this lncRNA is at the crossroads of crosstalk between conflicting histone modifying complexes. While little is known about the regulation of lncRNAs, DBE-T may be an example of a positive feedback loop which may be a common theme for other lncRNAs – in other words, lncRNA expression may be regulated by targets of the target that the lncRNA itself regulates.

Similar to the positive feedback observed between DBE-T and ASHL1, HOTTIP lncRNA and WDR5 operate analogously. Similar to NeST, HOTTIP physically interacts with WDR5, and WDR5 forms a complex with MLL1, which is a H3K4 methyltransferase, triggering gene expression. HOTTIP maintains an appropriate level of the WDR5/MLL1 at a gene cluster, and its influence over the gene cluster dissipates as a function of distance from its site of transcription. Thus, this lncRNA interacts indirectly with a histone modifying complex, is involved in a feedback loop with its interacting partner, and activates expression of a cluster of genes as a function of distance from its site of expression.

ANRIL is a lncRNA transcribed immediately upstream of a cluster of genes important in human cell proliferation and is probably the most studied lncRNA to date because of its important role in cancer. ANRIL is transcribed on the AS strand of three intimately linked genes. It can bind to the transcript of the nearest gene at the locus, *INK4*, through complementary base pairing and can act at the promoter to recruit both PRC1 and PRC2 to repress transcription. ANRIL, while seemingly with a wider diversity of function than other polycomb recruitment lncRNAs, may actually foreshadow the function of other PRC-recruiting lncRNAs. Specifically, that they may have a wide variety of functions at a particular locus, and the only reason this has not yet been identified is because of experimental design strategies. We suspect many PRC-interacting lncRNAs will have many other functions that complement their effects. The multi-mechanistic function of ANRIL also showcases the idea that not all lncRNAs operate by recruiting large histone modifying complexes. Instead, recently identified lncRNA often operate by binding to the primary target or acting as a decoy of repressive effectors of the target.

lincMD1 and TINCR are two examples of non-PRC-recruiting lncRNAs with novel function to refine expression of a target. In contrast to lncRNAs COLDAIR, IRT1, NeST, BVHT, DBE-T, HOTTIP, and ANRIL, lincMD1, is a lncRNA that appears to be a by-product or remnant of microRNA processing (Ala et al., 2013). Specifically, this lncRNA can act as a decoy for the targets of microRNA produced from the same locus as lincMD1 (Cesana et al., 2011). TINCR also differs from all reported lncRNAs to date as it appears to bind to a 25 bp "TINCR-box" present in the RNA of different coding transcripts and influence levels of these transcripts in a STAU-dependent manner. STAU is an RNA guidance protein initially identified for its

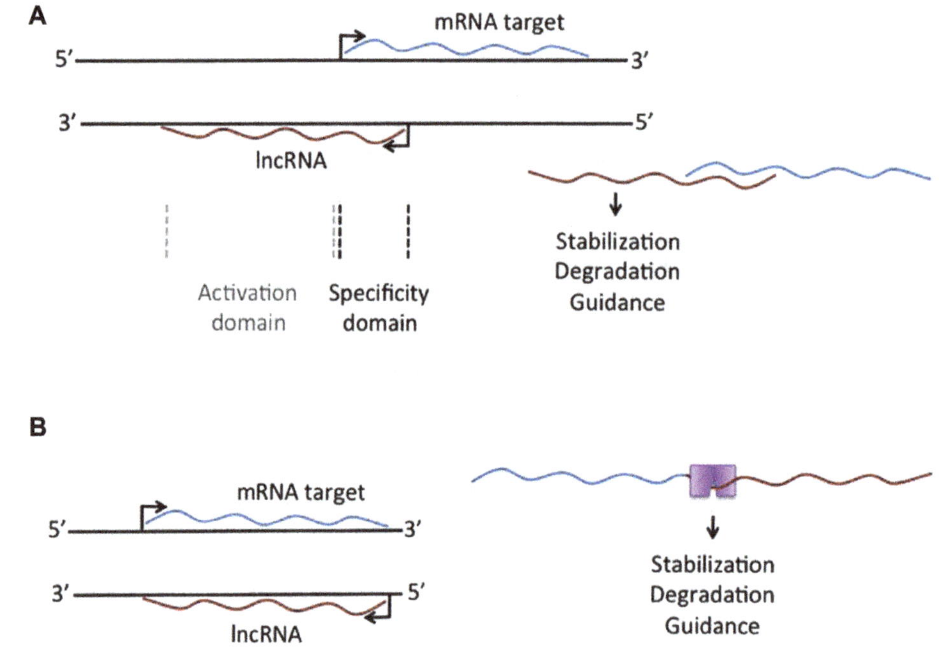

FIGURE 1 | Different mechanistic action of lncRNAs that overlap an mRNA target. (A) lncRNA expressed from the antisense strand may have an overlapping domain, providing specificity, and a non-overlapping activation domain. This activation domain could be a sequence that allows single-strand binding for different molecules, leading to stabilization or degradation of the RNA:RNA complex. **(B)** Complementary binding of overlapping mRNA and lncRNA could create a double-stranded binding site for protein binding, leading to selective degradation of stabilization of the RNA:RNA complex.

involvement in oocytes of *Drosophila.* All lncRNA described to date provide locus specificity for activating or repressive complexes to neighboring target genes, or interact directly with a target through sequence complementarity. TINCR on the other hand, appears to target specific RNA transcripts actively through an RNA sequence motif. lincMD1 also diverges drastically in that it is a by-product of pri-microRNA processing and acts to sponge the microRNAs from which it was initially processed. There may be many other pri-microRNA by-products that function similarly.

Another lncRNA that reportedly does not use large histone modifying complexes to alter a target, but instead operates through binding of a primary target, is AS-UCHL1. AS-UCHL1 has been shown recently to be important for proper targeting of sense transcript to polysomes, suggesting a stabilizing function for this lncRNA, demonstrated by a strong increase in UCHL1 protein with no difference in UCHL1 transcript on over-expression of AS-UCHL1. This principle of RNA stabilization to affect protein levels of targets may be a continuing theme for lncRNAs (e.g., Yoon et al., 2012). This lncRNA has a single target, binds it directly, and functions to increase protein of the primary target by stabilizing the mRNA. Besides this novel functional effect for a lncRNA, AS-UCHL1 action is driven by repeat elements within the AS transcript. Specifically, an orientation-specific SINEB2 repeat is required for the stabilizing function and protein synthesis activation of the sense strand. The overlapping portion of the AS gene with the sense gene thereby provides targeting information, while the SINEB2 region, which is not overlapped by the sense strand, confers protein synthesis activation (see **Figure 1A**).

The idea of repeat elements in the genome, acting through lncRNAs, has also been described with respect to Alu repeats, one of the most common repeats in the human genome. The description of overlapping Alu repeats, one in an AS strand and one in the 3′UTR of the sense strand, can lead to formation of a STAU1 binding site, which allows for STAU1 to stabilize base pairing and target the RNA duplex for degradation. Similar to all lncRNAs described here, these Alu-containing lncRNAs can regulate the levels of a transcript through an mRNA decay pathway. This was specifically demonstrated for *SERPINE1* and *FLJ21870* mRNAs between their 3′ UTR Alu element and the Alu element in a single lncRNA (see **Figure 1B**).

SYNTHESIS OF lncRNA FEATURES FROM DIFFERENT SPECIES

These examples support an important role for lncRNAs in the genome, and highlight the diverse function of lncRNAs, but also some similarities. First, there appears to be no relationship between the particular function of a lncRNA, its size, or how it is processed. This suggests that lncRNAs will represent a diverse range of characteristics. Second, neither transcriptional direction nor strand specificity appears to have an effect on function. The key element is that lncRNAs are produced either within their target gene or in the vicinity of target genes. Those lncRNAs produced from overlapping regions of their target gene are more likely to bind to the target, however, due to direct complementarity with the target. Whether these lncRNAs come from the same or the AS strand as a target appears not to have functional impact. Future experiments should document this for all newly described lncRNAs to determine whether this remains the case. These ideas may help guide issues of categorization of lncRNAs, and we propose a system that anchors lncRNAs in the target molecule. This may not prove useful for those lncRNA, like the intergenic lncRNAs, that do not appear to have nearby targets. Their function may prove to be completely different and independent from those lncRNAs expressed in relationship to mRNAs.

Most lncRNAs are modulators of a primary transcript suggesting that, evolutionarily, they arose after the primary transcript. For example, HOTTIP, either evolved with or after the *HOX* gene cluster that it regulates. There is little evidence for lncRNAs that operate in isolation (although the lincRNAs may be an exception, reflected by their distinct locations and conservation across species (Managadze et al., 2013), but rather form part of a transcriptional regulation complex of a specific target or a cluster of targets. This suggests that characterizing lncRNA might best be done grounded in the primary target rather than through effector status.

Many lncRNAs act though histone modifying complexes and appear to affect either a single target gene or a cluster of genes in a local region. They may require an intermediate binding partner for recruitment of the histone complex or interact directly with one of the proteins in these complexes. Determining whether lncRNAs bind directly to the target, interact directly with a histone modifying complex, or require a partner to bind histone modifying complex, will be important information as new lncRNAs are uncovered. Most lncRNAs do not share any sequence similarity (i.e., no indication yet of any conserved domain within lncRNAs) and it seems the position of lncRNAs in relation to the target(s) are of fundamental importance to their function. While there are many remarkable functions attributed to lncRNAs, we strongly suspect that the function of even these lncRNA will prove more diverse as they undergo further investigation.

lncRNAs IN CNS DEVELOPMENT

Functional and mechanistic data generated by studying ncRNAs in different molecular systems and species suggests lncRNAs likely play an important role in all cellular systems. As evidenced in the previous sections, lncRNAs most likely act as modifiers of a complementary RNA, interact with large histone complexes, interact with complementary DNA sequences, or act completely independently in the nucleus with no obvious partners required. Given this diverse potential, the complexity of the nervous system in any species might be partially due to the additional level of control over the cellular machinery by lncRNAs. lncRNAs may provide a means to tweak a cellular system at many levels and to operate rapidly in response to external signals whether axon guidance cues or environmental exposure. In line with these ideas, we synthesize recent reports of lncRNAs in the developing nervous system.

lncRNA IN NEURAL STEM CELLS

Some of the first experiments to underscore the importance of lncRNAs were done in mouse or human stem cells from. Stem cells used for research are either derived from the inner cell mass of a fertilized embryo (Thomson et al., 1998) or induced to pluripotency by the experimental increase of transcription factors normally present in early embryonic stages (Takahashi and Yamanaka, 2006) in terminally differentiated cells. These stem cells can be differentiated to a neural stem cell (NSC) fate and these NSCs can then give way to glia and neurons (Hu et al., 2010). In a wide ranging, exploratory analysis, Ng and colleagues (Ng et al., 2012) examined neuronal differentiation from human embryonic stem cells (hESCs). They used a two-step differentiation protocol from radial glial-like cells to largely dopaminergic cells, and then assessed global gene expression levels of pre-selected lncRNAs in radial-glial cells compared to dopaminergic-like cells. They identified 35 lncRNAs that were differentially expressed between progenitor and mature states, and then tested some of these for functionality. Following similar designs of non-neuronal studies of lncRNA, they assessed the association of differentially expressed lncRNA with SUZ12 and the neurogenesis repressor complex REST/NRSF (neural restrictive silencer factor; Naruse et al., 1999). In a study using just three lncRNAs, their data supported interaction of one lncRNA with REST and another lncRNA with SUZ12. While the SUZ12 interaction is consistent with previous lncRNA studies, the interaction with REST/NRSF is novel for lncRNAs in neurons, although it does associate with HOTAIR in non-neuronal cell types to repress expression of neuronal genes. This suggests that the lncRNA in the Ng et al. (2012) study may interact with REST to regulate neuronal gene expression.

While ES (embryonic stem cells)- and iPSC (induced pluripotent stem cell)-derived NSCs may not perfectly capture the developmental progression of the human brain, they provide an excellent model with which to screen for important factors as the cells develop from stem cells to electrically active neurons. A study monitoring iPSC-NSC differentiation accompanied by RNA sampling at different timepoints, contrasted with brain temporal lobe brain tissue RNA levels from the same donor has revealed a gradual increase in the expression of different lncRNAs as NSCs differentiate (Hjelm et al., 2013). This is supported by our own study, where we observed an increase in the neurodevelopmentally important intergenic lncRNA00299 as iPSC-NSCs differentiated (Talkowski et al., 2012). A recent report using adult NSCs in mice has further confirmed that lncRNAs increase as cells differentiate. Ramos et al. (2013) sorted stem cells of the sub-ventricular zone of mice and screened these cells for expression levels of different lncRNAs creating publically accessible expression maps for lncRNAs that may be relevant to glial-neuron specification in adult brain.

lncRNA IN DEVELOPING BRAIN

In the mammalian brain, lncRNAs have long been recognized as important in neurodevelopment, although they were traditionally referred to as AS transcripts. An example of this is the AS transcripts near the *Sox4* and *Sox11* loci produced during development of the mouse cerebral cortex (Ling et al., 2009). Sox proteins contain a high mobility group, and this refers to the ability of these proteins to bind and bend DNA. Using global gene expression analysis tools, Ling et al. (2011) showed that AS *Sox4* and *Sox11* transcripts are produced during proliferating and differentiating states, suggesting that the regulation of these important genes is by complementary lncRNAs. Recently, this same group documented a similar effect with respect to *Nrgn* and *Camk2n1* gene product in mouse cerebral corticogenesis. A recent study in adult brain also suggests that electrical activity in neurons stimulates lncRNA

(Barry et al., 2013). Given the importance of initial synaptic contacts and communication between cells, it stands to reason that there may be lncRNAs that respond to activity and independent lncRNAs that increase expression as differentiation proceeds, similar to data from *in vitro* NSC models. Of intense interest as well are the loci of the genome where AS transcripts are transcribed from the same genomic locations as brain-relevant genes and how some of these may be specific to human. For example, *BDNF* is transcribed from chromosome 11 and an AS transcript is produced from the opposite strand in humans but not in mouse (Aid et al., 2007; Pruunsild et al., 2007). The discovery of an ever increasing number of AS transcripts that may assist in regulation of genes fundamental to brain development will likely be forthcoming. Determining the exact role of these AS transcripts, their size, and binding dynamics will be important.

Recent data from our group suggest that lncRNAs may be important in neurodevelopmental disease (Talkowski et al., 2012); we showed that a nuclear, multi-exon lncRNA was disrupted in subjects with global developmental delays. This complements work from others, where ncRNAs have long been suspected of causing certain neurological problems, the best example of which may be Prader–Willi syndrome (PWS). PWS is characterized by intellectual disability, sleep disorders, and psychosis, and can be caused by deletion of 15q11-13 on the paternal chromosome. Genes in this region are suppressed on the maternal chromosome, meaning that paternally expressed genes likely provide the optimal dosage of expression. The minimal required locus within this ~10 Mb region implicates *116HG*, a lncRNA retained in the nucleus, as well as the small nucleolar RNA *SNORD116* (Sahoo et al., 2008). Both ncRNAs are the control of the imprinting control region, involving multiple overlap of genes – suggesting that transcription and splicing in this region are complex. Recently, Powell et al. (2013) reported the first experiments to determine the function of lncRNA 116HG. They found that 116HG forms RNA "clouds" specific to nuclei in mouse brain, and that these 116HG clouds change size and shape in predictable ways as the brain develops. Using RNA and DNA FISH mapping, they show that 116HG likely interacts with the paternal *UBE3a* locus, a gene found immediately upstream of the 116HG locus and known to be important in neurodevelopment. Their data further suggest that 116HG interacts with RBBP5, a subunit of the MLL complex, which acts as a transcriptional activator by methylation of H3K4. This model conforms nicely to what is known of lncRNA functions in other species; 116HG might associate with MLL complex and interact with histones at the *UBE3a* locus. How 116HG itself is regulated is unknown, but this will be clearly important to understand better the neurobiology of PWS.

lncRNAs likely have a role in many aspects of the cell, and brain development might be an area where their structure and function is particularly suited. This may suggest that many more lncRNAs await discovery in novel systems as well as in added layers of control for well known processes of neurodevelopment.

ACKNOWLEDGMENTS

Carl Ernst is supported by a Canada Research Chair and holds funding from the Banting Foundation of Toronto and the Natural Science and Engineering Research Council of Canada. Cynthia C. Morton acknowledges support from the NIH (GM061354).

REFERENCES

Aid, T., Kazantseva, A., Piirsoo, M., Palm, K., and Timmusk, T. (2007). Mouse and rat BDNF gene structure and expression revisited. *J. Neurosci. Res.* 85, 525–535. doi: 10.1002/jnr.21139

Ala, U., Karreth, F. A., Bosia, C., Pagnani, A., Taulli, R., Leopold, V., et al. (2013). Integrated transcriptional and competitive endogenous RNA networks are cross-regulated in permissive molecular environments. *Proc. Natl. Acad. Sci. U.S.A.* 110, 7154–7159. doi: 10.1073/pnas.1222509110

Barry, G., Briggs, J. A., Vanichkina, D. P., Poth, E. M., Beveridge, N. J., Ratnu, V. S., et al. (2013). The long non-coding RNA Gomafu is acutely regulated in response to neuronal activation and involved in schizophrenia-associated alternative splicing. *Mol. Psychiatry* doi: 10.1038/mp.2013.45 [Epub ahead of print].

Batista, P. J., and Chang, H. Y. (2013). Long noncoding RNAs: cellular address codes in development and disease. *Cell* 152, 1298–1307. doi: 10.1016/j.cell.2013.02.012

Birney, E., Stamatoyannopoulos, J. A., Dutta, A., Guigo, R., Gingeras, T. R., Margulies, E. H., et al. (2007). Identification and analysis of functional elements in 1% of the human genome by the ENCODE pilot project. *Nature* 447, 799–816. doi: 10.1038/nature05874

Bond, A. M., Vangompel, M. J., Sametsky, E. A., Clark, M. F., Savage, J. C., Disterhoft, J. F., et al. (2009). Balanced gene regulation by an embryonic brain ncRNA is critical for adult hippocampal GABA circuitry. *Nat. Neurosci.* 12, 1020–1027. doi: 10.1038/nn.2371

Bovolenta, M., Erriquez, D., Valli, E., Brioschi, S., Scotton, C., Neri, M., et al. (2012). The DMD locus harbours multiple long non-coding RNAs which orchestrate and control transcription of muscle dystrophin mRNA isoforms. *PLoS ONE* 7:e45328. doi: 10.1371/journal.pone.0045328

Cabianca, D. S., Casa, V., Bodega, B., Xynos, A., Ginelli, E., Tanaka, Y., et al. (2012). A long ncRNA links copy number variation to a polycomb/trithorax epigenetic switch in FSHD muscular dystrophy. *Cell* 149, 819–831. doi: 10.1016/j.cell.2012.03.035

Carninci, P., Kasukawa, T., Katayama, S., Gough, J., Frith, M. C., Maeda, N., et al. (2005). The transcriptional landscape of the mammalian genome. *Science* 309, 1559–1563. doi: 10.1126/science.1112014

Carrieri, C., Cimatti, L., Biagioli, M., Beugnet, A., Zucchelli, S., Fedele, S., et al. (2012). Long non-coding antisense RNA controls Uchl1 translation through an embedded SINEB2 repeat. *Nature* 491, 454–457. doi: 10.1038/nature11508

Cesana, M., Cacchiarelli, D., Legnini, I., Santini, T., Sthandier, O., Chinappi, M., et al. (2011). A long noncoding RNA controls muscle differentiation by functioning as a competing endogenous RNA. *Cell* 147, 358–369. doi: 10.1016/j.cell.2011.09.028

Crick, F. (1970). Central dogma of molecular biology. *Nature* 227, 561–563. doi: 10.1038/227561a0

Czech, B., and Hannon, G. J. (2011). Small RNA sorting: matchmaking for Argonautes. *Nat. Rev. Genet.* 12, 19–31. doi: 10.1038/nrg2916

Esteller, M. (2011). Non-coding RNAs in human disease. *Nat. Rev. Genet.* 12, 861–874. doi: 10.1038/nrg3074

Faghihi, M. A., Modarresi, F., Khalil, A. M., Wood, D. E., Sahagan, B. G., Morgan, T. E., et al. (2008). Expression of a noncoding RNA is elevated in Alzheimer's disease and drives rapid feed-forward regulation of beta-secretase. *Nat. Med.* 14, 723–730. doi: 10.1038/nm1784

Gomez, J. A., Wapinski, O. L., Yang, Y. W., Bureau, J. F., Gopinath, S., Monack, D. M., et al. (2013). The NeST long ncRNA controls microbial susceptibility and epigenetic activation of the interferon-gamma locus. *Cell* 152, 743–754. doi: 10.1016/j.cell.2013.01.015

Gong, C., and Maquat, L. E. (2011). lncRNAs transactivate STAU1-mediated mRNA decay by duplexing with 3′ UTRs via Alu elements. *Nature* 470, 284–288. doi: 10.1038/nature09701

Grote, P., Wittler, L., Hendrix, D., Koch, F., Wahrisch, S., Beisaw, A., et al. (2013). The tissue-specific lncRNA Fendrr is an essential regulator of heart and body wall development in the mouse. *Dev. Cell* 24, 206–214. doi: 10.1016/j.devcel.2012.12.012

Guttman, M., Russell, P., Ingolia, N. T., Weissman, J. S., and Lander, E. S. (2013). Ribosome profiling provides evidence that large noncoding RNAs do not encode proteins. *Cell* 154, 240–251. doi: 10.1016/j.cell.2013.06.009

Heo, J. B., and Sung, S. (2011). Vernalization-mediated epigenetic silencing by a long intronic noncoding RNA. *Science* 331, 76–79. doi: 10.1126/science.1197349

Hjelm, B. E., Salhia, B., Kurdoglu, A., Szelinger, S., Reiman, R. A., Sue, L. I., et al. (2013). In vitro-differentiated neural cell cultures progress towards donor-identical brain tissue. *Hum. Mol. Genet.* 22, 3534–3546. doi: 10.1093/hmg/ddt208

Hu, B. Y., Weick, J. P., Yu, J., Ma, L. X., Zhang, X. Q., Thomson, J. A., et al. (2010). Neural differentiation of human induced pluripotent stem cells follows developmental principles but with variable potency. *Proc. Natl. Acad. Sci. U.S.A.* 107, 4335–4340. doi: 10.1073/pnas.0910012107

Huarte, M., Guttman, M., Feldser, D., Garber, M., Koziol, M. J., Kenzelmann-Broz, D., et al. (2010). A large intergenic noncoding RNA induced by p53 mediates global gene repression in the p53 response. *Cell* 142, 409–419. doi: 10.1016/j.cell.2010.06.040

Jeon, Y., Sarma, K., and Lee, J. T. (2012). New and Xisting regulatory mechanisms of X chromosome inactivation. *Curr. Opin. Genet. Dev.* 22, 62–71. doi: 10.1016/j.gde.2012.02.007

Klattenhoff, C. A., Scheuermann, J. C., Surface, L. E., Bradley, R. K., Fields, P. A., Steinhauser, M. L., et al. (2013). Braveheart, a long noncoding RNA required for cardiovascular lineage commitment. *Cell* 152, 570–583. doi: 10.1016/j.cell.2013.01.003

Kretz, M., Siprashvili, Z., Chu, C., Webster, D. E., Zehnder, A., Qu, K., et al. (2013). Control of somatic tissue differentiation by the long non-coding RNA TINCR. *Nature* 493, 231–235. doi: 10.1038/nature11661

Lamond, A. I., and Gibson, T. J. (1990). Catalytic RNA and the origin of genetic systems. *Trends Genet.* 6, 145–149. doi: 10.1016/0168-9525(90)90148-Y

Lee, J. T. (2012). Epigenetic regulation by long noncoding RNAs. *Science* 338, 1435–1439. doi: 10.1126/science.1231776

Lin, M. F., Jungreis, I., and Kellis, M. (2011). PhyloCSF: a comparative genomics method to distinguish protein coding and non-coding regions. *Bioinformatics* 27, i275-282. doi: 10.1093/bioinformatics/btr209

Ling, K. H., Hewitt, C. A., Beissbarth, T., Hyde, L., Banerjee, K., Cheah, P. S., et al. (2009). Molecular networks involved in mouse cerebral corticogenesis and spatio-temporal regulation of Sox4 and Sox11 novel antisense transcripts revealed by transcriptome profiling. *Genome Biol.* 10, R104. doi: 10.1186/gb-2009-10-10-r104

Ling, K. H., Hewitt, C. A., Beissbarth, T., Hyde, L., Cheah, P. S., Smyth, G. K., et al. (2011). Spatiotemporal regulation of multiple overlapping sense and novel natural antisense transcripts at the Nrgn and Camk2n1 gene loci during mouse cerebral corticogenesis. *Cereb. Cortex* 21, 683–697. doi: 10.1093/cercor/bhq141

Managadze, D., Lobkovsky, A. E., Wolf, Y. I., Shabalina, S. A., Rogozin, I. B., and Koonin, E. V. (2013). The vast, conserved mammalian lincRNome. *PLoS Comput. Biol.* 9:e1002917. doi: 10.1371/journal.pcbi.1002917

Melo, C. A., Drost, J., Wijchers, P. J., Van De Werken, H., De Wit, E., Oude Vrielink, J. A., et al. (2013). eRNAs are required for p53-dependent enhancer activity and gene transcription. *Mol. Cell* 49, 524–535. doi: 10.1016/j.molcel.2012.11.021

Nakamura, A., Amikura, R., Mukai, M., Kobayashi, S., and Lasko, P. F. (1996). Requirement for a noncoding RNA in *Drosophila* polar granules for germ cell establishment. *Science* 274, 2075–2079. doi: 10.1126/science.274.5295.2075

Naruse, Y., Aoki, T., Kojima, T., and Mori, N. (1999). Neural restrictive silencer factor recruits mSin3 and histone deacetylase complex to repress neuron-specific target genes. *Proc. Natl. Acad. Sci. U.S.A.* 96, 13691–13696. doi: 10.1073/pnas.96.24.13691

Ng, S. Y., Johnson, R., and Stanton, L. W. (2012). Human long non-coding RNAs promote pluripotency and neuronal differentiation by association with chromatin modifiers and transcription factors. *EMBO J.* 31, 522–533. doi: 10.1038/emboj.2011.459

Novikova, I. V., Hennelly, S. P., Tung, C. S., and Sanbonmatsu, K. Y. (2013). Rise of the RNA machines: exploring the structure of long non-coding RNAs. *J. Mol. Biol.* 425, 3731–3746. doi: 10.1016/j.jmb.2013.02.030

Powell, W. T., Coulson, R. L., Crary, F. K., Wong, S. S., Ach, R. A., Tsang, P., et al. (2013). A Prader–Willi locus lncRNA cloud modulates diurnal genes and energy expenditure. *Hum. Mol. Genet.* doi: 10.1093/hmg/ddt281 [Epub ahead of print].

Pruunsild, P., Kazantseva, A., Aid, T., Palm, K., and Timmusk, T. (2007). Dissecting the human BDNF locus: bidirectional transcription, complex splicing, and multiple promoters. *Genomics* 90, 397–406. doi: 10.1016/j.ygeno.2007.05.004

Qureshi, I. A., and Mehler, M. F. (2012). Emerging roles of non-coding RNAs in brain evolution, development, plasticity and disease. *Nat. Rev. Neurosci.* 13, 528–541. doi: 10.1038/nrn3234

Rahim, G., Araud, T., Jaquier-Gubler, P., and Curran, J. (2012). Alternative splicing within the elk-1 5′ untranslated region serves to modulate initiation events downstream of the highly conserved upstream open reading frame 2. *Mol. Cell. Biol.* 32, 1745–1756. doi: 10.1128/MCB.06751-11

Ramos, A. D., Diaz, A., Nellore, A., Delgado, R. N., Park, K. Y., Gonzales-Roybal, G., et al. (2013). Integration of genome-wide approaches identifies lncRNAs of adult neural stem cells and their progeny in vivo. *Cell Stem Cell* 12, 616–628. doi: 10.1016/j.stem.2013.03.003

Rinn, J. L., and Chang, H. Y. (2012). Genome regulation by long non-coding RNAs. *Annu. Rev. Biochem.* 81, 145–166. doi: 10.1146/annurev-biochem-051410-092902

Rinn, J. L., Kertesz, M., Wang, J. K., Squazzo, S. L., Xu, X., Brugmann, S. A., et al. (2007). Functional demarcation of active and silent chromatin domains in human HOX loci by noncoding RNAs. *Cell* 129, 1311–1323. doi: 10.1016/j.cell.2007.05.022

Sahoo, T., Del Gaudio, D., German, J. R., Shinawi, M., Peters, S. U., Person, R. E., et al. (2008). Prader–Willi phenotype caused by paternal deficiency for the HBII-85 C/D box small nucleolar RNA cluster. *Nat. Genet.* 40, 719–721. doi: 10.1038/ng.158

Sana, J., Faltejskova, P., Svoboda, M., and Slaby, O. (2012). Novel classes of non-coding RNAs and cancer. *J. Transl. Med.* 10, 103. doi: 10.1186/1479-5876-10-103

Santoro, F., Mayer, D., Klement, R. M., Warczok, K. E., Stukalov, A., Barlow, D. P., et al. (2013). Imprinted Igf2r silencing depends on continuous Airn lncRNA expression and is not restricted to a developmental window. *Development* 140, 1184–1195. doi: 10.1242/dev.088849

Sharp, P. A. (1985). On the origin of RNA splicing and introns. *Cell* 42, 397–400. doi: 10.1016/0092-8674(85)90092-3

Takahashi, K., and Yamanaka, S. (2006). Induction of pluripotent stem cells from mouse embryonic and adult fibroblast cultures by defined factors. *Cell* 126, 663–676. doi: 10.1016/j.cell.2006.07.024

Talkowski, M. E., Maussion, G., Crapper, L., Rosenfeld, J. A., Blumenthal, I., Hanscom, C., et al. (2012). Disruption of a large intergenic noncoding RNA in subjects with neurodevelopmental disabilities. *Am. J. Hum. Genet.* 91, 1128–1134. doi: 10.1016/j.ajhg.2012.10.016

Thomson, J. A., Itskovitz-Eldor, J., Shapiro, S. S., Waknitz, M. A., Swiergiel, J. J., Marshall, V. S., et al. (1998). Embryonic stem cell lines derived from human blastocysts. *Science* 282, 1145–1147. doi: 10.1126/science.282.5391.1145

van Werven, F. J., Neuert, G., Hendrick, N., Lardenois, A., Buratowski, S., Van Oudenaarden, A., et al. (2012). Transcription of two long noncoding RNAs mediates mating-type control of gametogenesis in budding yeast. *Cell* 150, 1170–1181. doi: 10.1016/j.cell.2012.06.049

Wan, G., Mathur, R., Hu, X., Liu, Y., Zhang, X., Peng, G., et al. (2013). Long non-coding RNA ANRIL (CDKN2B-AS) is induced by the ATM-E2F1 signaling pathway. *Cell. Signal.* 25, 1086–1095. doi: 10.1016/j.cellsig.2013.02.006

Wan, Y., Kertesz, M., Spitale, R. C., Segal, E., and Chang, H. Y. (2011). Understanding the transcriptome through RNA structure. *Nat. Rev. Genet.* 12, 641–655. doi: 10.1038/nrg3049

13

Toward a unified biological hypothesis for the BDNF Val66Met-associated memory deficits in humans: a model of impaired dendritic mRNA trafficking

Gabriele Baj[1], Davide Carlino[2,3], Lucia Gardossi[4] and Enrico Tongiorgi[1]*

[1] *Department of Life Sciences, Brain Centre for Neurosciences, University of Trieste, Trieste, Italy*
[2] *Brain Centre for Neurosciences, University of Trieste, Trieste, Italy*
[3] *Department of Medical, Surgical, and Health Sciences, Psychiatric Clinic, University of Trieste, Trieste, Italy*
[4] *Dipartimento di Scienze Chimiche e Farmaceutiche, University of Trieste, Trieste, Italy*

***Correspondence:**
Enrico Tongiorgi, Department of Life Sciences, University of Trieste, Via L. Giorgieri, 5 Building Q, 34127 Trieste, Italy
e-mail: tongi@units.it

Brain-derived neurotrophic factor (BDNF) represents promotesa key molecule for the survival and differentiation of specific populations of neurons in the central nervous system. BDNF also regulates plasticity-related processes underlying memory and learning. A common single nucleotide polymorphism (SNP) rs6265 has been identified on the coding sequence of human *BDNF* located at 11p13. The SNP rs6265 is a single base mutation with an adenine instead of a guanine at position 196 (G196A), resulting in the amino acid substitution Val66Met. This polymorphism only exists in humans and has been associated with a plethora of effects ranging from molecular, cellular and brain structural modifications in association with deficits in social and cognitive functions. To date, the literature on Val66Met polymorphism describes a complex and often conflicting pattern of effects. In this review, we attempt to provide a unifying model of the Val66Met effects. We discuss the clinical evidence of the association between Val66Met and memory deficits, as well as the molecular mechanisms involved including the reduced transport of BDNF mRNA to the dendrites as well as the reduced processing and secretion of BDNF protein through the regulated secretory pathway.

Keywords: neurotrophins, BDNF, memory deficits, post-traumatic stress disorder, hippocampus atrophy, dendritic mRNA trafficking, regulated protein secretion

BDNF AND ITS Val66Met POLYMORPHISM

The development and plasticity of the nervous system are deeply influenced by a family of neurotrophic factors called neurotrophins. In mammals, the neurotrophin family consists of four related proteins: Nerve growth factor (NGF; Levi-Montalcini and Booker, 1960), Brain-derived neurotrophic factor (BDNF; Barde et al., 1982), neurotrophin-3 (NT3; Hohn et al., 1990) and neurotrophin-4 (NT4; Berkemeier et al., 1991). Since its discovery three decades ago, BDNF has been firmly implicated in the development of the nervous system in vertebrates and in the support of survival and differentiation of specific populations of neurons in the central nervous system (Bibel and Barde, 2000; Binder and Scharfman, 2004; Chao et al., 2006). More recently, BDNF has also emerged in several functions in the adult brain in which it is involved in the main plasticity-related processes including memory and learning (Tyler et al., 2002; Yamada et al., 2002).

A common single nucleotide polymorphism (SNP) rs6265 has been identified on the coding sequence of human *BDNF* located at 11p13 (Hall et al., 2003). This single base mutation, presenting an adenine instead of a guanine at position 196 (G196A), results in the amino acid substitution Val66Met (Hall et al., 2003). The polymorphism Val66Met only exists in humans and has been associated with a plethora of effects ranging from detrimental molecular, cellular and brain structural modifications associated with social and cognitive dysfunction (Dincheva et al., 2012). While the literature on this polymorphism is rapidly increasing, there is little consensus on the pattern of results. In this review, we discuss how this single nucleotide variant affects molecular mechanisms of memory formation and maintenance and summarize clinical evidence on the association between Val66Met SNP and memory deficits. In conclusion, we hypothesize a biological mechanism underlying the memory deficits associated with the Val66Met polymorphism of BDNF.

CLINICAL ASPECTS OF Val66Met

The first association between the BDNF Val66Met SNP and a clinical phenotype was reported in schizophrenic patients, their relatives and healthy controls (Egan et al., 2003). This study showed a specific effect on cognitive functions where the Met BDNF allele reduced the delayed recall of episodic memory in all three groups but had no influence on other cognitive domains or *intelligence quotient* (IQ). Both Egan et al. (2003) and Hariri et al. (2003) demonstrated that Met BDNF carriers displayed reduced hippocampal engagement during encoding and retrieval of a spatial task with respect to Val/Val homozygotes. Two independent MRI investigations (Pezawas et al., 2004; Szeszko et al., 2005) extended the findings of Egan and Hariri (Egan et al., 2003; Hariri et al., 2003) by demonstrating in affectively ill individuals that Val/Met heterozygotes displayed lower hippocampal volumes than their Val/Val counterparts. In addition, there was volume reduction of the gray matter in the dorsolateral prefrontal cortex (DLPFC), which is implicated in learning and memory processes involving also the hippocampus (Pezawas et al., 2004; Hwang et al., 2006; Benjamin et al., 2010). Further, in a community of elderly Caucasian individuals, BDNF Val/Val homozygotes performed significantly better compared to both Val/Met heterozygous and Met/Met homozygous individuals on a delayed recall task and an alphabet-coding task (ACT), a measure of processing speed. Another study also noted a decrease in hippocampal volume in Met BDNF allele carries, although it did not reach statistical significance (Goldberg et al., 2008) while other authors confirmed the hippocampal volume reduction (Benjamin et al., 2010; Teh et al., 2012; Tost et al., 2013).

In a fascinating study, Schofield et al. (2009) found a significant increase in hippocampal activity in Met BDNF carriers compared to BDNF Val/Val homozygotes during the auditory oddball task. The auditory oddball task has been asserted to measure auditory attention and auditory capacity. It is a task in which the subject must detect a relevant ("oddball") stimulus, which is presented infrequently and randomly within a train of task-irrelevant stimuli. Schofield and colleagues argued that Met BDNF carriers might require greater hippocampal activation than their BDNF Val/Val counterparts to process auditory stimuli presented during the task, inadvertently depleting resources for prefrontal processing of stimuli. Similar circuit dysregulation has been observed previously. A protective role for the Met allele has also been proposed for enhanced verbal reasoning in the elderly (Harris et al., 2006), systemic lupus erythematosus (Oroszi et al., 2006), multiple sclerosis patients (Zivadinov et al., 2007; Cerasa et al., 2010), and Parkinson's disease (Foltynie et al., 2005), although others have challenged these findings (Liguori et al., 2007; Miyajima et al., 2008).

Additional studies found either a gender-specific association (Echeverria et al., 2005) or no association in schizophrenic and bipolar patients (Strauss et al., 2004; Tramontina et al., 2009) as well as in other clinical conditions characterized by cognitive impairment such as multiple sclerosis (Cerasa et al., 2010), HIV-associated neurocognitive disorders (Levine et al., 2012) and Parkinson's disease (Guerini et al., 2009). The disagreement between results could be due to a number of differences in study design (Hong et al., 2011). First, these studies are typically characterized by relatively small sample size. Second, there is a noticeable variability in the phenotypes analyzed in these researches. Specifically, it is striking how many different tasks have been used in the field to assess arguably similar (or the same) cognitive functions. In animal studies, BDNF is studied especially in the hippocampus and therefore, the impact of the Val66Met polymorphism could have been more evident in studies focused on hippocampus-specific tasks. More orchestrated approaches to behavioral phenotyping for cognition in the design of genetic and allelic association studies on BDNF are highly needed and are likely to be fruitful. Otherwise, using the mere Mini Mental State Examination (MMSE), the effects of BDNF Val66Met polymorphism on cognition may not emerge (Zivadinov et al., 2007; Forlenza et al., 2010).

More recently, Mandelman and Grigorenko (2012) conducted a meta-analysis examining the relationship between BDNF Val66Met and cognition, but did not find significant associations between the Val66Met polymorphism and any of the phenotypes that were included. The obvious additional candidates for sources of such between-study heterogeneity are demographic characteristics such as gender, age and phase of illness, ethnicity, physical exercise, cardio-vascular health status and diagnosis (Verhagen et al., 2010; Lu et al., 2012; Martinho et al., 2012; Nagata et al., 2012; Smith et al., 2012). Furthermore, it is possible that the genetic variant may be associated with some intermediate phenotypes, which, in turn, could be related to some but not all of the cognitive phenotypes examined. Given the multitude of BDNF protein isoforms and the diversity of its transcripts in different brain areas, it is conceivable that cognitive phenotypes should be grouped not by their behavioral similarities, but by similarities in the brain activation pathways that underlie these phenotypes.

BIOLOGICAL ASPECTS OF THE Val66Met MUTATION

It is important to highlight that the vast majority of the results reported in the literature and summarized here, were obtained by investigations on homozygous BDNF Met/Met (Met BDNF) mice while the double mutated Met/Met allele is quite rare in humans. Indeed, data from human population are collected mostly from individuals with BDNF Val/Met (single allele mutation) or BDNF Val/Val (Mandelman and Grigorenko, 2012). The biological effects of the SNP rs6265 in the BDNF coding sequence have been studied using several different model systems, starting from cell culture, to animal models or in correlation with diseases affecting the human population. In their initial work, using an *in vitro* neuronal culture system, Egan and coworkers (Egan et al., 2003) demonstrated that Met BDNF in fusion with GFP and transfected in neurons is produced at levels similar to the control Val BDNF-GFP. The knock-in animal model for the Met BDNF allele created by Chen et al. (2006; Dincheva et al., 2012) confirmed that the total BDNF production is not affected by the polymorphism. A recent detailed analysis of BDNF protein in specific brain areas of BDNF Val and Met knock-in mice showed slight but significant reduction of BDNF in the hippocampus (HPC) and in prefrontal (PFC) cortex while no variations were found at the level of amygdala and striatum (Bath et al., 2012; Yu et al., 2012).

Chen and coworkers were first to describe the functional importance of the Met BDNF allele in the brain. They reported alterations in hippocampal anatomy of the Met BDNF knock-in model. The total hippocampal volume of Met BDNF mice was found to be reduced and dentate gyrus (DG) neurons were measured as significantly smaller in total volume and dendritic complexity while the soma size was not affected (Chen et al., 2008). Similar neuronal volumetric reduction was reported also on ventromedial prefrontal cortex cells, while neurons from the striatum were not affected (Yu et al., 2009). The same mouse model was used to establish the effect of Met polymorphism in prefrontal cortex (PFC) where a specific atrophy of apical dendrites of layer 5 pyramidal cells was found (Liu et al., 2012). Furthermore, the induction of synaptogenesis in PFC by ketamine administration on brain slices was almost abolished indicating impaired synaptic formation/maturation. Met BDNF mice presented alterations in the generation of LTP at the CA3-CA1 synapse. Ninan and colleagues were able to dissect out the mechanism of this deficit revealing that NMDA receptor-dependent LTP was specifically affected while the non-NMDA receptor neurotransmission (i.e., mGluR) was normal (Ninan et al., 2010; Pattwell et al., 2012). Yet, it remains unclear if these BDNF effects on the NMDA receptor are either due to a deficit in the basal regulation of the NMDA receptor trafficking and expression or to an altered acute activity-dependent release of BDNF, in Met BDNF mice. Additional evidence of deficits in neuronal activation was observed in ventromedial prefrontal cortex of Met BDNF mice following fear extinction tests: the Met substitution results in reduced cellular activation (–50%), visualized by cFOS positivitity (Yu et al., 2009).

EFFECTS OF Val66Met ON BDNF SECRETION AND TRAFFICKING

Apart from of the evidence of a normal (total brain) or reduced (HPC and PFC) production of BDNF in presence of the Val66Met mutation, several studies reported a differential subcellular distribution of BDNF upon Met substitution. BDNF was shown to regulate synaptic strength in a site-restricted manner and therefore, its subcellular distribution is very important for the functional role of this neurotrophin (Steward and Schuman, 2001; Horch and Katz, 2002; Lu, 2003; Alonso et al., 2004; Horch, 2004). Given the morphological structure of neurons, highly polarized cells with a relatively small soma and long dendritic/axonal processes, the BDNF protein must overcome several challenges to exert its functions correctly. In particular, just like many other synaptic proteins, BDNF is synthesized in response to synaptic stimuli and needs to be delivered at synapses located in specific subcellular districts where it can modify their structure and function (Tongiorgi, 2008; Edelmann et al., 2013). Jiang and Schuman reviewed the three different models that explain how a protein can be selectively delivered at activated synapses (Jiang and Schuman, 2002). In one model, the protein is synthesized in the soma and becomes tagged for a specific group of synapses, where it is subsequently transported. The second model describes the protein as being transported to dendrites without a specific target but is then captured by "tagged" synapses which have been appropriately stimulated. In the third model, which is in contrast with the previous two, it is the mRNA that is transported to dendrites and where it is locally translated by the protein synthesis machinery localized in proximity to synapses. Most researchers in the field agree that these three mechanisms might coexist within the same cell and may even regulate the very same protein.

BDNF represents one clear example of coexistence of multiple mechanisms for dendritic trafficking. The mRNAs encoding for BDNF were found to be transported and actively accumulate along dendrites following stimulation of electrical activity (Tongiorgi et al., 1997, 2004; Jakawich et al., 2010; Baj et al., 2013). BDNF was also shown to be transported to dendrites in large, dense-core secretory vesicles (Edelmann et al., 2013). Accordingly, recent studies have investigated the possible impact of the Met polymorphism on these BDNF sorting mechanisms.

In neurons, Met BDNF seems to be preferentially produced and accumulated within the somatic district and only partially transported in the proximal area of the primary dendrites whereas Val BDNF is produced and transported also in secondary and tertiary dendrites (Egan et al., 2003; Chen et al., 2004, 2006). The abnormal distribution of Met BDNF in cell body and dendrites of hippocampal neurons appears to be accompanied by a significant reduction in the secretion of this neurotrophin. In particular, secretion of BDNF induced in response to electrical stimuli is compromised (30% decrease) while the constitutive secretion of BDNF is not affected (Egan et al., 2003; Chen et al., 2004, 2006). BDNF sorting into the regulated secretory pathway appears to be mediated by two mechanisms. The first, involves interaction of the sorting receptor carboxypeptidase-E (CPE) with a tetrad of aminoacids (I16, E18, I105, D106) in the mature BDNF region forming a specific recognition motif, which is not affected by the Val66Met mutation (Lou et al., 2005). The second sorting mechanism, regards the interaction of sortilin with the pro-BDNF region comprised between aminoacid 44 and 102, wherein the Val66Met occurs (Chen et al., 2005). The current view is that substitution of the Val66 with a Met causes less efficient interaction with sortilin and therefore Met BDNF proteins show decreased targeting to the regulated secretory pathway (Dincheva et al., 2012). Further evidence in support of this hypothesis was provided by experiments in which Met BDNF was able to impair the trafficking of BDNF Val following formation of BDNF Val and Met BDNF heterodimers (Chen et al., 2004).

The combinatorial effect of normal BDNF production but decreased dendritic localization and secretion is correlated with the reduction of BDNF in SecII-positive secretory vesicles and Synaptophysin-positive synapses (Egan et al., 2003). The different biochemical pathways involved in the production and sorting of BDNF was also specified by Del Toro et al. (2006), who not only confirmed the reduction of Met BDNF in SecII vesicles but also found an accumulation of Met BDNF in Giantin-positive Golgi vesicle. Moreover, they found no visible differences between Val BDNF and Met BDNF in calnexin-positive vesicles while the total number of vesicles containing BDNF was strongly reduced by about 30% (Del Toro et al., 2006). It is important to note that the polymorphism at position 66 in the BDNF pro-domain does not affect the cleavage of the pro-BDNF form and the generation of the mature form of BDNF (Chen et al., 2004).

Transport dynamics of Met BDNF and Val BDNF through the secretory pathway are comparable. Indeed, the calculation of the mean translocation velocity of secretory vesicles in dendrites showed no differences between the two alleles (Del Toro et al., 2006), suggesting that the transported BDNF protein may probably account for the constitutive secretion of this neurotrophin which is also not affected by the Val66Met polymorphism. However, the physiological role of BDNF is only partially supported by this basal BDNF secretion. In fact, Bath et al. (2008) showed that TrkB activation in neurogenic regions, revealed by the intensity of pTrkB staining on total TrkB (in SVZ and OB), was reduced by 30% in Met BDNF brains. These data point out that activity-dependent BDNF secretion is required for a physiological activation of TrkB to sustain neurogenesis and neuroblast survival. Accordingly, a recent characterization of the neurogenesis at the level of the DG in Met BDNF mice, calculated as density of BrdU positive cells, found a considerable reduction (ca. –25%) of proliferating neural progenitors (Bath et al., 2012).

EFFECTS OF BDNF Val66Met: AN mRNA PERSPECTIVE

Among the studies that report a reduced amount of dendritic BDNF protein in presence of the Val66Met polymorphism, one study compared the trafficking of BDNF mRNA with the wild type sequence (Val allele: G196) or with the mutation (Met allele: A196). The results revealed significantly reduced dendritic trafficking of BDNF mRNA bearing the A196 mutation (Chiaruttini et al., 2009). This effect was found to be due to a reduced affinity of the RNA-binding protein Translin for the mutated allele. The RNA-binding domain of Translin contains a pocket made by three aminoacids (His90, Arg21, Glu89), which specifically recognizes a guanosine by forming three hydrogen bonds (**Figures 1A,C**). The transition from Guanine to Adenine at position 196 reduces the number of possible hydrogen bonds to just one, with a dramatic reduction in binding stability of Translin to the mutated BDNF mRNA (**Figure 1B**). Since the minimal tertiary configuration with which Translin can bind to an mRNA is a dimer, the authors hypothesized that there could be a second Guanine on BDNF mRNA that binds to the other Translin participating to the dimer (**Figure 1D**). Indeed, a second Guanine was identified at position 177 and when mutated to an Adenine, it abolished the Translin-mediated trafficking in dendrites exactly like the Guanine at position 196 (Chiaruttini et al., 2009). This mutation of Guanine at position 179 has no effect. Translin is involved in the trafficking of other transcripts associated with synaptic functions such as the mRNA encoding for CamK-II and the small non-coding RNA BC-1, implicated in regulating protein translation in dendrites (Li et al., 2008; Jaendling and McFarlane, 2010). In addition, Translin can associate with a similar protein called TRAX to form a complex known as C3PO, which is highly conserved from fungi to humans (Tian et al., 2011; Ye et al., 2011). This complex is a component of the RISC complex and is required for the mechanisms of RNA interference, which regulates mRNA stability and translation at synapses. Remarkably, Translin KO-mice have several behavioral abnormalities (Li et al., 2008; Jaendling and McFarlane, 2010). Thus, abnormal interactions of an mRNA with Translin may have profound effects on its trafficking, stability and translation (Wu et al., 2011).

A recent study in humans found a strong correlation, defined by linkage disequilibrium, among rs6265 and two other SNPs in the 3′UTR of BDNF (rs11030100 and rs11030099). All these SNPs transform BDNF mRNA into a target for the miR-26s family providing a supplementary component involved in regulation of mRNA stability and translation (Caputo et al., 2011). These pieces of evidence are consistent with findings in BDNF Met/Met mice and suggest that the reduced availability of BDNF mRNA in dendrites of Met BDNF neurons leads to a deficiency of local protein translation and may contribute to deficits in activity-dependent release of Met BDNF from post-synaptic terminals.

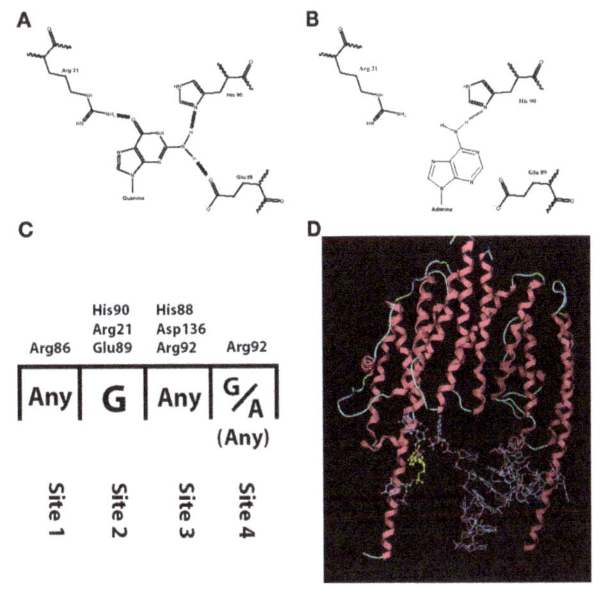

FIGURE 1 | Interaction between translin and human BDNF mRNA is hampered by the mutation G196A. (A) Translin amino acids His-90, Arg-21, Glu-89 form three hydrogen bonds with G196 (normal human BDNF) but only one with A196 (mutated BDNF, in **B**). **(C)** The nucleotide: amino acid selectivity for each of the four binding sites on translin. **(D)** Three-dimensional modeling of translin dimer binding to human BDNF mRNA.

TOWARD A UNIFIED BIOLOGICAL HYPOTHESIS FOR THE Val66Met DEFICIT

At the current state of research, a number of studies support the view that the BDNF Val66Met polymorphism, with one or two copies of the Met BDNF allele, can lead to altered performance of learning and memory functions, especially with impairments in hippocampal and cortical processes, suggestive of reduced neuroplasticity. To better understand how a small change in the BDNF sequence may determine so many effects, we believe that it is worth trying to assemble the large body of data available in the literature into a unified model at the cellular, anatomical, and behavioral level.

From a cellular standpoint, converging studies suggest that neuroplasticity deficits can be accounted mainly by impairment of activity-dependent translation and release of the Met BDNF from post-synaptic sites. We now describe how this deficit can result from a combination of reduced transport of Met BDNF mRNA in dendrites and an altered processing of the mutated protein through the regulated secretory pathway.

The deficit in mRNA trafficking found in BDNF bearing the Met allele (G196A at mRNA level), can be put into the broader perspective of the so called “spatial code model of BDNF transcripts” (Chiaruttini et al., 2008; Tongiorgi, 2008). BDNF is encoded by multiple mRNAs generated by alternative splicing. Eleven non-coding 5′UTR exons are alternatively spliced to a common downstream exon containing the coding region and a 3′UTR with two polyadenylation signals (An et al., 2008) which produce 11 different transcripts in rodents and 14 in humans, each with two 3′ tails (Aid et al., 2007; Pruunsild et al., 2007). Following a series of studies demonstrating the differential subcellular distribution of the different BDNF transcripts (Pattabiraman et al., 2005; Chiaruttini et al., 2008), a hypothesis was put forward that the multiple transcripts encoding the very same protein are used to generate a spatial code for expressing BDNF at restricted subcellular locations, leading to localized effects (Tongiorgi, 2008). This model was recently completed by real-time PCR analysis of all rodent transcripts *in vivo* (Baj et al., 2013). In untreated rats, various BDNF transcripts were detected in dendritic fields of the hippocampus (exons 6 and 7 in CA1; exons 1, 6, and 9a in CA3; and exons 5, 6, 7, and 8 in DG). However, due to the very low levels of most of these transcripts, the exon 6 was found to be the main transcript present in dendrites at resting conditions. Strong neuronal activation by pilocarpine-induced *status epilepticus* caused an increase in all hippocampal dendritic laminae of BDNF transcripts encoding exons 2, 4, and 6 and also of BDNF exons 3 and 9a in DG molecular layer, whereas the other transcripts were restricted to the soma (Baj et al., 2013). Importantly, overexpression or silencing of the four most abundant brain BDNF transcripts, encoding exon 1, 2, 4, or 6, led to differential effects on dendritic morphology *in vitro*, with exons 1 and 4 affecting only the proximal dendritic domains and exon 2 and 6 being able to shape the distal dendritic district (Baj et al., 2011). This differential effect was demonstrated to be due to localized expression of BDNF and activation of TrkB receptor in the same subcellular domains where the specific mRNAs are localized (Baj et al., 2011). The conversion of the G196 to A196 impairs the transport of BDNF mRNA into dendritic domains (Chiaruttini et al., 2009), and therefore, the mutated BDNF mRNAs can only be translated in the soma. Since the change G196A is located within the coding region, it is present in all BDNF mRNA variants but its effects on mRNA trafficking are only evident for those BDNF transcripts that are actively transported to dendrites (such as exon 2 or 6). However, activity-dependent secretion is also impaired, causing entrapment of the Met BDNF protein within the Golgi apparatus and the observed increase in the soma and depletion from distal dendrites (Egan et al., 2003; Chen et al., 2004) (**Figure 2**). In this model, the converging misplacement of BDNF mRNA and protein can affect the ability of BDNF to maintain dendritic branching in the periphery of the neuron and support plasticity at synapses in distal dendrites which represent, notably, the large majority of glutamatergic excitatory synapses (**Figures 2 A,B**).

Anatomically and physiologically, the available findings are consistent with morphological and functional hippocampal atrophy and subsequent cortico-cortical disconnection syndrome, which involves the disruption of neural networks between the anterior and posterior cerebral areas (Delbeuck et al., 2003; Nagata et al., 2010). Met carriers, rather than subjects with the Val/Val phenotype, may benefit from a protective role on executive function through hippocampus cortical atrophy or other subcortical tract changes, as reported in previous studies on elderly people. This observation may in part, explain the reason why this polymorphism despite its deleterious effects remains highly diffused in the world population. Interestingly, as pointed out by Li et al. (2010), the negative effects of the BDNF polymorphism on episodic memory are most likely observed when associative and executive demands are high. These observations are in line with the hypothesis that the magnitude of genetic effects on cognition is greater when brain resources are reduced, as with old age. Accordingly, the logical consequence of the strong effects of Met BDNF allele on mental performance are pronounced in the elderly, with poor or no impact on reproduction and therefore, transmission of the allele to the following generations can

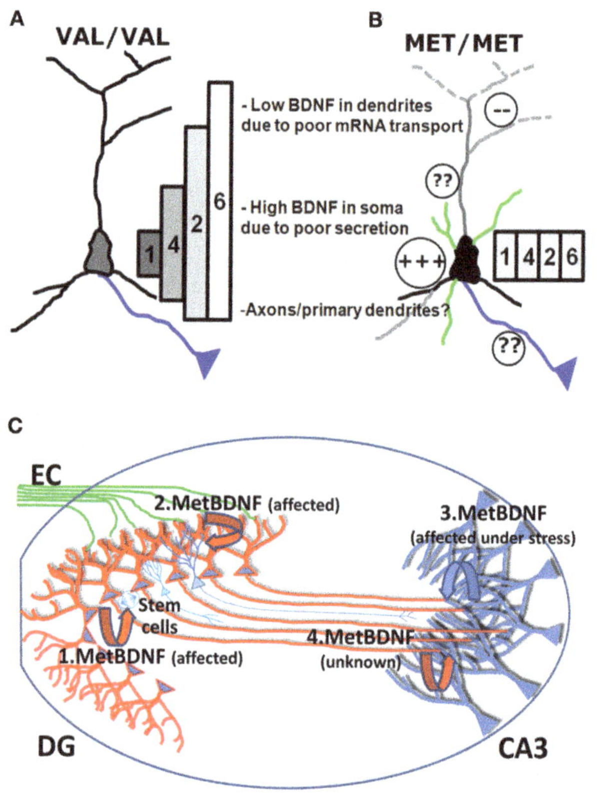

FIGURE 2 | Model of the biological effects of Val and Met BDNF alleles in the hippocampus. (A) In Val/Val neurons, the four most abundant brain BDNF transcripts, encoding exon 1, 2, 4, or 6, are differentially localized in dendrites, forming a gradient with exon 1 being restricted to the soma and exons 2, 4, and 6 in dendrites at increasing distances according to the gradient 4<2<6. Exons 1 and 4 can affect only the morphology of proximal dendrites while exon 2 and 6 are able to shape the distal dendrites. **(B)** In Met/Met neurons, transport of BDNF mRNA in distal dendritic districts is impaired and all BDNF mRNA isoforms accumulate in the cell soma. Met BDNF protein increases in the soma, also due to poor secretion. It is unclear if axonal transport and secretion can be affected by the Met mutation. **(C)** BDNF released from DG granule cells is shown in red, while BDNF released from CA3 neurons is shown in blue. (1.BDNF) Secretion of Met BDNF from the soma of DG granule cells is altered and provides insufficient local trophic support for survival and differentiation of DG subgranular neural stem cells. (2.BDNF) Release of Met BDNF from dendrites of DG neurons is affected and cannot provide sufficient autocrine support to dendrites which show reduced arborization. (3.BDNF) Local production and release of Met BDNF from apical dendrites of CA3 neurons is reduced and cannot provide the target-derived trophic support to promote innervation of CA3 from newly formed mossy fibers and autocrine support to CA3 dendrites. (4.BDNF) It is unknown if anterogradely transported Met BDNF in mossy fibers can support dendrites from CA3 neurons and if anterograde transport of BDNF is affected by the Met mutation.

be ensured thus contributing to the maintenance of this mutation through the population. In addition, carriers of the Met allele are more likely to develop post-traumatic stress disorder after traumatic experiences.

In **Figure 2C** we summarize how deficits in Met BDNF secretion and transport, observed at the cellular level, may affect the hippocampal neuronal circuit. According to this model, since various studies reported that survival and differentiation of DG subgranular neural stem cells are reduced, secretion of Met BDNF from the soma of DG granule cells is altered and provides insufficient local trophic support (**Figure 2**). In addition, since DG dendrites were shown to have reduced arborization, the model predicts that release of Met BDNF from dendrites of DG neurons is affected and cannot provide sufficient autocrine support. Following stressful situations, Met/Met individuals may undergo more easily to atrophy of apical dendrites of CA3 neurons. According to our model, the local production and release of Met BDNF from apical dendrites of CA3 neurons is reduced and thus cannot provide target-derived trophic support to promote the innervation of CA3 from newly formed mossy fibers and autocrine support to CA3 dendrites. It is currently unclear if Met BDNF present in mossy fibers can support dendrites from CA3 neurons and if anterograde transport of BDNF is affected by the Met mutation. One plausible hypothesis is that BDNF released from mossy fiber axons may stimulate dendritic targeting of BDNF mRNA in CA3 apical dendrites, thus contributing to a local autocrine loop which supports the maintenance of apical dendritic arborization.

In conclusion, the view proposed here does not claim to be the ultimate interpretation of the available data but intends to be a stimulus for the scientific community to develop a consensus model on the biological mechanisms of one of the most fascinating human mutations affecting cellular functions, brain morphology, and cognition.

ACKNOWLEDGMENTS

This study was supported by FRA—Università di Trieste 2012. Gabriele Baj was recipient of a fellowship from Sanofi Aventis Foundation.

REFERENCES

Aid, T., Kazantseva, A., Piirsoo, M., Palm, K., Timmusk, T. (2007). Mouse and rat BDNF gene structure and expression revisited. *J. Neurosci. Res.* 85, 525–535. doi: 10.1002/jnr.21139

Alonso, M., Medina, J. H., and Pozzo-Miller, L. (2004). ERK1/2 activation is necessary for BDNF to increase dendritic spine density in hippocampal CA1 pyramidal neurons. *Learn. Mem.* 11, 172–178. doi: 10.1101/lm.67804

An, J. J., Gharami, K., Liao, G. Y., Woo, N. H., Lau, A. G., Vanevski, F., et al. (2008). Distinct role of long 3' UTR BDNF mRNA in spine morphology and synaptic plasticity in hippocampal neurons. *Cell* 134, 175–187. doi: 10.1016/j.cell.2008.05.045

Baj, G., Del Turco, D., Schlaudraff, J., Torelli, L., Deller, T., and Tongiorgi, E. (2013). Regulation of the spatial code for BDNF mRNA isoforms in the rat hippocampus following pilocarpine-treatment: a systematic analysis using laser microdissection and quantitative real-time PCR. *Hippocampus* 23, 413–423. doi: 10.1002/hipo.22100

Baj, G., Leone, E., Chao, M. V., Tongiorgi, E. (2011). Spatial segregation of BDNF transcripts enables BDNF to differentially shape distinct dendritic compartments. *Proc. Natl. Acad. Sci. U.S.A.* 108, 16813–16818. doi: 10.1073/pnas.1014168108

Barde, Y. A., Edgar, D., Thoenen, H. (1982). Purification of a new neurotrophic factor from mammalian brain. *EMBO J.* 1, 549–553.

Bath, K. G., Jing, D. Q., Dincheva, I., Neeb, C. C., Pattwell, S. S., Chao, M. V., et al. (2012). BDNF Val66Met impairs fluoxetine-induced enhancement of adult hippocampus plasticity. *Neuropsychopharmacology* 37, 1297–1304. doi: 10.1038/npp.2011.318

Bath, K. G., Mandairon, N., Jing, D., Rajagopal, R., Kapoor, R., Chen, Z. Y., et al. (2008). Variant brain-derived neurotrophic factor Val66Met alters adult olfactory bulb neurogenesis and spontaneous olfactory discrimination. *J. Neurosci.* 28, 2383–2393. doi: 10.1523/JNEUROSCI.4387-07.2008

Benjamin, S., McQuoid, D. R., Potter, G. G., Payne, M. E., MacFall, J. R., Steffens, D. C., et al. (2010). The brain-derived neurotrophic factor Val66Met polymorphism, hippocampal volume, and cognitive function in geriatric depression. *Am. J. Geriatr. Psychiatry* 18, 323–331. doi: 10.1097/JGP.0b013e3181cabd2b

Berkemeier, L. R., Winslow, J. W., Kaplan, D. R., Nikolics, K., Goeddel, D. V., and Rosenthal, A. (1991). Neurotrophin-5: a novel neurotrophic factor that activates trk and trkB. *Neuron* 7, 857–866. doi: 10.1016/0896-6273(91)90287-A

Bibel, M., and Barde, Y. A. (2000). Neurotrophins: key regulators of cell fate and cell shape in the vertebrate nervous system. *Genes Dev.* 14, 2919–2937. doi: 10.1101/gad.841400

Binder, D. K., and Scharfman, H. E. (2004). Brain-derived neurotrophic factor. *Growth Factors* 22, 123–131. doi: 10.1080/08977190410001723308

Caputo, V., Sinibaldi, L., Fiorentino, A., Parisi, C., Catalanotto, C., Pasini, A., et al. (2011). Brain derived neurotrophic factor (BDNF) expression is regulated by microRNAs miR-26a and miR-26b allele-specific binding. *PLoS ONE* 6:e28656. doi: 10.1371/journal.pone.0028656

Cerasa, A., Tongiorgi, E., Fera, F., Gioia, M. C., Valentino, P., Liguori, M., et al. (2010). The effects of BDNF Val66Met polymorphism on brain function in controls and patients with multiple sclerosis: an imaging genetic study. *Behav. Brain Res.* 207, 377–386. doi: 10.1016/j.bbr.2009.10.022

Chao, M. V., Rajagopal, R., and Lee, F. S. (2006). Neurotrophin signalling in health and disease. *Clin. Sci.* 110, 167–173. doi: 10.1042/CS20050163

Chen, Z. Y., Bath, K., McEwen, B., Hempstead, B., and Lee, F. (2008). Impact of genetic variant BDNF (Val66Met) on brain structure and function. *Novartis Found. Symp.* 289, 180–188. discussion: 188–195. doi: 10.1002/9780470751251.ch14

Chen, Z. Y., Ieraci, A., Teng, H., Dall, H., Meng, C. X., Herrera, D. G., et al. (2005). Sortilin controls intracellular sorting of brain-derived neurotrophic factor to the regulated secretory pathway. *J. Neurosci.* 25, 6156–6166. doi: 10.1523/JNEUROSCI.1017-05.2005

Chen, Z. Y., Jing, D., Bath, K. G., Ieraci, A., Khan, T., Siao, C. J., et al. (2006). Genetic variant BDNF (Val66Met) polymorphism alters anxiety-related behaviour. *Science* 314, 140–143. doi: 10.1126/science.1129663

Chen, Z. Y., Patel, P. D., Sant, G., Meng, C. X., Teng, K. K., Hempstead, B. L., et al. (2004). Variant brain-derived neurotrophic factor (BDNF) (Met66) alters the intracellular trafficking and activity-dependent secretion of wild-type BDNF in neurosecretory cells and cortical neurons. *J. Neurosci.* 24, 4401–4411. doi: 10.1523/JNEUROSCI.0348-04.2004

Chiaruttini, C., Sonego, M., Baj, G., Simonato, M., Tongiorgi, E. (2008). BDNF mRNA splice variants display activity-dependent targeting to distinct hippocampal laminae. *Mol. Cell. Neurosci.* 37, 11–19. doi: 10.1016/j.mcn.2007.08.011

Chiaruttini, C., Vicario, A., Li, Z., Baj, G., Braiuca, P., Wu, Y., et al. (2009). Dendritic trafficking of BDNF mRNA is mediated by translin and blocked by the G196A (Val66Met) mutation. *Proc. Natl. Acad. Sci. U.S.A.* 106, 16481–16486. doi: 10.1073/pnas.0902833106

Delbeuck, X., Van Der Linden, M., and Collette, F. (2003). Alzheimer's disease as a disconnection syndrome. *Neuropsychol. Rev.* 13, 79–92. doi: 10.1023/A:1023832305702

Del Toro, D., Canals, J. M., Gines, S., Kojima, M., Egea, G., and Alberch, J. (2006). Mutant huntingtin impairs the post-Golgi trafficking of brain-derived neurotrophic factor but not its Val66Met polymorphism. *J. Neurosci.* 26, 12748–12757. doi: 10.1523/JNEUROSCI.3873-06.2006

Dincheva, I., Glatt, C. E., Lee, F. S. (2012). Impact of the BDNF Val66Met polymorphism on cognition: implications for behavioral genetics. *Neuroscientist* 18, 439–451. doi: 10.1177/1073858411431646

Echeverria, D., Woods, J. S., Heyer, N. J., Rohlman, D. S., Farin, F. M., Bittner, A. C., et al. (2005). Chronic low-level mercury exposure, BDNF polymorphism, and associations with cognitive and motor function. *Neurotoxicol. Teratol.* 27, 781–796. doi: 10.1016/j.ntt.2005.08.001

Edelmann, E., Leßmann, V., Brigadski, T. (2013). Pre- and postsynaptic twists in BDNF secretion and action in synaptic plasticity. *Neuropharmacology.* doi: 10.1016/j.neuropharm.2013.05.043. [Epub ahead of print].

Egan, M. F., Kojima, M., Callicott, J. H., Goldberg, T. E., Kolachana, B. S., Bertolino, A., et al. (2003). The BDNF Val66met polymorphism affects activity-dependent secretion of BDNF and human memory and hippocampal function. *Cell* 112, 257–269. doi: 10.1016/S0092-8674(03)00035-7

Foltynie, T., Lewis, S. G., Goldberg, T. E., Blackwell, A. D., Kolachana, B. S., Weinberger, D. R., et al. (2005). The BDNF Val66Met polymorphism has a gender specific influence on planning ability in Parkinson's disease. *J. Neurol.* 252, 833–838. doi: 10.1007/s00415-005-0756-5

Forlenza, O. V., Diniz, B. S., Teixeira, A. L., Ojopi, E. B., Talib, L. L., Mendonca, V. A., et al. (2010). Effect of brain-derived neurotrophic factor Val66Met polymorphism and serum levels on the progression of mild cognitive impairment. *World J. Biol. Psychiatry* 11, 774–780. doi: 10.3109/15622971003797241

Goldberg, T. E., Iudicello, J., Russo, C., Elvevag, B., Straub, R., Egan, M. F., et al. (2008). BDNF Val66Met polymorphism significantly affects d' in verbal recognition memory at short and long delays. *Biol. Psychol.* 77, 20–24. doi: 10.1016/j.biopsycho.2007.08.009

Guerini, F. R., Beghi, E., Riboldazzi, G., Zangaglia, R., Pianezzola, C., Bono, G., et al. (2009). BDNF Val66Met polymorphism is associated with cognitive impairment in Italian patients with Parkinson's disease. *Eur. J. Neurol.* 16, 1240–1245. doi: 10.1111/j.1468-1331.2009.02706.x

Hall, D., Dhilla, A., Charalambous, A., Gogos, J. A., and Karayiorgou, M. (2003). Sequence variants of the brain-derived neurotrophic factor (BDNF) gene are strongly associated with obsessive-compulsive disorder. *Am. J. Hum. Genet.* 73, 370–376. doi: 10.1086/377003

Hariri, A. R., Goldberg, T. E., Mattay, V. S., Kolachana, B. S., Callicott, J. H., Egan, M. F., et al. (2003). Brain-derived neurotrophic factor Val66met polymorphism affects human memory-related hippocampal activity and predicts memory performance. *J. Neurosci.* 23, 6690–6694.

Harris, S. E., Fox, H., Wright, A. F., Hayward, C., Starr, J. M., Whalley, L. J., et al. (2006). The brain-derived neurotrophic factor Val66Met polymorphism is associated with age-related change in reasoning skills. *Mol. Psychiatry* 11, 505–513. doi: 10.1038/sj.mp.4001799

Hohn, A., Leibrock, J., Bailey, K., and Barde, Y. A. (1990). Identification and characterization of a novel member of the nerve growth factor/brain-derived neurotrophic factor family. *Nature* 344, 339–341. doi: 10.1038/344339a0

Hong, C. J., Liou, Y. J., Tsai, S. J. (2011). Effects of BDNF polymorphisms on brain function and behavior in health and disease. *Brain Res. Bull.* 86, 287–297. doi: 10.1016/j.brainresbull.2011.08.019

Horch, H. W. (2004). Local effects of BDNF on dendritic growth. *Rev. Neurosci.* 15, 117–129. doi: 10.1515/REVNEURO.2004.15.2.117

Horch, H. W., Katz, L. C. (2002). BDNF release from single cells elicits local dendritic growth in nearby neurons. *Nat. Neurosci.* 5, 1177–1184. doi: 10.1038/nn927

Hwang, J. P., Tsai, S. J., Hong, C. J., Yang, C. H., Lirng, J. F., and Yang, Y. M. (2006). The Val66Met polymorphism of the brain-derived neurotrophic-factor gene is associated with geriatric depression. *Neurobiol. Aging* 27, 1834–1837. doi: 10.1016/j.neurobiolaging.2005.10.013

Jaendling, A., McFarlane, R. J. (2010). Biological roles of translin and translin-associated factor-X: RNA metabolism comes to the fore. *Biochem. J.* 429, 225–234. doi: 10.1042/BJ20100273

Jakawich, S. K., Nasser, H. B., Strong, M. J., McCartney, A. J., Perez, A. S., Rakesh, N., et al. (2010). Local presynaptic activity gates homeostatic changes in presynaptic function driven by dendritic BDNF synthesis. *Neuron* 68, 1143–1158. doi: 10.1016/j.neuron.2010.11.034

Jiang, C., and Schuman, E. M. (2002). Regulation and function of local protein synthesis in neuronal dendrites. *Trends Biochem. Sci.* 27, 506–513. doi: 10.1016/S0968-0004(02)02190-4

Levi-Montalcini, R., and Booker, B. (1960). Destruction of the sympathetic ganglia in mammals by an antiserum to a nerve-growth protein. *Proc. Natl. Acad. Sci. U.S.A.* 46, 384–391. doi: 10.1073/pnas.46.3.384

Levine, A. J., Sinsheimer, J. S., Bilder, R., Shapshak, P., and Singer, E. J. (2012). Functional polymorphisms in dopamine-related genes: effect on neurocognitive functioning in HIV+ adults. *J. Clin. Exp. Neuropsychol.* 34, 78–91. doi: 10.1080/13803395.2011.623118

Li, S. C., Chicherio, C., Nyberg, L., Von Oertzen, T., Nagel, I. E., Papenberg, G., et al. (2010). Ebbinghaus revisited: influences of the BDNF Val66Met polymorphism on backward serial recall are modulated by human aging. *J. Cogn. Neurosci.* 22, 2164–2173. doi: 10.1162/jocn.2009.21374

Li, Z., Wu, Y., Baraban, J. M. (2008). The Translin/Trax RNA binding complex: clues to function in the nervous system. *Biochim. Biophys. Acta* 1779, 479–485. doi: 10.1016/j.bbagrm.2008.03.008

Liguori, M., Fera, F., Gioia, M. C., Valentino, P., Manna, I., Condino, F., et al. (2007). Investigating the role of brain-derived neurotrophic factor in relapsing-remitting multiple sclerosis. *Genes Brain Behav.* 6, 177–183. doi: 10.1111/j.1601-183X.2006.00245.x

Liu, R. J., Lee, F. S., Li, X. Y., Bambico, F., Duman, R. S., and Aghajanian, G. K. (2012). Brain-derived neurotrophic factor Val66Met allele impairs basal and ketamine-stimulated synaptogenesis in prefrontal cortex. *Biol. Psychiatry* 71, 996–1005. doi: 10.1016/j.biopsych.2011.09.030

Lou, H., Kim, S. K., Zaitsev, E., Snell, C. R., Lu, B., Loh, Y. P. (2005). Sorting and activity-dependent secretion of BDNF require interaction of a specific motif with the sorting receptor carboxypeptidase-E. *Neuron* 45, 245–255. doi: 10.1016/j.neuron.2004.12.037

Lu, B. (2003). BDNF and activity-dependent synaptic modulation. *Learn. Mem.* 10, 86–98. doi: 10.1101/lm.54603

Lu, W., Zhang, C., Yi, Z., Li, Z., Wu, Z., and Fang, Y. (2012). Association between BDNF Val66Met polymorphism and cognitive performance in antipsychotic-naive patients with schizophrenia. *J. Mol. Neurosci.* 47, 505–510. doi: 10.1007/s12031-012-9750-4

Mandelman, S. D., and Grigorenko, E. L. (2012). BDNF Val66Met and cognition: all, none, or some. A meta-analysis of the genetic association. *Genes Brain Behav.* 11, 127–136. doi: 10.1111/j.1601-183X.2011.00738.x

Martinho, E. Jr., Michelon, L., Ayres, A. M., Scazufca, M., Menezes, P. R. Schaufelberger, M.S et al. (2012). BDNF gene polymorphism, cognition and symptom severity in a Brazilian population-based sample of first-episode psychosis subjects. *Rev. Bras. Psiquiatr.* 34(Suppl. 2), S219–S225. doi: 10.1016/j.rbp.2012.06.004

Miyajima, F., Ollier, W., Mayes, A., Jackson, A., Thacker, N., Rabbitt, P., et al. (2008). Brain-derived neurotrophic factor polymorphism Val66Met influences cognitive abilities in the elderly. *Genes Brain Behav.* 7, 411–417. doi: 10.1111/j.1601-183X.2007.00363.x

Nagata, T., Shinagawa, S., Nukariya, K., Yamada, H., and Nakayama, K. (2012). Association between BDNF polymorphism (Val66Met) and executive function in patients with amnestic mild cognitive impairment or mild Alzheimer disease. *Dement. Geriatr. Cogn. Disord.* 33, 266–272. doi: 10.1159/000339358

Nagata, T., Shinagawa, S., Ochiai, Y., Aoki, R., Kasahara, H., Nukariya, K., et al. (2010). Association between executive dysfunction and hippocampal volume in Alzheimer's disease. *Int. Psychogeriatr.*, 1–8. doi: 10.1017/S1041610210002164

Ninan, I., Bath, K. G., Dagar, K., Perez-Castro, R., Plummer, M. R., Lee, F. S., et al. (2010). The BDNF Val66Met polymorphism impairs NMDA receptor-dependent synaptic plasticity in the hippocampus. *J. Neurosci.* 30, 8866–8870. doi: 10.1523/JNEUROSCI.1405-10.2010

Oroszi, G., Lapteva, L., Davis, E., Yarboro, C. H., Weickert, T., Roebuck-Spencer, T., et al. (2006). The Met66 allele of the functional Val66Met polymorphism in the brain-derived neurotrophic factor gene confers protection against neurocognitive dysfunction in systemic lupus erythematosus. *Ann. Rheum. Dis.* 65, 1330–1335. doi: 10.1136/ard.2006.051623

Pattabiraman, P. P., Tropea, D., Chiaruttini, C., Tongiorgi, E., Cattaneo, A., Domenici, L. (2005). Neuronal activity regulates the developmental expression and subcellular localization of cortical BDNF mRNA isoforms *in vivo*. *Mol.*

Cell. Neurosci. 28, 556–570. doi: 10.1016/j.mcn.2004.11.010

Pattwell, S. S., Bath, K. G., Perez-Castro, R., Lee, F. S., Chao, M. V., and Ninan, I. (2012). The BDNF Val66Met polymorphism impairs synaptic transmission and plasticity in the infralimbic medial prefrontal cortex. *J. Neurosci.* 32, 2410–2421. doi: 10.1523/JNEUROSCI.5205-11.2012

Pezawas, L., Verchinski, B. A., Mattay, V. S., Callicott, J. H., Kolachana, B. S., Straub, R. E., et al. (2004). The brain-derived neurotrophic factor Val66met polymorphism and variation in human cortical morphology. *J. Neurosci.* 24, 10099–10102. doi: 10.1523/JNEUROSCI.2680-04.2004

Pruunsild, P., Kazantseva, A., Aid, T., Palm, K., Timmusk, T. (2007). Dissecting the human BDNF locus: bidirectional transcription, complex splicing, and multiple promoters. *Genomics* 90, 397–406. doi: 10.1016/j.ygeno.2007.05.004

Schofield, P. R., Williams, L. M., Paul, R. H., Gatt, J. M., Brown, K., Luty, A., et al. (2009). Disturbances in selective information processing associated with the BDNF Val66Met polymorphism: evidence from cognition, the P300 and fronto-hippocampal systems. *Biol. Psychol.* 80, 176–188. doi: 10.1016/j.biopsycho.2008.09.001

Smith, G. N., Thornton, A. E., Lang, D. J., Macewan, G. W., Ehmann, T. S., Kopala, L. C. et al. (2012). Hippocampal volume and the brain-derived neurotrophic factor Val66Met polymorphism in first episode psychosis. *Schizophr. Res.* 134, 253–259. doi: 10.1016/j.schres.2011.11.022

Steward, O., and Schuman, E. M. (2001). Protein synthesis at synaptic sites on dendrites. *Annu. Rev. Neurosci.* 24, 299–325. doi: 10.1146/annurev.neuro.24.1.299

Strauss, J., Barr, C. L., George, C. J., King, N., Shaikh, S., Devlin, B., et al. (2004). Association study of brain-derived neurotrophic factor in adults with a history of childhood onset mood disorder. *Am. J. Med. Genet. B Neuropsychiatr. Genet.* 131B, 16–19. doi: 10.1002/ajmg.b.30041

Szeszko, P. R., Lipsky, R., Mentschel, C., Robinson, D., Gunduz-Bruce, H., Sevy, S., et al. (2005). Brain-derived neurotrophic factor Val66met polymorphism and volume of the hippocampal formation. *Mol. Psychiatry* 10, 631–636. doi: 10.1038/sj.mp.4001656

Teh, C. A., Lee, T. S., Kuchibhatla, M., Ashley-Koch, A., MacFall, J., Krishnan, R., et al. (2012). Bipolar disorder, brain-derived neurotrophic factor (BDNF) Val66Met polymorphism and brain morphology. *PLoS ONE* 7:e38469. doi: 10.1371/journal.pone.0038469

Tian, Y., Simanshu, D. K., Ascano, M., Diaz-Avalos, R., Park, A. Y., Juranek, S. A., et al. (2011). Multimeric assembly and biochemical characterization of the Trax-translin endonuclease complex. *Nat. Struct. Mol. Biol.* 18, 658–664. doi: 10.1038/nsmb.2069

Tongiorgi, E. (2008). Activity-dependent expression of brain-derived neurotrophic factor in dendrites: facts and open questions. *Neurosci. Res.* 61, 335–346. doi: 10.1016/j.neures.2008.04.013

Tongiorgi, E., Armellin, M., Giulianini, P. G., Bregola, G., Zucchini, S., Paradiso, B., et al. (2004). Brain-derived neurotrophic factor mRNA and protein are targeted to discrete dendritic laminas by events that trigger epileptogenesis. *J. Neurosci.* 24, 6842–6852. doi: 10.1523/JNEUROSCI.5471-03.2004

Tongiorgi, E., Righi, M., and Cattaneo, A. (1997). Activity-dependent dendritic targeting of BDNF and TrkB mRNAs in hippocampal neurons. *J. Neurosci.* 17, 9492–9505.

Tost, H., Alam, T., Geramita, M., Rebsch, C., Kolachana, B., Dickinson, D., et al. (2013). Effects of the BDNF Val(66)met polymorphism on white matter microstructure in healthy adults. *Neuropsychopharmacology* 38, 525–532. doi: 10.1038/npp.2012.214

Tramontina, J. F., Yates, D., Magalhaes, P. V., Trentini, C., Sant'anna, M. K., Fries, G. R., et al. (2009). Brain-derived neurotrophic factor gene Val66met polymorphism and executive functioning in patients with bipolar disorder. *Rev. Bras. Psiquiatr.* 31, 136–140. doi: 10.1590/S1516-44462009000200010

Tyler, W. J., Alonso, M., Bramham, C. R., and Pozzo-Miller, L. D. (2002). From acquisition to consolidation: on the role of brain-derived neurotrophic factor signaling in hippocampal-dependent learning. *Learn. Mem.* 9, 224–237. doi: 10.1101/lm.51202

Verhagen, M., Van Der Meij, A., Van Deurzen, P. A., Janzing, J. G., Arias-Vasquez, A., Buitelaar, J. K., et al. (2010). Meta-analysis of the BDNF Val66Met polymorphism in major depressive disorder: effects of gender and ethnicity. *Mol. Psychiatry* 15, 260–271. doi: 10.1038/mp.2008.109

Wu, Y. C., Williamson, R., Li, Z., Vicario, A., Xu, J., Kasai, M. et al. (2011). Dendritic trafficking of brain-derived neurotrophic factor mRNA: regulation by translin-dependent and -independent mechanisms. *J. Neurochem.* 116, 1112–1121. doi: 10.1111/j.1471-4159.2010.07166.x

Yamada, K., Mizuno, M., and Nabeshima, T. (2002). Role for brain-derived neurotrophic factor in learning and memory. *Life Sci.* 70, 735–744. doi: 10.1016/S0024-3205(01)01461-8

Ye, X., Huang, N., Liu, Y., Paroo, Z., Huerta, C., Li, P., et al. (2011). Structure of C3PO and mechanism of human RISC activation. *Nat. Struct. Mol. Biol.* 18, 650–657. doi: 10.1038/nsmb.2032

Yu, H., Wang, D. D., Wang, Y., Liu, T., Lee, F. S., and Chen, Z. Y. (2012). Variant brain-derived neurotrophic factor Val66Met polymorphism alters vulnerability to stress and response to antidepressants. *J. Neurosci.* 32, 4092–4101. doi: 10.1523/JNEUROSCI.5048-11.2012

Yu, H., Wang, Y., Pattwell, S., Jing, D., Liu, T., Zhang, Y., et al. (2009). Variant BDNF Val66Met polymorphism affects extinction of conditioned aversive memory. *J. Neurosci.* 29, 4056–4064. doi: 10.1523/JNEUROSCI.5539-08.2009

Zivadinov, R., Weinstock-Guttman, B., Benedict, R., Tamano-Blanco, M., Hussein, S., Abdelrahman, N., et al. (2007). Preservation of gray matter volume in multiple sclerosis patients with the Met allele of the rs6265 (Val66Met) SNP of brain-derived neurotrophic factor. *Hum. Mol. Genet.* 16, 2659–2668. doi: 10.1093/hmg/ddm189

14

Circulating microRNAs in Alzheimer's disease: the search for novel biomarkers

Véronique Dorval[1,2], Peter T. Nelson[3] and Sébastien S. Hébert[1,2]**

[1] Axe Neurosciences, Centre de Recherche du Centre Hospitalier Universitaire de Québec (Centre Hospitalier de l'Université Laval), Québec, QC, Canada
[2] Département de Psychiatrie et de Neurosciences, Université Laval, Québec, QC, Canada
[3] Sanders-Brown Center on Aging, University of Kentucky, Lexington, KY, USA

***Correspondence:**
Sébastien S. Hébert and Véronique Dorval, Axe Neurosciences, Centre de Recherche du Centre Hospitalier Universitaire de Québec (Centre Hospitalier de l'Université Laval), 2705 Boulevard Laurier, RC-9800, QC G1V 4G2, Canada
e-mail: sebastien.hebert@neurosciences.ulaval.ca; veronique.dorval@crchul.ulaval.ca

Alzheimer's disease (AD) is the most common neurodegenerative disease in the elderly. While advancements have been made in understanding the genetic and molecular basis of AD, the clinical diagnosis of AD remains difficult, and post-mortem confirmation is often required. Furthermore, the onset of neurodegeneration precedes clinical symptoms by approximately a decade. Consequently, there is a crucial need for an early and accurate diagnosis of AD, which can potentially lead to strategies that can slow down or stop the progression of neurodegeneration and dementia. Recent advances in the non-coding RNA field have shown that microRNAs (miRNAs) can function as powerful biomarkers in human diseases. Studies are emerging suggesting that circulating miRNAs in the cerebrospinal fluid and blood serum have characteristic changes in AD patients. Whether miRNAs can be used in AD diagnosis, alone or in combination with other AD biomarkers (e.g., amyloid and tau), warrants further investigation.

Keywords: microRNA, Alzheimer's disease, biomarker, diagnosis, mild cognitive impairment

INTRODUCTION

Alzheimer's disease (AD) is a prevalent, devastating, and progressive neurodegenerative disorder. Epidemiological studies predict that over 35 million people worldwide will be affected by 2050, thus significantly increasing social and economical burdens. There is no cure at hand, and only a few medications aimed at slowing down memory deficits and clinical symptoms are available, with limited benefits. Consequently, there is an urgent need for the identification of biomarkers that will allow the detection of AD at early (prodromal) stages, potentially leading to novel diagnostic or therapeutic strategies.

Pathologically, AD is characterized by the gradual, widespread loss of neurons, synapses, and neuropil, culminating in ~40% loss of brain mass in end-stage disease (West et al., 1994; Gomez-Isla et al., 1996). There are two main AD pathological hallmarks: extracellular amyloid (senile) plaques and intracellular neurofibrillary tangles (NFTs; Hyman et al., 2012). The amyloid plaques comprise aggregated amyloid-beta (Aβ) peptides that are generated by sequential cleavage of amyloid precursor protein (APP) by β-secretase/BACE1 and the presenilin (PSEN)-containing γ-secretase complex (Kang et al., 1987; Wolfe, 2006). The NFTs result from the abnormal aggregation of hyperphosphorylated microtubule-associated protein, tau. The reason for tau aggregation into tangles remains under investigation, but may result from an imbalance in the delicate regulation of tau kinases and phosphatases. Whereas approximately 1–5% of AD cases can be explained genetically by mutations in APP or PSEN genes, the exact cause(s) of sporadic AD remains obscure. Most experts agree, however, that sporadic AD is caused by a combination of genes and environmental factors (multifactorial), perhaps exacerbated by oxidative stress and inflammation.

Biomarkers are used to measure or indicate the effects or progress of a disease or condition. A subtype of biomarkers relates to specific and traceable biochemical molecules or compounds found in body fluids. Detection of these substances may indicate disease states or allow correlations with the progression or the susceptibility to a disease or a given treatment. They can be measured in, for instance, saliva, sweat, breath, blood/serum, urine, and cerebrospinal fluid (CSF). The collection of these biological fluids is significantly less invasive than biopsies, an important and practical issue when studying neurodegenerative disorders like AD.

Accumulating evidence suggests that circulating biomarkers may be used in AD diagnosis, the most common being Aβ peptides (Aβ40 and Aβ42, the latter being more prone to aggregation) and tau/phospho-tau (Thr181 being one of the common phospho-epitopes). While this area of research continues to progress (Tarawneh and Holtzman, 2010; Holtzman, 2011), large variability exists in the literature, hampering or delaying their routine use in the clinic (Ingelson et al., 1999). Moreover, their potential use as prodromal AD biomarkers remains uncertain. Therefore, most experts agree that additional biomarkers are required for an accurate and early diagnosis of AD vs. other potential causes of dementia. In this review, we discuss recent studies suggesting that miRNAs could function as novel, non-invasive biomarkers in AD.

miRNAs AS BIOMARKERS

The miRNAs are a class of small (~22 nt) non-protein-coding RNAs crucially involved in the post-transcriptional regulation of gene expression. They are important for multiple biological processes such as development, proliferation, inflammation, and apoptosis (Xu et al., 2004; Pasquinelli et al., 2005; Thounaojam et al., 2013). The biogenesis and role(s) of the miRNA pathway have been recently and thoroughly reviewed by Treiber et al. (2012, and references therein). In brief, miRNAs function by binding with partial complementarity to messenger RNA (mRNA) sequences, mainly in the 3′ untranslated region (3′UTR). This targeting leads to either degradation or translational repression of the mRNA template(s), causing an overall downregulation in protein output. The miRNAs can target several disease-related genes involved in neurodegeneration (Delay et al., 2012; Abe and Bonini, 2013).

The precise mechanism(s) involved in miRNA release from cells remain largely unknown, but may involve the ceramide-dependent secretory machinery (Kosaka et al., 2010). Alternatively, there may be a passive leakage from necrotic or apoptotic cells (Zernecke et al., 2009). In any case, these small RNAs are highly stable in body fluids such as plasma and CSF (Mraz et al., 2009), making them attractive biomarkers. There are several factors involved in modulating (distant) circulating miRNAs. These small RNAs are transported in free forms, exosomes, liposomes, or high-density lipoproteins, which protect them from degradation (Vickers et al., 2011; Hu et al., 2012; the stable packaging, processing, and functionality of miRNAs in biofluids is a fascinating and important area of research mostly beyond what is addressed in the current review). While some miRNAs are ubiquitously expressed, others are present in specific cells or tissues, including the central nervous system (CNS; Landgraf et al., 2007). Furthermore, bioinformatics studies suggest that miRNA abundance is directly correlated with mRNA target activity (Dorval et al., 2012).

Interestingly, miRNAs have been described as epigenetic contributors to age-related cognitive changes (Kosik et al., 2012). It has been suggested that dysregulation of these miRNA-dependent epigenetic functions in vulnerable brain regions may lead to cognitive impairments. Accordingly, the past few years have witnessed an explosion of papers linking miRNAs to disease states, and current research efforts establish that miRNA expression profiles are altered in a variety of pathogenic conditions. This is particularly recognized in the cancer field (Sayed and Abdellatif, 2011). Interestingly, the various changes in miRNA levels are observable not only in cells/tissues directly related to disease (e.g., tumors vs. adjacent tissues), but often in the periphery or distant biological systems (e.g., tumors vs. blood). It is noteworthy that most peripheral miRNAs are also found in the brain, albeit at various levels (Hebert et al., 2013).

CIRCULATING miRNA BIOMARKERS IN AD

CEREBROSPINAL FLUID

Cerebrospinal fluid is a clear fluid that flows within the ventricles and around the surface of the brain and spinal cord. One primary function of CSF is to circulate nutrients within the CNS and, in turn, to act as a waste remover. The CSF is an attractive source of biomarkers as it is in direct and constant contact with the extracellular space of the brain, and can reflect biochemical and/or physiological changes that occur inside the brain.

In a pioneer study by Cogswell et al. (2008), the group performed a large-scale expression analysis of miRNAs in control and AD CSF. About 201 (out of 242 tested) miRNAs were detected above background levels, as measured by quantitative reverse transcriptase-polymerase chain reaction (qRT-PCR) using TaqMan probes (Applied Biosystems). They identified 60 miRNAs, including let-7i, that were significantly altered in AD CSF (Braak V stage) when compared to healthy elderly controls (Braak I stage; $n = 10$ per group, $P < 0.05$). Using biological pathway enrichment algorithms, the group observed an association between misregulated miRNAs and the immune system, including pathways such as *innate immunity* (e.g., miR-146b) and *T cell activation and differentiation* (e.g., miR-181a, miR-142-5p). Putative targets for these miRNAs include IRAK1, TRAF6 (Lindsay, 2008), and Bcl-2 family members (Ouyang et al., 2012). The authors suggested that abnormally expressed miRNAs in the CSF were likely derived from immune cells. This was the first study demonstrating that miRNAs can be detected in the CSF (even when initially frozen) and are altered in neurodegenerative disease conditions.

van Harten et al. (2011) confirmed that it was technically feasible to perform genome-wide expression analyses of circulating miRNAs in control and AD CSF. The authors used two stem-loop qRT-PCR methods, including: (1) an individual miRNA TaqMan qRT-PCR and (2) a Megaplex modified microarray. Using this latter approach, the authors detected 667 miRNAs from one control and one AD subject (note that more than 2,000 human miRNAs are currently registered in the miRNA database – www.mirbase.org). The authors specifically quantified and validated changes in neuronal miR-802, a suppressor of caveolin-1 (Lin et al., 2011), in the CSF of control ($n = 8$) and AD ($n = 14$) patients. Clinical tests, combined with Aβ42, t-tau, and p-tau-181 measurements in the CSF were globally consistent with the diagnosis of either group. Whether other miRNAs were misregulated in AD conditions was not evaluated.

Only recently have two critical questions been addressed in relation to CSF miRNAs in neurodegenerative diseases. The first natural question relates to *why* miRNAs are stably present in this biofluid. After all, RNAs are notoriously unstable in solution, and yet there presence has been reliably affirmed. Thus, there is a tantalizing possibility that the miRNAs in solution – and in biochemical packaging as described above – may be playing a role in the CNS. In an elegant study, Lehmann et al. (2012) demonstrated that circulating miRNAs, and in particular let-7b, could exacerbate brain damage and neurodegeneration by binding directly to the Toll-like receptor 7 (TLR7). As measured by miRNA qRT-PCR, AD CSF ($n = 13$) contained significantly higher levels of let-7b when compared to controls ($n = 11$). Here, AD patients were selected, in part, on the basis of Aβ42 and t-tau levels. Unfortunately, no correlation between these AD markers and let-7b levels was provided. However, this study demonstrates that miRNAs in CNS are bioactive, and may have paracrine/hormonal-like functions, which, if generally true, provides a novel and potentially incredibly important context for miRNA function (and pathological impact) in the brain.

A second key question is more practical, and was addressed by Alexandrov et al. (2012): is there a correlation between Aβ peptides and miRNA levels in the CSF? In this study, the patient groups consisted of six AD and six age-matched controls. Consistent with previous studies using enzyme-linked immunosorbent assay (ELISA), they reported a decrease in Aβ40 and Aβ42 in AD CSF, although this observation did not reach statistical significance ($P \sim 0.06$). Interestingly, the authors measured higher (greater than 100-fold) levels of total miRNAs (total mass) when compared to Aβ peptides, and this, both in control and AD CSF. Fluorescence-based miRNA microarrays indicated that the pro-inflammatory miRNAs miR-9, miR-125b, miR-146a, and miR-155 were significantly increased in AD CSF. These observations were further validated by a highly sensitive light-emitting diode (LED)-based Northern dot-blot analysis. This increase of specific miRNAs was extended to *in vitro* paradigms, where primary human neuronal/glial cells treated with AD-derived extracellular fluid lead to an increase of the same set of miRNAs. Significant negative correlations were observed between Aβ42 peptides and miR-137 ($r = -0.75$, $P = 0.003$), miR-181c ($r = -0.57$, $P = 0.037$), miR-9 ($r = -0.7$, $P = 0.007$), miR-29a ($r = -0.64$, $P = 0.01$), and miR-29b-1 ($r = -0.569$, $P = 0.037$), and this, in both control and AD patients. Based on these observations, it is tempting to speculate that miRs, alone or in combination with known AD biomarkers, could provide a better assessment of AD diagnosis.

BLOOD

Blood circulates in the principal vascular system, composed of arteries and veins, to carry oxygen to and carbon dioxide from tissues. The combination of lymphocytes, monocytes, and macrophages composes the peripheral blood mononuclear cells (PBMCs) population. These blood cells are critical components in the immune system.

Schipper (2007) assessed miRNA levels in blood mononuclear cells (BMCs) derived from sporadic AD and age-matched controls ($n = 16$ per group), using a microarray chip containing 462 human miRNAs. Several miRNAs were identified to be significantly altered in AD BMCs. A large number of miRNAs, including miR-34a, miR-181b, and let-7f, were validated by miRNA qRT-PCR. Interestingly, miR-34 targets include p53 (He et al., 2007), Notch (Bu et al., 2013), and Bcl-2 (Cole et al., 2008). The let-7 targets the oncogene Ras protein, thus promoting tumorigenesis (Johnson et al., 2005). Inversely, let-7 expression is regulated by the oncogenic Myc protein (Chang et al., 2008), suggesting a regulatory feedback loop. Together, these observations highlight the importance of these miRNAs in cell/tissue homeostasis.

Geekiyanage and Chan (2011) showed by miRNA qRT-PCR a decrease in miR-137, miR-181c, miR-9, and miR-29a/b levels in the neocortical region of controls ($n = 7$) and AD subjects ($n = 7$), which negatively correlated with Aβ42 levels in post-mortem brain tissues. In a follow-up study, using the same technical approach, the group reported that the same miRNAs were also present in the blood, albeit at lower basal levels (Geekiyanage et al., 2012). They were found to be downregulated in the blood serum of mild cognitive impairment (MCI; $n = 7$) and "probable" AD patients ($n = 7$) when compared controls ($n = 7$).

Villa et al. (2013) provided further evidence that dysregulation of peripheral miRNAs might contribute to AD development. In isolated PBMCs, they first showed that the transcription factor Sp1 was regulated at a post-transcriptional level by miR-29b. Interestingly, Sp1 regulates the expression of AD-related genes such as APP (La Fauci et al., 1989) and tau (Heicklen-Klein and Ginzburg, 2000). In a cohort of 393 AD patients and 412 healthy controls, the group observed an inverse relationship between Sp1 mRNA and miR-29b levels in PBMCs ($p = 0.002$). To our knowledge, this is the first report suggesting that changes in miRNA levels (e.g., miR-29b) and its/their target(s) (e.g., Sp1) may serve as cooperative biomarkers for AD diagnosis. Whether a genuine interaction between both molecules occurs in the blood remains to be validated.

Very recently, Bekris et al. (2013) reported in an elegant 3-phase study including post-mortem brain arrays and qRT-PCR validation that plasma miR-15a correlated with neuritic plaque score and Braak stages in AD. This particular miRNA was predicted to modulate 9 AD-relevant genes, including APP (Liu et al., 2012) and tau (Hebert et al., 2010). The authors concluded that pathologically-altered brain miRNAs might be detected in CSF or plasma during life, providing further proof of principle that miRNAs are relevant clinical biomarkers of AD pathology.

CIRCULATING miRNA BIOMARKERS IN MILD COGNITIVE IMPAIRMENT, AND CHALLENGES IN PATHOLOGICAL SPECIFICITY

Mild cognitive impairment is a term often conflated with indicating early clinical manifestation of AD, and many do indeed progress to full-blown AD clinically, although many other pathologies than AD underlie the clinical state of MCI (see below). Nevertheless, it is essential to develop tools that can accurately discriminate between normal aging, MCI, AD, and likely other cognitive disabilities. An attractive approach has recently been proposed, namely miRNA "pairs." This concept uses, following single qRT-PCR TaqMan assays, bioinformatics to analyze the ratios of all measured miRNAs, and select the most promising pair(s) of biomarkers (Sheinerman et al., 2012). In a pilot study, 13 miRNA pairs allowed to discriminate between AD and age-matched controls, as well as between MCI and age-matched controls ($n = 10$ per group), and this, with up to 90% accuracy. The proposed sets of miRNAs could detect pre-symptomatic MCI 1–5 years before the diagnosis in 70% of cases. Finally, the same pairs of miRNAs have been able to discriminate between aged and young healthy controls ($n = 20$ per group).

There are two practical issues that are important to the clinical relevance of any biomarker: sensitivity and specificity. The issue of sensitivity is basic and relates to the fact that by the time AD is manifest as MCI, it may be too late for (at least some) therapeutic interventions. The Aβ/tau CSF studies have now shown that, as expected (Nelson et al., 2009), up to one-third of non-demented subjects harbor some AD-type pathology (Nelson et al., 2012). It is increasingly appreciated that these are the patients that should be targeted for biomarker studies as well as clinical trials.

Aspects of biomarker specificity are perhaps paramount, and often under-appreciated. Although MCI is often used to indicate an early stage of AD, MCI was originally defined according to neuropsychological features (Portet et al., 2006), which have been recognized to entail "multiple sources of heterogeneity." As such, it is quite usual for MCI to be associated with brain pathologies other than AD: dementia with Lewy bodies (DLB), vascular pathologies, hippocampal sclerosis (HS-Aging), frontotemporal lobar dementia (FTLD), and other conditions may cause or contribute to MCI, as to dementia (Jicha et al., 2006). This highlights an important aspect of AD-related biomarkers: they are not only used in predicting whom will become demented, but also for specifying which subtype of dementia will be predominant; the importance of this specificity for clinical trials is obvious. Novel insights, relevant to this consideration, were obtained by deep sequencing miRNAs from brains of individuals with multiple different pathological diseases (AD, DLB, FTLD, and HS-Aging; Hebert et al., 2013). Although not a biomarker study *per se*, this showed that some miRNAs (particularly miR-132-5p) are downregulated in neurodegenerative diseases non-specifically. In the future, it is hoped that more specific miRNA "fingerprints" may help to distinguish the individual subtypes of neurodegenerative diseases before their earliest manifestations.

CONCLUSION AND PERSPECTIVES

To date, most researchers have relied on the combination of Aβ peptides, total-tau, and phospho-tau (Thr181) ratios to provide the best discriminative values for individuals with or without AD. However, in most cases, large variability and differences between studied groups did not reach statistical significance, leaving inconsistencies. Without excluding the amyloid and tau biological markers, a combination of biomarkers may provide a better tool for AD diagnosis, therefore improving their clinical usefulness. Known examples include structural (e.g., hippocampal shrinkage), functional (e.g., glucose metabolism), and molecular imaging (e.g., fluorescent Pittsburgh compound B; Chintamaneni and Bhaskar, 2012).

Small non-coding RNAs, and in particular miRNAs, have come a long way in the past two decades. As discussed herein, circulating miRNAs provide an exciting and emerging research area in the biomarker field. As of now, long lists of miRNAs potentially misregulated in disease conditions have been reported, although finding overlaps is challenging (note that this is also the case for miRNA profiling studies in the brain). However, some AD-specific miRNAs were "consistently" identified, including some let-7 family members (let-7f, let-7b, and let-7i), miR-9, miR-181, and miR-29 (Maes et al., 2009). These miRNAs seem involved in processes previously associated with AD, that is to say inflammation and immunological response. Perhaps expectedly, several miRNAs and their functions as biomarkers have been patented or in the process thereof (see, e.g., www.freepatentsonline.com or www.patentlawlinks.com). Although very attractive, the applicability of miRNAs as diagnostic tools into the clinic for AD (or MCI) will require extensive validation and follow-up studies in larger cohorts of patients. This is important as AD is a heterogeneous, multifactorial disease, with often display overlapping pathologies (e.g., Aβ deposits and Lewy bodies; Gomperts et al., 2008) and/or co-morbid diagnoses (e.g., diabetes, stroke). Obviously, the ultimate goal is to provide a sensitive, reproducible, and accurate detection of AD neuropathological changes prior to the onset of the disease and the appearance of the clinical symptoms. To this end, future studies will require better neuropathological validations as well as, ultimately, far greater sample sizes for robust statistical power.

In conclusion, circulating miRNAs are amongst the promising next generation of biomarkers for AD, and ultimately the discrimination between neurodegenerative diseases. They may be small molecules, but miRNAs certainly provide a big potential for the diagnosis of human diseases.

Note: While this work was in progress, a report has been published with regards to a circulating miRNA signature in AD patients (Leidinger et al., 2013).

ACKNOWLEDGMENTS

This work was supported by the Alzheimer's Society of Canada and the Canadian Institutes of Health Research.

REFERENCES

Abe, M., and Bonini, N. M. (2013). MicroRNAs and neurodegeneration: role and impact. *Trends Cell Biol.* 23, 30–36. doi: 10.1016/j.tcb.2012.08.013

Alexandrov, P. N., Dua, P., Hill, J. M., Bhattacharjee, S., Zhao, Y., and Lukiw, W. J. (2012). microRNA (miRNA) speciation in Alzheimer's disease (AD) cerebrospinal fluid (CSF) and extracellular fluid (ECF). *Int. J. Biochem. Mol. Biol.* 3, 365–373.

Bekris, L. M., Lutz, F., Montine, T. J., Yu, C. E., Tsuang, D., Peskind, E. R., et al. (2013). MicroRNA in Alzheimer's disease: an exploratory study in brain, cerebrospinal fluid and plasma. *Biomarkers* 18, 455–466. doi: 10.3109/1354750X.2013.814073

Bu, P., Chen, K. Y., Chen, J. H., Wang, L., Walters, J., Shin, Y. J., et al. (2013). A microRNA miR-34a-regulated bimodal switch targets notch in colon cancer stem cells. *Cell Stem Cell* 12, 602–615. doi: 10.1016/j.stem.2013.03.002

Chang, T. C., Yu, D., Lee, Y. S., Wentzel, E. A., Arking, D. E., West, K. M., et al. (2008). Widespread microRNA repression by Myc contributes to tumorigenesis. *Nat. Genet.* 40, 43–50. doi: 10.1038/ng.2007.30

Chintamaneni, M., and Bhaskar, M. (2012). Biomarkers in Alzheimer's disease: a review. *ISRN Pharmacol.* 2012, 984786. doi: 10.5402/2012/984786

Cogswell, J. P., Ward, J., Taylor, I. A., Waters, M., Shi, Y., Cannon, B., et al. (2008). Identification of miRNA changes in Alzheimer's disease brain and CSF yields putative biomarkers and insights into disease pathways. *J. Alzheimers Dis.* 14, 27–41.

Cole, K. A., Attiyeh, E. F., Mosse, Y. P., Laquaglia, M. J., Diskin, S. J., Brodeur, G. M., et al. (2008). A functional screen identifies miR-34a as a candidate neuroblastoma tumor suppressor gene. *Mol. Cancer Res.* 6, 735–742. doi: 10.1158/1541-7786.MCR-07-2102

Delay, C., Mandemakers, W., and Hebert, S. S. (2012). MicroRNAs in Alzheimer's disease. *Neurobiol. Dis.* 46, 285–290. doi: 10.1016/j.nbd.2012.01.003

Dorval, V., Smith, P. Y., Delay, C., Calvo, E., Planel, E., Zommer, N., et al. (2012). Gene network and pathway analysis of mice with conditional ablation of dicer in post-mitotic neurons. *PLoS ONE* 7: e44060. doi: 10.1371/journal.pone.0044060

Geekiyanage, H., and Chan, C. (2011). MicroRNA-137/181c regulates serine palmitoyltransferase and in turn amyloid beta, novel targets in sporadic Alzheimer's disease. *J. Neurosci.* 31, 14820–14830. doi: 10.1523/JNEUROSCI.3883-11.2011

Geekiyanage, H., Jicha, G. A., Nelson, P. T., and Chan, C. (2012). Blood serum miRNA: non-invasive biomarkers for Alzheimer's disease. *Exp. Neurol.* 235, 491–496. doi: 10.1016/j.expneurol.2011.11.026

Gomez-Isla, T., Price, J. L., Mckeel, D. W., Jr., Morris, J. C., Growdon, J. H., and Hyman, B. T. (1996). Profound loss of layer II entorhinal cortex neurons occurs in very mild Alzheimer's disease. *J. Neurosci.* 16, 4491–4500.

Gomperts, S. N., Rentz, D. M., Moran, E., Becker, J. A., Locascio, J. J., Klunk, W. E., et al. (2008). Imaging amyloid deposition in Lewy body diseases. *Neurology* 71, 903–910. doi: 10.1212/01.wnl.0000326146.60732.d6

He, L., He, X., Lim, L. P., De Stanchina, E., Xuan, Z., Liang, Y., et al. (2007). A microRNA component of the p53 tumour suppressor network. *Nature* 447, 1130–1134. doi: 10.1038/nature05939

Hebert, S. S., Papadopoulou, A. S., Smith, P., Galas, M. C., Planel, E., Silahtaroglu, A. N., et al. (2010). Genetic ablation of Dicer in adult forebrain neurons results in abnormal tau hyperphosphorylation and neurodegeneration. *Hum. Mol. Genet.* 19, 3959–3969. doi: 10.1093/hmg/ddq311

Hebert, S. S., Wang, W. X., Zhu, Q., and Nelson, P. T. (2013). A study of small RNAs from cerebral neocortex of pathology-verified Alzheimer's disease, dementia with Lewy bodies, hippocampal sclerosis, frontotemporal lobar dementia, and non-demented human controls. *J. Alzheimers Dis.* 35, 335–348. doi: 10.3233/JAD-122350

Heicklen-Klein, A., and Ginzburg, I. (2000). Tau promoter confers neuronal specificity and binds Sp1 and AP-2. *J. Neurochem.* 75, 1408–1418. doi: 10.1046/j.1471-4159.2000.0751408.x

Holtzman, D. M. (2011). CSF biomarkers for Alzheimer's disease: current utility and potential future use. *Neurobiol. Aging* 32(Suppl.1), S4–S9. doi: 10.1016/j.neurobiolaging.2011.09.003

Hu, G., Drescher, K. M., and Chen, X. M. (2012). Exosomal miRNAs: biological properties and therapeutic potential. *Front. Genet.* 3:56. doi: 10.3389/fgene.2012.00056

Hyman, B. T., Phelps, C. H., Beach, T. G., Bigio, E. H., Cairns, N. J., Carrillo, M. C., et al. (2012). National Institute on Aging–Alzheimer's Association guidelines for the neuropathologic assessment of Alzheimer's disease. *Alzheimers Dement.* 8, 1–13. doi: 10.1016/j.jalz.2011.10.007

Ingelson, M., Blomberg, M., Benedikz, E., Wahlund, L. O., Karlsson, E., Vanmechelen, E., et al. (1999). Tau immunoreactivity detected in human plasma, but no obvious increase in dementia. *Dement. Geriatr. Cogn. Disord.* 10, 442–445. doi: 10.1159/000017187

Jicha, G. A., Parisi, J. E., Dickson, D. W., Johnson, K., Cha, R., Ivnik, R. J., et al. (2006). Neuropathologic outcome of mild cognitive impairment following progression to clinical dementia. *Arch. Neurol.* 63, 674–681. doi: 10.1001/archneur.63.5.674

Johnson, S. M., Grosshans, H., Shingara, J., Byrom, M., Jarvis, R., Cheng, A., et al. (2005). RAS is regulated by the let-7 microRNA family. *Cell* 120, 635–647. doi: 10.1016/j.cell.2005.01.014

Kang, J., Lemaire, H. G., Unterbeck, A., Salbaum, J. M., Masters, C. L., Grzeschik, K. H., et al. (1987). The precursor of Alzheimer's disease amyloid A4 protein resembles a cell-surface receptor. *Nature* 325, 733–736. doi: 10.1038/325733a0

Kosaka, N., Iguchi, H., Yoshioka, Y., Takeshita, F., Matsuki, Y., and Ochiya, T. (2010). Secretory mechanisms and intercellular transfer of microRNAs in living cells. *J. Biol. Chem.* 285, 17442–17452. doi: 10.1074/jbc.M110.107821

Kosik, K. S., Rapp, P. R., Raz, N., Small, S. A., Sweatt, J. D., and Tsai, L. H. (2012). Mechanisms of age-related cognitive change and targets for intervention: epigenetics. *J. Gerontol. A Biol. Sci. Med. Sci.* 67, 741–746. doi: 10.1093/gerona/gls110

La Fauci, G., Lahiri, D. K., Salton, S. R., and Robakis, N. K. (1989). Characterization of the 5′-end region and the first two exons of the beta-protein precursor gene. *Biochem. Biophys. Res. Commun.* 159, 297–304. doi: 10.1016/0006-291X(89)92437-6

Landgraf, P., Rusu, M., Sheridan, R., Sewer, A., Iovino, N., Aravin, A., et al. (2007). A mammalian microRNA expression atlas based on small RNA library sequencing. *Cell* 129, 1401–1414. doi: 10.1016/j.cell.2007.04.040

Lehmann, S. M., Kruger, C., Park, B., Derkow, K., Rosenberger, K., Baumgart, J., et al. (2012). An unconventional role for miRNA: let-7 activates Toll-like receptor 7 and causes neurodegeneration. *Nat. Neurosci.* 15, 827–835. doi: 10.1038/nn.3113

Leidinger, P., Backes, C., Deutscher, S., Schmitt, K., Muller, S. C., Frese, K., et al. (2013). A blood based 12-miRNA signature of Alzheimer disease patients. *Genome Biol.* 14, R78. doi: 10.1186/gb-2013-14-7-r78 [Epub ahead of print].

Lin, D. H., Yue, P., Pan, C., Sun, P., and Wang, W. H. (2011). MicroRNA 802 stimulates ROMK channels by suppressing caveolin-1. *J. Am. Soc. Nephrol.* 22, 1087–1098. doi: 10.1681/ASN.2010090927

Lindsay, M. A. (2008). MicroRNAs and the immune response. *Trends Immunol.* 29, 343–351. doi: 10.1016/j.it.2008.04.004

Liu, W., Liu, C., Zhu, J., Shu, P., Yin, B., Gong, Y., et al. (2012). MicroRNA-16 targets amyloid precursor protein to potentially modulate Alzheimer's-associated pathogenesis in SAMP8 mice. *Neurobiol. Aging* 33, 522–534. doi: 10.1016/j.neurobiolaging.2010.04.034

Maes, O. C., Chertkow, H. M., Wang, E., and Schipper, H. M. (2009). MicroRNA: implications for Alzheimer disease and other human CNS disorders. *Curr. Genomics* 10, 154–168. doi: 10.2174/138920209788185252

Mraz, M., Malinova, K., Mayer, J., and Pospisilova, S. (2009). MicroRNA isolation and stability in stored RNA samples. *Biochem. Biophys. Res. Commun.* 390, 1–4. doi: 10.1016/j.bbrc.2009.09.061

Nelson, P. T., Alafuzoff, I., Bigio, E. H., Bouras, C., Braak, H., Cairns, N. J., et al. (2012). Correlation of Alzheimer disease neuropathologic changes with cognitive status: a review of the literature. *J. Neuropathol. Exp. Neurol.* 71, 362–381. doi: 10.1097/NEN.0b013e31825018f7

Nelson, P. T., Braak, H., and Markesbery, W. R. (2009). Neuropathology and cognitive impairment in Alzheimer disease: a complex but coherent relationship. *J. Neuropathol. Exp. Neurol.* 68, 1–14. doi: 10.1097/NEN.0b013e3181919a48

Ouyang, Y. B., Lu, Y., Yue, S., and Giffard, R. G. (2012). miR-181 targets multiple Bcl-2 family members and influences apoptosis and mitochondrial function in astrocytes. *Mitochondrion* 12, 213–219. doi: 10.1016/j.mito.2011.09.001

Pasquinelli, A. E., Hunter, S., and Bracht, J. (2005). MicroRNAs: a developing story. *Curr. Opin. Genet. Dev.* 15, 200–205. doi: 10.1016/j.gde.2005.01.002

Portet, F., Ousset, P. J., Visser, P. J., Frisoni, G. B., Nobili, F., Scheltens, P., et al. (2006). Mild Cognitive Impairment (MCI) in medical practice: a critical review of the concept and new diagnostic procedure. Report of the MCI Working Group of the European Consortium on Alzheimer's Disease. *J. Neurol. Neurosurg. Psychiatry* 77, 714–718. doi: 10.1136/jnnp.2005.085332

Sayed, D., and Abdellatif, M. (2011). MicroRNAs in development and disease. *Physiol. Rev.* 91, 827–887. doi: 10.1152/physrev.00006.2010

Schipper, H. M. (2007). Biomarker potential of heme oxygenase-1 in Alzheimer's disease and mild cognitive impairment. *Biomark. Med.* 1, 375–385. doi: 10.2217/17520363.1.3.375

Sheinerman, K. S., Tsivinsky, V. G., Crawford, F., Mullan, M. J., Abdullah, L., and Umansky, S. R. (2012). Plasma microRNA biomarkers for detection of mild cognitive impairment. *Aging (Albany NY)* 4, 590–605.

Tarawneh, R., and Holtzman, D. M. (2010). Biomarkers in translational research of Alzheimer's disease. *Neuropharmacology* 59, 310–322. doi: 10.1016/j.neuropharm.2010.04.006

Thounaojam, M. C., Kaushik, D. K., and Basu, A. (2013). MicroRNAs in the brain: it's regulatory role in neuroinflammation. *Mol. Neurobiol.* 47, 1034–1044. doi: 10.1007/s12035-013-8400-3

Treiber, T., Treiber, N., and Meister, G. (2012). Regulation of microRNA biogenesis and function. *Thromb. Haemost.* 107, 605–610. doi: 10.1160/TH11-12-0836

van Harten, A., Mulders, J., Çevik, C., Kester, M., Scheltens, P., Flier, W. V. D., et al. (2011). "MicroRNA analysis in the spinal fluid of Alzheimer patients: a methodological feasibility study," in *Circulating Nucleic Acids in Plasma and Serum*, ed. P. B. Gahan (Dordrecht: Springer Netherlands), 275–282.

Vickers, K. C., Palmisano, B. T., Shoucri, B. M., Shamburek, R. D., and Remaley, A. T. (2011). MicroRNAs are transported in plasma and delivered to recipient cells by high-density lipoproteins. *Nat. Cell Biol.* 13, 423–433. doi: 10.1038/ncb2210

Villa, C., Ridolfi, E., Fenoglio, C., Ghezzi, L., Vimercati, R., Clerici, F., et al. (2013). Expression of the transcription factor Sp1 and its regulatory hsa-miR-29b in peripheral blood mononuclear cells from patients with Alzheimer's disease. *J. Alzheimers Dis.* 35, 487–494. doi: 10.3233/JAD-122263

West, M. J., Coleman, P. D., Flood, D. G., and Troncoso, J. C. (1994). Differences in the pattern of hippocampal neuronal loss in normal ageing and Alzheimer's disease. *Lancet* 344, 769–772. doi: 10.1016/S0140-6736(94)92338-8

Wolfe, M. S. (2006). The gamma-secretase complex: membrane-embedded proteolytic ensemble. *Biochemistry* 45, 7931–7939. doi: 10.1021/bi060799c

Xu, P., Guo, M., and Hay, B. A. (2004). MicroRNAs and the regulation of cell death. *Trends Genet.* 20, 617–624. doi: 10.1016/j.tig.2004.09.010

15

Cause or effect: misregulation of microRNA pathways in neurodegeneration

Eduardo Gascon* and Fen-Biao Gao*

Department of Neurology, University of Massachusetts Medical School, Worcester, MA, USA

Correspondence:
Eduardo Gascon and Fen-Biao Gao, Department of Neurology, University of Massachusetts Medical School, Worcester, MA 01605, USA.
e-mail: eduardo.gascon@umassmed.edu; fen-biao.gao@umassmed.edu

During normal aging or neurodegenerative diseases, neuronal survival and function depend on protein homeostasis, which is regulated by multiple mechanisms, including the microRNA (miRNA) pathway. In different cells types, the absence of Dicer, a key miRNA processing enzyme, leads to neurodegeneration through cell-autonomous and non-cell-autonomous mechanisms. Loss of certain miRNAs also causes neurodegeneration in some model organisms. On the other hand, miRNA expression is misregulated in patients with different neurodegenerative diseases. Thus, the miRNA pathway appears to be essential in the pathogenesis of several age-dependent neurodegenerative conditions; however, our understanding of the underlying mechanism remains rudimentary. The precise causal relationships between specific miRNAs and neurodegeneration in humans need to be further investigated.

Keywords: Alzheimer's disease, ALS, FTD, microRNAs, neurodegeneration, C9ORF72, CHMP2B, TDP-43

INTRODUCTION

Alzheimer's disease (AD), frontotemporal dementia (FTD), Parkinson's disease (PD), and other neurodegenerative disorders are a major health problem in both developed and developing countries (Hampel et al., 2011; Reitz et al., 2011; Wittchen et al., 2011). Since no effective treatments are available, it is unlikely that the adverse societal effects of these disorders will be substantially alleviated in the near future. These disorders are characterized by progressive neuronal dysfunction that initially affects selected groups of neurons in specialized neuronal circuits. A number of cellular and molecular mechanisms lead to neuronal demise. Among them, neurotoxicity induced by misfolding, mislocalization, or abnormally elevated concentrations of particular protein species seems to be a common theme (Jucker and Walker, 2011; Lee et al., 2012; Selkoe et al., 2012). Mutations in several genes that are apparently functionally unrelated can cause the same neurodegenerative disease. Multiple environmental factors (e.g., viral infections and exposure to certain toxins) might also contribute to the development of a neurodegenerative disease (Ahmed and Wicklund, 2011; Gao and Hong, 2011). Much remained to be learned about how the complex interactions of environmental and genetic factors initially lead to the misregulation of protein homeostasis and subsequently to neuronal dysfunction.

MicroRNAs (miRNAs) are small non-coding RNAs that post-transcriptionally regulate gene expression by degrading their target mRNAs or repressing their translation (Ambros et al., 2003; Bartel, 2004; Ghildiyal and Zamore, 2009). Initially discovered in *Caenorhabditis elegans* (Lee et al., 1993), miRNAs have been found in plants, invertebrates, mammals, and humans (Bartel, 2009). Two major features of miRNAs indicate their potential contributions to neurodegenerative diseases. First, miRNAs can simultaneously regulate many target transcripts, and up to 50% of all coding genes may be regulated by miRNAs (Krol et al., 2010). Thus, miRNAs are central regulators of genetic networks. Second, miRNAs ensure stable protein levels under variable conditions and are therefore essential for the robustness of biological processes (Herranz and Cohen, 2010). Given the importance of protein homeostasis and the diversity of cellular pathways potentially leading to neurodegeneration, it has been hypothesized that miRNAs might contribute to neurodegenerative diseases (Eacker et al., 2009; Lau and de Strooper, 2010; Sonntag, 2010).

In this review, we briefly describe the biogenesis of miRNAs and their potential involvement in the evolution of the human brain. Then we will discuss accumulating evidence that miRNAs are important contributors to neurodegenerative diseases. Indeed, some observations suggest that miRNA alterations can disrupt protein homeostasis and may be at the root of neurodegenerative processes (**Figure 2**). Conversely, other data strongly suggest that altered miRNA networks are a consequence of abnormal neuronal physiology (**Figure 3**). Examples in specific neurodegenerative diseases will be presented.

miRNA BIOGENESIS

Although several alternative mechanisms also exist (Yang and Lai, 2011), the canonical pathway for miRNA biogenesis involves a primary transcript generated by RNA polymerase II (Lee et al., 2004). The primary miRNA is cleaved by a nuclear complex containing Drosha and DGCR8, giving rise to a hairpin precursor molecule (pre-miRNA) of 70–100 nt (Lee et al., 2003; Han et al., 2004). This pre-miRNA is exported to the cytoplasm, where Dicer, a RNA-III nuclease, catalyzes the final cleavage in the maturation process, resulting in an imperfect RNA duplex (Hutvagner et al., 2001). One strand (guide strand) is loaded into an RNA-induced silencing complex (RISC) to bind the target mRNA; the other strand (passenger strand) is usually destroyed (Chendrimada et al., 2005; Bartel, 2009). miRNAs control gene expression at the post-transcriptional level through imperfect base pairing with specific sequences, located mostly in the 3′UTRs of mRNAs. After recognition, miRNA–target interactions often result in mRNA degradation or inhibition of mRNA translation (Krol et al., 2010; **Figure 1**).

ROLE OF miRNA IN BRAIN EVOLUTION

Neurodegenerative diseases are considered devastating disorders because they often impair cognitive and executive functions. Understanding how the human brain acquired such functions is a challenging task that might also provide important insights into the mechanisms of neurodegeneration. It was noted four decades ago that genetic differences among species do not account for brain divergence and that brain evolution could have been driven by changes in gene expression levels (King and Wilson, 1975). Early comparative transcriptome analyses confirmed this hypothesis and revealed more human-specific than chimpanzee-specific expression changes in the prefrontal cortex (PFC); no such changes were observed in blood, liver, or other tissues (Enard et al., 2002; Caceres et al., 2003). These findings supported the idea that brain-specific changes in gene expression levels shaped the evolution of the nervous system.

To determine whether miRNAs participated in this process, miRNA expression profiles in human and chimpanzee brains were compared (Berezikov et al., 2006). Many of the newly identified brain miRNAs were expressed only in humans, and many were restricted to primates. Another study reported that 10–35% of miRNAs were expressed in the human brain but not in chimpanzee or macaque brains (Hu et al., 2011). More importantly,

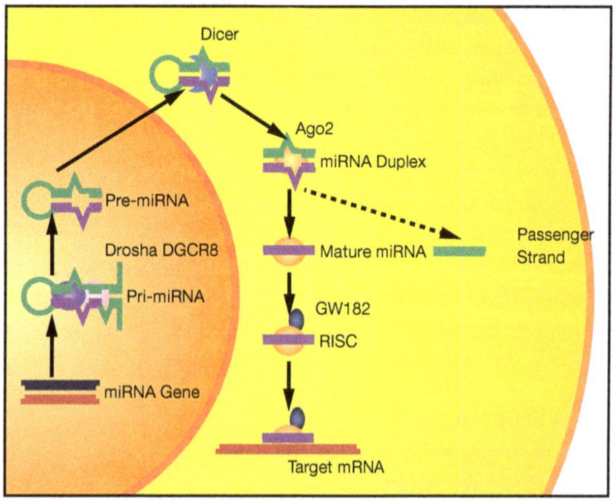

FIGURE 1 | The canonical miRNA biogenesis pathway. miRNAs are produced from long Pol II transcripts (pri-miRNA). A nuclear complex containing Drosha (purple oval) and DGCR8 (pink oval) cleaves the primary transcript and generates a precursor miRNA (pre-miRNA). After nuclear export, pre-miRNA is further processed by Dicer (blue croissant). Then, Ago2 (yellow oval) binds to the complex formed by miRNA duplex and Dicer. Ago2 induces Dicer dissociation and the release of the passenger strand from the complex. Finally, other proteins, such as GW182 (dark blue), associate with Ago2 and form a RISC complex that recognizes and then silences (by mRNA degradation and/or translation inhibition) a target mRNA.

developmental profiles of brain miRNAs and their target genes showed the fastest rates of human-specific evolutionary change (Somel et al., 2011). Although experimental evidence is still lacking, it is an attractive hypothesis that rapidly evolving miRNAs in the human brain are essential for neuronal function and maintenance.

GLOBAL LOSS OF miRNAs CAUSES NEURODEGENERATION

Genetic disruption of miRNAs biogenesis pathways has been used to probe the potential link between miRNAs and neurodegeneration. In mouse cerebellar Purkinje cells, conditional knockout of Dicer leads to age-dependent cerebellar degeneration and ataxia (Schaefer et al., 2007). Cell-type specific deletion of Dicer in striatal, retinal, spinal, and cortical neurons produced similar results (Cuellar et al., 2008; Damiani et al., 2008; Davis et al., 2008; Haramati et al., 2010). Dicer deletion also altered the phosphorylation pattern of tau before neuronal cell loss (Hebert et al., 2010), indicating that some mechanisms of neurodegeneration might be controlled through miRNAs. Of note, neurodegeneration in the absence of Dicer could also result from the toxic accumulation of pre-miRNAs or from the loss of other Dicer functions unrelated to miRNAs biogenesis. However, reduced production of a small proportion of miRNAs as a result of DGCR8 haploinsufficiency also leads to neuronal dysfunction (Stark et al., 2008; Fenelon et al., 2011; Schofield et al., 2011), supporting the notion that neurodegeneration could indeed arise from loss of miRNAs.

Glial cell defects may also profoundly influence neuronal survival (Ilieva et al., 2009; Prinz et al., 2011). Indeed, neurodegeneration ensues after targeted deletion of Dicer in astrocytes (Tao et al., 2011), oligodendrocytes (Shin et al., 2009), and Schwann cells (Pereira et al., 2010; Verrier et al., 2010; Wu et al., 2012). Thus, alteration of miRNA networks has the potential to disrupt neuronal function in a cell-autonomous or non-cell-autonomous manner and lead to neurodegenerative phenotypes.

LOSS OF INDIVIDUAL miRNAs LEADS TO NEURODEGENERATION

Genetic analyses have revealed essential roles for specific miRNAs in long-term neuronal survival, for example in mice lacking *miR-124-1* (Sanuki et al., 2011), one of the most well studied miRNAs in neuronal development (Gao, 2010). The mouse genome has three *miR-124* loci (*miR-124-1*, *-2*, and *-3*). Deletion of *miR-124-1*, the dominant source of this miRNA, increases apoptosis in the hippocampus and retina, causing a significant decrease in brain size. Furthermore, in the tail-suspension test, adult mutant mice exhibit a front and hind limb clasping response, a common phenotype in mouse models of neurodegenerative disorders. This effect seems to be mediated by regulation of the transcription factor Lhx2 (Sanuki et al., 2011). It is unclear whether the neurodegeneration also reflects the absence of miR-124's well-documented developmental functions (Cao et al., 2007; Visvanathan et al., 2007; De Pietri Tonelli et al., 2008; Cheng et al., 2009; Maiorano and Mallamaci, 2009). This question could be answered by using a conditional knockout approach.

Another miRNA that might be involved in neuronal survival is miR-8 (Karres et al., 2007), which is not brain-specific and has a complex pattern of expression. Mutant flies lacking *miR-8* have limb and wing defects and increased apoptosis in the central nervous system. Moreover, concomitant decrease of the transcriptional repressor atrophin in *miR-8* mutant flies partially rescued the phenotypes (Karres et al., 2007). It remains to be determined whether miR-8 contributes to neurodegeneration in mammals.

miRNAs IN NEURODEGENERATIVE DISEASES

Microarray studies have shown that the brain expresses a wide range of miRNAs, suggesting that these small RNA molecules participate in nervous system physiology (Lagos-Quintana et al., 2002; Miska et al., 2004; Lim et al., 2005; Manakov et al., 2009). More importantly, profiling studies revealed profound changes in several miRNAs (i.e., miR-9, miR-29 cluster, miR-107, miR-125b, and miR-128) in patient brains such as that of AD patients (Kim et al., 2007; Lukiw, 2007; Hebert et al., 2008; Johnson et al., 2008; Wang et al., 2008, 2011; Nunez-Iglesias et al., 2010). Although these global changes in miRNA expression under pathological conditions should be interpreted cautiously, they support the notion that dysregulation of miRNAs networks is a common theme in neurodegenerative diseases.

ALZHEIMER'S DISEASE

A pathological hallmark of AD, the most prevalent neurodegenerative disease, is the accumulation of plaques formed by short β-amyloid (Aβ) peptides, commencing in the hippocampus, and spreading progressively throughout the brain (Ballard et al., 2011; Selkoe et al., 2012). It is likely the accumulation is caused by both increased production and impaired clearance of Aβ. Aβ peptides exert toxic effects and elicit an inflammatory response. Both may contribute to disrupt neuronal homeostasis and alter the integrity of neuronal networks involved in learning, memory, and other cognitive functions. Aβ peptides are generated through proteolytic cleavage of the amyloid precursor protein (APP) by γ-secretase and β-site APP-cleaving enzyme 1 (BACE1; O'Brien and Wong, 2011).

Amyloid precursor protein and BACE1 each contain several miRNA target sites in their 3′UTRs. Several miRNAs have been reported to repress APP expression, including miR-16, miR-101, miR-106a, and miR-520c (Patel et al., 2008; Liu et al., 2010; Long and Lahiri, 2011). miR-137 and miR-181c regulate serine palmitoyltransferase, which modulates the Aβ level (Geekiyanage and Chan, 2011). Interestingly, polymorphisms in miRNA binding sites in the 3′UTR of APP gene could influence the binding efficiency of these miRNAs. Thus, miRNAs might fine tune APP expression, which may enhance or limit the risk of AD (Delay et al., 2011).

A significant decrease in neuronal miR-107 expression and a parallel increase in BACE1 have been observed in AD patients (even those at the earliest stages of AD; Wang et al., 2008, 2011; Nelson and Wang, 2010). The 3′UTR of BACE1 mRNA has functional binding sites for miR-29 (Hebert et al., 2008), miR-107 (Wang et al., 2008), and miR-124 (Fang et al., 2012). Interestingly, miR-107 also controls the expression of other proteins relevant to AD pathology, such as cofilin (Yao et al., 2010), an actin-binding protein that accumulates in cytoplasmic inclusions known as Hirano bodies (Hirano, 1994). Thus, a single miRNA deregulation could activate multiple potentially pathogenic cascades upstream of Aβ accumulation.

Another possibility is that alterations in miRNAs in AD brain are a consequence of amyloid deposits. For instance, miR-106b is aberrantly expressed in APPswe/PSE9 mice (Wang et al., 2010a), and miR-146a levels are increased in AD brains and in several mouse models of AD (Li et al., 2011). *In vitro* exposure of hippocampal neurons to Aβ peptides preferentially decreases the levels of mature miRNAs (only a small fraction of miRNAs were upregulated; Schonrock et al., 2010). miRNAs were similarly deregulated in the hippocampus of APP23 mice at the onset of plaque formation. Overall, these studies suggest that miRNA deregulation is an essential pathogenic mechanism that is induced by Aβ aggregation and contributes to the progression and severity of AD.

POLYGLUTAMINE DISEASES

Polyglutamine (polyQ) diseases are a group of nine neurodegenerative disorders caused by an unstable CAG expansion in the coding region of their respective associated genes (Orr and Zoghbi, 2007). Apart from this common feature, polyQ diseases have distinct clinical presentations, and the proteins involved in these diseases have no structural/functional homology (**Table 1**). Huntington disease (HD), the most frequent polyQ disease, is characterized by the progressive loss of striatal neurons and motor impairment (typically resulting in the involuntary writhing movements called chorea) and is often associated with cognitive and behavioral deficits (Shoulson and Young, 2011).

The causal mutation in HD is an expanded repetition of the CAG trinucleotide in the first exon of the gene encoding huntingtin (HTT; Gilliam et al., 1987), a large protein (3300 amino acids) whose functions remains mostly unknown. HTT associates with Ago2 in P-bodies, and HTT depletion impairs miRNA-mediated gene silencing (Savas et al., 2008). Expanded HTT may sequester RNA processing factors in the cytoplasm (Jiang et al., 2011).

miRNAs were implicated in HD pathogenesis by two lines of evidence. First, the levels of repressor element 1 silencing transcription (REST) factor, a major pathogenic pathway in HD (Buckley et al., 2010), is elevated in HD neurons, resulting in repression of hundreds of key neuronal genes (Zuccato et al., 2003, 2007; Johnson et al., 2010b). Canonical and non-canonical REST binding motifs have been mapped in close proximity to 22 miRNA sites in the human genome, including several miRNAs that are abundant in neurons (Bruce et al., 2004; Jothi et al., 2008; Yu et al., 2011). Could abnormal REST deregulate miRNAs network in HD patients? REST and its cofactor coREST possess functional target sites for miR-9 and miR-9*, respectively (Packer et al., 2008), and miR-9 and miR-9* (together with miR-7, miR-124, miR-132, and other miRNAs) are downregulated in HD patients (Johnson et al., 2008; Packer et al., 2008; Marti et al., 2010). miRNA deregulation in HD was confirmed by profiling studies in animal models, although there was a high degree of variability among the models (Lee et al., 2011). These observations strongly suggest that altered miRNA transcription is a major event in HD pathogenesis.

Besides HD, there are eight other polyQ diseases: dentatorubral–pallidoluysian atrophy, spinal and bulbar muscular atrophy, and spinocerebellar ataxia (SCA) 1, 2, 3, 6, 7, and 17 (Gatchel and Zoghbi, 2005). In fly models of SCA3, reduction of

Table 1 | List of miRNAs and the neurodegenerative diseases to which they have been associated.

miRNA	Disease	Type of evidence	Mechanism	Reference
miR-7	PD	*In vitro* reporter assay Overexpression *in vitro*	Regulation of α-synuclein	Doxakis (2010), Junn et al. (2009)
miR-8		miR-8 flies	Upregulation of atrophin	Karres et al. (2007)
miR-9/9*	HD	Profiling studies *In vitro* reporter assays	REST–coREST	Marti et al. (2010), Packer et al. (2008)
miR-9/9*	ALS	Profiling in mouse mode	Neurofilament expression	Haramati et al. (2010)
miR-16	AD	Profiling in mouse model Overexpression *in vitro*/*in vivo*	Regulation of APP levels	Liu et al. (2010)
miR-19	SCA1	*In vitro* reporter assays *In vitro* gain of function	Regulation of ataxin-1	Lee et al. (2008)
miR-29	AD	Profiling in patients *In vitro* reporter assay Overexpression/blocking *in vitro*	Regulation of BACE1 levels	Hebert et al. (2008)
miR-29b	FTD	*In vitro* reporter assay Overexpression *in vitro*	Regulation of progranulin	Jiao et al. (2010)
miR-34		SCA3 overexpression in fly	Protective role	Liu et al. (2012)
miR-101	AD	*In vitro* reporter assays Overexpression/blocking *in vitro*	Regulation of APP levels	Long and Lahiri (2011)
miR-101	SCA1	*In vitro* reporter assays *In vitro* gain of function	Regulation of ataxin-1	Lee et al. (2008)
miR-106a	AD	*In vitro* reporter assays Overexpression *in vitro*	Regulation of APP levels	Patel et al. (2008)
miR-106b	AD	Expression in mouse model	TGF-β	Wang et al. (2010a)
miR-107	AD	Profiling in patients *In vitro* reporter assay *In situ* hybridization in patients	Regulation of BACE1 levels	Wang et al. (2008)
miR-107	AD	*In vitro* reporter assay Levels in mouse models	Regulation of cofilin	Yao et al. (2010)
miR-107	FTD	*In vitro* overexpression	Regulation of progranulin	Wang et al. (2010b)
miR-124		miR-124-1 knockout mouse	Altered expression of Lhx2	Sanuki et al. (2011)
miR-124	AD	Overexpression *in vitro*/*in vivo*	Regulation of BACE1 levels	Fang et al. (2012)
miR-130	SCA1	*In vitro* reporter assays *In vitro* gain of function	Regulation of ataxin-1	Lee et al. (2008)
miR-133b	PD	Profiling in patients	???	Kim et al. (2007)
miR-137	AD	Profiling in patients *In vitro* reporter assay Blocking *in vitro*	Regulation of Aβ levels	Geekiyanage and Chan (2011)
miR-144	SCA1	Profiling in patients *In vitro* reporter assay	Regulation of ataxin-1	Persengiev et al. (2011)
miR-146a	AD	Profiling in patients Expression in cell lines and mouse models	Downstream of Aβ	Li et al. (2011)
miR-153	PD	*In vitro* reporter assay Overexpression *in vitro*	Regulation of α-synuclein	Doxakis (2010)
miR-181c	AD	Profiling in patients *In vitro* reporter assay Blocking *in vitro*	Regulation of Aβ levels	Geekiyanage and Chan (2011)
miR-520c	AD	*In vitro* reporter assays Overexpression *in vitro*	Regulation of APP levels	Patel et al. (2008)
miR-659	FTD–ALS	Human polymorphism	Regulation of progranulin	Rademakers et al. (2008)

miRNA processing after knockout of Dicer1 markedly enhances the toxicity induced by mutant ataxin-3 (Bilen et al., 2006). In parallel genetic screens, a single miRNA, *bantam*, was identified as a potent downstream modulator of both polyQ and tau toxicity in flies (Bilen et al., 2006). In a recent paper, miR-34 was shown to be protective against expanded SCA3 (Liu et al., 2012). Subsequent work suggested that miR-19, miR-101, and miR-130 are important for the post-translational regulation of ataxin-1 (Lee et al., 2008). Inhibition of those miRNAs enhanced the cytotoxicity of polyQ-expanded ataxin-1 in human cells. Moreover, miR-144, a highly conserved miRNA, also regulated ataxin-1 expression and appeared to be associated with aging. Ataxin-1 levels are higher in the cerebellum and cortex of SCA1 patients than in healthy aged brains (Persengiev et al., 2011). On the other hand, ataxin-2 might be required for miRNA function (McCann et al., 2011), further supporting the intimate association between miRNAs and polyQ diseases.

PARKINSON'S DISEASE

Parkinson's disease is a neurodegenerative disorder that primarily affects movement. Clinical symptoms include bradykinesia (decreased ability to start and continue movements), resting tremor, and rigidity. These symptoms are due to the relatively selective loss of dopaminergic neurons in the substantia nigra (Dauer and Przedborski, 2003) and reflect the impairment of neuronal networks important for regulating motor function. The past two decades have witnessed significant advances in the identification of distinct genetic loci at which pathogenic mutations are associated with parkinsonism (for review, see Lesage and Brice, 2009; Zimprich, 2011). Most research is focused on genes that have been conclusively linked to PD pathogenesis, including those encoding α-synuclein, leucine-rich repeat kinase 2, PTEN-induced putative kinase 1, parkin, and DJ-1.

Profiling studies of PD brains have revealed abnormalities in miRNA content (Kim et al., 2007; Minones-Moyano et al., 2011). One of the miRNAs found to be downregulated in these studies, miR-133b, plays a major role in the development of midbrain dopaminergic neurons by regulating the transcription factor Pitx3 (Kim et al., 2007). miR-7 and miR-153 control the expression of α-synuclein (Junn et al., 2009; Doxakis, 2010). Since intracellular levels of this protein appear to be critical in mediating its toxicity, deregulation of those miRNAs might lead to increased toxic levels of α-synuclein.

AMYOTROPHIC LATERAL SCLEROSIS AND FRONTOTEMPORAL DEMENTIA

Amyotrophic lateral sclerosis (ALS) is a progressive, lethal, degenerative disorder characterized by the selective death of motor neurons in the brain and spinal cord (Pasinelli and Brown, 2006; Ferraiuolo et al., 2011). ALS shares many clinical, pathological, and molecular features with FTD, the second most common early-onset dementia (Ferrari et al., 2011). Clinically, FTD progresses from an insidious onset of behavioral changes, impaired frontal executive functions, and language deficits to more severe cognitive defects and, finally, to generalized dementia (Boxer and Miller, 2005). In familial FTD cases, the mutated locus has been identified in the genes encoding tau (Hong et al., 1998; Hutton et al., 1998), VCP (Watts et al., 2004), CHMP2B (Skibinski et al., 2005), progranulin (Baker et al., 2006; Cruts et al., 2006), and C9ORF72 (DeJesus-Hernandez et al., 2011; Renton et al., 2011). Among them, CHMP2B (Parkinson et al., 2006; Cox et al., 2010), VCP (Johnson et al., 2010a), and C9ORF72 (DeJesus-Hernandez et al., 2011; Renton et al., 2011; Stewart et al., 2012) have also been implicated in ALS. Moreover, the RNA-binding proteins TDP-43 and FUS have been strongly implicated in both FTD and ALS (Arai et al., 2006; Neumann et al., 2006, 2009; Sreedharan et al., 2008; Vance et al., 2009).

As described above, Dicer deletion in spinal motor neurons mimics most of the clinical (e.g., progressive paralysis) and pathological (e.g., astrocytosis and signs of axonopathy) features of ALS (Haramati et al., 2010). Whether the miRNA pathway is involved in the molecular pathogenesis of FTD/ALS caused by *C9ORF72* repeat expansion is unknown. However, a potential involvement for the miRNA pathway in other forms of FTD or ALS comes from limited studies on TDP-43. TDP-43 is mutated in a subset of ALS patients (Sreedharan et al., 2008), and the expression levels of some miRNAs are affected in *TDP-43* mutant flies (Buratti et al., 2010). Biochemical interactions between TDP-43 and Drosha, a key miRNA processing enzyme (Han et al., 2004), have been observed (Gregory et al., 2004; Ling et al., 2010). These findings raise the possibility that TDP-43 may play a role in miRNA processing; however, the mechanism remains to be elucidated and whether endogenous Drosha and TDP-43 physically interact needs to be demonstrated.

Some miRNAs are emerging as important contributors to ALS pathogenesis. The muscle-specific miR-206 is upregulated upon nerve injury and is required for regeneration of neuromuscular synapses. Moreover, miR-206 deficiency accelerates disease progression in a mouse model of ALS (Williams et al., 2009). miR-9/9*, an evolutionarily conserved and multifunctional miRNA (Yuva-Aydemir et al., 2011), is also potentially involved in ALS. Profiling of miRNA expression in motor neurons harboring an *SMN1mut* allele found in pediatric spinal motor atrophy revealed decreases of more than 90% in miR-9 and miR-9* levels (Haramati et al., 2010). More importantly, changes in the expression levels of the neurofilament subunits likely contribute to the disease, and miR-9 is an upstream regulator of the neurofilament mRNAs.

Several other miRNAs might be linked to FTD–ALS through different mechanisms. For example, miR-29b and miR-107 regulate progranulin levels (Jiao et al., 2010; Wang et al., 2010b). Since progranulin haploinsufficiency can cause FTD[114,115], excessive levels of those miRNAs might decrease progranulin levels and be a risk factor for the disease. Consistent with these observations, a genetic polymorphism in the 3′UTR of the progranulin gene is associated with a higher risk of FTD–ALS, and multiple miRNAs are misregulated in FTD with TDP-43 pathology (Rademakers et al., 2008; Kocerha et al., 2011). This genetic variant (rs5848) affects the miR-659 binding site, resulting in more efficient binding and, consequently, decreased progranulin levels. It is not known whether translational regulation by miRNAs is a common mechanism in FTD caused by progranulin deficiency or in other neurodegenerative diseases (Rollinson et al., 2011).

SUMMARY

Although much progress has been made in our understanding of how miRNAs control gene expression at the post-transcriptional level during development, their contributions

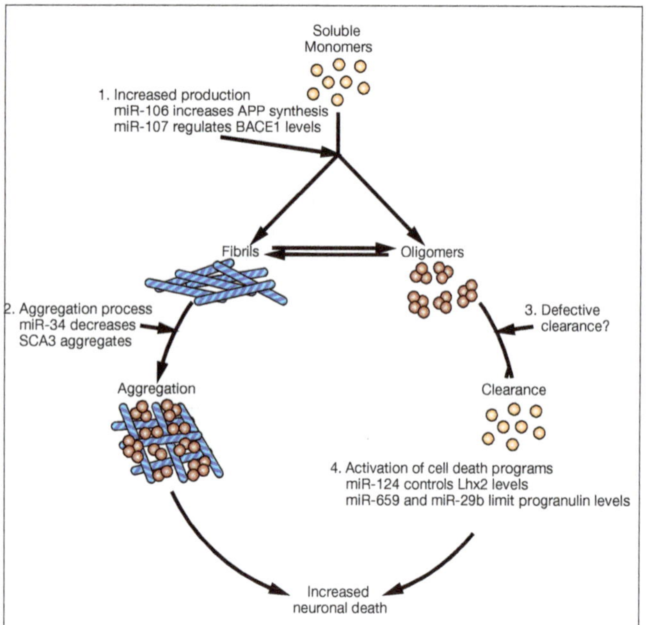

FIGURE 2 | Potential mechanisms of neurodegeneration induced by altered miRNA networks. Schematic representation of cellular pathways that could be affected downstream of miRNAs. (1) Defects in miRNAs could increase the levels of aggregation-prone proteins either directly (i.e., miR-106a and APP in AD) or indirectly (i.e., miR-107 acting through BACE1 or miR-137 acting through serine palmitoyltransferase in AD). (2) miRNAs could control the expression of proteins involved in proper folding or quality control, increasing the risk of protein aggregation. (3) miRNAs could impair the removal of aggregated proteins and therefore increase their levels and toxicity. (4) Finally, altered miRNAs might result in neuronal cell death due to increased levels of certain transcription factors (i.e., miR-124 controls neuronal survival by limiting the expression of Lhx2) or the imbalance between pro-survival and pro-apoptotic signals (i.e., in FTD, polymorphism rs5848 results in more efficient binding of miR-659 and decreased levels of the pro-survival factor progranulin).

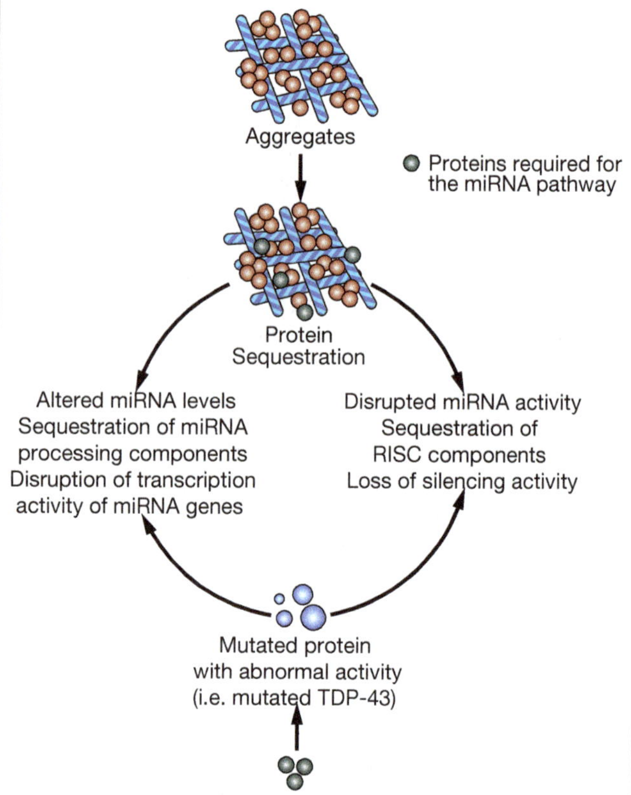

FIGURE 3 | Potential mechanisms of miRNA deregulation downstream of protein aggregation/dysfunction. Two basic mechanisms could affect miRNA regulatory networks: (1) alteration of miRNA levels through deregulation of transcription (i.e., miR-9 downstream of HTT and REST) or processing (i.e., TDP-43 altering Drosha function) and (2) interference with RISC activity (i.e., ataxin-2 seems to be required for optimal miRNA silencing).

to neurodegenerative disease remain poorly understood. Many fundamental questions need to be addressed. Is miRNA pathway disruption a downstream consequence or a cause of neurodegeneration (**Figure 2** vs. **Figure 3**)? Are miRNAs essential for the proper regulation of aggregation-prone proteins or do they control additional pathogenic pathways? Are individual miRNAs especially important in particular neurodegenerative diseases? Which miRNA target or targets are relevant for the disease? To address these questions, it will be essential to generate novel experimental models, such as conditional knockouts in which developmental defects can be circumvented, allowing assessment of the functions of specific miRNAs in the adult brain. We expect many exciting findings will be made in years to come.

ACKNOWLEDGMENTS

We thank S. Ordway for editorial assistance, A. Wilson for help with the figures, and Gao lab members for discussions. This work is supported by the National Institutes of Health (RO1NS066586 and RO1NS057553 to Fen-Biao Gao).

REFERENCES

Ahmed, A., and Wicklund, M. P. (2011). Amyotrophic lateral sclerosis: what role does environment play? *Neurol. Clin.* 29, 689–711.

Ambros, V., Lee, R. C., Lavanway, A., Williams, P. T., and Jewell, D. (2003). MicroRNAs and other tiny endogenous RNAs in *C. elegans*. *Curr. Biol.* 13, 807–818.

Arai, T., Hasegawa, M., Akiyama, H., Ikeda, K., Nonaka, T., Mori, H., Mann, D., Tsuchiya, K., Yoshida, M., Hashizume, Y., and Oda, T. (2006). TDP-43 is a component of ubiquitin-positive tau-negative inclusions in frontotemporal lobar degeneration and amyotrophic lateral sclerosis. *Biochem. Biophys. Res. Commun.* 351, 602–611.

Baker, M., Mackenzie, I. R., Pickering-Brown, S. M., Gass, J., Rademakers, R., Lindholm, C., Snowden, J., Adamson, J., Sadovnick, A. D., Rollinson, S., Cannon, A., Dwosh, E., Neary, D., Melquist, S., Richardson, A., Dickson, D., Berger, Z., Eriksen, J., Robinson, T., Zehr, C., Dickey, C. A., Crook, R., McGowan, E., Mann, D., Boeve, B., Feldman, H., and Hutton, M. (2006). Mutations in progranulin cause tau-negative frontotemporal dementia linked to chromosome 17. *Nature* 442, 916–919.

Ballard, C., Gauthier, S., Corbett, A., Brayne, C., Aarsland, D., and Jones, E. (2011). Alzheimer's disease. *Lancet* 377, 1019–1031.

Bartel, D. P. (2004). MicroRNAs: genomics, biogenesis, mechanism, and function. *Cell* 116, 281–297.

Bartel, D. P. (2009). MicroRNAs: target recognition and regulatory functions. *Cell* 136, 215–233.

Berezikov, E., Thuemmler, F., van Laake, L. W., Kondova, I., Bontrop, R., Cuppen, E., and Plasterk, R. H. (2006). Diversity of microRNAs in human and chimpanzee brain. *Nat. Genet.* 38, 1375–1377.

Bilen, J., Liu, N., Burnett, B. G., Pittman, R. N., and Bonini, N. M. (2006). MicroRNA pathways modulate polyglutamine-induced neurodegeneration. *Mol. Cell* 24, 157–163.

Boxer, A. L., and Miller, B. L. (2005). Clinical features of frontotemporal dementia. *Alzheimer Dis. Assoc. Disord.* 19(Suppl. 1), S3–S6.

Bruce, A. W., Donaldson, I. J., Wood, I. C., Yerbury, S. A., Sadowski, M. I., Chapman, M., Gottgens, B., and Buckley, N. J. (2004). Genome-wide analysis of repressor element 1 silencing transcription factor/neuron-restrictive silencing factor (REST/NRSF) target genes. *Proc. Natl. Acad. Sci. U.S.A.* 101, 10458–10463.

Buckley, N. J., Johnson, R., Zuccato, C., Bithell, A., and Cattaneo, E. (2010). The role of REST in transcriptional and epigenetic dysregulation in Huntington's disease. *Neurobiol. Dis.* 39, 28–39.

Buratti, E., De Conti, L., Stuani, C., Romano, M., Baralle, M., and Baralle, F. (2010). Nuclear factor TDP-43 can affect selected microRNA levels. *FEBS J.* 277, 2268–2281.

Caceres, M., Lachuer, J., Zapala, M. A., Redmond, J. C., Kudo, L., Geschwind, D. H., Lockhart, D. J., Preuss, T. M., and Barlow, C. (2003). Elevated gene expression levels distinguish human from non-human primate brains. *Proc. Natl. Acad. Sci. U.S.A.* 100, 13030–13035.

Cao, X., Pfaff, S. L., and Gage, F. H. (2007). A functional study of miR-124 in the developing neural tube. *Genes Dev.* 21, 531–536.

Chendrimada, T. P., Gregory, R. I., Kumaraswamy, E., Norman, J., Cooch, N., Nishikura, K., and Shiekhattar, R. (2005). TRBP recruits the Dicer complex to Ago2 for microRNA processing and gene silencing. *Nature* 436, 740–744.

Cheng, L. C., Pastrana, E., Tavazoie, M., and Doetsch, F. (2009). miR-124 regulates adult neurogenesis in the subventricular zone stem cell niche. *Nat. Neurosci.* 12, 399–408.

Cox, L. E., Ferraiuolo, L., Goodall, E. F., Heath, P. R., Higginbottom, A., Mortiboys, H., Hollinger, H. C., Hartley, J. A., Brockington, A., Burness, C. E., Morrison, K. E., Wharton, S. B., Grierson, A. J., Ince, P. G., Kirby, J., and Shaw, P. J. (2010). Mutations in CHMP2B in lower motor neuron predominant amyotrophic lateral sclerosis (ALS). *PLoS ONE* 5, e9872. doi:10.1371/journal.pone.0009872

Cruts, M., Gijselinck, I., van der Zee, J., Engelborghs, S., Wils, H., Pirici, D., Rademakers, R., Vandenberghe, R., Dermaut, B., Martin, J. J., van Duijn, C., Peeters, K., Sciot, R., Santens, P., De Pooter, T., Mattheijssens, M., Van den Broeck, M., Cuijt, I., Vennekens, K., De Deyn, P. P., Kumar-Singh, S., and Van Broeckhoven, C. (2006). Null mutations in progranulin cause ubiquitin-positive frontotemporal dementia linked to chromosome 17q21. *Nature* 442, 920–924.

Cuellar, T. L., Davis, T. H., Nelson, P. T., Loeb, G. B., Harfe, B. D., Ullian, E., and McManus, M. T. (2008). Dicer loss in striatal neurons produces behavioral and neuroanatomical phenotypes in the absence of neurodegeneration. *Proc. Natl. Acad. Sci. U.S.A.* 105, 5614–5619.

Damiani, D., Alexander, J. J., O'Rourke, J. R., McManus, M., Jadhav, A. P., Cepko, C. L., Hauswirth, W. W., Harfe, B. D., and Strettoi, E. (2008). Dicer inactivation leads to progressive functional and structural degeneration of the mouse retina. *J. Neurosci.* 28, 4878–4887.

Dauer, W., and Przedborski, S. (2003). Parkinson's disease: mechanisms and models. *Neuron* 39, 889–909.

Davis, T. H., Cuellar, T. L., Koch, S. M., Barker, A. J., Harfe, B. D., McManus, M. T., and Ullian, E. M. (2008). Conditional loss of Dicer disrupts cellular and tissue morphogenesis in the cortex and hippocampus. *J. Neurosci.* 28, 4322–4330.

De Pietri Tonelli, D., Pulvers, J. N., Haffner, C., Murchison, E. P., Hannon, G. J., and Huttner, W. B. (2008). miRNAs are essential for survival and differentiation of newborn neurons but not for expansion of neural progenitors during early neurogenesis in the mouse embryonic neocortex. *Development* 135, 3911–3921.

DeJesus-Hernandez, M., Mackenzie, I. R., Boeve, B. F., Boxer, A. L., Baker, M., Rutherford, N. J., Nicholson, A. M., Finch, N. A., Flynn, H., Adamson, J., Kouri, N., Wojtas, A., Sengdy, P., Hsiung, G. Y., Karydas, A., Seeley, W. W., Josephs, K. A., Coppola, G., Geschwind, D. H., Wszolek, Z. K., Feldman, H., Knopman, D. S., Petersen, R. C., Miller, B. L., Dickson, D. W., Boylan, K. B., Graff-Radford, N. R., and Rademakers, R. (2011). Expanded GGGGCC hexanucleotide repeat in noncoding region of C9ORF72 causes chromosome 9p-linked FTD and ALS. *Neuron* 72, 245–256.

Delay, C., Calon, F., Mathews, P., and Hebert, S. S. (2011). Alzheimer-specific variants in the 3′UTR of amyloid precursor protein affect microRNA function. *Mol. Neurodegener.* 6, 70.

Doxakis, E. (2010). Post-transcriptional regulation of alpha-synuclein expression by mir-7 and mir-153. *J. Biol. Chem.* 285, 12726–12734.

Eacker, S. M., Dawson, T. M., and Dawson, V. L. (2009). Understanding microRNAs in neurodegeneration. *Nat. Rev. Neurosci.* 10, 837–841.

Enard, W., Khaitovich, P., Klose, J., Zollner, S., Heissig, F., Giavalisco, P., Nieselt-Struwe, K., Muchmore, E., Varki, A., Ravid, R., Doxiadis, G. M., Bontrop, R. E., and Paabo, S. (2002). Intra- and interspecific variation in primate gene expression patterns. *Science* 296, 340–343.

Fang, M., Wang, J., Zhang, X., Geng, Y., Hu, Z., Rudd, J. A., Ling, S., Chen, W., and Han, S. (2012). The miR-124 regulates the expression of BACE1/beta-secretase correlated with cell death in Alzheimer's disease. *Toxicol. Lett.* 209, 94–105.

Fenelon, K., Mukai, J., Xu, B., Hsu, P. K., Drew, L. J., Karayiorgou, M., Fischbach, G. D., Macdermott, A. B., and Gogos, J. A. (2011). Deficiency of Dgcr8, a gene disrupted by the 22q11.2 microdeletion, results in altered short-term plasticity in the prefrontal cortex. *Proc. Natl. Acad. Sci. U.S.A.* 108, 4447–4452.

Ferraiuolo, L., Kirby, J., Grierson, A. J., Sendtner, M., and Shaw, P. J. (2011). Molecular pathways of motor neuron injury in amyotrophic lateral sclerosis. *Nat. Rev. Neurol.* 7, 616–630.

Ferrari, R., Kapogiannis, D., Huey, E. D., and Momeni, P. (2011). FTD and ALS: a tale of two diseases. *Curr. Alzheimer Res.* 8, 273–294.

Gao, F. B. (2010). Context-dependent functions of specific microRNAs in neuronal development. *Neural Dev.* 1, 5:25. doi: 10.1186/1749-8104-5-25

Gao, H. M., and Hong, J. S. (2011). Gene-environment interactions: key to unraveling the mystery of Parkinson's disease. *Prog. Neurobiol.* 94, 1–19.

Gatchel, J. R., and Zoghbi, H. Y. (2005). Diseases of unstable repeat expansion: mechanisms and common principles. *Nat. Rev. Genet.* 6, 743–755.

Geekiyanage, H., and Chan, C. (2011). MicroRNA-137/181c regulates serine palmitoyltransferase and in turn amyloid beta, novel targets in sporadic Alzheimer's disease. *J. Neurosci.* 31, 14820–14830.

Ghildiyal, M., and Zamore, P. D. (2009). Small silencing RNAs: an expanding universe. *Nat. Rev. Genet.* 10, 94–108.

Gilliam, T. C., Tanzi, R. E., Haines, J. L., Bonner, T. I., Faryniarz, A. G., Hobbs, W. J., MacDonald, M. E., Cheng, S. V., Folstein, S. E., Conneally, P. M., Wexler, N. S., and Gusella, J. F. (1987). Localization of the Huntington's disease gene to a small segment of chromosome 4 flanked by D4S10 and the telomere. *Cell* 50, 565–571.

Gregory, R. I., Yan, K. P., Amuthan, G., Chendrimada, T., Doratotaj, B., Cooch, N., and Shiekhattar, R. (2004). The Microprocessor complex mediates the genesis of microRNAs. *Nature* 432, 235–240.

Hampel, H., Prvulovic, D., Teipel, S., Jessen, F., Luckhaus, C., Frolich, L., Riepe, M. W., Dodel, R., Leyhe, T., Bertram, L., Hoffmann, W., and Faltraco, F. (2011). The future of Alzheimer's disease: the next 10 years. *Prog. Neurobiol.* 95, 718–728.

Han, J., Lee, Y., Yeom, K. H., Kim, Y. K., Jin, H., and Kim, V. N. (2004). The Drosha-DGCR8 complex in primary microRNA processing. *Genes Dev.* 18, 3016–3027.

Haramati, S., Chapnik, E., Sztainberg, Y., Eilam, R., Zwang, R., Gershoni, N., McGlinn, E., Heiser, P. W., Wills, A. M., Wirguin, I., Rubin, L. L., Misawa, H., Tabin, C. J., Brown, R. J., Chen, A., and Hornstein, E. (2010). miRNA malfunction causes spinal motor neuron disease. *Proc. Natl. Acad. Sci. U.S.A.* 107, 13111–13116.

Hebert, S. S., Horre, K., Nicolai, L., Papadopoulou, A. S., Mandemakers, W., Silahtaroglu, A. N., Kauppinen, S., Delacourte, A., and De Strooper, B. (2008). Loss of microRNA cluster miR-29a/b-1 in sporadic Alzheimer's disease correlates with increased BACE1/beta-secretase expression. *Proc. Natl. Acad. Sci. U.S.A.* 105, 6415–6420.

Hebert, S. S., Papadopoulou, A. S., Smith, P., Galas, M. C., Planel, E., Silahtaroglu, A. N., Sergeant, N., Buee, L., and De Strooper, B. (2010). Genetic ablation of Dicer in adult forebrain neurons results in abnormal tau hyperphosphorylation and neurodegeneration. *Hum. Mol. Genet.* 19, 3959–3969.

Herranz, H., and Cohen, S. M. (2010). MicroRNAs and gene regulatory networks: managing the impact of noise in biological systems. *Genes Dev.* 24, 1339–1344.

Hirano, A. (1994). Hirano bodies and related neuronal inclusions. *Neuropathol. Appl. Neurobiol.* 20, 3–11.

Hong, M., Zhukareva, V., Vogelsberg-Ragaglia, V., Wszolek, Z., Reed, L., Miller, B. I., Geschwind, D. H., Bird, T. D., McKeel, D., Goate, A., Morris, J. C., Wilhelmsen, K. C., Schellenberg, G. D., Trojanowski, J. Q., and Lee, V. M. (1998). Mutation-specific functional impairments in distinct tau isoforms of hereditary FTDP-17. *Science* 282, 1914–1917.

Hu, H. Y., Guo, S., Xi, J., Yan, Z., Fu, N., Zhang, X., Menzel, C., Liang, H., Yang, H., Zhao, M., Zeng, R., Chen, W., Paabo, S., and Khaitovich, P. (2011). MicroRNA expression and regulation in human, chimpanzee, and macaque brains. *PLoS Genet.* 7, e1002327. doi:10.1371/journal.pgen.1002327

Hutton, M., Lendon, C. L., Rizzu, P., Baker, M., Froelich, S., Houlden, H., Pickering-Brown, S., Chakraverty, S., Isaacs, A., Grover, A., Hackett, J., Adamson, J., Lincoln, S., Dickson, D., Davies, P., Petersen, R. C., Stevens, M., de Graaff, E., Wauters, E., van Baren, J., Hillebrand, M., Joosse, M., Kwon, J. M., Nowotny, P., Che, L. K., Norton, J., Morris, J. C., Reed, L. A., Trojanowski, J., Basun, H., Lannfelt, L., Neystat, M., Fahn, S., Dark, F., Tannenberg, T., Dodd, P. R., Hayward, N., Kwok, J. B., Schofield, P. R., Andreadis, A., Snowden, J., Craufurd, D., Neary, D., Owen, F., Oostra, B. A., Hardy, J., Goate, A., van Swieten, J., Mann, D., Lynch, T., and Heutink, P. (1998). Association of missense and 5′-splice-site mutations in tau with the inherited dementia FTDP-17. *Nature* 393, 702–705.

Hutvagner, G., McLachlan, J., Pasquinelli, A. E., Balint, E., Tuschl, T., and Zamore, P. D. (2001). A cellular function for the RNA-interference enzyme Dicer in the maturation of the let-7 small temporal RNA. *Science* 293, 834–838.

Ilieva, H., Polymenidou, M., and Cleveland, D. W. (2009). Non-cell autonomous toxicity in neurodegenerative disorders: ALS and beyond. *J. Cell Biol.* 187, 761–772.

Jiang, Y. J., Che, M. X., Yuan, J. Q., Xie, Y. Y., Yan, X. Z., and Hu, H. Y. (2011). Interaction with polyglutamine-expanded huntingtin alters cellular distribution and RNA processing of huntingtin yeast two-hybrid protein A (HYPA). *J. Biol. Chem.* 286, 25236–25245.

Jiao, J., Herl, L. D., Farese, R. V., and Gao, F. B. (2010). MicroRNA-29b regulates the expression level of human progranulin, a secreted glycoprotein implicated in frontotemporal dementia. *PLoS ONE* 5, e10551. doi:10.1371/journal.pone.0010551

Johnson, J. O., Mandrioli, J., Benatar, M., Abramzon, Y., Van Deerlin, V. M., Trojanowski, J. Q., Gibbs, J. R., Brunetti, M., Gronka, S., Wuu, J., Ding, J., McCluskey, L., Martinez-Lage, M., Falcone, D., Hernandez, D. G., Arepalli, S., Chong, S., Schymick, J. C., Rothstein, J., Landi, F., Wang, Y. D., Calvo, A., Mora, G., Sabatelli, M., Monsurro, M. R., Battistini, S., Salvi, F., Spataro, R., Sola, P., Borghero, G., Galassi, G., Scholz, S. W., Taylor, J. P., Restagno, G., Chio, A., and Traynor, B. J. (2010a). Exome sequencing reveals VCP mutations as a cause of familial ALS. *Neuron* 68, 857–864.

Johnson, R., Richter, N., Jauch, R., Gaughwin, P. M., Zuccato, C., Cattaneo, E., and Stanton, L. W. (2010b). The human accelerated region 1 noncoding RNA is repressed by REST in Huntington's disease. *Physiol. Genomics* 41, 269–274.

Johnson, R., Zuccato, C., Belyaev, N. D., Guest, D. J., Cattaneo, E., and Buckley, N. J. (2008). A microRNA-based gene dysregulation pathway in Huntington's disease. *Neurobiol. Dis.* 29, 438–445.

Jothi, R., Cuddapah, S., Barski, A., Cui, K., and Zhao, K. (2008). Genome-wide identification of in vivo protein-DNA binding sites from ChIP-Seq data. *Nucleic Acids Res.* 36, 5221–5231.

Jucker, M., and Walker, L. C. (2011). Pathogenic protein seeding in Alzheimer disease and other neurodegenerative disorders. *Ann. Neurol.* 70, 532–540.

Junn, E., Lee, K. W., Jeong, B. S., Chan, T. W., Im, J. Y., and Mouradian, M. M. (2009). Repression of alpha-synuclein expression and toxicity by microRNA-7. *Proc. Natl. Acad. Sci. U.S.A.* 106, 13052–13057.

Karres, J. S., Hilgers, V., Carrera, I., Treisman, J., and Cohen, S. M. (2007). The conserved microRNA miR-8 tunes atrophin levels to prevent neurodegeneration in *Drosophila*. *Cell* 131, 136–145.

Kim, J., Inoue, K., Ishii, J., Vanti, W. B., Voronov, S. V., Murchison, E., Hannon, G., and Abeliovich, A. (2007). A MicroRNA feedback circuit in midbrain dopamine neurons. *Science* 317, 1220–1224.

King, M. C., and Wilson, A. C. (1975). Evolution at two levels in humans and chimpanzees. *Science* 188, 107–116.

Kocerha, J., Kouri, N., Baker, M., Finch, N., DeJesus-Hernandez, M., Gonzalez, J., Chidamparam, K., Josephs, K. A., Boeve, B. F., Graff-Radford, N. R., Crook, J., Dickson, D. W., and Rademakers, R. (2011). Altered microRNA expression in frontotemporal lobar degeneration with TDP-43 pathology caused by progranulin mutations. *BMC Genomics* 12, 527. doi:10.1186/1471-2164-12-527

Krol, J., Loedige, I., and Filipowicz, W. (2010). The widespread regulation of microRNA biogenesis, function and decay. *Nat. Rev. Genet.* 11, 597–610.

Lagos-Quintana, M., Rauhut, R., Yalcin, A., Meyer, J., Lendeckel, W., and Tuschl, T. (2002). Identification of tissue-specific microRNAs from mouse. *Curr. Biol.* 12, 735–739.

Lau, P., and de Strooper, B. (2010). Dysregulated microRNAs in neurodegenerative disorders. *Semin. Cell Dev. Biol.* 21, 768–773.

Lee, E. B., Lee, V. M., and Trojanowski, J. Q. (2012). Gains or losses: molecular mechanisms of TDP43-mediated neurodegeneration. *Nat. Rev. Neurosci.* 13, 38–50.

Lee, R. C., Feinbaum, R. L., and Ambros, V. (1993). The *C. elegans* heterochronic gene lin-4 encodes small RNAs with antisense complementarity to lin-14. *Cell* 75, 843–854.

Lee, S. T., Chu, K., Im, W. S., Yoon, H. J., Im, J. Y., Park, J. E., Park, K. H., Jung, K. H., Lee, S. K., Kim, M., and Roh, J. K. (2011). Altered microRNA regulation in Huntington's disease models. *Exp. Neurol.* 227, 172–179.

Lee, Y., Ahn, C., Han, J., Choi, H., Kim, J., Yim, J., Lee, J., Provost, P., Radmark, O., Kim, S., and Kim, V. N. (2003). The nuclear RNase III Drosha initiates microRNA processing. *Nature* 425, 415–419.

Lee, Y., Kim, M., Han, J., Yeom, K. H., Lee, S., Baek, S. H., and Kim, V. N. (2004). MicroRNA genes are transcribed by RNA polymerase II. *EMBO J.* 23, 4051–4060.

Lee, Y., Samaco, R. C., Gatchel, J. R., Thaller, C., Orr, H. T., and Zoghbi, H. Y. (2008). miR-19, miR-101 and miR-130 co-regulate ATXN1 levels to potentially modulate SCA1 pathogenesis. *Nat. Neurosci.* 11, 1137–1139.

Lesage, S., and Brice, A. (2009). Parkinson's disease: from monogenic forms to genetic susceptibility factors. *Hum. Mol. Genet.* 18, R48–R59.

Li, Y. Y., Cui, J. G., Hill, J. M., Bhattacharjee, S., Zhao, Y., and Lukiw, W. J. (2011). Increased expression of miRNA-146a in Alzheimer's disease transgenic mouse models. *Neurosci. Lett.* 487, 94–98.

Lim, L. P., Lau, N. C., Garrett-Engele, P., Grimson, A., Schelter, J. M., Castle, J., Bartel, D. P., Linsley, P. S., and Johnson, J. M. (2005). Microarray analysis shows that some microRNAs downregulate large numbers of target mRNAs. *Nature* 433, 769–773.

Ling, S. C., Albuquerque, C. P., Han, J. S., Lagier-Tourenne, C., Tokunaga, S., Zhou, H., and Cleveland, D. W. (2010). ALS-associated mutations in TDP-43 increase its stability and promote TDP-43 complexes with FUS/TLS. *Proc. Natl. Acad. Sci. U.S.A.* 107, 13318–13323.

Liu, N., Landreh, M., Cao, K., Abe, M., Hendriks, G. J., Kennerdell, J. R., Zhu, Y., Wang, L. S., and Bonini, N. M. (2012). The microRNA miR-34 modulates ageing and neurodegeneration in *Drosophila*. *Nature* 482, 519–523.

Liu, W., Liu, C., Zhu, J., Shu, P., Yin, B., Gong, Y., Qiang, B., Yuan, J., and Peng, X. (2010). MicroRNA-16 targets amyloid precursor protein to potentially modulate Alzheimer's-associated pathogenesis in SAMP8 mice. *Neurobiol. Aging* 33, 522–534.

Long, J. M., and Lahiri, D. K. (2011). MicroRNA-101 downregulates Alzheimer's amyloid-beta precursor protein levels in human cell cultures and is differentially expressed. *Biochem. Biophys. Res. Commun.* 404, 889–895.

Lukiw, W. J. (2007). Micro-RNA speciation in fetal, adult and Alzheimer's disease hippocampus. *Neuroreport* 18, 297–300.

Maiorano, N. A., and Mallamaci, A. (2009). Promotion of embryonic cortico-cerebral neuronogenesis by miR-124. *Neural Dev.* 4, 40.

Manakov, S. A., Grant, S. G., and Enright, A. J. (2009). Reciprocal regulation of microRNA and mRNA profiles in neuronal development and synapse formation. *BMC Genomics* 10, 419. doi:10.1186/1471-2164-10-419

Marti, E., Pantano, L., Banez-Coronel, M., Llorens, F., Minones-Moyano, E., Porta, S., Sumoy, L., Ferrer, I., and Estivill, X. (2010). A myriad of miRNA variants in control and Huntington's disease brain regions detected by massively parallel sequencing. *Nucleic Acids Res.* 38, 7219–7235.

McCann, C., Holohan, E. E., Das, S., Dervan, A., Larkin, A., Lee, J. A., Rodrigues, V., Parker, R., and Ramaswami, M. (2011). The Ataxin-2 protein is required for microRNA function and synapse-specific long-term olfactory habituation.

Proc. Natl. Acad. Sci. U.S.A. 108, E655–E662.

Minones-Moyano, E., Porta, S., Escaramis, G., Rabionet, R., Iraola, S., Kagerbauer, B., Espinosa-Parrilla, Y., Ferrer, I., Estivill, X., and Marti, E. (2011). MicroRNA profiling of Parkinson's disease brains identifies early downregulation of miR-34b/c which modulate mitochondrial function. *Hum. Mol. Genet.* 20, 3067–3078.

Miska, E. A., Alvarez-Saavedra, E., Townsend, M., Yoshii, A., Sestan, N., Rakic, P., Constantine-Paton, M., and Horvitz, H. R. (2004). Microarray analysis of microRNA expression in the developing mammalian brain. *Genome Biol.* 5, R68.

Nelson, P. T., and Wang, W. X. (2010). MiR-107 is reduced in Alzheimer's disease brain neocortex: validation study. *J. Alzheimers Dis.* 21, 75–79.

Neumann, M., Rademakers, R., Roeber, S., Baker, M., Kretzschmar, H. A., and Mackenzie, I. R. (2009). A new subtype of frontotemporal lobar degeneration with FUS pathology. *Brain* 132, 2922–2931.

Neumann, M., Sampathu, D. M., Kwong, L. K., Truax, A. C., Micsenyi, M. C., Chou, T. T., Bruce, J., Schuck, T., Grossman, M., Clark, C. M., McCluskey, L. F., Miller, B. L., Masliah, E., Mackenzie, I. R., Feldman, H., Feiden, W., Kretzschmar, H. A., Trojanowski, J. Q., and Lee, V. M. (2006). Ubiquitinated TDP-43 in frontotemporal lobar degeneration and amyotrophic lateral sclerosis. *Science* 314, 130–133.

Nunez-Iglesias, J., Liu, C. C., Morgan, T. E., Finch, C. E., and Zhou, X. J. (2010). Joint genome-wide profiling of miRNA and mRNA expression in Alzheimer's disease cortex reveals altered miRNA regulation. *PLoS ONE* 5, e8898. doi:10.1371/journal.pone.0008898

O'Brien, R. J., and Wong, P. C. (2011). Amyloid precursor protein processing and Alzheimer's disease. *Annu. Rev. Neurosci.* 34, 185–204.

Orr, H. T., and Zoghbi, H. Y. (2007). Trinucleotide repeat disorders. *Annu. Rev. Neurosci.* 30, 575–621.

Packer, A. N., Xing, Y., Harper, S. Q., Jones, L., and Davidson, B. L. (2008). The bifunctional microRNA miR-9/miR-9* regulates REST and CoREST and is downregulated in Huntington's disease. *J. Neurosci.* 28, 14341–14346.

Parkinson, N., Ince, P. G., Smith, M. O., Highley, R., Skibinski, G., Andersen, P. M., Morrison, K. E., Pall, H. S., Hardiman, O., Collinge, J., Shaw, P. J., and Fisher, E. M. (2006). ALS phenotypes with mutations in CHMP2B (charged multivesicular body protein 2B). *Neurology* 67, 1074–1077.

Pasinelli, P., and Brown, R. H. (2006). Molecular biology of amyotrophic lateral sclerosis: insights from genetics. *Nat. Rev. Neurosci.* 7, 710–723.

Patel, N., Hoang, D., Miller, N., Ansaloni, S., Huang, Q., Rogers, J. T., Lee, J. C., and Saunders, A. J. (2008). MicroRNAs can regulate human APP levels. *Mol. Neurodegener.* 3, 10.

Pereira, J. A., Baumann, R., Norrmen, C., Somandin, C., Miehe, M., Jacob, C., Luhmann, T., Hall-Bozic, H., Mantei, N., Meijer, D., and Suter, U. (2010). Dicer in Schwann cells is required for myelination and axonal integrity. *J. Neurosci.* 30, 6763–6775.

Persengiev, S., Kondova, I., Otting, N., Koeppen, A. H., and Bontrop, R. E. (2011). Genome-wide analysis of miRNA expression reveals a potential role for miR-144 in brain aging and spinocerebellar ataxia pathogenesis. *Neurobiol. Aging* 32, 2316.e17–2316.e27.

Prinz, M., Priller, J., Sisodia, S. S., and Ransohoff, R. M. (2011). Heterogeneity of CNS myeloid cells and their roles in neurodegeneration. *Nat. Neurosci.* 14, 1227–1235.

Rademakers, R., Eriksen, J. L., Baker, M., Robinson, T., Ahmed, Z., Lincoln, S. J., Finch, N., Rutherford, N. J., Crook, R. J., Josephs, K. A., Boeve, B. F., Knopman, D. S., Petersen, R. C., Parisi, J. E., Caselli, R. J., Wszolek, Z. K., Uitti, R. J., Feldman, H., Hutton, M. L., Mackenzie, I. R., Graff-Radford, N. R., and Dickson, D. W. (2008). Common variation in the miR-659 binding-site of GRN is a major risk factor for TDP43-positive frontotemporal dementia. *Hum. Mol. Genet.* 17, 3631–3642.

Reitz, C., Brayne, C., and Mayeux, R. (2011). Epidemiology of Alzheimer disease. *Nat. Rev. Neurol.* 7, 137–152.

Renton, A. E., Majounie, E., Waite, A., Simon-Sanchez, J., Rollinson, S., Gibbs, J. R., Schymick, J. C., Laaksovirta, H., van Swieten, J. C., Myllykangas, L., Kalimo, H., Paetau, A., Abramzon, Y., Remes, A. M., Kaganovich, A., Scholz, S. W., Duckworth, J., Ding, J., Harmer, D. W., Hernandez, D. G., Johnson, J. O., Mok, K., Ryten, M., Trabzuni, D., Guerreiro, R. J., Orrell, R. W., Neal, J., Murray, A., Pearson, J., Jansen, I. E., Sondervan, D., Seelaar, H., Blake, D., Young, K., Halliwell, N., Callister, J. B., Toulson, G., Richardson, A., Gerhard, A., Snowden, J., Mann, D., Neary, D., Nalls, M. A., Peuralinna, T., Jansson, L., Isoviita, V. M., Kaivorinne, A. L., Holtta-Vuori, M., Ikonen, E., Sulkava, R., Benatar, M., Wuu, J., Chio, A., Restagno, G., Borghero, G., Sabatelli, M., Heckerman, D., Rogaeva, E., Zinman, L., Rothstein, J. D., Sendtner, M., Drepper, C., Eichler, E. E., Alkan, C., Abdullaev, Z., Pack, S. D., Dutra, A., Pak, E., Hardy, J., Singleton, A., Williams, N. M., Heutink, P., Pickering-Brown, S., Morris, H. R., Tienari, P. J., and Traynor, B. J. (2011). A hexanucleotide repeat expansion in C9ORF72 is the cause of chromosome 9p21-linked ALS-FTD. *Neuron* 72, 257–268.

Rollinson, S., Rohrer, J. D., van der Zee, J., Sleegers, K., Mead, S., Engelborghs, S., Collinge, J., De Deyn, P. P., Mann, D. M., Van Broeckhoven, C., and Pickering-Brown, S. M. (2011). No association of PGRN 3'UTR rs5848 in frontotemporal lobar degeneration. *Neurobiol. Aging* 32, 754–755.

Sanuki, R., Onishi, A., Koike, C., Muramatsu, R., Watanabe, S., Muranishi, Y., Irie, S., Uneo, S., Koyasu, T., Matsui, R., Cherasse, Y., Urade, Y., Watanabe, D., Kondo, M., Yamashita, T., and Furukawa, T. (2011). miR-124a is required for hippocampal axogenesis and retinal cone survival through Lhx2 suppression. *Nat. Neurosci.* 14, 1125–1134.

Savas, J. N., Makusky, A., Ottosen, S., Baillat, D., Then, F., Krainc, D., Shiekhattar, R., Markey, S. P., and Tanese, N. (2008). Huntington's disease protein contributes to RNA-mediated gene silencing through association with Argonaute and P bodies. *Proc. Natl. Acad. Sci. U.S.A.* 105, 10820–10825.

Schaefer, A., O'Carroll, D., Tan, C. L., Hillman, D., Sugimori, M., Llinas, R., and Greengard, P. (2007). Cerebellar neurodegeneration in the absence of microRNAs. *J. Exp. Med.* 204, 1553–1558.

Schofield, C. M., Hsu, R., Barker, A. J., Gertz, C. C., Blelloch, R., and Ullian, E. M. (2011). Monoallelic deletion of the microRNA biogenesis gene Dgcr8 produces deficits in the development of excitatory synaptic transmission in the prefrontal cortex. *Neural Dev.* 6, 11.

Schonrock, N., Ke, Y. D., Humphreys, D., Staufenbiel, M., Ittner, L. M., Preiss, T., and Gotz, J. (2010). Neuronal microRNA deregulation in response to Alzheimer's disease amyloid-beta. *PLoS ONE* 5, e11070. doi:10.1371/journal.pone.0011070

Selkoe, D., Mandelkow, E., and Holtzman, D. (2012). Deciphering Alzheimer disease. *Cold Spring Harb. Perspect. Med.* 2, a011460.

Shin, D., Shin, J. Y., McManus, M. T., Ptacek, L. J., and Fu, Y. H. (2009). Dicer ablation in oligodendrocytes provokes neuronal impairment in mice. *Ann. Neurol.* 66, 843–857.

Shoulson, I., and Young, A. B. (2011). Milestones in Huntington disease. *Mov. Disord.* 26, 1127–1133.

Skibinski, G., Parkinson, N. J., Brown, J. M., Chakrabarti, L., Lloyd, S. L., Hummerich, H., Nielsen, J. E., Hodges, J. R., Spillantini, M. G., Thusgaard, T., Brandner, S., Brun, A., Rossor, M. N., Gade, A., Johannsen, P., Sorensen, S. A., Gydesen, S., Fisher, E. M., and Collinge, J. (2005). Mutations in the endosomal ESCRTIII-complex subunit CHMP2B in frontotemporal dementia. *Nat. Genet.* 37, 806–808.

Somel, M., Liu, X., Tang, L., Yan, Z., Hu, H., Guo, S., Jiang, X., Zhang, X., Xu, G., Xie, G., Li, N., Hu, Y., Chen, W., Paabo, S., and Khaitovich, P. (2011). MicroRNA-driven developmental remodeling in the brain distinguishes humans from other primates. *PLoS Biol.* 9, e1001214. doi:10.1371/journal.pbio.1001214

Sonntag, K. C. (2010). MicroRNAs and deregulated gene expression networks in neurodegeneration. *Brain Res.* 1338, 48–57.

Sreedharan, J., Blair, I. P., Tripathi, V. B., Hu, X., Vance, C., Rogelj, B., Ackerley, S., Durnall, J. C., Williams, K. L., Buratti, E., Baralle, F., de Belleroche, J., Mitchell, J. D., Leigh, P. N., Al-Chalabi, A., Miller, C. C., Nicholson, G., and Shaw, C. E. (2008). TDP-43 mutations in familial and sporadic amyotrophic lateral sclerosis. *Science* 319, 1668–1672.

Stark, K. L., Xu, B., Bagchi, A., Lai, W. S., Liu, H., Hsu, R., Wan, X., Pavlidis, P., Mills, A. A., Karayiorgou, M., and Gogos, J. A. (2008). Altered brain microRNA biogenesis contributes to phenotypic deficits in a 22q11-deletion mouse model. *Nat. Genet.* 40, 751–760.

Stewart, H., Rutherford, N. J., Briemberg, H., Krieger, C., Cashman, N., Fabros, M., Baker, M., Fok, A., DeJesus-Hernandez, M., Eisen, A., Rademakers, R., and Mackenzie, I. R. (2012). Clinical and pathological features of amyotrophic lateral sclerosis caused by mutation in the C9ORF72 gene on chromosome 9p. *Acta Neuropathol.* 123, 409–417.

Tao, J., Wu, H., Lin, Q., Wei, W., Lu, X. H., Cantle, J. P., Ao, Y., Olsen, R. W., Yang, X. W., Mody, I., Sofroniew, M. V., and Sun, Y. E.

(2011). Deletion of astroglial Dicer causes non-cell-autonomous neuronal dysfunction and degeneration. *J. Neurosci.* 31, 8306–8319.

Vance, C., Rogelj, B., Hortobagyi, T., De Vos, K. J., Nishimura, A. L., Sreedharan, J., Hu, X., Smith, B., Ruddy, D., Wright, P., Ganesalingam, J., Williams, K. L., Tripathi, V., Al-Saraj, S., Al-Chalabi, A., Leigh, P. N., Blair, I. P., Nicholson, G., de Belleroche, J., Gallo, J. M., Miller, C. C., and Shaw, C. E. (2009). Mutations in FUS, an RNA processing protein, cause familial amyotrophic lateral sclerosis type 6. *Science* 323, 1208–1211.

Verrier, J. D., Semple-Rowland, S., Madorsky, I., Papin, J. E., and Notterpek, L. (2010). Reduction of Dicer impairs Schwann cell differentiation and myelination. *J. Neurosci. Res.* 88, 2558–2568.

Visvanathan, J., Lee, S., Lee, B., Lee, J. W., and Lee, S. K. (2007). The microRNA miR-124 antagonizes the anti-neural REST/SCP1 pathway during embryonic CNS development. *Genes Dev.* 21, 744–749.

Wang, H., Liu, J., Zong, Y., Xu, Y., Deng, W., Zhu, H., Liu, Y., Ma, C., Huang, L., Zhang, L., and Qin, C. (2010a). miR-106b aberrantly expressed in a double transgenic mouse model for Alzheimer's disease targets TGF-beta type II receptor. *Brain Res.* 1357, 166–174.

Wang, W. X., Wilfred, B. R., Madathil, S. K., Tang, G., Hu, Y., Dimayuga, J., Stromberg, A. J., Huang, Q., Saatman, K. E., and Nelson, P. T. (2010b). miR-107 regulates granulin/progranulin with implications for traumatic brain injury and neurodegenerative disease. *Am. J. Pathol.* 177, 334–345.

Wang, W. X., Huang, Q., Hu, Y., Stromberg, A. J., and Nelson, P. T. (2011). Patterns of microRNA expression in normal and early Alzheimer's disease human temporal cortex: white matter versus gray matter. *Acta Neuropathol.* 121, 193–205.

Wang, W. X., Rajeev, B. W., Stromberg, A. J., Ren, N., Tang, G., Huang, Q., Rigoutsos, I., and Nelson, P. T. (2008). The expression of microRNA miR-107 decreases early in Alzheimer's disease and may accelerate disease progression through regulation of beta-site amyloid precursor protein-cleaving enzyme 1. *J. Neurosci.* 28, 1213–1223.

Watts, G. D., Wymer, J., Kovach, M. J., Mehta, S. G., Mumm, S., Darvish, D., Pestronk, A., Whyte, M. P., and Kimonis, V. E. (2004). Inclusion body myopathy associated with Paget disease of bone and frontotemporal dementia is caused by mutant valosin-containing protein. *Nat. Genet.* 36, 377–381.

Williams, A. H., Valdez, G., Moresi, V., Qi, X., McAnally, J., Elliott, J. L., Bassel-Duby, R., Sanes, J. R., and Olson, E. N. (2009). MicroRNA-206 delays ALS progression and promotes regeneration of neuromuscular synapses in mice. *Science* 326, 1549–1554.

Wittchen, H. U., Jacobi, F., Rehm, J., Gustavsson, A., Svensson, M., Jonsson, B., Olesen, J., Allgulander, C., Alonso, J., Faravelli, C., Fratiglioni, L., Jennum, P., Lieb, R., Maercker, A., van Os, J., Preisig, M., Salvador-Carulla, L., Simon, R., and Steinhausen, H. C. (2011). The size and burden of mental disorders and other disorders of the brain in Europe 2010. *Eur. Neuropsychopharmacol.* 21, 655–679.

Wu, D., Raafat, A., Pak, E., Clemens, S., and Murashov, A. K. (2012). Dicer-microRNA pathway is critical for peripheral nerve regeneration and functional recovery in vivo and regenerative axonogenesis in vitro. *Exp. Neurol.* 233, 555–565.

Yang, J. S., and Lai, E. C. (2011). Alternative miRNA biogenesis pathways and the interpretation of core miRNA pathway mutants. *Mol. Cell* 43, 892–903.

Yao, J., Hennessey, T., Flynt, A., Lai, E., Beal, M. F., and Lin, M. T. (2010). MicroRNA-related cofilin abnormality in Alzheimer's disease. *PLoS ONE* 5, e15546. doi:10.1371/journal.pone.0015546

Yu, H. B., Johnson, R., Kunarso, G., and Stanton, L. W. (2011). Coassembly of REST and its cofactors at sites of gene repression in embryonic stem cells. *Genome Res.* 21, 1284–1293.

Yuva-Aydemir, Y., Simkin, A., Gascon, E., and Gao, F. B. (2011). MicroRNA-9: functional evolution of a conserved small regulatory RNA. *RNA Biol.* 8, 557–564.

Zimprich, A. (2011). Genetics of Parkinson's disease and essential tremor. *Curr. Opin. Neurol.* 24, 318–323.

Zuccato, C., Belyaev, N., Conforti, P., Ooi, L., Tartari, M., Papadimou, E., MacDonald, M., Fossale, E., Zeitlin, S., Buckley, N., and Cattaneo, E. (2007). Widespread disruption of repressor element-1 silencing transcription factor/neuron-restrictive silencer factor occupancy at its target genes in Huntington's disease. *J. Neurosci.* 27, 6972–6983.

Zuccato, C., Tartari, M., Crotti, A., Goffredo, D., Valenza, M., Conti, L., Cataudella, T., Leavitt, B. R., Hayden, M. R., Timmusk, T., Rigamonti, D., and Cattaneo, E. (2003). Huntingtin interacts with REST/NRSF to modulate the transcription of NRSE-controlled neuronal genes. *Nat. Genet.* 35, 76–83.

16

Genome-wide approaches to dissect the roles of RNA binding proteins in translational control: implications for neurological diseases

***Katannya Kapeli*[1] *and Gene W. Yeo*[1,2,3,4] ***

[1] Department of Cellular and Molecular Medicine, University of California San Diego, La Jolla, CA, USA
[2] UCSD Stem Cell Program, University of California San Diego, La Jolla, CA, USA
[3] UCSD Moores Cancer Center, University of California San Diego, La Jolla, CA, USA
[4] Institute for Genomic Medicine, University of California San Diego, La Jolla, CA, USA

Correspondence:
Gene W. Yeo, Department of Cellular and Molecular Medicine, Stem Cell Program, Institute for Genomic Medicine, University of California, 2880 Torrey Pines Scenic Drive, La Jolla, CA 92037, USA.
e-mail: geneyeo@ucsd.edu

Translational control of messenger RNAs (mRNAs) is a key aspect of neurobiology, defects of which can lead to neurological diseases. In response to stimuli, local translation of mRNAs is activated at synapses to facilitate long-lasting forms of synaptic plasticity, the cellular basis for learning, and memory formation. Translation, as well as all other aspects of RNA metabolism, is controlled in part by RNA binding proteins (RBPs) that directly interact with mRNAs to form mRNA-protein complexes. Disruption of RBP function is becoming widely recognized as a major cause of neurological diseases. Thus understanding the mechanisms that govern the interplay between translation control and RBP regulation in both normal and diseased neurons will provide new opportunities for novel diagnostics and therapeutic intervention. As a means of studying translational control, genome-wide methods are emerging as powerful tools that have already begun to unveil mechanisms that are missed by single-gene studies. Here, we describe the roles of RBPs in translational control, review genome-wide approaches to examine translational control, and discuss how the application of these approaches may provide mechanistic insight into the pathogenic underpinnings of RBPs in neurological diseases.

Keywords: translation, neurological disease, RNA binding proteins, ribosome profiling, CLIP

INTRODUCTION

Analogous to DNA, which is organized and packed via strong associations with histones in the nucleus, precursor, and mature messenger RNAs (mRNAs) never exist as "naked" ribonucleic acid sequences. After transcription in the nucleus, RNA binding proteins (RBPs) recognize *cis*-regulatory RNA elements within precursor mRNA sequence to form messenger ribonucleoprotein (mRNP) complexes. Again, analogous to DNA-binding proteins such as transcription factors that regulate gene expression by binding to DNA elements in the promoters of genes, RBPs regulate the fate of target RNAs by interacting with specific sequences or RNA secondary structural features within the transcribed RNA molecule. These *cis*-regulatory RNA elements can be found in the 5′ and 3′ untranslated regions (UTRs), introns, and exons of all protein-coding genes. RNA elements in 5′ and 3′ UTRs are frequently involved in targeting RNA to specific cellular compartments, affecting 3′ end formation, controlling RNA stability, and regulating mRNA translation. RNA elements in introns and exons are known to function as splicing enhancers or silencers to control the process of precursor mRNA splicing (Jensen et al., 2009).

A genome-wide survey of 323 mouse RBPs by *in situ* hybridization in the developing brain yielded the surprising result that two-thirds of those RBPs are expressed in a cell type specific manner (McKee et al., 2005). Compared to other cells in the body, the complex structure and specialization of neurons explains the need for having many RBPs to maintain proper neural function. Consistent with the crucial roles of RBPs in regulating RNA homeostasis in the nervous system, mutations that impair RBP function have been linked to severe neurological diseases such as Fragile X syndrome (FXS), Fragile X-associated tremor/ataxia syndrome (FXTAS), Amyotrophic lateral sclerosis (ALS), Frontotemporal lobar dementia (FTLD), Spinal muscular atrophy (SMA), and Myotonic dystrophy (Lukong et al., 2008). To understand the impact of mutations within RBPs in neurodegeneration, we need to elucidate the normal activities of RBPs in neurons. It is well-known that RBPs are intimately involved with the regulation of alternative splicing, a process by which numerous isoforms are generated from a single genetic loci, and is in fact, more prevalent in the nervous system than in any other cell types (Yeo et al., 2004; Wang et al., 2008a). RBPs are required to protect mRNAs during their transport from the soma to distal axonal and dendritic locations, and once at these locations, RBPs mediate local *de novo* synthesis of proteins (translation). Local translation at or near axonal and dendritic synapses is the underlying mechanism of synaptic plasticity (Sutton and Schuman, 2006), which refers to the ability of synapses to undergo long-lasting biochemical and morphological changes in response to stimuli (Richter and Klann, 2009). As a result, local translation is critical for cognition and memory. Local synaptic translation is also critical for axon guidance and nerve regeneration (Willis et al., 2005). Accordingly, pharmacological inhibition of protein synthesis prevents some forms of synaptic plasticity in cultured neurons and attenuates long-term memory in mice (Scharf et al., 2002; Kelleher et al., 2004; Banko et al., 2005; Sutton and Schuman, 2006).

Given the significance of RBP biology and mRNA translation in controlling neuron structure and function, advances in sequencing and microarray technology have sparked the development of genome-wide methods that enable the neuroscience community to dissect the roles that RBPs play in controlling mRNA translation in the brain. Here we review how RBPs associate with different mRNP complexes to regulate translation, summarize emerging genome-wide methods that enable an unbiased examination of translation on a global scale, and discuss how genome-wide studies using these methods have and will continue to aid our understanding of translational control in normal and pathological neurobiology.

MESSENGER RNP COMPLEXES AND TRANSLATIONAL CONTROL

From synthesis to destruction, mRNAs are coated with RBPs that sequester mRNA into mRNP complexes and ultimately influence their cellular fate. These mRNP complexes, as depicted in **Figure 1**, are polysomes, RNA granules, RNA particles, stress granules (SGs), processing bodies (P-bodies), and RNA-induced silencing complexes (RISCs). This section provides a brief description of these complexes and introduces RBPs with roles in translation that associate with these complexes (for more details, see Kiebler and Bassell, 2006; Sossin and DesGroseillers, 2006; Erickson and Lykke-Andersen, 2011).

Polysome complexes are the centers of protein production and are present in the cell body, axon, and dendrites of a neuron (Steward and Levy, 1982; Giuditta et al., 2002). RBPs such as Lin28 have been shown to decorate polysomes and promote translation (Balzer and Moss, 2007). It is important to note, however, that polysome-associated mRNAs can be translationally repressed. Several groups have reported instances where translational inhibition of certain proteins did not correspond to a decrease in ribosome number on the encoding mRNAs (Olsen and Ambros, 1999; Braat et al., 2004; Nottrott et al., 2006; Petersen et al., 2006). Such observations likely captured an event called ribosomal stalling, where ribosomes temporarily or permanently stop elongating along transcripts. The RBPs FMRP and Staufen have been shown to induce ribosomal stalling (Thomas et al., 2009; Darnell et al., 2011). MicroRNA-loaded RISCs (miRISCs), which also associate with polysomes, can repress translation by promoting ribosomal pausing (Maroney et al., 2006; Nottrott et al., 2006; Petersen et al., 2006).

For local translation to occur at synapses, mRNAs must be transported from the soma to synapses. RNA particles and RNA granules function to traffic mRNAs to designated subcellular compartments. These structures are complexes of mRNAs and interacting RBPs, motor proteins, and adaptor proteins that tether the RBPs to motor complexes. While in transit, transcripts are both protected from degradation and are translationally repressed until the appropriate signals are received. The RBP Zip-code binding protein 1 (ZBP1) is a well-known regulator of mRNA transport and translational repression (Huttelmaier et al., 2005). Other RBPs such as Staufen, FMRP, and Pumilio also suppress translation of

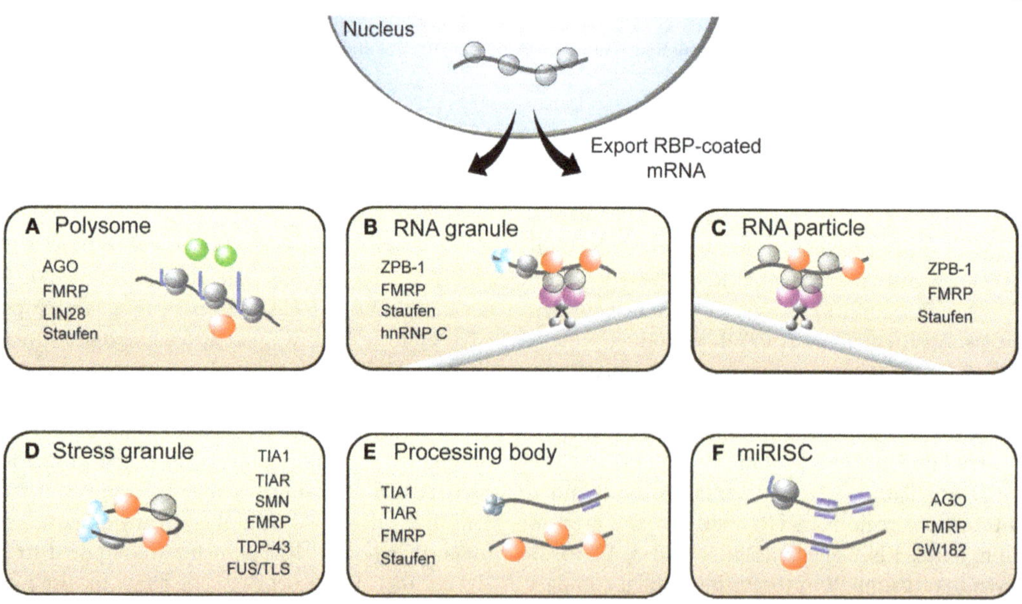

FIGURE 1 | Messenger RNAs associate with several RNP structures that influence their translational state. (A) Polysomes, sites of translation, contain RBPs that activate (green spheres) or repress (red spheres) translation. Following synthesis and processing, mRNA is exported from the nucleus and transported throughout the cell along microtubules via **(B)** RNA granules and **(C)** RNA particles. Repressor RBPs (red spheres) are present within RNA particles to ensure that mRNAs are not translated during transit. Messenger RNAs within RNA granules are associated with translation initiation machinery (light blue spheres) including ribosomes, suggesting that translation has commenced but is halted during transit. The translational fate of mRNA is dictated in part by the RBPs bound to them. If targeted by repressor RBPs or miRISCs (blue squares), mRNAs will associate with **(D)** stress granules, **(E)** processing bodies, or **(F)** miRISC structures resulting in either degradation or translational repression. Some RBPs present in neuronal RNP complexes are listed.

transcripts in RNA particles and RNA granules (Kiebler and DesGroseillers, 2000; Wang et al., 2010). An important distinction between these two transport complexes is that RNA granules contain ribosomes while RNA particles do not (Sossin and DesGroseillers, 2006). The presence of ribosomes within RNA granules suggests that the translation of associated mRNAs is blocked at the step of translational elongation.

RNA binding proteins are present within several other mRNP complexes that contain translationally repressed mRNAs. In response to cellular stress, TIA1, TIAR, G3BP, and other RBPs aggregate to form SGs, which contain untranslated mRNAs (Buchan and Parker, 2009; Kedersha and Anderson, 2009). SGs are proposed to safeguard specific mRNAs from destruction during cellular stress and, upon relief of the stress signal, disassemble to allow translationally repressed mRNAs to re-enter translation. P-bodies are another type of mRNP structure containing non-translating mRNAs that are destined for degradation (Coller and Parker, 2005; Teixeira et al., 2005) or are eventually released to re-associate with polysomes (Brengues et al., 2005). While the composition of P-bodies is not fully characterized, they generally contain decapping enzymes, exonucleases, translational repressors, microRNA (miRNA) silencing machinery, and translation-regulating RBPs, including CPEB, Staufen, and eIF4E (Parker and Sheth, 2007). Lastly, guided by miRNAs, RISCs repress translation of target mRNAs at the stage of translation initiation or elongation (Valencia-Sanchez et al., 2006). RBPs such as FMRP have been shown to associate with miRISC (Caudy et al., 2002; Witold, 2005) and can either promote or antagonize the repressive actions of miRISC (Brodersen and Voinnet, 2009). These translation-silencing mRNPs have been observed to interact with one another, and some components of SGs overlap with P-bodies (Kedersha et al., 2005) and miRISCs with P-bodies (Liu et al., 2005; Edbauer et al., 2010); however, the mechanisms that mediate these interactions remain to be established.

The interactions between RBPs and mRNAs are dynamic, allowing mRNAs to move from one mRNP to another in a controlled, bidirectional manner. This is an important feature of mRNA regulation because it ensures that the post-transcriptional fate of mRNAs is responsive to intracellular and extracellular signals. Signaling pathways, which are stimulated by various intracellular and extracellular cues, largely influence the mRNP distribution, and thus translational status, of mRNAs by regulating the expression and/or function of RBPs. For example, activation of the mTOR signaling results in the phosphorylation of FMRP (Narayanan et al., 2008); this post-translational modification affects the ability of FMRP to regulate translation (Ceman et al., 2003) and associate with RISC complexes (Cheever and Ceman, 2009). In response to signal-induced synaptic activation, the RBP Staufen was shown to activate translation by redistributing target mRNAs from RNA granules to translating polyribosomes (Krichevsky and Kosik, 2001).

GENOME-WIDE APPROACHES TO STUDY TRANSLATION

Advancements in technologies have significantly improved our ability to study translation at a genome-wide scale. Highly parallel techniques such as microarrays and high-throughput sequencing (deep sequencing) have revolutionized approaches to gene discovery, offering unbiased approaches and may be modified to investigate specific aspects of RNA regulation. In this section, we highlight studies that have utilized such technologies to investigate translational control at the genome-wide level both within neural and non-neural contexts (summarized in **Table 1**).

GENE EXPRESSION PROFILING USING MICROARRAYS

With the ability to examine gene expression on a global scale, microarray studies have provided evidence that diverse populations of mRNAs are localized at synapses. Martin and colleagues (Poon et al., 2006) were one of the first groups to examine synaptically localized mRNAs in rat neurons by mechanically separating axonal and dendritic processes from the cell body and performing microarray analysis on the isolated mRNA. Strikingly, they found that a significant proportion of synaptic mRNAs encoded translation factors and regulators, and proposed that this may be a general mechanism to enhance the capacity for local translation at synapses. Zhong et al. (2006) performed microarray studies on rat brain mRNA, which led to the discovery that the repertoire of synaptic mRNAs is more diverse than previously thought. The group not only identified transcripts that encoded translation factors and regulators, but also transcripts that encoded receptor and channel proteins, signaling molecules, cytoskeleton, and adhesion proteins, membrane trafficking proteins, and molecules involved in protein degradation. Additional studies have examined the synaptic transcriptome within other contexts, such as brain-derived nerve growth stimulation (Schratt et al., 2004) or neurons displaying molecular signatures of Alzheimer's disease (Williams et al., 2009). Interestingly, results from these and other microarray studies displayed little overlap, suggesting that a large number of mRNAs can be sequestered at synapses but that their localization largely depends on the cellular context.

POLYSOME PROFILING

A widely held view is that mRNA expression correlates closely with expression of the protein it encodes; this is certainly true in most instances, but is not always the case (Anderson and Seilhamer, 1997; Gygi et al., 1999). Indeed, the lack of correlation between mRNA and protein expression is expected given that multiple mechanisms are in place to control translation of mRNA. In this regard, microarray studies provide limited insight into the translational status of mRNA. An alternative approach called polysome profiling exploits the observation that, in general, polysome-associated mRNAs are translationally active. By separating polysome, monosome, and other mRNP complexes by centrifugation through a sucrose gradient, well-translated mRNAs can easily be distinguished from poorly translated mRNAs (**Figure 2A**). Morris and colleagues were one of the pioneering groups that used polysome profiling to examine the translation state of the transcriptome (Zong et al., 1999). Specifically, total cytoplasmic extracts from cultured human fibroblasts were layered onto sucrose gradients and centrifuged to separate the different mRNP complexes. Transcripts residing in high-density fractions containing polysome species were examined by microarray analysis to identify well-translated mRNAs (**Figure 2A**), while

Table 1 | Genome-wide methods to study translation.

	RNA isolation methodology	Novelty/advantages	Limitations	Reference
Polysome profiling	Purification of polysome-associated mRNAs by centrifugation through a sucrose gradient	Original method to examine translation status of transcriptome	Labor intensive; scaling issues; does not differentiate between active and stalled ribosomes	Zong et al. (1999)
TRAP	Immunoprecipitation (IP) of EFGP-L10a-associated mRNAs from mouse brain tissue	Examines polysome-associated mRNAs within a specific cell type *in vivo*	Each bacTRAP mouse line is limited to surveying one cell type; EGFP antibodies are costly relative to anti-HA antibody; does not differentiate between active and stalled ribosomes	Heiman et al. (2008), Doyle et al. (2008)
RiboTag	IP of Rlp22-HA-associated mRNAs from mouse tissue	Examines polysome-associated mRNAs within a specific cell type *in vivo*; takes advantage of Cre recombinase-expressing mouse lines to expand the range of cell types that can be investigated; commercial anti-HA antibody is less costly than in-house EGFP (see TRAP)	Does not differentiate between active and stalled ribosomes	Sanz et al. (2009)
Ribosome profiling	Nuclease digestion of polysome complexes, followed by centrifugation through a sucrose gradient or cushion to purify ribosome-mRNA complexes; ribosome-protected fragments are deep sequenced	Determines ribosome position and translation efficiency for individual mRNAs; reveals novel translational regulatory features (e.g., uORFs, start and termination sites, ribosome stall position)	May be difficult to apply to mouse models	Ingolia et al. (2009), Ingolia et al. (2011)
CLIP	UV-mediated crosslinking of mRNA-protein complexes, followed by nuclease digestion and IP of RBP of interest to recover RBP-protected mRNA fragments	Demonstrated the feasibility of crosslinking mRNA and protein using UV irradiation, which results in covalent bonds	Generated a limited dataset with a high false positive rate; low crosslinking efficiency	Ule et al. (2003)
CLIP-seq or HITS-CLIP	CLIP coupled with deep sequencing	Identifies direct RBP binding sites at nucleotide resolution	Low crosslinking efficiency	Licatalosi et al. (2008)
iCLIP	HITS-CLIP with modifications whereby a 5′ adapter and random barcode is attached to cDNA molecules after reverse transcription; the former modification allows for circularization of the cDNA	Introduction of a random barcode enables identification and quantification of unique cDNA products; cDNA circularization allows for the capture and sequencing of truncated cDNAs usually lost with standard CLIP, revealing crosslinking sites at nucleotide resolution	Low crosslinking efficiency	König et al. (2010)
PAR-CLIP	Photoreactive ribonucleoside analogs (e.g., 4SU or 6-SG) are incorporated into mRNA; nuclease digestion and IP of RBP of interest isolates RBP-protected mRNA fragments	Use of 4SU or 6-SG increases crosslinking efficiency; exact crosslinking sites are revealed after sequencing by T to C transitions in the cDNA prepared from RBP-bound mRNA	Some RBPs may not be amenable to PAR-CLIP	Hafner et al. (2010), Castello et al. (2012)
iPAR-CLIP	PAR-CLIP method applied to *C. elegans* exposed to 4SU	First demonstration of CLIP in a non-cell line system; allows for physiologically relevant, context-dependent studies of protein-RNA interactions in *C. elegans*	Technique yet to be applied to other *in vivo* models	Jungkamp et al. (2011)

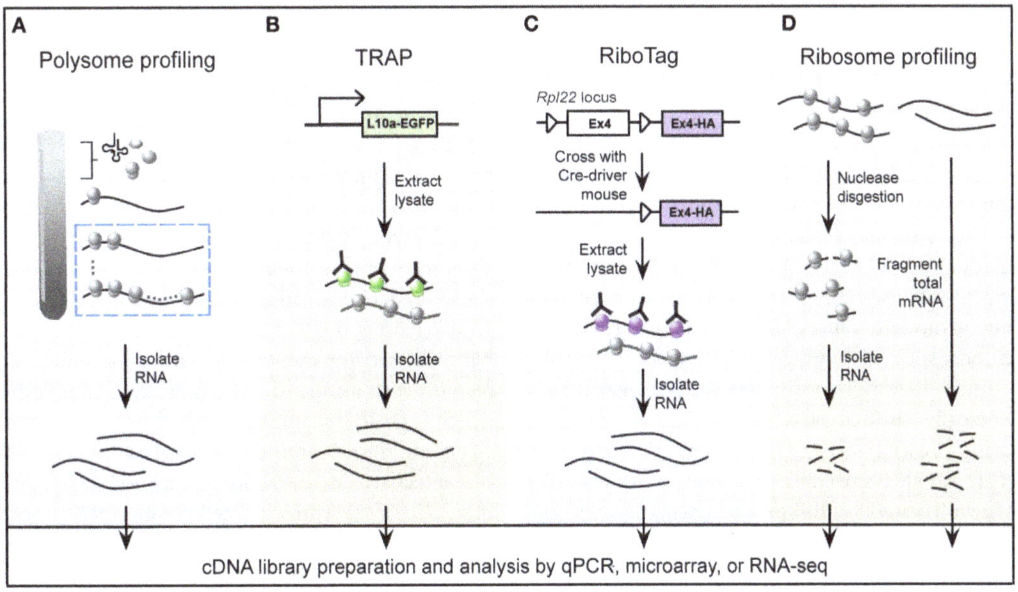

FIGURE 2 | Schematic of genome-wide methods to study polysome-associated mRNAs *in vitro* and *in vivo*. (A) With polysome profiling, cytoplasmic lysates from cells are layered onto a sucrose gradient and undergo centrifugation to separate tRNAs, 40S, 60S, and 80S ribosomes, and polysomes. Messenger RNAs from fractions corresponding to polysomes (dashed blue box) are isolated and identified by various approaches. **(B)** Engineered bacTRAP mice drive expression of EGFP-tagged L10a, a ribosomal protein found in polysomes (green ribosomes), from promoters that are activated in specific cells of the central nervous system. EGFP-L10a-mRNA complexes are immunopurified from brain tissue from bacTRAP mice, and associated mRNAs are identified by various techniques. **(C)** The RiboTag mouse carries an Rpl22 allele with a floxed wild-type C-terminal Exon4 followed by a HA-tagged Exon4. When the RiboTag mouse is crossed with a mouse expressing Cre-recombinase in a cell-type specific manner, Cre-recombinase activates expression of HA-tagged Rpl22, which incorporates into polysomes (purple ribosomes). Homogenized tissues from the offspring are subjected to co-immunoprecipitation using antibodies against HA, and associated mRNAs are identified by various techniques. **(D)** Using ribosome profiling to identify ribosome occupancy on mRNAs, cycloheximide-treated lysates from cultured cells are digested by micrococcal nucleases to remove mRNA sequences that are not bound by ribosomes (left). The resulting monosome complexes are purified by ultracentrifugation through a sucrose gradient or cushion. Ribosome-protected fragments are recovered and deep sequenced. In parallel, total mRNA from a similar preparation of cycloheximide-treated lysate is fragmented and deep sequenced (right), and serves as a normalizing control.

transcripts sequestered within the low-density fractions were also examined to identify poorly translated mRNAs.

Polysome profiling has been readily applied to various cellular contexts and cell types, including neurons (Johannes et al., 1999; Preiss et al., 2003; Rajasekhar et al., 2003; Schratt et al., 2004; Iguchi et al., 2006). A study that nicely demonstrated the utility of polysome profiling examined the widespread inhibition of translation in response to cellular stress, a process that appears to involve movement of mRNAs from polysome complexes to P-bodies. The movement of transcripts from polysomes to P-bodies is thought to be a general phenomenon in eukaryotic cells, yet evidence to support this view is based on single-gene studies. To address whether this view is in fact a general phenomenon or limited to a subset of mRNAs, Arribere et al. (2011) used polysome profiling in combination with translation inhibitors to measure the translational activity and ribosome occupancy upon glucose withdrawal and at different times following glucose re-addition. This study illustrated the power of genome-wide studies: Arribere and colleagues were able to provide mechanistic insight to translational control based on an examination of the entire transcriptome, not just representative transcripts. Notably, they showed that a substantial portion of mRNAs, many of which encode survival factors, is actively translated during stress; this finding disputed the prevailing notion that translational inhibition is widespread during stress. They also found that re-entry of pre-existing mRNAs, presumably from P-bodies, into polysomes is restricted to a subset of mRNAs rather than a general phenomenon as initially proposed (Brengues et al., 2005; Teixeira et al., 2005; Brengues and Parker, 2007; Hoyle et al., 2007).

As with any technique, polysome profiling has several limitations. From a technical perspective, this technique is labor intensive and difficult to scale-up (Larsson and Nadon, 2008). Furthermore, polysome profiling should be performed with uniform cell populations due to the heterogenic nature of transcriptomes between cell types; this presents a significant challenge in performing polysome profiling *in vivo*. A final consideration concerns the underlying assumption of the technique that mRNAs bound by multiple ribosomes are translationally active. Both active and stalled ribosomes have been shown to co-sediment during isolation of polysome complexes through sucrose gradients (Sivan et al., 2007), indicating that polysome profiling does not completely distinguish translationally active from repressed mRNAs. Complementary molecular studies that directly measure *de novo* protein synthesis would be useful to discriminate translating from non-translating polysome-associated mRNAs. Despite these limitations, polysome profiling has been extremely successful in identifying translationally active mRNAs.

TARGETING RIBOSOMES: IMMUNOPRECIPITATION-BASED METHODS TO ISOLATE POLYSOMES *IN VIVO*

Gene expression patterns vary greatly between cell types, thereby requiring that genome-wide studies be performed using homogeneous cell populations. The difficulty of performing polysome profiling *in vivo* is due to the challenge of extracting homogeneous cell populations in sufficient quantities without affecting the transcriptome (Okaty et al., 2011). The mammalian brain is an especially challenging model because of its immense heterogeneity. Enrichment methods, such as fluorescence-activated cell sorting or laser-capture microdissection, accurately separate genetically and/or morphologically distinct cells; however, manipulation of purified cells during isolation steps may alter the transcriptional profile (Okaty et al., 2011). These challenges must be addressed in order to achieve the full benefits of this powerful technique *in vivo*, giving us a better understanding of translational control within a physiologically relevant setting. Two groups independently tackled this issue by genetically altering mice to express epitope-tagged versions of ribosome protein subunits (Heiman et al., 2008; Sanz et al., 2009). All translated mRNAs are, at one point, associated with ribosomes; thus affinity purification of epitope-tagged ribosomes would allow for isolation of polysome-associated mRNAs. Heiman and colleagues termed their methodology translating ribosome affinity purification, or TRAP, which involves a series of bacterial artificial chromosome transgenic mice, called bacTRAP mice. In these genetically modified mice, expression of EGFP-tagged ribosomal protein L10a is driven by defined promoters that are activated in specific cell types of the central nervous system (CNS; **Figure 2B**; Heiman et al., 2008). Heiman and colleagues used bacTRAP mice that expressed EGFP-L10a from the Drd1a receptor or Drd2 receptor to isolate polysome-associated mRNA from striatonigral or striatopallidal cells, respectively, of the mouse striatum. Cells derived from the brain tissue of these mice were lysed in cycloheximide-spiked buffer to halt elongating ribosomes. Polysome-associated RNA corresponding to the specific cell type was isolated by immunoprecipitating for EGFP-L10a and examined by microarray analysis. The group demonstrated the power of TRAP technology by identifying polysome-associated mRNAs unique to four neuronal populations that are intermixed and morphologically indistinguishable. In an accompanying paper, Doyle et al. (2008) demonstrated the generality of TRAP with a comprehensive study of polysome profiles for 24 additional CNS cell types.

Using a strategy similar to TRAP, Sanz and colleagues engineered a mouse line called RiboTag that contains three HA tags inserted into the locus of *Rlp22*, a gene that encodes a ribosome protein present within polysomes (**Figure 2C**; Sanz et al., 2009). Expression of the $Rlp22^{HA}$ allele depends on Cre recombination such that in the absence of Cre recombinase only endogenous (wild-type) *Rlp22* is expressed. The group crossed RiboTag mice with several neuron-specific Cre recombinase-expressing mice. Brain tissues from the resulting offspring were used to immunopurify HA-tagged polysomes and recover associated mRNAs from specific cell populations expressing Cre recombinase. They demonstrated that the RiboTag system indeed reliably purifies mRNAs associated with ribosomes in a cell type specific manner.

The RiboTag strategy has several advantages over TRAP. First, the RiboTag mouse can be crossed to any Cre recombinase-expressing mouse line, allowing for polysome profiling of a significantly greater variety of cell types. In contrast, the TRAP system requires engineering a separate bacTRAP mouse line for each cell type of interest. Second, $Rlp22^{HA}$ is expressed at levels similar to wild-type Rlp22, thereby maintaining the appropriate stoichiometry of ribosomal subunits and kinetics of translation. The TRAP system instead expresses the EGFP-L10a transgene from an exogenous promoter, which may result in different levels of EGFP-L10a compared to wild-type L10a. Third, the TRAP method recommends using in-house monoclonal antibodies against EGFP, which is costly for multiple experiments. Alternatively, the RiboTag method uses commercial anti-HA antibodies that are far more cost effective. Finally, RiboTag technology has the ability to generate $Rlp22^{HA}$-expressing cells (e.g., $Rpl22^{HA}$-expressing mouse embryo fibroblasts) that are capable of proliferating indefinitely in culture dishes, providing an abundant and renewable model for obtaining ribosome-associated mRNAs by immunoprecipitation rather than by the laborious process of sucrose gradient centrifugation. This ability is severely limited with the TRAP system since many of the current bacTRAP mice activate EGFP-L10a expression in neurons, which do not proliferate *in vitro*. Overall, the TRAP and RiboTag technologies provide an efficient and rapid method of isolating polysome-associated mRNA from a single cell type *in vivo*.

RIBOSOME PROFILING

Many studies that have been designed to investigate global translation typically use genome-wide measurements of mRNA and/or protein expression as indicators of protein synthesis. However, transcripts are subject to multiple levels of translational control rendering mRNA expression an imperfect substitute for protein synthesis. Global protein expression, as measured by mass spectrometry, is also a poor proxy of translation since protein stability also contributes to changes in protein expression. The most precise measurement of translation is direct quantification of protein synthesis. To measure protein synthesis at a genome-wide level, Weissmann and colleagues developed a technique called ribosome profiling (**Figure 2D**) to map the precise positions of ribosomes within the transcriptome (Ingolia et al., 2009, 2012). Ribosome profiling involves nuclease digestion of cell extracts to degrade mRNA that is unprotected by ribosomes, leaving approximately 28 nucleotide RNA fragments. Individual ribosome-RNA complexes are isolated by centrifugation through a sucrose gradient or cushion, followed by a series of purification steps to recover the mRNA fragments, such as fragment size selection by gel electrophoresis and rRNA removal by subtractive hybridization. Ribosome-protected mRNA fragments (ribosome footprints) are identified by high-throughput sequencing, revealing the locations of ribosomes along transcripts at nucleotide-level resolution.

Ingolia et al. (2009) first demonstrated the utility of ribosome profiling for studying translation in *Saccharomyces cerevisiae*. The authors showed that ribosome profiling could quantitatively measure translational efficiency as defined by the ratio of ribosome footprint density to total mRNA. Ribosome density proved to be a much better predictor of protein production than measurements of mRNA levels. Ribosome density also correlated well, though not perfectly, with protein expression, suggesting that in some contexts mass spectrometry does not accurately measure global protein synthesis. The correlative differences between protein expression and translation efficiency observed by Ingolia et al. (2009) may be attributed to protein degradation and such differences could be exploited to investigate protein stability at a genome-wide level.

In addition to being an optimal tool for genome-wide measurements of protein synthesis, ribosome profiling is capable of discovering novel regulatory mechanisms of translation. While ribosome footprints are expected to map to coding regions of transcripts, a ribosome profiling study in yeast found that a small fraction of footprints (1.2%) mapped to non-coding regions (Ingolia et al., 2009). The majority of this 1.2% footprint fraction mapped to the 5′ UTRs of transcripts, leading to the discovery of 153 upstream open reading frames (uORFs) of which fewer than 30 had previously been described. Because ribosome profiling maps ribosome footprints at a resolution such that a three-base codon periodicity is observed, positional data can be used to identify frame shifts, non-canonical start codons, and stop codon readthrough, all of which would predict novel protein isoforms. As proof-of-principle, Ingolia et al. (2009) uncovered a total of 143 non-canonical (non-AUG) start sites in the yeast transcriptome; this was a substantial contribution to our knowledge of translational initiation sites in yeast, having previously known of only two examples of non-AUG initiation sites (Chang and Wang, 2004; Tang et al., 2004). Sites of premature translational termination can also be identified with this technique. Subsequent ribosome profiling studies have been implemented in other model organisms, including bacterial and mammalian systems (Guo et al., 2010; Ingolia et al., 2011; Oh et al., 2011; Hsieh et al., 2012). Not surprisingly, these studies also produced a long list of candidate initiation start sites, alternative reading frames, and uORFs. A comparison of these features across model systems or among different cell types will be informative in determining whether translational control mechanisms are general or specific to individual organisms and cell types.

Translation is a dynamic process and the kinetics of ribosome movement along transcripts is an important aspect of translational control. While the temporally static nature of ribosome profiling may appear ineffective for the study of translational kinetics, Ingolia et al. (2011) demonstrated otherwise by combining ribosome profiling with run-off elongation assays. First, cells were treated with harringtonine to block additional rounds of translation initiation (Fresno et al., 1977). Cycloheximide was then applied to cells for various time points afterward, freezing all actively translating ribosomes. Ribosome positions were then determined by ribosome profiling and the change in ribosome positioning over the course of cycloheximide treatment was used to generate a moving picture of ribosomes. In log phase growing yeast, the rate of translation appeared to be independent of mRNA class, mRNA length, or whether mRNAs encoded secreted or cytoplasmic proteins. This adaptation of ribosome profiling will prove useful in studying the rate of translation in the context of cellular stress or pathological states where global translation is aberrant (Shenton et al., 2006).

Another critical aspect of translational kinetics is ribosomal pausing. Pausing or stalling of elongating ribosomes is a mechanism of translational repression and may be caused by steric hindrance (due to secondary structure of the transcript or by the exiting nascent peptide), recruitment of low abundance tRNAs to rare codons, or RBPs such as FMRP (Lovett and Rogers, 1996; McNulty et al., 2003; Darnell et al., 2011). Ingolia et al. (2011) reasoned that the ribosome density for a given codon should be commensurate with the average ribosome dwell time, i.e., slower ribosome movement across a codon (longer dwell time) should result in more footprint counts at that codon. Based on this inference, the group used ribosome profiling to identify thousands of ribosome pause sites at the resolution of individual codons in the transcriptome of mouse embryonic stem cells (ESCs). Ribosomal pausing was initially described more than 20 years ago (Wolin and Walter, 1988), yet the mechanisms underlying this process is not well understood. Recently, ribosome profiling in bacteria was used to show that a Shine–Dalgarno-like feature in mRNA facilitates ribosome pausing (Li et al., 2012), demonstrating that this technique provides an efficient method to elucidate the molecular mechanisms of ribosomal pausing.

Similar to polysome profiling, ribosome profiling is technically challenging and labor intensive. Another consideration regarding ribosome profiling, as pointed out by Morris (2009), relates to the use of cycloheximide to halt elongating ribosomes prior to cell lysate preparation. Although characterized as an inhibitor of elongation, cycloheximide has also been shown to block translation initiation at similar concentrations used for ribosome profiling (Obrig et al., 1971) but not at lower concentrations (Lodish, 1971). If the exposure of cycloheximide to cells is low such that translation elongation, but not initiation, is negatively affected, this may allow ribosomes to accumulate at the 5′ end of transcripts and result in spurious ribosome footprinting patterns. This is especially relevant to applications of ribosome footprinting where drug delivery and exposure are difficult to control, for example with studies using mice. Yet despite these drawbacks, ribosome profiling remains a powerful tool to identify novel uORFs, initiation start sites, termination sites, and alternative reading frames, and provide insights on ribosome movement or lack thereof (ribosomal pausing) will most certainly lead to the discovery of new and unexpected modes of translational control.

TARGETING RBPs: IMMUNOPRECIPITATION-BASED METHODS TO ISOLATE RBP-mRNA COMPLEXES

Key to understanding the role of RBPs in translational control, and RNA metabolism in general, is identifying their mRNA targets. Initial genome-wide attempts to identify RBP targets employed a technique called RNA immunoprecipitation (RIP). Much like its DNA counterpart chromatin immunoprecipitation, RIP involves formaldehyde crosslinking of proteins to RNA, followed by immunoprecipitation of the protein-RNA complex. Protein-bound transcripts are then used to make a cDNA library that is subjected to microarray analysis. A major concern with RIP is the low signal to noise ratio: RNA tends to be "sticky" making the technique vulnerable to extracting non-physiological binding partners (Mili and Steitz, 2004). The Darnell group introduced an alternate approach to RBP target identification termed CLIP for crosslinking and immunoprecipitation (Ule et al., 2003), which is reviewed in greater detail elsewhere (Darnell, 2010a; König et al., 2012). Instead of using formaldehyde, CLIP uses UV irradiation to cement protein-RNA interactions by creating covalent bonds between proteins and RNA that are within distances of a few angstroms. Unprotected RNA is removed by partial digestion with RNase (in the original protocol) or micrococcal nuclease, which is easily inactivated by EGTA to avoid spurious, continual RNA digestion throughout the procedure (Yeo et al., 2009; Zisoulis et al., 2010; Polymenidou et al., 2011). Protein-RNA complexes are recovered by immunoprecipitation with antibodies against the protein of interest. Transcript fragments of approximately 60–100 nucleotides in length are released from proteins and are further processed for sequencing. Several aspects of this technique make it well suited to study RBP-RNA interactions. First, the direct interaction between a RBP and its target mRNA is faithfully preserved, since crosslinking only occurs with RBPs and mRNA that are within angstrom distances. Second, crosslinking of direct RBP-mRNA interactions via strong covalent bonds allows these complexes to be purified under stringent conditions, further reducing background signal. Third, UV irradiation does not preserve protein–protein interactions, thereby avoiding the possibility of indirect protein-mRNA interactions.

The first application of CLIP sought to identify RNA targets of Nova, a neuronal KH-type RBP, which is implicated in paraneoplastic neurologic degeneration (Darnell, 2010b). In this study, Ule, Jensen, and colleagues identified 34 candidate mRNA targets, most of which are involved in neuron function (Ule et al., 2003). The use of low-throughput sequencing to identify Nova targets, however, generated a limited dataset and made it difficult to discern authentic from spurious mRNA targets (50% false positive rate). To remedy this, the same group performed a subsequent study in which CLIP was combined with high-throughput sequencing (HITS-CLIP). This strategy generated a more robust dataset of Nova targets, confirming, and refining their previous assertions about Nova as a splicing regulator (Licatalosi et al., 2008). Further modifications to the standard CLIP protocol have improved the crosslinking efficiency. One such modification called Photoactivatable-Ribonucleoside-Enhanced Crosslinking and Immunoprecipitation, or PAR-CLIP, uses photoreactive ribonucleoside analogs [e.g., 4-thiouridine (4SU) or 6-thioguanosine], which are incorporated into nascent mRNAs in live cells (Hafner et al., 2010) or whole organisms (Jungkamp et al., 2011). Photoreactive ribonucleoside analogs crosslink with proteins more efficiently than endogenous ribonucleotides, thereby enhancing the signal to noise ratio. During cDNA preparation of labeled mRNA, crosslinked sites are prone to thymidine to cytidine or guanosine to adenosine transitions (when 4-thiouridine or 6-thioguanosine is used, respectively), revealing exact locations of nucleotide-protein interactions. This feature was initially exploited to identify individual ribonucleotides of a small nucleolar RNA that interacted with RBPs (Granneman et al., 2009) and has since been applied to other RBPs (Hafner et al., 2010). It is important to note that not all RBPs may be amenable to PAR-CLIP, such as CUG triplet repeat RNA binding protein (CELF1; Castello et al., 2012). A further modification of CLIP, called individual nucleotide resolution CLIP or iCLIP, provided an alternative approach to locating the exact crosslinking position (König et al., 2010). Ule and colleagues took advantage of the fact that reverse transcriptase arrests at sites of nucleotide-peptide crosslinking (a peptide remnant of the RBP remains after proteinase K digestion). The resulting truncated cDNAs are normally lost during the standard CLIP library preparation, but the iCLIP protocol was designed to recover and sequence these truncated cDNAs to identify exact crosslinking sites. Using CLIP or its variants, genome-wide protein-RNA interaction maps have been assembled for numerous RBPs, including Nova (Ule et al., 2003), RBFOX2 (Yeo et al., 2009), Argonaute proteins (Chi et al., 2009; Hafner et al., 2010; Zisoulis et al., 2010; Leung et al., 2011), TDP-43 (Polymenidou et al., 2011; Tollervey et al., 2011), FMRP (Darnell et al., 2011), and hnRNP proteins (Katz et al., 2010; König et al., 2010; Huelga et al., 2012).

The ability to systematically uncover RBP target sites within the transcriptome has provided insights to the role of RBPs in translational control. Using *in vivo* PAR-CLIP (iPAR-CLIP), Jungkamp et al. (2011) proposed several models by which GLD-1, a conserved germline-specific RBP in *C. elegans*, functioned as a translational repressor. Highly conserved 5′UTR GLD-1 binding sites were discovered near the start codon of target transcripts; this was unexpected, since prior studies indicated that GLD-1 primarily targets the 3′ UTR. Following extensive biochemical validation of their iPAR-CLIP results, Jungcamp and colleagues proposed that GLD-1 dimers, as the protein is known to form, bind to the 5′ and 3′ UTRs to promote circularization of mRNA to block translational initiation (Jungkamp et al., 2011). Alternatively, binding of GLD-1 near start codons at the 5′ UTR may prevent assembly of the ribosome.

Another study examined the mechanisms of translational control by FMRP using CLIP. Darnell et al. (2011) identified FMRP binding sites within the polysome-associated fraction of the transcriptome by HITS-CLIP and found that a significant portion of gene targets were involved in neuronal synaptic plasticity and synaptic-related signaling pathways. The majority of FMRP binding sites (66%) resided within coding sequences. This was unexpected given that translational control mechanisms usually involve binding of RBPs within UTRs. Furthermore, the distribution of FMRP appeared to be uniform along transcripts. Darnell and colleagues extended these observations to demonstrate that FMRP directly stalls ribosomes on polysome-associated transcripts, thereby suppressing translation. Given that many of these FMRP targets are involved in synaptic transmission, it was hypothesized that FMRP functions to represses translation of associated mRNAs during transit from the soma to synapses, to prevent premature translation and/or degradation. Upon release of FMRP from its targets, presumably by signal activation, the transcripts, already loaded with ribosomes, would be rapidly translated. Given that the interplay between different RBPs is important for RNA transport and local synaptic translation, the application of CLIP to neuronal RBPs will be instrumental in defining their individual and combinatorial contributions to translational control.

CLIP is accompanied by limitations that are either unique to particular version or inherent to all (Darnell, 2010a; Ascano et al.,

2012). The ability of CLIP to capture a representative population of RNA targets is influenced by the crosslinking efficiency between RNA and the protein of interest, nuclease digestion conditions (Kishore et al., 2011), and the specificity and reactivity of the antibody used to isolate the RBP-RNA complex; these factors are inherent to all CLIP versions. Crosslinking efficiency will vary depending on the RNA target sequence and the type of amino acids available for crosslinking. Nucleic acids are generally more reactive with cysteine, lysine, phenylalanine, tryptophan, and tyrosine residues, and this residue preference appears to be preserved with 4SU-labeled RNA (Meisenheimer and Koch, 1997; Meisenheimer et al., 2000). 4SU is generally thought to enhance crosslinking efficiency compared to unmodified uridine (Hafner et al., 2010), although this is not always the case (Kishore et al., 2011). A drawback of PAR-CLIP is the difficulty in applying this method to whole animal models other than *C. elegans*, where uniform exposure of 4SU may not be feasible. Until a method for efficient delivery of ribonucleoside analogs to other animal models (namely mice) is established, HITS-CLIP is generally required for *in vivo* studies.

CHALLENGES AND CONSIDERATIONS FOR GENOME-WIDE STUDIES IN NEURONS

The application of genome-wide studies to neurobiology is accompanied by challenges intrinsic to the neural model system and genome-wide method being employed. The mammalian CNS is complex with hundreds of morphologically distinct cell types, each expressing a unique transcriptome. This presents several challenges when performing genome-wide studies: purifying homogenous cell populations, isolating cells in a manner that does not alter gene expression, and obtaining sufficient quantities of cells for analysis. With regard to the latter issue, some of the techniques discussed here require up to several tens of micrograms of RNA for analysis, necessitating a large input of cells. Cultured primary neurons derived from rodent tissue are a well established model for studying neurons; however, they do not obviate some of these challenges, as they may not be completely homogenous and do not expand *in vitro*. Furthermore, establishing and maintaining cultured primary neurons are difficult and require extensive training of the researcher. Neural progenitor cells (NPCs), either differentiated from stem cells or isolated from rodents or humans, are an alternative *in vitro* model system. NPCs can be expanded and differentiated into multiple neuronal cell types, and developing an efficient differentiation process should yield a fairly homogenous cell population. Importantly, NPCs that were differentiated from induced pluripotent stem cells (iPSCs) derived from patient fibroblasts have been shown to recapitulate the molecular phenotypes of corresponding neurological diseases (Thonhoff et al., 2009; Ming et al., 2011), providing valuable models to investigate global translation in neurodegeneration using the methods described herein. The labor intensive, lengthy, and costly nature of establishing iPSCs and differentiating them into NPCs is a major drawback of this system.

Similar to cell culture models, animal models used for genome-wide analyses are fraught with the challenges mentioned above; however, some of these challenges may be circumvent with the use current technologies. To remedy the issue of complex cell heterogeneity, the RiboTag and TRAP technologies were created, both of which expressed tagged versions of ribosomes in a cell type specific manner to isolate mRNA. While the intended application of RiboTag and TRAP technologies is for *in vivo* polysome profiling, conceivably they may be adapted for *in vivo* ribosome profiling studies. The application of HITS-CLIP in heterogenic mouse brain tissue is standard (Licatalosi et al., 2008; Polymenidou et al., 2011), and in theory the technology required to perform HITS-CLIP in a cell type specific manner *in vivo* is obvious. Using transgenic mice that express a tagged version of the RBP of interest in a cell type specific manner, HITS-CLIP could easily be implemented *in vivo* using an antibody that recognizes the tag. Naturally, adaptation of RiboTag or TRAP technologies for ribosome profiling or transgenic mice for HITS-CLIP will require optimization steps to ensure a seamless integration.

Lastly, bioinformatic challenges inherent to genome-wide studies largely center on the issue of interpreting the massive amounts of raw sequencing data. A convergent problem in analyzing sequence reads derived from CLIP and ribosome profiling data is the identification of precise binding sites of RBPs or ribosome footprints within the RNA transcripts. Thus far, the analysis of CLIP data relies on inherent assumptions about the binding kinetics of RBPs with their preferred RNA substrates. The RBPs that have been studied thus far, for example Nova, RBFOX2, Argonaute, and hnRNP family members (Ule et al., 2003; Yeo et al., 2009; Hafner et al., 2010; Zisoulis et al., 2010; Leung et al., 2011; Huelga et al., 2012), tend to interact with several binding sites within a given substrate very strongly. Computational approaches to identify precise binding sites (or clusters of reads), known as cluster- or peak-finding algorithms, have therefore assumed that the majority of reads within an RNA that are below an expected threshold are not true binding sites and represent experimental noise or artifacts. However, other RBPs, such as TDP-43 or FUS/TLS may act in "scanning" mode on some RNA substrates (unpublished observations), interacting with many sites with low affinity, rather than a few sites with high affinity, with nevertheless important biological effects. Most algorithms are conservative in identifying these low affinity sites within RNA substrates. Ribosome footprints fall into this second category of many locations of roughly equal occupancy with mRNA substrates. Thus while ribosome profiling data can be used to measure quantitative differences in ribosome occupancy within RNAs, precise footprint sites such as paused ribosomes are not effectively identified by existing peak-finding algorithms. Another glaring concern with analyzing RNA and ribosome binding data is the dearth of statistical tools to measure whether we have saturated the potential number of binding sites observed (one strategy is suggested in Polymenidou et al., 2011), or techniques to measure the rate of false positives and negatives. Lastly, not all RNA binding sites are functional. It is not unthinkable that while highly expressed RBPs interact with many RNA substrates, many, if not most of these sites may not actually have an impact in regulating the life cycle of the RNA molecule. Integrating other genome-wide assays that reveal the dependence of the RNA on the RBP is important for assigning functions to these binding sites. Thus, computational approaches to integrate these information with CLIP data, and also comparing how multiple RBPs affect the same molecules will be crucial in going forward (Huelga et al., 2012).

Table 2 | List of RBPs involved in translation and implicated in neurological diseases.

RBP	Function	Disease	Reference
FMRP	Repressor	FXS	Darnell et al. (2011)
hnRNP A2/B1	Activator	ALS, FTLD	Kwon et al. (1999)
hnRNP C	Activator	AD	Lee et al. (2010)
IGHMBP2	Regulator	SMA	Grohmann et al. (2001), de Planell-Saguer et al. (2009)
Musashi	Repressor	AD	Okano et al. (2002), Perry et al. (2012)
SMN	Putative repressor	SMA	Piazzon et al. (2008)
TDP-43	Repressor	ALS, FTLD	Lagier-Tourenne et al. (2010)

AD, Alzheimer's disease; ALS, Amyotrophic lateral sclerosis; FTLD, Frontotemporal lobar dementia; FXS, Fragile X syndrome; SMA, Spinal muscular atrophy.

RNA BINDING PROTEINS AND TRANSLATIONAL CONTROL IN NEUROLOGICAL DISEASES

RNA binding proteins have received considerable attention for their roles in neurodegeneration (Lukong et al., 2008; Liu-Yesucevitz et al., 2011). This is not surprising given that many RNA processing events, including local translation, are important for neuronal function (Klann and Dever, 2004; Sutton and Schuman, 2006). Of the handful of RBPs that are associated with neurological diseases, only several have been implicated in the regulation of translation (**Table 2**), most of which were determined by single-gene studies. Presently, there is a lack of genome-wide studies examining the translation functions of RBPs in normal or pathological contexts of neurons, with the exception of a few studies related to FMRP; this leaves much room for investigation. In this section, we review our current knowledge of these RBPs (listed in **Table 2**) with regard to translational control and neurological diseases.

FMRP is one of the more extensively studied RBPs with regard to translational regulation and neurodegeneration. It is essential for proper synaptic function, as loss of FMRP in mice results in aberrant pre- and post-synaptic plasticity (Deng et al., 2011). FMRP is present throughout the neuron (soma, dendrites, and axon) with the majority of FMRP (85–90%) being associated with polysomes (Zhang and Darnell, 2011). Misregulation of FMRP is linked with FXS, a condition characterized by impaired cognitive, physical, emotional, and sensory function (Bagni and Greenough, 2005; Bassell and Warren, 2008). Consistent with this, a genome-wide analysis of FMRP RNA targets using HITS-CLIP revealed that many FMRP targets encode proteins that are implicated in autism spectrum disorders (Darnell et al., 2011). Mutations within the FMRP gene that lead to reduced expression are frequently observed in patients with FXS (De Boulle et al., 1993; Snow et al., 1993). As discussed in this review, FMRP downregulates target gene expression by blocking translation in part by stalling ribosomes. FMRP has also been shown to interact with Ago2 and recruit miRISCs to its mRNA targets, providing another mechanism by which FMRP mediates gene silencing (Muddashetty et al., 2011). Edbauer et al. (2010) demonstrated that FMRP associates with miR-125b and miR-132 in the mouse brain, interactions of which greatly influence dendritic spine morphology and synaptic physiology of hippocampal neurons. In this study, a limited set of miRNAs was examined and it is likely that additional miRNAs interact with FMRP. Genome-wide methods tailored to studying miRNA-protein interactions are well established (reviewed in Wilbert and Yeo, 2011) and would provide an unbiased approach toward identifying miRNAs that interact with FMRP as well as other RBPs.

The evolutionarily conserved Musashi family of RBPs, herein referred to as Musashi, has strong links to both translational control and neurobiology (Okano et al., 2002). Preferentially expressed in the mammalian nervous system, Musashi is a key contributor to maintaining the proliferative capacity of neural stem cells, such that its loss leads to a reduction in the formation of neurospheres, or cell-cultured neural stem cells (Sakakibara et al., 2002). Musashi is thought to maintain progenitor self-renewal by translationally repressing inhibitors of stem cell proliferation. One of these targets, *CDKN1A*, encodes the anti-proliferative cell-cycle inhibitor p21WAF (Battelli et al., 2006). *NUMB* is another Musashi target that is a negative regulator of the Notch signaling pathway. Since Notch signaling is crucial for neural stem cell maintenance (Hitoshi et al., 2002), translational silencing of *NUMB* by Musashi augments this proliferative process (Imai et al., 2001). The mechanism by which Musashi suppresses translation, at least for *NUMB*, involves disrupting the eIF4G binding-Poly(A)-binding protein interaction, thereby preventing assembly of the 80S ribosomal complex (Kawahara et al., 2008). If Musashi is characteristic of most RBPs, then it has hundreds or thousands targets in addition to *NUMB* and *CDKN1A*. Using the iterative *in vitro* selection process SELEX, the RNA binding sequence of Musashi has been identified (Imai et al., 2001). This sequence information can be useful in predicting targets, yet still requires single-gene studies to validate authentic Musashi targets. The application of HITS-CLIP to define Musashi targets and locate discrete binding sites would prove invaluable toward understanding the function of Musashi. This knowledge may also be useful to investigate a potential role of Musashi in Alzheimer's disease (AD). A link between Musashi and AD was established with the observation that Musashi expression was reduced in patients with AD (Perry et al., 2012). Consistent with Musashi being a key regulator of neural stem cells, downregulation of Musashi correlated with a reduction in neural stem cells, the latter event of which is often observed in AD. Whether misexpression of Musashi is a determinant or downstream effect of AD is uncertain. As will be a common theme among the RBPs addressed in this section, a comprehensive understanding of Musashi function as revealed by genome-wide studies will likely identify therapeutic targets implicated in AD.

TAR DNA-binding protein-43 (TDP-43) is another RBP that has recently been recognized as an important contributor to neurological diseases (Lagier-Tourenne et al., 2010). It is well established as a regulator of transcriptional repression and alternative splicing, but may also have a role in translational repression (Wang et al., 2008c). In hippocampal neurons, TDP-43 appears to reside within RNA granules and P-bodies – storage sites of repressed mRNAs; this observation is consistent with the finding that TDP-43 acts as a translational repressor in an *in vitro* assay (Wang et al., 2008b). A pathological link between TDP-43 and neurodegeneration was initially established with the observation that TDP-43 is present within brain cell inclusions of patients with ALS or FTLD (Arai et al., 2006; Neumann et al., 2006). Dominant mutations in the gene encoding TDP-43 (*TARDBP*) that cause mislocalization of the protein were subsequently identified in cohorts of patients with ALS and FTLD (Gitcho et al., 2008; Kabashi et al., 2008; Sreedharan et al., 2008; Van Deerlin et al., 2008; Yokoseki et al., 2008). The aberrant activities of disease-associated forms of TDP-43 have been recapitulated in transgenic rats displaying neurological impairments and in reprogrammed pluripotent stem cells, confirming the pathogenicity of these mutations (Bilican et al., 2012; Huang et al., 2012). Findings from genome-wide studies have advanced our understanding of the pathogenic activities TDP-43 by revealing that TDP-43 modulates the levels and alternative splicing of many of its RNA targets, a substantial portion of which encode proteins involved in neuronal development and function (Polymenidou et al., 2011; Tollervey et al., 2011). A connection between the translational silencing and alternative splicing functions of TDP-43 may exist, as demonstrated for the TDP-43 target S6 kinase 1 (S6K1) Aly/REF-like target (SKAR). TDP-43 appears to control alternative splicing of SKAR, which results in the expression of a SKAR isoform that can no longer activate the translation-stimulating SK61-dependent signaling pathway (Fiesel et al., 2012). Therefore loss of TDP-43 resulted in upregulation of the SKAR isoform that increases SK61 activity and consequently stimulated global translation. Whether the translational functions of TDP-43 require its splicing activities for other TDP-43 targets remains elusive. But more pressingly, the role of TDP-43 in global translation remains unknown and will best be addressed by employing the genome-wide methods discussed here.

Two members of the heterogeneous nuclear ribonucleoproteins (hnRNPs) family of RBPs, hnRNP A2/B1, and hnRNP C, have been recognized as translational regulators with links to neurological diseases. As key regulators of RNA metabolism, hnRNP proteins are widely expressed in various tissues, including the mammalian brain (Kamma et al., 1995, 1999). In neuroblastoma cells, hnRNP C was shown to enhance the translation of mRNA that encode amyloid precursor protein (APP), a protein that, when aberrantly processed, is the major constituent of cerebral amyloid plaques found in patients with AD (Lee et al., 2010). By associating with the same region of APP mRNA as FMRP, hnRNP C prevents FMRP from binding to and silencing the translation of APP mRNA. Another hnRNP family member, hnRNP A2/B1, promotes translation through several mechanisms. It has been show to mediate transport of specific mRNAs to distant dendrites where they are translated to produce a local supply of proteins that are required for synaptic plasticity (Gao et al., 2008). HnRNP A2/B1 was also shown to control the translation of mRNAs encoding myelin basic protein and c-Myc through mechanisms that are not well defined (Kwon et al., 1999; Shi et al., 2011). Similar to TDP-43, hnRNP A2/B1 is present within brain cell inclusions of some patients with FXTAS, suggesting that the pathogenic functions of hnRNP A2/B1 involve its mislocalization (Iwahashi et al., 2006).

Several other neurological disease-related RBPs are suggested to have roles in translation, yet direct experimental evidence to support such roles is lacking. One example is survival of motor neuron (SMN), an RBP that is mutated in patients with SMA (Wirth, 2000). SMA is a fatal autosomal recessive disorder characterized by degeneration of lower motor neurons that ultimately results in paralysis with muscular atrophy (Lunn and Wang, 2008). SMN is involved in the assembly of the spliceosome, a complex that carries out gene splicing, and is also required in the formation of SGs (Liu et al., 1997; Pellizzoni et al., 1998; Hua and Zhou, 2004). In addition, SMN is found in complexes with FMRP, and through this association, SMN is hypothesized to mediate translation (Piazzon et al., 2008). Loss-of-function mutations in the gene encoding the RBP immunoglobulin μ-binding protein 2 (IGHMBP2) are also associated with SMA (Grohmann et al., 2001). Biochemical analysis of IGHMBP2 determined that it physically associates with regulators of tRNA transcription and ribosome biogenesis, suggesting strong ties to translational regulation (de Planell-Saguer et al., 2009). The mechanism by which defects in SMN and IGHMBP2 cause SMA or whether either protein has a defined role in translation remains unknown.

CONCLUSION

Although progress has been made in understanding translational regulation in neurons, many questions remain unaddressed. Specifically, our knowledge of RBP function in translation is very limited. We have yet to discover direct targets of many neuron-related RBPs and determine how these RBPs influence ribosome kinetics and the distribution of mRNAs with different mRNP complexes. We also do not know whether misregulation of certain RBPs, an event often associated with neurological diseases, affects global translation and, if so, whether aberrant translational regulation is an underlying mechanism leading to disease. Genome-wide methods are well suited to address these questions, and can provide highly detailed information that will reveal novel translational control mechanisms. The tool set of genome-wide methods to study translational control is extensive, and improvements to both the molecular and computational components of these techniques are ongoing. Since genome-wide studies have the capacity to produce an overwhelming amount of data, it is important that key findings be independently validated by molecular techniques. Such complementation between genome-wide and molecular studies will undoubtedly provide insights to the basis for translational regulation in neurons.

ACKNOWLEDGMENTS

We thank Amy Pasquinelli (University of California, San Diego), Karen Wehner, Nicholas Guydosh, and Rachel Green (Johns Hopkins University), and members of the Yeo lab, specifically Julia Nussbacher, Yan Song, and Gabriel Pratt for their comments on the manuscript. This work was partially supported by grants from the US National Institutes of Health (HG004659, NS075449, and GM084317) and the California Institute of Regenerative Medicine (RB1-01413 and RB3-05009) to Gene W. Yeo. Gene W. Yeo is an Alfred P. Sloan Research Fellow.

REFERENCES

Anderson, L., and Seilhamer, J. (1997). A comparison of selected mRNA and protein abundances in human liver. *Electrophoresis* 18, 533–537.

Arai, T., Hasegawa, M., Akiyama, H., Ikeda, K., Nonaka, T., Mori, H., et al. (2006). TDP-43 is a component of ubiquitin-positive tau-negative inclusions in frontotemporal lobar degeneration and amyotrophic lateral sclerosis. *Biochem. Biophys. Res. Commun.* 351, 602–611.

Arribere, J. A., Doudna, J. A., and Gilbert, W. V. (2011). Reconsidering movement of eukaryotic mRNAs between polysomes and P bodies. *Mol. Cell* 44, 745–758.

Ascano, M., Hafner, M., Cekan, P., Gerstberger, S., and Tuschl, T. (2012). Identification of RNA-protein interaction networks using PAR-CLIP. *Wiley Interdiscip. Rev. RNA* 3, 159–177.

Bagni, C., and Greenough, W. T. (2005). From mRNP trafficking to spine dysmorphogenesis: the roots of fragile X syndrome. *Nat. Rev. Neurosci.* 6, 376–387.

Balzer, E., and Moss, E. G. (2007). Localization of the developmental timing regulator Lin28 to mRNP complexes, P-bodies and stress granules. *RNA Biol.* 4, 16–25.

Banko, J. L., Poulin, F., Hou, L., DeMaria, C. T., Sonenberg, N., and Klann, E. (2005). The translation repressor 4E-BP2 is critical for eIF4F complex formation, synaptic plasticity, and memory in the hippocampus. *J. Neurosci.* 25, 9581–9590.

Bassell, G. J., and Warren, S. T. (2008). Fragile X syndrome: loss of local mRNA regulation alters synaptic development and function. *Neuron* 60, 201–214.

Battelli, C., Nikopoulos, G. N., Mitchell, J. G., and Verdi, J. M. (2006). The RNA-binding protein Musashi-1 regulates neural development through the translational repression of p21WAF-1. *Mol. Cell. Neurosci.* 31, 85–96.

Bilican, B., Serio, A., Barmada, S. J., Nishimura, A. L., Sullivan, G. J., Carrasco, M., et al. (2012). Mutant induced pluripotent stem cell lines recapitulate aspects of TDP-43 proteinopathies and reveal cell-specific vulnerability. *Proc. Natl. Acad. Sci. U.S.A.* 109, 5803–5808.

Braat, A. K., Yan, N., Arn, E., Harrison, D., and Macdonald, P. M. (2004). Localization-dependent oskar protein accumulation; control after the initiation of translation. *Dev. Cell* 7, 125–131.

Brengues, M., and Parker, R. (2007). Accumulation of polyadenylated mRNA, Pab1p, eIF4E, and eIF4G with P-bodies in *Saccharomyces cerevisiae. Mol. Biol. Cell* 18, 2592–2602.

Brengues, M., Teixeira, D., and Parker, R. (2005). Movement of eukaryotic mRNAs between polysomes and cytoplasmic processing bodies. *Science* 310, 486–489.

Brodersen, P., and Voinnet, O. (2009). Revisiting the principles of microRNA target recognition and mode of action. *Nat. Rev. Mol. Cell Biol.* 10, 141–148.

Buchan, J. R., and Parker, R. (2009). Eukaryotic stress granules: the ins and outs of translation. *Mol. Cell* 36, 932–941.

Castello, A., Fischer, B., Eichelbaum, K., Horos, R., Beckmann, B. M., Strein, C., et al. (2012). Insights into RNA biology from an atlas of mammalian mRNA-binding proteins. *Cell* 149, 1393–1406.

Caudy, A. A., Myers, M., Hannon, G. J., and Hammond, S. M. (2002). Fragile X-related protein and VIG associate with the RNA interference machinery. *Genes Dev.* 16, 2491–2496.

Ceman, S., O'Donnell, W. T., Reed, M., Patton, S., Pohl, J., and Warren, S. T. (2003). Phosphorylation influences the translation state of FMRP-associated polyribosomes. *Hum. Mol. Genet.* 12, 3295–3305.

Chang, K.-J., and Wang, C.-C. (2004). Translation initiation from a naturally occurring non-AUG codon in *Saccharomyces cerevisiae. J. Biol. Chem.* 279, 13778–13785.

Cheever, A., and Ceman, S. (2009). Phosphorylation of FMRP inhibits association with Dicer. *RNA* 15, 362–366.

Chi, S. W., Zang, J. B., Mele, A., and Darnell, R. B. (2009). Argonaute HITS-CLIP decodes microRNA-mRNA interaction maps. *Nature* 460, 479–486.

Coller, J., and Parker, R. (2005). General translational repression by activators of mRNA decapping. *Cell* 122, 875–886.

Darnell, J. C., Van Driesche, S. J., Zhang, C., Hung, K. Y. S., Mele, A., Fraser, C. E., Stone, E. F., et al. (2011). FMRP stalls ribosomal translocation on mRNAs linked to synaptic function and autism. *Cell* 146, 247–261.

Darnell, R. B. (2010a). HITS-CLIP: panoramic views of protein-RNA regulation in living cells. *Wiley Interdiscip. Rev. RNA* 1, 266–286.

Darnell, R. B. (2010b). RNA regulation in neurologic disease and cancer. *Cancer Res. Treat.* 42, 125–129.

De Boulle, K., Verkerk, A. J., Reyniers, E., Vits, L., Hendrickx, J., Van Roy, B., et al. (1993). A point mutation in the FMR-1 gene associated with fragile X mental retardation. *Nat. Genet.* 3, 31–35.

de Planell-Saguer, M., Schroeder, D. G., Rodicio, M. C., Cox, G. A., and Mourelatos, Z. (2009). Biochemical and genetic evidence for a role of IGHMBP2 in the translational machinery. *Hum. Mol. Genet.* 18, 2115–2126.

Deng, P.-Y., Sojka, D., and Klyachko, V. A. (2011). Abnormal presynaptic short-term plasticity and information processing in a mouse model of fragile X syndrome. *J. Neurosci.* 31, 10971–10982.

Doyle, J. P., Dougherty, J. D., Heiman, M., Schmidt, E. F., Stevens, T. R., Ma, G., et al. (2008). Application of a translational profiling approach for the comparative analysis of CNS cell types. *Cell* 135, 749–762.

Edbauer, D., Neilson, J. R., Foster, K. A., Wang, C.-F., Seeburg, D. P., Batterton, M. N., et al. (2010). Regulation of synaptic structure and function by FMRP-associated microRNAs miR-125b and miR-132. *Neuron* 65, 373–384.

Erickson, S. L., and Lykke-Andersen, J. (2011). Cytoplasmic mRNP granules at a glance. *J. Cell Sci.* 124, 293–297.

Fiesel, F. C., Weber, S. S., Supper, J., Zell, A., and Kahle, P. J. (2012). TDP-43 regulates global translational yield by splicing of exon junction complex component SKAR. *Nucleic Acids Res.* 40, 2668–2682.

Fresno, M., Jiménez, A., and Vázquez, D. (1977). Inhibition of translation in eukaryotic systems by harringtonine. *Eur. J. Biochem.* 72, 323–330.

Gao, Y., Tatavarty, V., Korza, G., Levin, M. K., and Carson, J. H. (2008). Multiplexed dendritic targeting of α calcium calmodulin-dependent protein kinase II, neurogranin, and activity-regulated cytoskeleton-associated protein RNAs by the A2 pathway. *Mol. Biol. Cell* 19, 2311–2327.

Gitcho, M. A., Baloh, R. H., Chakraverty, S., Mayo, K., Norton, J. B., Levitch, D., et al. (2008). TDP-43 A315T mutation in familial motor neuron disease. *Ann. Neurol.* 63, 535–538.

Giuditta, A., Kaplan, B. B., van Minnen, J., Alvarez, J., and Koenig, E. (2002). Axonal and presynaptic protein synthesis: new insights into the biology of the neuron. *Trends Neurosci.* 25, 400–404.

Granneman, S., Kudla, G., Petfalski, E., and Tollervey, D. (2009). Identification of protein binding sites on U3 snoRNA and pre-rRNA by UV cross-linking and high-throughput analysis of cDNAs. *Proc. Natl. Acad. Sci. U.S.A.* 106, 9613–9618.

Grohmann, K., Schuelke, M., Diers, A., Hoffmann, K., Lucke, B., Adams, C., et al. (2001). Mutations in the gene encoding immunoglobulin mu-binding protein 2 cause spinal muscular atrophy with respiratory distress type 1. *Nat. Genet.* 29, 75–77.

Guo, H., Ingolia, N. T., Weissman, J. S., and Bartel, D. P. (2010). Mammalian microRNAs predominantly act to decrease target mRNA levels. *Nature* 466, 835–840.

Gygi, S. P., Rochon, Y., Franza, B. R., and Aebersold, R. (1999). Correlation between protein and mRNA abundance in yeast. *Mol. Cell. Biol.* 19, 1720–1730.

Hafner, M., Landthaler, M., Burger, L., Khorshid, M., Hausser, J., Berninger, P., et al. (2010). Transcriptome-wide identification of RNA-binding protein and microRNA target sites by PAR-CLIP. *Cell* 141, 129–141.

Heiman, M., Schaefer, A., Gong, S., Peterson, J. D., Day, M., Ramsey, K. E., et al. (2008). A translational profiling approach for the molecular characterization of CNS cell types. *Cell* 135, 738–748.

Hitoshi, S., Alexson, T., Tropepe, V., Donoviel, D., Elia, A. J., Nye, J. S., et al. (2002). Notch pathway molecules are essential for the maintenance, but not the generation, of mammalian neural stem cells. *Genes Dev.* 16, 846–858.

Hoyle, N. P., Castelli, L. M., Campbell, S. G., Holmes, L. E. A., and Ashe, M. P. (2007). Stress-dependent relocalization of translationally primed mRNPs to cytoplasmic granules that are kinetically and spatially distinct from P-bodies. *J. Cell Biol.* 179, 65–74.

Hsieh, A. C., Liu, Y., Edlind, M. P., Ingolia, N. T., Janes, M. R., Sher, A., et al. (2012). The translational landscape of mTOR signalling steers cancer initiation and metastasis. *Nature* 485, 55–61.

Hua, Y., and Zhou, J. (2004). Survival motor neuron protein facilitates assembly of stress granules. *FEBS Lett.* 572, 69–74.

Huang, C., Tong, J., Bi, F., Zhou, H., and Xia, X.-G. (2012). Mutant TDP-43 in motor neurons promotes the onset and progression of ALS in rats. *J. Clin. Invest.* 122, 107–118.

Huelga, S. C., Vu, A. Q., Arnold, J. D., Liang, T. Y., Liu, P. P., Yan,

B. Y., et al. (2012). Integrative genome-wide analysis reveals cooperative regulation of alternative splicing by hnRNP proteins. *Cell Rep.* 1, 167–178.

Huttelmaier, S., Zenklusen, D., Lederer, M., Dictenberg, J., Lorenz, M., Meng, X., et al. (2005). Spatial regulation of I[beta]I-actin translation by Src-dependent phosphorylation of ZBP1. *Nature* 438, 512–515.

Iguchi, N., Tobias, J. W., and Hecht, N. B. (2006). Expression profiling reveals meiotic male germ cell mRNAs that are translationally up- and down-regulated. *Proc. Natl. Acad. Sci. U.S.A.* 103, 7712–7717.

Imai, T., Tokunaga, A., Yoshida, T., Hashimoto, M., Mikoshiba, K., Weinmaster, G., et al. (2001). The neural RNA-binding protein Musashi1 translationally regulates mammalian numb gene expression by interacting with its mRNA. *Mol. Cell. Biol.* 21, 3888–3900.

Ingolia, N. T., Brar, G. A., Rouskin, S., McGeachy, A. M., and Weissman, J. S. (2012). The ribosome profiling strategy for monitoring translation in vivo by deep sequencing of ribosome-protected mRNA fragments. *Nat. Protoc.* 7, 1534–1550.

Ingolia, N. T., Ghaemmaghami, S., Newman, J. R. S., and Weissman, J. S. (2009). Genome-wide analysis in vivo of translation with nucleotide resolution using ribosome profiling. *Science* 324, 218–223.

Ingolia, N. T., Lareau, L. F., and Weissman, J. S. (2011). Ribosome profiling of mouse embryonic stem cells reveals the complexity and dynamics of mammalian proteomes. *Cell* 147, 789–802.

Iwahashi, C. K., Yasui, D. H., An, H.-J., Greco, C. M., Tassone, F., Nannen, K., et al. (2006). Protein composition of the intranuclear inclusions of FXTAS. *Brain* 129, 256–271.

Jensen, C. J., Oldfield, B. J., and Rubio, J. P. (2009). Splicing, cis genetic variation and disease. *Biochem. Soc. Trans.* 37, 1311.

Johannes, G., Carter, M. S., Eisen, M. B., Brown, P. O., and Sarnow, P. (1999). Identification of eukaryotic mRNAs that are translated at reduced cap binding complex eIF4F concentrations using a cDNA microarray. *Proc. Natl. Acad. Sci. U.S.A.* 96, 13118–13123.

Jungkamp, A.-C., Stoeckius, M., Mecenas, D., Grün, D., Mastrobuoni, G., Kempa, S., et al. (2011). In vivo and transcriptome-wide identification of RNA binding protein target sites. *Mol. Cell* 44, 828–840.

Kabashi, E., Valdmanis, P. N., Dion, P., Spiegelman, D., McConkey, B. J., Vande Velde, C., et al. (2008). TARDBP mutations in individuals with sporadic and familial amyotrophic lateral sclerosis. *Nat. Genet.* 40, 572–574.

Kamma, H., Horiguchi, H., Wan, L., Matsui, M., Fujiwara, M., Fujimoto, M., et al. (1999). Molecular characterization of the hnRNP A2/B1 proteins: tissue-specific expression and novel isoforms. *Exp. Cell Res.* 246, 399–411.

Kamma, H., Portman, D. S., and Dreyfuss, G. (1995). Cell type-specific expression of hnRNP proteins. *Exp. Cell Res.* 221, 187–196.

Katz, Y., Wang, E. T., Airoldi, E. M., and Burge, C. B. (2010). Analysis and design of RNA sequencing experiments for identifying isoform regulation. *Nat. Methods* 7, 1009–1015.

Kawahara, H., Imai, T., Imataka, H., Tsujimoto, M., Matsumoto, K., and Okano, H. (2008). Neural RNA-binding protein Musashi1 inhibits translation initiation by competing with eIF4G for PABP. *J. Cell Biol.* 181, 639–653.

Kedersha, N., and Anderson, P. (2009). Regulation of translation by stress granules and processing bodies. *Prog. Mol. Biol. Transl. Sci.* 90, 155–185.

Kedersha, N., Stoecklin, G., Ayodele, M., Yacono, P., Lykke-Andersen, J., Fritzler, M. J., et al. (2005). Stress granules and processing bodies are dynamically linked sites of mRNP remodeling. *J. Cell Biol.* 169, 871–884.

Kelleher, R. J., Govindarajan, A., Jung, H.-Y., Kang, H., and Tonegawa, S. (2004). Translational control by MAPK signaling in long-term synaptic plasticity and memory. *Cell* 116, 467–479.

Kiebler, M. A., and Bassell, G. J. (2006). Neuronal RNA granules: movers and makers. *Neuron* 51, 685–690.

Kiebler, M. A., and DesGroseillers, L. (2000). Molecular insights into mRNA transport and local translation in the mammalian nervous system. *Neuron* 25, 19–28.

Kishore, S., Jaskiewicz, L., Burger, L., Hausser, J., Khorshid, M., and Zavolan, M. (2011). A quantitative analysis of CLIP methods for identifying binding sites of RNA-binding proteins. *Nat. Methods* 8, 559–564.

Klann, E., and Dever, T. E. (2004). Biochemical mechanisms for translational regulation in synaptic plasticity. *Nat. Rev. Neurosci.* 5, 931–942.

König, J., Zarnack, K., Luscombe, N. M., and Ule, J. (2012). Protein–RNA interactions: new genomic technologies and perspectives. *Nat. Rev. Genet.* 13, 77–83.

König, J., Zarnack, K., Rot, G., Curk, T., Kayikci, M., Zupan, B., et al. (2010). iCLIP reveals the function of hnRNP particles in splicing at individual nucleotide resolution. *Nat. Struct. Mol. Biol.* 17, 909–915.

Krichevsky, A. M., and Kosik, K. S. (2001). Neuronal RNA Granules: A Link between RNA Localization and Stimulation-Dependent Translation. *Neuron* 32, 683–696.

Kwon, S., Barbarese, E., and Carson, J. H. (1999). The cis-acting RNA trafficking signal from myelin basic protein mRNA and its cognate trans-acting ligand hnRNP A2 enhance cap-dependent translation. *J. Cell Biol.* 147, 247–256.

Lagier-Tourenne, C., Polymenidou, M., and Cleveland, D. W. (2010). TDP-43 and FUS/TLS: emerging roles in RNA processing and neurodegeneration. *Hum. Mol. Genet.* 19, R46–64.

Larsson, O., and Nadon, R. (2008). Gene expression – time to change point of view? *Biotechnol. Genet. Eng. Rev.* 25, 77–92.

Lee, E. K., Kim, H. H., Kuwano, Y., Abdelmohsen, K., Srikantan, S., Subaran, S. S., et al. (2010). hnRNP C promotes APP translation by competing with FMRP for APP mRNA recruitment to P bodies. *Nat. Struct. Mol. Biol.* 17, 732–739.

Leung, A. K. L., Young, A. G., Bhutkar, A., Zheng, G. X., Bosson, A. D., Nielsen, C. B., et al. (2011). Genome-wide identification of Ago2 binding sites from mouse embryonic stem cells with and without mature microRNAs. *Nat. Struct. Mol. Biol.* 18, 237–244.

Li, G.-W., Oh, E., and Weissman, J. S. (2012). The anti-Shine-Dalgarno sequence drives translational pausing and codon choice in bacteria. *Nature* 484, 538–541.

Licatalosi, D. D., Mele, A., Fak, J. J., Ule, J., Kayikci, M., Chi, S. W., et al. (2008). HITS-CLIP yields genome-wide insights into brain alternative RNA processing. *Nature* 456, 464–469.

Liu, J., Valencia-Sanchez, M. A., Hannon, G. J., and Parker, R. (2005). MicroRNA-dependent localization of targeted mRNAs to mammalian P-bodies. *Nat. Cell Biol.* 7, 719–723.

Liu, Q., Fischer, U., Wang, F., and Dreyfuss, G. (1997). The spinal muscular atrophy disease gene product, SMN, and its associated protein SIP1 are in a complex with spliceosomal snRNP proteins. *Cell* 90, 1013–1021.

Liu-Yesucevitz, L., Bassell, G. J., Gitler, A. D., Hart, A. C., Klann, E., Richter, J. D., et al. (2011). Local RNA translation at the synapse and in disease. *J. Neurosci.* 31, 16086–16093.

Lodish, H. F. (1971). Alpha and beta globin messenger ribonucleic acid. Different amounts and rates of initiation of translation. *J. Biol. Chem.* 246, 7131–7138.

Lovett, P. S., and Rogers, E. J. (1996). Ribosome regulation by the nascent peptide. *Microbiol. Rev.* 60, 366–385.

Lukong, K. E., Chang, K., Khandjian, E. W., and Richard, S. (2008). RNA-binding proteins in human genetic disease. *Trends Genet.* 24, 416–425.

Lunn, M. R., and Wang, C. H. (2008). Spinal muscular atrophy. *Lancet* 371, 2120–2133.

Maroney, P. A., Yu, Y., Fisher, J., and Nilsen, T. W. (2006). Evidence that microRNAs are associated with translating messenger RNAs in human cells. *Nat. Struct. Mol. Biol.* 13, 1102–1107.

McKee, A. E., Minet, E., Stern, C., Riahi, S., Stiles, C. D., and Silver, P. A. (2005). A genome-wide in situ hybridization map of RNA-binding proteins reveals anatomically restricted expression in the developing mouse brain. *BMC Dev. Biol.* 5, 14. doi:10.1186/1471-213X-5-14

McNulty, D. E., Claffee, B. A., Huddleston, M. J., Porter, M. L., Cavnar, K. M., and Kane, J. F. (2003). Mistranslational errors associated with the rare arginine codon CGG in *Escherichia coli*. *Protein Expr. Purif.* 27, 365–374.

Meisenheimer, K. M., and Koch, T. H. (1997). Photocross-linking of nucleic acids to associated proteins. *Crit. Rev. Biochem. Mol. Biol.* 32, 101–140.

Meisenheimer, K. M., Meisenheimer, P. L., and Koch, T. H. (2000). Nucleoprotein photo-cross-linking using halopyrimidine-substituted RNAs. *Meth. Enzymol.* 318, 88–104.

Mili, S., and Steitz, J. A. (2004). Evidence for reassociation of RNA-binding proteins after cell lysis: implications for the interpretation of immunoprecipitation analyses. *RNA* 10, 1692–1694.

Ming, G.-L., Brüstle, O., Muotri, A., Studer, L., Wernig, M., and Christian, K. M. (2011). Cellular reprogramming: recent advances in modeling neurological diseases. *J. Neurosci.* 31, 16070–16075.

Morris, D. R. (2009). Ribosomal footprints on a transcriptome landscape. *Genome Biol.* 10, 215.

Muddashetty, R. S., Nalavadi, V. C., Gross, C., Yao, X., Xing, L., Laur, et al. (2011). Reversible inhibition of PSD-95 mRNA translation by miR-125a, FMRP phosphorylation, and mGluR signaling. *Mol. Cell* 42, 673–688.

Narayanan, U., Nalavadi, V., Nakamoto, M., Thomas, G., Ceman, S., Bassell, G. J., et al. (2008). S6K1 Phosphorylates and regulates fragile X mental retardation protein (FMRP) with the neuronal protein synthesis-dependent mammalian target of rapamycin (mTOR) signaling cascade. *J. Biol. Chem.* 283, 18478–18482.

Neumann, M., Sampathu, D. M., Kwong, L. K., Truax, A. C., Micsenyi, M. C., Chou, T. T., et al. (2006). Ubiquitinated TDP-43 in frontotemporal lobar degeneration and amyotrophic lateral sclerosis. *Science* 314, 130–133.

Nottrott, S., Simard, M. J., and Richter, J. D. (2006). Human let-7a miRNA blocks protein production on actively translating polyribosomes. *Nat. Struct. Mol. Biol.* 13, 1108–1114.

Obrig, T. G., Culp, W. J., McKeehan, W. L., and Hardesty, B. (1971). The mechanism by which cycloheximide and related glutarimide antibiotics inhibit peptide synthesis on reticulocyte ribosomes. *J. Biol. Chem.* 246, 174–181.

Oh, E., Becker, A. H., Sandikci, A., Huber, D., Chaba, R., Gloge, F., et al. (2011). Selective ribosome profiling reveals the cotranslational chaperone action of trigger factor in vivo. *Cell* 147, 1295–1308.

Okano, H., Imai, T., and Okabe, M. (2002). Musashi: a translational regulator of cell fate.

Okaty, B. W., Sugino, K., and Nelson, S. B. (2011). Cell type-specific transcriptomics in the brain. *J. Neurosci.* 31, 6939–6943.

Olsen, P. H., and Ambros, V. (1999). The lin-4 regulatory RNA controls developmental timing in *Caenorhabditis elegans* by blocking LIN-14 protein synthesis after the initiation of translation. *Dev. Biol.* 216, 671–680.

Parker, R., and Sheth, U. (2007). P bodies and the control of mRNA translation and degradation. *Mol. Cell* 25, 635–646.

Pellizzoni, L., Kataoka, N., Charroux, B., and Dreyfuss, G. (1998). A novel function for SMN, the spinal muscular atrophy disease gene product, in pre-mRNA splicing. *Cell* 95, 615–624.

Perry, E. K., Johnson, M., Ekonomou, A., Perry, R. H., Ballard, C., and Attems, J. (2012). Neurogenic abnormalities in Alzheimer's disease differ between stages of neurogenesis and are partly related to cholinergic pathology. *Neurobiol. Dis.* 47, 155–162.

Petersen, C. P., Bordeleau, M.-E., Pelletier, J., and Sharp, P. A. (2006). Short RNAs repress translation after initiation in mammalian cells. *Mol. Cell* 21, 533–542.

Piazzon, N., Rage, F., Schlotter, F., Moine, H., Branlant, C., and Massenet, S. (2008). In vitro and in cellulo evidences for association of the survival of motor neuron complex with the fragile X mental retardation protein. *J. Biol. Chem.* 283, 5598–5610.

Polymenidou, M., Lagier-Tourenne, C., Hutt, K. R., Huelga, S. C., Moran, J., Liang, T. Y., et al. (2011). Long pre-mRNA depletion and RNA missplicing contribute to neuronal vulnerability from loss of TDP-43. *Nat. Neurosci.* 14, 459–468.

Poon, M. M., Choi, S.-H., Jamieson, C. A. M., Geschwind, D. H., and Martin, K. C. (2006). Identification of process-localized mRNAs from cultured rodent hippocampal neurons. *J. Neurosci.* 26, 13390–13399.

Preiss, T., Baron-Benhamou, J., Ansorge, W., and Hentze, M. W. (2003). Homodirectional changes in transcriptome composition and mRNA translation induced by rapamycin and heat shock. *Nat. Struct. Biol.* 10, 1039–1047.

Rajasekhar, V. K., Viale, A., Socci, N. D., Wiedmann, M., Hu, X., and Holland, E. C. (2003). Oncogenic Ras and Akt signaling contribute to glioblastoma formation by differential recruitment of existing mRNAs to polysomes. *Mol. Cell* 12, 889–901.

Richter, J. D., and Klann, E. (2009). Making synaptic plasticity and memory last: mechanisms of translational regulation. *Genes Dev.* 23, 1–11.

Sakakibara, S., Nakamura, Y., Yoshida, T., Shibata, S., Koike, M., Takano, H., et al. (2002). RNA-binding protein Musashi family: roles for CNS stem cells and a subpopulation of ependymal cells revealed by targeted disruption and antisense ablation. *Proc. Natl. Acad. Sci. U.S.A.* 99, 15194–15199.

Sanz, E., Yang, L., Su, T., Morris, D. R., McKnight, G. S., and Amieux, P. S. (2009). Cell-type-specific isolation of ribosome-associated mRNA from complex tissues. *Proc. Natl. Acad. Sci. U.S.A.* 106, 13939–13944.

Scharf, M. T., Woo, N. H., Lattal, K. M., Young, J. Z., Nguyen, P. V., and Abel, T. (2002). Protein synthesis is required for the enhancement of long-term potentiation and long-term memory by spaced training. *J. Neurophysiol.* 87, 2770–2777.

Schratt, G. M., Nigh, E. A., Chen, W. G., Hu, L., and Greenberg, M. E. (2004). BDNF regulates the translation of a select group of mRNAs by a mammalian target of rapamycin-phosphatidylinositol 3-kinase-dependent pathway during neuronal development. *J. Neurosci.* 24, 7366–7377.

Shenton, D., Smirnova, J. B., Selley, J. N., Carroll, K., Hubbard, S. J., Pavitt, G. D., et al. (2006). Global translational responses to oxidative stress impact upon multiple levels of protein synthesis. *J. Biol. Chem.* 281, 29011–29021.

Shi, Y., Frost, P., Hoang, B., Benavides, A., Gera, J., and Lichtenstein, A. (2011). IL-6-induced enhancement of c-Myc translation in multiple myeloma cells: critical role of cytoplasmic localization of the rna-binding protein hnRNP A1. *J. Biol. Chem.* 286, 67–78.

Sivan, G., Kedersha, N., and Elroy-Stein, O. (2007). Ribosomal slow-down mediates translational arrest during cellular division. *Mol. Cell. Biol.* 27, 6639–6646.

Snow, K., Doud, L. K., Hagerman, R., Pergolizzi, R. G., Erster, S. H., and Thibodeau, S. N. (1993). Analysis of a CGG sequence at the FMR-1 locus in fragile X families and in the general population. *Am. J. Hum. Genet.* 53, 1217–1228.

Sossin, W. S., and DesGroseillers, L. (2006). Intracellular trafficking of RNA in neurons. *Traffic* 7, 1581–1589.

Sreedharan, J., Blair, I. P., Tripathi, V. B., Hu, X., Vance, C., Rogelj, B., et al. (2008). TDP-43 mutations in familial and sporadic amyotrophic lateral sclerosis. *Science* 319, 1668–1672.

Steward, O., and Levy, W. (1982). Preferential localization of polyribosomes under the base of dendritic spines in granule cells of the dentate gyrus. *J. Neurosci.* 2, 284–291.

Sutton, M. A., and Schuman, E. M. (2006). Dendritic protein synthesis, synaptic plasticity, and memory. *Cell* 127, 49–58.

Tang, H.-L., Yeh, L.-S., Chen, N.-K., Ripmaster, T., Schimmel, P., and Wang, C.-C. (2004). Translation of a yeast mitochondrial tRNA synthetase initiated at redundant non-AUG codons. *J. Biol. Chem.* 279, 49656–49663.

Teixeira, D., Sheth, U., Valencia-Sanchez, M. A., Brengues, M., and Parker, R. (2005). Processing bodies require RNA for assembly and contain nontranslating mRNAs. *RNA* 11, 371–382.

Thomas, M. G., Tosar, L. J. M., Desbats, M. A., Leishman, C. C., and Boccaccio, G. L. (2009). Mammalian Staufen 1 is recruited to stress granules and impairs their assembly. *J. Cell. Sci.* 122, 563–573.

Thonhoff, J. R., Ojeda, L., and Wu, P. (2009). Stem cell-derived motor neurons: applications and challenges in amyotrophic lateral sclerosis. *Curr. Stem Cell Res. Ther.* 4, 178–199.

Tollervey, J. R., Curk, T., Rogelj, B., Briese, M., Cereda, M., Kayikci, M., et al. (2011). Characterizing the RNA targets and position-dependent splicing regulation by TDP-43. *Nat. Neurosci.* 14, 452–458.

Ule, J., Jensen, K. B., Ruggiu, M., Mele, A., Ule, A., and Darnell, R. B. (2003). CLIP identifies Nova-regulated RNA networks in the brain. *Science* 302, 1212–1215.

Valencia-Sanchez, M. A., Liu, J., Hannon, G. J., and Parker, R. (2006). Control of translation and mRNA degradation by miRNAs and siRNAs. *Genes Dev.* 20, 515–524.

Van Deerlin, V. M., Leverenz, J. B., Bekris, L. M., Bird, T. D., Yuan, W., Elman, L. B., et al. (2008). TARDBP mutations in amyotrophic lateral sclerosis with TDP-43 neuropathology: a genetic and histopathological analysis. *Lancet Neurol.* 7, 409–416.

Wang, D. O., Martin, K. C., and Zukin, R. S. (2010). Spatially restricting gene expression by local translation at synapses. *Trends Neurosci.* 33, 173–182.

Wang, E. T., Sandberg, R., Luo, S., Khrebtukova, I., Zhang, L., Mayr, C., et al. (2008a). Alternative isoform regulation in human tissue transcriptomes. *Nature* 456, 470–476.

Wang, I.-F., Wu, L.-S., Chang, H.-Y., and Shen, C.-K. J. (2008b). TDP-43, the signature protein of FTLD-U, is a neuronal activity-responsive factor. *J. Neurochem.* 105, 797–806.

Wang, I.-F., Wu, L.-S., and Shen, C.-K. J. (2008c). TDP-43: an emerging new player in neurodegenerative diseases. *Trends. Mol. Med.* 14, 479–485.

Wilbert, M. L., and Yeo, G. W. (2011). Genome-wide approaches in the study of microRNA biology. *Wiley Interdiscip. Rev. Syst. Biol. Med.* 3, 491–512.

Williams, C., Mehrian Shai, R., Wu, Y., Hsu, Y.-H., Sitzer, T., Spann, B., et al. (2009). Transcriptome analysis of synaptoneurosomes identifies neuroplasticity genes overexpressed in incipient Alzheimer's disease. *PLoS ONE* 4, e4936. doi:10.1371/journal.pone.0004936

Willis, D., Li, K. W., Zheng, J.-Q., Chang, J. H., Smit, A., Kelly, T., et al. (2005). Differential transport and local translation of cytoskeletal, injury-response, and neurodegeneration protein mRNAs in axons. *J. Neurosci.* 25, 778–791.

Wirth, B. (2000). An update of the mutation spectrum of the survival motor neuron gene (SMN1) in autosomal recessive spinal muscular atrophy (SMA). *Hum. Mutat.* 15, 228–237.

Witold, F. (2005). RNAi: the nuts and bolts of the RISC machine. *Cell* 122, 17–20.

Wolin, S. L., and Walter, P. (1988). Ribosome pausing and stacking during translation of a eukaryotic mRNA. *EMBO J.* 7, 3559–3569.

Yeo, G., Holste, D., Kreiman, G., and Burge, C. B. (2004). Variation in alternative splicing across human tissues. *Genome Biol.* 5, R74.

Yeo, G. W., Coufal, N. G., Liang, T. Y., Peng, G. E., Fu, X.-D., and Gage, F. H. (2009). An RNA code for the FOX2 splicing regulator revealed by mapping RNA-protein interactions in stem cells. *Nat. Struct. Mol. Biol.* 16, 130–137.

Yokoseki, A., Shiga, A., Tan, C.-F., Tagawa, A., Kaneko, H., Koyama, A., et al. (2008). TDP-43 mutation in familial amyotrophic lateral sclerosis. *Ann. Neurol.* 63, 538–542.

Zhang, C., and Darnell, R. B. (2011). Mapping in vivo protein-RNA interactions at single-nucleotide resolution from HITS-CLIP data. *Nat. Biotechnol.* 29, 607–614.

Zhong, J., Zhang, T., and Bloch, L. M. (2006). Dendritic mRNAs encode diversified functionalities in hippocampal pyramidal neurons. *BMC Neurosci.* 7, 17. doi:10.1186/1471-2202-7-17

Zisoulis, D. G., Lovci, M. T., Wilbert, M. L., Hutt, K. R., Liang, T. Y., Pasquinelli, A. E., et al. (2010). Comprehensive discovery of endogenous argonaute binding sites in *Caenorhabditis elegans*. *Nat. Struct. Mol. Biol.* 17, 173–179.

Zong, Q., Schummer, M., Hood, L., and Morris, D. R. (1999). Messenger RNA translation state: the second dimension of high-throughput expression screening. *Proc. Natl. Acad. Sci. U.S.A.* 96, 10632–10636.

17

microRNAs in nociceptive circuits as predictors of future clinical applications

Michaela Kress[1]*, Alexander Hüttenhofer[2], Marc Landry[3], Rohini Kuner[4], Alexandre Favereaux[3], David Greenberg[5], Josef Bednarik[6], Paul Heppenstall[7], Florian Kronenberg[2], Marzia Malcangio[8], Heike Rittner[9], Nurcan Üçeyler[9], Zlatko Trajanoski[2], Peter Mouritzen[10], Frank Birklein[11], Claudia Sommer[9] and Hermona Soreq[12]

[1] *Department of Physiology and Medical Physics, Division of Physiology, Medical University Innsbruck, Innsbruck, Austria*
[2] *Medical University Innsbruck, Innsbruck, Austria*
[3] *UMR 5297, Interdisciplinary Institute for Neuroscience, Centre National de la Recherche Scientifique, University of Bordeaux, Bordeaux, France*
[4] *University Heidelberg, Heidelberg, Germany*
[5] *Hebrew University of Jerusalem, Jerusalem, Israel*
[6] *University Hospital Brno, Brno, Czech Republic*
[7] *The European Molecular Biology Laboratory, Monterotondo, Italy*
[8] *King's College London, London, UK*
[9] *University Hospital Würzburg, Würzburg, Germany*
[10] *Exiqon A/S, Vedbaek, Denmark*
[11] *University Hospital Mainz, Mainz, Germany*
[12] *Laboratory of Molecular Neuroscience, Department of Biological chemistry, Hebrew University of Jerusalem, Jerusalem, Israel*

***Correspondence:**
Michaela Kress, Department of Physiology and Medical Physics, Division of Physiology, Medical University Innsbruck, Fritz-Pregl-Street 3, 6020 Innsbruck, Austria
e-mail: michaela.kress@i-med.ac.at

Neuro-immune alterations in the peripheral and central nervous system play a role in the pathophysiology of chronic pain, and non-coding RNAs – and microRNAs (miRNAs) in particular – regulate both immune and neuronal processes. Specifically, miRNAs control macromolecular complexes in neurons, glia and immune cells and regulate signals used for neuro-immune communication in the pain pathway. Therefore, miRNAs may be hypothesized as critically important master switches modulating chronic pain. In particular, understanding the concerted function of miRNA in the regulation of nociception and endogenous analgesia and defining the importance of miRNAs in the circuitries and cognitive, emotional and behavioral components involved in pain is expected to shed new light on the enigmatic pathophysiology of neuropathic pain, migraine and complex regional pain syndrome. Specific miRNAs may evolve as new druggable molecular targets for pain prevention and relief. Furthermore, predisposing miRNA expression patterns and inter-individual variations and polymorphisms in miRNAs and/or their binding sites may serve as biomarkers for pain and help to predict individual risks for certain types of pain and responsiveness to analgesic drugs. miRNA-based diagnostics are expected to develop into hands-on tools that allow better patient stratification, improved mechanism-based treatment, and targeted prevention strategies for high risk individuals.

Keywords: chronic pain, biomarker, polymorphism, miRNA-based diagnostics, miRNA expression patterns, miRNA polymorphisms, antagomir, miRNA-based analgesic

INTRODUCTION

Human chronic pain disorders are bio-psycho-social diseases, which are difficult to treat due to their diversity. Chronic pain syndromes that develop after nerve damage, trauma or surgery are characterized by persistent and severe pain; they induce anxiety and depression and greatly impair patients' quality of life. One out of five Europeans suffers from chronic pain with most reporting that they endure it for more than two years (Breivik et al., 2006; Baker et al., 2010). Due to direct and follow-up costs they constitute a heavy burden for the health system (Phillips, 2006).

Of the painful neuropathies, the most frequent, painful diabetic polyneuropathy is a common complication of diabetes mellitus occurring in up to 20% of patients (Sommer, 2003; Sadosky et al., 2008). Good glycemic control can reduce the incidence of diabetic polyneuropathy but not painful diabetic polyneuropathy (PDPN) for which only symptomatic therapy of low to moderate efficacy is available to date (Vincent et al., 2011). Cellular mechanisms are emerging that include the classical changes of the diabetic milieu (Bierhaus and Nawroth, 2012; Bierhaus et al., 2012) however various studies have also identified signatures of neuroinflammation as critical components of painful diabetic polyneuropathy (Pabreja et al., 2011; Vincent et al., 2011). Pathological neuro-immune communication has also been associated with painful neuropathy that occurs in up to 50% of patients with traumatic peripheral nerve injury as a consequence of accidents, warfare or surgical procedures (Myers et al., 2006; Ciaramitaro et al., 2010; Birch et al., 2012). Also the neurogenic complex regional pain syndrome (CRPS) occurring as a complication of bone fracture, tissue injury or surgical interventions has a neuro-inflammatory component (Parkitny et al., 2013).

In the majority of cases symptoms grossly resolve, however in 30% of patients pain symptoms persist or even intensify (Marinus et al., 2011). The beneficial effect of therapy with glucocorticosteroids in the acute phase of CRPS supports pathophysiological mechanisms associated with neuro-immune dysfunction (Üceyler et al., 2007a; Fischer et al., 2010; Marinus et al., 2011). Thus, converging evidence suggests that neuro-immune alterations in the peripheral and central nervous system play a major role in the general pathophysiology of neurogenic and neuropathic pain (McMahon and Malcangio, 2009; Kuner, 2010). Non-coding RNAs (ncRNAs), including microRNAs (miRNAs) and Piwi-binding piRNAs, are intimately associated with normal cellular as well as pathological processes (Mattick, 2004; Hüttenhofer et al., 2005; Hüttenhofer and Schattner, 2006). In this review we will focus on miRNAs since they are most extensively studied so far.

Various diseases, including neuropathic pain disorders, reveal unique miRNA expression signatures that can be exploited as diagnostic and prognostic markers. Recent reports on miRNA modulation of both neuronal and immune processes further predict therapeutic potential for manipulating disease-modified miRNAs in diseases affecting both the immune system and brain function, such as neuropathic pain disorders, Alzheimer's disease, Parkinson's disease, multiple sclerosis, and anxiety-related disorders (Soreq and Wolf, 2011; O'Connor et al., 2012).

miRNAs that function within both the nervous and the immune systems possibly act as "negotiators" between these two interacting compartments (**Figure 1**). These "neurimmiRs" primarily target transcription factor genes or other regulatory genes, which enables simultaneous modulation of both immune and neuronal processes including cognition through direct or indirect alterations of neuron–glia or brain-to-body signaling (Soreq and Wolf, 2011). Thus, a given miRNA controls multiple cellular pathways, and miRNAs can act as "master switches" of the transcriptome or proteome, regulating multiple gene products and orchestrating multiple pathways including genes that encode cellular enzymes, trophic factors, receptor proteins, and ion channels many of which are individually pursued as drug targets.

Pain conditions have been suggested to deregulate the expression of miRNAs in pain pathways from primary afferent nociceptors to brain areas associated with emotional components of pain perception (Bai et al., 2007; Aldrich et al., 2009; Kusuda et al., 2011; Imai et al., 2011; Poh et al., 2011; von Schack et al., 2011). miRNAs are frequently deregulated and expressed at aberrant levels in diseased tissue, and first evidence suggests that this applies to neurogenic pain in CRPS (Orlova et al., 2011). Altered miRNA expression is frequently a consequence of genetic mutations, which may also cause loss or gain of function (Mishra and Bertino, 2009). This may account for inter-individual variation of pain sensitivity. However, the functional consequences of polymorphisms in miRNA genes and/or their binding sites, the downstream targets of miRNAs and the mechanisms by which miRNAs regulate circuitries and processes modulating nociception and endogenous analgesia are as yet unresolved.

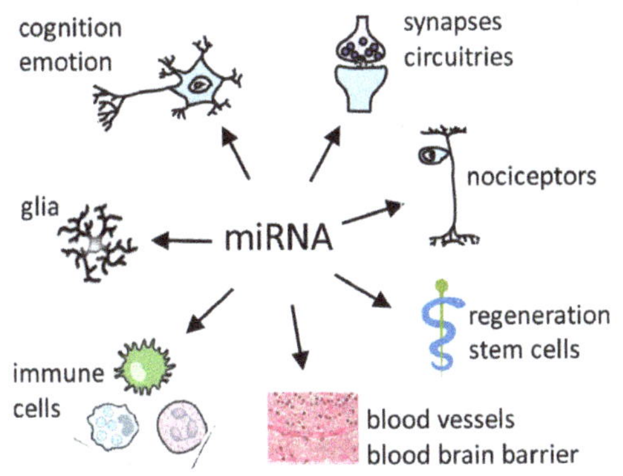

FIGURE 1 | Targets of miRNA regulation networks.

Therapeutic miRNA regulation has been thoroughly studied and widely established in cancer research but its impact and the therapeutic prospects of miRNAs in the pain field are largely unexplored. Manipulation of miRNAs offers the possibility to control multiple targets including neuro-immune interactions, nociceptive processing and cognitive pathways. Both miRNAs and their isomiRNA versions are likely to each interact with many different targets, which may lead to downstream changes either due to the direct suppression of these targets or because of regulatory effects of those targets. Such downstream effects may be rather elaborate and are defined by some researchers "off-target" effects. However, we find that this definition may be misleading as it assumes that the physiological role of each miRNA is limited to the suppression of its direct targets. It is expected that miRNAs and miRNA derivatives will have few, if any, sequence-specific "off-target" effects. Thus, miRNA based diagnostics and therapeutics may have superior advantages by targeting multiple pain-associated genes and miRNA-based drugs may be the most appropriate therapy for the prevention or treatment of neuropathic pain.

BIOMARKERS FOR NEUROPATHIC AND NEUROGENIC PAIN SYNDROMES

Painful diabetic polyneuropathy is the most frequent painful neuropathy occurring in up to 20% of diabetic patients (Sommer, 2003; Sadosky et al., 2008). CRPS is an extremely painful condition that occurs in some patients after bone or tissue injury and peripheral nerve injury (traumatic neuropathy) and results in chronic neuropathic pain in many of these patients. These well-characterized albeit aetiologically diverse (metabolic, inflammatory, traumatic) neuropathic/neurogenic pain syndromes cover a spectrum of mechanisms underlying chronic pain. Nevertheless, the medical need for these syndromes is prevalent, and each of them is prototypic for an entire group of pain disorders.

It is unclear why diabetic neuropathy or traumatic neuropathy are painful in some instances and painless in others or why some patients develop CRPS after bone fracture, and why some recover from CRPS and others do not (Marinus et al., 2011). Thus, as yet unknown factors determine whether a given disorder entails chronic neuropathic pain. A first approach to be able to predict the individual risk of pain chronification was to use sensory phenotypes as surrogate markers for possible underlying mechanisms. Quantitative sensory testing (QST) is now well established but is still insufficient to disentangle specific pathophysiological mechanisms of chronic pain (Baron et al., 2012). One of the major hindrances in translating such findings into better therapy of neuropathic and neurogenic pain syndromes is the complexity of their pathophysiology. It is well known that alterations in many processes including ion channels, inflammatory mediators, neurotrophic factors, synaptic plasticity, and de- and regeneration, are involved, and that they even change during the course of the disease (Hehn et al., 2012). Therefore, a search for better and more specific diagnostic trait and state markers is one of the prerequisites for successful treatment in the future. Circulating miRNAs are detectable in body fluids including blood and cerebrospinal fluid and may be useful as novel biomarkers amenable to clinical diagnostic applications for various types of disease (Cogswell et al., 2008; Orlova et al., 2011; Ajit, 2012; Weiland et al., 2012; Machida et al., 2013). Therefore, it should be likewise promising to carefully assess which circulating miRNAs and novel ncRNAs are associated with neurogenic and neuropathic pain syndromes and may emerge as reliable diagnostic biomarkers for painful diabetic polyneuropathy, nerve injury pain, CRPS, headache and migraine.

NEW DRUGGABLE MOLECULAR TARGETS FOR PAIN TREATMENT

Treatment of painful diabetic polyneuropathy is far from satisfactory in many patients although this is the most intensely studied painful neuropathy in randomized controlled trials (RCTs). National and international guidelines differ in their recommendations about first and second line treatment choices. While pregabalin is favored by some (Bril et al., 2013), duloxetine or even tricyclic antidepressants are first choice in others (NICE-guideline; Attal et al., 2010; Dworkin et al., 2007). All of these drugs have adverse effects on diabetes. Furthermore, mean treatment effects comprise only two points of pain reduction on a 11-point Likert scale. In other types of neuropathy, like traumatic neuropathy or the frequent inflammatory types, there is little or no data at all from RCTs on pain treatment. Even worse, treatment of CRPS is neither standardized, nor satisfactory, nor based on multicentre RCTs. From single center studies with very limited patient numbers some evidence exists for anti-inflammatory treatment by corticosteroids or bisphosphonates in acute but not chronic stages, and for behavioral therapy for selected patients in chronic stages (de Tran et al., 2010). For the most frequently used invasive treatment modalities such as sympathetic blockers no RCT evidence of efficacy is available (Straube et al., 2010). Thus, more efficacious and specific medications are needed for both neurogenic and neuropathic pain syndromes.

Both, the novel and specific mode of action and the ability to function as master switches of entire signaling networks has triggered enthusiasm for miRNAs as promising therapeutic targets although relatively little is known about the mechanisms of cellular uptake, storage and mode of action of miRNA modulators (van Rooji and Olson, 2012). In several rodent pain models, deregulated expression of miRNAs was found in pain pathways from primary afferent nociceptors to brain areas associated with emotional components of pain perception (**Figure 2**; Bai et al., 2007; Aldrich et al., 2009; Kusuda et al., 2011; Imai et al., 2011; Poh et al., 2011; von Schack et al., 2011). First evidence supporting a future for analgesic miRNA treatment comes from mice intrathecally receiving miR-124, miR-103 or miR-23b which are reported to prevent and treat persistent inflammatory and neuropathic pain (Favereaux et al., 2011; Imai et al., 2011; Willemen et al., 2012). Despite the fact that these miRNA treatments reduced signatures of synaptic modification, neuroinflammation and microglial response, the full extent and the mechanisms of the analgesic effect are not understood to date (Favereaux et al., 2011; Willemen et al., 2012).

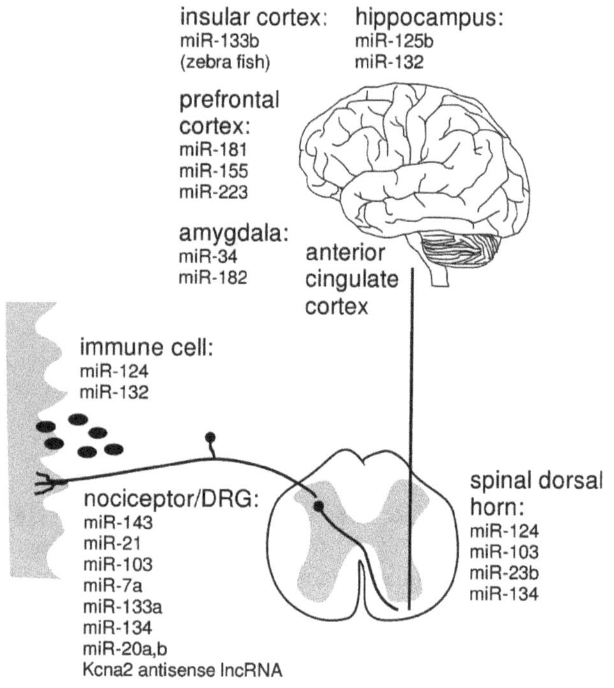

FIGURE 2 | miRNA that may be causally associated with maintained neuropathic pain in immune cells (Shaked et al., 2009; Soreq and Wolf, 2011; Ponomarev et al., 2013), nociceptors/DRG (Zhao et al., 2010, 2013; Tam Tam et al., 2011; Ni et al., 2012; Sakai and Suzuki, 2013), spinal cord (Brandenburger et al., 2012; Favereaux et al., 2011; Im et al., 2012; Ni et al., 2012; Willemen et al., 2012), insular cortex (Sanchez-Simon et al., 2010), amygdala (Meerson et al., 2010; Haramati et al., 2011; Griggs et al., 2013), prefrontal cortex (Poh et al., 2011), hippocampus (Edbauer et al., 2010).

CIRCUITRIES AND PROCESSES MODULATING NOCICEPTION AND ENDOGENOUS ANALGESIA

Various studies have identified signatures of neuroinflammation as critical components of diabetic polyneuropathy (Pabreja et al., 2011; Vincent et al., 2011) in addition to the cellular mechanisms that include the classical changes of the diabetic milieu (Bierhaus et al., 2012; Bierhaus and Nawroth, 2012). Pathological neuro-immune communication has been associated with painful neuropathy following traumatic peripheral nerve injury (Myers et al., 2006; Ciaramitaro et al., 2010; Birch et al., 2012). Moreover, CRPS occurring as a complication of bone fracture or tissue injury results from neurogenic inflammatory processes (Goebel, 2011). In humans, a systemic pro-inflammatory profile distinguishes painful from painless neuropathy, and a local pro-inflammatory profile is part of the pathophysiology of small fiber neuropathy (Üceyler et al., 2007b, 2010). Specialized peripheral neurons, the nociceptors sense inflammatory or neuropathic conditions and respond with increased excitability and sensitivity leading to persisting pain and hyperalgesia (Costigan and Woolf, 2000; Sommer and Kress, 2004; Berta et al., 2008; Üceyler et al., 2009). However, mice lacking receptors for pro-inflammatory mediators in their nociceptor neurons are frequently protected from certain signatures of pathological pain (Andratsch et al., 2009; Schweizerhof et al., 2009; Mair et al., 2011; Quarta et al., 2011). The deficiency in anti-inflammatory cytokines in patients with CRPS (Üceyler et al., 2007a) together with beneficial effect of therapy with glucocorticosteroids support pathophysiological mechanisms associated with neuro-immune dysfunction (Fischer et al., 2010).

Inflammatory processes are also activated in the spinal cord upon peripheral nerve injury and involve microglia activation and leakage at the blood nerve barrier along the entire neuraxis (McMahon and Malcangio, 2009; Beggs et al., 2010, 2012). Microglia activation occurs in diabetic neuropathy in rodents (Wodarski et al., 2009; Beggs et al., 2012; Talbot and Couture, 2012) and has been recognized to be critical for the maintenance of neuropathic pain via the release of pro-nociceptive mediators (Clark et al., 2007). Leakage of the blood nerve barrier or the blood spinal barrier is just emerging in the pathophysiology of neuropathic pain accompanied by changes in tight junction proteins (Echeverry et al., 2011). Tight junction proteins which are critically involved in maintaining the blood–brain barrier like claudin-1 are also new targets, e.g., of miR-155 (Qin et al., 2013).

Deregulated miRNAs can be a consequence or cause of local inflammatory processes such as regulation of nociceptor sensitisation by controlling phospholipase A2 activation (Sun et al., 2012). Analyses of expression profiles of dorsal root ganglia (DRG) containing nociceptor cell bodies reveal that particular miRNAs are deregulated in rodent pain models giving rise to deregulation of miRNA-targeted ion channel expression patterns and metabotropic receptor transcripts in peripheral neurons which presumably cause nociceptor dysfunction (Zhao et al., 2010; von Schack et al., 2011). miRNAs are universal regulators of differentiation, activation and polarization of microglia in normal and inflammatory conditions (Ponomarev et al., 2013). Microglia and macrophage activity is suppressed by specific miRNAs, e.g., miR-124, and it is therefore anticipated that miRNA regulation is critically involved in endogenous inhibition and resolution of inflammation by e.g., resolvins (Ponomarev et al., 2011; Recchiuti et al., 2011). Certain miRNAs are substantially suppressed in glucocorticoid-treated thymocytes by reduced expression of the key miRNA processing enzymes Dicer, Drosha, and DGCR8/Pasha (Smith et al., 2010). This observation is of great relevance since CRPS for example is regarded a prototype disorder of failed termination of inflammation (Birklein and Kingery, 2009). The spinal release of immune modulators affects both spinal synaptic processes and local inhibitory circuits, possibly by classical cytokine-prostaglandin signaling and dys-inhibition of e.g., glycinergic spinal control (Samad et al., 2001; Harvey et al., 2004). Plastic changes at synapses in the spinal dorsal horn promote neuropathic and neurogenic pain via mechanisms involving enhanced nociceptive transmission but also inhibition of spinal endogenous analgesic circuits (Hartmann et al., 2004; Harvey et al., 2004; Fossat et al., 2007; Sandkühler, 2007, 2009; Pernía-Andrade et al., 2009; Zeilhofer et al., 2009; Fossat et al., 2010; Laffray et al., 2012).

For a few miRNAs and long ncRNAs, downstream target proteins have been reported. For example, a conserved long ncRNA seems to modulate sensory neuron excitability by activation of a transcription factor MZF and downregulation of *Kcna2* potassium channel expression and this has been causally associated with neuropathic pain (Zhao et al., 2013). In addition, the functional consequences of miR-103 regulation of voltage-gated Cav1.2 calcium channels and intrinsic excitability of spinal projection neurons have been demonstrated (Favereaux et al., 2011). It is well accepted that certain hereditary forms of migraine are associated with polymorphisms of voltage-gated calcium channels Cav2.1 and Cav2.2 (Pietrobon and Striessnig, 2003). Novel evidence suggests that in particular endogenous pain control systems including GABAergic and opioidergic synaptic signals are down-regulated by miRNAs such as miR-134 or miR-181a (Ni et al., 2012; Sengupta et al., 2013). Some of them link miRNAs like let-7 or miR-339 to opioid tolerance (He et al., 2010; He and Wang, 2012; Wu et al., 2013). In analogy, miRNA neuronal dys-regulation should not only apply to neurogenic or neuropathic pain but very likely the same principles and pathways should apply to other pain syndromes like headaches and in particular hereditary and other forms of migraine.

COGNITIVE, EMOTIONAL AND BEHAVIORAL COMPONENTS OF PAIN

Neuropsychological alterations are present in 65 % of CRPS patients and in particular cognitive impairment and deficits of emotional decision-making may impact their quality of life especially in risky, emotional situations (Apkarian et al., 2004). Emotional deficits and functional alterations in corresponding brain regions are reported in chronic CRPS patients and pain-related fear is one of the strongest predictors of disability in chronic pain disorders (Geha et al., 2008; de Jong et al., 2011).

Specific areas in the brain are actively involved in pain perception and behavior in humans and rodents and structural brain changes are associated with sensory and emotional function in rodent long-term neuropathic pain. In particular, decreased volumes of primary somatosensory and frontal cortex, retrosplenial and entorhinal cortex, anterior cingulate cortex and insula are maintained for months (Seminowicz et al., 2009). Specifically, abnormalities in hippocampus volume are observed in human CRPS and the mouse spared nerve injury (SNI) model. Similar to CRPS patients, SNI mice show increased anxiety like behavior and abnormal contextual fear extinction and this is associated with reduced extracellular signal-regulated kinase (ERK) expression, decreased neurogenesis and altered synaptic plasticity (Kodama et al., 2007; Mutso et al., 2012). Mice with experimental neuropathic pain also show cognitive deficits in novel object recognition and this is associated with deregulation of glycinergic neurotransmission in the hippocampus (Kodama et al., 2011), and may relate to reported enhanced quantal neurotransmitter release in the anterior cingulate cortex of mice with neuropathic pain (Toyoda et al., 2009). Dopaminergic and glutamatergic inputs from amygdala, hippocampus and prefrontal cortex to the nucleus accumbens participate in the putative emotional control circuits and recent human brain activity studies have examined the nucleus accumbens in the emotional aspects of pain processing (Baliki et al., 2010). These reports further link chronic pain with emotional dysfunction, and maladaptive responses of the nucleus accumbens in neuropathic pain have recently been associated with deregulated miRNAs in this region (Imai et al., 2011).

Brain-specific miRNAs are emerging as regulators of cognition, neuronal plasticity and memory by manipulating synapse structure and function, and specific miRNAs not only control cognition and emotional processes but also neuro-immune communication in the brain (Bredy et al., 2011; Soreq and Wolf, 2011). Mental retardation has been associated with miR-125b, miR-132 and other miRNAs and this arises from effects on dendritic spine morphology and synaptic physiology at hippocampal neurons. AMPA-mediated miniature mEPSC amplitude and frequency are reduced by neuronal over-expression of miR-125b and increased by miR-132 and this is due to differential regulation of glutamate NR2A and NR2B receptor mRNA levels (Edbauer et al., 2010). Other glutamate receptor subunits in the brain are regulated by dopamine through miR-181a which has recently been associated with the pain system (Saba et al., 2012). miR-132 is a highly interesting brain specific miRNA since it is up-regulated by brain derived neurotrophic factor (BDNF) and other growth factors in cortical neurons and this results in an increased expression of synaptic proteins including glutamate receptors (NR2A, NR2B and GluA1), an effect that is attenuated by glucocorticoids (Kawashima et al., 2010; Numakawa et al., 2011). Hippocampal miR-132 mediates stress-inducible cognitive deficits through acetylcholinesterase as a downstream target and specifically in the amygdala miR-34 is associated with the repression of stress-induced anxiety (Haramati et al., 2011; Shaltiel et al., 2013). More generally, happiness, anxiety and depression seem to depend on miRNA expression levels. Specific miRNAs are deregulated in patients suffering from depression and anxiety, and in pre-clinical models of psychological stress (Meerson et al., 2010). Moreover, psychoactive agents, including antidepressants and mood stabilizers, utilize miRNAs as downstream effectors (O'Connor et al., 2012). This further links neuropathic pain to emotional disorders and to the clinical benefit of antidepressants for pain treatment (Dworkin et al., 2007).

PAIN PREDISPOSING GENETIC POLYMORPHISMS

There is evidence that chronic pain, pain sensitivity and responsiveness to analgesic opioids show a sufficient heritability to make these phenotypes highly interesting sources for genetic variability which has an influence on pain (Angst et al., 2012; Hocking et al., 2012; Nielsen et al., 2012). Altered miRNA expression is frequently a consequence of genetic mutations, which may also cause loss or gain of function (Mishra and Bertino, 2009). This may account for significant inter-individual variation in the response to painful stimuli and analgesic drugs. Polymorphisms of specific molecular targets may be associated with certain pain phenotypes and this has emerged for example for a specific calcium channel subunit in a *Drosophila* screen that is conserved in mice and humans (Mogil, 2012; Neely et al., 2010). Several meta-analyses are available of the genetics of pain and associated specific loss- or gain of function polymorphisms with altered pain perception (LaCroix-Fralish et al., 2011; Mogil, 2012). A recent genome-wide association (GWA) study revealed three susceptibility loci for common migraine in the general population, however, systematic association studies are unavailable for DPN and CRPS to date (Chasman et al., 2011). In general, genetic studies have helped to understand the role and downstream mechanisms of individual proteins in pain processing, but specific single nucleotide polymorphism (SNP) related pain disorders apply to small numbers of individuals only and so far do not explain the large variability regarding susceptibility to distinct pain disorders or the responsiveness to different pain therapies in the general population (Dworkin et al., 2007; Attal et al., 2010).

The functional consequences of polymorphisms in miRNA genes and/or their binding sites, the downstream targets of miRNAs and the mechanisms by which miRNAs regulate circuitries and processes modulating nociception and endogenous analgesia are entirely unaddressed. SNPs in miRNAs or their target sites are not only bioinformatically predicted to be associated with the pathogenesis of diseases but are also experimentally validated (Wu et al., 2008; Coassin et al., 2010). It is known that SNPs are less common in miRNAs or their target sites than in other parts of the genome which points to the importance of miRNAs for cellular processes. However, on the other hand SNPs in these sites can affect the expression of a large number of genes when the production of the miRNA is influenced by that particular SNP. Moreover, SNPs in target sites of miRNAs can either modulate/disrupt existing binding sites or create new binding sites for the miRNAs that may then influence gene expression. SNPs in these regions have become a major focus of research and some of them are expected to explain pathogenetic mechanisms in disease development (Glinsky, 2008; Haas et al., 2012). For example, miRNA expression is markedly different between normal tissues and tumor tissues although otherwise miRNA expression is strictly controlled. This might be explained by somatic mutations that are introduced during carcinogenesis. The investigation of genetic variants at miRNAs or their target sites and their association with various diseases is only in its infancy. Initial studies show that these RNA chains might also be involved in neurological diseases such as Parkinson's disease (Martins et al., 2011), Alzheimer's disease (Serpente et al., 2011) or frontotemporal lobar degeneration (Villa et al., 2011). The identification of SNPs in miRNA related regions of the genome might be advantageous over classical GWA study since individual ncRNAs may control and regulate whole networks and pathways involving a multitude of functional proteins. This may open a new avenue that may potentially improve our understanding of extensive inter-individual differences in patients.

TRANSLATION OF PRE-CLINICAL AND CLINICAL RESULTS INTO SOLUTIONS FOR THE BENEFIT OF PATIENTS

As stated above, one of the major hindrances in the way of translating such findings into better therapy of neuropathic and neurogenic pain syndromes is the complexity of their pathophysiology, which even changes during the course of disease. Based on and in analogy to recent developments in the oncology field, an improved understanding of the role of miRNAs in neuropathic pain might be highly useful for diagnostic and prognostic assessments. For example, aberrant expression or functional deregulation of miRNAs has been associated with the risk for and progression of malignancies and this knowledge is expected to advance the management of certain cancer types through the development of novel personalized miRNA-based diagnostics and therapies (Dreussi et al., 2012; Rossi and Calin, 2013). Increasing evidence indicates that certain miRNAs may be aberrantly expressed or deregulated in certain individuals after tissue injury or with diabetes. This may be associated with increased risk of pain chronification or even responsiveness to analgesic drugs (Ivanov et al., 2012). Therefore, miRNAs are expected to have potential for personalized pain medicine as biomarkers for risk assessment, drug selection and novel therapies.

Therapeutic miRNA regulation has been thoroughly studied and begins to be established in different types of cancer, and the first miRNA targeted drug has entered phase II clinical trials (Lindow and Kauppinen, 2012). In contrast, the potential therapeutic impact of miRNAs in the pain field is as yet largely unexplored. To date, therapeutic approaches have been restricted to rodent models and intrathecal administration and some inconsistencies have emerged; thus miRNA increases in a disease may be either a cause or a feedback reaction to the observed symptoms. For example, although miR-124 is up-regulated after chronic constrictive nerve injury (CCI), intrathecal administration of miR-124 can prevent and treat persistent inflammatory and neuropathic pain (Willemen et al., 2012). Likewise, miR-132 levels are increased in colon biopsies from patients with intestinal bowel disease which should predictably limit inflammation (Maharshak et al., 2013). Importantly, manipulation of miRNAs offers the possibility to control multiple targets including neuro-immune interactions, nociceptive processing and cognitive and affective pathways. Thus, miRNA based therapeutics may have superior advantages by targeting multiple pain-associated genes and miRNA-based drugs may be the most appropriate therapy for the prevention or treatment of neuropathic and neurogenic pain. At least, recent developments provide an optimistic perspective on the evolution of therapeutic ncRNAs despite the drawback of unresolved obstacles for successful delivery and unknown, however unlikely, off-target effects (Cho, 2012).

Manipulating miRNAs as a therapeutic tool presents significant theoretical and practical challenges that must be overcome before this approach becomes a reality. Specific examples involve two of the more straightforward approaches for miRNA modulation, miRNA mimics and antagomirs (**Figure 3**). miRNA mimics consist of over-expressing specific miRNAs that are reduced in the disease state. This mimic approach could be done by introducing synthetic oligonucleotides (natural or modified) or involve over-expression of such miRNAs from an introduced viral vector. Antagomirs are synthetic oligonucleotide sequences that are designed to be inversely oriented (antisense) to miRNAs that are over-expressed in the disease state and which can form Watson-Crick base pairing with the target miRNA. This can either inactivate a miRNA or result in its degradation. Similar to the miRNA mimics, these therapeutic and research tools can also consist of synthetic or modified nucleic acid sequences or be overexpressed from viral vectors.

Alternative methodologies used in experimental settings include miRNA sponges, which are exogenous DNA repeats of the target sequence and can serve to soak up excess copies of the excess miRNA (Ebert et al., 2007). The miRNA sponges may be produced under the regulation of RNA Polymerase III promoters and can generate high amounts of specific target sequences. Another novel yet promising approach involves target protection. In this application, modified antisense oligonucleotides such as LNA or morpholinos are prepared that will be complementary to a specific sequence in the target gene messenger RNA. These are added to the cells, where they bind to the target sequence, block its down-regulation by the miRNA complex and ensure sufficient expression of the target mRNA (Choi et al., 2007). Enhanced and prolonged miRNA suppression and simultaneous targeting of multiple miRNAs can be achieved by inhibitors carrying clustered hairpins based on the "Tough decoy" (TuD) design which offer the advantage of standardized suppression of families or clusters of miRNAs and can be combined with recombinant adenovirus vectors (Haraguchi et al., 2009; Xie et al., 2012; Bak et al., 2013; Hollensen et al., 2013).

An important difficulty that may be predicted for developing neuronal miRNA therapeutics is delivery, since targeting to the brain involves the significant hurdle of crossing the blood–brain barrier. Nevertheless, therapeutic efficacy of certain approaches such as the use of LNA antagomirs has been demonstrated even in primate models, and certain neuronal miRNA therapeutic approaches are now in preclinical development. These studies cover several creative approaches that have been developed to overcome the delivery problem. Thus, ~20-mer miRNA-size oligonucleotides are indeed unlikely to cross the blood–brain barrier. However, peripheral administration of oligonucleotide controllers of inflammation-regulating miRNAs would change the levels of cytokines, and cytokines can penetrate and affect the brain. Such effects have been demonstrated for miR-132 (Shaked et al., 2009) and miR-212 (Hollander et al., 2010). Other means include direct infection of cerebral neurons with viral vectors that may be adapted for better tropism to neuronal cells (Barbash et al., 2013). Direct introduction of antisense oligonucleotides can alternatively be performed by intracerebroventricular or local stereotactic injection though these would be extremely problematic in pain syndromes. Yet more recent work described the use of rabies virus glycoprotein labeled nanoparticles to enable direct delivery of a miRNA mimic to neuronal cells (Hwang do et al., 2011).

CONCLUSION

Recently, specific miRNAs have been associated with pathological pain and the deregulation of ion channel expression in sensory neurons in rodent pain models (Zhao et al., 2010; Favereaux et al., 2011; Li et al., 2011). Pain conditions have been suggested to deregulate the expression of miRNAs in pain pathways from primary afferent nociceptors to brain areas associated with emotional components of pain perception (Bai et al., 2007; Aldrich et al., 2009; Imai et al., 2011; Kusuda et al., 2011; Poh et al., 2011; von Schack et al., 2011; Arai et al., 2013; Genda et al., 2013; Sakai and Suzuki, 2013). Unique signatures of miRNAs are associated with altered innate immune signaling and secreted miRNAs are even considered

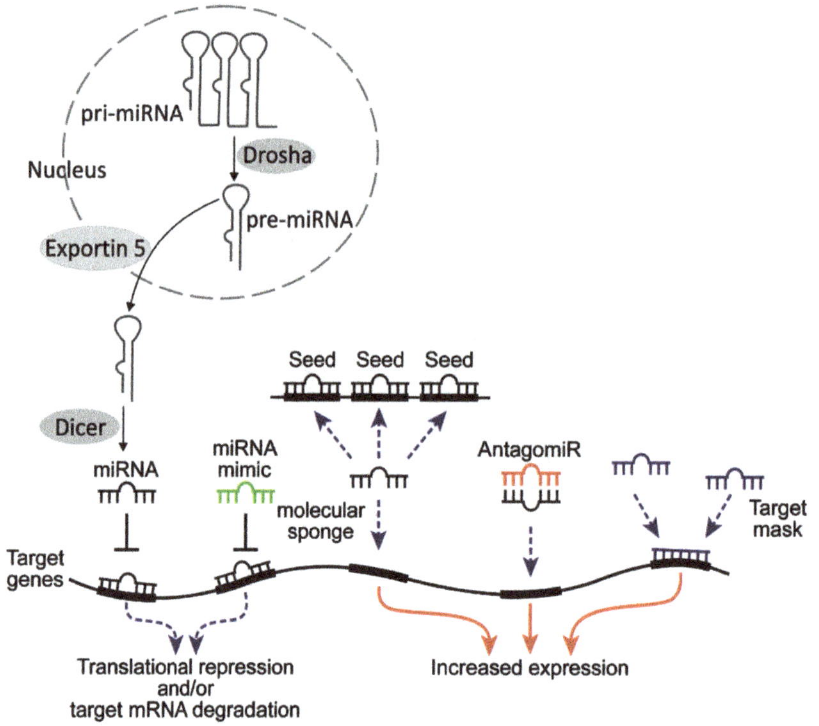

FIGURE 3 | Endogenous miRNAs are generated from primary (pri-) miRNAs via cleavage by the RNAse Drosha into pre-miRNAs in the nucleus. They are exported into the cytosol by Exportin 5 and there are cleaved into active miRNAs by the RNAse Dicer. Depending on the degree of homology, miRNAs trigger translational repression or degradation of target mRNAs (for review see He and Hannon, 2004; Bartel, 2009). Therapeutic manipulations of miRNAs may involve various methods. Host tissue miRNAs (gray) bind to complementary sequences, which are often located in the 3′-untranslated region (3′-UTR) of the target genes. This leads to translational repression, often accompanied by degradation. Mimicking this process, miRNA mimics (green) with similar sequences to those of miRNAs may be designed to target the same mRNAs. Such mimics are synthetic oligonucleotides that are chemically protected against nucleolytic degradation. Alternative routes include molecular "sponges," with several binding sites of a certain miRNA; antagomiRs (red), complementary oligonucleotides to the host miRNA which bind to it and limit its function, and target masks, which bind to part of target miRNAs and compete with their function. Thus, tools exist both for inducing gain of function (red arrows) or loss of function (dashed blue arrows).

a new form of neuroimmune communication and control immune cell activity as well as neuron function (Peng et al., 2010; Bredy et al., 2011; Chen et al., 2012; Ponomarev et al., 2013). miRNAs act at the neuro-immune interface which controls neuronal plasticity and memory but also are linked to the etiology of anxiety and mood disorders (Bredy et al., 2011; Soreq and Wolf, 2011; O'Connor et al., 2012; Shaltiel et al., 2013). Such deficits in the interaction of immune cells and neurons together with cognitive and emotional alterations in patients with neuropathic or neurogenic pain syndromes are hypothesized to converge on miRNA deregulated mechanisms along the entire neuraxis, and alterations in miRNA expression may account for the variation of susceptibility to certain types of pain or even for the responsiveness to analgesics and opioid tolerance (Parsons et al., 2008). Understanding the role of miRNAs in pain mechanisms is suggested to provide great benefit for clinical diagnostic and therapeutic applications.

ACKNOWLEDGMENTS

This work is supported by the European Commission (GA N 602133 - ncRNAPain).

REFERENCES

Ajit, S. K. (2012). Circulating microRNAs as biomarkers, therapeutic targets, and signaling molecules. *Sensors (Basel)* 12, 3359–3369. doi: 10.3390/s120303359

Aldrich, B. T., Frakes, E. P., Kasuya, J., Hammond, D. L., and Kitamoto, T. (2009). Changes in expression of sensory organ-specific microRNAs in rat dorsal root ganglia in association with mechanical hypersensitivity induced by spinal nerve ligation. *Neuroscience* 164, 711–723. doi: 10.1016/j.neuroscience.2009.08.033

Andratsch, M., Mair, N., Constantin, C. E., Scherbakov, N., Benetti, C., Quarta, S., et al. (2009). A key role for gp130 expressed on peripheral sensory nerves in pathological pain. *J. Neurosci.* 29, 13473–13483. doi: 10.1523/JNEUROSCI.1822-09.2009

Angst, M. S., Phillips, N. G., Drover, D. R., Tingl, M., Ray, A., Swan, G. E., et al. (2012). Pain sensitivity and opioid analgesia: a pharmacogenomic twin study. *Pain* 153, 1397–1409. doi: 10.1016/j.pain.2012.02.022

Apkarian, A. V., Sossa, Y., Krauss, B. R., Thomas, P. S., Fredrickson, B. E., Levy, R. E., et al. (2004). Chronic pain patients are impaired on an emotional decision-making task. *Pain* 108, 129–136. doi: 10.1016/j.pain.2003.12.015

Arai, M., Genda, Y., Ishikawa, M., Shunsuke, T., Okabe, T., and Sakamoto, A. (2013). The miRNA and mRNA changes in rat hippocampi after chronic constriction injury. *Pain Med.* 14, 720–729. doi: 10.1111/pme.12066

Attal, N., Cruccu, G., Baron, R., Haanpää, M., Hansson, P., Jensen, T. S., et al. (2010). EFNS guidelines on the pharmacologcial treatment of neuropathic pain. *Eur. J. Neurol.*

17, 1113–e88. doi: 10.1111/j.1468-1331.2010.02999.x

Bai, G., Ambalavanar, R., Wei, D., and Dessem, D. (2007). Downregulation of selective microRNAs in trigeminal ganglion neurons following inflammatory muscle pain. *Mol. Pain* 3, 15. doi: 10.1186/1744-8069-3-15

Bak, R. O., Hollensen, A. K., Primo, M. N., Sørensen, C. D., and Mikkelsen, J. G. (2013). Potent microRNA suppression by RNA Pol II-transcribed 'Tough Decoy' inhibitors. *RNA* 19, 280–293 doi: 10.1261/rna.034850.112

Baker, M., Collet, B., Fischer, A., Hermann, V., Huygen, F. J. P., Trueman, P., et al. (2010). *Improving the Current and Future Management of Chronic Pain.* A European Consensus Report. Brussels: Pfizer.

Baliki, M. N., Geha, P. Y., Fields, H. L., and Apkarian, A. V. (2010). Predicting value of pain and analgesia: nucleus accumbens response to noxious stimuli changes in the presence of chronic pain. *Neuron* 66, 149–160. doi: 10.1016/j.neuron.2010.03.002

Barbash, S., Hanin, G., and Soreq, H. (2013). Stereotactic injection of microRNA-expressing lentiviruses to the mouse hippocampus CA1 region and assessment of the behavioral outcome. *J. Vis. Exp.* doi: 10.3791/50170

Baron, R., Forster, M., and Binder, A. (2012). Subgrouping of patients with neuropathic pain according to pain-related sensory abnormalities: a first step to a stratified treatment approach. *Lancet Neurol.* 11, 999–1005. doi: 10.1016/S1474-4422(12)70189-8

Bartel, D. P. (2009). MicroRNAs: target recognition and regulatory functions. *Cell* 136, 2015–2233 doi: 10.1016/j.cell.2009.01.002

Beggs, S., Liu, X. J., Kwan, C., and Salter, M. W. (2010). Peripheral nerve injury and TRPV1-expressing primary afferent C-fibres cause opening of hte blood–brain barrier. *Mol. Pain* 6, 74. doi: 10.1186/1744-8069-6–74.

Beggs, S., Trang, T., and Salter, M. W. (2012). P2X4R(+) microglia drive neuropathic pain. *Nat. Neurosci.* 15, 1068–1073. doi: 10.1038/nn.3155

Berta, T., Poirot, O., Pertin, M., Ji, R. R., Kellenberger, S., and Decosterd, I. (2008). Transcriptional and functional profiles of voltage-gated Na+ channels in injured and non-injured DRG neurons in the SNI model of neuropathic pain. *Mol. Cell. Neurosci.* 37, 196–208. doi: 10.1016/j.mcn.2007.09.007

Bierhaus, A., Fleming, T., Stoyanov, S., Leffler, A., Babes, A., Neacsu, C., et al. (2012). Methylglyoxal modification of Na(v)1.8 facilitates nociceptive neuron firing and causes hyperalgesia in diabetic neuropathy. *Nat. Med.* 18, 926–933. doi: 10.1038/nm.2750

Bierhaus, A., and Nawroth, P. P. (2012). Critical evaluation of mouse models used to study pain and loss of pain perception in diabetic neuropathy. *Exp. Clin. Endocrinol. Diabetes* 120, 188–190. doi: 10.1055/s-0032-1304567

Birch, R., Misra, P., Stewart, M. P., Eardley, W. G., Ramasamy, A., Brown, K., et al. (2012). Nerve injuries sustained during warfare: part I – epidemiology. *J. Bone Joint Surg. Br.* 94, 523–528. doi: 10.1302/0301-620X.94B4.28483

Birklein, F., and Kingery, W. S. (2009). Complex regional pain syndrome: a loss of inhibition? *Pain* 142, 177–178. doi: 10.1016/j.pain.2009.01.029

Brandenburger, T., Castoldi, M., Brendel, M., Grievink, H., Schlösser, L., Werdehausen, R., et al. (2012). Expression of spinal cord microRNAs in a rat model of chronic neuropathic pain. *Neurosci. Lett.* 506, 281–286. doi: 10.1016/j.neulet.2011.11.023

Bredy, T. W., Lin, Q., Wei, W., Baker-Andresen, D., and Mattick, J. S. (2011). MicroRNA regulation of neural plasticity and memory. *Neurobiol. Learn. Mem.* 96, 89–94. doi: 10.1016/j.nlm.2011.04.004

Breivik, H., Collett, B., Ventafridda, V., Cohen, R., and Gallacher, D. (2006). Survey of chronic pain in Europe: prevalence, impact on daily life, and treatment. *Eur. J. Pain* 10, 287–333. doi: 10.1016/j.ejpain.2005.06.009

Bril, V., England, J., Franklin, G. M., Backonja, M., Cohen, J., Del Toro, D., et al. (2013). Evidence-based guidelines: treatment of painful diabetic neuropathy: report of the American Academy of Neurology, the American Association of Neuromuscular and Electrodiagnostic Medicine and the American Academy of Physical Medicine and Rehabilitation. *Neurology* 76, 1758–1765. doi: 10.1212/WNL.0b013e3182166ebe

Chasman, D. I., Schürks, M., Anttila, V., de Vries, B., Schminke, U., Launer, L. J., et al. (2011). Genome-wide association study reveals thress susceptibility loci for common migraine in the general population. *Nat. Genet.* 43, 695–698. doi: 10.1038/ng.856

Cho, W. C. (2012). Exploiting the therapeutic potential of microRNAs in human cancer. *Expert Opin. Ther. Targets* 16, 345–350. doi: 10.1517/14728222.2012.663354

Choi, W. Y., Giraldez, A. J., and Schier, A. F. (2007). Target protectors reveal dampening and balancing of Nodal agonist and antagonist by miR-430. *Science* 318, 271–274. doi: 10.1126/science.1147535

Ciaramitaro, P., Mondelli, M., Logullo, F., Grimaldi, S., Battiston, B., Sard, A., et al. (2010). Traumatic peripheral nerve injuries: epidemiological findings, neuropathic pain and quality of life in 158 patients. *J. Peripher. Nerv. Syst.* 15, 120–127. doi: 10.1111/j.1529-8027.2010.00260.x

Clark, A. K., Yip, P. K., Grist, J., Gentry, C., Staniland, A. A., Marchand, F., et al. (2007). Inhibition of spinal microglial cathepsin S for the reversal of neuropathic pain. *Proc. Natl. Acad. Sci. U.S.A.* 104, 10655–10660. doi: 10.1073/pnas.0610811104

Coassin, S., Brandstätter, A., and Kronenberg, F. (2010). Lost in the space of bioinformatic tools: a constantly updated survival guide for genetic epidemiology. The GenEpi Toolbox. *Atherosclerosis* 209, 321–335. doi: 10.1016/j.atherosclerosis.2009.10.026

Cogswell, J. P., Ward, J., Taylor, I. A., Waters, M., Shi, Y., Cannon, B., et al. (2008). Identification of miRNA changes in Alzheimer's disease brain and CSF yields putative biomarkers and insights into disease pathways. *Alzheimers Dis.* 14, 27–41.

Costigan, M., and Woolf, C. J. (2000). Pain: molecular mechanisms. *J. Pain* 3, 35–44. doi: 10.1054/jpai.2000.9818

de Jong, J. R., Vlaeyen, J. W., de Gelder, J. M., and Patijn, J. (2011). Pain-related fear, perceived harmfulness of activities, and functional limitations in complex regional pain syndrome type I. *J. Pain* 12, 1209–1218. doi: 10.1016/j.jpain.2011.06.010

de Tran, Q. H., Duon, G., Bertini, P., and Finlayson, R. J. (2010). Treatment of complex regional pain syndrome: a review of the evidence. *Can. J. Anaesth.* 57, 149–166. doi: 10.1007/s12630-009-9237-0

Dreussi, E., Biason, P., Toffoli, G., and Ceccin, E. (2012). miRNA pharmacogenomics: the new frontier for personalized medicine in cancer? *Pharmacogenomics* 13, 1635–1650. doi: 10.2217/pgs.12.147

Dworkin, R. H., O'Connor, A. B., Backonja, M., Farrar, J. T., Finnerup, N. B., Jensen, T. S., et al. (2007). Pharmacological management of neuropathic pain: evidence-based recommendations. *Pain* 132, 237–251. doi: 10.1016/j.pain.2007.08.033

Ebert, M. S., Neilson, J. R., and Sharp, P. A. (2007). MicroRNA sponges: competitive inhibitors of small RNAs in mammalian cells. *Nat. Methods* 4, 721–726. doi: 10.1038/nmeth1079

Echeverry, S., Shi, X. Q., Rivest, S., and Zhang, J. (2011). Peripheral nerve injury alters blood-spinal cord barrier functional and molecular integrity through a selective inflammatory pathway. *J. Neurosci.* 31, 10819–10828. doi: 10.1523/JNEUROSCI.1642-11.2011

Edbauer, D., Neilson, J. R., Foster, K. A., Wang, C.-F., Seeburg, D. P., Batterton, M. N., et al. (2010). Regulation of synaptic structure and function by FMRP-associated microRNAs miR-125b and miR-132. *Neuron* 65, 373–384. doi: 10.1016/j.neuron.2010.01.005

Favereaux, A., Thoumine, O., Bouali-Benazzouz, R., Roques, V., Papon, M. A., Salam, S. A., et al. (2011). Bidirectional integrative regulation of Cav1.2 calcium channel by microRNA miR-103: role in pain. *EMBO J.* 30, 3830–3841. doi: 10.1038/emboj.2011.249

Fischer, S. G. L., Zuumond, W. W. A., Birklein, F., Loer, S. A., and Perez, R. S. G. M. (2010). Anti-inflammatory treatment of complex regional pain syndrome. *Pain* 151, 251–256. doi: 10.1016/j.pain.2010.07.020

Fossat, P., Dobremez, E., Bouali-Benazzouz, R., Favereaux, A., Bertrand, S., Kilk, K., et al. (2010). Knock-down of L calcium channel subtypes: differential effects on neuropathic pain. *J. Neurosci.* 30, 1073–1085. doi: 10.1523/JNEUROSCI.3145-09.2010

Fossat, P., Sibon, I., LeMasson, G., Landry, M., and Nagy, F. (2007). L-type calcium channels and NMDA receptors: a determinant duo for short-term nociceptive plasticity. *Eur. J. Neurosci.* 25, 127–135. doi: 10.1111/j.1460-9568.2006.05256.x

Geha, P. Y., Baliki, M. N., Harden, R. N., Bauer, W. R., Parrish, T. B., and Apkarian, A. V. (2008). The brain in chronic CRPS pain: abnormal gray-white matter interactions in emotional and authnomic regions. *Neuron* 60, 570–581. doi: 10.1016/j.neuron.2008.08.022

Genda, Y., Arai, M., Ishikawa, M., Tanaka, S., Okabe, T., and Sakamoto, A. (2013). microRNA changes in the dorsal horn of the spinal cord of rats with chronic constriction injury: A TaqMan® Low Density Array study. *Int. J. Mol. Med.* 31, 129–137. doi: 10.3892/ijmm.2012.1163

Glinsky, G. V. (2008). An SNP-guided microRNA map of fifteen common human disorders identifies a consensus disease phenocode aiming at prinicpal components of the nuclear import pathway. *Cell Cycle* 7, 2570–2583. doi: 10.4161/cc.7.16.6524

Goebel, A. (2011). Complex regional pain syndrome in adults. *Rheumatology* 50, 1739–1750. doi: 10.1093/rheumatology/ker202

Griggs, E. M., Young, E. J., Rumbaugh, G., and Miller, C. A. (2013). MicroRNA-182 regulates amygdala-dependent memory formation. *J. Neurosci.* 33, 1734–1740. doi: 10.1523/JNEUROSCI.2873-12.2013

Haas, U., Sczakiel, G., and Laufer, S. D. (2012). MicroRNA-mediated regulation of gene expression is affected by disease-asssociated SNPs within the 3'UTR via altered RNA structure. *RNA Biol.* 9, 924–937. doi: 10.4161/rna.20497

Haraguchi, T., Ozaki, Y., and Iba, H. (2009). Vectors expressing efficient RNA decoys achieve the long-term suppression of specific microRNA activity in mammalian cells. *Nucleic Acid Res.* 37, e43. doi: 10.1093/nar/gkp040

Haramati, S., Navon, I., Issler, O., Ezra-Nevo, G., Gil, S., Zwang, R., et al. (2011). microRNAs as repressors of stress-induced anxiety: the case of amygdalar miR-34. *J. Neurosci.* 31, 14191–14203. doi: 10.1523/JNEUROSCI.1673-11.2011

Hartmann, B., Ahmadi, S., Heppenstall, P. A., Lewin, G. R., Schott, C., Borchardt, T., et al. (2004). The AMPA receptor subunits GluR-A and GluR-B reciprocally modulate spinal synaptic plasticity and inflammatory pain. *Neuron* 44, 637–650. doi: 10.1016/j.neuron.2004.10.029

Harvey, R. J., Depner, U. B., Wassle, H., Ahmadi, S., Heindl, C., Reinold, H., et al. (2004). GlyR alpha3: an essential target for spinal PGE2-mediated inflammatory pain sensitization. *Science* 304, 884–887. doi: 10.1126/science.1094925

He, L., and Hannon, G. J. (2004). MicroRNAs: small RNAs with a big role in gene regulation. *Nat. Rev. Genet.* 5, 522–531. doi: 10.1038/nrg1379

He, Y., and Wang, Z. J. (2012). Let-7 microRNAs and opoid tolerance. *Front. Genet.* 3:110 doi: 10.3389/fgene.2012.000110

He, Y., Yang, C., Kirkmire, C. M., and Wang, Z. J. (2010). Regulation of opoid tolerace by let-7 family microRNA targeting the mu opioid receptor. *J. Neurosci.* 30, 10251–10258. doi: 10.1523/JNEUROSCI.2419-10.2010

Hehn, C. A., Baron, R., and Woolf, C. J. (2012). Deconstructing the neuropathic pain phenotype to reveal neural mechanisms. *Neuron* 73, 638–652. doi: 10.1016/j.neuron.2012.02.008

Hocking, L. J., Generation, S., Morris, A. D., Dominiczak, A. F., Porteous, D. J., and Smith, B. H. (2012). Heritability of chronic pain in 2195 extended families. *Eur. J. Pain* 16, 1053–1063. doi: 10.1002/j.1532-2149.2011.00095.x

Hollander, J. A., Im, H. I., Amelio, A. L., Kocerha, J., Bali, P., Lu, Q., et al. (2010). Striatal microRNA controls cocaine intake through CREB signalling. *Nature* 466, 197–202. doi: 10.1038/nature09202

Hollensen, A. K., Bak, R. O., Haslund, D., and Mikkelsen, J. G. (2013). Suppression of microRNAs by dual targeting and clustered Tough Decoy inhibitors. *RNA Biol.* 10, 406–414. doi: 10.4161/rna.23543

Hüttenhofer, A., and Schattner, P. (2006). The principles of guiding by RNA: chimeric RNA–protein enzymes. *Nat. Rev. Genet.* 7, 475–482. doi: 10.1038/nrg1855

Hüttenhofer, A., Schattner, P., and Polacek, N. (2005). Non-coding RNAs: hope or hype? *Trends Genet.* 21, 289–297. doi: 10.1016/j.tig.2005.03.007

Hwang do, W., Son, S., Jang, J., Youn, H., Lee, S., Lee, D., et al. (2011). A brain-targeted rabies virus glycoprotein-disulfide linked PEI nanocarrier for delivery of neurogenic microRNA. *Biomaterials* 32, 4968–4975. doi: 10.1016/j.biomaterials.2011.03.047

Im, Y. B., Choi, J. I., Cho, H. T., Kwon, O. H., and Kang, S. K. (2012). Molecular targeting of NOX4 for neuropathic pain after traumatic injury of the spinal cord. *Cell Death Dis.* 3, e426. doi: 10.1038/cddis.2012.168

Imai, S., Saeki, M., Yanase, M., Horiuchi, H., Abe, M., Narita, M., et al. (2011). Change in microRNAs associated with neuronal adaptive responses in the nucleus accumbens under neuropathic pain. *J. Neurosci.* 31, 15294–15299. doi: 10.1523/JNEUROSCI.0921-11.2011

Ivanov, M., Kacevska, M., and Ingelman-Sundberg, M. (2012). Epigenomics and interindividual differences in drug response. *Clin. Pharmacol. Ther.* 92, 727–736. doi: 10.1038/clpt.2012.152

Kawashima, H., Numakawa, T., Kumumaru, E., Adachi, N., Mizuno, H., Ninomiya, M., Kunugi, H., and Hashido, K. (2010). Glucocorticoid attenuates brain-derived neurotrophic factor-dependent upregulation of glutamate receptors via the suppression of microRNA-132 expression. *Neuroscience* 165, 1301–1311. doi: 10.1016/j.neuroscience.2009.11.057

Kodama, D., Ono, H., and Tanabe, M. (2007). Altered hippocampal long-term potentiation after peripheral nerve injury in mice. *Eur. J. Pharmacol.* 574, 127–132. doi: 10.1016/j.ejphar.2007.07.054

Kodama, D., Ono, H., and Tanabe, M. (2011). Increased hippocampal glycine uptake and cognitive dysfunction after peripheral nerve injury. *Pain* 152, 809–817. doi: 10.1016/j.pain.2010.12.029

Kuner, R. (2010). Central mechanisms fo pathological pain. *Nat. Med.* 16, 1258–1266. doi: 10.1038/nm.2231

Kusuda, R., Cadetti, F., Ravanelli, M. I., Sousa, T. A., Zanon, S., De Lucca, F. L., et al. (2011). Differential expression of microRNAs in mouse pain models. *Mol. Pain* 7, 17. doi: 10.1186/1744-8069-7-17

LaCroix-Fralish, M. L., Austin, J.-S., Zheng, F. Y., Levitin, D. J., and Mogil, J. S. (2011). Patterns of pain: meta-analysis of microarray studies of pain. *Pain* 152, 1888–1898. doi: 10.1016/j.pain.2011.04.014

Laffray, S., Bouali-Benazzouz, R., Papon, M. A., Favereaux, A., Jiang, Y., Holm, T., et al. (2012). Impairment of GABAB receptor dimer by endogenous 14-3-z in chronic pain conditions. *EMBO J.* 31, 3239–3251. doi: 10.1038/emboj.2012.161

Lindow, M., and Kauppinen, S. (2012). Discovering the first micrRNA-targeted drug. *J. Cell Biol.* 199, 407–412. doi: 10.1083/jcb.201208082

Machida, A., Ohkubo, T., and Yokota, T. (2013). Circulating microRNAs in the cerebrospinal fluid of patients with brain diseases. *Methods Mol. Biol.* 1024, 203–209. doi: 10.1007/978-1-62703-453-1_16

Maharshak, N., Shenhar-Tsarfaty, S., Aroyo, N., Orpaz, N., Guberman, I., Canaani, J., et al. (2013). Micro-RNA-132 modulates cholinergic signaling and inflammation in human inflammatory bowel disease. *Inflamm. Bowel Dis.* 19, 1346–1353. doi: 10.1097/MIB.0b013e318281f47d

Mair, N., Benetti, C., Andratsch, M., Leitner, M. G., Constantin, C. E., Camprubí-Robles, M., et al. (2011). Genetic evidence for involvement of neuronally expressed S1P1receptor in nociceptor sensitization and inflammatory pain. *PLoS ONE* 6:e17268. doi: 10.1371/journal.pone.0017268

Marinus, J., Moseley, G. L., Birklein, F. B., Baron, R., Maihöfner, C., Kingery, W. S., et al. (2011). Clinical features and pathophysiology of complex regional pain syndrome. *Lancet Neurol.* 10, 637–648. doi: 10.1016/S1474-4422(11)70106-5

Martins, M., Rosa, A., Guedes, L. C., Fonseca, B. V., Gotovac, K., Violante, S., et al. (2011). Convergence of miRNA expression profiling, a-synuclein interacton and GWAS in Parkinson's disease. *PLoS ONE* 6:e25443. doi: 10.1371/journal.pone.0025443

Mattick, J. S. (2004). RNA regulation: a new genetics? *Nat. Rev. Genet.* 5, 316–323. doi: 10.1038/nrg1321

McMahon, S. B., and Malcangio, M. (2009). Current challenges in glia-pain biology. *Neuron* 64, 46–54. doi: 10.1016/j.neuron.2009.09.033

Meerson, A., Cacheaux, L., Goosens, K. A., Sapolsky, R. M., Soreq, H., and Kaufer, D. (2010). Changes in brain microRNAs contribute to cholinergic stress reactions. *J. Mol. Neurosci.* 40, 47–55. doi: 10.1007/s12031-009-9252-1

Mishra, P. J., and Bertino, J. R. (2009). MicroRNA polymorphisms: the future of pharmacogenomics, molecular epidemiology and individualized medicine. *Pharmacogenomics* 10, 399–416. doi: 10.2217/14622416.10.3.399

Mogil, J. S. (2012). Pain genetics: past, present and future. *Trends Genet.* 28, 258–266. doi: 10.1016/j.tig.2012.02.004

Mutso, A. A., Radzicki, D., Baliki, M. N., Huang, L., Banisadr, G., Centeno, M. V., et al. (2012). Abnormalities in hippocampal functioning with persistent pain. *J. Neurosci.* 32, 5747–5756. doi: 10.1523/JNEUROSCI.0587-12.2012

Myers, R. R., Campana, W. M., and Shubayev, V. I. (2006) The role of neuroinflammation in neuropathic pain: mechanisms and therapeutic targets. *Drug Discov. Today* 11, 8–20. doi: 10.1016/S1359-6446(05)03637-8

Neely, G. G., Hess, A., Costigan, M., Keene, A. C., Goulas, S., Langeslag, M., et al. (2010). A genome-wide *Drosophila* screen for heat nociception identifies a2d3 as evolutionarily conserved pain gene. *Cell* 143, 628–638. doi: 10.1016/j.cell.2010.09.047

Ni, J., Gao, Y., Gong, S., Guo, S., Hisamitsu, T., and Jiang, X. (2012). Regulation of m-opioid type 1 receptors my microRNA134 in dorsal root gnaglion neurons following peripheral inflammation. *Eur. J. Pain* doi: 10.1002/j.1532-2149.2012.00197.x

Nielsen, C. C., Knudsen, G. P., and Steingrimsdottir, O. A. (2012). Twin studies of pain. *Clin. Genet.* 82, 331–340. doi: 10.1111/j.1399-0004.2012.01938.x

Numakawa, T., Yamamoto, N., Chiba, S., Richards, M., Ooshima, Y., Kishi, S., et al. (2011). Growth factors stimulate expression of neuronal and glial miR-132. *Neurosci. Lett.* 505, 242–247. doi: 10.1016/j.neulet.2011.10.025

O'Connor, R. M., Dinan, T. G., and Cryan, J. F. (2012). Little things on which happiness depends: micrRNAs as novel therapeutic targets for the treatment of axiety and depression. *Mol. Psychiatry* 17, 359–376. doi: 10.1038/mp.2011.162

Orlova, I. A., Alexander, G. M., Qureshi, R. A., Sacan, A., Graziano, A., Barret, J. E., et al. (2011). MicroRNA modulation in complex regional pain syndrome. *J. Transl. Med.* 9, 195–204. doi: 10.1186/1479-5876-9-195

Pabreja, K., Dua, K., Sharma, S., Padi, S. S. V., and Kulkarni, S. K. (2011). Minocycline attenuates the development of diabetic neuropathic pain: possible anti-inflammatory and anti-oxidant mechanisms. *Eur. J. Pharmacol.* 661, 15–21. doi: 10.1016/j.ejphar.2011.04.014

Parkitny, L., McAuley, J. H., Di Pietro, F., Stanton, T. R., O'Connell, N. E., Marinus, J., et al. (2013). Inflammation in complex regional pain syndrome: a systematic review and meta-analysis. *Neurology* 80, 106–117. doi: 10.1212/WNL.0b013e31827b1aa1

Parsons, M. J., Grimm, C. H., Paya-Cano, J. L., Sugden, K., Nietfeld, W., Lehrach, H., et al. (2008). Using hippocampal microRNA expression diffeneces between mouse inbred strains to characterise miRNA function. *Mamm. Genome* 19, 552–560. doi: 10.1007/s00335-008-9116-y

Peng, X., Gralinski, L., Armour, C. D., Ferris, M. T., Thomas, M. J., and Proll S., et al. (2010). Unique signatures of long noncoding RNA expression in response to virus infection and altered innated immune signaling. *mBio.* 1, e00206–e00210. doi: 10.1128/mBio.00206-10

Pernía-Andrade, A. J., Kato, A., Witschi, R., Nyilas, R., Katona, I., Freund, T. F., et al. (2009). Spinal endocannabinoids and CB1 receptors mediate C-fiber-induced heterosynaptic pain sensitization. *Science* 325, 764. doi: 10.1126/science.1171870

Phillips, C. J. (2006). Economic burden of chronic pain. *Exp. Rev. Pharmacoecon. Outcomes Res.* 6, 591–601. doi: 10.1586/14737167.6.5.591

Pietrobon, D., and Striessnig, J. (2003). Neurobiology of migraine. *Nat. Rev. Neurosci.* 4, 386–398. doi: 10.1038/nrn1102

Poh, K. W., Yeo, J. F., and Ong, W. Y. (2011). MicroRNA changes in the mouse prefrontal cortex after inflammatory pain. *Eur. J. Pain* 801, e1–e12.

Ponomarev, E. D., Veremeyko, T., Barteneva, N., Krichevsky, A. M., and Weiner, H. L. (2011). MicroRNA-124 promotes microglia quiescence and suppresses EAE by deactivating macrophages via the C/EPB-a-PU.1 pathway. *Nat. Med.* 17, 64–70. doi: 10.1038/nm.2266

Ponomarev, E. D., Veremeyko, T., and Weiner, H. L. (2013). MicroRNAs are universal regulators of differentiation, activation, and polarization of microglia and macrophages in normal and diseased CNS. *Glia* 61, 91–103. doi: 10.1002/glia.22363

Qin, W., Ren, Q., Liu, T., Huang, Y., and Wang, J. (2013). MicroRNA-155 is a novel suppressor of ovarian cancer-initiating cells that targets CLDN1. *FEBS Lett.* 587, 1434–1439. doi: 10.1016/j.febslet.2013.03.023

Quarta, S., Vogl, C., Constantin, C. E., Üceyler, N., Sommer, C., and Kress, M. (2011). Genetic evidence for an essential role of neuronally expressed IL-6 signal transducer gp130 in the induction and maintenance of experimentally induced mechanical hypersensitivity in vivo and in vitro. *Mol. Pain* 7, 73. doi: 10.1186/1744-8069-7-73

Recchiuti, A., Krishnamoorthy, S., Fredman, G., Chiang, N., and Serhan, C. N. (2011). MicroRNAs in resolution of acute inflammation: identification of novel resolvin D1-miRNA circuits. *FASEB J.* 25, 544–560. doi: 10.1096/fj.10-169599

Rossi, S., and Calin, G. A. (2013). Bioinformatics, non-coding RNAs and its possible application in personalized medicine. *Adv. Exp. Med. Biol.* 774, 21–37. doi: 10.1007/978-94-007-5590-1_2

Saba, R., Störchel, P. H., Aksoy-Aksel, A., Kepura, F., Lippi, G., Plant, T. D., et al. (2012). Dopamine-regulated microRNA miR-181a controls GluA2 surface expression in hippocampal neurons. *Mol. Cell. Biol.* 32, 619–632. doi: 10.1128/MCB.05896-11

Sadosky, A., McDermott, A. M., Brandenburg, N. A., and Strauss, M. (2008). A review of the epidemiology of painful diabetic peripheral neuropathy, postherpetic neuroalgia, and less commonly studied neuropathic pain conditions. *Pain Pract.* 8, 56. doi: 10.1111/j.1533-2500.2007.00164.x

Sakai, A., and Suzuki, H. (2013). Nerve injury-induced upregulation of miR-21 in the primary sensory neurons contributes to neuropathic pain in rats. *Biochem. Biophys. Res. Commun.* 435, 176–181. doi: 10.1016/j.bbrc.2013.04.089

Samad, T. A., Moore, K. A., Sapirstein, A., Billet, S., Allchorne, A., Poole, S., et al. (2001). Interleukin-1b-mediated induction of Cox-2 in the CNS contributes to inflammatory pain hypersensitivty. *Nature* 410, 471–475. doi: 10.1038/35068566

Sanchez-Simon, F. M., Zhang, X. X., Loh, H. H., Law, P. Y., and Rodriguez, R. E. (2010). Morphine regulates dopaminergic neuron differentiation via miR-133b. *Mol. Pharmacol.* 78, 942. doi: 10.1124/mol.110.066837

Sandkühler, J. (2007). Understanding LTP in pain pathways. *Mol. Pain* 3, 9. doi: 10.1186/1744-8069-3-9

Sandkühler, J. (2009). Models and mechanisms of hyperalgesia and allodynia. *Physiol. Rev.* 89, 707–758. doi: 10.1152/physrev.00025.2008

Schweizerhof, M., Stösser, S., Kurejova, M., Njoo, C, Gangadharan, V., Agarwal, N., et al. (2009). Hematopoietic colony-stimulating factors mediate tumor-nerve interactions and bone cancer pain. *Nat. Med.* 15, 802–807. doi: 10.1038/nm.1976

Seminowicz, D. A., Laferriere, A. L., Millecamps, M., Yu, J. S. C., Codderre, T. J., and Bushnell, M. C. (2009). MRI structural brain changes associated with sensory and emotional function in a rat model of long-term neuropathic pain. *Neuroimage* 47, 1007–1014. doi: 10.1016/j.neuroimage.2009.05.068

Sengupta, J. N., Pochiraju, S., Kannampalli, P., Bruckert, M., Addya, S., Yadav, P., et al. (2013). MicroRNA-mediated GABA(Aalpha-1) receptor subunit downregulation in adult spinal cord following neonatal cystitis-induced chronic visceral pain in rats. *Pain* 154, 59–70. doi: 10.1016/j.pain.2012.09.002

Serpente, M., Fenoglio, C., Villa, C., Cortini, F., Cantoni, C., Ridolfi, E., et al. (2011). Role of OLR1 and its regulating hsa-miR369-3p in Alzheimer's disease: genetics and expression analysis. *J. Alzheimers Dis.* 26, 787–793.

Shaked, I., Meerson, A., Wolf, Y., Avni, R., Greenberg, D., Gilboa-Geffen, A., et al. (2009). MicroRNA-132 potentiates cholinergic anti-inflammatory signaling by targeting acetylcholinesterase. *Immunity* 31, 965–973. doi: 10.1016/j.immuni.2009.09.019

Shaltiel, G., Hanan, M., Wolf, Y., Barbash, S., Kovalev, E., Shoham, S., et al. (2013). Hippocampal microRNA-132 mediates stress-inducible cognitive deficits through its acetylcholinesterase target. *Brain Struct. Funct.* 218, 59–72. doi: 10.1007/s00429-011-0376-z

Smith, L. K., Shah, R. R., and Cidlowski, J. A. (2010). Glucocorticoids modulate microRNA expression and processing during lymphocyte apoptosis. *J. Biol. Chem.* 285, 36698–36708. doi: 10.1074/jbc.M110.162123

Sommer, C. (2003). Painful neuropathies. *Curr. Opin. Neurol.* 16, 623–628. doi: 10.1097/00019052-200310000-00009

Sommer, C., and Kress, M. (2004). Recent findings on how proinflammatory cytokines cause pain: peripheral mechanisms in inflammatory and neuropathic hyperalgesia. *Neurosci. Lett.* 361, 184–187. doi: 10.1016/j.neulet.2003.12.007

Soreq, H., and Wolf, Y. (2011). NeurimmiRs: microRNAs in the neuroimmune interface. *Trends Mol. Med.* 17, 548–555. doi: 10.1016/j.molmed.2011.06.009

Straube, S., Derry, S., Moore, R. A., and McQuay, H. J. (2010). Cervico-thoracic or lumbar sympathectomy for neuropathic pain and complex regional pain syndrome. *Cochrane Database Syst. Rev.* CD002918. doi: 10.1002/14651858.CD002918.pub2

Sun, Y., Li, X. Q., Sahbaie, P., Shi, X. Y., Li, W. W., Liang, D. Y., et al. (2012). miR-203 regulates nociceptive sensitization after incision by controlling phospholipase A2 activating protein expression. *Anesthesiology* 117, 626–638. doi: 10.1097/ALN.0b013e31826571aa

Talbot, S., and Couture, R. (2012). Emerging role of microglial kinin B1 receptor in diabetic pain neuropathy. *Exp. Neurol.* 234, 373–381. doi: 10.1016/j.expneurol.2011.11.032

Tam Tam, S., Bastian, I., Zhou, X. F., Vander Hoek, M., Michael, M. Z., Gibbins, I. L., et al. (2011). MicroRNA-143 expression in dorsal root ganglion neurons. *Cell Tissue Res.* 346, 163–173. doi: 10.1007/s00441-011-1263-x

Toyoda, H., Zhao, M. G., and Zhou, M. (2009). Enhnaced quantal release of excitatory transmitter in anterior cingulate cortex of adult mice with chronic pain. *Mol. Pain* 5, 4. doi: 10.1186/1744-8069-5-4.

Üceyler, N., Eberle, T., Rolke, R., Birklein, F., and Sommer, C. (2007a). Differential expression patterns of cytokines in complex regional pain syndrome. *Pain* 132, 195–205. doi: 10.1016/j.pain.2007.07.031

Üceyler, N., Rogausch, J. P., Toyka, K. V., and Sommer, C. (2007b). Differential expression of cytokines in painful and painless neuropathies. *Neurology* 69, 42–49. doi: 10.1212/01.wnl.0000265062.92340.a5

Üceyler, N., Göbel, K., Meuth, S. G., Ortler, S., Stoll, G., Sommer, C., et al. (2010). Deficiency of the negative immune regulator B7-H1 enhances inflammation and neuropathicpain after chronic constriction injury of mouse sciatic nerve. *Exp. Neurol.* 222, 153–160. doi: 10.1016/j.expneurol.2009.12.026

Üceyler, N., Schäfers, M., and Sommer, C. (2009). Modeof action of cytokines on nociceptive neurons. *Exp. Brain Res.* 196, 67–78. doi: 10.1007/s00221-009-1755-z

van Rooji, E., and Olson, E. N. (2012). MicroRNA therapeutics for cardiovascular disease: opportunities and obstacles. *Nat. Rev. Drug Discov.* 11, 860–872. doi: 10.1038/nrd3864

Villa, C., Fenoglio, C., De Riz, M., Clerici, F., Marcone, A., Benussi, L., et al. (2011). Role of hnRNP-A1 and miR-590-3p in neuronal death: genetics and expression analysis in patients with Alzheimer disease and frontotemporal lobar degeneration. *Rejuvenation Res.* 14, 275–281. doi: 10.1089/rej.2010.1123

Vincent, A. M., Callaghan, B. C., Smith, A. L., and Feldman, E. L. (2011). Diabetic neuropathy: cellular mechanisms as therapeutic targets. *Nat. Rev. Neurol.* 7, 573–583. doi: 10.1038/nrneurol.2011.137

von Schack, D., Agostino, M. J., Murray, B. S., Li, Y., Reddy, P. S., Chen, J., et al. (2011). Dynamic changes in the microRNA expression profile reveal mutliple regulatory mechanisms in the spinal nerve ligation model of neuropathic pain. *PLoS ONE* 6:e17670. doi: 10.1371/journal.pone.0017670

Weiland, M., Gao, X. H., Zhou, L., and Mi, Q. S. (2012). Small RNAs have a large impact: circulating microRNAs as biomarkers for human diseases. *RNA Biol.* 9, 850–859. doi: 10.4161/rna.20378

Willemen, H. L., Huo, X. J., Mao-Ying, Q. L., Zijlstra, J., Heijnen, C. J., and Kavelaars, A. (2012). MicroRNA-124 as a novel treatment for persistent hyperalgesia. *J. Neuroinflammation* 9, 143. doi: 10.1186/1742-2094-9–143

Wodarski, R., Clark, A. K., Grist, J., Marchand, F., and Malcangio, M. (2009). Gabapentin reverses microglial activation in the spinal cord of streptozotocin-induced diabetic rats. *Eur. J. Pain* 13, 807–811. doi: 10.1016/j.ejpain.2008.09.010

Wu, M., Jolicoeur, N., Li, Z., Zhang, L., Fortin, Y., L'Abbe, D., et al. (2008). Genetic variations of microRNAs in human cancer and their effects on teh expression of miRNAs. *Carcinogenesis* 29, 1710–1716. doi: 10.1093/carcin/bgn073

Wu, Q., Hwang, C. K., Zheng, H., Wagley, Y., Lin, H. Y., Kim, D. K., et al. (2013). MicroRNA 339 down-regulates m-opioid receptor at the post-transcriptional level in response to opioid treatment. *FASEB J.* 27, 522–535. doi: 10.1096/fj.12-213439

Xie, J., Ameres, S. L., Friedline, R., Hung, J.-H., Zhang, J., Zhang, Y., et al. (2012). Long-term, efficient inhibition of microRNA function in mice using rAAV vectors. *Nat. Methods* 9, 403–409. doi: 10.1038/nmeth.1903

Zeilhofer, H. U., Witschi, R., and Hösl, K. (2009). Subtype-selective GABAA receptor mimetics - novel antihyperalgesic agents? *J. Mol. Med.* 87, 465–469. doi: 10.1007/s00109-009-0454-3

Zhao, J., Lee, M. C., Momin, A., Cendan, C. M., Shepperd, S. T., Baker, M. D., et al. (2010). Small RNAscontrol sodium channel expression, nociceptor excitability and pain thresholds. *J. Neurosci.* 30, 10860–10871. doi: 10.1523/JNEUROSCI.1980-10.2010

Zhao, X., Tan, Z., Zhang, H., Atianjoh, F. E., Zhao, J. Y., Liang, L., et al. (2013). A long noncoding RNA contributes to neurpathic pain by silencing Kcna2 in primary afferent neurons. *Nat. Neurosci.* 16, 1024–1031. doi: 10.1038/nn.3438

18

Emerging roles of alternative pre-mRNA splicing regulation in neuronal development and function

Adam D. Norris and John A. Calarco*

FAS Center for Systems Biology, Harvard University, Cambridge, MA, USA

Correspondence:
John A. Calarco, FAS Center for Systems Biology, Harvard University, 52 Oxford Street, Room 457.40, Cambridge, MA 02138, USA.
e-mail: jcalarco@fas.harvard.edu

Alternative pre-mRNA splicing has the potential to greatly diversify the repertoire of transcripts in multicellular organisms. Increasing evidence suggests that this expansive layer of gene regulation plays a particularly important role in the development and function of the nervous system, one of the most complex organ systems found in nature. In this review, we highlight recent studies that continue to emphasize the influence and contribution of alternative splicing regulation to various aspects of neuronal development in addition to its role in the mature nervous system.

Keywords: alternative splicing, genomics, nervous system, RNA processing, gene regulation

INTRODUCTION

The nervous system is a uniquely complex structure, composed of diverse classes of neuronal cells found in close proximity to one another. These many classes of neurons must form precise synaptic connections with other neurons that can often be separated by large distances, establishing the circuitry governing the ability to sense, interpret, and appropriately respond to stimuli. It is of great interest to understand how the developmental steps of neurogenesis, migration, pathfinding, synapse formation, and maintenance are controlled with such precision. Defects in any of these processes in humans lead to numerous cognitive and motor disabilities (Mitchell, 2011). It is also of importance to better grasp the mechanisms governing the modulation of synaptic strength and plasticity in the mature nervous system, which plays a key role in sensory adaptation, learning and memory, and other behaviors.

Diverse spatio-temporal gene regulatory mechanisms have proven vital for the control of patterning in the nervous system, including regulation of mRNA synthesis by transcription factors (West and Greenberg, 2011), the dynamic alteration of chromatin states through modifying enzymes (Ooi and Wood, 2008; Yoo et al., 2009), turnover or translational repression by microRNAs (Meza-Sosa et al., 2012), RNA decay pathways, and regulation by post-translational modifications (Fukushima et al., 2009). Alternative splicing, the process in which multiple mRNA isoforms can be generated through differential splice site selection in precursor transcripts, is an additional important mechanism of gene regulation, contributing to transcriptomic and proteomic diversity in metazoans (Nilsen and Graveley, 2010). Greater than 95% of human genes undergo alternative splicing (Pan et al., 2008; Wang et al., 2008), and disruption of splicing contributes to a number of genetic diseases (Chakarova et al., 2002; Briese et al., 2005; Winkler et al., 2005; Mordes et al., 2006; Wang and Cooper, 2007). The nervous system exhibits particularly high levels of alternative splicing (Yeo et al., 2004; Grosso et al., 2008). Indeed, a recent large scale study of human tissues found that the cerebellum exhibited the highest degree of alternative splicing among 11 tested tissue samples, containing 50% more differentially expressed alternative exons than the next highest tissue (the testes; de la Grange et al., 2010). These results suggest that regulated splicing can serve as a potential mechanism for generating the high levels of molecular and cellular diversity observed in the nervous system (Lipscombe, 2005; Li et al., 2007).

The fidelity and efficiency of splicing depends on the action of five small nuclear RNAs (snRNAs) functioning as components of ribonucleoprotein particles called snRNPs, in conjunction with up to hundreds of additional auxiliary proteins (Wahl et al., 2009). This elaborate and highly dynamic complex known as the spliceosome regulates splicing with single nucleotide precision (Will and Luhrmann, 2011). Decades of research have begun to elucidate the "splicing code," the complete set of *cis*-acting RNA features (for example, sequence motifs, exon and intron length, secondary structure) and *trans*-acting splicing factors that dictate where and in what context differential splicing will occur in transcripts (Wang and Burge, 2008; Barash et al., 2010). The combination of detailed biochemical experiments with more recent genome-wide approaches and computational analyses have revealed diverse mechanisms by which alternative splicing can occur, and have been described in greater detail in several recent excellent reviews (Blencowe, 2006; Chen and Manley, 2009; Licatalosi and Darnell, 2010; Han et al., 2011; McManus and Graveley, 2011; Irimia and Blencowe, 2012).

In this focused review, we examine the role of alternative splicing during neuronal development and in response to neuronal activity. Although a large number of alternative isoforms derived from important neuronal genes have been reported in the literature, special emphasis is given here to recent findings illuminating the role of specific *trans*-acting splicing factors and select target splicing events they regulate in the biogenesis and function of neurons.

ALTERNATIVE SPLICING IN NEURONAL DEVELOPMENT AND MAINTENANCE

Alternative splicing plays an important role in generating diversity and specificity in the developing and mature nervous system. In this section we describe several recent examples demonstrating the role of alternative splicing in neuronal differentiation, migration and pathfinding, and synapse formation and function (**Figure 1**).

THE RS DOMAIN-CONTAINING SPLICING FACTOR nSR100/SRRM4 CONTROLS NEUROGENESIS

The neural-specific SR-related protein of 100 kDa (nSR100/SRRM4) was identified in a computational and expression based survey of genes encoding RS domain-containing proteins (Calarco et al., 2009). In a similar manner as other members of the SR and SR-related family of proteins, nSR100 was found to regulate alternative splicing decisions. Microarray profiling experiments in mouse neuroblastoma cells and tissues revealed that depletion of nSR100 results in increased skipping of alternatively spliced exons normally included in the brain, suggesting that it mainly acts to promote the inclusion of alternative exons. A significant fraction of genes containing these regulated exons are known to be important for regulating neuronal differentiation, raising the possibility that specific splice variants modulated by nSR100 could contribute to this process. Consistent with this notion, nSR100 was found to play a critical role in neuronal differentiation and neurite extension *in vitro* as well as nervous system and sensory organ development in zebrafish embryos *in vivo* (Calarco et al., 2009). The specific mechanism by which nSR100 regulates alternative splicing in the nervous system remains to be elucidated. However, nSR100 was shown to be required for proper inclusion of a target neural-specific exon using *in vitro* splicing extracts, indicating that it plays a direct biochemical role in promoting exon inclusion (Calarco et al., 2009). It was also found that the introns flanking alternative exons regulated by nSR100 are enriched in pyrimidine rich motifs (Calarco et al., 2009). The majority of these motifs are likely recognized by the polypyrimidine tract binding protein PTBP1 and its tissue-specific paralog PTBP2 (also called neural- or brain-enriched n/brPTB). Consistent with a link between these regulators in modulating neural-specific alternative splicing, many nSR100-dependent alternative splicing events are also regulated by PTBP1 and PTBP2.

More recently, it was discovered that nSR100 indirectly controls the steady-state abundance of a network of transcripts in neuronal cells distinct from the population of mRNAs that it regulates at the level of splicing. Depletion of nSR100 in mouse neuroblastoma cells led to decreased levels of hundreds of transcripts, and a subset of these changes were shown to be dependent on repressor element 1 silencing transcription factor (REST, also known as NRSF), a transcriptional repressor of genes involved in neurogenesis (Raj et al., 2011). In neuronal cells, REST transcripts include an additional exon that results in the introduction of a stop codon and production of a truncated protein lacking domains required for its repressive activity. Raj et al. (2011) found that nSR100 plays a critical role in promoting the inclusion of this alternative exon, suggesting that the expression of nSR100 in neurons contributes to the reduced activity of REST upon differentiation to the neural lineage. Importantly however, REST was also found to directly repress nSR100 transcription in non-neuronal cells and thus indirectly inhibit neural-specific alternative splicing. This negative feedback loop between two gene regulatory levels was found to be important for developmental outcomes in the nervous system, as inhibiting nSR100 expression in mouse brain disrupted cortical neurogenesis, preventing neuronal precursor cells from committing to a neuronal fate (Raj et al., 2011). These results are in agreement with

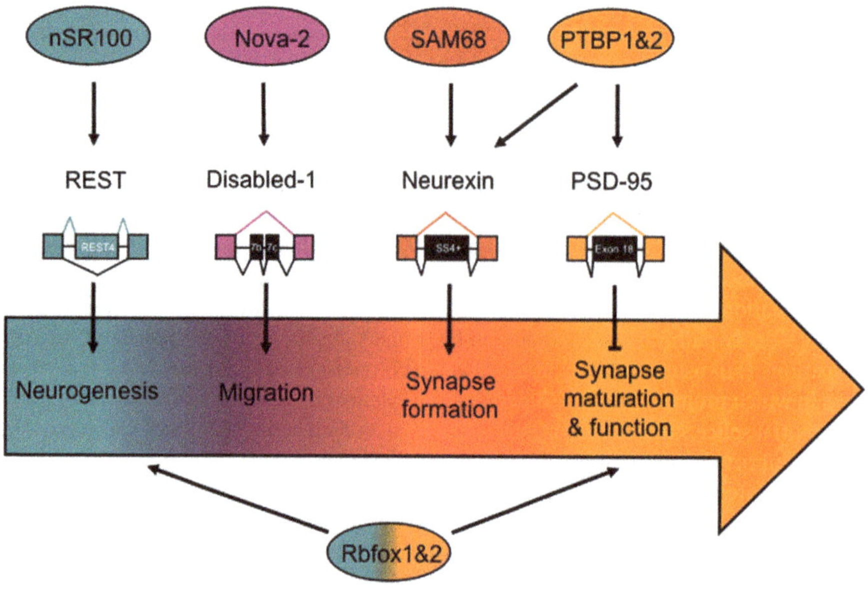

FIGURE 1 | Notable alternative splicing events important for nervous system development, and the factors implicated in regulating their splicing patterns. Developmental stages are represented by the arrow, beginning with neurogenesis, and ending with synapse maturation and function. Splicing factors and target transcripts are correspondingly color coded to the developmental stage in which they are known to be important.

previous studies showing that loss of REST de-represses neuronal transcripts in non-neuronal tissues, while REST overexpression inhibits the expression of transcripts in neuronal tissues, which in one study was shown to result in axon pathfinding errors in chick embryos (Chen et al., 1998; Paquette et al., 2000).

THE SPLICING FACTOR NOVA-2 CONTROLS NEURONAL MIGRATION

The Nova RNA binding proteins were among the first tissue-specific regulators of alternative splicing to be identified. Initial studies in knockout mice indicated that Nova-1 plays a critical role in the maintenance of brainstem and spinal cord neurons, likely through the regulation of alternative splicing in these cells (Jensen et al., 2000). Subsequent studies utilizing splicing-sensitive microarray profiling in mice lacking Nova proteins identified a network of brain-specific splicing events coordinated by these factors (Ule et al., 2005). Importantly, transcripts with Nova-regulated exons encoded proteins that were significantly enriched in functions associated with the synapse (Ule et al., 2005). Integrating this alternative splicing regulatory network with genome-wide cross-linking and immunoprecipitation (CLIP; Licatalosi et al., 2008; Zhang et al., 2010) studies identifying *in vivo* Nova binding sites in the transcriptome has provided two key advances. First, these datasets have enabled the formulation of "RNA maps" correlating cognate Nova YCAY *cis*-element locations in pre-mRNA transcripts with effects on splicing regulation, leading to mechanistic insights into how the Nova proteins modulate alternative splicing (Ule et al., 2006). Second, these networks identify target isoforms and pathways that likely contribute to aspects of neuronal physiology. In agreement with this latter point, subsequent phenotypic exploration of the role of target isoforms in Nova knockout mice has identified a role for Nova in modulating slow synaptic inhibition and in neuromuscular junction formation (Huang et al., 2005; Ule et al., 2005; Ruggiu et al., 2009).

Recently, Nova-2 was demonstrated to be important for the migration of late-born cortical and Purkinje neurons in mice (Yano et al., 2010). These migration defects were due largely to the aberrant splicing of transcripts from a single gene, the Reelin signaling adaptor Disabled-1. Reelin signaling is an important pathway regulating neuronal migration in the cortex, cerebellum, and hippocampus. Binding of the Reelin ligand to its receptors ApoER2 and Vldlr leads to phosphorylation of Disabled-1, which recruits various adaptor proteins that mediate cytoskeletal rearrangements and appropriate neuronal migration and positioning (Bar et al., 2000; Ayala et al., 2007). Disabled-1 is thus a critical effector protein in the Reelin signaling pathway.

Nova-2 suppresses the inclusion of Disabled-1 exons 7b and 7c (7bc+), which encode an additional 33 amino acid peptide of unknown function, the inclusion of which could produce a protein isoform with dominant negative activity. Expression of Disabled-1 (7bc+) in the E14.5 mouse cortex was sufficient to cause migration defects similar to the Nova-2 knockout, while Disabled-1 (7bc−) substantially rescued the Nova-2 knockout migration defects (Yano et al., 2010). Thus, Nova-2 controls the sensitivity of neurons to the Reelin signaling pathway, presumably by affecting the balance of Disabled-1 (7bc−), which is activated by the Reelin signaling pathway, and Disabled-1 (7bc+), which, through a currently unknown mechanism, impairs the Reelin signaling pathway. The alternative splicing of Disabled-1 is developmentally regulated, with Nova2-dependent suppression of Disabled-1 (7bc+) highest during the critical window of migration for late-born neurons (E14.5–E16.5), suggesting Nova2-mediated alternative splicing of Disabled-1 as a mechanism to control neuronal sensitivity to Reelin signaling throughout development (Yano et al., 2010).

PTBP1 AND PTBP2 REGULATE SYNAPSE FORMATION AND MAINTENANCE

PTBP1 and PTBP2 display mutually exclusive patterns of expression in the developing brain, with PTBP1 found in glial and non-neuronal cells, and PTBP2 in neurons (Boutz et al., 2007). This non-overlapping pattern of expression is established by an elegant cross-regulatory network where PTBP1 normally suppresses the inclusion of an exon in PTBP2 transcripts, leading to a non-functional isoform degraded by the nonsense-mediated mRNA decay (NMD) pathway (Boutz et al., 2007; Spellman et al., 2007). In neurons however, PTBP1 is silenced by miR-124, a neuron-specific microRNA, leading to the de-repression of PTBP2 (Makeyev et al., 2007). The consequences of modulating the relative levels of PTBP1 and PTBP2 in neuronal cells have been initially revealed through splicing-sensitive microarray profiling of mouse neuroblastoma cells depleted of these factors (Boutz et al., 2007). Analogous to the Nova-regulated alternative splicing network, PTBP1- and PTBP2-dependent alternative splicing events are frequently found in transcripts expressed from genes with known roles in neuronal differentiation and physiology.

A role for PTBP1 and PTBP2 in regulating the expression of PSD-95, an important scaffolding protein essential for synaptic maturation and plasticity of excitatory neurons, has recently been identified. Overexpression of PTBP1 and PTBP2 in cultured hippocampal neurons was shown to repress synaptic activity, dendritic spine formation, and reduce levels of PSD-95 transcripts (Zheng et al., 2012). This reduced mRNA abundance is caused by PTBP1 and PTBP2 binding to a pyrimidine rich *cis*-element upstream of PSD-95 exon 18, leading to increased exon skipping and the production of a transcript containing a premature termination codon that is targeted for degradation by the NMD pathway. Importantly, the increased expression of PSD-95 in developing neurons in the cortex was found to correlate with three distinct phases of PTBP1 and PTBP2 expression. At the neural progenitor stage, when PTBP1 levels are high, PSD-95 expression is at its lowest. In embryonic neurons, the weaker repressor PTBP2 is more highly expressed while PTBP1 expression is lost, leading to intermediate levels of PSD-95. Finally, in post-natal cortical neurons, PTBP2 is no longer expressed, allowing PSD-95 abundance to reach its highest levels (Zheng et al., 2012). These results indicate that the sequential changes in relative expression of PTBP1 and PTBP2 can allow for distinct splicing regulatory programs to be established at different stages in neuronal maturation.

ALTERNATIVE SPLICING OF NEUREXINS AND NEUROLIGINS IN SYNAPSE FORMATION AND MAINTENANCE

A number of studies in recent years have demonstrated the importance of alternative splicing of neurexins and their binding partners neuroligins in establishing and/or maintaining synapses

(Boucard et al., 2005; Chih et al., 2006; Graf et al., 2006). Neurexins and neuroligins function as adhesion proteins across the synaptic cleft, and increasing evidence suggest that these factors are central organizing proteins at both glutamatergic and GABAergic synapses in the brain (Craig and Kang, 2007). The Neurexin gene loci are highly complex, with the capacity of generating thousands of potential transcript variants pre-synaptically in mammals through the use of alternative promoters and alternative splicing (Boucard et al., 2005; Chih et al., 2006). The post-synaptic neuroligins also undergo alternative splicing, but to a lesser degree. Several key variants from each of these factors have been functionally characterized in cell culture, leading to the proposal of a trans-synaptic adhesive splicing "code" in which particular neurexin isoforms have specific affinity to particular neuroligin isoforms, and the isoforms utilized in neurons affect the functional properties of the synapse (Boucard et al., 2005; Chih et al., 2006; Graf et al., 2006). For instance, the addition of an alternative exon (B+) to neuroligin 1 decreased its ability to recruit GABAergic synaptic components but increased its glutamatergic synaptic recruitment. This change in activity was due to the reduced binding of neuroligin 1 (B+) isoforms to neurexin variants with splice site #4 selected (SS4+; Chih et al., 2006). Neuroligin (B+) bound neurexin (SS4−) strongly but exhibited only weak binding with neurexin (SS4+), while neuroligin (B−) had strong interaction with both neurexin (SS4+) and (SS4−) isoforms. These results point toward a role of neurexin and neuroligin alternative splicing in shaping the strength and class of synapses (Chih et al., 2006).

Several factors involved in the splicing of neurexin transcripts have been identified. The first was PTBP2, which was demonstrated to suppress selection of SS4 in neurexin-2α (Resnick et al., 2008). More recently it has been demonstrated that the KH domain RNA binding protein SAM68 regulates selection of the SS4 in neurexin 1 and neurexin 3, and that it does so in a neuronal activity-dependent fashion (further discussed below; Iijima et al., 2011).

Rbfox-1/A2BP1 AND Rbfox-2/Rbm9 PLAY A ROLE IN NEURONAL DEVELOPMENT AND FUNCTION

Members of the Rbfox family of RNA binding proteins display enriched or highly specific expression patterns in the neuromuscular system, and regulate alternative splicing decisions through interactions with the highly conserved *cis*-element (U)GCAUG (Underwood et al., 2005; Zhang et al., 2008; Zhou and Lou, 2008; Sun et al., 2012). Focused biochemical studies and several genome-wide analyses have demonstrated that the Rbfox proteins can function as activators or repressors of splicing, depending on the location of (U)GCAUG elements in target pre-mRNA transcripts (Jin et al., 2003; Zhang et al., 2008; Zhou and Lou, 2008; Sun et al., 2012). Together, these studies have begun to shed light on the relevant networks of transcripts modulated by these factors, although the role of the Rbfox proteins in nervous system development and function *in vivo* has remained somewhat unclear.

Two recent studies from the Black laboratory using Rbfox knockout mice have provided further insight toward the functional importance of these proteins in the nervous system (Gehman et al., 2011, 2012). Deletion of *Rbfox1* specifically in the nervous system of transgenic mice did not seem to have any effects on neuronal development or morphology in the brain. However, loss of *Rbfox1* did lead to spontaneous seizures, increased sensitivity to induced seizures, and increased excitability in neurons of the dentate gyrus. Integration of splicing-sensitive microarray profiling and CLIP-Seq datasets identified alternative splicing events differentially regulated in the brains of *Rbfox1*$^{-/-}$ mice, several of which were linked to genes known to be associated with epilepsy and others with roles in synaptic function (Gehman et al., 2011).

In contrast to loss of Rbfox1 in the nervous system, deletion of the gene encoding Rbfox2 in the nervous system led to pronounced defects in cerebellar development. *Rbfox2*$^{-/-}$ animals have much smaller cerebella than wild-type littermates, defects in Purkinje cell migration and dendritic arborization, and reduction in the migration and number of granule cells (Gehman et al., 2012). Again, splicing-sensitive microarray profiling experiments were performed, revealing alternative splicing events displaying significant changes upon loss of Rbfox2. Genes with affected exons were associated with neuronal development and function, and a subset of Rbfox2-dependent alternative splicing events were also regulated by Rbfox1, suggesting partial redundancy between the two factors. In agreement with these data, double knockout mice displayed far more severe phenotypes than those observed in either single knockout mutant (Gehman et al., 2012). Finally, in an attempt to separate a possible role for both Rbfox proteins in the mature nervous system from their collective role in development, transgenic animals were generated that deleted these two factors specifically in Purkinje cells. Intriguingly, these double knockout mice possess no gross morphological or developmental abnormalities, but display impaired motor skills and significant reductions in spontaneous firing frequency of Purkinje cells, demonstrating that the Rbfox proteins also play an important role in mature neural circuitry in addition to their contribution to development (Gehman et al., 2012).

NEURONAL ACTIVITY-DEPENDENT ALTERNATIVE SPLICING REGULATION

DEPOLARIZATION CAUSES CHANGES IN SPLICING OF NEURONAL TRANSCRIPTS

Calcium signaling has long been recognized to play an important role in various cellular processes such as muscle contraction and gene transcription, and in neurons it is critical for modulating neuronal activity and for learning and memory (West et al., 2001). Increasing evidence suggests that depolarization-induced calcium influx can also regulate alternative splicing in neurons. One of the early studies showing splicing differences in response to neuronal depolarization utilized the cholinergic agonist pilocarpine administered to the brains of rats. Chronic induction of depolarization with pilocarpine caused altered alternative splicing patterns in a number of key neuronal transcripts in the rat hippocampus and cortex, including tra2-beta, clathrin light chain B, NMDAR1, and *c-src* (Daoud et al., 1999).

A number of further studies using chemical treatments to induce or inhibit calcium signaling in neurons have revealed additional calcium-dependent alternative splicing events. For example, Ania-6, an RNA polymerase II-associated cyclin, exhibited increased inclusion of intron 6 upon glutamate stimulation, but decreased inclusion when stimulated by depolarizing concentrations of KCl. Increased intron inclusion leads to altered protein localization such that the longer isoform is found in nuclear speckles and is associated with hyperphosphorylated RNA Pol II (Berke et al., 2001; Sgambato et al., 2003). In a separate study, mature transcripts encoding SNAP25, a membrane-bound component of the SNARE complex essential for synaptic vesicle fusion, were found to include one of two mutually exclusive alternative exons (5a or 5b). Chronic depolarization of PC12 cells or of cerebellar granule cells by exposure to elevated extracellular K^+ resulted in altered splicing in which the abundance of the 5b isoform is increased (Hepp et al., 2001).

MODULATION OF NEURONAL ACTIVITY DURING CIRCADIAN RHYTHMS

Another potentially interesting physiological process that influences neuronal activity over sufficiently long timescales to involve new gene synthesis and RNA processing lies in the regulation of circadian rhythms. The connection between regulated splicing and circadian rhythms in the nervous system has been supported by a recent study using RNA-Seq in the *Drosophila* brain, which identified numerous splicing events regulated in response to circadian time or period, including splicing of key circadian genes (Hughes et al., 2012). Furthermore, Sanchez et al. (2010) identified the arginine methyl transferase PRMT5, which methylates arginine residues in the spliceosomal Sm proteins, in a screen for novel genes affecting circadian clock regulation in *Arabidopsis*. Mutations in PRMT5 were found to affect both transcription and alternative splicing of many transcripts, including several components of the circadian clock (Sanchez et al., 2010). The authors further demonstrated that a mutation in the *Drosophila prmt5* ortholog causes aberrant circadian-dependent behavior as well as altered mRNA splicing patterns. In both organisms, loss of PRMT5 led predominantly to increased intron retention. Taken together, these findings suggest that PRMT5 directly methylates splicing factors, though an alternative model in which PRMT5 leads to epigenetic changes cannot be ruled out (Sanchez et al., 2010). Continued exploration of the mechanisms controlling circadian regulation of alternative splicing will undoubtedly reveal novel insights.

MECHANISMS OF ACTIVITY-DEPENDENT ALTERNATIVE SPLICING

Although key neuronal transcripts undergoing depolarization-dependent alternative splicing have been discovered, our understanding of the mechanisms controlling this phenomenon is still in its infancy. In the sections that follow, we will highlight current progress in elucidating the role of *cis*-elements, chromatin states, and RNA binding protein modification as regulators of activity-dependent splicing (see **Figure 2** for an outline of examples).

mRNA ELEMENTS NECESSARY FOR ACTIVITY-DEPENDENT SPLICING

Several *cis*-elements regulating alternative splicing changes in response to neuronal depolarization have been identified. The BK (Big Potassium) channel encoded by the *Slo* gene in mammals is important for determining calcium and voltage sensitivity in neurons. The STREX exon of *Slo*, which contributes to enhanced neuronal sensitivity to Calcium when included in transcripts (Saito et al., 1997; Xie and Black, 2001), undergoes increased exon exclusion in response to KCl-mediated depolarization in cultured cells (Xie and Black, 2001). This depolarization-dependent alternative splicing required the Ca^{2+}/calmodulin-dependent protein kinase CaMKIV. The *cis*-elements that conferred responsiveness to CAMKIV signaling were identified, and came to be known as CaRREs (CaMKIV-responsive RNA elements; Xie and Black, 2001). Since then, a number of additional depolarization-dependent alternatively spliced transcripts have been shown to contain CaRREs and to be responsive to CaMKIV (Xie et al., 2005; Lee et al., 2007). Candidate RNA binding proteins that act downstream of CaMKIV signaling and bind to the CaRREs had remained elusive until recent years, when the heterogeneous ribonucleoprotein hnRNP L was identified to interact with CaRRE1 at the

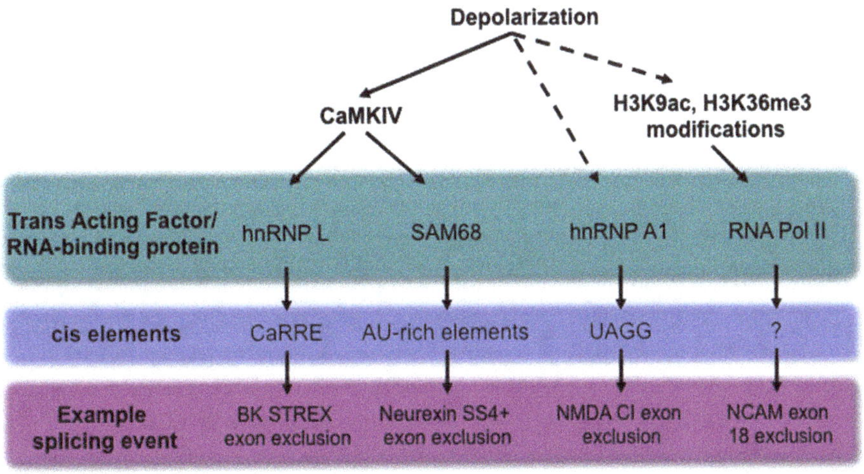

FIGURE 2 | Mechanisms of depolarization-dependent alternative splicing, including known trans-acting factors, the *cis*-elements with which they interact, and a representative alternative splicing event regulated by each specific pathway. Established mechanisms are shown as solid arrows, while activities with unknown intermediates are shown with dashed lines.

upstream 3′ splice site of the STREX exon (Yu et al., 2009). Knockdown of hnRNP L led to increased inclusion of the STREX exon prior to depolarization. However, upon treatment with KCl, cells with reduced hnRNP L levels had smaller STREX exon inclusion changes relative to control cells, but still showed some response to depolarization (Yu et al., 2009). These results suggest that other factors in addition to hnRNP L play a role in the observed depolarization-mediated effects. Indeed, a more recent study has now demonstrated that hnRNP L-like (hnRNP LL) acts redundantly with hnRNP L for the complete modulation of the STREX exon in response to depolarization (Liu et al., 2012).

An additional *cis*-element involved in activity-dependent splicing was uncovered in experiments demonstrating that K^+-induced alternative splicing of the CI cassette exon of the NMDA R1 receptor relied on the presence of two exonic UAGG silencing motifs. These motifs were previously identified as part of a multicomponent regulatory code involving 5′ splice site proximal GGGG elements in coordinating tissue-specific splicing regulation of the CI exon (Han et al., 2005). Introduction of the UAGG silencing motif into a constitutively spliced exon in an unrelated gene led to an increase in exon skipping, and importantly, further increased skipping in response to extracellular K^+-induced depolarization (An and Grabowski, 2007). Biochemical studies demonstrated that the association of the heterogeneous nuclear ribonucleoprotein hnRNP A1 with these UAGG motifs was increased in response to cellular excitation (An and Grabowski, 2007). Although it is currently unclear how this increased association is induced, these results suggest that a signaling cascade must exist that connects responses to depolarization at the plasma membrane and in the cytoplasm with subsequent effects on activity in the nucleus. Interestingly, hnRNP A1 has been demonstrated to alter its shuttling state between the nucleus and cytoplasm in response to osmotic shock (Allemand et al., 2005). It will be interesting to determine whether depolarization leads to increased shuttling of hnRNP A1 to the nucleus, which would offer a possible explanation for the elevated association with UAGG elements observed.

MODIFICATION OF CHROMATIN

Nucleosome positioning and chromatin modification have been recognized as important factors in memory formation and consolidation (Guan et al., 2002; Levenson et al., 2004). More recently, chromatin state and epigenetic marks, including post-translational histone-tail modifications and DNA methylation, have been found to play an important role in alternative splicing regulation (Hnilicova and Stanek, 2011; Luco and Misteli, 2011). Genome-wide analyses have indicated that nucleosomes have increased occupancy in exons compared to flanking intronic regions, and that local enrichment of certain histone modifications can facilitate alternative exon inclusion (see below; Nahkuri et al., 2009; Schwartz et al., 2009; Tilgner et al., 2009; Huff et al., 2010). Conversely, splicing has also been found to be important for the establishment of histone H3 lysine 36 methylation in intron-containing genes, suggesting a bi-directional communication (de Almeida et al., 2011; Kim et al., 2011). Additionally, and not mutually exclusive with the influence of chromatin state, the rate of RNA polymerase II (pol II) elongation during nascent transcript synthesis has also been found to affect alternative splicing (de la Mata et al., 2003). Two models have been proposed to possibly account for how pol II elongation can regulate alternative splicing. First, the recruitment model suggests that splicing factors can directly associate with pol II, likely via the C-terminal domain (CTD) of its largest subunit. These pol II-recruited splicing factors would then be available to recognize cognate *cis*-elements found in nascent pre-mRNA transcripts. Second, the kinetic model posits that chromatin structure influences the local rate of pol II transcription elongation, potentially exposing normally weak splice sites for extended periods of time, thereby allowing them to be more efficiently recognized by the spliceosome (Kornblihtt, 2007; Munoz et al., 2010).

Several recent studies have reinforced the hypothesis that local enrichment of distinct histone marks and DNA methylation status at alternative exons and flanking sequences can modulate pol II elongation rate and alternative splicing. First, Luco et al. (2010) have demonstrated using chromatin-immunoprecipitation assays that H3-K36me3 marks are enriched around a set of alternative splicing events regulated by PTBP1. The histone-tail binding protein MRG15, which specifically recognizes H3-K36me3, was also found to be in a physical complex with PTBP1, thus providing a link between histone modifications and the potential recruitment of splicing factors (Luco et al., 2010). In a separate study, Shukla et al. (2011) discovered that the DNA-binding protein CCCTC-binding factor (CTCF) can mediate local pausing of pol II and inclusion of weak alternative exons. These authors further revealed that the action of CTCF is inhibited by DNA methylation at these regulated exons, suggesting a mechanistic link between elongation rate, methylation, and CTCF binding in modulating alternative splicing. Finally, Close et al. (2012) have identified a novel polymerase-associated complex called DBIRD that was found to promote exon skipping. Depletion of components of the DBIRD complex was found to predominantly increase pol II occupancy surrounding regulated exons, leading to more inclusion. In a search for *cis*-elements associated with these DBIRD-sensitive exons, the authors identified enrichment of (A + T) rich sequences, which have been shown in previous studies to act as pol II elongation pause sites *in vitro* (Close et al., 2012).

An intriguing study has also implicated neuronal depolarization in the control of alternative splicing by affecting RNA Pol II transcription kinetics. Exon 18 of Neural Cell Adhesion Molecule (NCAM) transcripts undergoes developmentally regulated alternative splicing in which the exon-excluded isoform (NCAM140) is abundant in neuronal precursors while the exon-included isoform (NCAM180) is increasingly expressed throughout the process of neuronal differentiation (Pollerberg et al., 1985, 1986; Cunningham et al., 1987). Schor et al. (2009) demonstrated that NCAM exon 18 skipping increased in response to neuronal depolarization with KCl. This depolarization-mediated increase in exon skipping was not dependent on CaMKIV (Schor et al., 2009), but instead involved histone modification changes specifically in the vicinity of exon 18. Depolarization led to an increase in histone H3-K9 acetylation and H3-K36 tri-methylation exclusively in the region between exons 17 and 19, as well as a local increase in chromatin relaxation and accessibility. Furthermore, exon 18 inclusion could be artificially recapitulated by either using a mutant "slow" pol II or by applying the drug trichostatin which inhibits histone deacetylation. These results are consistent with a kinetic coupling model in which depolarization leads to specific local histone modifications in the region of the alternative exon causing local chromatin relaxation, in turn increasing the speed of pol II transit through the exon and facilitating increased exon 18 exclusion (Schor et al., 2009).

REGULATION OF RNA BINDING SPLICING FACTORS

The activity or levels of RNA binding proteins that regulate alternative splicing can be modified in response to neuronal activity. For instance, phosphorylation of hnRNP L on serine 513 by CaMKIV was found to play a crucial role in the differential regulation of STREX exon inclusion upon treatment of cells with KCl (Liu et al., 2012), providing a molecular link between signaling downstream of the stimulus and subsequent effects on alternative splicing. Two recent studies have implicated the RNA binding protein Rbfox-1/A2BP1 as an important splicing factor mediating activity-dependent alternative splicing (Lee et al., 2009; Amir-Zilberstein et al., 2012). First, transcripts encoding Rbfox-1 itself were identified as a target of depolarization-dependent splicing in mouse cells, where exon 19 was found to be excluded in response to depolarization. Exclusion of exon 19 led to the accumulation of a Rbfox-1 protein isoform targeted to the nucleus, where it re-activated the inclusion of target exons initially displaying more skipping upon depolarization (Lee et al., 2009). Thus, Rbfox-1 lacking exon 19 counteracted the effects of depolarization-dependent exon exclusion, suggesting a novel feedback-based mechanism for adapting to chronic neuronal depolarization. A second study implicated Rbfox-1 as an important downstream target of the hypothalamic transcription factor Orthopedia (Otp) in response to stress (Amir-Zilberstein et al., 2012). Rbfox-1 was demonstrated to be a transcriptional target of Otp, and Rbfox-1 transcript levels were upregulated in response to stress in mice. Rbfox-1 in turn was responsible for an increase in exon 14 inclusion in the pituitary adenylate cyclase-activating peptide (PACAP) receptor PAC1. Inclusion of PAC1 exon 14 led to a decrease in the levels of stress-induced corticotropin-releasing hormone (CRH). These results suggest that Rbfox-1-mediated inclusion of PAC1 exon 14 creates an isoform that helps terminate stress-induced transcription of CRH. In agreement with this model, zebrafish injected with morpholinos inhibiting PAC1 exon 14 inclusion exhibited abnormal "anxiety like" behavior and sustained expression of CRH transcription (Amir-Zilberstein et al., 2012).

Another recent study has provided insight into the mechanism controlling depolarization-dependent splicing of the neurexin SS4+ alternative isoform. As discussed above, alternative splicing of SS4 can modulate the affinity of neurexins for their post-synaptic ligands. Iijima et al. (2011) have now shown that SS4 selection can be suppressed in response to induced depolarization by various methods in cultured mouse neurons. This suppression of exon inclusion is dependent on CaMKIV and results in altered *trans*-synaptic signaling in response to depolarization. The STAR family RNA binding protein SAM68 was shown to be required for depolarization-dependent splicing of SS4, and to bind directly to AU-rich response elements in the neurexin pre-mRNA. Although SAM68 protein levels and localization were not affected by depolarization, a serine residue within a consensus CaMKIV recognition motif was found to be more highly phosphorylated following depolarization. These results suggest a model where neuronal depolarization affects CaMKIV due to increased intracellular calcium, leading to the phosphorylation of SAM68, which then alters neurexin splicing. Interestingly, the neurexin pre-mRNA does not contain recognizable CaRRE sequences, and a number of transcripts containing CaRRE sequences were not affected by loss of SAM68 (Iijima et al., 2011). Thus, it appears that CamKIV-dependent alternative splicing regulation depends on multiple downstream RNA binding proteins binding to distinct *cis*-elements.

PERSPECTIVES AND FUTURE DIRECTIONS

Neuronal depolarization can affect the splicing of many transcripts in the nervous system, but the mechanisms by which it does so still remain largely unknown. A major future challenge will be to identify the signaling cascades in addition to the CAMKIV pathway linking cellular excitation to alternative splicing via changes in the activity of splicing factors, chromatin state, and perhaps additional mechanisms. Moreover, experiments thus far revealing a role of depolarization in nervous system alternative splicing have relied on *in vitro* cell culture models requiring chronic depolarization for many hours to modulate the firing activity of neurons. It will now be important to understand how chronic neuronal depolarization affects alternative splicing regulation *in vivo*, and what consequences the affected isoforms have on neuronal physiology. Many depolarization-responsive alternative splicing events that have been identified are found in transcripts encoding channel proteins, neurotransmitter receptors, and other modulators of synaptic strength (Xie and Black, 2001; Lee et al., 2007; Iijima et al., 2011). Phenomena such as synaptic gain control and homeostasis, where synapses can alter their sensitivity in response to chronic hyper- or hypo-stimulation (Burrone and Murthy, 2003), are thought to occur over the course of hours. This time-frame overlaps well with the temporal dynamics of depolarization-induced alternative splicing changes observed *in vitro*. While it may be a technically challenging feat, genome-wide analyses of splicing changes in organisms maintained under differing stimuli or behavioral paradigms inducing such synaptic gain control or homeostatic maintenance would provide further insight into the mechanisms and relevance of neuronal activity in regulating alternative splicing *in vivo*.

The fact that the activity of master regulators of gene expression such as transcription factors can be modified by alternative splicing has blurred the lines of how differentiation programs in cell lineages are established (Gabut et al., 2011; Raj et al., 2011). In a broader sense, these results raise an interesting implication, namely, that regulation of alternative splicing events by RNA binding proteins can play a causal rather than simply consequential role in developmental transitions and activity states of neurons. Future experiments establishing the contribution of splicing and other RNA binding regulators to the identity and fate of neuronal lineages represents an important goal in basic research but also in biomedical applications such as regenerative medicine. While a handful of neuronal-specific splicing regulators have been discovered, it is unlikely that the current repertoire of known factors is sufficient to account for the remarkable degree of splicing complexity observed in the nervous system. Hundreds of RNA binding proteins have been identified in metazoan genomes, and many of them remain uncharacterized. As such, identification and characterization of novel regulators of splicing will be important. Large scale RNAi and/or cDNA overexpression studies in cell culture models will be useful for identifying factors whose inhibition or overexpression affect splicing in the nervous system. Importantly, invertebrate model organisms provide a valuable platform for performing forward and reverse genetic screens to identify previously uncharacterized factors affecting splicing. In addition to identifying RNA binding proteins, such unbiased screens may also uncover novel classes of genes not previously known to regulate alternative splicing, such as non-coding RNAs and chromatin regulators (Luco et al., 2010; Tripathi et al., 2010). Furthermore, it will be important to understand splicing regulation in the nervous system in a more spatially resolved manner, for example, by identifying brain sub-region and neuronal-subtype specific alternative splicing events, the factors that control these events, and the effects they have on specification and function of individual classes of neurons.

Although progress has been made toward understanding splicing changes in the development of the nervous system, much less is known about the interplay between alternative splicing regulation and aging in the brain. It has been known for some time that aberrant splicing of the LMNA gene leads to an accelerated aging phenotype found in individuals with Hutchinson-Gilford Progeria Syndrome (HGPS; Todorova et al., 2003). Recent studies using patient cell lines and HGPS mouse models have identified candidate regulators involved in the cryptic splicing of LMNA (Lopez-Mejia et al., 2011). It is tempting to speculate that analysis of the aging brain will also implicate splicing factors in both aging-related splicing changes and the gradual deterioration of the nervous system. Indeed, a recent study has identified dynamic alterations in splicing during normal brain aging consistent with an increase in PTB-dependent splicing, as well as splicing changes in diseased brain consistent with decreased NOVA-dependent splicing (Tollervey et al., 2011). In the future, it will be important to more directly understand the mechanisms and the consequences of splicing in the aging brain.

Finally, mutations in several RNA binding proteins expressed in the nervous system have been associated with neurodevelopmental disorders (Grabowski and Black, 2001; Wang and Cooper, 2007; Morikawa and Manabe, 2010). Advances in genome-wide approaches to globally monitor transcripts bound by these proteins and their effects on aspects of mRNA metabolism, including alternative splicing, are beginning to shed light on underlying mechanisms of action and provide a more detailed understanding of disease etiology. Deeper investigation of the transcript networks regulated by these RNA binding proteins will hopefully provide promising new insight into the development of treatments for some of these disorders.

As demonstrated by some of the examples described above, alternative splicing has the potential to generate multiple protein isoforms, but can also modulate other properties of mRNA transcripts, including their stability. Two recent studies have suggested that tissue-specific alternative exons can frequently encode structurally disordered regions in proteins and have the potential to influence post-translational modification and protein–protein interaction interfaces (Buljan et al., 2012; Ellis et al., 2012). These observations collectively indicate that it remains an important goal to develop additional techniques and approaches that will facilitate the characterization of the functional consequences of alternative splicing events in biological pathways.

We are embarking on an exciting time where techniques for large scale analysis of nervous system transcriptome dynamics in distinct cellular subtypes, multiple developmental states, and in response to environmental cues, are constantly improving. These approaches are already revealing a previously unappreciated role for post-transcriptional gene regulatory mechanisms in almost all aspects of nervous system physiology. Continued research taking advantage of these techniques, coupled with emerging computational approaches and more traditional biochemical and molecular genetic assays, will produce a more comprehensive understanding of alternative splicing regulation. These integrative analyses should also shed further light on the interplay between alternative splicing and other layers of gene regulation in generating the constellation of neuronal subtypes and their diverse functional properties.

ACKNOWLEDGMENTS

We would like to thank Ben Blencowe, Mathieu Gabut, and Arneet Saltzman for helpful feedback and comments on the manuscript. Research from our group is supported by Harvard University and the NIH (grant 1DP5OD009153). We would also like to apologize to colleagues whose work we were unable to reference due to space limitations.

REFERENCES

Allemand, E., Guil, S., Myers, M., Moscat, J., Caceres, J. F., and Krainer, A. R. (2005). Regulation of heterogenous nuclear ribonucleoprotein A1 transport by phosphorylation in cells stressed by osmotic shock. *Proc. Natl. Acad. Sci. U.S.A.* 102, 3605–3610.

Amir-Zilberstein, L., Blechman, J., Sztainberg, Y., Norton, W. N. J., Reuveny, A., Borodovsky, N., Tahor, M., Bonkowsky, J. L., Bally-Cuif, L., Chen, A., and Levkowitz, G. (2012). Homeodomain protein otp and activity-dependent splicing modulate neuronal adaptation to stress. *Neuron* 73, 279–291.

An, P., and Grabowski, P. J. (2007). Exon silencing by UAGG motifs in response to neuronal excitation. *PLoS Biol.* 5, e36. doi:10.1371/journal.pbio.0050036

Ayala, R., Shu, T. Z., and Tsai, L. H. (2007). Trekking across the brain: the journey of neuronal migration. *Cell* 128, 29–43.

Bar, I., de Rouvroit, C. L., and Goffinet, A. M. (2000). The Reelin signaling pathway in mouse cortical development. *Eur. J. Morphol.* 38, 321–325.

Barash, Y., Calarco, J. A., Gao, W. J., Pan, Q., Wang, X. C., Shai, O., Blencowe, B. J., and Frey, B. J. (2010). Deciphering the splicing code. *Nature* 465, 53–59.

Berke, J. D., Sgambato, V., Zhu, P. P., Lavoie, B., Vincent, M., Krause, M., and Hyman, S. E. (2001). Dopamine and glutamate induce distinct striatal splice forms of Ania-6, an RNA polymerase II-associated cyclin. *Neuron* 32, 277–287.

Blencowe, B. J. (2006). Alternative splicing: new insights from global analyses. *Cell* 126, 37–47.

Boucard, A. A., Chubykin, A. A., Comoletti, D., Taylor, P., and Sudhof, T. C. (2005). A splice code for trans-synaptic cell adhesion mediated by binding of neuroligin 1 to alpha- and beta-neurexins. *Neuron* 48, 229–236.

Boutz, P. L., Stoilov, P., Li, Q., Lin, C. H., Chawla, G., Ostrow, K., Shiue, L., Ares, M., and Black, D. L. (2007). A post-transcriptional regulatory switch in polypyrimidine tract-binding proteins reprograms alternative splicing in developing neurons. *Genes Dev.* 21, 1636–1652.

Briese, M., Esmaeili, B., and Sattelle, D. B. (2005). Is spinal muscular atrophy the result of defects in motor neuron processes? *Bioessays* 27, 946–957.

Buljan, M., Chalancon, G., Eustermann, S., Wagner, G. P., Fuxreiter, M., Bateman, A., and Babu, M. M. (2012). Tissue-specific splicing of disordered segments that embed binding motifs rewires protein interaction networks. *Mol. Cell* 46, 871–883.

Burrone, J., and Murthy, V. N. (2003). Synaptic gain control and homeostasis. *Curr. Opin. Neurobiol.* 13, 560–567.

Calarco, J. A., Superina, S., O'Hanlon, D., Gabut, M., Raj, B., Pan, Q., Skalska, U., Clarke, L., Gelinas, D., van der Kooy, D., Zhen, M., Ciruna, B., and Blencowe, B. J. (2009). Regulation of vertebrate nervous system alternative splicing and development by an SR-related protein. *Cell* 138, 898–910.

Chakarova, C. F., Hims, M. M., Bolz, H., Abu-Safieh, L., Patel, R. J., Papaioannou, M. G., Inglehearn, C. F., Keen, T. J., Willis, C., Moore, A. T., Rosenberg, T., Webster, A. R., Bird, A. C., Gal, A., Hunt, D., Vithana, E. N., and Bhattacharya, S. S. (2002). Mutations in HPRP3, a third member of pre-mRNA splicing factor genes, implicated in autosomal dominant retinitis pigmentosa. *Hum. Mol. Genet.* 11, 87–92.

Chen, M., and Manley, J. L. (2009). Mechanisms of alternative splicing regulation: insights from molecular and genomics approaches. *Nat. Rev. Mol. Cell Biol.* 10, 741–754.

Chen, Z. F., Paquette, A. J., and Anderson, D. J. (1998). NRSF/REST is required in vivo for repression of multiple neuronal target genes during embryogenesis. *Nat. Genet.* 20, 136–142.

Chih, B., Gollan, L., and Scheiffele, P. (2006). Alternative splicing controls selective trans-synaptic interactions of the neuroligin-neurexin complex. *Neuron* 51, 171–178.

Close, P., East, P., Dirac-Svejstrup, A. B., Hartmann, H., Heron, M., Maslen, S., Chariot, A., Soding, J., Skehel, M., and Svejstrup, J. Q. (2012). DBIRD complex integrates alternative mRNA splicing with RNA polymerase II transcript elongation. *Nature* 484, 386–389.

Craig, A. M., and Kang, Y. (2007). Neurexin-neuroligin signaling in synapse development. *Curr. Opin. Neurobiol.* 17, 43–52.

Cunningham, B. A., Hemperly, J. J., Murray, B. A., Prediger, E. A., Brackenbury, R., and Edelman, G. M. (1987). Neural cell adhesion molecule: structure, immunoglobulin-like domains, cell surface modulation, and alternative RNA splicing. *Science* 236, 799–806.

Daoud, R., Da Penha Berzaghi, M., Siedler, F., Hubener, M., and Stamm, S. (1999). Activity-dependent regulation of alternative splicing patterns in the rat brain. *Eur. J. Neurosci.* 11, 788–802.

de Almeida, S. F., Grosso, A. R., Koch, F., Fenouil, R., Carvalho, S., Andrade, J., Levezinho, H., Gut, M., Eick, D., Gut, I., Andrau, J. C., Ferrier, P., and Carmo-Fonseca, M. (2011). Splicing enhances recruitment of methyltransferase HYPB/Setd2 and methylation of histone H3 Lys36. *Nat. Struct. Mol. Biol.* 18, 977–983.

de la Grange, P., Gratadou, L., Delord, M., Dutertre, M., and Auboeuf, D. (2010). Splicing factor and exon profiling across human tissues. *Nucleic Acids Res.* 38, 2825–2838.

de la Mata, M., Alonso, C. R., Kadener, S., Fededa, J. P., Blaustein, M., Pelisch, F., Cramer, P., Bentley, D., and Kornblihtt, A. R. (2003). A slow RNA polymerase II affects alternative splicing in vivo. *Mol. Cell* 12, 525–532.

Ellis, J. D., Barrios-Rodiles, M., Colak, R., Irimia, M., Kim, T., Calarco, J. A., Wang, X., Pan, Q., O'Hanlon, D., Kim, P. M., Wrana, J. L., and Blencowe, B. J. (2012). Tissue-specific alternative splicing remodels protein-protein interaction networks. *Mol. Cell* 46, 884–892.

Fukushima, N., Furuta, D., Hidaka, Y., Moriyama, R., and Tsujiuchi, T. (2009). Post-translational modifications of tubulin in the nervous system. *J. Neurochem.* 109, 683–693.

Gabut, M., Samavarchi-Tehrani, P., Wang, X., Slobodeniuc, V., O'Hanlon, D., Sung, H. K., Alvarez, M., Talukder, S., Pan, Q., Mazzoni, E. O., Nedelec, S., Wichterle, H., Woltjen, K., Hughes, T. R., Zandstra, P. W., Nagy, A., Wrana, J. L., and Blencowe, B. J. (2011). An alternative splicing switch regulates embryonic stem cell pluripotency and reprogramming. *Cell* 147, 132–146.

Gehman, L. T., Meera, P., Stoilov, P., Shiue, L., O'Brien, J. E., Meisler, M. H., Ares, M., Otis, T. S., and Black, D. L. (2012). The splicing regulator Rbfox2 is required for both cerebellar development and mature motor function. *Genes Dev.* 26, 445–460.

Gehman, L. T., Stoilov, P., Maguire, J., Damianov, A., Lin, C. H., Shiue, L., Ares, M., Mody, I., and Black, D. L. (2011). The splicing regulator Rbfox1 (A2BP1) controls neuronal excitation in the mammalian brain. *Nat. Genet.* 43, 706-U133.

Grabowski, P. J., and Black, D. L. (2001). Alternative RNA splicing in the nervous system. *Prog. Neurobiol.* 65, 289–308.

Graf, E. R., Kang, Y., Hauner, A. M., and Craig, A. M. (2006). Structure function and splice site analysis of the synaptogenic activity of the neurexin-1 beta LNS domain. *J. Neurosci.* 26, 4256–4265.

Grosso, A. R., Gomes, A. Q., Barbosa-Morais, N. L., Caldeira, S., Thorne, N. P., Grech, G., von Lindern, M., and Carmo-Fonseca, M. (2008). Tissue-specific splicing factor gene expression signatures. *Nucleic Acids Res.* 36, 4823–4832.

Guan, Z. H., Giustetto, M., Lomvardas, S., Kim, J. H., Miniaci, M. C., Schwartz, J. H., Thanos, D., and Kandel, E. R. (2002). Integration of long-term-memory-related synaptic plasticity involves bidirectional regulation of gene expression and chromatin structure. *Cell* 111, 483–493.

Han, J. H., Xiong, J., Wang, D., and Fu, X. D. (2011). Pre-mRNA splicing: where and when in the nucleus. *Trends Cell Biol.* 21, 336–343.

Han, K., Yeo, G., An, P., Burge, C. B., and Grabowski, P. J. (2005). A combinatorial code for splicing silencing: UAGG and GGGG motifs. *PLoS Biol.* 3, e158. doi:10.1371/journal.pbio.0030158

Hepp, R., Dupont, J. L., Aunis, D., Langley, K., and Grant, N. J. (2001). NGF enhances depolarization effects on SNAP-25 expression: induction of SNAP-25b isoform. *Neuroreport* 12, 673–677.

Hnilicova, J., and Stanek, D. (2011). Where splicing joins chromatin. *Nucleus* 2, 182–188.

Huang, C. S., Shi, S. H., Ule, J., Ruggiu, M., Barker, L. A., Darnell, R. B., Jan, Y. N., and Jan, L. Y. (2005). Common molecular pathways mediate long-term potentiation of synaptic excitation and slow synaptic inhibition. *Cell* 123, 105–118.

Huff, J. T., Plocik, A. M., Guthrie, C., and Yamamoto, K. R. (2010). Reciprocal intronic and exonic histone modification regions in humans. *Nat. Struct. Mol. Biol.* 17, 1495–1499.

Hughes, M. E., Grant, G. R., Paquin, C., Qian, J., and Nitabach, M. N. (2012). Deep sequencing the circadian and diurnal transcriptome of Drosophila brain. *Genome Res.* 22, 1266–1281.

Iijima, T., Wu, K. R., Witte, H., Hanno-Iijima, Y., Glatter, T., Richard, S., and Scheiffele, P. (2011). SAM68 regulates neuronal activity-dependent alternative splicing of neurexin-1. *Cell* 147, 1601–1614.

Irimia, M., and Blencowe, B. J. (2012). Alternative splicing: decoding an expansive regulatory layer. *Curr. Opin. Cell Biol.* 24, 323–332.

Jensen, K. B., Dredge, B. K., Stefani, G., Zhong, R., Buckanovich, R. J., Okano, H. J., Yang, Y. Y. L., and Darnell, R. B. (2000). Nova-1 regulates neuron-specific alternative splicing and is essential for neuronal viability. *Neuron* 25, 359–371.

Jin, Y., Suzuki, H., Maegawa, S., Endo, H., Sugano, S., Hashimoto, K., Yasuda, K., and Inoue, K. (2003). A vertebrate RNA-binding protein Fox-1 regulates tissue-specific splicing via the pentanucleotide GCAUG. *EMBO J.* 22, 905–912.

Kim, S., Kim, H., Fong, N., Erickson, B., and Bentley, D. L. (2011). Pre-mRNA splicing is a determinant of histone H3K36 methylation. *Proc. Natl. Acad. Sci. U.S.A.* 108, 13564–13569.

Kornblihtt, A. R. (2007). Coupling transcription and alternative splicing. *Adv. Exp. Med. Biol.* 623, 175–189.

Lee, J. A., Tang, Z. Z., and Black, D. L. (2009). An inducible change in Fox-1/A2BP1 splicing modulates the alternative splicing of downstream neuronal target exons. *Genes Dev.* 23, 2284–2293.

Lee, J. A., Xing, Y., Nguyen, D., Xie, J. Y., Lee, C. J., and Black, D. L. (2007). Depolarization and CaM kinase IV modulate NMDA receptor splicing through two essential RNA elements. *PLoS Biol.* 5, 281–294. doi:10.1371/journal.pbio.0050040

Levenson, J. M., O'Riordan, K. J., Brown, K. D., Trinh, M. A., Molfese, D. L., and Sweatt, J. D. (2004). Regulation of histone acetylation during memory formation in the hippocampus. *J. Biol. Chem.* 279, 40545–40559.

Li, Q., Lee, J. A., and Black, D. L. (2007). Neuronal regulation of alternative pre-mRNA splicing. *Nat. Rev. Neurosci.* 8, 819–831.

Licatalosi, D. D., and Darnell, R. B. (2010). RNA processing and its regulation: global insights into biological networks. *Nat. Rev. Genet.* 11, 75–87.

Licatalosi, D. D., Mele, A., Fak, J. J., Ule, J., Kayikci, M., Chi, S. W., Clark, T. A., Schweitzer, A. C., Blume, J. E., Wang, X., Darnell, J. C., and Darnell, R. B.

(2008). HITS-CLIP yields genome-wide insights into brain alternative RNA processing. *Nature* 456, 464–469.

Lipscombe, D. (2005). Neuronal proteins custom designed by alternative splicing. *Curr. Opin. Neurobiol.* 15, 358–363.

Liu, G., Razanau, A., Hai, Y., Yu, J., Sohail, M., Lobo, V. G., Chu, J., Kung, S. K., and Xie, J. (2012). A conserved serine of heterogeneous nuclear ribonucleoprotein L (hnRNP L) mediates depolarization-regulated alternative splicing of potassium channels. *J. Biol. Chem.* 287, 22709–22716.

Lopez-Mejia, I. C., Vautrot, V., De Toledo, M., Behm-Ansmant, I., Bourgeois, C. F., Navarro, C. L., Osorio, F. G., Freije, J. M., Stevenin, J., De Sandre-Giovannoli, A., Lopez-Otin, C., Lévy, N., Branlant, C., and Tazi, J. (2011). A conserved splicing mechanism of the LMNA gene controls premature aging. *Hum. Mol. Genet.* 20, 4540–4555.

Luco, R. F., and Misteli, T. (2011). More than a splicing code: integrating the role of RNA, chromatin and non-coding RNA in alternative splicing regulation. *Curr. Opin. Genet. Dev.* 21, 366–372.

Luco, R. F., Pan, Q., Tominaga, K., Blencowe, B. J., Pereira-Smith, O. M., and Misteli, T. (2010). Regulation of alternative splicing by histone modifications. *Science* 327, 996–1000.

Makeyev, E. V., Zhang, J. W., Carrasco, M. A., and Maniatis, T. (2007). The MicroRNA miR-124 promotes neuronal differentiation by triggering brain-specific alternative pre-mRNA splicing. *Mol. Cell* 27, 435–448.

McManus, C. J., and Graveley, B. R. (2011). RNA structure and the mechanisms of alternative splicing. *Curr. Opin. Genet. Dev.* 21, 373–379.

Meza-Sosa, K. F., Valle-Garcia, D., Pedraza-Alva, G., and Perez-Martinez, L. (2012). Role of microRNAs in central nervous system development and pathology. *J. Neurosci. Res.* 90, 1–12.

Mitchell, K. J. (2011). The genetics of neurodevelopmental disease. *Curr. Opin. Neurobiol.* 21, 197–203.

Mordes, D., Luo, X. Y., Kar, A., Kuo, D., Xu, L. L., Fushimi, K., Yu, G. W., Sternberg, P., and Wu, J. Y. (2006). Pre-mRNA splicing and retinitis pigmentosa. *Mol. Vis.* 12, 1259–1271.

Morikawa, T., and Manabe, T. (2010). Aberrant regulation of alternative pre-mRNA splicing in schizophrenia. *Neurochem. Int.* 57, 691–704.

Munoz, M. J., de la Mata, M., and Kornblihtt, A. R. (2010). The carboxy terminal domain of RNA polymerase II and alternative splicing. *Trends Biochem. Sci.* 35, 497–504.

Nahkuri, S., Taft, R. J., and Mattick, J. S. (2009). Nucleosomes are preferentially positioned at exons in somatic and sperm cells. *Cell Cycle* 8, 3420–3424.

Nilsen, T. W., and Graveley, B. R. (2010). Expansion of the eukaryotic proteome by alternative splicing. *Nature* 463, 457–463.

Ooi, L., and Wood, I. C. (2008). Regulation of gene expression in the nervous system. *Biochem. J.* 414, 327–341.

Pan, Q., Shai, O., Lee, L. J., Frey, B. J., and Blencowe, B. J. (2008). Deep surveying of alternative splicing complexity in the human transcriptome by high-throughput sequencing. *Nat. Genet.* 40, 1413–1415.

Paquette, A. J., Perez, S. E., and Anderson, D. J. (2000). Constitutive expression of the neuron-restrictive silencer factor (NRSF)/REST in differentiating neurons disrupts neuronal gene expression and causes axon pathfinding errors in vivo. *Proc. Natl. Acad. Sci. U.S.A.* 97, 12318–12323.

Pollerberg, E. G., Sadoul, R., Goridis, C., and Schachner, M. (1985). Selective expression of the 180-kD component of the neural cell adhesion molecule N-CAM during development. *J. Cell Biol.* 101, 1921–1929.

Pollerberg, G. E., Schachner, M., and Davoust, J. (1986). Differentiation state-dependent surface mobilities of two forms of the neural cell adhesion molecule. *Nature* 324, 462–465.

Raj, B., O'Hanlon, D., Vessey, J. P., Pan, Q., Ray, D., Buckley, N. J., Miller, F. D., and Blencowe, B. J. (2011). Cross-regulation between an alternative splicing activator and a transcription repressor controls neurogenesis. *Mol. Cell* 43, 843–850.

Resnick, M., Segall, A., Rozic-Kotliroff, G., Lupowitz, Z., and Zisapel, N. (2008). Alternative splicing of neurexins: a role for neuronal polypyrimidine tract binding protein. *Neurosci. Lett.* 439, 235–240.

Ruggiu, M., Herbst, R., Kim, N., Jevsek, M., Fak, J. J., Mann, M. A., Fischbach, G., Burden, S. J., and Darnell, R. B. (2009). Rescuing Z+ agrin splicing in Nova null mice restores synapse formation and unmasks a physiologic defect in motor neuron firing. *Proc. Natl. Acad. Sci. U.S.A.* 106, 3513–3518.

Saito, M., Nelson, C., Salkoff, L., and Lingle, C. J. (1997). A cysteine-rich domain defined by a novel exon in a slo variant in rat adrenal chromaffin cells and PC12 cells. *J. Biol. Chem.* 272, 11710–11717.

Sanchez, S. E., Petrillo, E., Beckwith, E. J., Zhang, X., Rugnone, M. L., Hernando, C. E., Cuevas, J. C., Herz, M. A. G., Depetris-Chauvin, A., Simpson, C. G., Brown, J. W., Cerdán, P. D., Borevitz, J. O., Mas, P., Ceriani, M. F., Kornblihtt, A. R., and Yanovsky, M. J. (2010). A methyl transferase links the circadian clock to the regulation of alternative splicing. *Nature* 468, 112–116.

Schor, I. E., Rascovan, N., Pelisch, F., Allo, M., and Kornblihtt, A. R. (2009). Neuronal cell depolarization induces intragenic chromatin modifications affecting NCAM alternative splicing. *Proc. Natl. Acad. Sci. U.S.A.* 106, 4325–4330.

Schwartz, S., Meshorer, E., and Ast, G. (2009). Chromatin organization marks exon-intron structure. *Nat. Struct. Mol. Biol.* 16, 990–995.

Sgambato, V., Minassian, R., Nairn, A. C., and Hyman, S. E. (2003). Regulation of ania-6 splice variants by distinct signaling pathways in striatal neurons. *J. Neurochem.* 86, 153–164.

Shukla, S., Kavak, E., Gregory, M., Imashimizu, M., Shutinoski, B., Kashlev, M., Oberdoerffer, P., Sandberg, R., and Oberdoerffer, S. (2011). CTCF-promoted RNA polymerase II pausing links DNA methylation to splicing. *Nature* 479, 74-U99.

Spellman, R., Llorian, M., and Smith, C. W. J. (2007). Crossregulation and functional redundancy between the splicing regulator PTB and its paralogs nPTB and ROD1. *Mol. Cell* 27, 420–434.

Sun, S., Zhang, Z., Fregoso, O., and Krainer, A. R. (2012). Mechanisms of activation and repression by the alternative splicing factors RBFOX1/2. *RNA* 18, 274–283.

Tilgner, H., Nikolaou, C., Althammer, S., Sammeth, M., Beato, M., Valcarcel, J., and Guigo, R. (2009). Nucleosome positioning as a determinant of exon recognition. *Nat. Struct. Mol. Biol.* 16, 996-U124.

Todorova, A., Halliger-Keller, B., Walter, M. C., Dabauvalle, M. C., Lochmuller, H., and Muller, C. R. (2003). A synonymous codon change in the LMNA gene alters mRNA splicing and causes limb girdle muscular dystrophy type 1B. *J. Med. Genet.* 40, e115.

Tollervey, J. R., Wang, Z., Hortobagyi, T., Witten, J. T., Zarnack, K., Kayikci, M., Clark, T. A., Schweitzer, A. C., Rot, G., Curk, T., Zupan, B., Rogelj, B., Shaw, C. E., and Ule, J. (2011). Analysis of alternative splicing associated with aging and neurodegeneration in the human brain. *Genome Res.* 21, 1572–1582.

Tripathi, V., Ellis, J. D., Shen, Z., Song, D. Y., Pan, Q., Watt, A. T., Freier, S. M., Bennett, C. F., Sharma, A., Bubulya, P. A., Blencowe, B. J., Prasanth, S. G., and Prasanth, K. V. (2010). The nuclear-retained noncoding RNA MALAT1 regulates alternative splicing by modulating SR splicing factor phosphorylation. *Mol. Cell* 39, 925–938.

Ule, J., Stefani, G., Mele, A., Ruggiu, M., Wang, X., Taneri, B., Gaasterland, T., Blencowe, B. J., and Darnell, R. B. (2006). An RNA map predicting Nova-dependent splicing regulation. *Nature* 444, 580–586.

Ule, J., Ule, A., Spencer, J., Williams, A., Hu, J. S., Cline, M., Wang, H., Clark, T., Fraser, C., Ruggiu, M., Zeeberg, B. R., Kane, D., Weinstein, J. N., Blume, J., and Darnell, R. B. (2005). Nova regulates brain-specific splicing to shape the synapse. *Nat. Genet.* 37, 844–852.

Underwood, J. G., Boutz, P. L., Dougherty, J. D., Stoilov, P., and Black, D. L. (2005). Homologues of the Caenorhabditis elegans Fox-1 protein are neuronal splicing regulators in mammals. *Mol. Cell. Biol.* 25, 10005–10016.

Wahl, M. C., Will, C. L., and Luhrmann, R. (2009). The spliceosome: design principles of a dynamic RNP machine. *Cell* 136, 701–718.

Wang, E. T., Sandberg, R., Luo, S. J., Khrebtukova, I., Zhang, L., Mayr, C., Kingsmore, S. F., Schroth, G. P., and Burge, C. B. (2008). Alternative isoform regulation in human tissue transcriptomes. *Nature* 456, 470–476.

Wang, G. S., and Cooper, T. A. (2007). Splicing in disease: disruption of the splicing code and the decoding machinery. *Nat. Rev. Genet.* 8, 749–761.

Wang, Z. F., and Burge, C. B. (2008). Splicing regulation: from a parts list of regulatory elements to an integrated splicing code. *RNA* 14, 802–813.

West, A. E., Chen, W. G., Dalva, M. B., Dolmetsch, R. E., Kornhauser, J. M., Shaywitz, A. J., Takasu, M. A., Tao, X., and Greenberg, M. E. (2001). Calcium regulation of neuronal gene expression. *Proc. Natl. Acad. Sci. U.S.A.* 98, 11024–11031.

West, A. E., and Greenberg, M. E. (2011). Neuronal activity-regulated gene transcription in synapse development and cognitive function. *Cold Spring Harb. Perspect. Biol.* 3, pii: a005744.

Will, C. L., and Luhrmann, R. (2011). Spliceosome structure and function. *Cold Spring Harb. Perspect. Biol.* 3, pii: a003707.

Winkler, C., Eggert, C., Gradl, D., Meister, G., Giegerich, M., Wedlich, D., Laggerbauer, B., and Fischer, U. (2005). Reduced U snRNP assembly causes motor axon degeneration in an animal model for spinal muscular atrophy. *Genes Dev.* 19, 2320–2330.

Xie, J. Y., and Black, D. L. (2001). A CaMK IV responsive RNA element mediates depolarization-induced alternative splicing of ion channels. *Nature* 410, 936–939.

Xie, J. Y., Jan, C., Stoilov, P., Park, J., and Black, D. L. (2005). A consensus CaMK IV-responsive RNA sequence mediates regulation of alternative exons in neurons. *RNA* 11, 1825–1834.

Yano, M., Hayakawa-Yano, Y., Mele, A., and Darnell, R. B. (2010). Nova2 regulates neuronal migration through an RNA switch in disabled-1 signaling. *Neuron* 66, 848–858.

Yeo, G., Holste, D., Kreiman, G., and Burge, C. B. (2004). Variation in alternative splicing across human tissues. *Genome Biol.* 5, R74.

Yoo, A. S., Staahl, B. T., Chen, L., and Crabtree, G. R. (2009). MicroRNA-mediated switching of chromatin-remodelling complexes in neural development. *Nature* 460, 642-U112.

Yu, J. K., Hai, Y., Liu, G. D., Fang, T. L., Kung, S. K. P., and Xie, J. Y. (2009). The heterogeneous nuclear ribonucleoprotein L is an essential component in the Ca^{2+}/calmodulin-dependent protein kinase IV-regulated alternative splicing through cytidine-adenosine repeats. *J. Biol. Chem.* 284, 1505–1513.

Zhang, C., Frias, M. A., Mele, A., Ruggiu, M., Eom, T., Marney, C. B., Wang, H., Licatalosi, D. D., Fak, J. J., and Darnell, R. B. (2010). Integrative modeling defines the Nova splicing-regulatory network and its combinatorial controls. *Science* 329, 439–443.

Zhang, C., Zhang, Z., Castle, J., Sun, S., Johnson, J., Krainer, A. R., and Zhang, M. Q. (2008). Defining the regulatory network of the tissue-specific splicing factors Fox-1 and Fox-2. *Genes Dev.* 22, 2550–2563.

Zheng, S., Gray, E. E., Chawla, G., Porse, B. T., O'Dell, T. J., and Black, D. L. (2012). PSD-95 is post-transcriptionally repressed during early neural development by PTBP1 and PTBP2. *Nat. Neurosci.* 15, 381-U202.

Zhou, H. L., and Lou, H. (2008). Repression of prespliceosome complex formation at two distinct steps by Fox-1/Fox-2 proteins. *Mol. Cell. Biol.* 28, 5507–5516.

19

Insights on the functional interactions between miRNAs and copy number variations in the aging brain

Stephan Persengiev, Ivanela Kondova and Ronald Bontrop*

Biomedical Primate Research Center, Rijswijk, Netherlands

Correspondence:
Ronald Bontrop, Biomedical Primate Research Center, Lange Kleiweg 139, 2288 GH Rijswijk, Netherlands
e-mail: bontrop@bprc.nl

MicroRNAs (miRNAs) are regulatory genetic elements that coordinate the expression of thousands of genes and play important roles in brain aging and neurodegeneration. DNA polymorphisms affecting miRNA biogenesis, dosage, and gene targeting may represent potentially functional variants. The consequences of single nucleotide polymorphisms affecting miRNA function were previously demonstrated by both experimental and computational methods. However, little is known about how copy number variations (CNVs) influence miRNA metabolism and regulatory networks. We discuss potential mechanisms of CNVs-mediated effects on miRNA function and regulation that might have consequences for brain aging. We argue that CNVs, which potentially can alter miRNA expression, regulation or target gene recognition, are possible functional variants and should be considered high priority candidates in genotype–phenotype mapping studies of brain-related disorders.

Keywords: miRNAs, CNV, brain aging, neurodegeneration, non-coding RNA

INTRODUCTION

The establishment of human cognitive abilities is a gradual process that takes place mostly in the period between birth and adulthood, although some developmental processes extend beyond this period (Sowell et al., 2004; Thompson et al., 2004; Zhan et al., 2013). During this time window, the brain undergoes dramatic molecular transformations, which are manifested both structurally and functionally (de Graaf-Peters and Hadders-Algra, 2006). Notably, shortly after the brain development is accomplished, the process of brain aging commences at early adulthood, which is revealed by the gradual decline of the brain ability to absorb and process the flow of information (Sowell et al., 2004; Peters et al., 2008; Salthouse, 2009; May, 2011; Zhan et al., 2013). However, more recent research has revealed that changes in brain circuits are not exclusively restricted to the early stages of brain development, and has supported the concept of continuous neuroplasticity throughout live (May, 2011; Taubert et al., 2012). Novel experience as a result environmental changes and new learning experience have been recognized as stimulating factors of brain function and underlying neuroanatomic networks. Experiments with animals have showed that mice living in active environment exhibited a reduced neuronal age-dependent degeneration and achieved a greater threshold for age-dependent deficits (Kempermann et al., 2002; Fryer et al., 2011).

The aging process is confronted by various neuroprotective mechanisms that are genetically programed and underlie the dynamics of the brain adaptive responses. The sole purpose of the multiple cellular and functional events that take place during brain aging is to maintain neural cells functionality and structural integrity. In cases where the neuroprotective mechanisms are overwhelmed by the accumulation of toxic products, the result is progressive neurodegeneration, as observed in Alzheimer's disease (AD), cerebellar ataxias, and Parkinson's disease (PD). The neuroprotective mechanisms can be augmented by dietary and behavioral modifications, but the genetic predisposition to accelerated aging is likely to be the main driving factor that triggers and maintains the advance of neurodegeneration.

miRNA MACHINERY REACTION TO THE BRAIN AGING

Non-coding RNAs and microRNAs (miRNAs) in particular, play an essential role in the regulation of a number of cell processes, including cell proliferation, development, differentiation, stress responses, blast transformation, and apoptosis. The rapid accumulation of knowledge in the field of miRNA research has revealed its role in regulating gene expression at transcriptional and post-transcriptional levels. Meanwhile, the role of miRNAs in senescence remains poorly understood. miRNAs regulate several pathways associated with the aging mechanisms, and recent genome-wide analysis of miRNA expression revealed age-related changes in their expression level (Kosik, 2006; Krichevsky et al., 2006; Cogswell et al., 2008; Hebert and De Strooper, 2009). These data have underscored the significance of miRNA in brain aging and neurodegeneration.

MicroRNA can affect pathways involved in aging, and miRNA profiling has shown significant alterations in their expression level. Importantly, recent data have shown the significance of miRNA in brain aging and neurodegeneration (Kosik, 2006; Krichevsky et al., 2006; Cogswell et al., 2008; Hebert and De Strooper, 2009). The genome-wide expression analysis of miRNAs in aging individuals revealed a general decline in miRNA levels that was linked to potential loss of control of genes that regulate the cell cycle progression and cell differentiation programing (Noren Hooten et al., 2010). Nine miRNAs (miR-103, miR-107, miR-128, miR-130a, miR-155, miR-24, miR-221, miR-496, and miR-1538) were identified to be significantly lower in the peripheral blood mononuclear cells of old individuals as compared to the young subjects were identified in this study.

The ability of miRNAs to regulate oxidative stress and cell death is displayed in relationship to the growth harmone/insulin-like growth factor (GH/IGF) pathway and several AD-related oxidative damaging proteins (Nakasa et al., 2008; Stanczyk et al., 2008; Wang et al., 2008). Oxidative DNA damage may occur due to free reactive oxygen species (ROS) binding to nucleic acids and thus preventing transcription and causing DNA damage (Cooke et al., 2003). miR-210 and miR-373 inhibit the expression of key DNA repair proteins following hypoxic stress (Crosby et al., 2009). p53, a critical factor for maintaining the genome integrity, is activated by DNA oxidative damage, which is partially due to the miR-29-induced repression of negative regulators of p53, p85a, and CDC42 (Park et al., 2009).

Apoptosis is an extremely important signaling events influenced by miRNAs, particularly in the context of aging and age-related diseases. Several members of the miR-34 family participate in the p53 network, which induces apoptosis, cell cycle arrest, and senescence (Chang et al., 2007; He et al., 2007). It appears that activation of apoptosis – through internal or external stimuli, leads to repression of miRNAs that would otherwise silence genes involved in activating the apoptosis cascade. The reciprocal action, once an apoptotic cascade is activated, is the upregulation of miRNAs targeting proliferative or cell-survival genes (Wang, 2007). These results illustrate the complexity of miRNA interactions and their contribution to the regulation of programed cell death mechanisms.

MicroRNAs play a role in the control of brain metabolism and subsequently the dynamic of miRNA expression levels reflects the cellular responses to aging progression and deterioration of neuronal functionality. Several miRNAs are selectively expressed in brain tissues (Landgraf et al., 2007) and the inactivation of miRNA processing enzyme Dicer was found to lead to rapid degeneration of Purkinje cells (Schaefer et al., 2007). The global signature of miRNA expression in the adult brain appears to be species-specific, as shown by several comparative studies carried out on different species (Lee et al., 2000; Fraser et al., 2005; Berezikov et al., 2006). Selected miRNAs have been shown to be involved in AD, spinocerebellar ataxias, PD, and other neurodegenerative pathologies (Lukiw, 2007, 2012; Cogswell et al., 2008; Nelson et al., 2008; Persengiev et al., 2012b; Dimmeler and Nicotera, 2013). Genome-wide screens of miRNAs and ncRNAs in the aging brain found that miRNA expression is differentially regulated in the cortex and cerebellum of humans and non-human primates. This observation is likely to reflect the temporal functional status of neuronal activity in the cortex and cerebellum. Despite the observation for the lack of unifying specific miRNA pattern associated with the brain aging, the ontological analysis of targeted genes revealed that they represent a relatively conserved group (Persengiev et al., 2011). Importantly, miR-144 was identified to be the sole miRNA that was consistently upregulated in the aging chimp and human cerebellum and cortex (Persengiev et al., 2011, 2012a). The mechanism underlying the selective increase of miR-144 transcripts is unknown at this point, but indicates that miR-144 might play a coordinating role in the post-transcriptional suppression of specific genes in the aging brain. The mechanisms that govern miRNA expression during brain development and aging are highly structured and largely unknown. Complex gene expression patterns are regulated at several levels, including regulation by *cis*-acting *trans*-regulatory factors or regulation on the basis of epigenetic modifications such as gene methylation and histone modifications that depend on the genomic landscape. Thus, the adaptive responses of the brain cells during the aging process, which is reflected by brain phenotypic changes and the associated pathologies, will depend on either the physical presence or accessibility of multiple regulatory elements.

COPY NUMBER VARIATIONS ASSOCIATED WITH miRNA GENES AND BRAIN ANOMALIES

Copy number variations (CNVs) in non-coding regions can have profound effects on human phenotype (Klopocki and Mundlos, 2011). CNVs most common outcome is altering the copy number of an entire gene that is predisposed to a dosage effect. In a different scenario, CNVs can result in position effects and cause long-distance effects as far as 1 Mb from the translocation breakpoints. CNVs have been associated with several neuropsychiatric disorders, such as autism, schizophrenia, and bipolar disorder (Cook and Scherer, 2008; Lee and Scherer, 2010). Furthermore, CNVs have been associated with PD and early onset AD, which support the possibility of the existence of CNVs-driven mechanism(s) in PD and AD pathogenesis (Toft and Ross, 2010; McNaughton et al., 2012).

Copy number variations have an impact on the miRNA-mediated post-transcription regulatory network as well. miRNAs preferentially regulate the centers of protein interaction and metabolic networks (Liang and Li, 2007; Baek et al., 2008) and CNVs of miRNA genes may fluctuate the dosage balance of signal transduction pathways, metabolic flux, or protein complexes (Veitia, 2004; Veitia et al., 2008), leading eventually to individuals of the same population or different populations having different susceptibility to diseases. Although a comprehensive investigation to evaluate the CNV-miRNAs health risks among human populations is still lacking, recent experimental studies have confirmed the role of CNV-causing dysregulation of miRNAs in disease occurrence (Volinia et al., 2010). High-frequency copy number abnormalities occur in miRNA-containing regions throughout the genome in a range of human diseases (Zhang et al., 2006; Guo et al., 2008; Rossi et al., 2008; Wong et al., 2008), which is associated with altered expression of multiple genes and pathways (Reddy et al., 2009; Whitman et al., 2010). Genome-wide association studies have confirmed such associations for dozens of protein-coding genes and showed that CNVs capture at least 18% of the total detected genetic variation in gene expression (Stranger et al., 2007). The expression of miRNA genes is modified by CNVs and there is a correlation between somatic CNV and the miRNA levels. Thus, the CNV of functionally relevant miRNAs can modulate or predispose to certain complex genetic diseases.

Copy number variations are segments of genomic DNA that are roughly 1 kb to 1 Mb in length that show variable numbers of copies in the genome due to deletions or duplications and may cause the impairment of neuronal structures. The co-localization of all miRNA loci with known CNV regions was analyzed by using bioinformatics tools (Marcinkowska et al., 2011). In total, 209 copy number variable miRNA genes (CNV-miRNAs) in CNV regions deposited in the Database of Genomic Variations (DGV) have been identified and validated. Eleven CNV-miRNAs in two sets of CNVs have been classified as highly polymorphic. The overall conclusions from this *in silico* study were that miRNA loci are underrepresented in highly polymorphic and well-validated CNV regions consistent with their essential biological functions. The potential importance and consequences of the miRNAs presence in detected CNV regions, however, has been recognized in several other studies, suggesting that rare CNV-miRNA variants might have significant functional impact (Morley and Montgomery, 2001; Sebat et al., 2004; McCarroll et al., 2008).

At this stage, little is known about CNV of miRNA genes that can cause reduced cognitive ability in normal individuals during aging. miRNA copy number change can cause aberrant miRNA expression and/or deregulation of their target genes in subjects with neurodegenerative disorders, intellectual disability, and congenital abnormalities. For instance, the potential role of CNVs in AD has been investigated and identified a number of genes overlapped by CNV calls (Heinzen et al., 2010; Swaminathan et al., 2012a,b). Case-control association revealed several loci containing CHRFAM7A, RELN, DOPEY2, CSMD1, HNRNPCL1, IMMP2L, SLC35F2, NRXN1, ERBB4, and HLA-DRA genes that are associated with AD. The NRX1 gene has been linked to AD, autism, and schizophrenia (Szatmari et al., 2007; Latella et al., 2009) and ERBB4 is likely to play a role in AD progression (Woo et al., 2010). Overall, there appears that gene duplications and deletions across AD cohorts might account for the differences in the individual susceptibility to the neurodegeneration progression.

Copy number variations were established to be a major contributor of the pathology of brain disorders, but almost all studies have focused on the protein-coding genes present in the CNV loci, while the impact of miRNAs present in these regions has been overlooked. In a more recent study the biological and functional significance of miRNAs present in CNV loci and their target genes has been addressed by using an array of computational tools (Vaishnavi et al., 2013). The study found that nearly 11% of the autism-associated CNV loci harbor miRNAs, most of which were not previously reported to be associated with autism. A systematic analysis of the CNV-miRNAs based on their interactions with the target genes enabled the authors to pinpoint 10 miRNAs, miR-590-3p, miR-944, miR-570, miR-34a, miR-124, miR-548f, miR-429, miR-200b, miR-195, and miR-497 as core factors. The newly identified autism-associated miRNAs were predicted to form a regulatory loop with transcription factors and their downstream target genes. In addition, miRNAs present in deleted and duplicated CNV loci may explain the difference in dosage of the crucial autism genes and can also affect core components of miRNA processing machinery through negative feedback loops. Interestingly, the most common genomic disorder in humans, the hemizygous deletion of a 1.5–3 Mb region of chromosome 22q11.2, which increases the risk of developing schizophrenia by approximately 25-fold includes *DGCR8* miRNA processing gene (Brzustowicz and Bassett, 2012). The exact mechanism by which this deletion increases risk is unknown, but the observation strongly suggests that altered miRNAs metabolism may be a factor in the pathogenesis of schizophrenia. Overall, the findings support a possible role of copy number change in miRNA expression and processing with consequences affecting cognition, brain disorders, and/or CNV-mediated developmental delay.

EFFECTS OF COPY NUMBER VARIATIONS ON miRNA FUNCTION

Heritable information is transformed into cellular and organismal functions by the orderly expression of the entire set of genes in the genome. The complex process of gene expression regulation functions at several levels can be affected by structural alteration in the genomic architecture. Variations in the human genome occur on several levels. Originally, they were described as single-nucleotide changes within or outside of the coding sequence, or as microscopically visible alterations (CNVs) that affect parts of or even entire chromosomes. The effects include regulation in *cis* by promoters, enhancers, and repressors; regulation in *trans* by, e.g., transcription factors or miRNAs; or regulation on the basis of epigenetic modification such as DNA methylation. These genomic segmental differences reflect the dynamic nature of the genome and are believed to account for a large part of human phenotypic variability, including the predisposition to disease.

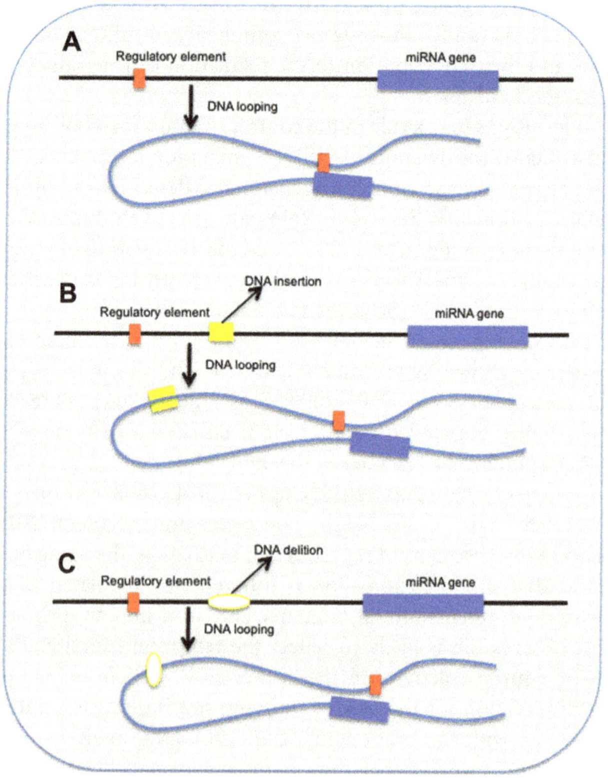

FIGURE 1 | Effect of genomic variations on long-range DNA interactions between remote regulatory elements and miRNA genes. **(A)** DNA looping allows factors associated with distant regulatory element(s) to bind miRNA consensus sequences and control gene expression. **(B,C)** Scenarios that depict the effect of DNA insertions and deletions on the repositioning of the remote regulatory element(s) and eventual loss of physiological control mechanism.

Selected genomic loci have been associated with non-coding pathogenic CNVs and their associated human disease phenotypes. CNVs were found to be distributed genome-wide that encompass non-coding sequences, thereby affecting the regulation of gene expression (Klopocki and Mundlos, 2011). More recently, a genome-wide scan identified 125 regions in which the same haplotypes are segregating in humans and chimpanzees, all with the exception of two encompassed non-coding regions (Leffler et al., 2013). In another study, a systematic search for DNA sequences missing in humans and present in chimpanzees, revealed that the identified sequences were almost exclusively from the non-coding regions of the genome (McLean et al., 2011). In addition, the study discovered that the absence of the penile bone in humans, which is present in chimpanzees, macaques, and mice, is due to the loss of a regulatory element that influences the expression of the androgen receptor gene. It is likely that these approaches will identify many more species-specific changes that relate to changes in phenotype.

Polymorphisms in miRNA genes can affect the expression of many downstream-regulated genes (Georges et al., 2007; Borel and Antonarakis, 2008). Single nucleotide mutations (SNPs) are most common form of polymorphism that affects the function of miRNAs, e.g., the structure of miRNA precursors, the efficiency of miRNA biogenesis and miRNA-target recognition. A series of *in silico* and experimental studies have revealed many SNPs located in different parts of miRNA genes (Duan et al., 2007; de Jong et al., 2013). The occurrence of SNPs in predominantly in the regions surrounding miRNA-coding elements, while sequences of mature miRNAs featured as the most conserved (Saunders et al., 2007). Functional analysis demonstrated that rare mutations naturally occurring within pre-miRNA sequences affect miRNA biogenesis and impair miRNA-mediated gene silencing (Duan et al., 2007; Sun et al., 2009). Recently, large genome-wide association study has demonstrated that SNPs located outside (>14 kb) of pre-miRNA sequences can modulate miRNA expression both as *cis-* and *trans*-regulators, as well (Borel et al., 2011). miRNA target sites are also conserved genetic elements and SNPs with potential to either disrupt or create new miRNA target sites are underrepresented in both experimentally validated and computationally predicted miRNA target sites (Chen and Rajewsky, 2006; Saunders et al., 2007; de Jong et al., 2013). Analysis of CNVs in the human and chimpanzee genomes demonstrates the potentially greater role of CNVs in evolutionary change than single base-pair sequence variation (Cheng et al., 2005). Comparisons of the human and chimpanzee genomes revealed that there are more than twice as many nucleotides involved in CNVs as there are in changes to individual nucleotides, 2.7% compared to 1.2%. Furthermore, the data revealed that while the majority of CNVs were shared between the human and chimpanzee genomes, approximately one-third of the CNVs observed in the human genome were unique and therefore acquired later in evolution. Additional studies have further revealed that CNVs are often linked to genetic diseases apparent in humans (Stankiewicz and Lupski, 2002). However, little is known about CNVs interactions with miRNAs.

Copy number variations have the propensity to alter the general organization of the chromatin in the affected chromosome regions that may have significant functional impact. Recent findings emphasized that nuclear architecture and chromatin organization play important role in the regulation of gene expression (Stankiewicz and Lupski, 2002), and that these components are essential epigenetic mechanisms for both the normal physiology as well as in the pathogenesis of a number of human maladies (Parada et al., 2004a). Portions of DNA, known as DNA loops, protrude from euchromatic portions of chromosomes, and the genes on these segments may localize to transcriptionally active chromatin centers that contain intergenic or intragenic miRNA genes (Osborne et al., 2004). Chromosome looping that enables remote segments of DNA from the same chromosome or from different chromosomes to interact and to modify the expression of distant genes presents a plausible mechanism that links the global misregulation of miRNA expression in AD and other neurodegenerative diseases to CNVs (**Figure 1**). As a consequence of

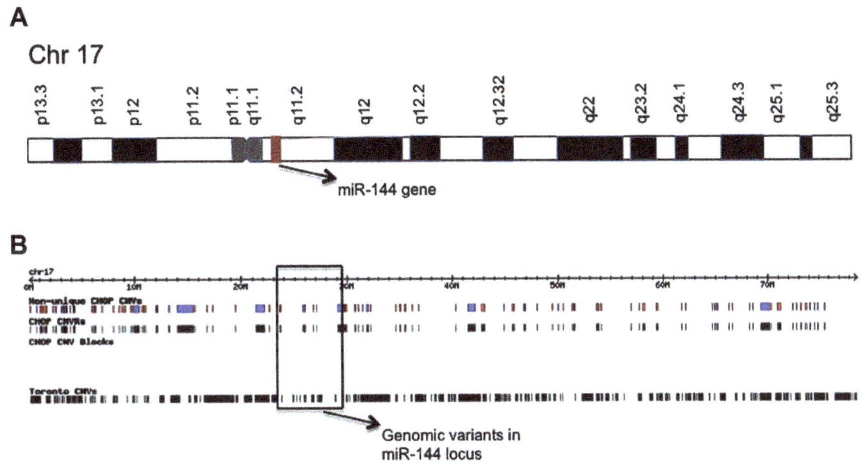

FIGURE 2 | Schematic of chromosome 17 showing the location of miR-144 gene (A); and distribution map of genetic variants identified in chromosome 17 (B). The location of miR-144 gene is shown and CNVs in its respective genomic region are marked.

CNV-induced chromatin reorganization, accessibility of miRNA binding elements within 3′ untranslated region (UTR) of target genes, miRNA promoters availability, as well as the expression of long ncRNAs that serve as sponges for miRNAs might be dramatically altered (Sanyal et al., 2012; Memczak et al., 2013). CNVs that are in close proximity of these loops may also trigger recombination and chromatin rearrangements (Parada et al., 2004b).

Interestingly, the aging-specific miR-144 is located on chromosome 17 in a region reported to be polymorphic, including several inversions and duplications, according to CNV database (**Figure 2**; **Table 1**). The significance of CNVs in the vicinity

Table 1 | Genomic variations in the vicinity of miR-144 genomic location on chromosome 17 according to the Database for Genomic Variants.

Locus	Landmark	Variation type	Cytoband	Position (Mb)	Known genes in the locus
chr17:27013684-27014304	chr17:27,013,684..27,014,304	InDel	17q11.2	27.0	
chr17:27107800-27123735	chr17:27,107,800..27,123,735	Copy number	17q11.2	27.1	
	chr17:27,120,270..27,121,891	Copy number	17q11.2	27.1	
chr17:27122880-27122983	chr17:27,122,880..27,122,983	InDel	17q11.2	27.1	
chr17:27130078-27131878	chr17:27,130,078..27,131,878	Copy number	17q11.2	27.1	
	chr17:27,130,682..27,131,776	Copy number	17q11.2	27.1	
chr17:27130696-27131659	chr17:27,130,930..27,131,420	InDel	17q11.2	27.1	
	chr17:27,130,738..27,131,656	InDel	17q11.2	27.1	
	chr17:27,130,736..27,131,659	InDel	17q11.2	27.1	
	chr17:27,130,696..27,131,638	InDel	17q11.2	27.1	
chr17:27245834-27562095	chr17:27,459,989..27,461,612	Copy number	17q11.2	27.5	UTP6
	chr17:27,412,804..27,436,507	Copy number	17q11.2	27.4	SUZ12
	chr17:27,465,972..27,469,974	Copy number	17q11.2	27.5	LRRC37B
	chr17:27,245,834..27,562,095	Copy number	17q11.2	27.2	SH3GL1P1
	chr17:27,466,732..27,471,357	Copy number	17q11.2	27.5	ARGFXP2
	chr17:27,333,922..27,335,931	Copy number	17q11.2	27.3	RHOT1
chr17:27384860-27385274	chr17:27,384,860..27,385,274	InDel	17q11.2	27.4	LRRC37B
chr17:27460863-27461165	chr17:27,460,863..27,461,165	InDel	17q11.2	27.5	
chr17:27614844-27619890	chr17:27,614,844..27,619,890	Copy number	17q11.2	27.6	RHBDL3
chr17:27621887-27622597	chr17:27,621,887..27,622,597	InDel	17q11.2	27.6	RHBDL3
chr17:27627845-27628095	chr17:27,627,845..27,628,095	InDel	17q11.2	27.6	RHBDL3
chr17:27633422-27634030	chr17:27,633,422..27,634,030	InDel	17q11.2	27.6	RHBDL3
chr17:27668824-27669757	chr17:27,668,824..27,669,757	InDel	17q11.2	27.7	RHBDL3
	chr17:27,669,594..27,669,594	InDel	17q11.2	27.7	
chr17:27788363-27788659	chr17:27,788,363..27,788,659	InDel	17q11.2	27.8	
chr17:27837365-27838765	chr17:27,837,365..27,838,765	Copy number	17q11.2	27.8	CDK5R1
chr17:27917975-27917975	chr17:27,917,975..27,917,975	InDel	17q11.2	27.9	MYO1D
chr17:28279105-28280814	chr17:28,279,105..28,280,814	Copy number	17q11.2	28.3	TMEM98
chr17:28341799-28342792	chr17:28,341,799..28,342,792	InDel	17q11.2	28.3	
chr17:28501812-28502008	chr17:28,501,828..28,502,008	InDel	17q11.2	28.5	ACCN1
	chr17:28,501,812..28,502,002	InDel	17q11.2	28.5	
chr17:28620758-28620884	chr17:28,620,758..28,620,884	InDel	17q11.2	28.6	ACCN1
chr17:28630652-28631318	chr17:28,630,652..28,631,318	InDel	17q11.2	28.6	ACCN1
chr17:28643047-28645208	chr17:28,643,047..28,645,208	Copy number	17q11.2	28.6	ACCN1
chr17:28670843-28673962	chr17:28,670,843..28,673,962	Copy number	17q11.2	28.7	ACCN1
chr17:28708062-28708198	chr17:28,708,062..28,708,198	InDel	17q11.2	28.7	ACCN1
chr17:28779244-28781640	chr17:28,779,244..28,781,640	Copy number	17q11.2	28.8	ACCN1

miR-144 gene is encoded on the minus strand and contains two exons on Chr:17q11.2; 23,396,926-27,188,636 locus.

of miR-144 gene is unclear at this point, but long-range regulatory chromatin interactions play an important role in gene regulation. Both intrachromosomal and interchromosomal long-range associations have been demonstrated, and DNA binding factors have been implicated in the maintenance of these interactions (Cremer et al., 2000; Branco and Pombo, 2006). Several distant DNA segments may interact with a single gene and influence its expression pattern. Monoallelically expressed genes, most notably imprinted genes, are frequently found to be regulated by these long-range interactions. In support of this concept, FLT3-internal tandem duplications (ITDs) on chromosome 13, an adverse prognostic marker in specific aging individuals, were found to affect negatively the expression of GATA-3 transcription factor and miR-144 (Whitman et al., 2010). Members of GATA transcription factor family are believed to play a role in the control miR-144 transcription. GATA-4 transcription factor been reported to be critical regulator of miR-144 expression and is supposed to be the responsible gene for the congenital heart defects (CHDs) in the chromosomal 8p23 deletion syndrome, a complex malformation syndrome with clinical symptoms manifested by facial anomalies, microcephaly, mental retardation, and CHDs (Guida et al., 2010; Zhang et al., 2010). These findings emphasize the importance of studying the geography and architecture of the nucleus as an important factor in the regulation of miRNA expression.

CONCLUSIONS AND OUTLOOK

The existing CNVs in the human genome cover approximately 360 Mb, or 12% of the human genome, as reported by the CNV Project database (http://www.sanger.ac.uk/research/areas/humangenetics/cnv/). CNVs encompass more nucleotide content per genome than SNPs, underscoring CNVs' significance to genetic diversity. A genome-wide map of CNVs shows that no region of the genome is exempt, and that between 6% and 19% of each individual's chromosomes exhibit CNVs (Redon et al., 2006).

The mechanisms that operate during the progress of brain aging and associated neurodegenerative diseases are complex and their malfunction is rarely due to the failure of a few cell death or neuronal differentiation genes. Because susceptibility to premature aging and cognitive decline is a result of the malfunction of numerous genes, miRNAs dysregulation that inevitably would alter the expression of multiple genes might provide the basis for neuronal cell deterioration.

Multiple factors participate in the control of miRNA expression. Here, we discuss the emerging role of CNVs in miRNA regulation and the potential impacts on brain aging and neurodegeneration. Our simple notion is that the long-range interactions between DNA segments affected by CNVs might directly modify miRNA expression pattern, and as consequence miRNA-mediated inhibition of genes that are important for maintaining neuron homeostasis. We argue that CNVs-miRNA interactions are an important part of increased brain susceptibility to external and internal stress during the aging process. A more complete understanding of CNVs effect on the global nuclear geography and chromatin organization in the vicinity of miRNA-encoding regions will allow defining the chromosome regions that represent risk factors for the brain anomalies. Therefore, the challenge now is to annotate CNVs, which potentially can alter miRNA expression and determine whether they are functional variants and should be considered high-priority candidates in genotype–phenotype mapping studies of brain-related disorders.

REFERENCES

Baek, D., Villen, J., Shin, C., Camargo, F. D., Gygi, S. P., and Bartel, D. P. (2008). The impact of microRNAs on protein output. *Nature* 455, 64–71. doi: 10.1038/nature07242

Berezikov, E., Thuemmler, F., van Laake, L. W., Kondova, I., Bontrop, R., Cuppen, E., et al. (2006). Diversity of microRNAs in human and chimpanzee brain. *Nat. Genet.* 38, 1375–1377. doi: 10.1038/ng1914

Borel, C., and Antonarakis, S. E. (2008). Functional genetic variation of human miRNAs and phenotypic consequences. *Mamm. Genome* 19, 503–509. doi: 10.1007/s00335-008-9137-6

Borel, C., Deutsch, S., Letourneau, A., Migliavacca, E., Montgomery, S. B., Dimas, A. S., et al. (2011). Identification of *cis-* and *trans-*regulatory variation modulating microRNA expression levels in human fibroblasts. *Genome Res.* 21, 68–73. doi: 10.1101/gr.109371.110

Branco, M. R., and Pombo, A. (2006). Intermingling of chromosome territories in interphase suggests role in translocations and transcription-dependent associations. *PLoS Biol.* 4:e138. doi: 10.1371/journal.pbio.0040138

Brzustowicz, L. M., and Bassett, A. S. (2012). miRNA-mediated risk for schizophrenia in 22q11.2 deletion syndrome. *Front. Genet.* 3:291. doi: 10.3389/fgene.2012.00291

Chang, T. C., Wentzel, E. A., Kent, O. A., Ramachandran, K., Mullendore, M., Lee, K. H., et al. (2007). Transactivation of miR-34a by p53 broadly influences gene expression and promotes apoptosis. *Mol. Cell* 26, 745–752. doi: 10.1016/j.molcel.2007.05.010

Chen, K., and Rajewsky, N. (2006). Natural selection on human microRNA binding sites inferred from SNP data. *Nat. Genet.* 38, 1452–1456. doi: 10.1038/ng1910

Cheng, Z., Ventura, M., She, X., Khaitovich, P., Graves, T., Osoegawa, K., et al. (2005). A genome-wide comparison of recent chimpanzee and human segmental duplications. *Nature* 437, 88–93. doi: 10.1038/nature04000

Cogswell, J. P., Ward, J., Taylor, I. A., Waters, M., Shi, Y., Cannon, B., et al. (2008). Identification of miRNA changes in Alzheimer's disease brain and CSF yields putative biomarkers and insights into disease pathways. *J. Alzheimers Dis.* 14, 27–41.

Cook, E. H. Jr., and Scherer, S. W. (2008). Copy-number variations associated with neuropsychiatric conditions. *Nature* 455, 919–923. doi: 10.1038/nature07458

Cooke, M. S., Evans, M. D., Dizdaroglu, M., and Lunec, J. (2003). Oxidative DNA damage: mechanisms, mutation, and disease. *FASEB J.* 17, 1195–1214. doi: 10.1096/fj.02-0752rev

Cremer, T., Kreth, G., Koester, H., Fink, R. H., Heintzmann, R., Cremer, M., et al. (2000). Chromosome territories, interchromatin domain compartment, and nuclear matrix: an integrated view of the functional nuclear architecture. *Crit. Rev. Eukaryot. Gene Expr.* 10, 179–212. doi: 10.1615/CritRevEukarGeneExpr.v10.i2.60

Crosby, M. E., Kulshreshtha, R., Ivan, M., and Glazer, P. M. (2009). MicroRNA regulation of DNA repair gene expression in hypoxic stress. *Cancer Res.* 69, 1221–1229. doi: 10.1158/0008-5472.CAN-08-2516

de Graaf-Peters, V. B., and Hadders-Algra, M. (2006). Ontogeny of the human central nervous system: what is happening when? *Early Hum. Dev.* 82, 257–266. doi: 10.1016/j.earlhumdev.2005.10.013

de Jong, V. M., Zaldumbide, A., van der Slik, A. R., Persengiev, S. P., Roep, B. O., and Koeleman, B. P. (2013). Post-transcriptional control of candidate risk genes for type 1 diabetes by rare genetic variants. *Genes Immun.* 14, 58–61. doi: 10.1038/gene.2012.38

Dimmeler, S., and Nicotera, P. (2013). MicroRNAs in age-related diseases. *EMBO Mol. Med.* 5, 180–190. doi: 10.1002/emmm.201201986

Duan, R., Pak, C., and Jin, P. (2007). Single nucleotide polymorphism associated with mature miR-125a alters the processing of pri-miRNA. *Hum. Mol. Genet.* 16, 1124–1131. doi: 10.1093/hmg/ddm062

Fraser, H. B., Khaitovich, P., Plotkin, J. B., Paabo, S., and Eisen, M. B. (2005).

Aging and gene expression in the primate brain. *PLoS Biol.* 3:e274. doi: 10.1371/journal.pbio.0030274

Fryer, J. D., Yu, P., Kang, H., Mandel-Brehm, C., Carter, A. N., Crespo-Barreto, J., et al. (2011). Exercise and genetic rescue of SCA1 via the transcriptional repressor Capicua. *Science* 334, 690–693. doi: 10.1126/science.1212673

Georges, M., Coppieters, W., and Charlier, C. (2007). Polymorphic miRNA-mediated gene regulation: contribution to phenotypic variation and disease. *Curr. Opin. Genet. Dev.* 17, 166–176. doi: 10.1016/j.gde.2007.04.005

Guida, V., Lepri, F., Vijzelaar, R., De Zorzi, A., Versacci, P., Digilio, M. C., et al. (2010). Multiplex ligation-dependent probe amplification analysis of GATA4 gene copy number variations in patients with isolated congenital heart disease. *Dis. Markers* 28, 287–292. doi: 10.1155/2010/530360

Guo, C., Sah, J. F., Beard, L., Willson, J. K., Markowitz, S. D., and Guda, K. (2008). The noncoding RNA, miR-126, suppresses the growth of neoplastic cells by targeting phosphatidylinositol 3-kinase signaling and is frequently lost in colon cancers. *Genes Chromosomes Cancer* 47, 939–946. doi: 10.1002/gcc.20596

He, L., He, X., Lim, L. P., de Stanchina, E., Xuan, Z., Liang, Y., et al. (2007). A microRNA component of the p53 tumour suppressor network. *Nature* 447, 1130–1134. doi: 10.1038/nature05939

Hebert, S. S., and De Strooper, B. (2009). Alterations of the microRNA network cause neurodegenerative disease. *Trends Neurosci.* 32, 199–206. doi: 10.1016/j.tins.2008.12.003

Heinzen, E. L., Need, A. C., Hayden, K. M., Chiba-Falek, O., Roses, A. D., Strittmatter, W. J., et al. (2010). Genome-wide scan of copy number variation in late-onset Alzheimer's disease. *J. Alzheimers Dis.* 19, 69–77.

Kempermann, G., Gast, D., and Gage, F. H. (2002). Neuroplasticity in old age: sustained fivefold induction of hippocampal neurogenesis by long-term environmental enrichment. *Ann. Neurol.* 52, 135–143. doi: 10.1002/ana.10262

Klopocki, E., and Mundlos, S. (2011). Copy-number variations, noncoding sequences, and human phenotypes. *Annu. Rev. Genomics Hum. Genet.* 12, 53–72. doi: 10.1146/annurev-genom-082410-101404

Kosik, K. S. (2006). The neuronal microRNA system. *Nat. Rev. Neurosci.* 7, 911–920. doi: 10.1038/nrn2037

Krichevsky, A. M., Sonntag, K. C., Isacson, O., and Kosik, K. S. (2006). Specific microRNAs modulate embryonic stem cell-derived neurogenesis. *Stem Cells* 24, 857–864. doi: 10.1634/stemcells.2005-0441

Landgraf, P., Rusu, M., Sheridan, R., Sewer, A., Iovino, N., Aravin, A., et al. (2007). A mammalian microRNA expression atlas based on small RNA library sequencing. *Cell* 129, 1401–1414. doi: 10.1016/j.cell.2007.04.040

Latella, G., Vetuschi, A., Sferra, R., Catitti, V., D'Angelo, A., Zanninelli G., et al. (2009). Targeted disruption of Smad3 confers resistance to the development of dimethylnitrosamine-induced hepatic fibrosis in mice. *Liver Int.* 29, 997–1009. doi: 10.1111/j.1478-3231.2009.02011.x

Lee, C., and Scherer, S. W. (2010). The clinical context of copy number variation in the human genome. *Expert Rev. Mol. Med.* 12, e8. doi: 10.1017/S1462399410001390

Lee, C. K., Weindruch, R., and Prolla, T. A. (2000). Gene-expression profile of the ageing brain in mice. *Nat. Genet.* 25, 294–297. doi: 10.1038/77046

Leffler, E. M., Gao, Z., Pfeifer, S., Segurel, L., Auton, A., Venn, O., et al. (2013). Multiple instances of ancient balancing selection shared between humans and chimpanzees. *Science* 339, 1578–1582. doi: 10.1126/science.1234070

Liang, H., and Li, W. H. (2007). MicroRNA regulation of human protein interaction network. *RNA* 13, 1402–1408. doi: 10.1261/rna.634607

Lukiw, W. J. (2007). Micro-RNA speciation in fetal, adult and Alzheimer's disease hippocampus. *Neuroreport* 18, 297–300. doi: 10.1097/WNR.0b013e3280148e8b

Lukiw, W. J. (2012). Evolution and complexity of micro RNA in the human brain. *Front. Genet.* 3:166. doi: 10.3389/fgene.2012.00166

Marcinkowska, M., Szymanski, M., Krzyzosiak, W. J., and Kozlowski, P. (2011). Copy number variation of microRNA genes in the human genome. *BMC Genomics* 12:183. doi: 10.1186/1471-2164-12-183

May, A. (2011). Experience-dependent structural plasticity in the adult human brain. *Trends Cogn. Sci.* 15, 475–482. doi: 10.1016/j.tics.2011.08.002

McCarroll, S. A., Kuruvilla, F. G., Korn, J. M., Cawley, S., Nemesh, J., Wysoker, A., et al. (2008). Integrated detection and population-genetic analysis of SNPs and copy number variation. *Nat. Genet.* 40, 1166–1174. doi: 10.1038/ng.238

McLean, C. Y., Reno, P. L., Pollen, A. A., Bassan, A. I., Capellini, T. D., Guenther, C., et al. (2011). Human-specific loss of regulatory DNA and the evolution of human-specific traits. *Nature* 471, 216–219. doi: 10.1038/nature09774

McNaughton, D., Knight, W., Guerreiro, R., Ryan, N., Lowe, J., Poulter, M., et al. (2012). Duplication of amyloid precursor protein (APP), but not prion protein (PRNP) gene is a significant cause of early onset dementia in a large UK series. *Neurobiol. Aging* 33, 426.e413–426.e421.

Memczak, S., Jens, M., Elefsinioti, A., Torti, F., Krueger, J., Rybak, A., et al. (2013). Circular RNAs are a large class of animal RNAs with regulatory potency. *Nature* 495, 333–338. doi: 10.1038/nature11928

Morley, K. I., and Montgomery, G. W. (2001). The genetics of cognitive processes: candidate genes in humans and animals. *Behav. Genet.* 31, 511–531. doi: 10.1023/A:1013337209957

Nakasa, T., Miyaki, S., Okubo, A., Hashimoto, M., Nishida, K., Ochi, M., et al. (2008). Expression of microRNA-146 in rheumatoid arthritis synovial tissue. *Arthritis Rheum.* 58, 1284–1292. doi: 10.1002/art.23429

Nelson, P. T., Wang, W. X., and Rajeev, B. W. (2008). MicroRNAs (miRNAs) in neurodegenerative diseases. *Brain Pathol.* 18, 130–138. doi: 10.1111/j.1750-3639.2007.00120.x

Noren Hooten, N., Abdelmohsen, K., Gorospe, M., Ejiogu, N., Zonderman, A. B., and Evans, M. K. (2010). microRNA expression patterns reveal differential expression of target genes with age. *PLoS ONE* 5:e10724. doi: 10.1371/journal.pone.0010724

Osborne, C. S., Chakalova, L., Brown, K. E., Carter, D., Horton, A., Debrand, E., et al. (2004). Active genes dynamically colocalize to shared sites of ongoing transcription. *Nat. Genet.* 36, 1065–1071. doi: 10.1038/ng1423

Parada, L. A., McQueen, P. G., and Misteli, T. (2004a). Tissue-specific spatial organization of genomes. *Genome Biol.* 5, R44. doi: 10.1186/gb-2004-5-7-r44

Parada, L. A., Sotiriou, S., and Misteli, T. (2004b). Spatial genome organization. *Exp. Cell Res.* 296, 64–70. doi: 10.1016/j.yexcr.2004.03.013

Park, K. J., Lee, S. H., Lee, C. H., Jang, J. Y., Chung, J., Kwon, M. H., et al. (2009). Upregulation of Beclin-1 expression and phosphorylation of Bcl-2 and p53 are involved in the JNK-mediated autophagic cell death. *Biochem. Biophys. Res. Commun.* 382, 726–729. doi: 10.1016/j.bbrc.2009.03.095

Persengiev, S., Kondova, I., and Bontrop, R. E. (2012a). Functional annotation of small noncoding RNAs target genes provides evidence for a deregulated ubiquitin-proteasome pathway in spinocerebellar ataxia type 1. *J. Nucleic Acids* 2012, 672536. doi: 10.1155/2012/672536

Persengiev, S. P., Kondova, I. I., and Bontrop, R. E. (2012b). The impact of microRNAs on brain aging and neurodegeneration. *Curr. Gerontol. Geriatr. Res.* 2012, 359–369. doi: 10.1155/2012/359369

Persengiev, S., Kondova, I., Otting, N., Koeppen, A. H., and Bontrop, R. E. (2011). Genome-wide analysis of miRNA expression reveals a potential role for miR-144 in brain aging and spinocerebellar ataxia pathogenesis. *Neurobiol. Aging* 32, e2317–e2327. doi: 10.1016/j.neurobiolaging.2010.03.014

Peters, A., Sethares, C., and Luebke, J. I. (2008). Synapses are lost during aging in the primate prefrontal cortex. *Neuroscience* 152, 970–981. doi: 10.1016/j.neuroscience.2007.07.014

Reddy, S. D., Pakala, S. B., Ohshiro, K., Rayala, S. K., and Kumar, R. (2009). MicroRNA-661, a c/EBPalpha target, inhibits metastatic tumor antigen 1 and regulates its functions. *Cancer Res.* 69, 5639–5642. doi: 10.1158/0008-5472.CAN-09-0898

Redon, R., Ishikawa, S., Fitch, K. R., Feuk, L., Perry, G. H., Andrews, T. D., et al. (2006). Global variation in copy number in the human genome. *Nature* 444, 444–454. doi: 10.1038/nature05329

Rossi, S., Sevignani, C., Nnadi, S. C., Siracusa, L. D., and Calin, G. A. (2008). Cancer-associated genomic regions (CAGRs) and noncoding RNAs: bioinformatics and therapeutic implications. *Mamm. Genome* 19, 526–540. doi: 10.1007/s00335-008-9119-8

Salthouse, T. A. (2009). When does age-related cognitive decline begin? *Neurobiol. Aging* 30, 507–514. doi: 10.1016/j.neurobiolaging.2008.09.023

Sanyal, A., Lajoie, B. R., Jain, G., and Dekker J. (2012). The long-range interaction landscape of gene promoters. *Nature* 489, 109–113. doi: 10.1038/nature11279

Saunders, M. A., Liang, H., and Li, W. H. (2007). Human polymorphism at microRNAs and microRNA target sites. *Proc. Natl. Acad. Sci. U.S.A.* 104, 3300–3305. doi: 10.1073/pnas.0611347104

Schaefer, A., O'Carroll, D., Tan, C. L., Hillman, D., Sugimori, M., Llinas, R., et al. (2007). Cerebellar neurodegeneration in the absence of microRNAs. *J. Exp. Med.* 204, 1553–1558. doi: 10.1084/jem.20070823

Sebat, J., Lakshmi, B., Troge, J., Alexander, J., Young, J., Lundin, P., et al. (2004). Large-scale copy number polymorphism in the human genome. *Science* 305, 525–528. doi: 10.1126/science.1098918

Sowell, E. R., Thompson, P. M., and Toga, A. W. (2004). Mapping changes in the human cortex throughout the span of life. *Neuroscientist* 10, 372–392. doi: 10.1177/1073858404263960

Stanczyk, J., Pedrioli, D. M., Brentano, F., Sanchez-Pernaute, O., Kolling, C., Gay, R. E., et al. (2008). Altered expression of microRNA in synovial fibroblasts and synovial tissue in rheumatoid arthritis. *Arthritis Rheum.* 58, 1001–1009. doi: 10.1002/art.23386

Stankiewicz, P., and Lupski, J. R. (2002). Genome architecture, rearrangements and genomic disorders. *Trends Genet.* 18, 74–82. doi: 10.1016/S0168-9525(02)02592-1

Stranger, B. E., Forrest, M. S., Dunning, M., Ingle, C. E., Beazley, C., Thorne, N., et al. (2007). Relative impact of nucleotide and copy number variation on gene expression phenotypes. *Science* 315, 848–853. doi: 10.1126/science.1136678

Sun, G., Yan, J., Noltner, K., Feng, J., Li, H., Sarkis, D. A., et al. (2009). SNPs in human miRNA genes affect biogenesis and function. *RNA* 15, 1640–1651. doi: 10.1261/rna.1560209

Swaminathan, S., Huentelman, M. J., Corneveaux, J. J., Myers, A. J., Faber, K. M., Foroud, T., et al. (2012a). Analysis of copy number variation in Alzheimer's disease in a cohort of clinically characterized and neuropathologically verified individuals. *PLoS ONE* 7:e50640. doi: 10.1371/journal.pone.0050640

Swaminathan, S., Shen, L., Kim, S., Inlow, M., West, J. D., Faber, K. M., et al. (2012b). Analysis of copy number variation in Alzheimer's disease: the NIALOAD/NCRAD Family Study. *Curr. Alzheimer Res.* 9, 801–814. doi: 10.2174/156720512802455331

Szatmari, P., Paterson, A. D., Zwaigenbaum, L., Roberts, W., Brian, J., Liu, X. Q., et al. (2007). Mapping autism risk loci using genetic linkage and chromosomal rearrangements. *Nat. Genet.* 39, 319–328. doi: 10.1038/ng1985

Taubert, M., Villringer, A., and Ragert, P. (2012). Learning-related gray and white matter changes in humans: an update. *Neuroscientist* 18, 320–325. doi: 10.1177/1073858411419048

Thompson, P. M., Hayashi, K. M., Sowell, E. R., Gogtay, N., Giedd, J. N., Rapoport, J. L., et al. (2004). Mapping cortical change in Alzheimer's disease, brain development, and schizophrenia. *Neuroimage* 23(Suppl. 1), S2–S18. doi: 10.1016/j.neuroimage.2004.07.071

Toft, M., and Ross, O. A. (2010). Copy number variation in Parkinson's disease. *Genome Med.* 2, 62.

Vaishnavi, V., Manikandan, M., Tiwary, B. K., and Munirajan, A. K. (2013). Insights on the functional impact of microRNAs present in autism-associated copy number variants. *PLoS ONE* 8:e56781. doi: 10.1371/journal.pone.0056781

Veitia, R. A. (2004). Gene dosage balance in cellular pathways: implications for dominance and gene duplicability. *Genetics* 168, 569–574. doi: 10.1534/genetics.104.029785

Veitia, R. A., Bottani, S., and Birchler, J. A. (2008). Cellular reactions to gene dosage imbalance: genomic, transcriptomic and proteomic effects. *Trends Genet.* 24, 390–397. doi: 10.1016/j.tig.2008.05.005

Volinia, S., Galasso, M., Costinean, S., Tagliavini, L., Gamberoni, G., Drusco, A., et al. (2010). Reprogramming of miRNA networks in cancer and leukemia. *Genome Res.* 20, 589–599. doi: 10.1101/gr.098046.109

Wang, E. (2007). MicroRNA, the putative molecular control for mid-life decline. *Ageing Res. Rev.* 6, 1–11. doi: 10.1016/j.arr.2007.02.004

Wang, W. X., Rajeev, B. W., Stromberg, A. J., Ren, N., Tang, G., Huang, Q., et al. (2008). The expression of microRNA miR-107 decreases early in Alzheimer's disease and may accelerate disease progression through regulation of beta-site amyloid precursor protein-cleaving enzyme 1. *J. Neurosci.* 28, 1213–1223. doi: 10.1523/JNEUROSCI.5065-07.2008

Whitman, S. P., Maharry, K., Radmacher, M. D., Becker, H., Mrozek, K., Margeson, D., et al. (2010). FLT3 internal tandem duplication associates with adverse outcome and gene- and microRNA-expression signatures in patients 60 years of age or older with primary cytogenetically normal acute myeloid leukemia: a cancer and leukemia group B study. *Blood* 116, 3622–3626. doi: 10.1182/blood-2010-05-283648

Wong, T. S., Liu, X. B., Wong, B. Y., Ng, R. W., Yuen, A. P., and Wei, W. I. (2008). Mature miR-184 as potential oncogenic microRNA of squamous cell carcinoma of tongue. *Clin. Cancer Res.* 14, 2588–2592. doi: 10.1158/1078-0432.CCR-07-0666

Woo, R. S., Lee, J. H., Yu, H. N., Song, D. Y., and Baik, T. K. (2010). Expression of ErbB4 in the apoptotic neurons of Alzheimer's disease brain. *Anat. Cell Biol.* 43, 332–339. doi: 10.5115/acb.2010.43.4.332

Zhan, J., Dinov, I. D., Li, J., Zhang, Z., Hobel, S., Shi, Y., et al. (2013). Spatial-temporal atlas of human fetal brain development during the early second trimester. *Neuroimage* 82C, 115–126. doi: 10.1016/j.neuroimage.2013.05.063

Zhang, L., Huang, J., Yang, N., Greshock, J., Megraw, M. S., Giannakakis, A., et al. (2006). microRNAs exhibit high frequency genomic alterations in human cancer. *Proc. Natl. Acad. Sci. U.S.A.* 103, 9136–9141. doi: 10.1073/pnas.0508889103

Zhang, X., Wang, X., Zhu, H., Zhu, C., Wang, Y., Pu, W. T., et al. (2010). Synergistic effects of the GATA-4-mediated miR-144/451 cluster in protection against simulated ischemia/reperfusion-induced cardiomyocyte death. *J. Mol. Cell. Cardiol.* 49, 841–850. doi: 10.1016/j.yjmcc.2010.08.007

20

RNA pathogenesis via Toll-like receptor-activated inflammation in expanded repeat neurodegenerative diseases

Robert I. Richards[1]*, Saumya E. Samaraweera[1], Clare L. van Eyk[1], Louise V. O'Keefe[1] and Catherine M. Suter[2]

[1] Discipline of Genetics and Centre for Molecular Pathology, School of Molecular and Biomedical Science, The University of Adelaide, Adelaide, SA, Australia
[2] Victor Chang Cardiac Research Institute, Darlinghurst, NSW, Australia

***Correspondence:**
Robert I. Richards, Discipline of Genetics and Centre for Molecular Pathology, School of Molecular and Biomedical Science, The University of Adelaide, North Terrace, Adelaide, SA 5000, Australia
e-mail: robert.richards@adelaide.edu.au

Previously, we hypothesized that an RNA-based pathogenic pathway has a causal role in the dominantly inherited unstable expanded repeat neurodegenerative diseases. In support of this hypothesis we, and others, have characterized $rCAG.rCUG_{100}$ repeat double-strand RNA (dsRNA) as a previously unidentified agent capable of causing pathogenesis in a *Drosophila* model of neurodegenerative disease. *Dicer*, *Toll*, and autophagy pathways have distinct roles in this *Drosophila* dsRNA pathology. *Dicer* dependence is accompanied by cleavage of $rCAG.rCUG_{100}$ repeat dsRNA down to $r(CAG)_7$ 21-mers. Among the "molecular hallmarks" of this pathway that have been identified in *Drosophila*, some [i.e., $r(CAG)_7$ and elevated tumor necrosis factor] correlate with observations in affected people (e.g., Huntington's disease and amyotrophic lateral sclerosis) or in related animal models (i.e., autophagy). The *Toll* pathway is activated in the presence of repeat-containing dsRNA and toxicity is also dependent on this pathway. How might the endogenously expressed dsRNA mediate *Toll*-dependent toxicity in neuronal cells? Endogenous RNAs are normally shielded from *Toll* pathway activation as part of the mechanism to distinguish "self" from "non-self" RNAs. This typically involves post-transcriptional modification of the RNA. Therefore, it is likely that $rCAG.rCUG_{100}$ repeat dsRNA has a characteristic property that interferes with or evades this normal mechanism of shielding. We predict that repeat expansion leads to an alteration in RNA structure and/or form that perturbs RNA modification, causing the unshielded repeat RNA (in the form of its *Dicer*-cleaved products) to be recognized by *Toll*-like receptors (TLRs), with consequent activation of the *Toll* pathway leading to loss of cell function and then ultimately cell death. We hypothesize that the proximal cause of expanded repeat neurodegenerative diseases is the TLR recognition (and resultant innate inflammatory response) of repeat RNA as "non-self" due to their paucity of "self" modification.

Keywords: RNA pathogenesis, Toll-like receptor, innate inflammation, expanded repeat diseases, neurodegeneration

INTRODUCTION

Since the first discovery of trinucleotide repeat expansion as the basis for many important human genetic diseases (Kremer et al., 1991; La Spada et al., 1991; Yu et al., 1991; Richards and Sutherland, 1992), there has been a vast amount of research in this area (*PubMed* search "trinucleotide repeat disorders" gives >3,700 results). Much of this research is aimed at identifying the mechanism of pathogenesis underlying diseases caused by this form of mutation. Individual diseases can follow either dominant or recessive mode of inheritance indicating distinct pathogenic pathways. Repeat sequences that are expanded in copy number are the basis for ~20 dominantly inherited neurodegenerative diseases, including Huntington's disease (HD). Despite some of the responsible genes being identified as long as 20 years ago, the identity and nature of the disease-causing pathogenic pathway remains a gap in knowledge for these diseases, i.e., no definitive molecular pathway from the mutation to the clinical symptoms has yet been identified. For at-risk individuals in families affected with dominantly inherited late-onset neurodegenerative diseases due to expanded repeats, the majority opt not to have the definitive pre-symptomatic diagnostic test. Their preference is to live with the uncertainty of not knowing, than the certainty of getting the disease, as no treatments are yet available. Therefore, determining the pathogenic pathway and identifying therapeutic targets for intervention is an urgent priority for reducing the impact of these devastating diseases. This understanding is essential for rational approaches to delay onset, slow progression, or ultimately effect cure.

MOLECULAR PATHWAY FROM REPEAT EXPANSION TO DISEASE

There are common properties exhibited by the various repeat expansions that give rise to human disease. The vast majority of these diseases originate from an existing repeat sequence that exhibits copy number variation in the human population. In each case, the disease alleles arise when copy number increases beyond a critical threshold. The repeat composition varies, but most are trinucleotide repeats. In some diseases, the repeat expands to the point where gene expression at the expanded repeat locus is either substantially reduced or lost altogether, resulting in loss-of-function of the repeat-harboring gene. Typically such diseases are inherited in a recessive manner. Many repeat loci, however, give rise to dominantly inherited diseases in a manner that is not gene-dose dependent (i.e., two mutant alleles are no worse, and may be even better than one – see Carroll et al., 2013). This suggests that gain-of-function is the mechanism rather than haploinsufficiency. Repeat copy number in many cases is a major determinant of age at onset of clinical symptoms (referred to as "anticipation") indicating that the repeat itself is a rate-limiting determinant of the pathogenic pathway. However, since pathology typically involves cell death and there are many ways in which cells die, identification of the disease-causing "toxic agent" has been problematic.

IS THERE A COMMON PATHOGENIC AGENT?

The unstable expanded repeat diseases (**Figure 1**) typically manifest as neurodegenerative and/or muscular diseases, some with a high degree of clinical overlap, despite affecting distinct proteins and unrelated loci. Where expanded repeats are translated, they generally code for polyglutamine; however, the proteins in which they are located are all unrelated in the remainder of their amino-acid sequence. Therefore, much attention has been focused on expanded polyglutamine as the common basis of pathology (McLeod et al., 2005; van Eyk et al., 2011, 2012). Some of these diseases, however, have repeat expansions located within untranslated RNAs and/or arise from repeat sequences that cannot encode polyglutamine (**Figure 1**; Richards, 2001; La Spada and Taylor, 2010). Despite these significant differences in the location of the repeat in this family of diseases they exhibit overlapping symptoms resulting from neuronal loss of function and/or neurodegeneration. In addition, in most cases the polyglutamine and "untranslated" diseases have similar disease allele copy number repeat thresholds (*HD* CAG > 36, *SCA17* CAG > 47, *FXTAS* CGG > 55, *HDL2* CUG > 44, *DM1* CUG > 50, *SCA12* CAG > 66). This suggests that there may be a common pathogenic agent or agents in the translated and untranslated repeat diseases.

RNA MAY BE PATHOGENIC IN TRANSLATED REPEAT DISEASES

While there is growing consensus that RNA plays a causal role in "non-coding repeat expansion disorders," its contribution when the repeat is located in coding regions (specifically polyglutamine disorders) is more controversial (Fiszer and Krzyzosiak, 2013). Yet even here there is evidence that RNA is key. For example, intermediate copy number CAG alleles of *SCA2* that are below the threshold required to encode aggregate forming polyglutamine, increase the risk of amyotrophic lateral sclerosis (ALS; Elden et al., 2010). Furthermore, interruption of CAG repeat with CAA dramatically mitigates polyglutamine toxicity in a *Drosophila* model of *SCA3* (Li et al., 2008).

It is possible that multiple pathways (at least one of which is RNA mediated) contribute to progression of expanded repeat neurodegenerative diseases. In support of this possibility, ALS and *SCA7* both appear to involve two cell types (nerve cells and glial cells; Furrer et al., 2011; Polymenidou and Cleveland, 2011). Astrocytes and glial cells have both been shown to affect their neighboring neurons in individuals with repeat expansions, leading Ilieva et al. (2009) to hypothesize that the onset of the disease is determined in the nerve cell, and the progression of the disease determined in adjacent glial(-like) cells. But

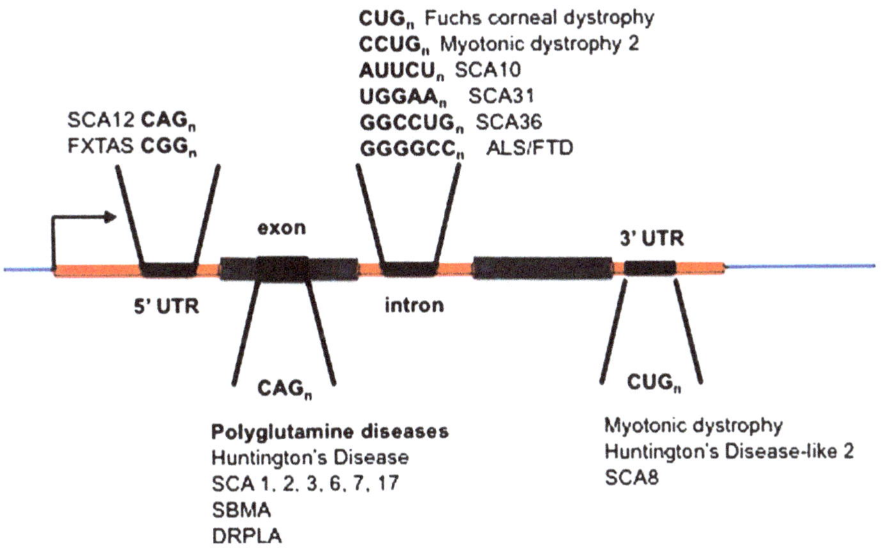

FIGURE 1 | Location of expanded repeats in disease genes. SCA, spinocerebellar ataxia (multiple loci numbered); FXTAS, fragile X tremor ataxia syndrome; ALS, amyotrophic lateral sclerosis; FTLD, frontotemporal lobar dementia; SBMA, spinobulbar muscular atrophy; DRPLA, dentatorubral-pallidoluysian atrophy.

importantly, there is consistent evidence implicating RNA as of principle importance as the originating causal event that initiates pathology.

MECHANISMS OF RNA-INITIATED PATHOLOGY

What precedents and potential mechanisms are there for RNA to initiate pathogenesis in human diseases? See **Figure 2**.

SINGLE-STRANDED RNA TOXICITY

Precedence for expanded repeat RNA being a disease-causing entity in its own right first came from the *DM1* and *DM2* repeat expansions that both give rise to myotonic dystrophy (Ranum and Day, 2004). The repeat expansions in these diseases are similar, but importantly, not identical (CUG vs. CCUG) and are located in untranslated regions (3′UTR or intron) of two otherwise unrelated genes (*DMPK* and *ZNF9*). In muscle cells, RNAs from expanded alleles of either repeat are able to bind and sequester alternative splicing factors (muscleblind and CUG-BP) and in so doing, perturb the splicing pathways of proteins for which alternative splicing is a necessary step for their complete range of functions (Mankodi et al., 2002; Ranum and Day, 2004). It is now generally accepted that RNA is the common pathogenic agent in these diseases most likely through its impact on alternative splicing, although this has recently been challenged with evidence that GSK3β mediates at least some aspects of the RNA-based pathology in myotonic dystrophy (Jones et al., 2012) and in a *Drosophila* model (van Eyk et al., 2011).

Evidence for a more widespread role for RNA in neurodegenerative diseases has been steadily accumulating. *SCA31* and *SCA36* are due to large expansions of *de novo* 5 bp TGGAA repeat and an existing 6 bp GGCCTG repeat, respectively – both located within introns of different genes (Sato et al., 2009; Kobayashi et al., 2011). An expanded GGGGCC repeat has recently been found to cause a substantial proportion of cases of ALS and frontotemporal lobar dementia (FTLD; DeJesus-Hernandez et al., 2011; Renton et al., 2011). As indicated by others (Orr, 2011) "The location of this repeat within an intron of the *C9ORF72* gene along with some evidence for alternative splicing of *C9ORF72* transcripts brings in to play a prominent aspect of non-coding repeat expansion disorders – the role of RNA metabolism in pathogenesis."

REPEAT ASSOCIATED NON-AUG TRANSLATION

The hairpin structure of expanded repeat RNA is such that it can enable the initiation of translation in the absence of the normal requirement of an AUG start codon (Zu et al., 2011). Although this mechanism involves conversion of the RNA into peptides, thereby rendering the RNA no longer "untranslated," the phenomenon can occur to RNA sequences that do not normally appear in protein-coding sequences, i.e., RNA from introns or 5′ or 3′ untranslated regions of mRNAs. The resultant translated

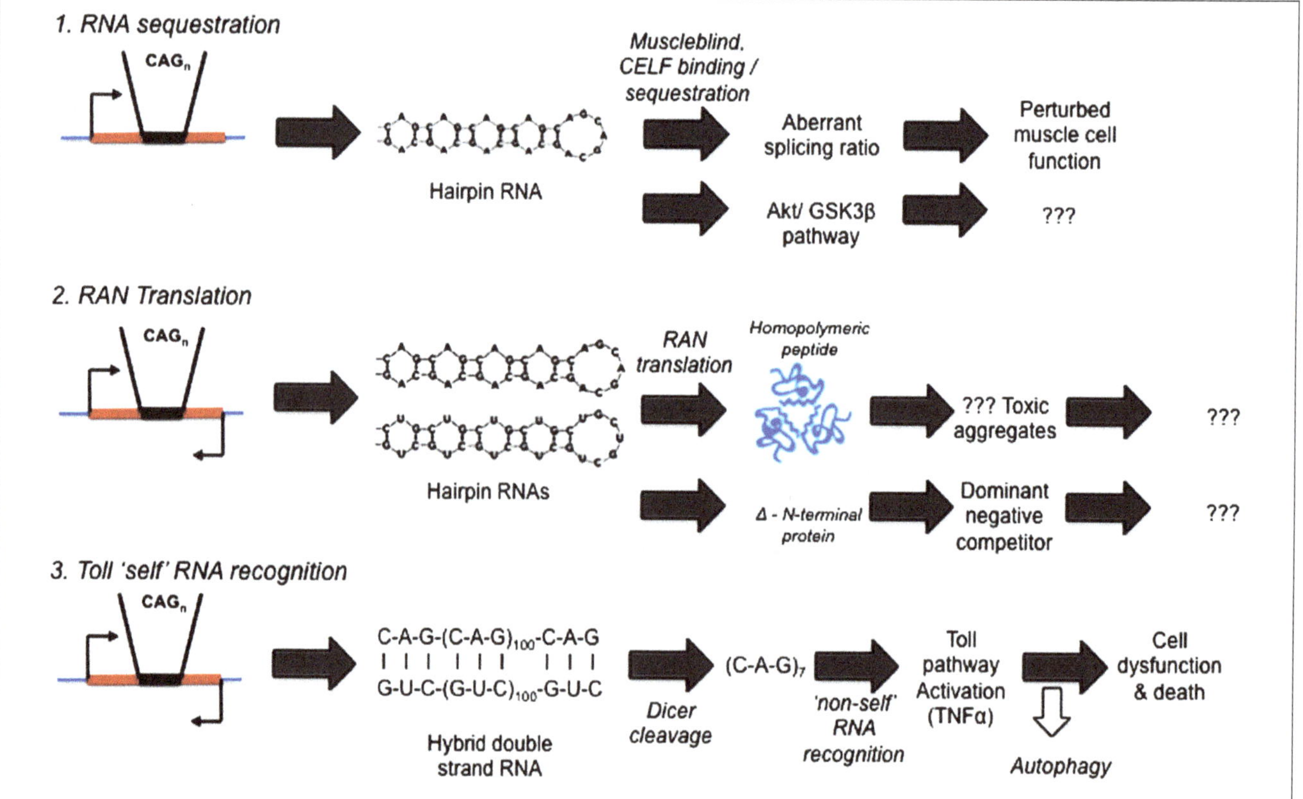

FIGURE 2 | Competing hypotheses of expanded repeat disease pathogenic pathways involving RNA. (1) RNA sequestration – via alternative splicing (Mankodi et al., 2002; Ranum and Day, 2004) or Akt/GSK3β pathway (van Eyk et al., 2011; Jones et al., 2012; Lawlor et al., 2012). (2) RAN (repeat associated non-AUG) Translation (Zu et al., 2011; Ash et al., 2013; Mori et al., 2013; Todd et al., 2013). (3) *Toll* "self" RNA recognition (Lawlor et al., 2011; Yu et al., 2011; Samaraweera et al., (2013)).

polypeptides can initiate from within the repeat sequence and in any reading frame, therefore, a single strand containing repeat RNA sequence can encode three different polypeptide sequences. Since expanded repeat sequences are typically located in regions of bi-directional transcription (Batra et al., 2010), the resultant transcripts from both strands potentially enable the production of six different peptide sequences, any of which may be toxic to the cell. Such polypeptides have now been detected in pathology specimens from individuals affected with a number of different expanded repeat diseases including DM1, fragile X syndrome (FRAXA) and ALS/FTLD (Zu et al., 2011; Ash et al., 2013; Mori et al., 2013; Todd et al., 2013). Of particular note, two recent publications (Ash et al., 2013; Mori et al., 2013) have identified repeat associated non-AUG (RAN)-translation of the GGGGCC expanded repeats that cause ALS/FTLD into polypeptides that also form aggregates in affected tissues. However, these aggregates are confined to nerve cells and are absent from adjacent glial cells that are also involved in the pathology. On the other hand, the absence of visible aggregates does not prove the absence of toxic peptides.

These RAN translation results have suggested that an aggregate polypeptide analogous to polyglutamine could be neurotoxic in diseases where the causative repeat expansion cannot encode polyglutamine. However, this is doubtful in the cases of ALS and FTLD because of the observations that mutations in either of two RNA-binding proteins, FUS and TDP-43, can also cause disease (Rutherford et al., 2008; Van Langenhove et al., 2010). In individuals affected due to these mutations, no such expanded polyGly-Pro polypeptide is evident, therefore while polyGly-Pro may lead to subtle differences in pathology (Ash et al., 2013; Mori et al., 2013), it appears to play a modifying role at most.

Furthermore, inhibition of an RNA lariat debranching enzyme has recently been shown to suppress TDP-43 toxicity in ALS disease models (Armakola et al., 2012). These observations reinforce the view that RNA has a central role to play in this disease. While the role of such polypeptides in disease pathogenesis is unclear, for example, whether their aggregation may actually be protective rather than pernicious, they are a curious set of products driven by the unusual structure of expanded repeat RNAs. One possibility is that rather than the homopolymeric polypeptides themselves being toxic, the initiation of translation within the repeats could give rise to *N*-terminal truncated proteins devoid of upstream functional domains that could then act as dominant negative competitors for the full-length functionally intact proteins.

DOUBLE-STRANDED EXPANDED REPEAT RNA IS PATHOGENIC

Drosophila models of expanded repeat diseases have been described that specifically investigate the intrinsic toxicity of both translated and untranslated expanded repeat sequences (Lawlor et al., 2011; van Eyk et al., 2011, 2012; Samaraweera et al., 2013). In one study (Lawlor et al., 2011), a single line of *Drosophila* expressing untranslated CAG was identified with a marked degenerative phenotype (whereas multiple other random insertion lines of the same transgene had no such phenotype). Upon detailed characterization, this degenerative phenotype line was found to have the repeat transgene inserted into an endogenous gene (*cheerio*) in the opposite orientation to normal transcription. Transcripts containing expanded repeats would, therefore, originate from both strands via bi-directional transcription. This finding coincided with numerous reports in the literature that expanded repeat disease loci are typically transcribed from both DNA strands (see Batra et al., 2010). Therefore, this *Drosophila* line mimicked a previously uncharacterized property of these disease genes. Bi-directional transcription was subsequently modeled in a controlled manner by co-expression from two different transgenes of expanded $rCAG_{\sim 100}$ together with $rCUG_{\sim 100}$ [giving rise to $rCAG.rCUG_{\sim 100}$ or double-strand RNA (dsRNA)] to produce repeat-containing dsRNA (Lawlor et al., 2011). Flies expressing dsRNA showed *Dicer*-dependent toxicity. Additionally dsRNA expression throughout the nervous system caused an age-dependent neurodegenerative phenotype. An abundance of $r(CAG)_7$ also implicated specific *Dicer* processing of the $rCAG.rCUG_{\sim 100}$ dsRNA as a pathogenic pathway in this model (Lawlor et al., 2011). Similar findings have also been reported in an independent *Drosophila* model (Yu et al., 2011). There are, with all animal models, caveats. In order to manifest a phenotype in the time frame of laboratory experiments, these *Drosophila* (and other animal) models employ copy numbers well in excess of those that cause pathology (after several decades) in some of these diseases. This is thought to be due to an inverse relationship between repeat copy number and age-at-onset, the basis of which could be somatic repeat instability over time (see **Figure 2** in Richards, 2001 and Swami et al., 2009). Furthermore, the level of expression of the repeat RNAs required to give an early phenotype in animal models (Lawlor et al., 2011; Yu et al., 2011) may be well in excess of that of the endogenous human disease gene. Importantly, however, examination of HD patient samples (Bañez-Coronel et al., 2012) revealed the presence of the same $r(CAG)_7$ cleavage product seen in the *Drosophila* models, providing evidence in support of the activity of this pathway in HD pathogenesis.

In an effort to identify further components of expanded repeat RNA pathogenesis in *Drosophila*, microarray analyses of *Drosophila* expressing $rCAG.rCUG_{\sim 100}$ dsRNA have been undertaken (Samaraweera et al., 2013). Changes in transcription profiles revealed candidate pathways for mediating the resultant pathogenesis. Alterations in transcripts common to several pathways were detected, including components of inflammation and innate immunity. Hallmarks of immune activation, including elevated plasma tumor necrosis factor (TNF), appear prior to clinical symptoms of dominantly inherited expanded repeat human diseases (Moreau et al., 2005; Björkqvist et al., 2008). Therefore, the *Drosophila* model expressing $rCAG.rCUG_{\sim 100}$ dsRNA was utilized to test two key elements of immune activation – the *Toll* and autophagy pathways for their contribution to expanded repeat RNA pathogenesis. *Toll* signaling pathway was identified as essential for dsRNA pathogenesis and autophagy was found to reduce toxicity in this model (Samaraweera et al., 2013). Furthermore, multiple reports implicate glial cells in the pathology of expanded repeat diseases. Neurons are dependent upon glial cell function that includes the destruction and removal of the carcasses of dead neurons. The $rCAG.rCUG_{\sim 100}$ dsRNA was found to impact nerve cell function even when exclusively expressed in glial cells (Samaraweera et al., 2013), providing evidence that dsRNA pathology in

Drosophila is, like the human expanded repeat diseases, non-cell autonomous (Ilieva et al., 2009; Furrer et al., 2011).

The requirement for *Toll* signaling pathway in this *Drosophila* model is intriguing. *Toll*-like receptors (TLRs) function in normal biology to protect an organism from infection by viruses and bacteria. They recognize foreign pathogen molecules including DNA and RNA through specific receptors (such as endosomal *TLR3*) and can distinguish these nucleic acids (as "non-self") from the endogenous nucleic acids ("self"). Therefore, while the *rCAG.rCUG~100* dsRNA is being expressed endogenously in this *Drosophila* model, it is being recognized by the *Toll* signaling pathway as foreign or "non-self" – a recognition that then activates innate inflammatory regulatory pathways, ultimately leading to cell death.

PATHOGENIC MUTATIONS IN PROTEINS THAT FUNCTIONALLY INTERACT WITH RNA

While it can be difficult to ascribe specific functions to RNA in pathogenic pathways, there are some noteworthy instances of disease-causing mutations in proteins that functionally interact with RNAs. By implication, the RNAs that these proteins normally act upon are, therefore, likely contributors to and/or mediators of the relevant pathogenic process.

RNA-BINDING MOTIFS – THE RNAS THAT HAVE THEM AND THE PROTEINS THAT RECOGNIZE THEM

Recent discoveries regarding the importance of RNA–protein recognition in disease pathogenesis have led to a renewed interest in the role that these interactions play in biological processes. While they have long been recognized as key regulators of gene expression, only a small fraction have been functionally characterized. A recent compendium of RNA-binding motifs (Ray et al., 2013) highlighted both the significance and scope of these interactions. The human genome encodes at least 400 known or predicted RNA-binding proteins with a diverse array of RNA sequence-binding motifs. Indeed the number of such human RNA-binding proteins appears to be much higher than this, with 860 identified in HeLa cells alone (Castello et al., 2012). The scope and specificity of RNA recognition is determined both by the number and variety of RNA-binding proteins and by the number and variety of RNA-sequence motifs that they bind.

FRAGILE X SYNDROME IS DUE TO LOSS OF RNA-BINDING PROTEIN FUNCTION

Fragile X syndrome is a striking example of the role of an RNA-binding protein in human disease. FRAXA is due to the expanded CGG repeat that is responsible for the *FRAXA* rare, folate-sensitive chromosomal fragile site (Kremer et al., 1991), located in the 5′UTR of the *FMR1* gene (Verkerk et al., 1991). Expansion of the repeat beyond ~230 copies results in inactivation of the gene and consequent loss of encoded FMRP (fragile X mental retardation protein) function (Pieretti et al., 1991). The FMRP is an RNA-binding protein with KH- and RGG-binding motifs (Ashley et al., 1993). The loss of function of this protein is responsible for the clinical symptoms as rare cases of point mutation or deletion of the *FMR1* gene have similar clinical symptoms. Indeed one of these pathogenic point mutations is at a highly conserved amino acid in a KH domain of FMRP highlighting the significance of the role of RNA interaction in FMRP function (De Boulle et al., 1993). The FMRP has an impact on the translation of the mRNAs with which it interacts (Darnell et al., 2001) and, therefore, its absence leads to the dysregulation of the translation of these specific mRNAs. This is thought to be the proximal cause of the symptoms of FRAXA.

THE INTRIGUING PATHOGENESIS OF AICARDI–GOUTIÈRES SYNDROME

Aicardi–Goutières syndrome (AGS) is a genetically heterogeneous disorder that is due (at least in a substantial proportion of cases) to the mutation of various nucleic acid-metabolizing enzymes, including various subunits of ribonuclease H2 or the RNA-editing enzyme ADAR1 (see Crow and Rehwinkel, 2009 and OMIM #225750). AGS is characterized, in its more severe forms, by severe neurological dysfunction in infancy that includes progressive microcephaly, spasticity, dystonic posturing, profound psychomotor retardation, and often death in early childhood (OMIM #225750) (**Figure 3**). In its milder forms, these neurological symptoms are diminished or even absent, but peripheral symptoms outside the nervous system are common to the phenotypic spectrum and include thrombocytopenia, hepatosplenomegaly, and elevated hepatic transaminases along with intermittent fever. Chilblains are also a typical feature. Together these symptoms demonstrate phenotypic overlap both with systemic lupus erythematosus and with the sequelae of congenital infection (Crow and Rehwinkel, 2009). The disease, therefore, appears to be due to defects in the processes that remove and/or modify endogenous nucleic acids. These endogenous unmodified nucleic acids then accumulate and are sensed as "non-self" by TLRs, that, in turn, activate innate inflammatory regulatory pathways. This bears a striking resemblance to mechanisms we have identified as responsible for dsRNA pathogenesis in the *Drosophila* model of expanded repeat neurodegenerative diseases.

HYPOTHESIS

EXPANDED REPEAT RNAS AS PATHOGENIC AGENTS BY *TOLL* "SELF" RNA RECOGNITION

Repeat RNA sequences represent a pivotal point of potential weakness in processes that utilize RNA–protein recognition, as the repeat RNA sequence will harbor either a paucity or excess of sequence-binding motifs. Expansion of repeat RNA sequences, therefore, clearly has the potential to give rise to too much or too little of an interaction that is a rate-limiting factor in a crucial biological process. RNA modification is one process that is sequence motif-dependent and known to be key to the distinction between "self" and "non-self" by components of the innate immune pathways. Indeed, it has been shown that exogenous "non-self" RNAs require *in vitro* modification in order to escape innate immune recognition and activation when transferred *in vivo* (Warren et al., 2007; Pan, 2013). The exposure of the innate immune activators to unmodified nucleic acids, including RNA, appears to be the proximal cause of AGS. We, therefore, hypothesize that this provides a clear molecular mechanism for the ability of expanded repeat RNA sequences, through their paucity of RNA modification, to initiate pathogenesis in the dominantly inherited, expanded repeat neurodegenerative diseases (**Figure 4**).

Aicardi-Goutieres Syndrome** (AGS) – **Mutations in RNA Metabolizing Genes in Bold

Locus – mutated gene

AGS1 – TREX1	*– DNA 3' to 5' exonuclease, prevents autoimmunity caused by endogenous retroelements.*
AGS2 – RNASEH2B	**– subunit B of the human ribonuclease H2 enzyme complex – ribonuclease H2 cleaves ribonucleotides from RNA:DNA duplexes.**
AGS3 – RNASEH2C	**– subunit C of the human ribonuclease H2 enzyme complex.**
AGS4 – RNASEH2A	**– subunit A of the human ribonuclease H2 enzyme complex.**
AGS5 – SAMHD1	*– converts deoxynucleoside triphosphates to constituent deoxynucleoside and inorganic triphosphate. Restriction factor that renders human dendritic and myeloid cells largely refractory to HIV-1 infection.*
AGS6 – ADAR1	**– converts adenosine to inosine in double strand RNA**

FIGURE 3 | Pathogenic mutations in Aicardi–Goutières syndrome. Mutations in genes in at least six distinct loci are able give rise to the constellation of symptoms that defines Aicardi–Goutières syndrome. Four of these (AGS2, AGS3, AGS4, and AGS6) are in genes that encode RNA-metabolizing proteins. The remaining two that have been identified (AGS1 and AGS5) are also in enzymes that have roles in nucleic acid metabolism. Deficiencies in any one of these enzymes are thought to result in the accumulation of endogenous nucleic acids that are sensed as "non-self" by *Toll*-like receptors, that in turn activate innate inflammatory pathways (Crow and Rehwinkel, 2009).

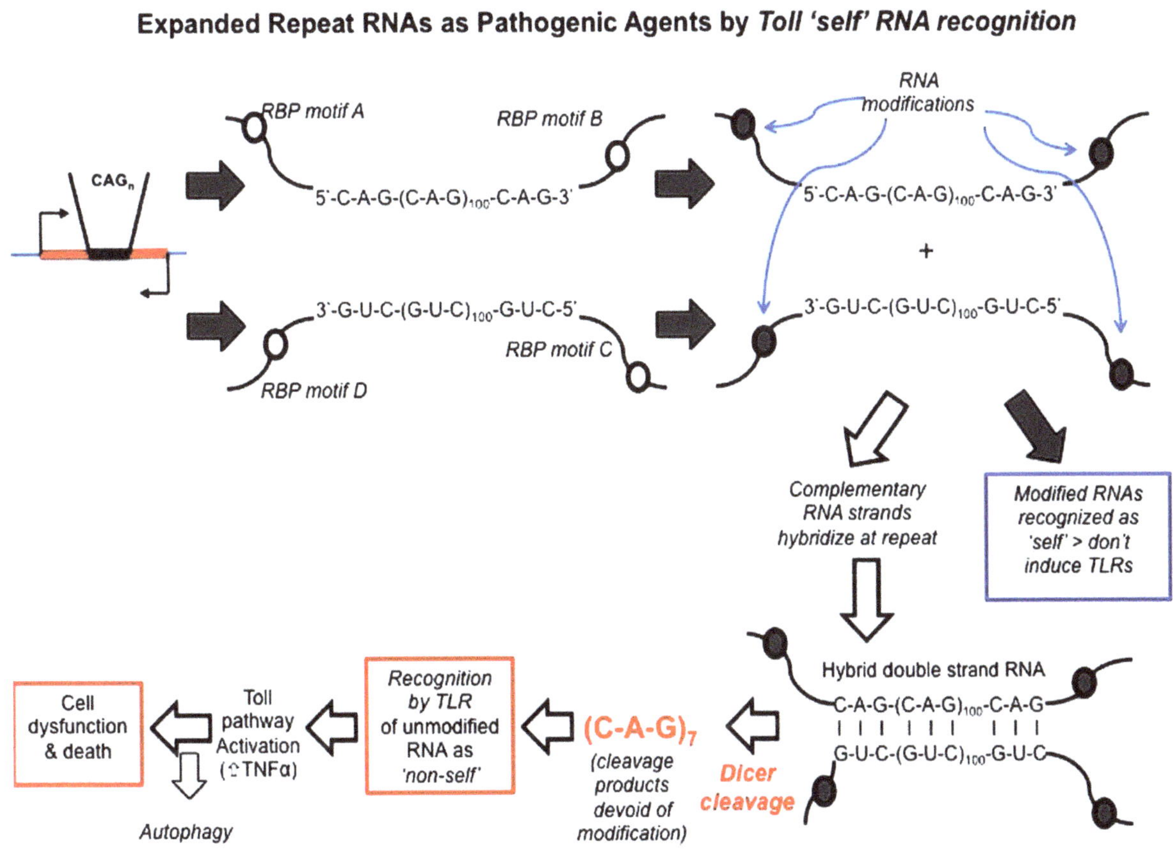

FIGURE 4 | Hypothesis: expanded repeat neurodegenerative diseases are caused by the TLR recognition (and resultant innate inflammatory response) of repeat RNA as "non-self" due to their paucity of "self" modification that is exposed upon Dicer processing of double-strand RNA. Open circles represent sequence motifs for RNA modifying proteins; filled circles represent the modification of RNA at these specific sequence motifs (e.g., by methylation or A > I editing). Dicer is required for pathology in the *Drosophila* model and cleaves long high copy number repeat RNA down to 21mers [mainly *r(CAG)*$_7$ mers; Lawlor et al., 2011]. These *r(CAG)*$_7$ mers are, therefore, unmodified and recognized by TLRs as "non-self." *Toll*-like receptor pathways (most probably the endosomal TLR3 receptor) are required for pathology (Samaraweera et al., 2013), through activation of the innate inflammatory pathway. Autophagy reduces pathology, possibly by metabolizing *r(CAG)*$_7$ mers.

ACTIVITY OF *TOLL* "SELF" RNA RECOGNITION IN NEURODEGENERATIVE DISEASES

Double strand expanded repeat RNA pathology has been modeled in *Drosophila*. What evidence is there that this pathway of TLR recognition of expanded repeat RNA and subsequent activation of the innate inflammatory cascade is active in the human dominantly inherited neurodegenerative diseases due to expansion of repeat sequences?

One of the key steps in dsRNA pathology is the generation of $r(CAG)_7$ 21mers from the much greater copy number double strand repeat RNA by *Dicer*. This $r(CAG)_7$ 21mer has been identified in the brain RNA of individuals affected with HD (Bañez-Coronel et al., 2012). The activity of *Dicer* is crucial to the observed pathology in the *Drosophila* model (Lawlor et al., 2011) and, therefore, it would appear that this step is a likely proximal event in the observed phenotype. The appearance of $r(CAG)_7$ 21mers in HD brain is therefore an important "molecular hallmark" of this pathway and a key indicator of its activity in the human disease. Another, albeit less direct, indicator of this pathway is seen in the increased activity of components of the innate inflammatory response mechanism in human diseases associated with expanded repeats. Elevated TNF is seen in the *Drosophila* model as one read-out of innate immune activation (Samaraweera et al., 2013) and both TNF and various interleukins (i.e., IL-4, IL-5, IL-6, IL-8, and IL-10) have been found to be elevated in people affected with the repeat expansion responsible for HD even before clinical manifestation of the disease (Björkqvist et al., 2008). Another indicator of innate immune activation in HD is the abnormal peripheral chemokine profile that has been observed in HD (Wild et al., 2011). Various reports indicate activation of innate adaptive immunity via TLR signaling in ALS (Casula et al., 2011; Sta et al., 2011) – a disease that has recently been found, at least in a proportion of instances, to be also due to an expanded repeat (DeJesus-Hernandez et al., 2011; Renton et al., 2011).

CONCLUSION

A growing body of literature indicates a consistent association between innate immunity, neuroinflammation and neurodegeneration (Shastri et al., 2013). Where there are exogenous causes (e.g., trauma or infection), activation of the TLR pathway can be attributed to an external agent (e.g., bacterial lipopolysaccharide or viral RNA); however, a causal basis for this relationship has not been clear when there is an endogenous basis to the disease, e.g., expansion of a repeat sequence beyond a pathogenic threshold. Recognition by the *Toll* receptor pathway of expanded repeat RNA as "non-self" and consequent activation of the innate immune inflammatory cascade provides a mechanism and a common pathogenic pathway for the neurodegenerative diseases due to expanded repeats. This new understanding, once proven in the relevant human diseases, will provide new targets for intervention and ultimately, we hope, therapeutic targets for drugs to delay onset and/or alleviate disease progression.

AUTHOR CONTRIBUTIONS

Robert I. Richards drafted the initial version of the manuscript, including the hypothesis, then edited in the additions and changes made by the other co-authors. Saumya E. Samaraweera and Clare L. van Eyk provided original unpublished data on which the manuscript and its hypothesis is based, as well as contributions to the development of the hypothesis, the text and figures. Louise V. O'Keefe contributed to the development of the hypothesis and additional text and revision of the manuscript. Catherine M. Suter contributed to information on RNA-binding proteins, the development of the hypothesis and content of the text and figures.

ACKNOWLEDGMENTS

This work is funded in part by a Project Grant (627183) from the National Health and Medical Research Council of Australia and a post-doctoral fellowship to Clare L. van Eyk from the National Ataxia Foundation (USA). Robert I. Richards wishes to thank Dan Kastner, Ivona Aksentijevich, and Massimo Gadina (NIH) for valuable discussions and Sarah Robertson, Amanda Choo, Danielle Fornarino, and Cheng Shoou Lee for helpful and constructive criticism of drafts of this manuscript.

REFERENCES

Armakola, M., Higgins, M. J., Figley, M. D., Barmada, S. J., Scarborough, E. A., Diaz, Z., et al. (2012). Inhibition of RNA lariat debranching enzyme suppresses TDP-43 toxicity in ALS disease models. *Nat. Genet.* 44, 1302–1309. doi: 10.1038/ng.2434

Ash, P. E., Bieniek, K. F., Gendron, T. F., Caulfield, T., Lin, W. L., Dejesus-Hernandez, M., et al. (2013). Unconventional translation of C9ORF72 GGGGCC expansion generates insoluble polypeptides specific to c9FTD/ALS. *Neuron* 77, 1–8. doi: 10.1016/j.neuron.2013.02.004

Ashley, C. T., Jr., Wilkinson, K. D., Reines, D., and Warren, S. T. (1993). FMR1 protein: conserved RNP family domains and selective RNA-binding. *Science* 262, 563–566. doi: 10.1126/science.7692601

Bañez-Coronel, M., Porta, S., Kagerbauer, B., Mateu-Huertas, E., Pantano, L., Ferrer, I., et al. (2012). A pathogenic mechanism in Huntington's disease involves small CAG-repeated RNAs with neurotoxic activity. *PLoS Genet.* 8:e1002481. doi: 10.1371/journal.pgen.1002481

Batra, R., Charizanis, K., and Swanson, M. S. (2010). Partners in crime: bidirectional transcription in unstable microsatellite disease. *Hum. Mol. Genet.* 19, R77–R82. doi: 10.1093/hmg/ddq132

Björkqvist, M., Wild, E. J., Thiele, J., Silvestroni, A., Andre, R., Lahiri, N., et al. (2008). A novel pathogenic pathway of immune activation detectable before clinical onset in Huntington's disease. *J. Exp. Med.*, 205, 1869–1877. doi: 10.1084/jem.20080178

Carroll, J. M., Quaid, K. A., Stone, K., Jones, R., Schubert, F., and Griffith, C. B. (2013). Two is better than one: a case of homozygous myotonic dystrophy type 1. *Am. J. Med. Genet.* 161, 1763–1767. doi: 10.1002/ajmg.a.35967

Castello, A., Fischer, B., Eichelbaum, K., Horos, R., Beckmann, B. M., Strein, C., et al. (2012). Insights into RNA biology from an atlas of mammalian mRNA-binding proteins. *Cell* 149, 1393–1406. doi: 10.1016/j.cell.2012.04.031

Casula, M., Iyer, A. M., Spliet, W. G., Anink, J. J., Steentjes, K., Sta, M., et al. (2011). Toll-like receptor signaling in amyotrophic lateral sclerosis spinal cord tissue. *Neuroscience* 179, 233–243. doi: 10.1016/j.neuroscience.2011.02.001

Crow, Y. J., and Rehwinkel, J. (2009). Aicardi–Goutières syndrome and related phenotypes: linking nucleic acid metabolism with autoimmunity. *Hum. Mol. Genet.* 18, R130–R136. doi: 10.1093/hmg/ddp293

Darnell, J. C., Jensen, K. B., Jin, P., Brown, V., Warren, S. T., and Darnell, R. B. (2001). Fragile X mental retardation protein targets G quartet mRNAs important for neuronal function. *Cell* 107, 489–499. doi: 10.1016/S0092-8674(01)00566-9

De Boulle, K., Verkerk, A. J. M. H., Reyniers, E., Vits, L., Hendrickx, J., Van Roy, B., et al. (1993). A point mutation in the FMR-1 gene associated with fragile X mental retardation. *Nat. Genet.* 3, 31–35. doi: 10.1038/ng0193-31

DeJesus-Hernandez, M., Mackenzie, I. R., Boeve, B. F., Boxer, A. L., Baker, M., Rutherford, N. J., et al. (2011). Expanded GGGGCC hexanucleotide repeat in non-coding region of C9ORF72 causes chromosome 9p-linked FTD and ALS. *Neuron* 72, 245–256. doi: 10.1016/j.neuron.2011.09.011

Elden, A. C., Kim, H. J., Hart, M. P., Chen-Plotkin, A. S., Johnson, B. S., Fang, X., et al. (2010). Ataxin-2 intermediate-length polyglutamine expansions are associated with increased risk for ALS. *Nature* 466, 1069–1075. doi: 10.1038/nature09320

Fiszer, A., and Krzyzosiak, W. J. (2013). RNA toxicity in polyglutamine disorders: concepts, models, and progress of research. *J. Mol. Med.* 91, 683–691. doi: 10.1007/s00109-013-1016-2

Furrer, S. A., Mohanachandran, M. S., Waldherr, S. M., Chang, C., Damian, V. A., Sopher, B. L., et al. (2011). Spinocerebellar ataxia type 7 cerebellar disease requires the coordinated action of mutant ataxin-7 in neurons and glia, and displays non-cell-autonomous Bergmann glia degeneration. *J. Neurosci.* 31, 16269–16278. doi: 10.1523/JNEUROSCI.4000-11.2011

Ilieva, H., Polymenidou, M., and Cleveland, D. W. (2009). Non-cell autonomous toxicity in neurodegenerative disorders: ALS and beyond. *J. Cell Biol.* 187, 761–772. doi: 10.1083/jcb.200908164

Jones, K., Wei, C., Iakova, P., Bugiardini, E., Schneider-Gold, C., Meola, G., et al. (2012). GSK3β mediates muscle pathology in myotonic dystrophy. *J. Clin. Invest.* 122, 4461–4472. doi: 10.1172/JCI64081

Kobayashi, H., Abe, K., Matsuura, T., Ikeda, Y., Hitomi, T., Akechi, Y., et al. (2011). Expansion of intronic GGCCTG hexanucleotide repeat in NOP56 causes SCA36, a type of spinocerebellar ataxia accompanied by motor neuron involvement. *Am. J. Hum. Genet.* 89, 121–130. doi: 10.1016/j.ajhg.2011.05.015

Kremer, E., Pritchard, M., Lynch, M., Yu, S., Holman, K., Warren, S., et al. (1991). DNA instability at the fragile X maps to a trinucleotide repeat sequence p(CCG)n. *Science* 252, 1711–1714. doi: 10.1126/science.1675488

La Spada, A. R., and Taylor, J. P. (2010). Repeat expansion disease: progress and puzzles in disease pathogenesis. *Nat. Rev. Genet.* 11, 247–258. doi: 10.1038/nrg2748

La Spada, A. R., Wilson, E. M., Lubahn, D. B., Harding, A. E., and Fischbeck, K. H. (1991). Androgen receptor gene mutations in X-linked spinal and bulbar muscular atrophy. *Nature* 352, 77–79. doi: 10.1038/352077a0

Lawlor, K. T., O'Keefe, L. V., Samaraweera, S., van Eyk, C., McLeod, C. J., Maloney, C., et al. (2011). Double stranded RNA is pathogenic in *Drosophila* models of expanded repeat neurodegenerative diseases. *Hum. Mol. Genet.* 20, 3757–3768. doi: 10.1093/hmg/ddr292

Lawlor, K. T., O'Keefe, L. V., Samaraweera, S. E., van Eyk, C. L., and Richards, R. I. (2012). Ubiquitous expression of CUG or CAG trinucleotide repeat RNA causes common morphological defects in a *Drosophila* model of RNA-mediated pathology. *PLoS ONE* 7:e38516. doi: 10.1371/journal.pone.0038516

Li, L.-B., Yu, Z., Teng, X., and Bonini, N. M. (2008). RNA toxicity is a component of ataxin-3 degeneration in *Drosophila*. *Nature* 453, 1107–1111. doi: 10.1038/nature06909

Mankodi, A., Takahashi, M. P., Jiang, H., Beck, C. L., Bowers, W. J., Moxley, R. T., et al. (2002). Expanded CUG repeats trigger aberrant splicing of ClC-1 chloride channel pre-mRNA and hyperexcitability of skeletal muscle in myotonic dystrophy. *Mol. Cell* 10, 35–44. doi: 10.1016/S1097-2765(02)00563-4

McLeod, C., O'Keefe, L., and Richards, R. I. (2005). The pathogenic agent in *Drosophila* models of 'polyglutamine' diseases. *Hum. Mol. Genet.* 14, 1041–1048. doi: 10.1093/hmg/ddi096

Moreau, C., Devos, D., Brunaud-Danel, V., Defebvre, L., Perez, T., Destée, A., et al. (2005). Elevated IL-6 and TNF-alpha levels in patients with ALS: inflammation or hypoxia? *Neurology* 65, 1958–1960. doi: 10.1212/01.wnl.0000188907.97339.76

Mori, K., Weng, S. M., Arzberger, T., May, S., Rentzsch, K., Kremmer, E., et al. (2013). The C9orf72 GGGGCC repeat is translated into aggregating dipeptide-repeat proteins in FTLD/ALS. *Science* 339, 1335–1338. doi: 10.1126/science.1232927

Orr, H. T. (2011). FTD and ALS: genetic ties that bind. *Neuron* 72, 189–190. doi: 10.1016/j.neuron.2011.10.001

Pan, T. (2013). N6-methyl-adenosine modification in messenger and long non-coding RNA. *Trends Biochem. Sci.* 38, 204–209. doi: 10.1016/j.tibs.2012.12.006

Pieretti, M., Zhang, F., Fu, Y.-H., Warren, S. T., Oostra, B. A., Caskey, C. T., et al. (1991). Absence of expression of the FMR-1 gene in fragile X syndrome. *Cell* 66, 817–822. doi: 10.1016/0092-8674(91)90125-I

Polymenidou, M., and Cleveland, D. W. (2011). The seeds of neurodegeneration: prion-like spreading in ALS. *Cell* 147, 498–508. doi: 10.1016/j.cell.2011.10.011

Ranum, L. P., and Day, J. W. (2004). Myotonic dystrophy: RNA pathogenesis comes into focus. *Am. J. Hum. Genet.* 74, 793–804. doi: 10.1086/383590

Ray, D., Kazan, H., Cook, K. B., Weirauch, M. T., Najafabadi, H. S., Li, X., et al. (2013). A compendium of RNA-binding motifs for decoding gene regulation. *Nature* 499, 172–177. doi: 10.1038/nature12311

Renton, A. E., Majounie, E., Waite, A., Simón-Sánchez, J., Rollinson, S., Gibbs, J. R., et al. (2011). A hexanucleotide repeat expansion in C9ORF72 is the cause of chromosome 9p21-linked ALS–FTD. *Neuron* 72, 257–268. doi: 10.1016/j.neuron.2011.09.010

Richards, R. I. (2001). Dynamic mutations: a decade of unstable expanded repeats in human disease *Hum. Mol. Genet.* 10, 2187–2194. doi: 10.1093/hmg/10.20.2187

Richards R. I., and Sutherland, G. R. (1992). Dynamic mutations: a new class of mutations causing human disease. *Cell* 70, 709–712. doi: 10.1016/0092-8674(92)90302-S

Rutherford, N. J., Zhang, Y. J., Baker, M., Gass, J. M., Finch, N. A., Xu, Y. F., et al. (2008). Novel mutations in TARDBP (TDP-43) in patients with familial amyotrophic lateral sclerosis. *PLoS Genet.* 4:e1000193. doi: 10.1371/journal.pgen.1000193

Samaraweera, S. E., O'Keefe, L. V., Price, G. R., Venter, D. J., and Richards, R. I. (2013). Distinct roles for Toll and autophagy pathways in double-stranded RNA toxicity in a *Drosophila* model of expanded repeat neurodegenerative diseases. *Hum. Mol. Genet.* 22, 2811–2819. doi: 10.1093/hmg/ddt130

Sato, N., Amino, T., Kobayashi, K., Asakawa, S., Ishiguro, T., Tsunemi, T., et al. (2009). Spinocerebellar ataxia type 31 is associated with "inserted" penta-nucleotide repeats containing (TGGAA)n. *Am. J. Hum. Genet.* 85, 544–557. doi: 10.1016/j.ajhg.2009.09.019

Shastri, A., Bonfati, D. M., and Kishore, U. (2013). Innate immunity and neuroinflammation. *Mediators Inflamm.* 2013, 342931. doi: 10.1155/2013/342931

Sta, M., Sylva-Steenland, R. M., Casula, M., de Jong, J. M., Troost, D., Aronica, E., et al. (2011). Innate and adaptive immunity in amyotrophic lateral sclerosis: evidence of complement activation. *Neurobiol. Dis.* 42, 211–220. doi: 10.1016/j.nbd.2011.01.002

Swami, M., Hendricks, A. E., Gillis, T., Massood, T., Mysore, J., Myers, R. H., et al. (2009). Somatic expansion of the Huntington's disease CAG repeat in the brain is associated with an earlier age of disease onset. *Hum. Mol. Genet.* 18, 3039–3047. doi: 10.1093/hmg/ddp242

Todd, P. K., Oh, S. Y., Krans, A., He, F., Sellier, C., Frazer, M., et al. (2013). CGG repeat-associated translation mediates neurodegeneration in fragile X tremor ataxia syndrome. *Neuron* 78, 440–455. doi: 10.1016/j.neuron.2013.03.026

van Eyk, C., McLeod, C. J., O'Keefe, L. V., and Richards, R. I. (2012). Comparative Toxicity of polyglutamine, polyalanine and polyleucine tracts in *Drosophila* models of expanded repeat disease. *Hum. Mol. Genet.* 21, 536–547. doi: 10.1093/hmg/ddr487

van Eyk, C. L., O'Keefe, L. V., Lawlor, K. T., Samaraweera, S. E., McLeod, C. J., Price, G. R., et al. (2011). Perturbation of the Akt/Gsk3-β signalling pathway is common to *Drosophila* expressing expanded untranslated CAG, CUG and AUUCU repeat RNAs. *Hum. Mol. Genet.* 20, 2783–2794. doi: 10.1093/hmg/ddr177

Van Langenhove, T., van der Zee, J., Sleegers, K., Engelborghs, S., Vandenberghe, R., Gijselinck, I., et al. (2010). Genetic contribution of FUS to frontotemporal lobar degeneration. *Neurology* 74, 366–371. doi: 10.1212/WNL.0b013e3181ccc732

Verkerk, A. J., Pieretti, M., Sutcliffe, J. S., Fu, Y. H., Kuhl, D. P., Pizzuti, A., et al. (1991). Identification of a gene (FMR-1) containing a CGG repeat coincident with a breakpoint cluster region exhibiting length variation in fragile X syndrome. *Cell* 65, 905–914. doi: 10.1016/0092-8674(91)90397-H

Warren, L., Manos, P. D., Ahfeldt, T., Loh, Y. H., Li, H., Lau, F., et al. (2007). Highly efficient reprogramming to pluripotency and directed differentiation of human cells with synthetic modified mRNA. *Cell Stem Cell* 7, 618–630. doi: 10.1016/j.stem.2010.08.012

Wild, E., Magnussen, A., Lahiri, N., Krus, U., Orth, M., Tabrizi, S. J., et al. (2011). Abnormal peripheral chemokine profile in Huntington's disease. *PLoS Curr.* 3:RRN1231. doi: 10.1371/currents.RRN1231

Yu, S., Kremer, E., Pritchard, M., Lynch, M., Nancarrow, J., Baker, E., et al. (1991). The fragile X genotype is characterized by an unstable region of DNA. *Science* 252, 1179–1182. doi: 10.1126/science.252.5009.1179

Yu, Z., Teng, X., and Bonini, N. M. (2011). Triplet repeat-derived siRNAs enhance RNA-mediated toxicity in a *Drosophila* model for myotonic dystrophy. *PLoS Genet.* 7:e1001340. doi: 10.1371/journal.pgen.1001340

Zu, T., Gibbens, B., Doty, N. S., Gomes-Pereira, M., Huguet, A., Stone, M. D., et al. (2011). Non-ATG-initiated translation directed by microsatellite expansions. *Proc. Natl. Acad. Sci. U.S.A.*, 108, 260–265. doi: 10.1073/pnas.1013343108

21

Small RNA sequencing-microarray analyses in Parkinson leukocytes reveal deep brain stimulation-induced splicing changes that classify brain region transcriptomes

*Lilach Soreq[1], Nathan Salomonis[2], Michal Bronstein[3], David S. Greenberg[4], Zvi Israel[5], Hagai Bergman[1,6] and Hermona Soreq[4,6]**

[1] Department of Medical Neurobiology, Hadassah Faculty of Medicine, The Hebrew University of Jerusalem, Jerusalem, Israel
[2] Division of Genomic Medicine, The Gladstone Institute of Cardiovascular Disease, San Francisco, CA, USA
[3] The Center for Genomic Technologies, The Institute of Life Sciences, The Hebrew University of Jerusalem, Jerusalem, Israel
[4] Department of Biological Chemistry, The Life Sciences Institute, The Hebrew University of Jerusalem, Jerusalem, Israel
[5] Department of Neurosurgery, The Centre for Functional and Restorative Neurosurgery, Hadassah University Hospital, Jerusalem, Israel
[6] The Edmond and Lili Safra Center for Brain Sciences, The Hebrew University of Jerusalem, Jerusalem, Israel

***Correspondence:**
*Hermona Soreq, Department of Biological Chemistry, The Alexander Silberman Institute of Life Sciences, Edmond Safra Campus - Givat Ram, The Hebrew University of Jerusalem, Jerusalem, 91904, Israel.
e-mail: hermona.soreq@mail.huji.ac.il*

MicroRNAs (miRNAs) are key post transcriptional regulators of their multiple target genes. However, the detailed profile of miRNA expression in Parkinson's disease, the second most common neurodegenerative disease worldwide and the first motor disorder has not been charted yet. Here, we report comprehensive miRNA profiling by next-generation small-RNA sequencing, combined with targets inspection by splice-junction and exon arrays interrogating leukocyte RNA in Parkinson's disease patients before and after deep brain stimulation (DBS) treatment and of matched healthy control volunteers (HC). RNA-Seq analysis identified 254 miRNAs and 79 passenger strand forms as expressed in blood leukocytes, 16 of which were modified in patients pre-treatment as compared to HC. 11 miRNAs were modified following brain stimulation 5 of which were changed inversely to the disease induced changes. Stimulation cessation further induced changes in 11 miRNAs. Transcript isoform abundance analysis yielded 332 changed isoforms in patients compared to HC, which classified brain transcriptomes of 47 PD and control independent microarrays. Functional enrichment analysis highlighted mitochondrion organization. DBS induced 155 splice changes, enriched in ubiquitin homeostasis. Cellular composition analysis revealed immune cell activity pre and post treatment. Overall, 217 disease and 74 treatment alternative isoforms were predictably targeted by modified miRNAs within both 3′ and 5′ untranslated ends and coding sequence sites. The stimulation-induced network sustained 4 miRNAs and 7 transcripts of the disease network. We believe that the presented dynamic networks provide a novel avenue for identifying disease and treatment-related therapeutic targets. Furthermore, the identification of these networks is a major step forward in the road for understanding the molecular basis for neurological and neurodegenerative diseases and assessment of the impact of brain stimulation on human diseases.

Keywords: deep brain stimulation, high throughput sequencing, leukocytes, miRNAs, Parkinson's disease, splice junction microarrays, substantia nigra

INTRODUCTION

Parkinson's disease (PD) is the second most prevalent neurodegenerative disease worldwide and the first movement disorder. PD is mainly expressed in progressive, deliberating motor symptoms, initiating with tremor at rest, slowness of movement and rigidity which develops into flexed posture, loss of postural reflexes and a freezing phenomenon (Fahn and Elton, 1987). Notably, the cardinal motor symptoms appear only years after initiation of the disease process. As the disease presents other central signs such as depression, sleep disorders (Bernal-Pacheco et al., 2012) and dementia (Goldman and Litvan, 2011) as well as peripheral symptoms (e.g., digestive and olfactory impairments) (Braak and Del Tredici, 2008), it is becoming increasingly clear that PD is not a pure motor disorder. In fact, it may initiate many years prior to the appearance of the motor symptoms (Hawkes et al., 2010). However, current methods do not allow early or accurate diagnosis and moreover, the genes and mechanisms that lead to the massive dopaminergic neuronal loss in the damaged PD brain are still largely obscure. The growing prevalence of PD worldwide introduces a strong incentive to find both disease biomarkers and future therapeutic targets.

MicroRNAs (miRNAs) are key post transcriptional regulators. Since their discovery 20 years ago (Lee et al., 1993), it has been well demonstrated that they play essential roles in development (Reinhart et al., 2000) as well as in human diseases, for example in cancer (Mraz et al., 2009; Li et al., 2013; Pu et al., 2013), as well as Alzheimer's disease and other central nervous system disorders (Maes et al., 2009). A growing body of evidence now suggests that peripheral miRNA changes may co-occur with the onset of PD (Doxakis, 2010; Martins et al., 2011). MiRNAs mediate post-transcriptional regulation of protein-coding genes by primarily binding to short sequence motifs found typically in the 3′ untranslated region of target mRNAs. Depending on the degree of sequence complementation, this leads to translational inhibition, mRNA destabilization or degradation. Since their discovery, miRNAs have emerged as crucial regulators of gene expression, and a single miRNA can concurrently regulate hundreds of target mRNAs. However, despite their biologic importance, determining global miRNA profiles and functional miRNA-target interactions remained a major challenge. Correspondingly, so far only about 20% of the known human miRNAs were found as constituting biological networks of functionally-associated molecules (Satoh and Tabunoki, 2011).

PD-related miRNAs have been characterized in midbrain dopaminergic neurons (Kim et al., 2007; Minones-Moyano et al., 2011), and were found to post-transcriptionally regulate the key PD genes LRRK2 (Gehrke et al., 2010) and SNCA (Doxakis, 2010). Also, down-regulation of miR-34b/c was found to modulate mitochondrial function in PD (Martins et al., 2011) and deletion of hsa-let-7 caused toxic effects similar to those of pathogenic LRRK2 (Gehrke et al., 2010) in a Drosophila PD model. However, post-mortem brain tissue is not readily accessible, and entails RNA quality concerns (Atz et al., 2007). Blood leukocytes, however, are readily obtained, and the RNA can be obtained at high quality from them. Recently, miRNAs were characterized in whole blood of PD patients (Martins et al., 2011) by microarrays and quantitative real time PCR (Margis and Rieder, 2011). RNA-Seq provides an unbiased, high resolution, accurate means for profiling miRNAs (Pritchard et al., 2012) and thus for investigating the miRNA mediated pathogenesis of complex diseases such as PD. Purported advantages of small RNA sequencing for this purpose include generation of data unlimited by genome annotations or probe affinities. Analyses of genome-wide small RNA transcriptome provide an accurate and comprehensive approach of globally characterizing miRNA expression in disease under different conditions. Yet, the global profile of miRNAs was not characterized yet by comprehensive RNA high throughput sequencing of PD blood cells.

Alternative splicing (AS) dramatically increases the complexity of the human transcriptome. Recent studies identified possible interactions between miRNAs and spliced targets (Salomonis et al., 2010; Irimia and Blencowe, 2012). Yet, while gene-expression changes have been characterized in PD blood cells using conventional microarrays (Scherzer et al., 2007; Soreq et al., 2008), such studies are blinded to alternative isoform regulation, which has been associated with at least 95% of the human genes (Pan et al., 2008). Deep brain stimulation (DBS) treatment dramatically improves the debilitating PD motor symptoms which are the cardinal disease symptoms (Bergman et al., 1990; Kingwell, 2011). Recently, we found disease-related transcription changes in PD leukocytes and demonstrated that DBS neurosurgery largely reversed the observed transcriptional changes (Soreq et al., 2012b). We have identified that a large number of predicted miRNA binding sites located in alternative exon regions.

Here, we performed a comparative miRNA, splicing and gene expression analysis of PD patients' blood leukocytes pre- and post-DBS with the electrical stimulus being on and following 1 h of stimulation cessation using high throughput RNA sequencing in conjunction with exon and splice junction microarrays of the same cells. Along with these samples, mRNA samples from matched healthy control volunteers were analysed.

We found significant differences in miRNA expression levels in PD, partially reversed by DBS, which co-occurred with high confidence changes in gene isoform expression levels and cell lineage composition. Our findings provide new insights for the molecular mechanisms underlying peripheral reactions to neurological and neurodegenerative diseases and opening new avenues for future development of both new diagnostic and therapeutic tools. We provide these data sets and analyses as a resource for understanding miRNA expression and splicing in Parkinson's patient's leukocytes pre- and post-treatment.

MATERIALS AND METHODS

This study was authorized and approved by the Ethics committee of the human review board (Hadassah University Hospital, Ein-Kerem, approval number 07.09.07-6) in accordance with the Declaration of Helsinki principles. Following oral agreement all participants signed informed consent prior to inclusion in the study.

PATIENT RECRUITMENT, DBS NEUROSURGERY AND CLINICAL EVALUATION

Overall, seven PD male patients nominated for bilateral STN-DBS neurosurgery and six healthy age-matched male controls healthy control volunteers (HC) were recruited to the study. Clinical description appears under (Soreq et al., 2012b). Briefly, volunteers were assessed for their clinical background and state and fulfilled detailed medical history questionnaires. Patients with other medical conditions were excluded, including depression and past and current DSM Axis I and II psychological disorders (SM), chronic inflammatory disease, coagulation irregularities, previous malignancies or cardiac events, or any surgical procedure up to one year pre-DBS. All patients went through bilateral STN-DBS electrode implantation (Medtronics, USA) and were under dopamine replacement therapy (DRT) both pre- and post-DBS (on significantly reduced dosage post-DBS both on and off stimulation), the last medication administered at least 5 h pre-sampling. Patients further exhibited similar total white and red blood cell counts pre- and post-DBS. Blood samples were collected from each patient at three time points: (1) one day pre-DBS upon hospitalization, with medication; (2) post-DBS (range 6–18 weeks), when reaching optimal clinical state as evaluated by a neurologist and on a lower DRT dose, ON stimulus and (3) OFF stimulus, following 1 h while being OFF electrical stimulation (counted from stage 2). The clinical severity of disease symptoms was assessed by a neurologist for Unified PD Rating Scale (UPDRS-III) (Fahn and Elton, 1987) section 3 (motor section) in the tested stated (pre-DBS, post-DBS on and off stimulation). Controls were recruited among Hadassah hospital staff and researchers at the Edmond J. Safra Campus (Jerusalem). Exclusion criteria for the HC volunteers included smoking, chronic inflammatory diseases, drug or alcohol use, major depression, previous cardiac events and past year hospitalizations. The only group of other control volunteers that was included in this study is of healthy control volunteers, as samples from "sham" operated healthy (or patient) volunteers cannot be obtained or authorized by ethics committee. However, in addition to the healthy controls, we tested the leukocyte RNA from PD patients post-DBS also following 1 h of electrical stimulation cessation, to enable discrimination of the electrical stimulation effect from the effect of the presence of the stimulator in the brain *per-se*.

BLOOD SAMPLE COLLECTION AND RNA EXTRACTION

Blood collection was conducted between 10AM–14PM. Collected venous blood (9 ml blood using 4.5 ml EDTA (anti-coagulant tubes) was immediately filtered using the LeukoLock fractionation and stabilization kit™ (Ambion, Applied Biosystems, Inc., Foster City, CA) and incubated in RNALater (Ambion). Stabilized filters and serum samples were stored at −80°C. RNA extraction followed the manufacturers' alternative protocol instructions. Briefly, cells were flushed (TRI-Reagent™, Ambion) into 1-bromo-3-chloropropane-(BCP) containing 15 ml tubes and centrifuged. 0.5 and 1.25 volume water and ethanol were added to the aqueous phase. Samples were filtered through spin cartridges, stored in pre-heated 150 μl EDTA; RNA was quantified and examined on Bioanalyzer 2100 (Agilent, Santa Clara, CA, USA) and frozen in −80°C.

RNA-Seq SREK LIBRARY PREPARATION AND SEQUENCING

Libraries for next generation sequencing (NGS) were prepared from total leukocyte RNA with the manufacturers' SREK (Small RNA Expression Kit) protocol (Applied Biosystems, Foster City, CA, USA). A total of 12 libraries were prepared: from RNA of blood leukocytes from three healthy volunteers and three PD patients in three states (pre-DBS, post-DBS on stimulation and following 1 h of stimulation cessation). The DNA Libraries were tested using a DNA 1000 Lab Chip on Bioanalyzer 2100. Briefly, the total RNA samples were hybridized with Adaptor Mix A which is a set of oligonucleotides with a single-stranded degenerate sequence at one end and a defined sequence required for SOLiD sequencing at the other end. The Adaptor Mix A constrains the orientation of the RNA in the ligation reaction such that hybridization with it yields template for SOLiD sequencing from the 5′ end of the sense strand. After hybridization, the adaptors are ligated to the small RNA molecules using Ligation Enzyme Mix, which is a mixture of an RNA Ligase and other components. Ligation requires an RNA molecule with a 5′-monophosphate and a 3′-hydroxyl end; therefore, most small RNAs can participate in this reaction, and intact mRNA molecules with a 5′ cap structure are excluded. Next, the small RNA population with ligated adaptors of each sample was reverse transcribed, to generate cDNA libraries. Treatment with RNase H followed, to digest the RNA from RNA/cDNA duplexes and to reduce the concentration of unligated adaptors and adaptor by-products. The cDNA libraries were amplified using bar-coded primer sets and 15–18 cycles of PCR. The amplified cDNA libraries were cleaned up and size selected from gel—PCR products ~105–150 bp were isolated, corresponding to inserts derived from the small RNA population. These contained each 90 base pairs of primers and adapter sequences. The amplified cDNA libraries generated with the SOLiD Small RNA Expression Kit were thereafter ready for attachment to beads at the emulsion PCR step of the SOLiD sample preparation workflow. The slides were analyzed on a SOLiD system V3.5 (Applied Biosystems). Applied Biosystems SOLiD sequencer generated 50-base read sequences as.csfasta files and the corresponding quality control (.qual) and statistic (.stat) files.

MAPPING TO miRBase AND TO HUMAN REFERENCE GENOME

A total of 166,746,207 sequencing reads were obtained. All the reads (50 bp long) were subjected to trimming of the tag end terminal base pairs to remove these 15 bases. The trimmed (35 bases) reads were subjected to further trimming of the P1 start adapter as well as the SOLiD miRNA reverse primer sequences using CLC genomics workbench V4.0 [CLC bio, Cambridge, MA (Sakharkar et al., 2004a)] through local Smith–Watermann alignment. Overall, 50.92% of the reads contained adapter sequence and/or the SOLiD small RNA reverse primer. The 5′ terminal nucleotide was removed as well. This left 71,939,833 reads (average length of 21 bases). The minimum sampling count threshold was set to 1. The remaining reads obtained from each patient's library were first aligned against the human miRNA genome (miRBase release 15, which contained 940 Homo sapiens sequences). On average, 55% (range 44.9–61.3%) of the sequences of all libraries were mapped (**Figure S1**) and 50.94% (range 43.8–54.8%) of miRBase genes were detected (**Figure S1**). Of the mapped reads, on average 42.03% (range 39.6–45.9%) had perfect match to the aligned genes, 45.16% (range 42.5–55.3%)-one mismatch, 12.6% (range 0–18.6%)-2 mismatches and 0.18% (range 0.1–0.6%)-3 mismatches (**Figure S1**).

The full count information obtained from the alignment results is given under **Tables S4** and **S5**. Through the annotation and merge counts, only reads longer than 15 bases were analyzed. Match parameters included for mature length variants (IsomiRs)-additional 2 upstream and downstream bases, and 2 missing upstream and downstream bases. Alignment settings were not conducted on the color space. This allowed detection of mismatches up to the maximal allowed by the software (of 3 mismatches instead of 2). Annotated samples were grouped by both precursor and mature sequence identity.

Mapping to another repository of reference human genome database (Ambion, life technologies) which included 764 sequences of snoRNAs, tRNA, SINE/ALU sequences, LINEs and spliceosomal U genes (such as U1) yielded mapping to 293.3 (38.37%) filter database sequences on average across libraries. Only 151,136.8 (3.16%) reads of these that were not mapped to miRBase were mapped against this repository. Of them, 60.88% (range 42.2–68.9%) exhibited perfect match to the reference sequences, 30.94% (range 25.4–41.2%) with 1 mismatch, 6.24 (2.1–13.4%)-2 mismatches and 1.94% (0.5–3.3%)-3 mismatches (**Figure S1**). An average of 486 miRNAs per sample (range: 436–515) were detected in the analyzed libraries (**Figures S2A–E**:

selected examples, full raw digital count quantification under **Tables S4** and **S5**).

DIFFERENTIAL EXPRESSION ANALYSIS OF NEXT GENERATION SEQUENCING DETECTED miRNAs

Overall, a total of 335 mature and 79 mature-star (mature 3′/5′ forms) that exhibited count of at least 1 per million reads in at least one sample were analyzed for differential expression between the different experimental conditions. Differential expression analysis of the miRNAs was conducted using the Bioconductor R package biocLite. The EdgeR (Robinson et al., 2010) and BaySeq (Hardcastle and Kelly, 2010) components were used for differential expression. 251 miRNAs were detected in all the samples, but the statistical methodologies applied allowed to analyze the detected miRNAs that presented count of at least 1 per million reads in at least one sample using adjacent statistical approach, to avoid false negative results. First, a combined data matrix of all the miRNAs detected in any of the samples was created. Then, the detected miRNAs were filtered for low count values such that only these with count ≥ 1 in at least 3 of the 12 libraries remained and 351 miRNAs were analyzed.

AFFYMETRIX HJAY SPLICE JUNCTION MICROARRAY HYBRIDIZATIONS AND DATA PRE-PROCESSING

The Affymetrix human junction arrays (HJAY) were obtained through collaboration with the EURASNET consortium and were used to assess genome-wide changes in exon and junction expression. Briefly, these microarrays interrogate 335,663 human transcripts from ~25,000 Ensembl genes, 260,488 junctions and 360,569 exons. Using the program AltAnalyze, probe set level RMA [17] expression and DAGB *p*-values were obtained by calling Affymetrix Power Tools (APT) [17]. From this probe set-level data, sample gene expression values were calculated from the mean intensity of constitutive annotated exon and junction probe sets in AltAnalyze.

HJAY MICROARRAY PROFILING, DATABASE CONSTRUCTION AND ANALYSIS

HJAY profiling of blood leukocytes was conducted using exon array pre-prepared hybridization samples and Gene-Chip Whole Transcript Sense target Labeling assay kit (Affymetrix), as per manufacturers' instructions. The high-density HJAYs (Affymetrix) were washed and stained with streptavidin-phycoerythrin and signal amplification was done using a biotinylated anti-streptavidin antibody. The microarrays were scanned on an Affymetrix GeneChip Scanner 3000 7G scanner, according to the Affymetrix GeneChip Whole Transcript Sense Target Labeling Assay protocol for the GeneChip Exon 1.0 ST array (Lapuk et al., 2010). A total of 12 microarray samples were obtained. AltAnalyze probe set-to-exon associations were obtained by matching the annotated exon sequences (two exons for each junction probe sets) provided by Affymetrix to the reference Ensembl genome (version 62) for Affymetrix annotated gene symbols. In cases where probe sets aligned to an intron, a novel exon annotation indicating the relative intron position was assigned. Alternative exons, junctions, reciprocal probe sets and event-annotations (e.g., alternative cassette exon, alternative promoter) were obtained by comparing the exon-structure of mRNAs from Ensembl, the UCSC genome database and novel junctions assayed by the HJAY array, as previously described (Salomonis et al., 2010). Linear regression analysis was employed by an updated version of AltAnalyze to detect AS with a fold change of 2. $P < 0.05$ were considered significant. Additional methods details are provided at http://www.altanalyze.org/help_main.htm.

FUNCTIONAL PREDICTION ANALYSIS OF THE HJAY DETECTED SPLICED TRANSCRIPTS

Each detected alternative event or alternative promoter selection (e.g., cassette exon, alternative 3′ end, alternative 5′ end, intron retention, bleeding exon, alternative C-terminal exon) or alternative promoter (alternative N-terminal exon or alternative promoter) of a pair of reciprocal junctions detected by the linear regression analysis was examined for putative protein domains or motifs and for miRNA binding sites. Alternative event annotations were obtained from both from Ensembl and UCSC Genome Browser databases as previously described for AltAnalyze version 1.0 (Salomonis et al., 2009).

FUNCTIONAL ENRICHMENT ANALYSIS OF THE HJAY DETECTED SPLICED TRANSCRIPTS

Functional *post-hoc* analysis of the alternative isoforms detected by linear regression analysis conducted for each comparison (e.g., patients pre-treatment as compared with controls) using Gene Ontology (GO) Elite (Zambon et al., 2012) was called directly from the adopted AltAnalyze version using Ensembl database. Genes cut-off parameters included minimal 2-fold change; and *t*-test raw *P*-value < 0.05 with minimum number of 3 changed genes defined. GO terms (Ashburner et al., 2000) and WikiPathways (Pico et al., 2008) were ranked by a combination of *z*-score (cut-off: 1.96) and gene number. Over-representation analysis (ORA) was conducted with 2000 permutations.

BRAIN TRANSCRIPTOME MICROARRAY ANALYSIS

Raw expression data was obtained from the GEO dataset GSE8397 [citation: (Moran et al., 2006)], where Affymetrix HG_U133 array sets were used to determine the whole genome transcription profile of clinically documented and neuropathologically confirmed cases of sporadic Parkinson's disease as well as controls (and overall 47 samples). The .CEL files were normalized by sketch normalization and the expression of the 22,283 interrogated probe-sets (including 68 microarray internal control prone-sets) was quantified from 506,944 probes using the Probe Logarithmic Intensity Error estimation (PLIER) method (Affymetrix) while considering mismatch probes (Irizarry et al., 2003) (PLIER-MM).

EXON MICROARRAYS HYBRIDIZATION AND INITIAL PRE-PROCESSING

1 μg of total RNA was labeled using the Affymetrix exon array using whole transcripts sense targeting labeling assay according to the manufacturers' instructions; cDNA samples were hybridized to GeneChip® Exon_1.0_ST_Array (Affymetrix, Santa Clara, CA, USA) microarrays, and results were scanned (GeneChip scanner 30007G, 27 CEL files). Hybridizations were conducted in

7 batches. To prevent possible batch effect, each microarray hybridization included samples of patients from all experimental stages (pre, post-DBS on stimulation and post-DBS off stimulation) by random sampling. Quality assurance, normalization (quantile) and probe set summation (DABG iter-PLIER) were conducted using expression console (EC) 1.0 (Affymetrix, Santa Clara, CA). Only core level probe sets ($n = 284.241$) were included to summarize probe set level expression for 22,011 exon array transcript clusters (iter-PLIER, EC). Filtering out transcript clusters with no annotated gene symbols maintained 17,658 of 22,011 transcripts.

CELLULAR LINEAGE ANALYSIS OF THE EXON MICROARRAYS DATA

To identify cellular composition, the new module of AltAnalyze, LineageProfiler, was used. To derive correlation scores to different cell types and tissues (lineages), a database reflecting the most specifically-expressed genes or exons present in each particular lineage, relative to all lineages types examined (ranging from 50 to 150), was created. The resulting database was a small subset of the original, containing the most informative markers. The exon microarrays RNA-profile expression data of patients in three states (pre-DBS, post-DBS on and following 1 h of electrical stimulation cessation-off stimulus) and HC volunteers was compared to the profile of each lineage only for these markers to derive correlation coefficients and resulting Z scores based on the distribution of values for each user RNA-profile. Z scores to each lineage were calculated from the distribution of Pearson correlation coefficients, specifically for each sample or condition analyzed. Lineage differences between conditions were specifically evaluated via the AltAnalyze GO-Elite module using the database of lineage specific markers for examined differentially expressed genes. The results were visualized as association scores at the level of hierarchically clustered cell types and curated lineage networks.

miRNA: TARGET PREDICTIONS

Prediction of target gene sites for the miRNAs that exhibited differential expression in PD patients compared to control volunteers was conducted using the miRWalk repository which combines prediction data from 8 different prediction programs and adds inspection of all the gene regions including 5′ UTR and coding regions (Dweep et al., 2011). The analysis was conducted through a construction of a local MySQL database based on these predictions. For each miRNA that was detected as differentially expressed in the deep transcriptome of patients compared to control volunteers and post compared to pre-DBS, all the predicted targets were searched for and were filtered to these that were detected as alternatively spliced in the human junction microarrays for the corresponding tested condition. The final network of miRNA-spliced targets was analyzed and created through Cytoscape (Smoot et al., 2011).

ACCESSION CODES

All the data discussed in this publication have been deposited in NCBI's Gene Expression Omnibus (Edgar et al., 2002):

(1) High throughput SOLiD small RNA sequencing: http://www.ncbi.nlm.nih.gov/geo/query/acc.cgi?token=btahlkcwcuyyyvy&acc=GSE40915
(2) Splice junction microarrays: http://www.ncbi.nlm.nih.gov/geo/query/acc.cgi?token=hhmvrgcsyyciexi&acc=GSE37591
(3) Exon microarrays: http://www.ncbi.nlm.nih.gov/geo/query/acc.cgi?acc=GSE23676

RESULTS

DIFFERENTIAL EXPRESSION OF miRNAs IN PD COMPARED TO HEALTHY VOLUNTEERS

To delineate the expression of miRNAs in PD peripheral blood leukocytes, we employed SOLiD high throughput sequencing on small RNAs from peripheral blood leukocytes of PD patients followed by alignment to miRBase. The samples were obtained from three PD patients prior to DBS treatment, and upon symptoms stabilization following DBS neurosurgery, both on stimulation and following 1 hour off-electrical stimulation and as compared with matched healthy volunteers (clinical information under **Table S1**). The study design and overall analysis flow are given under **Figure 1**. The full mapping information and reads count are given under **Tables S2** and **S4**. Alignment to miRNABase detected 44.9–61.3% annotated reads (an average detection of 54.67% reads of each library) constituting 43.8–54.8% of miRBase (**Figure S1** and **Table S2**). For comparison, we also aligned all the sequenced libraries against a reference database which contained 764 human filter sequences of other non-coding RNAs including snoRNAs, scRNAs, 5S_RNAs, tRNAs, SINE and ALU repeats, LINEs, spliceosomal U sequences (number 1,2,3,4,5,6,8,13,14, and 17) and snRNAs. These alignments yielded an average of 8.9% aligned reads of the total number of reads (with a minimum of 0.2% and a maximum of 3.16%) (**Table S3**). The raw read counts of all the detected miRNA and miRNA* (also termed 3′/5′) mature forms (given under **Tables S4**, **S5** and **Figures S2A–E**) were filtered to include only those presenting at least 1 reads per million base pairs in at least three samples. Thus, 254 miRNAs, 2 recent miRNA-candidates which are highly likely tRNA fragments and 79 passenger strand miRNAs 3′/5′ forms (miRNA*) were analyzed (altogether 334). At the technical level, count data was discrete and skewed and was hence not well approximated by a normal distribution. To account for this, we applied a test based on the negative binomial distribution (a generalization of the Poisson model that recaptures biological variance correctly) (Robinson and Smyth, 2008). This test can effectively reflect the sample properties and detect differential expression. A linear modeling approach was applied for calculating fold changes and estimating standard errors, and was followed by an empirical Bayes smoothing (which provides more stable results for small sample sized data sets) to moderate the standard errors [16]. Also, sequenced libraries differ in the number of expressed transcripts. This fact could skew the differential expression calls due to over- and under-sampling (Robinson and Oshlack, 2010). To overcome this difficulty, we estimated tag-wise common dispersion to compute normalization factors for each library while accounting for library size. The common dispersion values enabled estimation of the bias using an empirical approach and were built into the library size. The computed concentrations were further subjected to an exact test to detect the top differentially

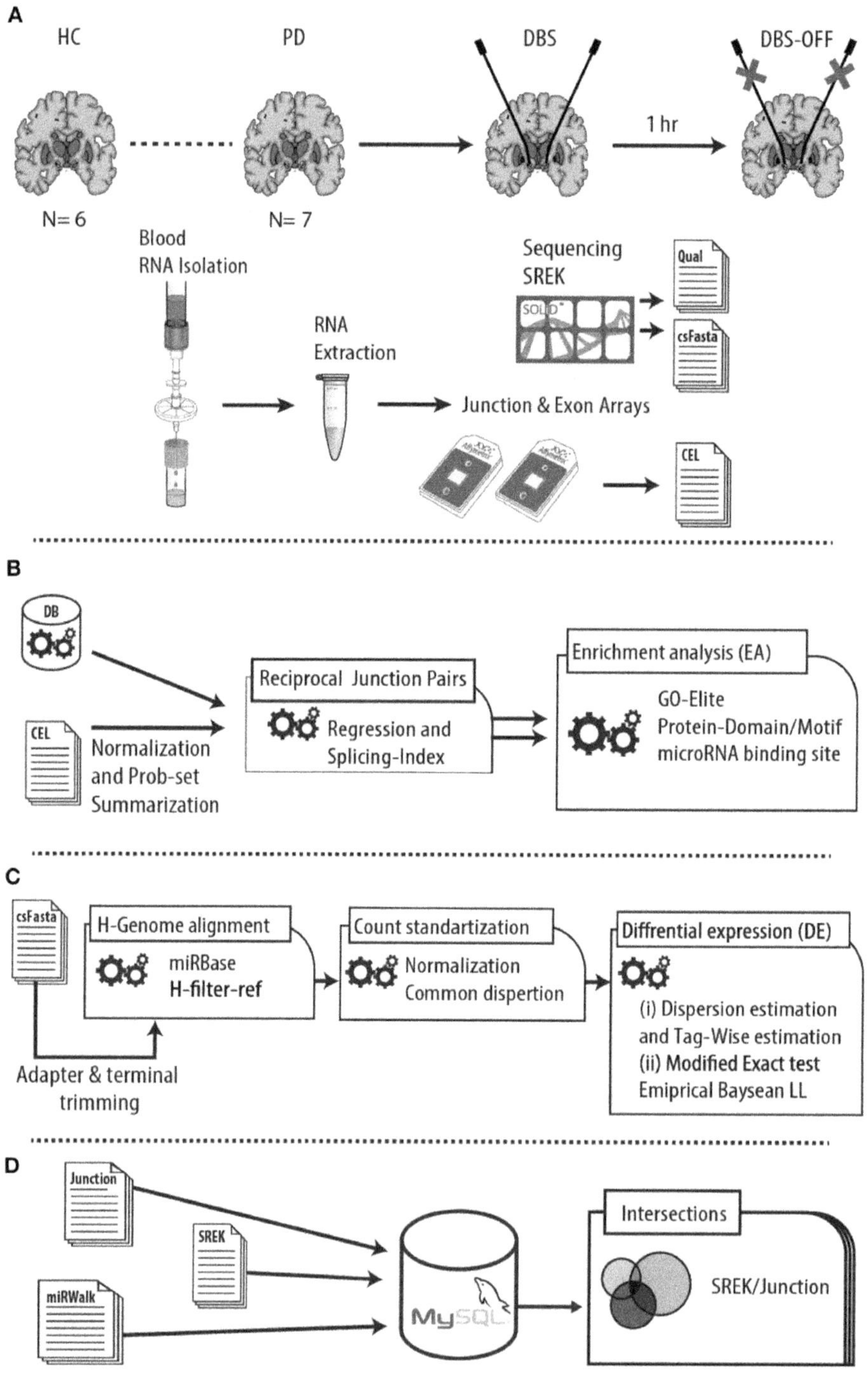

FIGURE 1 | Experimental design and analysis flow. (A) RNA was prepared from filtered blood leukocytes isolated from whole blood samples of PD patients 1 day prior to DBS (PD), and post-DBS upon clinical stabilization, both on (DBS-ON) and following 1 h off electrical stimulation (DBS-OFF). Age- and gender-matched male healthy volunteers served as a control group (HC). Small RNA libraries were prepared using a Small RNA Expression Kit (SREK) and constructed according to the manufacturer's instructions. cDNA of patients pre- and post-DBS on stimulation and controls was also applied to Affymetrix human exon and junction prototype microarrays (HJAY). **(B)** The junction array datasets were processed in AltAnalyze using a previously described regression approach (Sugnet et al., 2006) on junction pairs, followed by functional enrichment analysis. Protein domains and enriched motifs and miRNA binding sites were also identified. **(C)** Reads obtained by high throughput sequencing were trimmed from adapter and end terminal bases and were aligned to miRBase and the human filter reference sequences database. The counts were standardized for common tag-wise dispersion and tested for differential expression using both dispersion estimation and Bayesian learning methods. **(D)** Predictions of miRNA: target interactions with the spliced transcripts were conducted by using the miRWalk algorithm which combines 8 established databases with predictions beyond the 3′ untranslated region (3′-UTR).

expressed genes. This involved B statistic ranking (logarithm of the posterior odds that a gene is differentially expressed (Lonnstedt, 2002) through the EdgeR package (Robinson et al., 2010). Thus, the statistical approach applied here took into account both technical and biological variability of the digital count data.

Taking a B-statistic threshold of one, 16 out of the identified 332 miRNAs and one of the tRNA fragments previously named hsa-mir-1280 (Schopman et al., 2010), were detected as differentially expressed between PD patients and healthy control volunteers (**Figure 2A**, full list under **Table S6**). Of the detected miRNAs, 12 were predicted to bind AS regions detected as changed in the junction arrays. Overall, 6 miRNAs were down regulated and 10 up regulated in PD patients (**Figure 2B**), some of which already shown as relevant to PD or other neurodegenerative diseases. For example, hsa-miR-16-1 and hsa-miR-16-2, two precursors that yield the same mature hsa-miR-16 that was recently identified as changed in PD blood cells (Margis and Rieder, 2011), were decreased in PD leukocytes (log2 fold change: −1.11, *p*-value: 0.026, expression under **Figure 2B**). Inversely, the neuro-inflammatory regulator hsa-miR-20a that is down-regulated in human aging (Hackl et al., 2010) and is affected in Alzheimer's disease (Delay et al., 2011) was up regulated in patients (log fold change: 1.5, *p*-value: 0.002). Elevated hsa-miR-20a-mediated suppression of the transcription factor E2F1 can also impair transcription (Brock et al., 2009), whereas potentiated targeting by mir-20a of the BMPR2 protein, a pathogenic hallmark of pulmonary hypertension may compete with BMPR2 regulation by the miR-17/92 cluster (Huang et al., 2012) which was also identified as changed in our PD patients data (log fold change: −1.16, *p*-value: 13.28). Among the changed miRNAs was also the miR-17/92 cluster, which is regulated by the pro-inflammatory cytokine *interleukin (Il)-6* (Brock et al., 2009). Intriguing, miR-320, which was modified in PD patients as compared with matched healthy control volunteers is included in a miRNA signature of prion-induced neurodegeneration (Saba et al., 2008), perhaps reflecting a feedback response aimed to avoid disease symptoms that was potentiated by DBS. Inversely, the tumorigenic miR-18b was elevated in PD (Guo et al., 2008). Additionally, the recently identified primate-specific hsa-miR-1274b (log fold change: 1.07, *p*-value: −11.79) which closely resembles a known tRNA was also up regulated, predicting global aberrations in miRNA metabolism. The PD-increased hsa-miR-21 (log fold change: 1.3, *p*-value: 0.02) also participates in a miRNA-based gene dys-regulation pathway in Huntington's disease (Johnson et al., 2008), and potentiated suppression by hsa-miR-21 and hsa-miR-150 of *PTEN* and *Huntingtin* would intercept T- and B-cells proliferation (Krichevsky and Gabriely, 2008; Lundstrom, 2011). Also, the reduced hsa-miR-378 (log fold change: 2.25, *p*-value: 0.000109) predicts impaired control over mitochondrial metabolism (Carrer et al., 2012) and negatively regulated NK cell cytotoxicity (Wang et al., 2012). Predicted decrease in another target, the Alzheimer's disease *APP* gene encoding for the amyloid precursor protein (Hébert et al., 2009) may indicate that at least some of these miRNA changes reflect a protective feedback response.

DBS INDUCES STIMULUS-DEPENDENT REVERSAL OF PD LEUKOCYTE miRNA CHANGES

Patient leukocyte miRNAs were re-examined post DBS, upon symptoms stabilization. Similar analysis identified differential expression of 11 miRNAs, 8 down and 3 up-regulated as compared to the pre-DBS state ($p < 0.05$, B statistics, **Table S7** and **Figure 2C**) and the two tRNA fragments which were previously annotated as two new miRNAs (**Table S7**) (Schopman et al., 2010). Strikingly, 5 of the DBS-modified miRNAs were among those that were changed in PD compared to healthy volunteers but exhibited inverse direction of change post-DBS as compared with the PD-induced change (**Figures 2C,D**). For example, hsa-mir-20a and hsa-miR-378 increased in PD patients and decreased post-DBS, whereas the inflammatory-regulating hsa-miR-424 (Spinelli et al., 2013) showed no change in PD but it's passenger 3′ form (hsa-miR-18b-3p, previously called star form) exhibited down-regulation following DBS.

The DBS neurosurgery leads to greatly reduced medication dosage (which was also the case in our cohort), raising the possibility that the post-DBS changes were due to this reduced dose and/or to the surgical procedure itself. To challenge these questions, we also examined the small RNA transcriptome following 1-h stimulation cessation in the same group of patients. Disconnecting the electrical stimulation rapidly re-induces the motor disease symptoms that were apparent prior to the DBS treatment (such as resting tremor). Rapid and dynamic changes in miRNA expression were previously observed in several instances [e.g., (Perry et al., 2008; Pang et al., 2009)], but were never tested in live patients.

To isolate the effect of the electrical stimulation on the expression patterns of miRNAs in the blood we further analysed the small RNA transcriptome of patients' blood leukocytes following a 1 h of electrical stimulation cessation. Disconnecting the electrical stimulus notably induced in this short time lapse expression change in 11 miRNAs (six decreased and five increased) **Figures 2E,F** and **Table S8**). This indicated that the post-DBS miRNA changes were neither due to the lower medication dose nor to the surgical procedure, but rather reflected the retrieved motor symptoms under stimulus cessation. The modified leukocyte miRNAs also differed in their copy numbers (for example, hsa-miR-20a expressed higher as compared with hsa-miR-320, **Figure 2G**). One of the putative identified as changed miRNA (hsa-miR-1280) is a fragment of a tRNA (Griffiths-Jones et al., 2008). Disconnecting the electrical stimulus further reversed the DBS-mediated change in hsa-mir-143 (**Figure 2F**). Correspondingly, the post-DBS miRNA profile differed from that of healthy volunteers by only nine miRNAs (**Figure S3**) whereas 1 h OFF stimulation induced 4 miRNA count changes as compared with the healthy state including hsa-mir-320a and hsa-mir-320c (**Figure S3**). Taken together, these findings imply that miRNA profiles in both post-DBS and the OFF stimulus conditions differ from the pre-surgery disease state. The oncogenic hsa-mir-16 (Diniz et al., 2012), which was also observed in other PD blood cohort through RT-PCR as changed (Margis and Rieder, 2011) and the tumorigenic miR-18b (Guo et al., 2009).

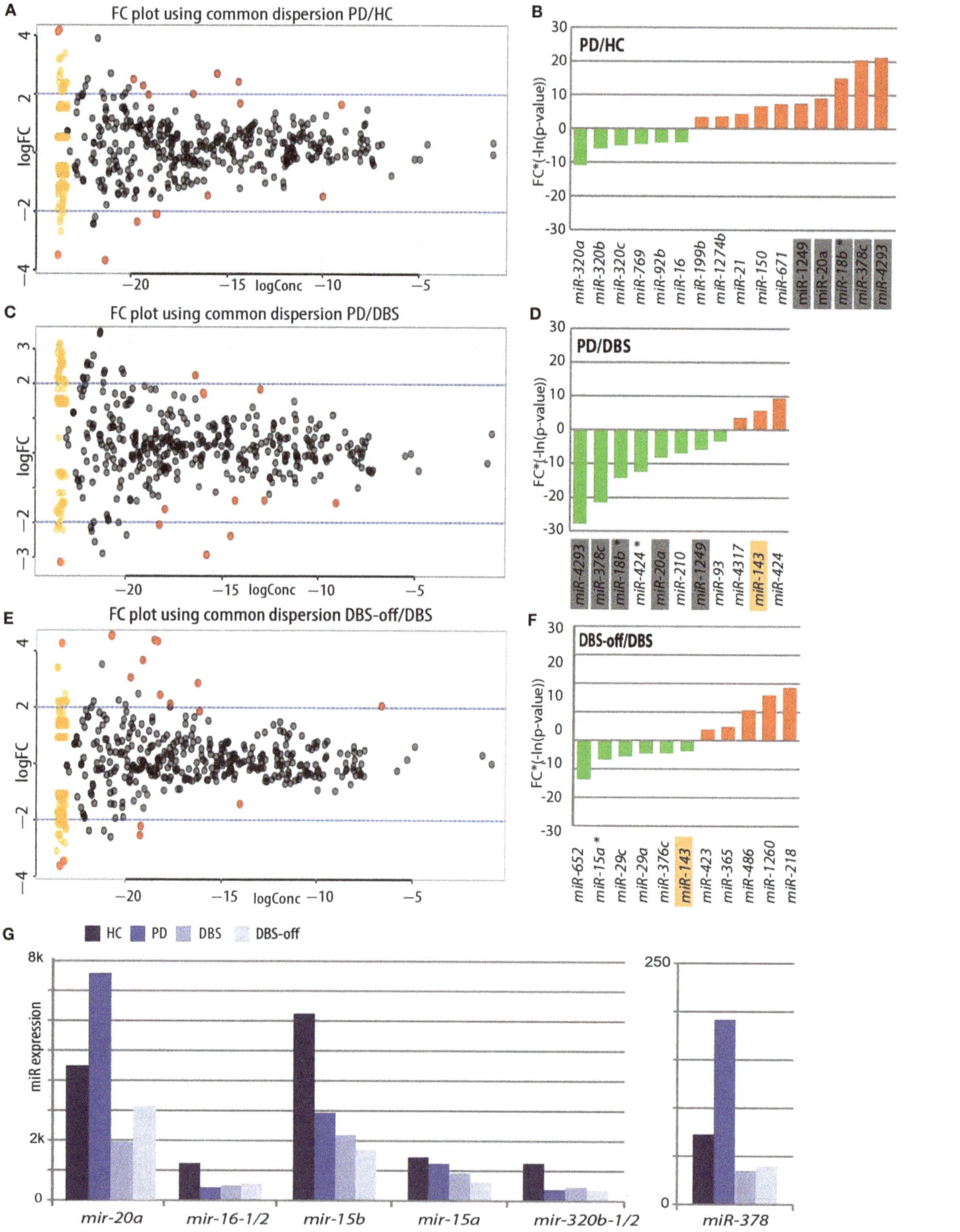

FIGURE 2 | Changed miRNA profiles in PD patients pre-and post-DBS and under stimulation cessation. (A,C,E). Log concentrations (*X*-axis) and log 2 fold change (FC) (*Y*-axis) of miRNAs detected as differentially expressed in **(A)** PD patient leukocytes pre- DBS as compared with healthy volunteers, **(C)** post-DBS on stimulation as compared to pre-DBS and **(E)** following 1 h off electrical stimulation. Red color denotes miRNAs that were detected as exhibiting a statistically significant change in each condition. The yellow

(Continued)

FIGURE 2 | Continued
dots represent all miRNAs that exhibited close to zero expression in one or both conditions (thus they cannot be plotted on a log scale). These condition-unique points in the plot are artificially represented by spreading them vertically as a "smear" on the left. **(B,D,F)** The fold change and *p*-value of miRNAs exhibiting down- or up-regulation in patients' pre-DBS compared to **(B)** healthy volunteers, **(D)** post-DBS on stimulation as compared to pre-DBS and **(E)** post-DBS off stimulation compared to on-stimulation **(F)**. The length of the bar denotes the statistical significance and the direction of change is denoted by positive or negative direction of the bar. *Y*-axis: -ln (*p*-value)*log fold change. Marked miRNA names: 5 (~30%) of the miRNAs that were detected as changed in PD patients were also detected as changed post-DBS and exhibited reverse direction of change compared to the disease state. **(G)** The non-normalized sequence read counts for selected changed miRNAs in the four tested states.

SPLICE JUNCTION MICROARRAY ANALYSIS IDENTIFIES PD-RELATED SPLICING CHANGES, INCLUDING IN miRNA BINDING SITES

The observed miRNA leukocyte changes predicted corresponding changes in the levels of target transcripts of modified miRNAs. Given that the great majority of human genes are subject to splicing regulation, based on our exon microarray-based detection of exon microarrays-based detection of differential expression changes in PD leukocytes (Soreq et al., 2012c), we hypothesized that differential splicing regulation also exist in the patient's blood leukocytes. Furthermore, alternative isoforms that occur through alternative polyadenylation can also result in the removal or addition of microRNA binding sites in the 3′ UTR. Supporting this notion, the full splicing picture is far more complex than the resolution level of the exon arrays technology, since each splicing event may simultaneously cover one or more than one exon. To accurately identify the complex scope of inclusion and exclusion events in particular transcripts, we applied here human splice junction arrays (HJAY) (Sugnet et al., 2006), which simultaneously interrogate all of the exons and all of the splice junctions of each gene, enabling to construct a more robust combinatorial splicing picture (**Figure 3A**).

To detect alternative exons regulated from these arrays, the expression of pairs of opposing (reciprocal) exon-junctions were compared using a previously described linear regression approach (Sugnet et al., 2006) in the program AltAnalyze (Salomonis et al., 2009) for ~25,000 detected transcripts. This analysis revealed 478 disease-associated inclusion and exclusion events that occurred in 332 distinct genes, along with exons containing predicted miRNA binding sites (**Table S9**). These predictions correctly classified the patients from control groups by both the inclusion changed junction probe-sets (**Figure 3B**) and the exclusion ones (**Figure 3C**). Of these changes, 142 events were predicted to be functionally relevant (i.e., modifying a protein domain or binding site effects). 31 of those predicted change in a miRNA binding site. Statistical significance enrichment analysis identified nine miRNAs as frequently targeted at detected spliced transcripts: hsa-miR-769-3p (predicted to target four transcripts), hsa-miR-378 (with six targets) as well as hsa-mir-320, hsa-miR-92b-5p, hsa-miR-16, hsa-miR-150, hsa-miR-671, hsa-miR-20a, and hsa-miR-18b (The full list and adjusted *p*-values are given under **Table S10**). The enriched differentially expressed miRNAs were predicted to target 14 spliced transcripts, including the reactive oxygen species modulator *ROMO1*, the RNA-binding motif *RBM6* transcript involved in splicing (Heath et al., 2010) and the splicing factor *hnRNPA2B1*. Three newly annotated miRNAs were only identified in the small RNA sequencing data. PD-mediated inclusion of exon 12 in the exon 12 of the hnRNP gene *hnRNPA2B1* was detected by a mutual change in three junction probe-set pairs sharing the same second probe-set (**Table S9**). Notably, *hnRNPA2B1* has a role in regulating the biogenesis of miRNAs and in up-regulating hsa-miR-18a (Guil and Caceres, 2007) and down-regulating hsa-let-7 (Eddington et al., 2007). Recently, a link was found between mutation in this hnRNP and the neurological disease Amyotrophic lateral sclerosis (ALS) (Kim et al., 2013). Also, we recently found that hnRNPA2B1 is essential for neuronal well-being in Alzheimer's disease (Berson et al., 2012), compatible with the stress-associated changes in AS (Meshorer et al., 2002). The domain that we found here as included in PD patients extends the c-terminus of the hnRNPA2B1 protein that is necessary for nuclear localization and is sufficient for sending cytosolic proteins to the nucleus in a phosphorylation-regulated manner (Allemand et al., 2005). Furthermore, hnRNPA2B1 mediates mRNA shuttling to distal processes in both neuronal cells and oligodendrocytes (Han et al., 2010), and blockade of its function by splice-isoform-specific antibodies impaired the assembly of RNA granules. Notably, the extended domain spans putative binding sites for several PD-modified miRNAs (including hsa-miR-320a, hsa-miR-320b, and hsa-miR-320c), together indicating that the observed inclusion events may entail functional relevance.

Post-hoc functional GO analysis through the GO-Elite module of AltAnalyze of the regression detected genes revealed enrichment in transcripts of cellular components of the ubiquitin ligase complex (GO:0000151, $p = 0.0165$) and biological processes of mitochondrion organization (GO:0007005, $p = 0.0185$) and regulation of translation (GO:0006417, $p = 0.015$) (**Table S11**). The *FOXP1* transcript showed disease-mediated exclusion in PD patients, which was further annotated as replacing the N-terminus in the 679 amino acid long FOXP1 protein by an alternative N-terminus that yields a shorter protein product of 573 residues.

Independent exon-level splicing-index (Gardina et al., 2006) analysis of the exon probe sets present on the junction microarrays detected 202 PD-inducible AS events pre-DBS (**Table S12**). The detected exon interrogating probe-sets successfully classified patients from controls by expression level (**Figure S4**). The interrogated spliced genes that included the changed probe-sets exhibited enrichment in AS and apoptotic nuclear changes (**Table S13**). Of the exon level detected disease regulated genes, 34 were among these detected by the linear regression analysis of reciprocal junction pairs. However, only four of the detected alternative exons were annotated as overlapping with alternative polyadenylation sites, in the 3′ UTR (**Table S12**). Of

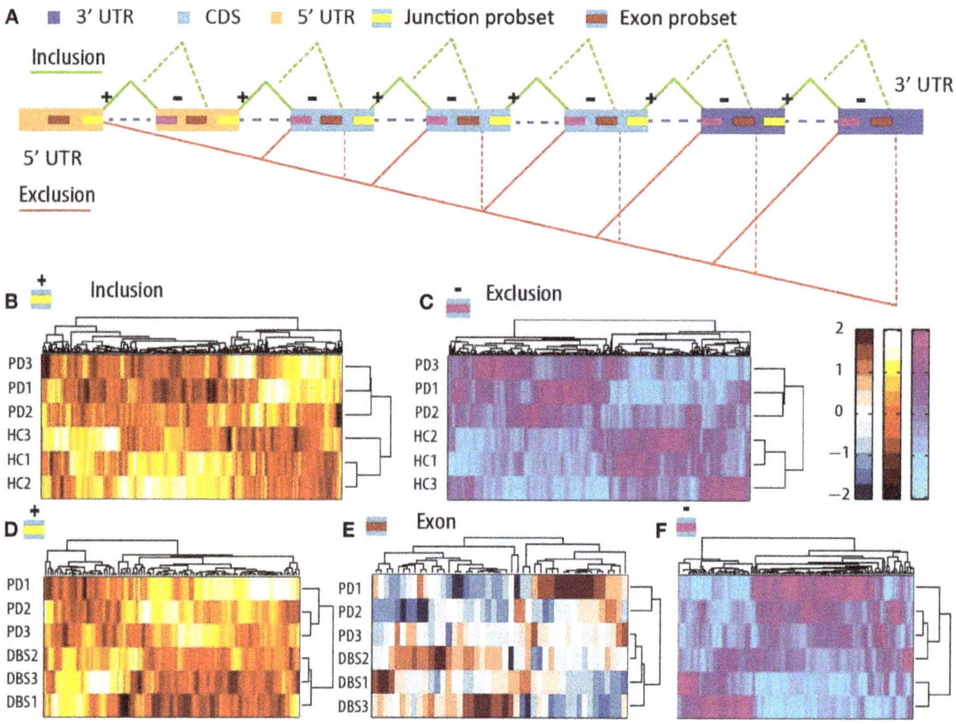

FIGURE 3 | Alternatively spliced exon-junction pairs classify PD and post-DBS samples correctly. (A) A schematic illustration of the interrogated gene regions and possible combinations of inclusion and exclusion of junction and exon probe-sets. Dashed lines: possible inter-exonic Probe Selection Region (PSR) combinations with exclusion of junction probe-sets. A splice change is defined where bi-directional changes occur in the inclusion and the exclusion probe-sets (or alternatively, in a PSR and exclusion probe-sets) concurrently between two tested states. **(B)** The global expression of inclusion junction array probe sets that were detected as changed classified patients from healthy volunteers successfully. **(C)** Classification by the exclusion junction probe-sets. **(D)** Classification following DBS classified post-DBS from pre-treatment state by inclusion detected probe-sets. **(E)** Internal exon classification of the post-DBS from pre-DBS states by PSRs. **(F)** Classification by the DBS-modified exclusion junction probe-sets. Distance method for all the clusterograms: city block.

the 202 identified events, 127 had functional predictions such as alternative terminals or non-sense-mediated decay regions (**Table S12**). Of these, 36 (~30%) were functionally predicted as miRNA binding sites (**Table S14**). Enrichment analysis detected 6 of the DBS-modified miRNAs as modified (having adjusted p-value < 0.05): hsa-miR-320 (a, b, and c) as predicted to bind 4 spliced transcripts including *hnRNPA2B1*, hsa-miR-378 (predicted to bind 6 spliced transcripts), hsa-miR-92b (predicted to bind 4 spliced transcripts), hsa-mir-150, hsa-miR-20a, and hsa-miR-18b (where hsa-miR-18b-3′ changed). Overall, the disease-modified miRNAs were predicted to target overall 11 spliced target transcripts, including *ROMO1* and *COL6A2* that were each predicted as targets of four different modified miRNAs (**Table S14**).

SPLICE JUNCTION MICROARRAYS IDENTIFY SPLICE ISOFORM CHANGES FOLLOWING DBS INCLUDING IN miRNA PREDICTED BINDING SITES

To identify splicing changes induced by DBS, we compared the splice junctions of patients post-DBS to these of pre-DBS state. Linear regression analysis was conducted through AltAnalyze software package (Salomonis et al., 2010) which was adopted for the human junction arrays. The analysis demonstrated a more limited AS induction of 155 inclusion and exclusion events in 117 genes (**Table S15**). Of these, 140 entailed functional predictions, 109 changing protein domain(s) and 31 modifying miRNA binding sites (**Table S15**). For ~1/3rd of the detected splicing events, an additional internal exon probe set reproduced the junction-level results (45 out of 155), indicating higher confidence exon expression changes. Each of these measurement types (inclusion junction, exclusion junction, or inclusion exon) properly segregated the post-DBS samples from pre-DBS ones: the inclusion junction probe-sets (**Figure 3D**), the exon probe-sets (**Figure 3E**) and most distinctly, the exclusion probe-sets (**Figure 3F**), which showed an inverse pattern of change to that observed for PD, completely segregating the post-DBS samples from pre-DBS ones. The linear regression detected genes were enriched in the molecular processes of RNA import into nucleus and MAPK phosphatase export from nucleus as well as in ubiquitin homeostasis (**Table S16**).

DBS induced potentiated reduction in the junction pair measuring exclusion of the PD-included region of *hnRNPA2B1*, however, with two additionally decreased internal exons. Thus, PD *hnRNPA2B1* transcripts showed lower expression of the internal exon probe-set as compared with control volunteers, and post-DBS state caused a yet lower level of this internal region;

and the opposite trend was found in the corresponding junction probes, reinforcing the conclusion that an exclusion event had occurred.

Exon level splicing-index analysis of PD patients post- compared to pre-DBS detected 131 AS events in 117 genes (**Table S17**), in 20 of them junctions were also detected as changed through the linear regression analysis of patients post- compared to pre-DBS. The detected exon-interrogating probe-sets successfully classified post- from pre-DBS samples (**Figure S5**). Intriguingly, 89 of the detected exons had functional transcript structure predictions. Two of these were annotated in Poly(A) sites (**Table S17**), and one of these predictions occured in the *Toll-like receptor (TLR)3*, a pro-inflammatory cytokine mediator (Butt et al., 2012). The NCBI web-based functional annotation tool DAVID (Database for Annotation, Visualization and Integrated Discovery) was used to investigate functional associations of splicing changes seen in the patients' blood leukocytes following DBS. Functional analysis using the DAVID EASE tool (Hosack et al., 2003) revealed that the DBS modified genes were functionally enriched in response to metal ion, AS, apoptosis, immune response and mitochondrion organization (**Table S18**). Enrichment analysis for miRNA binding sited in the detected genes detected 3 of the DBS-modified miRNAs: hsa-miR-20, hsa-mir-18, and hsa-miR-143 as highly predicted to bind AS targets detected by the exon level analysis (**Table S19**).

PD BRAIN REGION TRANSCRIPTOMES REVEAL DISEASE-ASSOCIATED EXPRESSION CHANGES IN THE ALTERNATIVELY SPLICED LEUKOCYTE TRANSCRIPTS

Expression profiling of the alternatively spliced transcripts identified by junction microarray analyses of PD leukocytes was independently examined in a large microarray dataset of PD Patient's brain samples (Moran et al., 2006). The external dataset was composed of 3′ microarray data obtained from post mortem brain samples from PD patients and matched healthy control volunteers and included three brain regions: the lateral and medial substantia nigra and the superior frontal gyrus. The expression of 707 probe sets interrogating genes that were found as alternatively spliced (**Table S20**) successfully classified patients from control samples in this independent data-set (**Figure 4**), reflecting brain region specificity of the tested transcriptomes. The brain regions were further segregated by the blink classifier which separated the frontal gyrus from the substantia nigra (**Figure 4**). In the medial substantia nigra, all the patients were correctly classified apart from the control volunteers. Two lateral substantia nigra control samples were misclassified with the patients, and two of eight patients were misclassified with the control volunteers based on the expression in the superior frontal gyrus. Thus, the vast majority of the samples were correctly classified as patients or control volunteers and overall over 91% accuracy of classification was obtained (**Figure 4**).

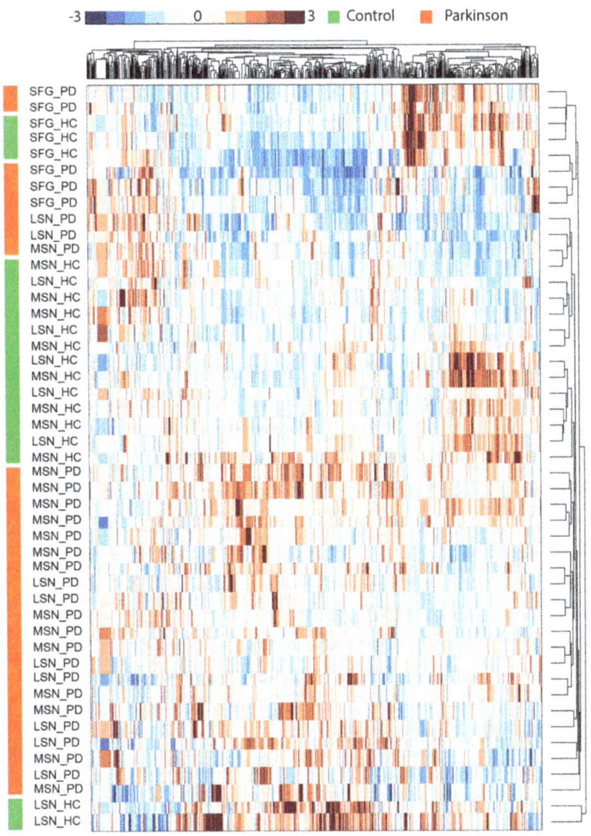

FIGURE 4 | Brain expression levels of spliced transcripts classify PD patients from control volunteers in a brain-region dependent manner. Shown is a hierarchical classification in the noted brain regions of those same transcripts found to be alternatively spliced in PD leukocytes. Note that the hierarchical classifier successfully classified patients from healthy control volunteers based on expression levels obtained by 3′ microarrays interrogating three post mortem brain regions: superior frontal gyrus (SFG), medial substantia nigra (MSN) and lateral substantia nigra (LSN). Each row corresponds to one probe set interrogating a gene that was detected as alternatively spliced in leukocytes from patients and each column represents one sample of brain transcriptome from either patient (red) or healthy volunteer (green) from the independent cohort (Moran et al., 2006). The frontal brain region was correctly segregated from both substantia nigra regions by the expression data, and overall 43 of the 47 samples (~91.5%) were correctly classified as healthy or Parkinsonian ones.

CELLULAR COMPOSITION OF PARKINSON PATIENTS LEUKOCYTES

To identify the most likely cell types and tissues represented in the samples analysed by small RNA deep sequencing and microarrays, we have conducted a cell lineage analysis of the exon microarray data of seven patients examined in three states each and the six control volunteers using a newly developed module of AltAnalyze software package called LineageProfiler program. A large human and mouse compendium of publically available microarray datasets (from the GEO public repository) were used to identify gene sets for each cell type or tissue with the highest correlations to an idealized binary expression profile. The analysed samples included these tested also by junction microarrays and small RNA deep sequencing. The results were visualized as association scores at the level of hierarchically clustered cell types and curated lineage networks (**Figure 5**). The created database reflects the most specifically-expressed exons present in the samples, relative to all lineages types examined. We identified substantial immune cell activity in the patient's blood leukocytes both pre and post- treatment. Furthermore, most healthy individual samples were classified with post-treated patients and pre-treated patients with the off stimulation state based on the expression data of the marker genes.

PD-SPLICED TRANSCRIPTS SHOW MULTIPLE GENE-WIDE BINDING SITES FOR PD-MODIFIED miRNAs

Recent evidence indicates miRNA targeting beyond the 3′-UTR (Orom et al., 2008; Lee et al., 2009). Therefore, we extended the search for putatively changed miRNA-AS pairs to both the coding region and the 5′-UTR using miRWalk (Dweep et al., 2011). The miRWalk database, which predicts miRNA binding sites on the complete sequence of each gene, includes thousands of predicted targets for each of the PD-differentially expressed miRNAs. Using this information, we constructed a network of potential links between the 13 PD-differentially expressed miRNAs and the 217 target transcripts that showed AS changes in PD (**Figure 6A**), which exhibited more inclusion than exclusion events (**Figure 6B**). A total of 560 putative connections emerged (an average of 2.6 miRNA binding sites/target gene), with 43% of those on the 3′ UTR, 38% in the coding sequence and the remaining 19% in the 5′ UTR of the AS target genes (**Figure 6C**). This could not be explained by the relative length of the different gene regions of the identified links, since the average fraction (out of the total length for all genes) is 24% for the 3′ UTR, 69% for the coding sequence and only 7% for the 5′ UTR (Mazumder et al., 2003; Sakharkar et al., 2004b). The network showed a remarkable asymmetry containing the highest number of spliced junctions (68) in predicted targets for miR-16 and the lowest (1) for hsa-miR-18b-3p (previously called hsa-miR-18b*). Supporting the notion of disease relevance, this network included several known PD-related genes such as the transcriptional regulator *FOXP1* (Kim et al., 2007), that promotes midbrain dopaminergic identity in stem cells (Konstantoulas et al., 2010), which was identified as alternatively spliced in this study and was predicted as targeted by the PD dys-regulated hsa-miR-320 cluster. We conclude that at least some of the PD-modified miRNAs may potentially interact with part of the AS-modified transcripts in PD leukocytes.

DBS-SPLICED TRANSCRIPTS SHOW MULTIPLE GENE-WIDE BINDING SITES FOR DBS-MODIFIED miRNAs

An independent network was constructed for the DBS-modified miRNAs and their DBS-spliced predicted targets (**Figure 7A**). Similar to the PD network, it included more AS predicted targets exhibiting inclusion (rather than exclusion) events (**Figure 7B**), with binding prediction again showing a decreasing order from the 3′ UTR (45%), coding sequence (35%) to the 5′ UTR (20%) (**Figure 7C**).The pair-sample DBS network was predictably thinner than the PD one as it avoided inter-individual variability and was based on smaller numbers of DBS-modified miRNAs and AS target transcripts. Nevertheless, this network as well showed asymmetry, with many more predicted AS targets for hsa-miR-424. The two networks shared four miRNAs that their predicted targets were modified both in PD and after DBS as compared with healthy volunteers. Of these, hsa-miR-378, and hsa-miR-20a were predicted to bind the same seven AS target genes, affecting the actin-associated protein VASP involved in axon guidance (Mohamed et al., 2012), the hemoglobin subunit gamma-2 *HBG2* gene, the retinoic acid receptor responder *RARRES3*, the androgen-regulated solute carrier *SLC14A1* (Vaarala et al., 2012), *TMEM69*, a dividing leukocytes biomarker (Solmi et al., 2006), the mitochondrial tRNA dimethyl-allyltransferase trit1 and the bZIP nuclear transcription factor. Other identified targets were the cAMP-responsive modulator *CREM*, which has known spliced isoform variants (Sanborn et al., 1997) and regulates target genes in different tissues (Fimia et al., 2000) and which associates with panic disorder (Domschke et al., 2003), depression and bipolar disorder (Crisafulli et al., 2012) and the schizophrenia-mutated GABA receptor modulator *DBI* which encodes a diazepam binding inhibitor (Niu et al., 2004).

We found three target genes (*HBG2*, *RARRES3*, and *VASP*) common to both networks to be regulated by the same miRNAs predicted to target the same junction probe-set pairs (**Figure 7A**), and in opposite directions from healthy controls to PD patients as compared with pre- to post-DBS. This is a remarkable overlap given that the DBS network was much narrower than the PD one and that the patient's pre-DBS changes were unpaired with the healthy volunteers.

The observed changes in miRNA-AS interactions in both PD and following DBS further raised the question if the predicted binding sites coincided with the modified splice junctions. Sequence comparisons of the miRNA seed regions with the junction probe sets that exhibited splicing changes identified such PD-and DBS-modified links in 34 (**Table S21**) and 33 events (**Table S22**) from 13 and 5 spliced transcripts, respectively. Given that this is only a small fraction of the modified transcripts in each case, this finding demonstrates that the great majority of the observed interactions did not involve active splicing events, while suggesting that direct miRNA-mediated splicing regulation is a viable, albeit exceptional case where the predicted binding site was located exactly in an alternatively spliced junction (**Tables S21** and **S22**). Exceptional PD-related examples of seemingly active miRNA-spliced transcripts predicted interactions include *USP13*, which promotes smooth (*SMO*) signaling by preventing its ubiquitination (Xia et al., 2012), *RGS3* which contributes to neural progenitor/stem cell regulation (Qiu et al., 2010) and *MGAT1*, involved in multiple sclerosis (Grigorian et al., 2012). The DBS-induced events occurred in the *NPC2* gene, which shares cholesterol interactions with *alpha-synuclein* (*SNCA*) (Liu et al., 2010), as well as in *HLX*, *C11orf5*, *CREM*, and the poly-A binding protein coding transcript *PABPC1L*.

DISCUSSION

Our study provides the first report of complete small RNA transcriptome sequencing of live human PD patients' blood leukocytes. Using a whole transcriptome high throughput sequencing technique in conjunction with analyses of both exon and junction microarrays interrogating RNA from the same cells of patients prior to brain stimulation treatment and post-treatment while being on and off electrical stimulation as well as of healthy volunteers, we characterized differential expression in miRNAs

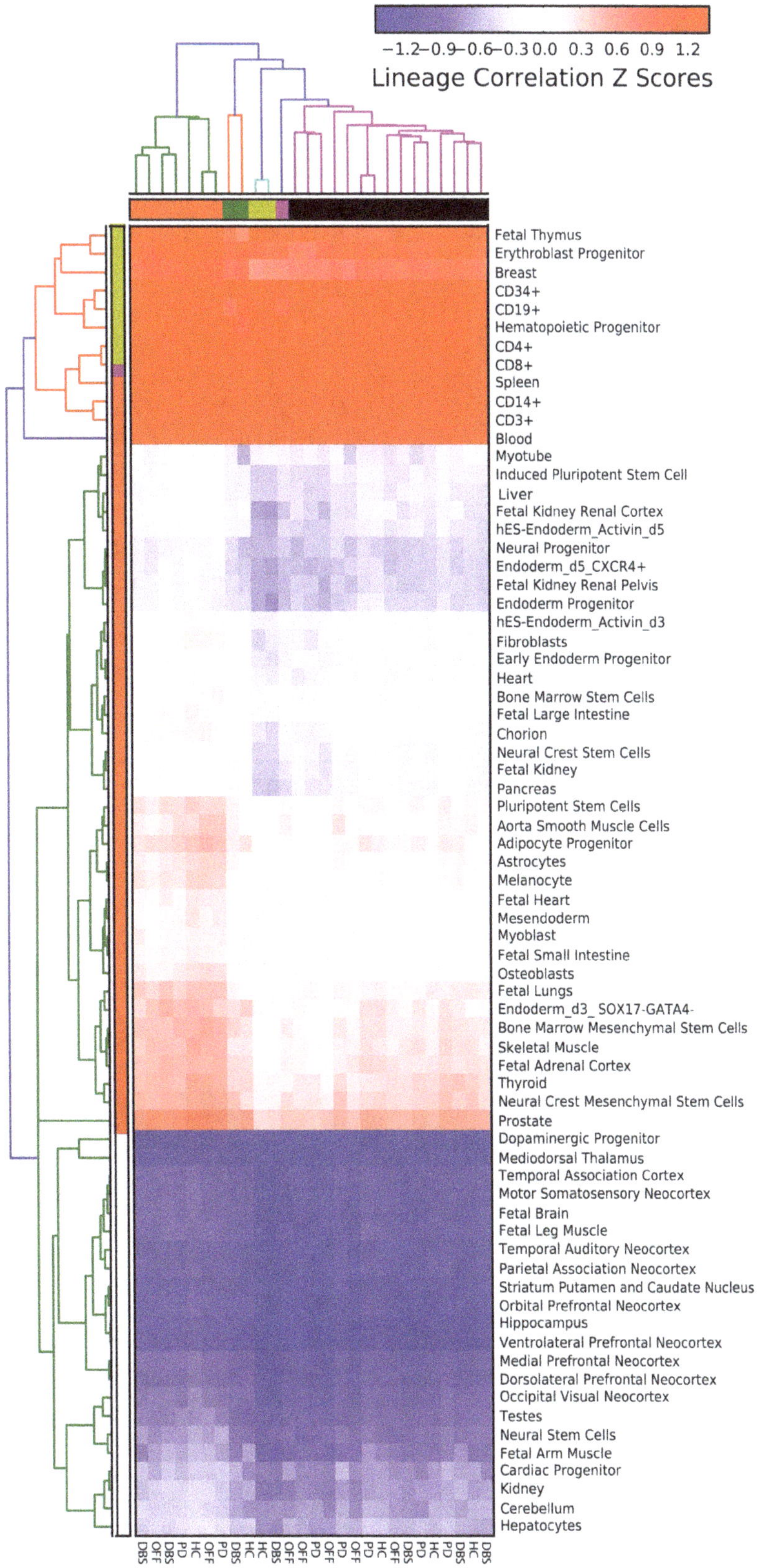

FIGURE 5 | Tissue and cell type origins of leukocyte-expressed AS-modified long transcripts. Putative cell and tissue specific markers subjected to alternative splicing in the leukocyte RNA samples from tested patients and controls were hierarchically classified by exon array gene level expression (columns) and markers (rows). The color code scale presents lineage correlation Z-scores.

predicted to bind target transcripts that underwent splicing modifications. We have successfully classified microarray brain samples of an independent patients cohort based on the expression levels of genes that were found as alternatively spliced in blood leukocytes, at high accuracy of over 91%. Moreover, the dual comparative analysis of patients pre- and post- treatment as well as compared to healthy volunteers yielded insights into both disease and treatment mechanisms. Practically all of the human genes are simultaneously subject to post-transcriptional regulation by either AS or miRNAs thus both miRNAs and spliced isoforns represent promising new candidates for PD research.

For the small RNA sequencing, we used SOLiD genome analyser, obtaining ~106 reads per sample, which is considered a sufficient coverage for small RNA profiling, with above 50% average mapping to the miRBase database, detecting about half of miRBase miRNAs in the samples in up to millions of copies for few of them. For comparison, alignment to a different

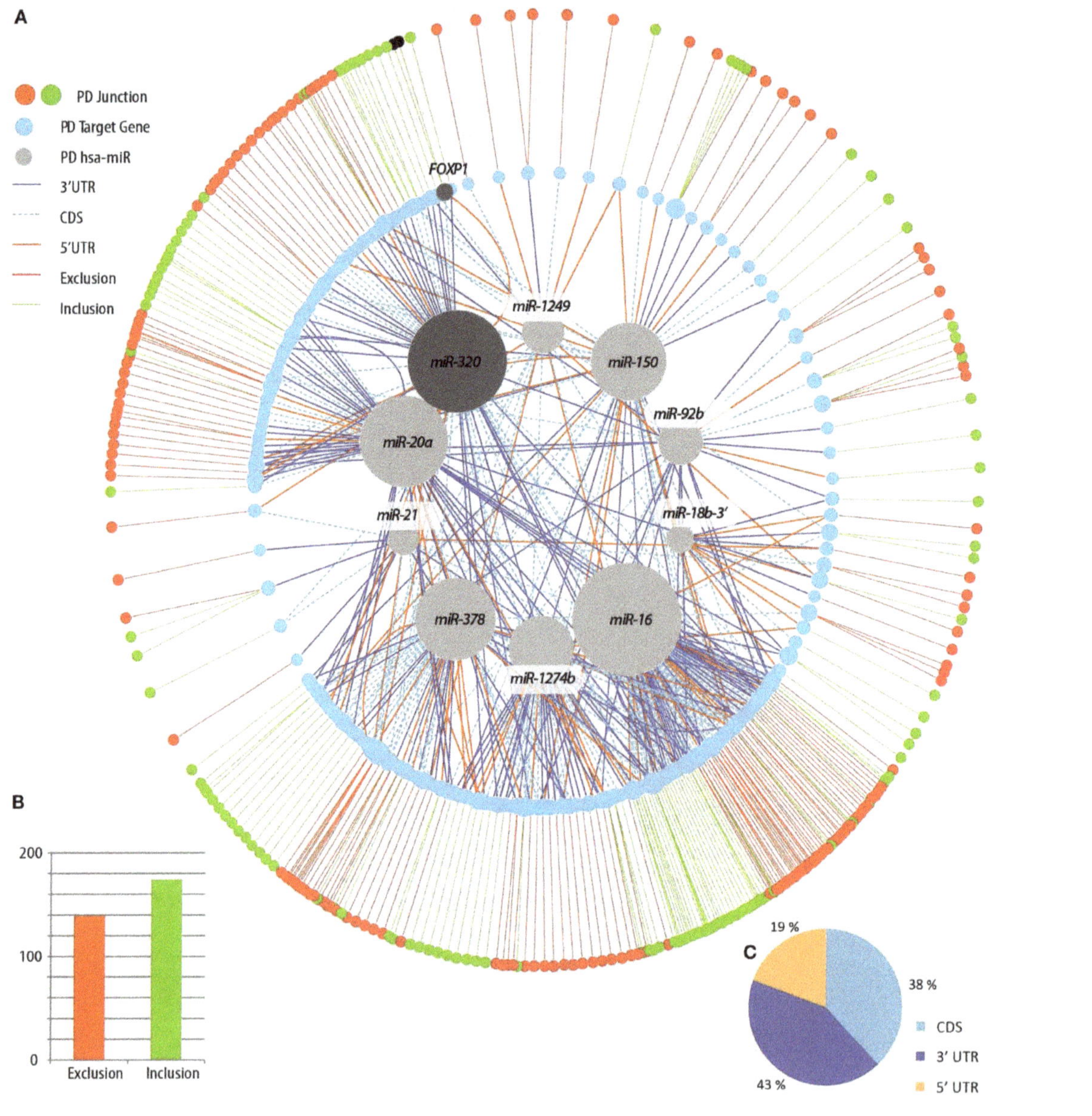

FIGURE 6 | A complex disease network of miRNA-spliced targets expanding beyond the 3′-UTR. (A) Circular tree layout of the network of connections between the miRNAs detected as differentially expressed in PD patients to their predicted target genes that exhibited AS change in splice junction microarrays. External circle: red node color indicates exclusion event in the predicted target splice junction and green—exclusion. Larger node size indicates higher number of connections. The color of edges represents the gene region in which the binding is predicted to occur (5′, 3′-UTR or coding sequence, CDS). Middle circle: the predicted target genes. Gray circle color highlights miRNAs targeting the PD-related gene *FOXP1*, previously linked to miRNA regulation. Curved lines in the inner circle denote two different regions predicted to be bound (e.g., 3′ UTR and coding sequence). **(B)** PD patients exhibited more inclusion events (green) as compared with exclusion (red). **(C)** The types of events predicted to be regulated by the changed miRNAs: 3′ un-translated region (UTR) (dark blue), 5′ UTR (orange) or coding sequence (CDS, light blue).

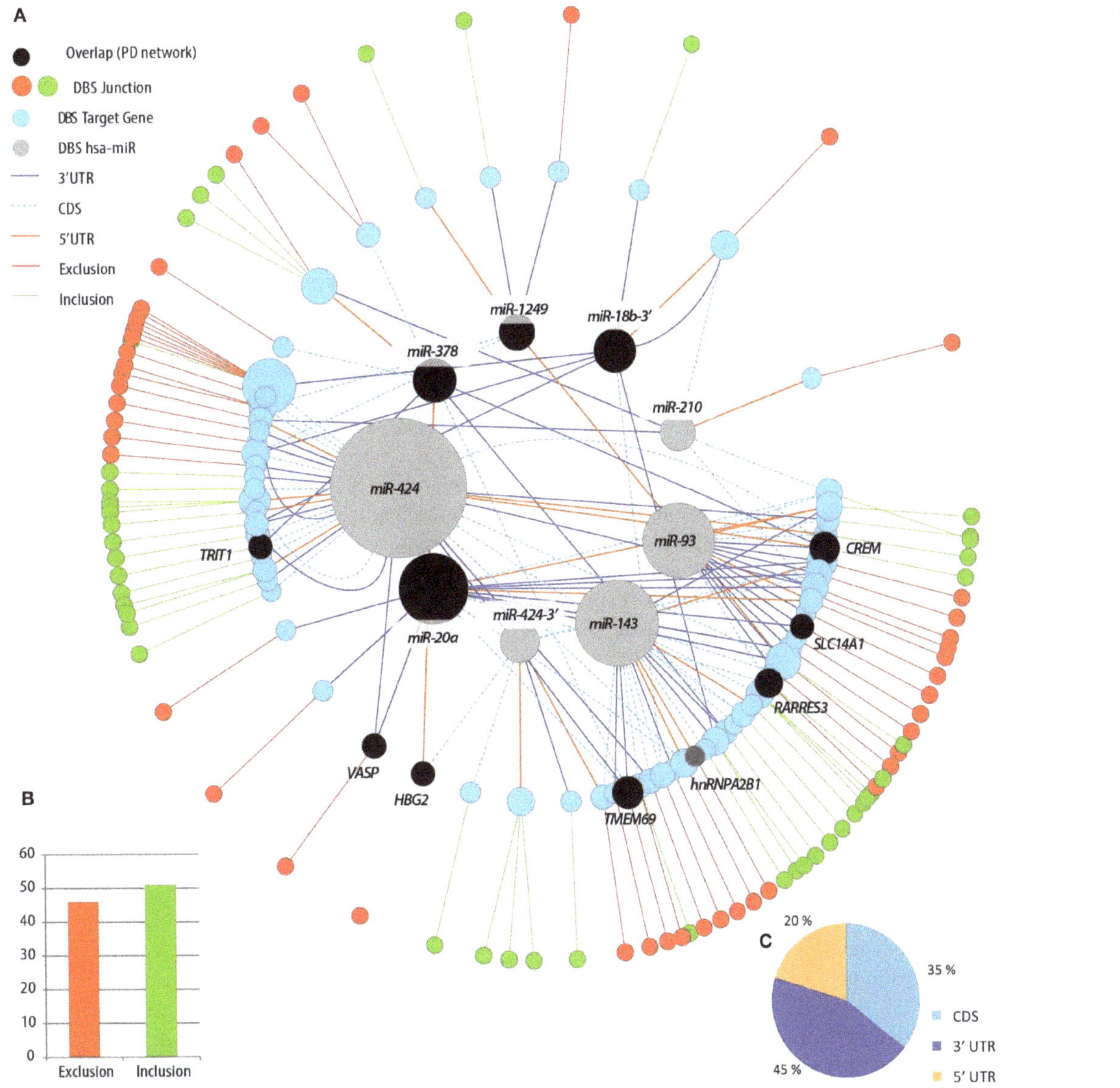

FIGURE 7 | The treatment induced network partially overlaps with the disease network. (A) A circular tree layout network of connections between miRNAs that exhibited differential expression following DBS neurosurgery as compared with the pre-DBS (non-stimulated) state, and their predicted target genes that exhibited alternative splicing changes in PD patients following the DBS treatment. External circle: red node color indicates exclusion event in the predicted target splice junction and green—exclusion. Larger node size indicates higher number of connections. The color of edges represents the gene region in which the binding is predicted to occur (5′, 3′-UTR or coding sequence, CDS). Middle circle: the predicted target genes. Curved lines in the inner circle denote two different regions predicted to be bound (e.g., 3′ UTR and coding sequence). Black miRNAs and dark blue target genes: overlap with the patients compared to healthy network presented in **Figure 6**. **(B)** The number of exon exclusion (red) and exon inclusion (green) events detected in the predicted target genes that were detected by linear regression analysis of reciprocal junction pairs interrogated by junction microarrays (*Y*-axis: number of detected events). **(C)** The types of events predicted to be regulated by DBS modified miRNAs: 3′ un-translated region (UTR) (dark blue), 5′ UTR (orange) or coding sequence (CDS, light blue).

reference sequence database of other (non-miRNA) non-coding human RNAs yielded only an average of ~9% alignment over all the sequenced libraries, supporting the notion that the miRNAs are composing the largest population of non-coding RNAs found in the studied human leukocyte samples. Concurrently, we employed splice junction and exon microarrays to generate parallel transcriptome datasets from the same cells in order to identify AS changes in the modified miRNAs predicted targets, as well as the cellular composition changes in patients and post-treatment. Testing the same patients before and after surgery further overcame individual variability. Further studies which will expand the size of the tested cohort will enable validation of our findings in larger and independent patient and healthy volunteer cohorts.

We identified significant changes in 16 miRNAs in PD and in 11 post-DBS treatment, 5 of which were among these changed in the disease, but inversely, reinforcing the disease and treatment relevance of these changes. The post-DBS change was further dependent on the stimulus itself, as 1-h off-stimulus completely changed the modified miRNA profiles as compared with the stimulated state. We also identified expression changes in two newly proposed miRNAs that were recently reported as tRNA fragments, possibly originated from the antisense copy of an endogenous retroviral element (Kawaji et al., 2008; De Hoon et al., 2010). The observed miRNA changes post DBS were compatible with miRNA expression being subject to rapid changes under modified environmental conditions (Brzuzan et al., 2012) and immunological challenges (Perry et al., 2008) and indicating possible relevance to the DBS-suppressed symptoms. To our knowledge, our study is the first to construct differential disease- and treatment-related networks in human cells from neurodegenerative disease patients.

Analysis of gene isoform differential expression, AS and modifications of miRNA binding sites revealed significant differences in splice isoform profiles between PD patients and healthy control volunteers blood leukocytes, as well as in post- compared to pre-DBS states, following the brain stimulation treatment. We identified enrichment of immune cellular composition by lineage cell profiling analysis of the exon microarrays, both pre- and post-DBS. Through database and network analyses we identified disease and treatment networks of miRNAs and spliced targets. Combining the data from differential expression of miRNAs with the changes in expression of their potential targets in a systematic way is a powerful tool to identify disease related miRNA-target gene networks. In this study we combined miRNA expression profile in PD patients pre- and post-treatment with data of splice junction microarrays from the same samples presenting a systems biology approach for studying blood leukocyte RNA expression patterns in neurological and neurodegenerative diseases.

Intriguingly, there were more predicted binding sites in the 3′ UTR as compared with the other regions, compatible with other's observations for more efficient regulation of the target genes through 3′-UTR binding. Whereas binding to the 3′ UTR induces stronger degradation of the target, auxiliary binding to the coding sequence may possibly increase this effect. Recent reports that target regulation through miRNAs may function also in coding regions, not only the 3′ UTR (Dolken et al., 2010; Hafner et al., 2010) are compatible with our predictions. Correspondingly, computational analyses of data produced from RNA-binding protein PAR-CLIP [Photoactivatable-Ribonucleoside-Enhanced Crosslinking and Immunoprecipitation, (Ule et al., 2003)] experiments showed that nearly 50% of the miRNA binding sites were located in the coding regions of the targets, compatible with the capacity of miRNAs to each regulate thousands of target mRNAs (Lewis et al., 2005; Friedman et al., 2009). Yet, the molecular mechanisms underlying the selection of interaction between distinct "seed" domains and single transcripts was not yet interrogated in a global manner in human diseases; therefore, the identification of more than a single miRNA-target putative interaction site opens yet more questions, for future studies.

The state specific enrichment for GO processes suggests disease- and treatment-specific functional involvement. The disease induced processes included ubiquitin ligase complex, mitochondrion organization and regulation of translation, compatible with our recent observations in brains of PD model mice (Soreq et al., 2012a). The DBS induced functional changes included MAPK phosphatase export from nucleus as well as in ubiquitin homeostasis, suggesting treatment influence on disease leukocyte pathways.

While the majority of small RNAs are found within the cells, a growing number of miRNAs are observed outside cells, including various body fluids and blood (Caby et al., 2005; Garcia et al., 2008). For example, small RNA deep sequencing identified miRNAs in exosomes, membrane-bound nanoparticles which are a newly described pathway of intercellular communications (Zhang and Grizzle, 2011), with a distinct signature during prion infection (Bellingham et al., 2012). Circulating miRNAs were recently suggested as new biomarkers in diagnosis, prognosis, and treatment of cancer (Allegra et al., 2012). Moreover, extracellular RNAs have immune-modulating properties (Altincicek et al., 2008) and can escape the blood-brain barrier to reach distal locations (Graner et al., 2009) or to tumors. This suggests a future venue to pursue.

Epigenetic chromatin changes were recently found in blood leukocytes of schizophrenia patients (Melas et al., 2012), where association was found between DNA methylation and antipsychotic drug treatment. As PD also entails hereditary component in about 10% of the cases, it will be highly interesting to pursue in the future for genome-wide DNA methylation changes in PD patients prior to, and following DBS, which may reveal dysregulated epigenome in PD and may further discover DNA methylation disease biomarkers in blood leukocytes.

Concurrent measurements of miRNAs and their mRNA targets were recently reported for two other human diseases: Myelodysplastic Syndromes (Beck et al., 2011) and multiple sclerosis (Angerstein et al., 2012) but these studies did not address splice isoform and AS changes of the targets. To our knowledge, this is the first report where quantitative information about miRNAs has been generated concurrently with particular splice variants in a genome wide scale for blood leukocytes of human patients prior to, and post brain stimulation treatment. Thus, our study provides a useful data set supplementing previous microarray datasets of transcripts, which lacked isoform-specific resolution, and provides a comparative resource for miRNA study in PD.

The PD network included the transcription factor *FOXP1*, implicated in a miRNA-mediated feedback loop controlling the survival of midbrain dopaminergic neurons (Kim et al., 2007), and shown to co-act with *PITX3* for maintaining neuronal fate in the midbrain. Specifically, *PITX3* promotes midbrain identity through regulation by this dopaminergic regulator (De Mena et al., 2010; Konstantoulas et al., 2010). These findings suggest relevance of the observed changes for both the PD phenotype and treatment efficacy, and foresee the use of similar network approaches for the study of other neurological and neurodegenerative diseases.

The identified miRNA changes in PD patient's blood leukocytes here may provide an important biomarker profile, reflecting remarkable dynamics and stimulus-dependent reversal of miRNA expression. Our data set provide a source for miRNA expression in blood leukocytes in healthy state and under neurodegenerative disease. Our analyses and findings suggest disease-associated dys-regulation and treatment-related reversal of the transcriptional mediated checks and balances that, under healthy state functioning, reflect balanced splicing events controlling leukocyte transcript profiles.

AUTHOR CONTRIBUTIONS

Conceived and designed the experiments: Lilach Soreq, Hermona Soreq, Hagai Bergman, Zvi Israel. Recruited study participants: Zvi Israel, Hagai Bergman, Lilach Soreq. Collected patient samples: Lilach Soreq. Produced RNA: Lilach Soreq. Conducted computational and Bioinformatic analyses: Lilach Soreq. Designed the software used in analysis: Nathan Salomonis, Lilach Soreq. Contributed analysis tools: Nathan Salomonis. Wrote the paper: Lilach Soreq and Hermona Soreq and all co-authors approved.

ACKNOWLEDGMENTS

We thank the patient and healthy volunteers that participated in this study. This work was supported by the Israeli Chief Scientist to Hagai Bergman and David S. Greenberg, the Thyssen Foundation to Hermona Soreq. and to Hagai Bergman, and the Rosetrees foundation and the European Network of excellence on alternative splicing (EURASNET) to Hermona Soreq. Lilach Soreq thanks HUJI sources for PhD fellowship support.

SUPPLEMENTARY MATERIAL

Figure S1 | The percentage of annotated reads and of the miRBase aligned reads. (A) Percent of annotated reads of the total number of RNA-Seq reads in all the 12 RNA-Seq libraries. **(B)** Percent of miRBase detected entries in the RNA-Seq libraries **(C)** Percent of sequence match between the reads and the miRBase sequences (perfect, 1 mismatch, 2 mismatches and 3 mismatches) **(D)** Percent of reads match to the human reference sequences database (which comprises of non protein-coding RNAs other than miRNAs, such as SINEs and ALU sequences).

Figure S2 | Expression map of selected miRNAs and miRNA* (3′/5′ forms) in blood leukocytes of Parkinson's patients pre- and post- treatment and of healthy control volunteers. Read counts of selected miRNAs that were detected through RNA-Seq alignment of small RNAs detected in blood leukocytes of PD patients pre- and post-DBS while being on and off electrical stimulation and of matched healthy control (HC) volunteers to the miRBase database. **(A)** Highly expressed miRNAs **(B)** Highest expressed, and **(C)** Count levels of miRNA* (3′/5′) forms presenting low expression **(D)** and high expression **(E)**. Y axis: the number of reads. Dark blue: the expression of miRNAs in HC volunteers, lighter blue: in PD patients pre-DBS, light blue: post-DBS while being on electrical stimulation (DBS) and lightest: in PD patients post-DBS following one hour of a complete electrical stimulation cessation (DBS-off).

Figure S3 | DBS differentially expressed miRNAs. Presented are miRNAs that were detected as changed in PD patients post-DBS as compared with HC volunteers (left), the same patients while being off electrical stimulation state as compared with healthy control volunteers (middle) and in off electrical stimulation state as compared with the pre-surgical state (right). Y axis: fold change multiplied by the ln (log base 10) of the significance p-value which was obtained by a negative binomial distribution test through a generalization of the Poisson model that recaptures biological variance correctly (Robinson and Smyth, 2008).

Figure S4 | Classification of patients pre-DBS from healthy control volunteers. The junction microarrays expression levels of blood leukocyte RNA exons that were detected as changed in PD patients as compared with HC volunteers correctly classified patient samples apart from the healthy samples. The unsupervised hierarchical classification was conducted on the expression signals of the exons detected by splicing-index analysis which was conducted by a modified version of AltAnalyze (Salomonis et al., 2009) that was adopted to the human splice junction (HJAY) arrays and is based on the measurement described for exon arrays under (Gardina et al., 2006). The classification was conducted with Euclidean distance measure on both the exons and the samples.

Figure S5 | Classification of patients post-DBS from pre-DBS state. The junction microarrays expression levels of blood leukocyte RNA exons that were detected as changed by splicing-index analysis in PD patients pre-DBS as compared with post-DBS (DBS) state (while being on electrical stimulation) correctly classified between the states. The classification was conducted by a hierarchical classifier with Euclidean distance measure on both the exons and the samples.

REFERENCES

Affymetrix Technical Notes. *Guide to Probe Logarithmic Intensity Error (PLIER) Estimation*. Available online at: http://www.affymetrix.com/support/technical/technotes/plier_technote.pdf

Allegra, A., Alonci, A., Campo, S., Penna, G., Petrungaro, A., Gerace, D., et al. (2012). Circulating microRNAs: new biomarkers in diagnosis, prognosis and treatment of cancer (Review). *Int. J. Oncol.* 41, 1897–1912.

Allemand, E., Guil, S., Myers, M., Moscat, J., Caceres, J. F., and Krainer, A. R. (2005). Regulation of heterogenous nuclear ribonucleoprotein A1 transport by phosphorylation in cells stressed by osmotic shock. *Proc. Natl. Acad. Sci. U.S.A.* 102, 3605–3610.

Altincicek, B., Stotzel, S., Wygrecka, M., Preissner, K. T., and Vilcinskas, A. (2008). Host-derived extracellular nucleic acids enhance innate immune responses, induce coagulation, and prolong survival upon infection in insects. *J. Immunol.* 181, 2705–2712.

Angerstein, C., Hecker, M., Paap, B. K., Koczan, D., Thamilarasan, M., Thiesen, H. J., et al. (2012). Integration of MicroRNA databases to study MicroRNAs associated with multiple sclerosis. *Mol. Neurobiol.* 45, 520–535.

Ashburner, M., Ball, C. A., Blake, J. A., Botstein, D., Butler, H., Cherry, J. M., et al. (2000). Gene ontology: tool for the unification of biology. *Nat. Genet.* 25, 25.

Atz, M., Walsh, D., Cartagena, P., Li, J., Evans, S., Choudary, P., et al. (2007). Methodological considerations for gene expression profiling of human brain. *J. Neurosci. Methods* 163, 295–309.

Beck, D., Ayers, S., Wen, J., Brandl, M. B., Pham, T. D., Webb, P., et al. (2011). Integrative analysis of next generation sequencing for small non-coding RNAs and transcriptional regulation in Myelodysplastic Syndromes. *BMC Med. Genomics* 4:19. doi: 10.1186/1755-8794-4-19

Bellingham, S. A., Coleman, B. M., and Hill, A. F. (2012). Small RNA deep sequencing reveals a distinct miRNA signature released in exosomes from prion-infected neuronal cells. *Nucleic Acids Res.* 40, 10937–10949.

Bergman, H., Wichmann, T., and Delong, M. R. (1990). Reversal of experimental parkinsonism by lesions of the subthalamic nucleus. *Science* 249, 1436–1438.

Bernal-Pacheco, O., Limotai, N., Go, C. L., and Fernandez, H. H. (2012). Nonmotor manifestations in Parkinson disease. *Neurologist* 18, 1–16.

Berson, A., Barbash, S., Shaltiel, G., Goll, Y., Hanin, G., Greenberg, D. S., et al. (2012). Cholinergic-associated loss of hnRNP-A/B in Alzheimer's disease impairs cortical splicing and cognitive function in mice. *EMBO Mol. Med.* 4, 730–742.

Braak, H., and Del Tredici, K. (2008). Invited article: nervous system pathology in sporadic Parkinson disease. *Neurology* 70, 1916–1925.

Brock, M., Trenkmann, M., Gay, R. E., Michel, B. A., Gay, S., Fischler, M., et al. (2009). Interleukin-6 modulates the expression of the bone morphogenic protein receptor type II through a novel STAT3–microRNA cluster 17/92 pathway. *Circ. Res.* 104, 1184–1191.

Brzuzan, P., Wozny, M., Wolinska, L., and Piasecka, A. (2012). Expression profiling *in vivo* demonstrates rapid changes in liver microRNA levels of whitefish (*Coregonus lavaretus*) following microcystin-LR exposure. *Aquat. Toxicol.* 122-123, 188–196.

Butt, A. Q., Ahmed, S., Maratha, A., and Miggin, S. M. (2012). 14-3-3{epsilon} and 14-3-3sigma inhibit TLR-mediated pro-inflammatory cytokine induction. *J. Biol. Chem.* 287, 38665–38679.

Caby, M. P., Lankar, D., Vincendeau-Scherrer, C., Raposo, G., and Bonnerot, C. (2005). Exosomal-like vesicles are present in human blood plasma. *Int. Immunol.* 17, 879–887.

Carrer, M., Liu, N., Grueter, C. E., Williams, A. H., Frisard, M. I., Hulver, M. W., et al. (2012). Control of mitochondrial metabolism and systemic energy homeostasis by microRNAs 378 and 378*. *Proc. Natl. Acad. Sci. U.S.A.* 109, 15330–15335.

Crisafulli, C., Shim, D. S., Andrisano, C., Pae, C. U., Chiesa, A., Han, C., et al. (2012). Case-control association study of 14 variants of CREB1, CREBBP and CREM on diagnosis and treatment outcome in major depressive disorder and bipolar disorder. *Psychiatry Res.* 198, 39–46.

De Hoon, M. J., Taft, R. J., Hashimoto, T., Kanamori-Katayama, M., Kawaji, H., Kawano, M., et al. (2010). Cross-mapping and the identification of editing sites in mature microRNAs in high-throughput sequencing libraries. *Genome Res.* 20, 257–264.

Delay, C., Calon, F., Mathews, P., and Hebert, S. S. (2011). Alzheimer-specific variants in the 3′UTR of Amyloid precursor protein affect microRNA function. *Mol. Neurodegener.* 6:70. doi: 10.1186/1750-1326-6-70

De Mena, L., Coto, E., Cardo, L. F., Diaz, M., Blazquez, M., Ribacoba, R., et al. (2010). Analysis of the Micro-RNA-133 and PITX3 genes in Parkinson's disease. *Am. J. Med. Genet. B Neuropsychiatr. Genet.* 153B, 1234–1239.

Diniz, M. G., Gomes, C. C., De Castro, W. H., Guimaraes, A. L., De Paula, A. M., Amm, H., et al. (2012). miR-15a/16-1 influences BCL2 expression in keratocystic odontogenic tumors. *Cell Oncol. (Dordr.)* 35, 285–291.

Dolken, L., Malterer, G., Erhard, F., Kothe, S., Friedel, C. C., Suffert, G., et al. (2010). Systematic analysis of viral and cellular microRNA targets in cells latently infected with human gamma-herpesviruses by RISC immunoprecipitation assay. *Cell Host Microbe* 7, 324–334.

Domschke, K., Kuhlenbaumer, G., Schirmacher, A., Lorenzi, C., Armengol, L., Dibella, D., et al. (2003). Human nuclear transcription factor gene CREM: genomic organization, mutation screening, and association analysis in panic disorder. *Am. J. Med. Genet. B Neuropsychiatr. Genet.* 117B, 70–78.

Doxakis, E. (2010). Post-transcriptional regulation of alpha-synuclein expression by mir-7 and mir-153. *J. Biol. Chem.* 285, 12726–12734.

Dweep, H., Sticht, C., Pandey, P., and Gretz, N. (2011). miRWalk–database: prediction of possible miRNA binding sites by "walking" the genes of three genomes. *J. Biomed. Inform.* 44, 839–847.

Eddington, K. M., Dolcos, F., Cabeza, R., Kr, R. K., and Strauman, T. J. (2007). Neural correlates of promotion and prevention goal activation: an fMRI study using an idiographic approach. *J. Cogn. Neurosci.* 19, 1152–1162.

Edgar, R., Domrachev, M., and Lash, A. E. (2002). Gene expression omnibus: NCBI gene expression and hybridization array data repository. *Nucleic Acids Res.* 30, 207–210.

Fahn, S. (2011). Classification of movement disorders. *Mov. Disord.* 26, 947–957.

Fahn, S., and Elton, R. (1987). "Members of the updrs Development Committee," in *Recent Developments in Parkinson's Disease*, Vol 2. eds S. Fahn, C. D. Marsden, D. B. Calne, and M. Goldstein (Florham Park, NJ: Macmillan Health Care Information), 153–163, 293–304.

Fimia, G. M., De Cesare, D., and Sassone-Corsi, P. (2000). A family of LIM-only transcriptional coactivators: tissue-specific expression and selective activation of CREB and CREM. *Mol. Cell. Biol.* 20, 8613–8622.

Friedman, R. C., Farh, K. K., Burge, C. B., and Bartel, D. P. (2009). Most mammalian mRNAs are conserved targets of microRNAs. *Genome Res.* 19, 92–105.

Garcia, J. M., Garcia, V., Pena, C., Dominguez, G., Silva, J., Diaz, R., et al. (2008). Extracellular plasma RNA from colon cancer patients is confined in a vesicle-like structure and is mRNA-enriched. *RNA* 14, 1424–1432.

Gardina, P. J., Clark, T. A., Shimada, B., Staples, M. K., Yang, Q., Veitch, J., et al. (2006). Alternative splicing and differential gene expression in colon cancer detected by a whole genome exon array. *BMC Genomics* 7:325. doi: 10.1186/1471-2164-7-325

Gehrke, S., Imai, Y., Sokol, N., and Lu, B. (2010). Pathogenic LRRK2 negatively regulates microRNA-mediated translational repression. *Nature* 466, 637–641.

Goldman, J. G., and Litvan, I. (2011). Mild cognitive impairment in Parkinson's disease. *Minerva Med.* 102, 441–459.

Graner, M. W., Alzate, O., Dechkovskaia, A. M., Keene, J. D., Sampson, J. H., Mitchell, D. A., et al. (2009). Proteomic and immunologic analyses of brain tumor exosomes. *FASEB J.* 23, 1541–1557.

Griffiths-Jones, S., Saini, H. K., Van Dongen, S., and Enright, A. J. (2008). miRBase: tools for microRNA genomics. *Nucleic Acids Res.* 36, D154–D158.

Grigorian, A., Mkhikian, H., Li, C. F., Newton, B. L., Zhou, R. W., and Demetriou, M. (2012). Pathogenesis of multiple sclerosis via environmental and genetic dysregulation of N-glycosylation. *Semin. Immunopathol.* 34, 415–424.

Guil, S., and Caceres, J. F. (2007). The multifunctional RNA-binding protein hnRNP A1 is required for processing of miR-18a. *Nat. Struct. Mol. Biol.* 14, 591–596.

Guo, J., Miao, Y., Xiao, B., Huan, R., Jiang, Z., Meng, D., et al. (2008). Differential expression of microRNA species in human gastric cancer versus non-tumorous tissues. *J. Gastroenterol. Hepatol.* 24, 652–657.

Guo, J., Miao, Y., Xiao, B., Huan, R., Jiang, Z., Meng, D., et al. (2009). Differential expression of microRNA species in human gastric cancer versus non-tumorous tissues. *J. Gastroenterol. Hepatol.* 24, 652–657.

Hackl, M., Brunner, S., Fortschegger, K., Schreiner, C., Micutkova, L., Mück, C., et al. (2010). miR-17, miR-19b, miR-20a, and miR-106a are down-regulated in human aging. *Aging Cell* 9, 291–296.

Hafner, M., Landthaler, M., Burger, L., Khorshid, M., Hausser, J., Berninger, P., et al. (2010). Transcriptome-wide identification of RNA-binding protein and microRNA target sites by PAR-CLIP. *Cell* 141, 129–141.

Han, S. P., Friend, L. R., Carson, J. H., Korza, G., Barbarese, E., Maggipinto, M., et al. (2010). Differential subcellular distributions and trafficking functions of hnRNP A2/B1 spliceoforms. *Traffic* 11, 886–898.

Hardcastle, T. J., and Kelly, K. A. (2010). baySeq: empirical Bayesian methods for identifying differential expression in sequence count data. *BMC Bioinformatics* 11:422. doi: 10.1186/1471-2105-11-422

Hawkes, C. H., Del Tredici, K., and Braak, H. (2010). A timeline for Parkinson's disease. *Parkinsonism Relat. Disord.* 16, 79–84.

Heath, E., Sablitzky, F., and Morgan, G. T. (2010). Subnuclear targeting of the RNA-binding motif protein RBM6 to splicing speckles and nascent transcripts. *Chromosome Res.* 18, 851–872.

Hébert, S. S., Horré, K., Nicolaï, L., Bergmans, B., Papadopoulou, A. S., Delacourte, A., et al. (2009). MicroRNA regulation of Alzheimer's Amyloid precursor protein expression. *Neurobiol. Dis.* 33, 422–428.

Hosack, D. A., Dennis, G. Jr., Sherman, B. T., Lane, H. C., and Lempicki, R. A. (2003). Identifying biological themes within lists of genes with EASE. *Genome Biol.* 4:R70. doi: 10.1186/gb-2003-4-10-r70

Huang, G., Nishimoto, K., Zhou, Z., Hughes, D., and Kleinerman, E. S. (2012). miR-20a encoded by the miR-17-92 cluster increases the metastatic potential of osteosarcoma cells by regulating Fas expression. *Cancer Res.* 72, 908–916.

Irimia, M., and Blencowe, B. J. (2012). Alternative splicing: decoding an expansive regulatory layer. *Curr. Opin. Cell Biol.* 24, 323–332.

Irizarry, R. A., Bolstad, B. M., Collin, F., Cope, L. M., Hobbs, B., and Speed, T. P. (2003). Summaries of affymetrix GeneChip probe level data. *Nucleic Acids Res.* 31, e15.

Johnson, R., Zuccato, C., Belyaev, N. D., Guest, D. J., Cattaneo, E., and Buckley, N. J. (2008). A microRNA-based gene dysregulation pathway in Huntington's disease. *Neurobiol. Dis.* 29, 438–445.

Kawaji, H., Nakamura, M., Takahashi, Y., Sandelin, A., Katayama, S., Fukuda, S., et al. (2008). Hidden layers of human small RNAs. *BMC Genomics* 9:157. doi: 10.1186/1471-2164-9-157

Kim, H. J., Kim, N. C., Wang, Y. D., Scarborough, E. A., Moore, J., Diaz, Z., et al. (2013). Mutations in prion-like domains in hnRNPA2B1 and hnRNPA1 cause multisystem proteinopathy and ALS. *Nature* 495, 467–473.

Kim, J., Inoue, K., Ishii, J., Vanti, W. B., Voronov, S. V., Murchison, E., et al. (2007). A MicroRNA feedback circuit in midbrain dopamine neurons. *Science* 317, 1220–1224.

Kingwell, K. (2011). Parkinson disease. Constant-current deep brain stimulation improves symptoms in Parkinson disease. *Nat. Rev. Neurol.* 8:119. doi: 10.1038/nrneurol.2012.14

Konstantoulas, C. J., Parmar, M., and Li, M. (2010). FoxP1 promotes midbrain identity in embryonic stem cell-derived dopamine neurons by regulating Pitx3. *J. Neurochem.* 113, 836–847.

Krichevsky, A. M., and Gabriely, G. (2008). miR-21: a small multi-faceted RNA. *J. Cell. Mol. Med.* 13, 39–53.

Lapuk, A., Marr, H., Jakkula, L., Pedro, H., Bhattacharya, S., Purdom, E., et al. (2010). Exon-level microarray analyses identify alternative splicing programs in breast cancer. *Mol. Cancer Res.* 8, 961.

Lee, I., Ajay, S. S., Yook, J. I., Kim, H. S., Hong, S. H., Kim, N. H., et al. (2009). New class of microRNA targets containing simultaneous5′-UTR and 3′-UTR interaction sites. *Genome Res.* 19, 1175–1183.

Lee, R. C., Feinbaum, R. L., and Ambros, V. (1993). The, *C.* elegans heterochronic gene lin-4 encodes small RNAs with antisense complementarity to lin-14. *Cell* 75, 843–854.

Lewis, B. P., Burge, C. B., and Bartel, D. P. (2005). Conserved seed pairing, often flanked by adenosines, indicates that thousands of human genes are microRNA targets. *Cell* 120, 15–20.

Li, J., Kong, X., Zhang, J., Luo, Q., Li, X., and Fang, L. (2013). MiRNA-26b inhibits proliferation by targeting PTGS2 in breast cancer. *Cancer Cell Int.* 13, 7.

Liu, J. P., Tang, Y., Zhou, S., Toh, B. H., McLean, C., and Li, H. (2010). Cholesterol involvement in the pathogenesis of neurodegenerative diseases. *Mol. Cell. Neurosci.* 43, 33–42.

Lonnstedt, I. S. T. (2002). Replicated microarray data. *Stat. Sinica* 12, 31–46.

Lundstrom, K. (2011). Micro-RNA in disease and gene therapy. *Curr. Drug Discov. Technol.* 8, 76–86.

Maes, O. C., Chertkow, H. M., Wang, E., and Schipper, H. M. (2009). MicroRNA: implications for Alzheimer disease and other human CNS disorders. *Curr. Genomics* 10, 154–168.

Margis, R., and Rieder, C. R. (2011). Identification of blood microRNAs associated to Parkinsonis disease. *J. Biotechnol.* 152, 96–101.

Martins, M., Rosa, A., Guedes, L. C., Fonseca, B. V., Gotovac, K., Violante, S., et al. (2011). Convergence of miRNA expression profiling, alpha-synuclein interacton and GWAS in Parkinson's disease. *PLoS ONE* 6:e25443. doi: 10.1371/journal.pone.0025443

Mazumder, B., Seshadri, V., and Fox, P. L. (2003). Translational control by the 3′-UTR: the ends specify the means. *Trends Biochem. Sci.* 28, 91–98.

Melas, P. A., Rogdaki, M., Osby, U., Schalling, M., Lavebratt, C., and Ekstrom, T. J. (2012). Epigenetic aberrations in leukocytes of patients with schizophrenia: association of global DNA methylation with antipsychotic drug treatment and disease onset. *FASEB J.* 26, 2712–2718.

Meshorer, E., Erb, C., Gazit, R., Pavlovsky, L., Kaufer, D., Friedman, A., et al. (2002). Alternative splicing and neuritic mRNA translocation under long-term neuronal hypersensitivity. *Science* 295, 508–512.

Minones-Moyano, E., Porta, S., Escaramis, G., Rabionet, R., Iraola, S., Kagerbauer, B., et al. (2011). MicroRNA profiling of Parkinson's disease brains identifies early downregulation of miR-34b/c which modulate mitochondrial function. *Hum. Mol. Genet.* 20, 3067–3078.

Mohamed, A. M., Boudreau, J. R., Yu, F. P., Liu, J., and Chin-Sang, I. D. (2012). The *Caenorhabditis elegans* Eph receptor activates NCK and N-WASP, and inhibits Ena/VASP to regulate growth cone dynamics during axon guidance. *PLoS Genet.* 8:e1002513. doi: 10.1371/journal.pgen.1002513

Moran, L. B., Duke, D. C., Deprez, M., Dexter, D. T., Pearce, R. K., and Graeber, M. B. (2006). Whole genome expression profiling of the medial and lateral substantia nigra in Parkinson's disease. *Neurogenetics* 7, 1–11.

Mraz, M., Pospisilova, S., Malinova, K., Slapak, I., and Mayer, J. (2009). MicroRNAs in chronic lymphocytic leukemia pathogenesis and disease subtypes. *Leuk. Lymphoma* 50, 506–509.

Niu, N., Rice, S. R., Heston, L. L., and Sobell, J. L. (2004). Multiple missense mutations in the diazepam binding inhibitor (DBI) gene identified in schizophrenia but lack of disease association. *Am. J. Med. Genet. B Neuropsychiatr. Genet.* 125B, 10–19.

Orom, U. A., Nielsen, F. C., and Lund, A. H. (2008). MicroRNA-10a binds the 5′UTR of ribosomal protein mRNAs and enhances their translation. *Mol. Cell* 30, 460–471.

Pan, Q., Shai, O., Lee, L. J., Frey, B. J., and Blencowe, B. J. (2008). Deep surveying of alternative splicing complexity in the human transcriptome by high-throughput sequencing. *Nat. Genet.* 40, 1413–1415.

Pang, M., Woodward, A. W., Agarwal, V., Guan, X., Ha, M., Ramachandran, V., et al. (2009). Genome-wide analysis reveals rapid and dynamic changes in miRNA and siRNA sequence and expression during ovule and fiber development in allotetraploid cotton (Gossypium hirsutum L.). *Genome Biol* 10, R122.

Perry, M. M., Moschos, S. A., Williams, A. E., Shepherd, N. J., Larner-Svensson, H. M., and Lindsay, M. A. (2008). Rapid changes in microRNA-146a expression negatively regulate the IL-1beta-induced inflammatory response in human lung alveolar epithelial cells. *J. Immunol.* 180, 5689–5698.

Pico, A. R., Kelder, T., Van Iersel, M. P., Hanspers, K., Conklin, B. R., and Evelo, C. (2008). WikiPathways: pathway editing for the people. *PLoS Biol.* 6:e184. doi: 10.1371/journal.pbio.0060184

Pritchard, C. C., Cheng, H. H., and Tewari, M. (2012). MicroRNA profiling: approaches and considerations. *Nat. Rev. Genet.* 13, 358–369.

Pu, X., Roth, J. A., Hildebrandt, M. A., Ye, Y., Wei, H., Minna, J. D., et al. (2013). MicroRNA-related genetic variants associated with clinical outcomes in early stage non-small cell lung cancer patients. *Cancer Res.* 73, 1867–1875.

Qiu, R., Wang, J., Tsark, W., and Lu, Q. (2010). Essential role of PDZ-RGS3 in the maintenance of neural progenitor cells. *Stem Cells* 28, 1602–1610.

Reinhart, B. J., Slack, F. J., Basson, M., Pasquinelli, A. E., Bettinger, J. C., Rougvie, A. E., et al. (2000). The 21-nucleotide let-7 RNA regulates developmental timing in *Caenorhabditis elegans*. *Nature* 403, 901–906.

Robinson, M. D., McCarthy, D. J., and Smyth, G. K. (2010). edgeR: a bioconductor package for differential expression analysis of digital gene expression data. *Bioinformatics* 26, 139–140.

Robinson, M. D., and Oshlack, A. (2010). A scaling normalization method for differential expression analysis of RNA-seq data. *Genome Biol.* 11, R25.

Robinson, M. D., and Smyth, G. K. (2008). Small-sample estimation of negative binomial dispersion, with applications to SAGE data. *Biostatistics* 9, 321–332.

Saba, R., Goodman, C. D., Huzarewich, R. L. C. H., Robertson, C., and

Booth, S. A. (2008). A miRNA signature of prion induced neurodegeneration. *PloS ONE* 3:e3652. doi: 10.1371/journal.pone.0003652

Sakharkar, M. K., Chow, V. T., Chaturvedi, I., Mathura, V. S., Shapshak, P., and Kangueane, P. (2004a). A report on single exon genes (SEG) in eukaryotes. *Front. Biosci.* 9:3262–3267. doi: 10.2741/1478

Sakharkar, M. K., Chow, V. T., and Kangueane, P. (2004b). Distributions of exons and introns in the human genome. *In Silico Biol.* 4, 387–393.

Salomonis, N., Nelson, B., Vranizan, K., Pico, A. R., Hanspers, K., Kuchinsky, A., et al. (2009). Alternative splicing in the differentiation of human embryonic stem cells into cardiac precursors. *PLoS Comput. Biol.* 5:e1000553. doi: 10.1371/journal.pcbi.1000553

Salomonis, N., Schlieve, C. R., Pereira, L., Wahlquist, C., Colas, A., Zambon, A. C., et al. (2010). Alternative splicing regulates mouse embryonic stem cell pluripotency and differentiation. *Proc. Natl. Acad. Sci. U.S.A.* 107, 10514–10519.

Sanborn, B. M., Millan, J. L., Meistrich, M. L., and Moore, L. C. (1997). Alternative splicing of CREB and CREM mRNAs in an immortalized germ cell line. *J. Androl.* 18, 62–70.

Satoh, J., and Tabunoki, H. (2011). Comprehensive analysis of human microRNA target networks. *BioData Min.* 4, 17.

Scherzer, C. R., Eklund, A. C., Morse, L. J., Liao, Z., Locascio, J. J., Fefer, D., et al. (2007). Molecular markers of early Parkinson's disease based on gene expression in blood. *Proc. Natl. Acad. Sci. U.S.A.* 104, 955–960.

Schopman, N. C., Heynen, S., Haasnoot, J., and Berkhout, B. (2010). A miRNA-tRNA mix-up: tRNA origin of proposed miRNA. *RNA Biol.* 7, 573–576.

Smoot, M. E., Ono, K., Ruscheinski, J., Wang, P. L., and Ideker, T. (2011). Cytoscape 2.8: new features for data integration and network visualization. *Bioinformatics* 27, 431–432.

Solmi, R., Ugolini, G., Rosati, G., Zanotti, S., Lauriola, M., Montroni, I., et al. (2006). Microarray-based identification and RT-PCR test screening for epithelial-specific mRNAs in peripheral blood of patients with colon cancer. *BMC Cancer* 6:250. doi: 10.1186/1471-2407-6-250

Soreq, L., Ben-Shaul, Y., Israel, Z., Bergman, H., and Soreq, H. (2012a). Meta-analysis of genetic and environmental Parkinson's disease models reveals a common role of mitochondrial protection pathways. *Neurobiol. Dis.* 45, 1018–1030.

Soreq, L., Bergman, H., Goll, Y., Greenberg, D. S., Israel, Z., and Soreq, H. (2012b). Deep brain stimulation induces rapidly reversible transcript changes in Parkinson's leucocytes. *J. Cell. Mol. Med.* 16, 1496–1507.

Soreq, L., Bergman, H., Israel, Z., and Soreq, H. (2012c). Exon arrays reveal alternative splicing aberrations in Parkinson's disease leukocytes. *Neurodegener. Dis.* 10, 203–206.

Soreq, L., Israel, Z., Bergman, H., and Soreq, H. (2008). Advanced microarray analysis highlights modified neuro-immune signaling in nucleated blood cells from Parkinson's disease patients. *J. Neuroimmunol.* 201–202, 227–236.

Spinelli, S. V., Diaz, A., D'Attilio, L., Marchesini, M. M., Bogue, C., Bay, M. L., et al. (2013). Altered microRNA expression levels in mononuclear cells of patients with pulmonary and pleural tuberculosis and their relation with components of the immune response. *Mol. Immunol.* 53, 265–269.

Sugnet, C. W., Srinivasan, K., Clark, T. A., O'Brien, G., Cline, M. S., Wang, H., et al.. (2006). Unusual intron conservation near tissue-regulated exons found by splicing microarrays. *PLoS Comput. Biol.* 2:e4. doi: 10.1371/journal.pcbi.0020004

Ule, J., Jensen, K. B., Ruggiu, M., Mele, A., Ule, A., and Darnell, R. B. (2003). CLIP identifies Nova-regulated RNA networks in the brain. *Science* 302, 1212–1215.

Vaarala, M. H., Hirvikoski, P., Kauppila, S., and Paavonen, T. K. (2012). Identification of androgen-regulated genes in human prostate. *Mol. Med. Rep.* 6, 466–472.

Wang, P., Gu, Y., Zhang, Q., Han, Y., Hou, J., Lin, L., et al. (2012). Identification of resting and type I IFN-activated human NK cell miRNomes reveals MicroRNA-378 and MicroRNA-30e as negative regulators of NK cell cytotoxicity. *J. Immunol.* 189, 211–221.

Xia, R., Jia, H., Fan, J., Liu, Y., and Jia, J. (2012). USP8 promotes smoothened signaling by preventing its ubiquitination and changing its subcellular localization. *PLoS Biol.* 10:e1001238. doi: 10.1371/journal.pbio.1001238

Zambon, A. C., Gaj, S., Ho, I., Hanspers, K., Vranizan, K., Evelo, C. T., et al. (2012). GO-Elite: a flexible solution for pathway and ontology over-representation. *Bioinformatics* 28, 2209–2210.

Zhang, H. G., and Grizzle, W. E. (2011). Exosomes and cancer: a newly described pathway of immune suppression. *Clin. Cancer Res.* 17, 959–964.

22

A molecular toolbox for rapid generation of viral vectors to up-or down-regulate neuronal gene expression *in vivo*

Melanie D. White, Ruth V. J. Milne and Matthew F. Nolan*

Centre for Integrative Physiology, University of Edinburgh, Edinburgh, Scotland, UK

Correspondence:
Matthew F. Nolan, Centre for Integrative Physiology, University of Edinburgh, Edinburgh EH8 9XD, Scotland, UK.
e-mail: mattnolan@ed.ac.uk

We introduce a molecular toolbox for manipulation of neuronal gene expression *in vivo*. The toolbox includes promoters, ion channels, optogenetic tools, fluorescent proteins, and intronic artificial microRNAs. The components are easily assembled into adeno-associated virus (AAV) or lentivirus vectors using recombination cloning. We demonstrate assembly of toolbox components into lentivirus and AAV vectors and use these vectors for *in vivo* expression of inwardly rectifying potassium channels (Kir2.1, Kir3.1, and Kir3.2) and an artificial microRNA targeted against the ion channel HCN1 (HCN1 miRNA). We show that AAV assembled to express HCN1 miRNA produces efficacious and specific *in vivo* knock-down of HCN1 channels. Comparison of *in vivo* viral transduction using HCN1 miRNA with mice containing a germ line deletion of *HCN1* reveals similar physiological phenotypes in cerebellar Purkinje cells. The easy assembly and re-usability of the toolbox components, together with the ability to up- or down-regulate neuronal gene expression *in vivo*, may be useful for applications in many areas of neuroscience.

Keywords: intronic miRNA, ion channel, lentivirus, AAV, RNAi, hippocampus, cerebellum, virus

INTRODUCTION

Tools for manipulation of gene expression in neurons *in vivo* are important for investigation of nervous system function and may be of therapeutic benefit. Using viruses to manipulate neuronal gene expression *in vivo* is increasingly popular due to the flexibility and economy of generating viruses compared with transgenic animals. Nevertheless, it can be time consuming to assemble new viral constructs containing different promoter, gene of interest, and reporter elements. Furthermore, many of these elements have not been widely validated for use in the central nervous system and their performance may vary between studies. A toolbox of standardized modular components that can be rapidly assembled into new viral constructs is therefore desirable. A critical component of such a toolbox will be methods to disrupt expression of targeted genes. One promising approach is gene knockdown by RNA interference (RNAi). However, current methods using RNAi in neurons are associated with off target effects and toxicity (Grimm et al., 2006; McBride et al., 2008; Ehlert et al., 2010; Martin et al., 2011). In non-neuronal cells intronic expression of miRNAs has been suggested to alleviate these problems, while simultaneously enabling stable, high level co-expression of reporter genes (Du et al., 2006). However, it is not clear if this approach will produce specific or selective gene knockdown when introduced into neurons *in vivo* using viruses.

To address the goal of establishing a molecular toolbox for *in vivo* up- and down-regulation of neuronal gene expression, we generated and validated a set of vectors containing cassettes encoding promoters, genes of interest, intronic miRNA constructs, and reporter proteins flanked by recombination sites. These cassettes can be assembled into various combinations and inserted into a viral backbone through a single recombination reaction. The system is extremely flexible, allowing rapid assembly of gene-reporter fusion constructs, bicistronic expression of genes, and expression of intronic interfering RNAs for gene knockdown. We show that viruses produced from these components give stable over-expression or efficient knockdown of genes in neurons *in vivo*, while simultaneously expressing reporter genes detectable by native fluorescence. By focusing on manipulation of the *HCN1* gene, which encodes an ion channel that has been relatively well characterized using knockout mice (e.g., Nolan et al., 2003, 2004, 2007; Chen et al., 2009, 2010; Huang et al., 2009), we go on to provide evidence that intronic expression of artificial miRNAs causes effective knockdown of gene expression and does not appear to be associated with off-target effects on physiological properties of neurons. Because this molecular toolbox enables identification and investigation of transduced cells by a variety of techniques, it may be useful for addressing a broad range of questions within neuroscience.

METHODS

VECTOR CONSTRUCTION

The principle behind the molecular toolbox is to allow rapid generation of new constructs through recombination cloning. To create the initial att-flanked PCR fragments for recombination, traditional digestion, and ligation methods were used for some constructs, whereas other were generated by PCR amplification or synthesized directly.

Restriction enzyme-based cloning

pEYFP-HCN1 was constructed by amplifying the *HCN1* sequence from pEGFP-HCN1 (gift from Bina Santoro, Columbia University, New York) and cloning it into the *Eco*RI and *Bam*HI sites of pEYFP-C1 (Clontech). pFCK-ChR2-mCherry was constructed by replacing the EGFP sequence in pFCK-ChR2-EGFP (from Ed Boyden, MIT, obtained via Addgene) with mCherry digested from pmCherry-N1 (Clontech) using *Age*I and *Bsr*GI. To enhance transgene expression we used a version of the woodchuck hepatitis virus post-transcriptional regulatory element optimized for safety (oPRE; Schambach et al., 2006). A lentivirus with oPRE cloned into the backbone (pLenti-oPRE) was generated by PCR amplifying the oPRE sequence from pMP71-GFP-opre (provided by Axel Schambach, Hannover Medical School, Hannover) and cloning it into the *Xho*I and *Xba*I sites of pLenti6/BLOCK-iT-DEST backbone (Invitrogen).

To create a Gateway-compatible adeno-associated virus (AAV) vector, pCAGW-ChR2-Venus (from Karel Svoboda, obtained via Addgene; Petreanu et al., 2009) was digested with *Afe*I and *Eco*RV to remove the CAGW promoter and the ChR2-Venus insert. Reading Frame B of the Gateway Vector Conversion System (Invitrogen) was ligated into the AAV backbone to create pAAV-Gateway.

Recombination cloning

While several technologies are available for recombination cloning (e.g., Ma et al., 1987; Bubeck et al., 1993; Oliner et al., 1993) here we use the Gateway system (Invitrogen). This recombination cloning system uses a three-step strategy to generate expression constructs (Hartley et al., 2000; Sasaki et al., 2004; Yahata et al., 2005). Cassettes encoding individual components of the constructs are amplified by PCR using primers containing att sites that allow recombination. The PCR cassettes are then recombined into a donor vector with matching att sites to generate an entry vector. By using alternative matching pairs of att sites different arrangements of entry vectors can be recombined into a destination vector to generate new expression constructs in a single reaction.

Generation of att-flanked cassettes. Cassettes encoding the toolbox components were generated by PCR amplification using primers containing the required att recombination site sequences in addition to 18–25 base pairs of template-specific sequence. Two hundred nanomolars of each primer was added to 1–2 μl of template DNA and 22.5 μl Platinum PCR SuperMix High Fidelity (Invitrogen). Thirty-five cycles of amplification were performed at 94°C for 30 s, 55°C for 30 s, and 68°C for 60 s.

Generation of entry vectors. To insert the att-flanked PCR cassettes into a vector, 50 fmol of purified att-flanked PCR products were added to 50 fmol of the appropriate pDONR vector (Invitrogen), 2 μl of BP Clonase II enzyme (Invitrogen), and TE Buffer pH 8.0 to a total of 10 μl. The recombination reactions were incubated at 25°C overnight generating entry vectors. One microliter of proteinase K solution (Invitrogen) was added and the reactions were incubated at 37°C for 10 min. Three microliters of each reaction was transformed into TOP10 cells (Invitrogen) and positive transformants were selected with kanamycin (Sigma-Aldrich).

Generation of viral destination vectors. To generate viral vectors, 10 fmol of each of the required entry vectors were added to 20 fmol of the destination vector (pLenti6/BLOCK-iT/DEST, pLenti-oPRE, or AAV-Gateway), 2 μl LR Clonase II Plus enzyme mix (Invitrogen), and TE Buffer pH 8.0 to a total volume of 10 μl. The reactions were incubated overnight at 25°C and treated with 1 μl proteinase K solution (Invitrogen) at 37°C for 10 min. The entire reaction was transformed into Stbl3 cells (Invitrogen) and positive transformants were selected using ampicillin (Invitrogen) and blasticidin (Autogen Bioclear).

General purpose cassettes. *CAMKIIα* short (0.4 kb) and long (1.3 kb) promoter variants (Dittgen et al., 2004) were amplified from mouse genomic DNA. Synapsin and enhanced synapsin promoters were amplified from pBSIISK-SYN-GFP-WPRE and pBSIISK-E/SYN-GFP-WPRE respectively (vectors provided by Hiroyuki Hioki, Kyoto University, Kyoto). The Netrin G1 putative promoter was amplified from mouse genomic DNA. A 1.1-kb sequence corresponding to the first 345 bp + 754 bp immediately upstream of the transcription start site of the Netrin G1 (*Ntng1*) transcript variant a, gene (NM_030699.2) was cloned. Fluorescent reporter cassettes were amplified from pEGFP-C1, pEYFP-C1, pECFP-C1, and pmCherry-C1 (Clontech). An IRES-EGFP cassette was amplified from pPRIG (provided by Patrick Martin, UNICE, Nice). A cassette containing oPRE was generated by flanking the oPRE sequence with the appropriate recombination sites to ensure it would always insert 3′ to any coding cassettes during recombination into a viral vector. EYFP-HCN1 cassettes were amplified from pEYFP-HCN1. Channelrhodopsin-mCherry was amplified from pFCK-ChR2-mCherry. CAMKII(0.4)-Halorhodopsin-EGFP was amplified from pFCK-Halo-EGFP (from Ed Boyden, MIT, obtained via Addgene; Han and Boyden, 2007). Following PCR, the amplicons were gel purified using a QIAquick gel extraction kit (Qiagen). To generate entry vectors, all cassettes were recombined into the appropriate donor vector.

Reporter constructs. To generate a reporter construct for visualization of neurons *in vivo*, pENTR-L1-CAMKII(0.4)-L4, pENTR-R4-EYFP-R3, and pENTR-L3-oPRE-L2 were recombined with pLenti6/BLOCK-iT-DEST to create pLenti6-CAMKII(0.4)-EYFP. To test the Netrin G1 promoter a reporter lentivirus was made by recombining pENTR-L1-NetrinG1-L4, pENTR-R4-EYFP-R3, and pENTR-L3-oPRE-L2 with pLenti6/BLOCK-iT-DEST to create pLenti6-NetrinG1-EYFP.

Ion channel constructs. The coding sequences of Kir2.1 (*Mus musculus* potassium inwardly rectifying channel, subfamily J, member 2 (*Kcnj2*) NM_008425) fused to a C-terminal mCherry sequence was synthesized by Mr Gene[1]. The sequence was flanked by L5-L2 att recombination sites to enable recombination cloning. pENTR-L1-CAMKII(0.4)-R5 and pENTR-L5-Kir2.1-mCherry-R2 were recombined with the pLenti-oPRE backbone to create the lentivector pLenti–CAMKII(0.4)-Kir2.1-mCherry. The coding sequences of Kir3.1 [*M. musculus* potassium inwardly rectifying channel, subfamily J, member 3 (*Kcnj3*) NM_008426.1] and Kir3.2 [*M. musculus* potassium inwardly rectifying channel, subfamily J, member 6 (*Kcnj6*), transcript variant Girk2A-1, NM_010606] fused to a C-terminal EGFP sequence and flanked by R4-R3

[1] http://mrgene.com

att sites were synthesized by Mr Gene. Entry vectors encoding the Kir3-EGFP constructs were recombined with pENTR-L1-CAMKII(0.4)-L4, pENTR-L3-oPRE-L2, and pLenti6/BLOCK-iT-DEST to create the lentivectors pLenti6–CAMKII(0.4)-Kir3.1-EGFP and pLenti6–CAMKII(0.4)-Kir3.2-EGFP.

miRNA constructs. BLOCK-iT miR RNAi sequences targeting mouse *HCN1* (Mmi510951, Mmi510952, and Mmi510953; Invitrogen) were cloned into the synthetic intron of pSM155 (provided by Guangwei Du, University of Texas Health Science Center, Houston) using procedures described in detail elsewhere (Du et al., 2006). Briefly, RNAi sequences were designed so that once annealed, the duplexes had four nucleotide overhangs compatible with the cohesive ends produced by *Bsm*BI digestion of pSM155. The RNAi duplexes were annealed and ligated into pSM155. A cassette containing the intronic miRNA upstream of EGFP was then amplified with B4r and B3r PCR primers and recombined to generate the entry vector pENTR-R4-HCN1miR-EGFP-R3. To create a negative control miRNA construct, a cassette containing a miRNA targeting luciferase upstream of an EGFP sequence was amplified from pSM155-Luc (also from Guangwei Du) with B4r and B3r primers and recombined to create the entry vector pENTR-R4-LucmiR-EGFP-R3. To generate a lentivirus expressing the HCN1 miRNA, pENTR-L1-ESYN-L4 was recombined with pENTR-R4-HCN1miR-EGFP-R3, pENTR-L3-oPRE-L2, and the pLenti6/BLOCK-iT-DEST backbone to create the lentivector pLenti6-ESYN-HCN1miR-EGFP. To generate AAV expressing miRNAs for experiments in **Figures 5–7**, pENTR-L1-ESYN-L4 and pENTR-L3-oPRE-L2 were recombined with pAAV-Gateway and pENTR-R4-HCN1miR-EGFP-R3 or pENTR-R4-LucmiR-EGFP-R3 to create pAAV-ESYN-HCN1miR-EGFP and pAAV-ESYN-LucmiR-EGFP.

PCR PRIMERS

CAMKII(1.3) attB1: GGGGACAAGTTTGTACAAAAAAGCAGGCTCATTATGGCCTTAGGTCACTT

CAMKII(0.4) attB1: GGGGACAAGTTTGTACAAAAAAGCAGGCTACTTGTGGACTAAGTTTGTTC

CAMKII attB5r: GGGGACAACTTTTGTATACAAAGTTGTTGCCCCCAGAACTAGGGGCCACT

CAMKII attB4: GGGGACAACTTTGTATAGAAAAGTTGGGTGCCCCAGAACTAGGGGCCACT

SYN attB1: GGGGACAAGTTTGTACAAAAAAGCATACCTGCAGAGGGCCCTGCGTAT

SYN attB5r: GGGGACAACTTTTGTATACAAAGTTGTCGCCGCAGCGCAGATGGTC

SYN attB4: GGGGACAACTTTGTATAGAAAAGTTGGGTGCGCCGCAGCGCAGATGGT

ESYN attB1: GGGGACAAGTTTGTACAAAAAAGCATACTAGTTATTAATAGTAATCAATT

NetrinG1 attB1: GGGGACAAGTTTGTACAAAAAAGCAGGCTCTCCTACTTGGACTTCAAC

NetrinG1 attB4: GGGGACAACTTTGTATAGAAAAGTTGGGTGCTTCCGAGGTCTCGGCAA

EXFP attB5: GGGGACAACTTTGTATACAAAAGTTGACCATGGTGAGCAAGGGCGAGGA

EXFP attB2: GGGGACCACTTTGTACAAGAAAGCTGGGTATTACTTGTACAGCTCGTCCA

EXFP attB4r: GGGGACAACTTTTCTATACAAAGTTGACCATGGTGAGCAAGGGCGAGGA

EXFP attB3r: GGGGACAACTTTATTATACAAAGTTGTTTACTTGTACAGCTCGTCCAT

IRES-EGFP attB3: GGGGACAACTTTGTATAATAAAGTTGGTTACTGGCCGAAGCCGCTT

oPRE attB3: GGGGACAACTTTGTATAATAAAGTTGGAGCATCTTACCGCCATTTATACC

oPRE attB5: GGGGACAACTTTGTATACAAAAGTTGGAGCATCTTACCGCCATTTATACC

oPRE attB2: GGGGACCACTTTGTACAAGAAAGCTGGGTACGACAACACCACGGAATT

miR EXFP attB5: GGGGACAACTTTGTATACAAAAGTTGCTAAGGTAGCCTTGCAGAAGTTGG

miR EXFP attB4r: GGGGACAACTTTTCTATACAAAGTTGCTAAGGTAGCCTTGCAGAAGTTGG

ChR2 attB5: GGGGACAACTTTGTATACAAAAGTTGATGCCACCATGGACTATGGCGGCG

ChR2 attB4r: GGGGACAACTTTTCTATACAAAGTTGCCACCATGGACTATGGCGGCG

HCN1 attB5: GGGGACAACTTTGTATACAAAAGTTGATATGGAAGGCGGCGGCAAACC

HCN1 attB4r: GGGGACAACTTTTCTATACAAAGTTGATATGGAAGGCGGCGGCAAACC

HCN1 attB3r: GGGGACAACTTTATTATACAAAGTTGTTAAATTCGAAGCAAAACGGGG

K2P9.1 attB4r: GGGGACAACTTTTCTATACAAAGTTGATATGAAGCGGCAGAACGTGCGT

K2P9.1 attB3r: GGGGACAACTTTATTATACAAAGTTGTTTAGATGGACTTGCGACGGAGGTG

HCN1F_*Eco*RI: AATTCTAATGGAAGGCGGCG

HCN1R_*Bam*HI: GATCTAAATTCGAAGCAAAACGG

CELL CULTURE AND TRANSFECTIONS

HEK293FT cells (Invitrogen) were maintained at 37°C, 5% CO_2 in DMEM with GlutaMAX I, 4,500 mg/l glucose and 110 mg/l sodium pyruvate (Gibco) supplemented with 50 U/ml each of penicillin and streptomycin (Sigma) and 10% v/v FCS (Invitrogen). DNA transfections of miRNA constructs were performed using FuGENE6 (Roche) according to manufacturer's instructions.

VIRUS PRODUCTION

Lentiviruses were generated by triple transfection of HEK293FT cells (Invitrogen) with the modified lentivector, psPAX2 (courtesy of Didier Trono, EPFL, Lausanne, obtained via Addgene) and the pLP/VSVG envelope plasmid (Invitrogen) using calcium phosphate transfection. The lentiviruses were harvested in serum-free medium after 3 days, filtered, and concentrated in primed Centricon Plus-80 filter devices (Millipore). Functional titer was determined by EGFP fluorescence after serial dilution on HEK293FT cells and was 0.5–2 × 10^8 TU/ml. Hybrid AAVs containing the ITR from AAV2 and Capsid from AAV1 were produced by Vectorbiolabs[2]. Titer was determined by measurement of vector genome copies and was 5×10^{13} GC/ml.

STEREOTAXIC INJECTION

All animal experiments were carried out according to the guidelines laid down by the University of Edinburgh's Animal Welfare Committee and in accordance with the UK Animals (Scientific Procedures) Act 1986. C57BL6/JOlaHsd mice aged 4–6 weeks were anesthetized with isoflurane and mounted into a stereotaxic frame. 0.2–1 μl of virus was injected at 0.1 μl/min using pre-calibrated pulled glass pipettes with a tip diameter of 6–10 μm following a previously described protocol (Cetin et al., 2006). The pipette was left in place for at least 5 min following the injection before being slowly raised. For the hippocampal injections, co-ordinates measured in millimeter from bregma were: −1.8 A/P, ±1.6 M/L, −1.3 D/V for CA1, −1.8 A/P, ±1.4 M/L, −1.6 D/V for dentate gyrus. Cerebellum injections targeted the vermis and were measured relative to lambda. Co-ordinates were −2 A/P and 0–1.5 M/L. Injections of ~200 nl of virus were made at $Z = -3, -2.5, -2, -1.5, -1$, and -0.5 either with the injecting arm angled back 10° and positioned over the midline, or angled back 30° and rotated 45° counter-clockwise to target more lobules. Following surgery, animals were returned to their home cages for at least 2 weeks before being used for analysis.

IMAGING

Mice were terminally anesthetized with sodium pentobarbital and perfused with 4% PFA. The brains were post-fixed and immersed in 30% sucrose overnight before 30 μm coronal sections were cut on a freezing microtome. Native fluorescence of reporter genes was imaged using Leica TCS-NT and Zeiss LSM510 confocal microscopes. For imaging of spines, data were acquired using a 1,024 × 1,024 pixel image size and a Zeiss Plan NeoFLUAR 1.4 NA 63× oil immersion lens. Image data, acquired at Nyquist sampling rates, were deconvolved using Huygens software (Scientific Volume Imaging, Netherlands), and the resulting three-dimensional images were analyzed using NIH ImageJ software.

FACS SORTING

Lentivirus expressing Kir2.1-mCherry was stereotaxically targeted to the dentate gyrus of 4-week-old C57BL6/JOlaHsd mice. Two weeks after injections the mice were culled and the brains rapidly removed and placed in modified oxygenated artificial cerebrospinal fluid (ACSF) of the following composition (mM): NaCl 86, NaH_2PO_4 1.2, KCl 2.5, $NaHCO_3$ 25, $CaCl_2$ 0.5, $MgCl_2$ 7, glucose 25, sucrose 75, at 6°C for 10 min. Four hundred micrometers coronal slices were cut through the hippocampus on a Vibratome 3000 sectioning system. The slices were transferred to oxygenated ACSF + 10 mM HEPES at 37°C and incubated for 10 min. The dorsal dentate gyrus from each slice was microdissected out under a dissecting microscope. All samples from a single brain were pooled together in 1 well of a 12-well plate with 900 μl oxygenated ACSF + HEPES. One hundred microliters papain (200 U/ml, Sigma-Aldrich) was added to each well and the samples were incubated at 37°C in 5% CO_2 for 45 min to dissociate the cells. The digestion reaction was stopped by the addition of 100 μl BSA (10 mg/ml; Sigma-Aldrich), 11 μl FCS (Invitrogen), 10 μl DNase (1,000 U/ml, Sigma-Aldrich), and 10 μl 5 mM Leupeptin (Sigma-Aldrich). The samples were triturated through a fire-polished Pasteur pipette until no clumps were visible. Two milliliters of cold ACSF + HEPES was added to each sample and they were passed through a 70-μm sieve (Becton Dickinson) to remove any remaining cell clumps. The cells were counted with a hemocytometer and resuspended in cold ACSF + HEPES to a density of 1×10^6 cells/ml. One microliter per milliliter of Live/Dead stain with a green fluorescent dye (Invitrogen) was added to each sample and they were incubated on ice for 30 min in the dark. The samples were centrifuged at 800 rpm and washed twice in 1 ml PBS + 1% FCS to remove unbound stain. Samples from an uninjected brain were prepared alongside the experimental brains to use as controls for the fluorescence-activated cell sorting (FACS) sorting and subsequent RT-PCR. A sample from the uninjected brain was heat-treated at 50°C to provide a positive control for calibration of the Live/Dead stain. The samples were sorted on a FACS Aria II (BD Biosciences, USA) with gating parameters set to collect live mCherry-fluorescent cells. The yield of positive cells isolated from the dentate gyrus of a single brain ranged from 5,249 to 14,624 cells.

REAL TIME PCR

Total mRNA was extracted using the Arcturus PicoPureAR RNA Isolation Kit (Applied Biosystems) in combination with the RNase-Free DNase Set (Qiagen) according to manufacturer's instructions. Taqman Gene expression assays were used for amplification of target genes. Specific assays used were Mm00484848_m1 (*Mcm6*), Mm01326464_m1 (NeuroD6), Mm01253033_m1 (*Gfap*), and Mm00446968_m1 (*Hprt1*). RT-PCR reactions were set up using the TaqManAR RNA-to-CT 1-Step Kit (Applied Biosystems). One microliter of mRNA template was added to 1 μl of the gene-specific probe and primers mix, 10 μl 2× TaqMan RT-PCR mix, 0.5 μl 40× TaqMan RT Enzyme mix, and 7.5 μl H_2O. Reverse transcription was performed at 48°C for 15 min followed by activation of AmpliTaq Gold DNA Polymerase at 95°C for 10 min and 40 cycles of amplification at 95°C for 15 s and 60°C for 1 min on an Mx3000P QPCR system (Stratagene). Quantification was performed using the comparative C_t method with normalization to *Hprt1*. Results were expressed as mean ± standard error of the mean and statistical analyses (ANOVA and Student's *t*-test) were performed using Smith's Statistical Package, version 2.80 or R (The R Foundation).

ELECTROPHYSIOLOGY

Cerebellar brain slices were prepared and whole-cell patch-recordings made from visually identified cerebellar Purkinje cells following previously described procedures (Nolan et al., 2003; Zonta et al., 2011). Briefly, sagittal cerebellar slices were cut in cold (4–6°C) modified oxygenated ACSF of the following composition (mM): NaCl 86, NaH_2PO_4 1.2, KCl 2.5, NaHCO3 25, $CaCl_2$

[2]www.vectorbiolabs.com

0.5, $MgCl_2$ 7, glucose 25, and sucrose 75. Slices were then maintained in standard oxygenated ACSF of the following composition: NaCl 124, NaH_2PO_4 1.2, KCl 2.5, NaHCO3 25, $CaCl_2$ 2, $MgCl_2$ 1, and glucose 20. Slices were initially kept for 10–20 min at 37 ± 1°C and then at room temperature for up to 6 h before making recordings. Patch-clamp recordings were obtained in standard ACSF using electrodes with resistance 2–5 MΩ. Whole-cell voltage-clamp recordings of hyperpolarization-activated currents were made at room temperature (18–22°C) in the following ACSF modified to reduce activation of other voltage-gated conductances: (mM) NaCl 115, NaH_2PO_4 1.2, KCl 5, $BaCl_2$ 1, $CdCl_2$ 0.1, 4-AP 1, $NiCl_2$ 1, TEA 5, $CaCl_2$ 2, $MgCl_2$ 1, NBQX 0.2, D-AP5 0.2, picrotoxin 0.5, and tetrodotoxin 0.001. Virally transduced cells were identified by EGFP fluorescence. All recordings from non-transduced (EGFP negative) cells were in slices also containing transduced (EGFP positive) cells. Recordings of action potentials and voltage responses to current steps were made in cell-attached voltage-clamp and whole-cell current-clamp configurations in standard ACSF at 34–36°C. Electrophysiological data were acquired using an Instrutech ITC-18 AD boards (HEKA Elektronik) and Axograph software. Analysis used custom written routines in IGOR pro (Wavemetrics) and R (R Foundation). Comparison of group data used analysis of variance (ANOVA and Student's t-test as appropriate). All data are expressed as mean $\pm$ standard error of the mean.

RESULTS

A MOLECULAR TOOLBOX FOR RAPID GENERATION OF VIRAL CONSTRUCTS

To enable rapid generation of new viral constructs we constructed a library of vectors containing cassettes compatible with the Gateway cloning system (**Figure 1**). The Gateway system uses site-specific recombination for directional cloning of PCR products and their subsequent subcloning into destination vectors. DNA fragments are amplified by PCR using primers that contain specific recombination sites (att sites). Entry vectors are generated using the BP Clonase enzyme to catalyze the insertion of the DNA fragment into a donor vector containing matching att sites. To generate the final construct the entry vector can then be recombined with a destination vector through the addition of LR Clonase. By including different att sites in each vector, att-flanked cassettes from up to four entry vectors can be directionally recombined into a destination vector in a single reaction (**Figures 1A,C**). Compared with conventional cloning approaches this reduces the number of

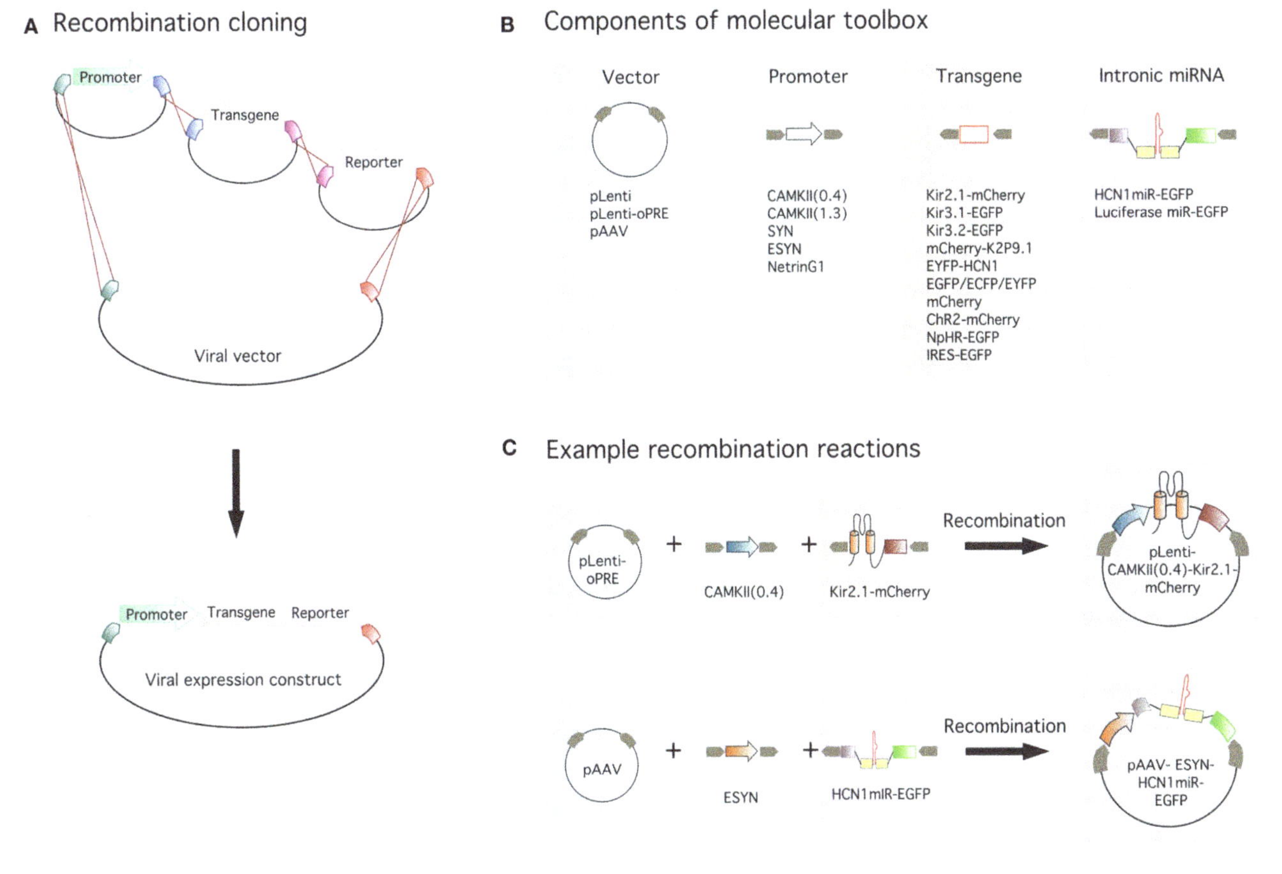

FIGURE 1 | Schematic illustrating modular cloning system. (A) Schematic illustrating recombination of entry vectors carrying cassettes flanked by att sites with a viral destination vector. In a single reaction recombination between matching att sites inserts the cassettes into the vector in the correct order, generating a viral expression construct. **(B)** Components of the molecular toolbox flanked with att sites ready for recombination. For simplicity, only the att-flanked regions of the entry vectors are illustrated in **(B,C)**. **(C)** Examples of recombination reactions to generate the lentiviral expression construct plenti-CAMKII(0.4)-Kir2.1-mCherry and the AAV expression construct pAAV-ESYN-HCN1miR-EGFP.

steps and time required to generate new constructs and allows the same components to be easily rearranged to form new viral vectors. The att-flanked cassettes that we describe here encode neuronal promoters, genes of interest, reporter genes, or intronic miRNAs (**Figure 1B**). The promoter cassettes are flanked with the appropriate att sites to ensure they insert as the first cassette in the viral backbone and therefore drive expression of downstream cassettes. The reporter cassettes are flanked with att sites permitting their expression either alone or as N- or C-terminal fusion proteins. Because we wanted to transduce neurons with either lentivirus or AAV and as there are currently no AAV vectors available that are compatible with the Gateway cloning system, we converted pACAGW-ChR2-Venus-AAV (Petreanu et al., 2009) into a Gateway-compatible destination vector by removing the CAG promoter and the ChR2-Venus insert and replacing them with a recombination cassette. Functionality of all the modular components of the toolbox was confirmed *in vitro* before using a subset to generate viruses for further *in vivo* validation (**Table 1**).

To evaluate expression of ion channels tagged with fluorescent reporter molecules, we first produced lentiviruses assembled from constructs encoding either an EYFP reporter alone, or Kir potassium channel-reporter fusion proteins, with expression driven by the short variant of the *CAMKIIα* promoter (CAMKII(0.4)-EYFP, CAMKII(0.4)-Kir2.1-mCherry, CAMKII(0.4)-Kir3.1-EGFP, and CAMKII(0.4)-Kir3.2-EGFP). To examine their expression in hippocampal neurons *in vivo*, we stereotaxically injected the viruses into CA1. Two weeks later, we sectioned the brains and imaged the native fluorescence in virally transduced cells. All four viruses expressed their transgenes at high levels within neurons. The subcellular localization of the reporter protein differed between

Table 1 | Molecular toolbox components.

Component	Symbol	Level of validation		
		In vitro expression	*In vivo* expression	Comment
PROMOTERS				
CAMKIIα short promoter	CAMKII(0.4)	+	+	
CAMKIIα long promoter	CAMKII(1.3)	+	+	
Synapsin	SYN	+	+	
Enhanced synapsin	ESYN	+	+	
Netrin G1	NetrinG1	+	+	Does not recapitulate the endogenous expression pattern of Netrin G1. Drives expression of transgenes in granule cells and interneurons of the dentate gyrus and pyramidal neurons and glial cells in CA1
TRANSGENES				
Ion channels				
Kir2.1 fused to mCherry	Kir2.1-mCherry	+	+	Localizes to the soma and proximal dendrites resulting in an inwardly rectifying potassium current confirmed by electrophysiological recordings (data not shown)
Kir3.1 fused to EGFP	Kir3.1-EGFP	+	+	Diffuse localization throughout soma and dendrites of CA1 pyramidal neurons but physiology not tested yet
Kir3.2 fused to EGFP	Kir3.2-EGFP	+	+	More punctate localization throughout soma and dendrites of CA1 pyramidal neurons but physiology not tested yet
K2P9.1 fused to mCherry	mCherry-K2P9.1	+	n/d	
HCN1 fused to EYFP	EYFP-HCN1	+	n/d	
Channelrhodopsin fused to mCherry	ChR2-mCherry	+	+	Diffuse localization throughout entire cell
Halorhodopsin fused to EGFP	NpHR-EGFP	+	n/d	
REPORTERS				
EGFP	EGFP	+	+	Diffuse localization throughout cytoplasm of entire cell
ECFP	ECFP	+	+	Diffuse localization throughout cytoplasm of entire cell
EYFP	EYFP	+	+	Diffuse localization throughout cytoplasm of entire cell
mCherry	mCherry	+	+	Diffuse localization throughout cytoplasm of entire cell
IRES-EGFP	IRES-EGFP	+	+	Diffuse localization throughout cytoplasm of entire cell
INTRONIC miRNAs				
HCN1 miRNA with EGFP	HCN1miR-EGFP	+	+	Knocks down HCN1 function without apparent off-target effects
Luciferase control miRNA with EGFP	LucmiR-EGFP	+	+	No apparent effects

The components of the toolbox that are in entry vectors ready for recombination to generate new viral vectors. The level of validation to date is indicated with observations regarding in vivo expression where appropriate.

the viruses, indicating differential trafficking of each ion channel (**Figure 2A**). The Kir2.1 channel was localized to the soma and proximal dendrites of CA1 pyramidal neurons, whereas Kir3.1 and Kir3.2 were present throughout the proximal and distal dendrites. Importantly, the differences in distribution of each ion channel reflect expression patterns for native channels *in vivo* (Miyashita and Kubo, 1997; Koyrakh et al., 2005) suggesting that this approach will be useful to study ion channel localization (cf. Piskorowski et al., 2011). Expression of Kir2.1-mCherry results in an inwardly rectifying potassium current confirmed by electrophysiological recordings from acute slices (data not shown) in agreement with previous findings using a similar construct in cultured neurons (Burrone et al., 2002). We are yet to assess the functional consequences of expression of Kir3.1-EGFP and Kir3.2-EGFP, but similar constructs yield functional channels in expression systems (Stevens et al., 1997).

Promoters other than *CAMKIIα* can also be successfully used to drive neuronal gene expression (Dittgen et al., 2004). To determine if a fragment of the putative *Netrin G1* promoter can drive gene expression *in vivo*, we assembled the lentivirus NetrinG1-EYFP and injected it into the dentate gyrus. Consistent with *in situ* hybridization data from the Allen Brain Atlas which suggests the presence of *Netrin G1* mRNA in the dentate gyrus, we found EYFP fluorescence in dentate gyrus granule cells transduced with NetrinG1-EYFP (**Figure 2B**). However, expression was not limited to granule cells, as we also observed occasional fluorescence in interneurons and glia (**Table 1**). The enhanced synapsin (ESYN) promoter also gave strong transgene expression in dentate gyrus granule cells transduced with the virus ESYN-LucmiR-EGFP (**Figure 2C**). The native fluorescence from this virus (and the other reporter viruses) was bright enough to allow imaging of dendritic spines (**Figure 2C**). Transgene expression was visible from 4 days post-injection, reached maximal intensity by 7–10 days post-injection and remained stable for at least 9–12 weeks (the longest time points we have analyzed for AAV and lentivirus respectively, data not shown).

ISOLATION OF NEURONS TRANSDUCED WITH VIRUSES *IN VIVO* FOR DOWNSTREAM ANALYSIS OF GENE EXPRESSION

Expression of virally encoded fluorescent proteins in addition to interfering RNAs or transgenes enables identification of transduced neurons and in principle might allow their isolation from non-transduced neurons. We therefore assessed the amenability to FACS of neurons transduced using viruses generated from the toolbox (**Figure 3**).

We stereotaxically injected CAMKII(0.4)-Kir2.1-mCherry into the dentate gyrus of mice aged 4–6 weeks (**Figure 3B**). Two weeks later the dentate gyrus was manually dissected out and dissociated (see Methods and **Figure 3A**). The cells were stained with a cell death marker and sorted by FACS to isolate a live Kir2.1-mCherry expressing population. We were able to harvest up to 14,624 positive cells from a single brain. RNA was extracted from these cells and screened by RT-PCR for markers of different hippocampal cell types to examine the purity of the sorted population. Samples of RNA from cells dissociated from either the entire dorsal hippocampus or just the dentate gyrus dissected from uninjected brains were used as controls.

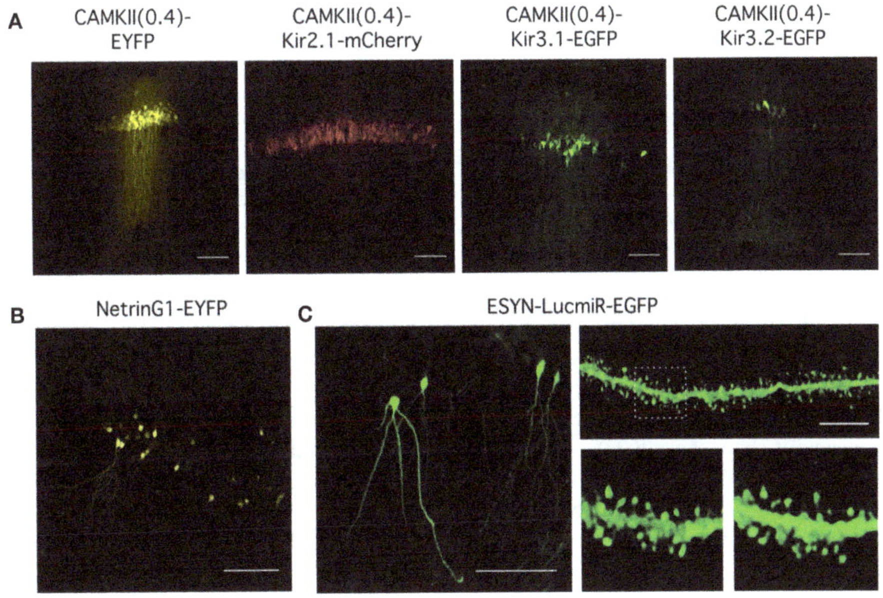

FIGURE 2 | Lentiviral expression of transgenes in hippocampal neurons. Examples of gene expression from lentiviruses injected into the hippocampus. **(A)** Expression of EYFP or Kir-reporter fusion proteins in CA1 pyramidal neurons of the hippocampus driven by the *CAMKIIα* short promoter. **(B)** Expression of EYFP in dentate gyrus granule cells driven by the Netrin G1 promoter. **(C)** EGFP expression in dentate gyrus granule cells driven by the ESYN promoter. Reporter protein expression is bright enough to enable imaging of dendritic spines (upper and lower right panels). Higher magnification spine images (lower right panel) correspond to different depths from a Z-stack of the region indicated by the white box (upper right panel). Scale bars = 100 μm for all images except the spine image, where the scale bar = 5 μm.

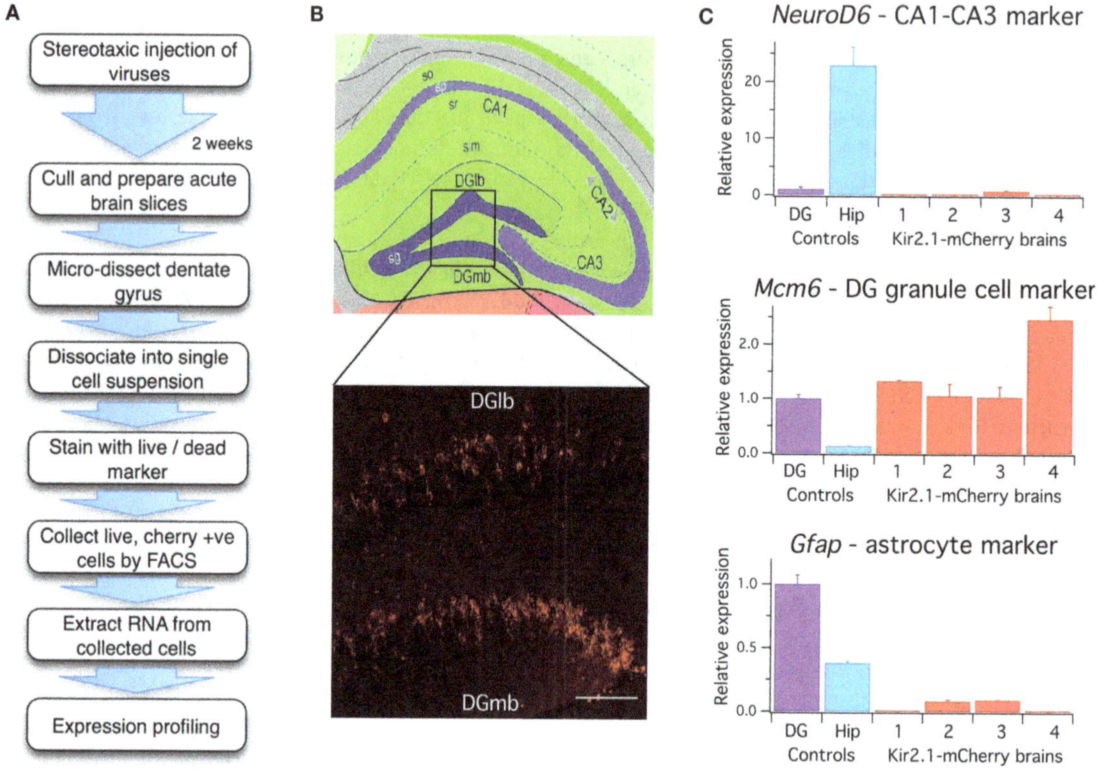

FIGURE 3 | Analysis of mRNA expression from dentate gyrus granule cells transduced *in vivo*. (A) Flow diagram of procedure for *ex vivo* isolation of virally transduced cells from the dentate gyrus. **(B)** *Upper panel*: Schematic of the hippocampus adapted from the Allen Mouse Brain Reference Atlas. Virus was injected into the dentate gyrus. *Lower panel*: Expression of Kir2.1-mCherry in granule cells (sg) of the dentate gyrus lateral blade (DGlb) and medial blade (DGmb). Scale bar = 100 μm. **(C)** RT-PCR quantification of gene expression in Kir2.1-mCherry positive cell populations isolated from four lentivirally injected brains compared to control cells dissociated from the entire hippocampus (Hip) or microdissected dentate gyrus (DG) of pooled uninjected brains ($n = 3$). All reactions were performed in triplicate and normalized to the housekeeping gene *Hprt1*. Gene expression levels in the Kir2.1-mCherry positive cells are expressed relative to control cells isolated from the dentate gyrus of uninjected brains. A one-way ANOVA revealed a significant difference in gene expression between the brain samples for *Neurod6* [$F(5,9) = 50.713$, $p = 3 \times 10^{-6}$], *Mcm6* [$F(5,9) = 8.204$, $p = 0.0036$], and *Gfap* [$F(5,9) = 154.442$, $p = 5 \times 10^{-9}$], *post hoc* analysis showed the dissected dentate gyrus and the mCherry positive cells populations isolated from Kir2.1-mCherry injected brains had significantly lower *Neurod6* expression than the entire hippocampus ($p < 0.05$ for DG, $p < 0.005$ for Kir2.1-mCherry brains, unpaired Student's *t*-test). Expression of the granule cell marker *Mcm6* was significantly higher in the Kir2.1-mCherry positive samples than in the entire hippocampus ($p < 0.05$, unpaired Student's *t*-test). *Gfap* expression was significantly higher in the dentate gyrus than the entire hippocampus ($p < 0.05$, unpaired Student's *t*-test), but the Kir2.1-mCherry positive samples expressed significantly less than both the dentate gyrus and the entire hippocampus ($p < 0.005$, unpaired Student's *t*-test). so, stratum oriens; sp, stratum pyramidale; sr, *stratum radiatum*; slm, stratum lacunosum-moleculare.

To evaluate the accuracy of cell isolation and the specificity of neuronal transduction we used RT-PCR of mRNA isolated from sorted cells to compare expression of anatomical and cell type markers. Expression of *Neurod6*, which is a marker for the CA1 and CA3 regions of the hippocampus (Lein et al., 2004), was over 20-fold higher in the entire hippocampus compared with the dissected dentate gyrus. *Neurod6* was undetectable in three out of four samples of sorted Kir2.1-mCherry positive cells and the remaining sample had extremely low expression of *Neurod6* mRNA that was significantly less than the entire hippocampus (**Figure 3C**). This indicates minimal contamination of the samples with cells from CA1 or CA3. Expression of *Mcm6*, which is a marker for dentate gyrus granule cells (Lein et al., 2004), was over seven-fold higher in the dentate gyrus compared with the entire hippocampus. All four Kir2.1-mCherry positive samples expressed *Mcm6* at a level equivalent to or higher than the dissected dentate gyrus and significantly higher than the entire hippocampus, suggesting they were enriched for dentate gyrus granule cells (**Figure 3C**). The dentate gyrus shows significantly increased mRNA expression of the glial marker *Gfap* compared to the entire hippocampus, possibly due to high levels of astrocytes in addition to *Gfap*-expressing neural progenitors in the subgranular zone (Liu et al., 2010). *Gfap* mRNA was only detectable in two of the four Kir2.1-mCherry positive samples and in both cases was at levels significantly less than both the dissected dentate gyrus and the entire hippocampus. The CAMKII(0.4) promoter should preclude expression of this construct in astrocytes but it may express in a small percentage of neural progenitors potentially explaining the low level of *Gfap* positivity (van Hooijdonk et al., 2009). Together, the RT-PCR results demonstrate that specific neuronal subpopulations can be transduced with viruses *in vivo* and then subsequently isolated for downstream analysis of gene expression.

REDUCTION OF NEURONAL GENE EXPRESSION BY VIRAL EXPRESSION OF INTRONIC miRNAs

Whilst our previous constructs altered neuronal gene expression through over-expression of transgenes, it is also useful to be able to reduce expression of endogenous genes. This can be achieved through the use of RNAi. However, early approaches for RNAi that used RNA Pol-III promoters to express high levels of short hairpin RNAs (shRNA) can be toxic *in vivo* (McBride et al., 2008; Ehlert et al., 2010). More recently, Pol-II promoters have been used to express artificial miRNA sequences. This reduces the off-target effects (McBride et al., 2008) but can result in poor expression levels of cocistronic reporter genes (Du et al., 2006). To address this issue, vectors were developed in which the miRNA is contained within a synthetic intron upstream of the reporter gene (Du et al., 2006). This allows for efficient expression of both the reporter and the miRNA from a Pol-II promoter (**Figure 4A**). We speculated that viruses expressing artificial intronic miRNAs and reporter genes would have improved specificity of gene knockdown in neuronal populations *in vivo*.

To test this possibility, we constructed AAVs targeting expression of the HCN1 ion channel and evaluated their efficacy in cerebellar Purkinje neurons. Oligonucleotide sequences designed against *HCN1* were cloned between the miRNA arms in the artificial intron of the pSM155 plasmid vector as previously described

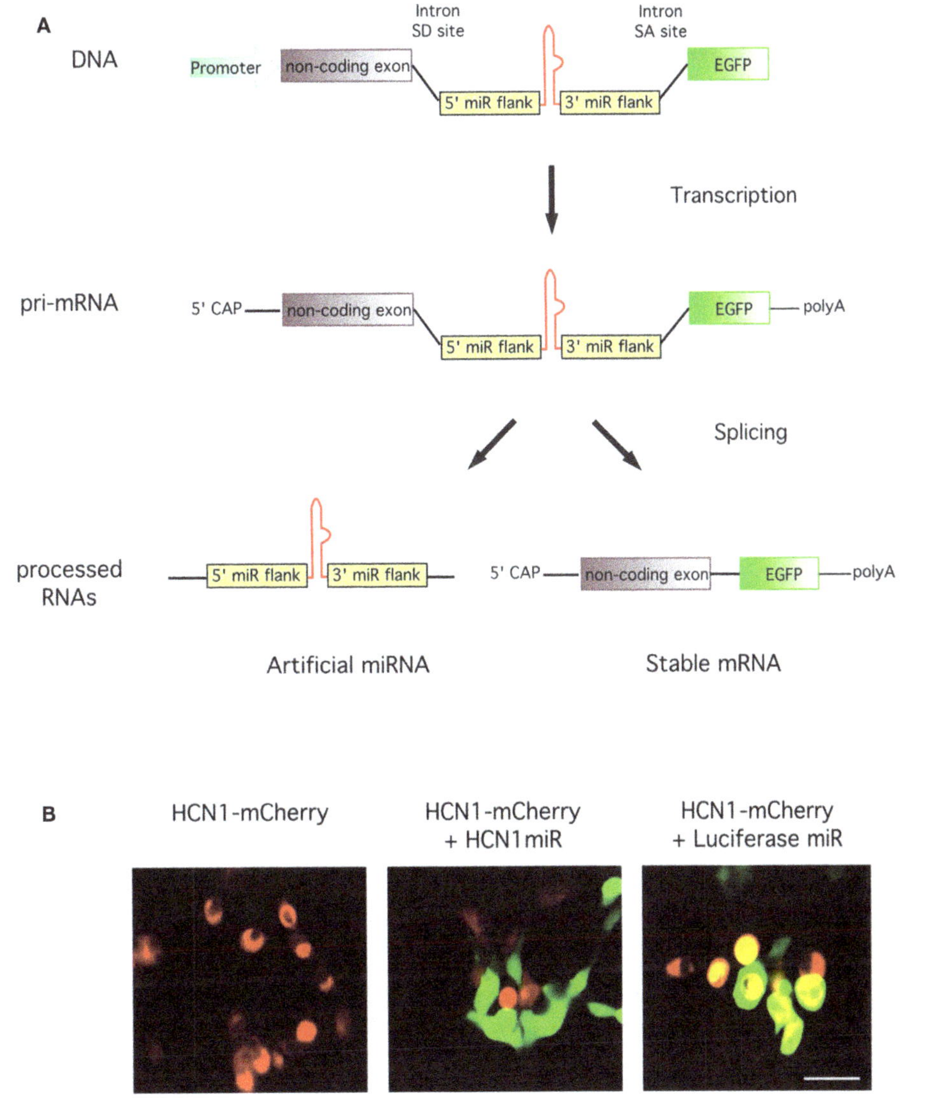

FIGURE 4 | Knockdown of a target gene and high co-expression of reporter gene following expression of constructs containing intronic artificial miRNAs. (A) Schematic illustrating processing of an artificial intronic miRNA construct. DNA transcription produces a pri-mRNA with an intronic miRNA. Cellular splicing excises the miRNA component ready for downstream processing, leaving a stable mRNA with a 5′CAP and polyA tail for the expression of EGFP. **(B)** *In vitro* testing of an intronic miRNA construct targeting expression of *HCN1*. HEK293FT cells were transfected with plasmids expressing HCN1-mCherry (red) alone or as a cotransfection with a vector encoding a miRNA targeting either *HCN1* (pSM155-HCN1miR) or luciferase (pSM155-LucmiR; green). Transfection with the HCN1 miRNA reduced expression of HCN1-mCherry whereas the luciferase miRNA did not. Scale bar = 50 μm.

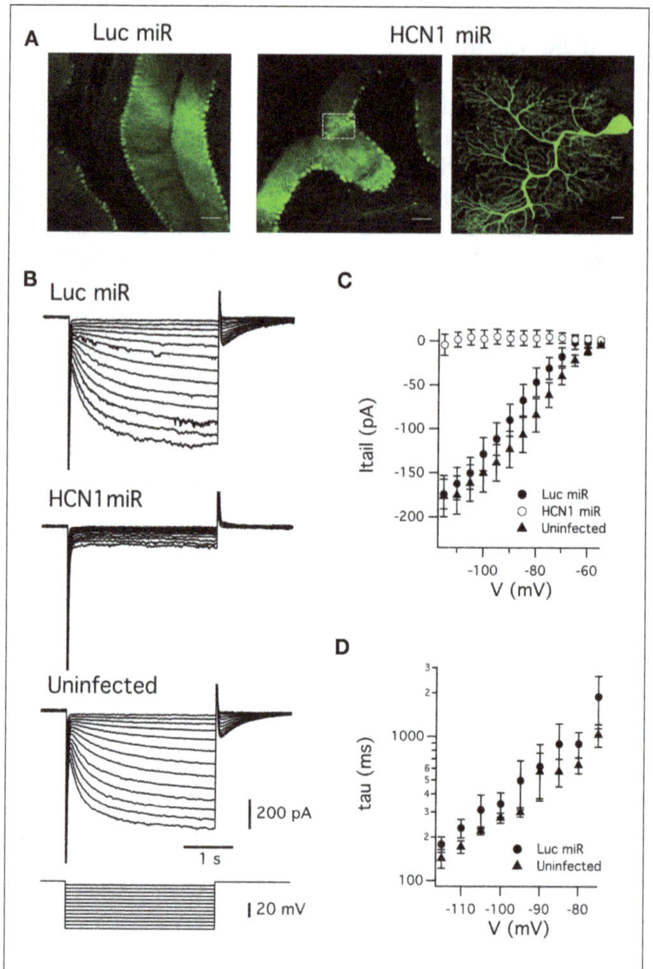

FIGURE 5 | Expression of miRNAs targeted against HCN1 abolishes hyperpolarization-activated currents recorded from cerebellar Purkinje cells. (A) Native fluorescence of EGFP from cerebellar Purkinje cells transduced with AAVs expressing miRNAs targeting either luciferase (Luc miRNA) or *HCN1* (HCN1 miR). The higher magnification image (right panel) corresponds to an HCN1miR expressing cell indicated by the white box (middle panel). Scale bars = 100 μm in left and middle panels, 10 μm in right panel. **(B)** Membrane current responses (upper three panels) to hyperpolarizing voltage steps (lower panel) recorded from a Purkinje cell infected with AAV expressing Luc miR (top), HCN1 miR (upper middle) or uninfected (lower middle). **(C)** Mean I_h tail current amplitude plotted as a function of test potential for each group of Purkinje cells. Tail current amplitudes differed between groups at membrane potentials ≤ -65 mV ($p < 0.05$, ANOVA). There was no significant difference between tail currents from Luc miR infected and uninfected neurons at any test potential ($p > 0.05$, unpaired Student's *t*-test), whereas tail currents from cells infected with HCN1 miR differed from the Luc miR group at potentials ≤ -75 mV and from the uninfected group at all test potentials ($p < 0.05$, unpaired Student's *t*-test). **(D)** Activation time constant plotted as a function of test potential for I_h recorded from uninfected Purkinje cells and from Purkinje cells infected with Luc miR. The time constant did not differ significantly between uninfected and Luc miR infected Purkinje cells at any test potential ($p > 0.05$, unpaired Student's *t*-test). Group sizes for voltage-clamp experiments were as follows: Luc miR ($n = 6$ cells, 3 mice); HCN1 miR ($n = 6$ cells, 3 mice); uninfected ($n = 6$ cells, 4 mice).

(Du et al., 2006). To check their effectiveness, the HCN1 miRNA vectors were cotransfected with an HCN1-mCherry plasmid into HEK293FT cells. The most effective HCN1 miRNA (Mmi510953) successfully reduced expression of HCN1-mCherry to almost undetectable levels whereas a negative control miRNA targeting luciferase (Luc miR) had no effect (**Figure 4B**). We therefore generated AAV vectors in which a cassette encoding the intronic HCN1 miRNA and EGFP reporter was inserted downstream of the ESYN promoter in a modified AAV backbone (AAV-Gateway). A negative control vector encoding Luc miR-EGFP was also constructed and viruses produced.

Injection of the AAVs into the cerebellum demonstrated widespread, high level expression of the EGFP reporter in Purkinje neurons within 2 weeks (**Figure 5A**). We evaluated the efficacy of the knockdown approach by measuring the hyperpolarization-activated current (I_h) recorded from cerebellar Purkinje cells in brain slices. Previous experiments with knockout mice suggest that HCN1 channels are the major ion channel type responsible for I_h recorded from cerebellar Purkinje cells (Nolan et al., 2003). Consistent with this idea, and with the efficacy of the artificial miRNA approach, I_h was abolished in Purkinje cells transduced with AAV expressing the HCN1 miRNA (**Figure 5B-C**). In contrast, Purkinje cells transduced with AAV expressing the miRNA targeted against luciferase had I_h with amplitude and kinetics comparable to non-transduced neurons (**Figure 5C-D**). Thus, efficacious suppression of HCN1 function by the viral expression of HCN1 miRNA is not due to non-specific effects of viral transduction or expression of miRNAs.

To further address the specificity of the miRNA approach, we took advantage of the ability of electrophysiological techniques to quantify key physiological properties of neurons. We compared measures of spontaneous action potential firing, action potential waveform, and basic integrative properties, in control and HCN1 miRNA transduced neurons from mice with global deletion of *HCN1* and wild-type littermate controls (**Figures 6 and 7**). We reasoned that if knockdown with HCN1 miRNA is specific, then the physiological properties of Purkinje cells transduced with AAV expressing HCN1 miRNA should be similar to Purkinje cells from mice with global deletion of *HCN1*. Differences between physiological properties due to viral transduction or off-target effects of the HCN1 miRNA should be apparent by a similar effect of transduction with the HCN1 miRNA AAV on Purkinje cells from wild-type mice and mice with global deletion of *HCN1*.

We first examined spontaneous action potentials generated by cerebellar Purkinje cells. In agreement with previous studies of global *HCN1* knockout mice (Nolan et al., 2003), we find that transduction with the HCN1 miRNA AAV does not affect the frequency of spontaneous action potentials recorded from Purkinje cells using either cell-attached or whole-cell configurations (**Figures 6A–C**). Properties of the action potential waveform were also similar in both uninfected and infected Purkinje cells from control mice and mice with global deletion of *HCN1* (**Figures 6A,D,E**). These data are consistent with the notion that HCN1 channels do not play a pacemaker role in cerebellar Purkinje cells (Raman and Bean, 1999; Nolan et al., 2003; Carter and Bean, 2011) and also indicate that transduction with AAV expressing intronic microRNAs and EGFP does not have off target effects on ion channels important for generation of action potentials.

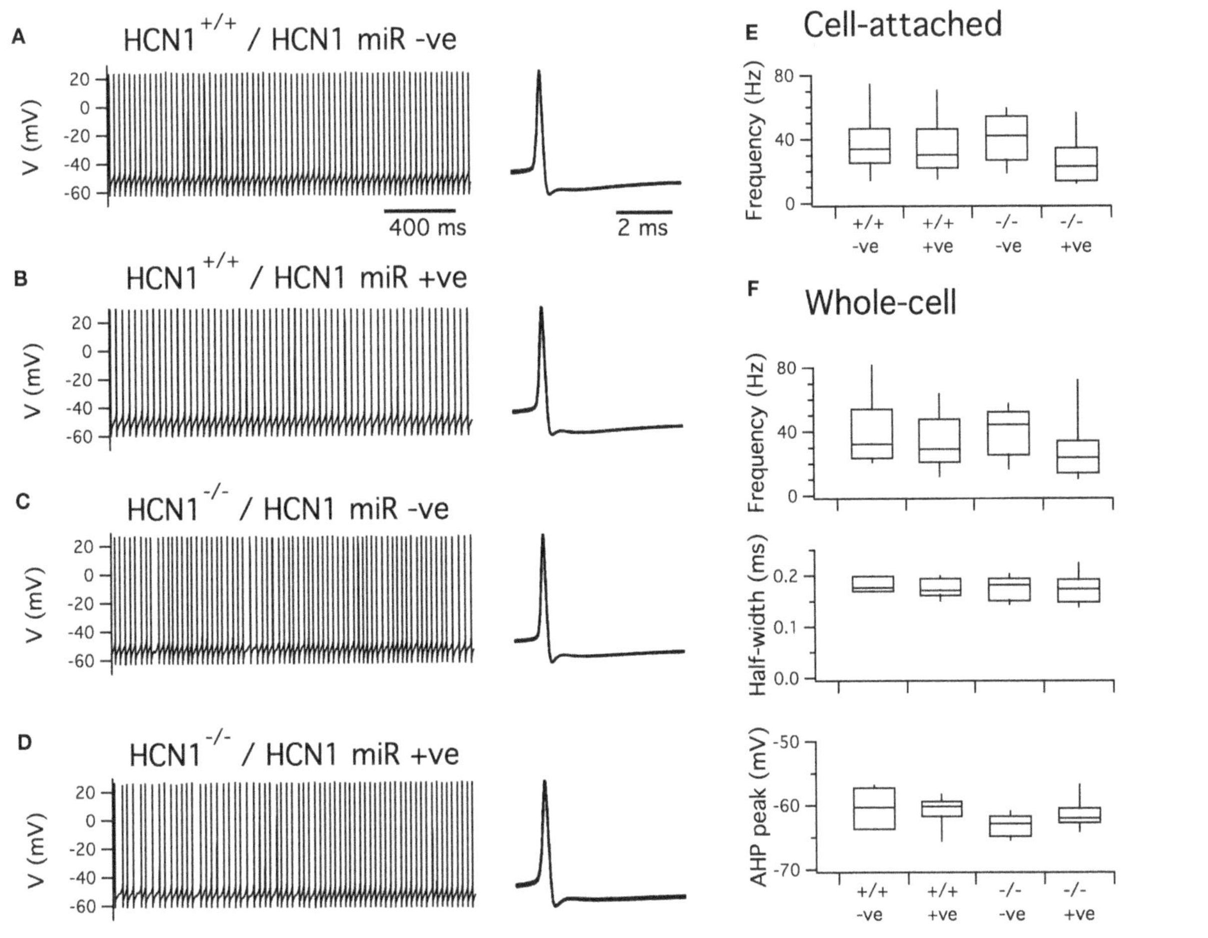

FIGURE 6 | Expression of HCN1 miR does not affect spontaneous firing of cerebellar Purkinje cells. (A–D). Examples of spontaneous action potentials recorded from cerebellar Purkinje cells in brain slices obtained from control mice **(A,B)** or global *HCN1* knockout mice **(C,D)**. Purkinje cells were either positive for EGFP indicating expression of HCN1 miRNA **(B,D)** or were EGFP negative indicating they did not express HCN1 miRNA **(A,C)**. **(E)** Frequency of spontaneous action potentials recorded from Purkinje cells in the cell-attached configuration. There was no significant effect of genotype ($p = 0.59$) or AAV infection ($p = 0.14$). **(F)** Frequency (upper panel), half-width (middle panel), and peak after-hyperpolarization (lower panel) for spontaneous action potentials recorded from Purkinje cells in the current-clamp configuration. There was no significant effect of AAV on any parameter ($p > 0.14$) and no significant effect of genotype on spike frequency or AHP duration. There was a significant effect of genotype on the AHP peak ($p = 0.034$). This effect is small and is inconsistent with previous comparisons suggesting it may simply reflect a chance outcome expected from repeated statistical testing. Statistical analysis uses ANOVA. Group sizes were as follows: HCN1+/+, HCN1 miR –ve ($n = 7$ cells, 3 mice); HCN1+/+, HCN1 miR+ve ($n = 9$ cells, 3 mice); HCN1+/+, HCN1 miR –ve ($n = 9$ cells, 3 mice); HCN1–/–, HCN1 miR+ve ($n = 9$ cells, 3 mice).

We next examined responses of Purkinje cells to injection of current steps though the recording electrode. In all experimental groups negative current steps switched off action potential firing. The relationship between action potential frequency and the amplitude of the current step was similar in Purkinje cells from wild-type and global *HCN1* knockout mice that were uninfected or that were infected with the HCN1 miRNA AAV (**Figures 7A–C**). In contrast, the relationship between the amplitude of current steps and the modal membrane potential was significantly modified by deletion of *HCN1* and also by knockdown of HCN1 following transduction with the HCN1 miRNA AAV (**Figures 7A,B,D**). Thus, whereas in Purkinje cells from wild-type mice the membrane potential remained close to spike threshold even during current steps that switched off spike firing, in Purkinje cells expressing HCN1 miRNA or from mice with deletion of *HCN1*, the membrane potential reached more hyperpolarized potentials. Importantly, there was no significant difference between any of the HCN1 knockdown or *HCN1* deletion groups in the relationship between current and membrane potential. These results are consistent with the idea that HCN1 channels are primarily activated in Purkinje cells by stimuli that inhibit spontaneous action potential firing and that they act to stabilize the membrane potential close to spike threshold (Nolan et al., 2003). They also indicate that the effects of AAV mediated expression of HCN1 miRNA are specific to HCN1 channels and do not involve changes in the properties of other membrane ion channels, or general pathological effects.

DISCUSSION

We have described a molecular toolbox of components to engineer lentiviral and AAV vectors for up- or down-regulation of neuronal

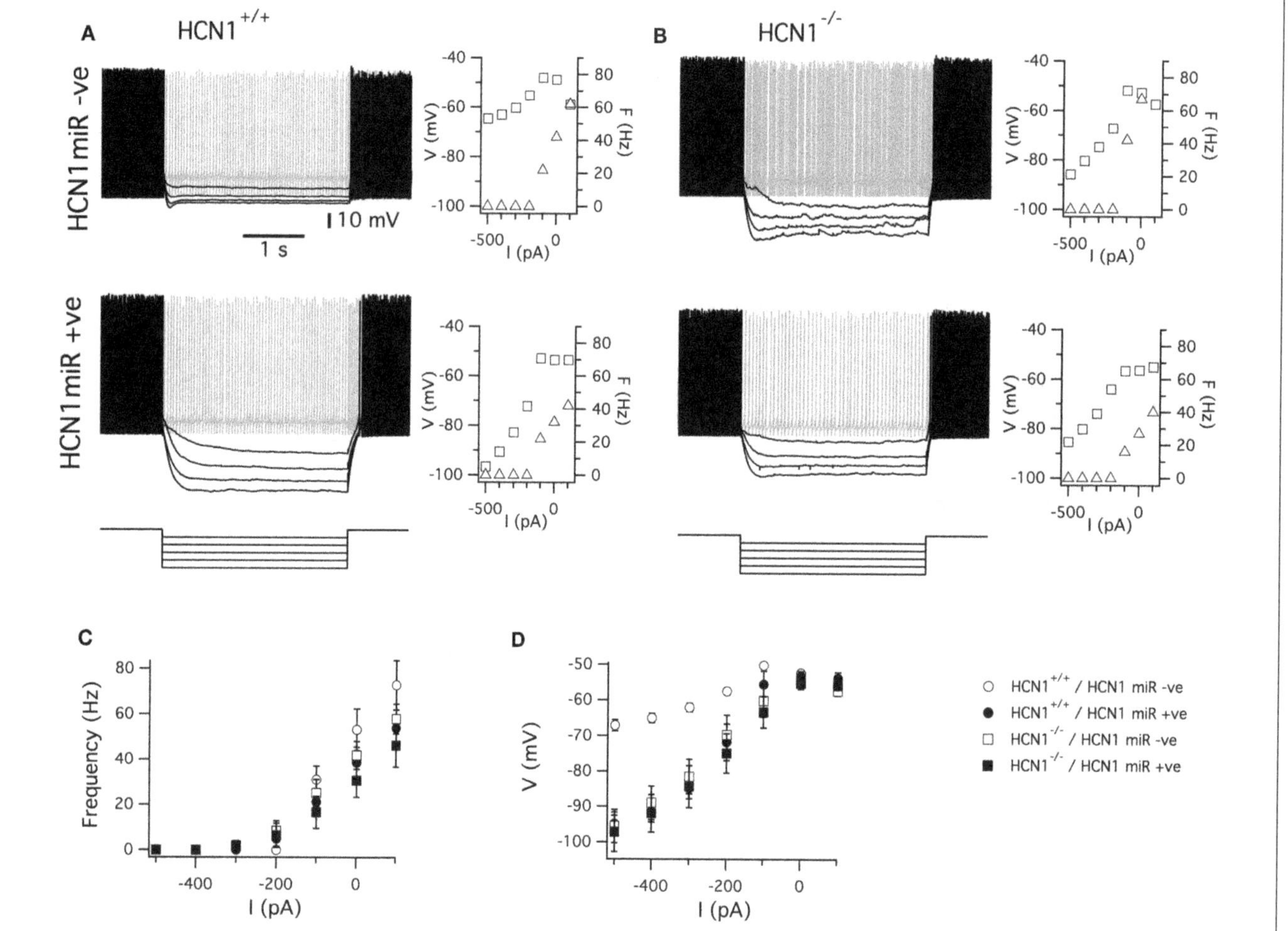

FIGURE 7 | HCN1 miR expression and global deletion of *HCN1* cause similar physiological phenotypes in cerebellar Purkinje cells. (A,B) Examples of responses of Purkinje cells from wild-type **(A)** and global *HCN1* knockout mice **(B)** to series of current steps. Upper traces are recordings from uninfected Purkinje cells. Lower traces are recordings from Purkinje cells infected with AAV HCN1 miRNA EGFP. Graphs plot spike frequency and modal membrane potential as a function of the amplitude of the test current step. **(C,D)** Plot of the average across all neurons in each group of the spike frequency **(C)** and modal membrane potential **(D)** as a function of the amplitude of the test current step. There was no significant effect ($p < 0.05$) of genotype or AAV expression on the spike frequency at any amplitude of current step. In contrast, the amplitude of the membrane potential change depended on genotype and AAV expression (for −500 pA current steps, $p = 0.0061$ for genotype, $p = 0.0044$ for AAV infection, and $p = 0.0050$ for interaction between AAV infection and genotype). Numbers of neurons and mice are as in **Figure 6**.

gene expression *in vivo*. The system is extremely flexible and allows rapid generation of viral constructs. Viruses produced from these vectors yield stable expression of transgenes at levels high enough to detect native fluorescence in neurons. This allows easy identification of virally transduced cells for downstream analysis of the effects of gene manipulation.

A MODULAR APPROACH TO VIRUS ASSEMBLY

Constructing viral vectors using traditional cloning methods is time consuming and inefficient as it requires restriction enzyme sites that must be compatible with insert sequences and the vector. Altering or rearranging elements within a vector often involves a new cloning strategy leading to unnecessary duplication of work. We have used a recombination cloning system to build a library of validated interchangeable components for the generation of viral vectors. New vectors can be quickly assembled using a single recombination reaction reducing both the time and expertise required. Entry vectors encoding the desired components must be generated initially, but they can then be utilized repeatedly in various combinations to create new expression constructs. By creating a Gateway-compatible AAV vector, the same entry vectors can now be used to produce both lentiviruses and AAV. Additional viral vectors, such as rabies or HSV could also be adapted to this system, further increasing the utility of the toolbox. In this way, the same components can be easily swapped between different viral systems, whilst minimizing the variability between them.

The toolbox could potentially be applied to diverse neuroscientific questions. Using different promoters we have expressed various reporter proteins, ion channels, and miRNA cassettes in neurons *in vivo*. The high level of native reporter fluorescence detectable from these viruses facilitates both imaging and electrophysiology-based techniques as well as *ex vivo* isolation of transduced cells. Using FACS to detect the reporter, we isolated relatively pure populations of transduced granule cells from the dentate gyrus. These cells could then be subjected to further *ex vivo* analysis of their DNA, RNA, or protein profiles to examine the consequences of the gene manipulation in more detail. This approach has the advantage of performing the manipulation in an *in vivo* context whilst allowing the use of a wide range of sensitive *in vitro* assays for the subsequent analysis of transduced cells. It is also possible to isolate many more transduced cells than is practical using other current techniques such as laser capture microscopy. It is important to note however, that the purity of the cell population isolated will depend on both the specificity of the transgene promoter and the accuracy of virus injection and tissue dissection.

In contrast to generation and maintenance of knockout and transgenic mice, which can be time consuming and expensive, the use of viruses to alter gene expression *in vivo* has several advantages. In particular, viruses can be rapidly produced and are easily targeted to well-defined brain regions at specific time points. However, there are potential disadvantages to consider when planning viral experiments. Transgene expression levels may vary unpredictably depending on how many viruses transduce each cell and where lentivirally expressed DNA integrates into the cell's genome. In addition, while careful injection of lentiviruses and AAV directly into the brain parenchyma results in little or no immune response (Chamberlin et al., 1998; Mazarakis et al., 2001; Abordo-Adesida et al., 2005; Lowenstein et al., 2007) any surgical procedure still has the potential to induce local inflammatory responses. This can be counteracted through the use of careful surgical techniques, viral production and purification methods (Scherr et al., 2002; Baekelandt et al., 2003) and injection of lower viral titers (Chamberlin et al., 1998; Abordo-Adesida et al., 2005; Lowenstein et al., 2007). Nevertheless, where appropriate residual effects must still be excluded through the use of experimental controls.

EFFICIENT AND SPECIFIC GENE KNOCKDOWN USING INTRONIC ARTIFICIAL microRNAs

Traditionally vectors for RNAi express shRNAs at high levels from strong Pol-III promoters. The shRNAs can result in toxicity by disrupting the biogenesis of natural miRNAs (Grimm et al., 2006), causing off-target silencing through accumulation of antisense RNAs (Birmingham et al., 2006) and stimulating innate immune responses to double stranded RNA (Samuel, 2001; Bridge et al., 2003; Fish and Kruithof, 2004). *In vivo*, these off-target effects can cause cell death and lethality (Grimm et al., 2006; Martin et al., 2011). Therefore, subsequent approaches have mimicked endogenous miRNAs by expressing artificial pri-miRNAs from Pol-II promoters. The pri-miRNAs are expressed at lower levels, but are more efficiently processed, minimizing their side effects (McBride et al., 2008). The use of Pol-II promoters also enables cocistronic expression of a reporter gene in a cell-specific and regulated manner. A weakness of this approach is that processing of the pri-miRNA can block translation of the cocistronic reporter gene as the resultant mRNA is unstable due to the lack of either a 5′ CAP or polyA tail (Du et al., 2006). Placing the miRNA within an artificial intron upstream of the reporter gene addresses this problem by allowing cellular splicing mechanisms to excise the miRNA whilst leaving an intact mRNA for co-expression of the reporter gene (Du et al., 2006). This approach has been used recently to generate a BAC transgenic mouse (Garbett et al., 2010), but viral delivery of artificial intronic miRNA to neurons has not been reported. Our results show that an AAV can be used to express an intronic miRNA targeting the HCN1 ion channel resulting in specific knockdown of I_h in cerebellar Purkinje neurons without apparent off-target effects.

The molecular toolbox we have developed allows for the rapid generation of lentiviral and AAV vectors. Nevertheless, it is still important to rigorously test each component in order to generate useful viruses. When expressed within a viral context, promoter sequences may not always yield the anticipated pattern of activity. Although our neuronal promoters gave the expected high levels of expression within principle neurons (Dittgen et al., 2004; Hioki et al., 2007; Nathanson et al., 2009a,b; van Hooijdonk et al., 2009), we also observed occasional expression in interneurons as has been reported previously (Nathanson et al., 2009a). Transgene expression driven by the *Netrin G1* promoter sequence did not recapitulate endogenous *Netrin G1* expression patterns and a *Netrin G2* promoter sequence failed to express *in vivo* despite high activity *in vitro* (data not shown). Ensuring viral transgene expression is restricted to very specific subpopulations of neurons may require additional strategies, such as using viral constructs incorporating FLEX switches in combination with Cre-expressing transgenic animals (Schnütgen et al., 2003; Atasoy et al., 2008). As new technologies for the control of gene expression become available, they can also be incorporated into the molecular toolbox. We anticipate these tools will be of widespread use for investigation of nervous system function and pathology.

ACKNOWLEDGMENTS

We thank Trudi Gillespi from the IMPACT Imaging facility at the University of Edinburgh for assistance with confocal microscopy and Bina Santoro, Hiroyuki Hioki, Patrick Martin, Axel Schambach, and Guangwei Du for generously providing plasmids. This work was funded by a Marie Curie Excellence Grant and grants from the BBSRC, the MRC, and NARSAD.

REFERENCES

Abordo-Adesida, E., Follenzi, A., Barcia, C., Sciascia, S., Castro, M. G., Naldini, L., and Lowenstein, P. R. (2005). Stability of lentiviral vector-mediated transgene expression in the brain in the presence of systemic antivector immune responses. *Hum. Gene Ther.* 16, 741–751.

Atasoy, D., Aponte, Y., Su, H. H., and Sternson, S. M. (2008). A FLEX switch targets Channelrhodopsin-2 to multiple cell types for imaging and long-range circuit mapping. *J. Neurosci.* 28, 7025–7030.

Baekelandt, V., Eggermont, K., Michiels, M., Nuttin, B., and Debyser, Z. (2003). Optimized lentiviral vector production and purification procedure prevents immune response after transduction of mouse brain. *Gene Ther.* 10, 1933–1940.

Birmingham, A., Anderson, E. M., Reynolds, A., Ilsley-Tyree, D., Leake, D., Fedorov, Y., Baskerville, S., Maksimova, E., Robinson, K., Karpilow, J., Marshall, W. S., and Khvorova, A. (2006). 3′ UTR seed matches, but not overall identity, are associated with RNAi off-targets. *Nat. Methods* 3, 199–204.

Bridge, A. J., Pebernard, S., Ducraux, A., Nicoulaz, A. L., and Iggo, R. (2003). Induction of an interferon response by RNAi vectors in mammalian cells. *Nat. Genet.* 34, 263–264.

Bubeck, P., Winkler, M., and Bautsch, W. (1993). Rapid cloning by homologous recombination *in vivo*. *Nucleic Acids Res.* 21, 3601–3602.

Burrone, J., O'Byrne, M., and Murthy, V. N. (2002). Multiple forms of synaptic plasticity triggered by selective suppression of activity in individual neurons. *Nature* 420, 414–418.

Carter, B. C., and Bean, B. P. (2011). Incomplete inactivation and rapid recovery of voltage-dependent sodium channels during high-frequency firing in cerebellar Purkinje neurons. *J. Neurophysiol.* 105, 860–871.

Cetin, A., Komai, S., Eliava, M., Seeburg, P. H., and Osten, P. (2006). Stereotaxic gene delivery in the rodent brain. *Nat. Protoc.* 1, 3166–3173.

Chamberlin, N. L., Du, B., de Lacalle, S., and Saper, C. B. (1998). Recombinant adeno-associated virus vector: use for transgene expression and anterograde tract tracing in the CNS. *Brain Res.* 793, 169–175.

Chen, X., Shu, S., and Bayliss, D. A. (2009). HCN1 channel subunits are a molecular substrate for hypnotic actions of ketamine. *J. Neurosci.* 29, 600–609.

Chen, X., Shu, S., Schwartz, L. C., Sun, C., Kapur, J., and Bayliss, D. A. (2010). Homeostatic regulation of synaptic excitability: tonic GABA(A) receptor currents replace *I*(h) in cortical pyramidal neurons of HCN1 knock-out mice. *J. Neurosci.* 30, 2611–2622.

Dittgen, T., Nimmerjahn, A., Komai, S., Licznerski, P., Waters, J., Margrie, T. W., Helmchen, F., Denk, W., Brecht, M., and Osten, P. (2004). Lentivirus-based genetic manipulations of cortical neurons and their optical and electrophysiological monitoring in vivo. *Proc. Natl. Acad. Sci. U.S.A.* 101, 18206–18211.

Du, G., Yonekubo, J., Zeng, Y., Osisami, M., and Frohman, M. A. (2006). Design of expression vectors for RNA interference based on miRNAs and RNA splicing. *FEBS J.* 273, 5421–5427.

Ehlert, E. M., Eggers, R., Niclou, S. P., and Verhaagen, J. (2010). Cellular toxicity following application of adeno-associated viral vector-mediated RNA interference in the nervous system. *BMC Neurosci.* 11, 20. doi: 10.1186/1471-2202-11-20

Fish, R. J., and Kruithof, E. K. (2004). Short-term cytotoxic effects and long-term instability of RNAi delivered using lentiviral vectors. *BMC Mol. Biol.* 5, 9. doi: 10.1186/1471-2199-5-9

Garbett, K. A., Horváth, S., Ebert, P. J., Schmidt, M. J., Lwin, K., Mitchell, A., Levitt, P., and Mirnics, K. (2010). Novel animal models for studying complex brain disorders: BAC-driven miRNA-mediated in vivo silencing of gene expression. *Mol. Psychiatry* 15, 987–995.

Grimm, D., Streetz, K. L., Jopling, C. L., Storm, T. A., Pandey, K., Davis, C. R., Marion, P., Salazar, F., and Kay, M. A. (2006). Fatality in mice due to oversaturation of cellular microRNA/short hairpin RNA pathways. *Nature* 441, 537–541.

Han, X., and Boyden, E. S. (2007). Multiple-color optical activation, silencing, and desynchronization of neural activity, with single-spike temporal resolution. *PLoS ONE* 2, e299. doi: 10.1371/journal.pone.0000299

Hartley, J. L., Temple, G. F., and Brasch, M. A. (2000). DNA cloning using in vitro site-specific recombination. *Genome Res.* 10, 1788–1795.

Hioki, H., Kameda, H., Nakamura, H., Okunomiya, T., Ohira, K., Nakamura, K., Kuroda, M., Furuta, T., and Kaneko, T. (2007). Efficient gene transduction of neurons by lentivirus with enhanced neuron-specific promoters. *Gene Ther.* 14, 872–882.

Huang, Z., Walker, M. C., and Shah, M. M. (2009). Loss of dendritic HCN1 subunits enhances cortical excitability and epileptogenesis. *J. Neurosci.* 29, 10979–10988.

Koyrakh, L., Luján, R., Colón, J., Karschin, C., Kurachi, Y., Karschin, A., and Wickman, K. (2005). Molecular and cellular diversity of neuronal G-protein-gated potassium channels. *J. Neurosci.* 25, 11468–11478.

Lein, E. S., Zhao, X., and Gage, F. H. (2004). Defining a molecular atlas of the hippocampus using DNA microarrays and high-throughput in situ hybridization. *J. Neurosci.* 24, 3879–3889.

Liu, Y., Namba, T., Liu, J., Suzuki, R., Shioda, S., and Seki, T. (2010). Glial fibrillary acidic protein-expressing neural progenitors give rise to immature neurons via early intermediate progenitors expressing both glial fibrillary acidic protein and neuronal markers in the adult hippocampus. *Neuroscience* 166, 241–251.

Lowenstein, P. R., Mandel, R. J., Xiong, W. D., Kroeger, K., and Castro, M. G. (2007). Immune responses to adenovirus and adeno-associated vectors used for gene therapy of brain diseases: the role of immunological synapses in understanding the cell biology of neuroimmune interactions. *Curr. Gene Ther.* 7, 347–360.

Ma, H., Kunes, S., Schatz, P. J., and Botstein, D. (1987). Plasmid construction by homologous recombination in yeast. *Gene* 58, 201–216.

Martin, J. N., Wolken, N., Brown, T., Dauer, W. T., Ehrlich, M. E., and Gonzalez-Alegre, P. (2011). Lethal toxicity caused by expression of shRNA in the mouse striatum: implications for therapeutic design. *Gene Ther.* doi: 10.1038/gt.2011.10. [Epub ahead of print].

Mazarakis, N. D., Azzouz, M., Rohll, J. B., Ellard, F. M., Wilkes, F. J., Olsen, A. L., Carter, E. E., Barber, R. D., Baban, D. F., Kingsman, S. M., Kingsman, A. J., O'Malley, K., and Mitrophanous, K. A. (2001). Rabies virus glycoprotein pseudotyping of lentiviral vectors enables retrograde axonal transport and access to the nervous system after peripheral delivery. *Hum. Mol. Genet.* 10, 2109–2121.

McBride, J. L., Boudreau, R. L., Harper, S. Q., Staber, P. D., Monteys, A. M., Martins, I., Gilmore, B. L., Burstein, H., Peluso, R. W., Polisky, B., Carter, B. J., and Davidson, B. L. (2008). Artificial miRNAs mitigate shRNA-mediated toxicity in the brain: implications for the therapeutic development of RNAi. *Proc. Natl. Acad. Sci. U.S.A.* 105, 5868–5873.

Miyashita, T., and Kubo, Y. (1997). Localization and developmental changes of the expression of two inward rectifying K(+)-channel proteins in the rat brain. *Brain Res.* 750, 251–263.

Nathanson, J. L., Jappelli, R., Scheeff, E. D., Manning, G., Obata, K., Brenner, S., and Callaway, E. M. (2009a). Short promoters in viral vectors drive selective expression in mammalian inhibitory neurons, but do not restrict activity to specific inhibitory cell-types. *Front. Neural Circuits* 3:19. doi: 10.3389/neuro.04.019.2009

Nathanson, J. L., Yanagawa, Y., Obata, K., and Callaway, E. M. (2009b). Preferential labeling of inhibitory and excitatory cortical neurons by endogenous tropism of adeno-associated virus and lentivirus vectors. *Neuroscience* 161, 441–450.

Nolan, M. F., Dudman, J. T., Dodson, P. D. and Santoro, B. (2007). 'HCN1 channels control resting and active integrative properties of stellate cells from layer II of the entorhinal cortex.', *J. Neurosci.*, 27(46), 12440–12451.

Nolan, M. F., Malleret, G., Dudman, J. T., Buhl, D. L., Santoro, B., Gibbs, E., Vronskaya, S., Buzsáki, G., Siegelbaum, S. A., Kandel, E. R., and Morozov, A. (2004). A behavioral role for dendritic integration: HCN1 channels constrain spatial memory and plasticity at inputs to distal dendrites of CA1 pyramidal neurons. *Cell* 119, 719–732.

Nolan, M. F., Malleret, G., Lee, K. H., Gibbs, E., Dudman, J. T., Santoro, B., Yin, D., Thompson, R. F., Siegelbaum, S. A., Kandel, E. R., and Morozov, A. (2003). The hyperpolarization-activated HCN1 channel is important for motor learning and neuronal integration by cerebellar Purkinje cells. *Cell* 115, 551–564.

Oliner, J. D., Kinzler, K. W., and Vogelstein, B. (1993). *In vivo* cloning of PCR products in *E. coli*. *Nucleic Acids Res.* 21, 5192–5197.

Petreanu, L., Mao, T., Sternson, S. M., and Svoboda, K. (2009). The subcellular organization of neocortical excitatory connections. *Nature* 457, 1142–1145.

Piskorowski, R., Santoro, B., and Siegelbaum, S. A. (2011). TRIP8b splice forms act in concert to regulate the localization and expression of HCN1 channels in CA1 pyramidal neurons. *Neuron* 70, 495–509.

Raman, I. M., and Bean, B. P. (1999). Ionic currents underlying spontaneous action potentials in isolated cerebellar Purkinje neurons. *J. Neurosci.* 19, 1663–1674.

Samuel, C. E. (2001). Antiviral actions of interferons. *Clin. Microbiol. Rev.* 14, 778–809; table of contents.

Sasaki, Y., Sone, T., Yoshida, S., Yahata, K., Hotta, J., Chesnut, J. D., Honda, T., and Imamoto, F. (2004). Evidence for high specificity and efficiency of multiple recombination signals in mixed DNA cloning by the Multisite Gateway system. *J. Biotechnol.* 107, 233–243.

Schambach, A., Bohne, J., Baum, C., Hermann, F. G., Egerer, L., von Laer, D., and Giroglou, T. (2006). Woodchuck hepatitis virus post-transcriptional regulatory element deleted from X protein and promoter sequences enhances retroviral vector titer and expression. *Gene Ther.* 13, 641–645.

Scherr, M., Battmer, K., Eder, M., Schüle, S., Hohenberg, H., Ganser, A., Grez, M., and Blömer, U. (2002). Efficient gene transfer into the CNS by lentiviral vectors purified by anion exchange chromatography. *Gene Ther.* 9, 1708–1714.

Schnütgen, F., Doerflinger, N., Calléja, C., Wendling, O., Chambon, P., and Ghyselinck, N. B. (2003). A directional strategy for monitoring

Cre-mediated recombination at the cellular level in the mouse. *Nat. Biotechnol.* 21, 562–565.

Stevens, E. B., Woodward, R., Ho, I. H., and Murrell-Lagnado, R. (1997). Identification of regions that regulate the expression and activity of G protein-gated inward rectifier K+ channels in *Xenopus* oocytes. *J. Physiol.* 503(Pt 3), 547–562.

van Hooijdonk, L. W., Ichwan, M., Dijkmans, T. F., Schouten, T. G., de Backer, M. W., Adan, R. A., Verbeek, F. J., Vreugdenhil, E., and Fitzsimons, C. P. (2009). Lentivirus-mediated transgene delivery to the hippocampus reveals sub-field specific differences in expression. *BMC Neurosci.* 10, 2. doi: 10.1186/1471-2202-10-2

Yahata, K., Kishine, H., Sone, T., Sasaki, Y., Hotta, J., Chesnut, J. D., Okabe, M., and Imamoto, F. (2005). Multi-gene gateway clone design for expression of multiple heterologous genes in living cells: conditional gene expression at near physiological levels. *J. Biotechnol.* 118, 123–134.

Zonta, B., Desmazieres, A., Rinaldi, A., Tait, S., Sherman, D. L., Nolan, M. F., and Brophy, P. J. (2011). A critical role for neurofascin in regulating action potential initiation through maintenance of the axon initial segment. *Neuron* 69, 945–956.

23

microRNA-431 regulates axon regeneration in mature sensory neurons by targeting the Wnt antagonist *Kremen1*

***Di Wu*[1] *and Alexander K. Murashov*[2] ***

[1] *Department of Neurobiology and Anatomy, Drexel University College of Medicine, Philadelphia, PA, USA*
[2] *The Harriet and John Wooten Laboratory for Alzheimer's and Neurodegenerative Diseases Research, Department of Physiology, School of Medicine, East Carolina University, Greenville, NC, USA*

Correspondence:
Alexander K. Murashov, The Harriet and John Wooten Laboratory for Alzheimer's and Neurodegenerative Diseases Research, Department of Physiology, School of Medicine, East Carolina University, Brody Building No. 6N-98, 600 Moye Boulevard, Greenville, NC 27834, USA
e-mail: murashoval@ecu.edu

MicroRNAs (miRNAs) are small, non-coding RNAs that function as key post-transcriptional regulators in neural development, brain function, and neurological diseases. Growing evidence indicates that miRNAs are also important mediators of nerve regeneration, however, the affected signaling mechanisms are not clearly understood. In the present study, we show that nerve injury-induced miR-431 stimulates regenerative axon growth by silencing *Kremen1*, an antagonist of Wnt/beta-catenin signaling. Both the gain-of-function of miR-431 and knockdown of *Kremen1* significantly enhance axon outgrowth in murine dorsal root ganglion neuronal cultures. Using cross-linking with AGO-2 immunoprecipitation, and 3′-untranslated region (UTR) luciferase reporter assay we demonstrate miR-431 direct interaction on the 3′-UTR of *Kremen1* mRNA. Together, our results identify miR-431 as an important regulator of axonal regeneration and a promising therapeutic target.

Keywords: miRNA, axon, regeneration, Wnt, *Kremen1*, miR-431, sensory neurons

INTRODUCTION

Axon loss is the hallmark of traumatic brain and spinal cord injury (SCI) as well as many neurodegenerative diseases including Alzheimer's (Coleman and Perry, 2002). A body of research is focused on understanding the mechanisms of axon degeneration and promoting axon regeneration, however, the molecular mechanisms of neural repair remain poorly understood (Fang and Bonini, 2012). Growing evidence indicates that microRNA (miRNA) pathway controls regulatory mechanism involved in neural repair and regeneration (Strickland et al., 2011; Wu et al., 2011, 2012; Yu et al., 2011a; Zhang et al., 2011; Zhou et al., 2012). miRNAs are short, non-coding RNAs that silence gene expression by imperfect binding to 3′-untranslated region (UTR) of mRNA (Bartel, 2004; Filipowicz et al., 2008). miRNAs ability to simultaneously regulate the expression of several genes suggests that they are critical regulators of complex transcriptional networks (McNeill and Van Vactor, 2012). In the nervous system, miRNAs have been implicated in neurodevelopment (Smith et al., 2010), neurogenesis (Shi et al., 2010), and neurological disorders (Hebert and De Strooper, 2007; Kim et al., 2007). Recent observations have identified a group of miRNAs which reside within the distal axonal domain of superior cervical ganglia neuron suggesting miRNA role in the maintenance of axonal structure and function (Natera-Naranjo et al., 2010). In addition, several miRNAs have been associated with axon regeneration in peripheral nervous system (PNS) neurons (Strickland et al., 2011; Yu et al., 2011a; Zhang et al., 2011; Zhou et al., 2012) and axon development in cortical neurons (Dajas-Bailador et al., 2012).

Recent studies from our laboratory have demonstrated that ablation of Dicer, a key enzyme required for miRNA biogenesis, markedly impairs the regenerative axon growth *in vivo* and *in vitro*, indicating that the intact Dicer-dependent miRNA pathway is critical for successful peripheral nerve regeneration (Wu et al., 2012). In the current study, we examine the mechanism of miRNA action in axon regeneration. Here we show that injury-induced miR-431 stimulates regenerative axon growth by silencing *Kremen1*, a negative regulator of Wnt/beta-catenin signaling pathway. Both the gain-of-function of miR-431 and loss-of-function of *Kremen1* significantly enhance regenerative axon growth in dissociated dorsal root ganglia (DRG) neuronal cultures. Using cross-linking with AGO-2 immunoprecipitation (CLIP), and 3′-UTR luciferase assay we demonstrate miR-431 direct interaction on the 3′-UTR of *Kremen1* mRNA. Collectively, our observations provide the first evidence for a role of miRNA in regulating Wnt/beta-catenin signaling pathway in nerve regeneration and identify miR-431 as an important regulator and a potential therapeutic target.

MATERIALS AND METHODS

ANIMALS

Eight-week-old CD-1 male mice were obtained from Charles River laboratories (Wilmington, MA, USA). The animal use protocol was approved by the institutional Animal Care and Use Committee of East Carolina University, an Association for Assessment and Accreditation of Laboratory Animal Care-accredited facility. Animals were housed individually under standard laboratory conditions, with a 12 h light/dark schedule and unlimited access to food and water.

CONDITIONING NERVE LESION

Before surgery, anesthesia was induced using an intraperitoneal ketamine (18 mg/ml)-xylazine (2 mg/ml) mixture (0.05 ml/10 g of body weight). The procedure followed a protocol described previously (Islamov et al., 2004). Exposure of the right sciatic nerve was performed with sterile surgical instruments through an incision on the middle thigh of the right hind limb. Approximately 5 mm of nerve was exposed from the sciatic notch to the trifurcation of the nerve. The exposed sciatic nerve was crushed in the mid-thigh for 15 s with a fine hemostat. The wounds were closed with 3M™ Vetbond™ Tissue Adhesive (3M, Saint Paul, MN, USA) and the animals were left to recover for 5 days.

DISSOCIATED DRG CULTURES

Mouse L4/5 DRG neurons were collected 5 days after a conditioning sciatic nerve crush from both the intact side and injured side. DRGs were dissociated with collagenase and 0.25% trypsin in Dulbecco's modified Eagle's medium (DMEM; Invitrogen, Carlsbad, CA, USA). The dissociated DRGs were plated on poly-L-lysine and laminin (Invitrogen), coated plates. DRGs were grown in DMEM/F12 containing 10% horse serum, L-glutamine, and N2 supplement (Gemini Bio-product, West Sacramento, CA, USA) at 37°C for 18 h. To inhibit glial cell growth cytosin β-D-arabinofuranoside (Arac, 10 μM) and 5,6-dichlorobenzimidazole riboside (DRB, 80 μM; Sigma, Saint Louis, MO, USA) or 50 nM 5-fluoro-2′-deoxyuridine (Sigma) were added to the growth medium.

PC12 CELL CULTURES

PC12 cells were cultured in DMEM containing 10% horse serum, 5% fetal bovine serum 2 mM glutamine, and penicillin and streptomycin (100 unitl/ml). The cells were plated on collagen-coated cell culture dishes. For nerve growth factor (NGF)-induced differentiation of PC12 cells, NGF (50 ng/ml) was added to cell culture medium to initiate neurite outgrowth. Medium was refreshed every 2–3 days.

TRANSFECTION OF miRNA MIMICS AND INHIBITORS

In order to determine the biological effects of each individual miRNA on regenerative axon growth, we performed functional analyses for injury-induced miRNAs. Gain-of-function experiments were performed with Ambion® Pre-miR™ miRNA Precursor Molecules (Ambion, Austin, TX, USA), which are also called miRNA mimics. With transfection reagent, these small, chemically modified double-stranded RNA molecules can be introduced into cells and be taken up into the RNA-induced silencing complex (RISC), mimicking endogenous mature miRNAs activity. Loss-of-function analyses were performed with Ambion® Anti-miR™ miRNA inhibitors. The miRNA inhibitors are chemically modified, single-stranded nucleic acids designed to specifically bind to complementary miRNAs. The binding between endogenous miRNA and miRNA inhibitors down-regulates endogenous miRNAs activity.

All miRNA mimics and miRNA inhibitors were obtained from Ambion. Transient transfections of DRGs were performed using Lipofectamine™ LTX and Plus Reagent (Invitrogen) according to the manufacturer's protocol. To extend the time window for effective transfection of miRNA precursors and inhibitors, as well as, initiation of miRNA machinery, we incubated DRG neurons with 20 μM of SP600125 for the first 24 h according to a protocol previously described (Davare et al., 2009). SP600125 is a specific inhibitor of JNK and reversibly inhibits axonogenesis (Davare et al., 2009). We then released the block on axonogenesis from the SP600125 by washing out SP600125 and change culture media. DRG neurons were then cultured for an additional 24 h to allow axon formation.

IMMUNOFLUORESCENT STAINING AND IMAGE ANALYSIS

The cells cultured on coverslips were fixed with 4% paraformaldehyde for 5 min and washed with phosphate buffered saline with Tween (PBST). After blocking with 10% goat serum for 1 h at room temperature, the samples were incubated with the indicated primary antibodies diluted at optimized concentrations at 4°C overnight. This was followed by incubation with secondary antibodies conjugated with FITC-, TX Red-, or Alexa Fluor® (Invitrogen). Negative controls included samples processed in parallel with non-immune serum or without primary antibodies. After mounting the slides with anti-fading media (Invitrogen), images were viewed with an Olympus IX81 fluorescent microscope and captured with CellSens Dimension software (Olympus America Inc., Center Valley, PA, USA). The images we acquired were all single plane fluorescent images.

Quantification of axon length and measurement of axon branches were performed following previously described lab protocol (Murashov et al., 2005). For each coverslip, 30 images were taken, and from each, 10–15 neurons, which were completely distinguishable from neighboring cells, were chosen for further analysis. The axon length was quantified by tracing the image of neurites with the ImageJ software (NIH, Bethesda, MD, USA). The longest axon for each neuron was measured and recorded. The number of neurite branches per neuron was also determined from each neuronal population manually. Only primary branches, which are routinely defined as neurites originating from the neuronal soma and are at least longer than two times the diameter of the cell body were counted (Liu et al., 2002).

IMMUNOBLOTTING ANALYSIS

Tissue samples were homogenized in ice-cold homogenization buffer (20 mM Tris, 2 mM EGTA, 2 mM EDTA, 6 mM β-mercaptoethanol, 1mM PMSF, and 10% Triton) containing protease inhibitor cocktail (Sigma), and centrifuged at 10,000 *g* for 10 min at 4°C. The supernatants were collected in fresh tubes and stored at −20°C. Proteins concentrations were quantified using the Bio-Rad reagent (Bio-Rad, Hercules, CA, USA) and samples for western blot analysis were prepared by boiling with standard stop buffer for 5 min. Equal amounts of solubilized proteins were loaded per lane on sodium dodecyl sulfate gels and separated by electrophoresis. The separated proteins were then transferred to immobilonP membranes (Millipore Corporation, Bedford, MA, USA).

Membranes were blocked in Odyssey blocking buffer (LI-COR, NE, USA) for 1 h at room temperature on a shaker, and then probed with a primary antibody in Odyssey blocking buffer at 4°C overnight. The membranes were washed three times with PBST, and then incubated with IRDye® conjugated secondary antibodies for 1 h at room temperature with gentle shaking. The fluorescent signals on membrane were detected with the Odyssey® Infrared Imaging System (LI-COR). Densitometry values were normalized to α-tubulin, to obtain the relative signal intensity.

LIST OF ANTIBODIES

Primary antibodies

Mouse monoclonal neuro-specific β III tubulin antibody (TUJ-1) Covance Research Products, Inc. (Denver, PA, USA). Goat polyclonal antibodies against *Kremen1* (R&D Systems, Minneapolis, MN, USA). Rabbit anti-GAP-43 polyclonal antibodies (Millipore, Billerica, MA, USA). Mouse monoclonal anti-α-tubulin antibodies Zymed (Zymed Laboratories, Carlsbad, CA, USA).

Secondary antibodies

IRDye 800CW goat anti-Mouse IgG, IRDye 680LT goat anti-Rabbit IgG, and IRDye 800CW donkey anti-goat IgG secondary antibodies (LI-COR Corporate, NE, USA). For fluorescence studies, secondary FITC-, TX Red-conjugated IgG (Jackson ImmunoResearch Laboratories, Inc., West Grove, PA, USA) or Alexa Fluor 594 donkey anti-goat from Invitrogen were applied.

Cross-linked immunoprecipitation (CLIP) analysis

Argonaute CLIP method to identify *in vivo* targets of miRNAs followed procedure described previously (Jaskiewicz et al., 2012). DRG neuronal cell cultures were transfected with 100 nM of miR-431 mimic or a scrambled miRNA mimic negative control. Two days post-transfection, the cells were rinsed once in PBS and then placed in UVP CL-1000 cross-linker (UVP, Upland, CA, USA) with the cover off. Cells were irradiated once for 400 mJ/cm^2 and once more for 200 mJ/cm^2 to establish protein-RNA reversible cross-linking. Cells were lysed in cell lysis buffer (100 mM KCl, 5 mM $MgCl_2$, 10 mM HEPES, PH 7.0, 0.5% NP-40, 1 mM DTT, 100 U/ml RNasin RNase inhibitor (Promega), 2 mM vanadyl-ribonucleoside complexes solution (Sigma)) supplemented with a mixture of protease inhibitors (Invitrogen). Cells were then detached with a cell scraper and lysate was transferred to a tube on ice. Cell lysates were centrifuged at 16,000 *g* for 15 min at 4°C and the supernatants (the protein lysates) were transferred to sterile tubes for further immunoprecipitation. Prior to the immunoprecipitation, protein G agarose beads (Sigma) were equilibrated by washing twice with a wash buffer (0.5% NP-40, 150 mM NaCl, 2 mM $MgCl_2$, 2 mM $CaCl_2$, 20 mM Tris, pH 7.5, 5 mM DTT, with protease inhibitor) containing 1 mg/ml yeast tRNA and 1 mg/ml BSA. After pre-clearing the protein lysate with equilibrated protein G-agarose beads, 5 μl of each sample was saved as an input fraction. The protein lysate was immunoprecipitated with specific mouse monoclonal antibodies against Ago-2 (Wako, Richmond, VA, USA) or control serum and bounded by protein G agarose beads with agitation at 4°C overnight. After precipitation, the beads were washed three times with washing buffer. Afterward, the bonds between RNA and protein were disrupted by heating at 50°C for 30 min. RNA was then extracted and purified using Trizol (Invitrogen) and used for qRT-PCR.

Luciferase assays

Luciferase assays were performed using the pMIR-REPORT™ miRNA expression reporter vector system (Ambion). pMIR-REPORT firefly luciferase (FL) plasmids were purified with Miniprep kit (Qiagen, Valencia, CA, USA) and digested with restriction enzymes *Spe*I and *Hind*III (New England BioLabs, Ipswich, MA, USA). Linearized vectors from the restriction digestion were retrieved by agarose gel electrophoresis and gel purification of DNA using Gel Extraction Kit (Omega Bio-Tek, Inc., Norcross, GA, USA). The 3′-UTR regions of mouse *Kremen1* gene were amplified from mouse *Kremen1* cDNA clones (Source Bioscience, Nottingham, UK). The primers were designed as: 5′-ATAACTAGTGCTCCGCTCCAAGCTCGAGTTTGC 3′ and 5′-GCGAAGCTTTCTCTTTTGTAAAAGTTAAGTACC 3′. Restriction enzyme sites for *Spe*I and *Hind*III were introduced into the PCR product to facilitate directional cloning. The 3′-UTR of *Kremen1* was inserted into downstream of FL gene in the pMIR-REPORT vector with T4 ligase (New England BioLabs), and subsequently transformed in DH5α competent cells (Invitrogen). Luciferase assays were performed using the Dual-Luciferase assay kit (Promega). PC12 cells (40,000) were cultured and co-transfected in 24-well plates with 400 ng of FL reporter construct, 100 nM miR-431 mimics or mimic negative controls, and 40 ng of pRL-TK control vector encoding renilla luciferase (RL; Promega). The transfection was performed with Lipofectamine 2000. Forty hours after transfection, the cells were harvested in passive lysis buffer and firefly and RL activities were measured in a Turner Biosystems 20/20n Luminometer (Turner Biosystems, Sunnyvale, CA, USA). The luciferase data is expressed as a ratio of FL to RL to normalize for transfection variability between samples. Luciferase experiments were repeated at least three independent times in triplicate.

miRNA and gene expression array analyses for DRG RNA

Total RNA for the microarray expression analysis was isolated from L4 to L5 DRGs, pooled from 10 mice at 4 days after sciatic nerve crush. Total RNA extraction was performed with miRVANA™ miRNA isolation kit following the manufacturer's instruction (Ambion). These pooled RNA samples were sent to UNC Lineberger Comprehensive Cancer Center Genomics Core for microarray analysis. After a quality control, they were hybridized to 8 × 15 miRNA one-color arrays (Agilent, Santa Clara, CA, USA). The same RNA samples were also hybridized to 4 × 44K mouse gene expression microarrays (Agilent) at the same Genomics Core. All microarray experiments were performed in duplicate and repeated twice. Normalization and further analyses of microarray data were performed with GeneSpring software (Agilent). Differentially expressed miRNAs were determined using a combination of t tests, with FDR correction of 0.1, and further defined by p-value < 0.05 after correction for multiple hypotheses. The analysis with GeneSpring allowed for identification of a different expression pattern of miRNAs in the crushed groups compared with the control groups. Statistically significant upregulated or downregulated miRNAs were then selected for further analysis. All microarray data have been submitted to GEO (access number pending).

Real-time PCR (RT-PCR)

Total RNA was isolated from L4-L5 DRGs using *mir*Vana™ miRNA Isolation Kit (Ambion). Total RNAs from DRG neuronal cell cultures were purified with RNAqueous Micro Scale RNA Isolation Kit (Ambion). RNA was quantified with a NanoDrop ND-1000 spectrophotometer (Thermo Scientific, Wilmington, DE, USA). Reverse transcription was performed with NCode™ VILO™ miRNA cDNA Synthesis Kit and SuperScript VILO cDNA Synthesis Kit (Invitrogen) for miRNA expression analysis and mRNA expression analysis, respectively. The real-time PCRs were carried out using EXPRESS SYBER® GreenER™ qPCR SuperMix Universal (Invitrogen) in triplicates for each cDNA sample on Applied Biosystems 7500 real-time PCR system (Applied Biosystems, Life Technologies, Carlsbad, CA, USA). Primers specific for each miRNA and mRNA were obtained from Invitrogen. As an internal control, primers for S12 (mitochondrial ribosome small subunit) were added for RNA template normalization, and the relative quantification of gene and miRNA expressions were calculated against S12 using a $2^{-\Delta\Delta CT}$ method. We routinely use S12 for qPCR studying axonal injuries. Other standard controls like beta-actin and GAPDH usually change in response to crush injury. All experiments were carried out three times independently.

List of primers

miR-21: 5′-TAGCTTATCAGACTGATGTTGA-3′
miR-431: 5′-CAGGCCGTCATGCAAA-3′
miR-744: 5′-GGGCTAGGGCTAACAGCA-3′
miR-124: 5′-GCGGTGAATGCCAAAAA-3′
miR-29a: 5′-TAGCACCATCTGAAATCGGTTA-3′
Kremen1: 5′-ACAGCCAACGGTGCAGATTAC-3′ and 5′-TGT TGTACGGATGCTGGAAAG-3′
GAP-43: 5′TGGTGTCAAGCCGGAAGATAA-3′ and 5′-GCTG GTGCATCACCCTTCT-3′
S-12: 5′-TGGCCCGGCCTTCTTTATG-3′ and 5′-CCTAAGCG GTGCATCTGGTT-3′

Statistical analysis

Data from multiple independent experiments were analyzed with GraphPad Prism version 5 for Windows (GraphPad Software, San Diego, CA, USA). The results were expressed as mean ± standard error of the mean in graphic and text representations. To determine the difference between three or more groups, a one-way analysis of variance (ANOVA) followed by Bonferroni's multiple comparison tests was utilized. For the analysis of two independent groups, Student's *t*-test was used. A *p*-value of less than 0.05 was considered statistically significant.

RESULTS

miRNAs ARE DIFFERENTIALLY EXPRESSED IN DRG UPON SCIATIC NERVE INJURY

We analyzed miRNA expressions in DRGs using microarrays at 4 days after sciatic nerve crush. DRGs were collected from both the pre-conditioned side, as well as the contralateral uninjured side. RNA from the contralateral uninjured side served as a control group. At 4 days post-injury, pre-conditioned DRG neurons show robust regenerative axon growth (Forman et al., 1980). RNA from the pre-conditioned DRG was considered the actively regenerating group. By comparing the miRNA expression pattern from pre-conditioned DRG and control DRG, miRNAs that were upregulated and down-regulated during the process of regeneration were determined. Several miRNAs demonstrated differential expression based on regenerative growth condition. Using 1.5-fold cut-off, statistical analyses revealed that 19 miRNA were differentially expressed in the pre-conditioned DRG compared to the non-conditioned contralateral DRG. Of those 19, 11 miRNAs had higher expression level in pre-conditioned group and the other eight miRNAs had lower expression level in DRG during regeneration (**Figure 1A**). miR-431, miR-714, miR-744, miR-877, miR-130b, miR-21, miR-323-3p, miR-325, miR-409-3p, miR-154*, and miR-681 were significantly increased 4 days post-sciatic nerve crush in pre-conditioned DRGs, while miR-190, miR-1, miR-33, miR-32, miR-153, miR-335-5p, miR-193, and miR-488 showed significantly decreased expression. The most upregulated miR-431 was selected for further analyses.

We validated the microarray data for miR-431 using real-time qPCR. We also included miR-744 and miR-21 as positive controls and miR-124 and miR-29a as non-regulated controls in our real-time PCR experiments. These last two miRNAs play various roles in neurodevelopment and maintenance of neuronal cell homeostasis (Cheng et al., 2009; Shioya et al., 2010); however, they did not show changes in their expression in our array data. In agreement with the microarray data, miR-431, miR-744, and miR-21 were significantly upregulated in regenerating neuronal cells. We detected 2.4-fold upregulation of miR-431, a twofold upregulation of miR-744, and a 2.5-fold upregulation of miR-21, respectively (**Figure 1B**). At the same time, RT-qPCR experiments showed that miR-29a and miR-124 did not change their expression during regeneration.

GAIN-OF-FUNCTION OF miR-431 INCREASES REGENERATIVE OUTGROWTH

To investigate the role of miR-431 in regenerative axon growth, we manipulated the level of miR-431 in dissociated DRG neurons. We observed a positive association between miR-431 expression and neurite outgrowth in dissociated DRG neuronal cell culture (**Figure 2A**). Increased mir-431 level was achieved by applying miR-431 mimic to DRG neuronal cell cultures at a final concentration of 100 nM. Overexpression of miR-431 significantly increased axon length. Additionally, blocking miR-431 activity with miR-431 inhibitor significantly inhibited neurite extension (no treatment control group: 100 ± 5%; miR-431 mimic group: 130 ± 6%; mimic negative group: 91 ± 4%; miR-431 inhibitor group: 75% ± 7%; inhibitor negative control: 90 ± 8%; **Figure 2B**). Moreover, manipulating miRNA-431 levels also affected axon branching, and led to a decrease in the number of branches per neuron due to transfection with miR-431 inhibitor (no treatment control group: 100 ± 9%; miR-431 mimic group: 110 ± 10%; mimic negative group: 82 ± 7%; miR-431 inhibitor group: 64% ± 6%; inhibitor negative control: 86 ± 10%; **Figure 2C**).

We next studied GAP-43 expression in DRG neurons with miR-431 mimic and inhibitor treatments, as a strong association

A

miRNA	Fold change
mmu-miR-431	2.611244 up
mmu-miR-714	2.1570964 up
mmu-miR-744	2.1376243 up
mmu-miR-877	2.081734 up
mmu-miR-130b	1.7698824 up
mmu-miR-21	1.7598088 up
mmu-miR-323-3p	1.6689959 up
mmu-miR-325	1.5642644 up
mmu-miR-409-3p	1.5461996 up
mmu-miR-154*	1.517914 up
mmu-miR-681	1.5049202 up
mmu-miR-190	2.063784 down
mmu-miR-1	1.7755992 down
mmu-miR-33	1.6966436 down
mmu-miR-32	1.6579413 down
mmu-miR-153	1.6068152 down
mmu-miR-335-5p	1.595247 down
mmu-miR-193	1.5837495 down
mmu-miR-488	1.5546329 down

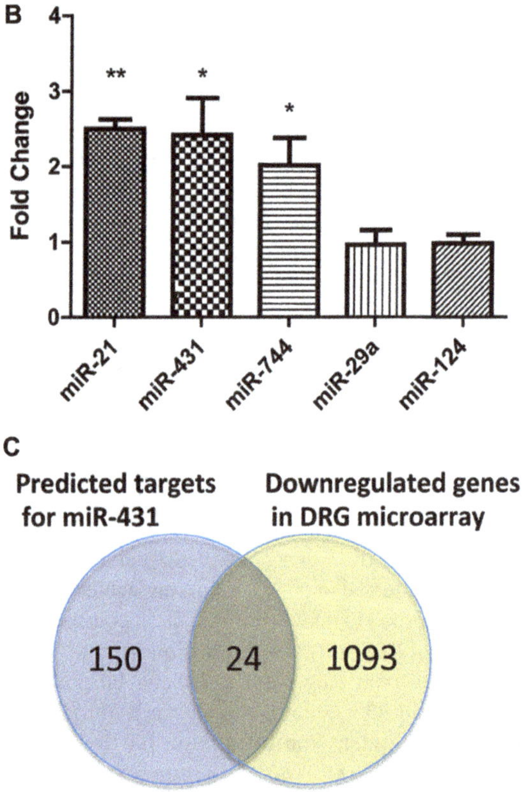

FIGURE 1 | Sciatic nerve injury induced changes in miRNA expression profile in DRG. (A) Total RNA for the microarray expression analysis was isolated from DRG 4 days after sciatic nerve crush. Agilent arrays were done in duplicates and repeated twice. Normalization and analyses were performed with GeneSpring software. miRNAs with a statistically significant upregulation or down-regulation over 1.5-fold were listed in the table. **(B)** Three miRNAs that were significantly upregulated were selected for further validation. Real-time qPCR for miRNA validated the relative changes in miRNA level. miRNA expression was normalized to reference *(Continued)*

FIGURE 1 | Continued
gene s12. The graph indicates a significant increase of miR-744, miR-431, and miR-21 in DRG after sciatic nerve crush, whereas the expression level of miR-124 and miR 29a did not change ($^*p < 0.05$, $^{**}p < 0.01$, $N = 3$). **(C)** Venn diagram of overlap in predicted miR-431 target genes and down-regulated genes in DRG after conditioning sciatic nerve lesion. The potential targets of miR-431 were chosen using three algorithms http://www.targetscan.org, http://www.microrna.org, and http://diana.cslab.ece.ntua.gr. Down-regulated genes were selected using fold change cut-offs of >2 and significance p-values of <0.05 expression based on microarray data for DRGs 4 days post-sciatic nerve injury. Overlap shows 24 genes having predicted binding site for miR-431 and significantly down-regulated expression level in DRG microarray. A one-way ANOVA followed by Bonferroni's multiple comparison tests was utilized. For the analysis of two independent groups, Student's t-test was used.

between neurite outgrowth and expression of GAP-43 has been reported in previous studies (Benowitz and Routtenberg, 1997). We observed significant increase in GAP-43 immunostaining caused by transfection with miR-431 (**Figure 2D**). GAP-43 mRNA level was further studied with RT-qPCR. **Figure 2E** clearly demonstrates a significant increase in GAP-43 mRNA in the cultures treated with 100 nM of miR-431 mimics, as compared to the group treated with the scrambled miRNA mimic control. This relates to immunofluorescent data demonstrating significant increase in axon outgrowth after overexpression of miR-431.

IDENTIFICATION OF miR-431 mRNA TARGETS

We used three databases[1] to generate a list of mRNAs with potential binding site for miR-431 in their 3′-UTR. The potential candidates were further selected based on evaluation of the gene expression microarray data for DRGs 4 days post-sciatic nerve injury (SNI). We hypothesized that an increased expression of miR-431 in preconditioned DRG, would negatively associate with expression of the target mRNAs in the same RNA samples. Using GeneSpring 10 software package (Agilent) we performed joint analysis of miRNA and gene expression data. This allowed us to narrow the list of potential targets to 24 genes. These 24 genes met both criteria, of having a predicted binding site for miR-431 in their 3′-UTR and significantly down-regulated expression level in DRG microarray (**Figure 1C**).

To investigate which genes may be regulated by miR-431, we initially screened potential targets in neuronal PC12 cells overexpressing miR-431. Transient overexpression of miR-431 was achieved using transfection of PC12 cells with miR-431 mimic. The expression of potential targets was studied with real-time RT-qPCR. The experiments revealed that only six genes (*Braf, Eif2s2, Kremen1, Msi2, Tnrc6b, Zkscan1*) were significantly down-regulated by miR-431 in PC12 cells (**Table 1**). We then applied the same approach to test these six genes with overexpression of miR-431 in primary DRG neurons. In the RT-qPCR experiments, overexpression of miR-431 led to significant suppression of the expression of only three genes including *Braf, Kremen1*, and *Zkscan1* (**Table 1**). Based on the literature data indicating that *Kremen1* is an antagonist of Wnt signaling pathway (Nakamura and Matsumoto, 2008), which

[1] http://www.targetscan.org, http://www.microrna.org, http://diana.cslab.ece.ntua.gr

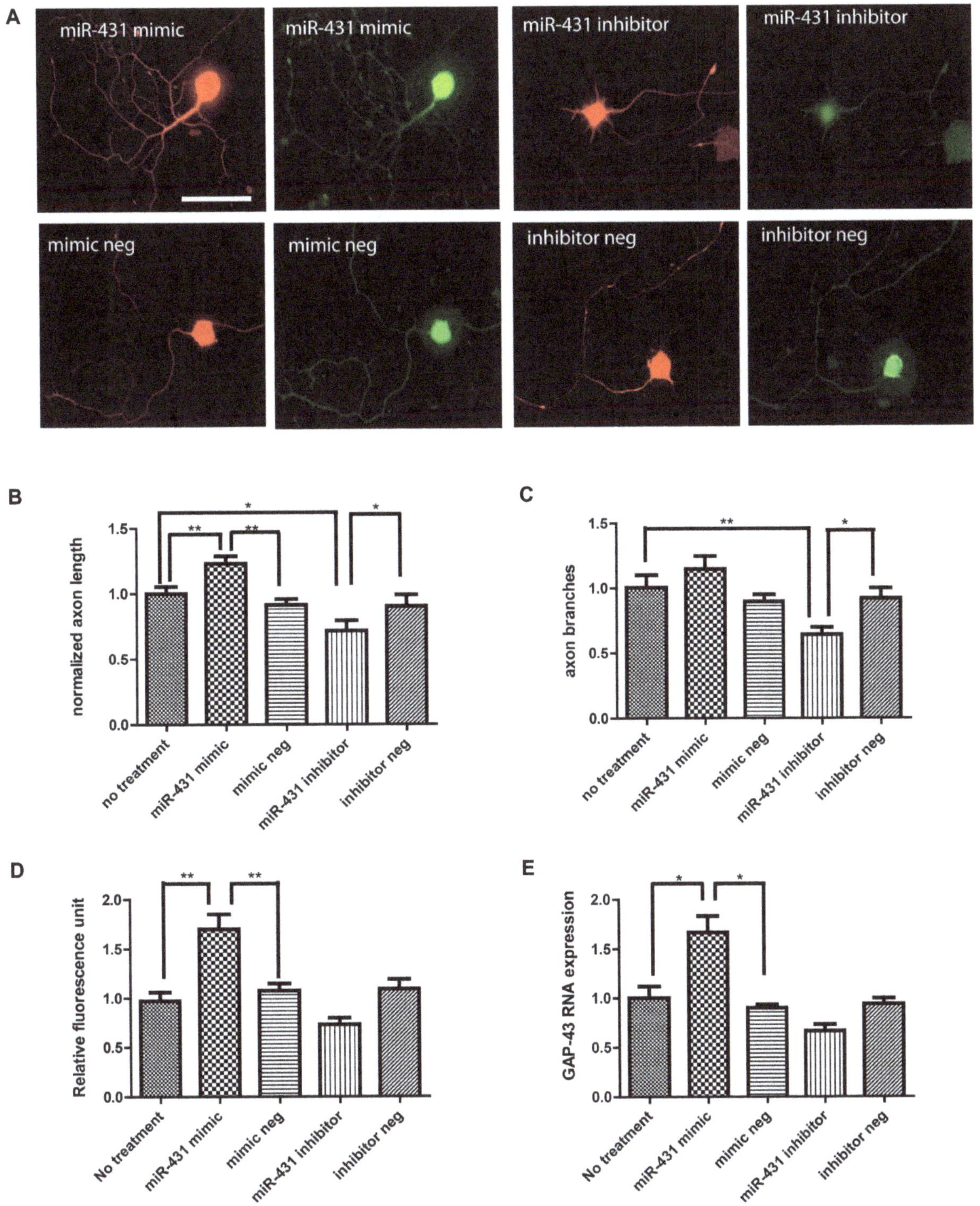

FIGURE 2 | miR-431 increases axon outgrowth in DRG neurons. Effects of miR-431 mimic and inhibitor on axon outgrowth. (A) Left panel shows the effect of the transfection of DRG neurons with miR-431 mimic. Right panel depicts the effect of transection with miR-431 inhibitor. Negative controls for miR-431 mimic and inhibitor are indicated on the lower images. Cells were stained with primary antibodies against neuronal β-tubulin and signals were visualized with TX-Red conjugated secondary antibody (scale bar: 50 μm). The expression of GAP-43, a marker for axon regeneration, was detected using an anti-GAP-43 antibody and visualized with FITC-conjugated secondary antibodies. The effect of miR-431 on axon length **(B)** and on axon branching **(C)** was quantified. Overexpression of miR-431 significantly increased axon extension, whereas suppression of miR-431 significantly blocked axon branching. The fluorescence signal intensity against GAP-43 was quantified in **(D)**. The significant increase in GAP-43 immunofluorescence reflects increase in regenerative axon growth. **(E)** Significant increase in GAP-43 expression on mRNA level quantified by RT-qPCR ($*p < 0.05$, $**p < 0.01$, $N = 50$). A one-way ANOVA followed by Bonferroni's multiple comparison tests was utilized.

Table 1 | Effect of miR-431 overexpression on levels of potential target genes in PC12 cells and primary DRG culture.

Gene	PC12 cells		DRG culture	
	Relative value	SEM	Relative value	SEM
Braf	**0.323**	**0.1245**	**0.5133**	**0.07055**
Cwf1912	5.833	0.8762		
Dlst	1.85	0.1531		
Eif2s2	**0.4367**	**0.06766**	0.8333	0.2028
Fgf12	0.68	0.1531		
Hip1	1.07	0.2608		
Kremen1	**0.3833**	**0.05364**	**0.6733**	**0.0393**
Luc712	0.69	0.1054		
Msi2	**0.4133**	**0.05364**	1.04	0.07024
Ncam1	1.103	0.2284		
Nudcd3	0.91	0.1002		
Slc30a10	1.163	0.02906		
Son	4.033	0.5044		
Tcf712	2.043	0.4937		
Tnrc6b	**0.58**	**0.0755**	0.9467	0.245
Vezt	1.253	0.1141		
Wnk3	0.9433	0.1601		
Zeb2	0.9733	0.1742		
Zkscan1	**0.2967**	**0.06009**	**0.45**	**0.06429**

Transient overexpression of miR-431 was achieved using transfection with its mimic. Relative mRNA levels of the potential target genes were evaluated by real-time RT-qPCR. Bold numbers indicate significant decrease in gene expression. Only three genes Kremen1, Braf, and Zkscan1 were significantly down-regulated in primary neuronal culture.

is critical for axonal remodeling (Purro et al., 2008), we focused our subsequent experiments on characterization of *Kremen1*–miR-431 interaction.

To investigate a direct interaction between target mRNAs and miR-431 in RISC, CLIP of the Ago-2 protein, the central component of the RISC was carried out. Applying miR-431 mimic to DRG neurons increased the expression level of miR-431 ~7.75-fold in DRG neuronal cell cultures (**Figure 3A**). Electrophoresis of CLIP samples confirmed the miR-431 induced association of *Kremen1* mRNA with RISC, suggesting *Kremen1* as the target gene for miR-431 (**Figure 3B**). **Figure 3B** shows the RT-PCR of *Kremen1* mRNA presented in the total RNA (input) and IP fractions from DRG cultures treated with miRNA mimic and the mimic negative control. In the total RNA samples from DRG cultures, overexpression of miR-431 reduced the amount of stable *Kremen1* mRNA when compared to the miRNA mimic negative control group. In the Ago-2 immunoprecipitated RNA samples, overexpression of miR-431 clearly increased the level of Ago-2 associated *Kremen1* mRNA. In the IP negative control group (non-immune serum), no detectable *Kremen1* mRNA was observed, confirming the specificity of the precipitation (**Figure 3B**).

LUCIFERASES REPORTER ASSAY CONFIRMS miR-431 TARGET *Kremen1* 3′UTR

Kremen1 has one binding site for miR-431 at its 3′-UTR, at the position 2530–2536 bp. It corresponds perfectly to nucleotides 2–7 of the mature miRNA in mouse, rat, and human. In addition, the seed target site is close the poly-A tail, which increases its accessibility. To confirm miR-431 direct interaction on *Kremen1* 3′ UTR, we established a *Kremen1* 3′UTR-FLs construct with the 3′-UTR of *Kremen1* inserted downstream of the FL gene. This construct allowed us to quantitatively evaluate the regulatory effect of miR-431 on the 3′-UTR of *Kremen1*. PC12 cells were transiently transfected with miR-431 mimics or mimic negative controls, *Kremen1* 3′UTR-FL construct, and RL plasmid DNA as internal control. As shown in **Figure 3F**, co-transfection of miR-431 mimic and *Kremen1* 3′UTR-FL construct resulted in significant decrease in FL activity. Luciferase activity reduced to 48% compared with the vector control, whereas co-transfection of mimic negative controls and *Kremen1* 3′UTR-FL construct did not affect the expression of FL gene (**Figure 3F**). Together, these data suggest that miR-431 actively modulates *Kremen1* protein and RNA expression within DRG neurons through association with *Kremen1* 3′UTR.

miR-431 MODULATES *Kremen1* EXPRESSION AT mRNA AND PROTEIN LEVELS IN PRIMARY NEURONAL CULTURES

To show that miR-431 regulates endogenous *Kremen1* in DRG neurons, we transfected cells with either miR-431 mimics, miR-431 inhibitors, mimic negative control, or inhibitor negative control. Since miRNA-mediated gene regulation can destabilize target mRNA and reduce the level of the target mRNA, we used RT-qPCR to determine the effect of miR-431 on *Kremen1*. We observed that transient transfection with miR-431 mimic, decreased the mRNA level of *Kremen1* to 30%. Application of miR-431 inhibitors significantly elevated the mRNA level of *Kremen1* (**Figure 3C**). These results demonstrated that miR-431 level is inversely correlated to *Kremen1* expression at mRNA level in DRG neurons.

We then performed proteomic analysis of *Kremen1* in DRG neurons. Whereas endogenous miR-431 was inhibited by transfection with miR-431 inhibitor, the expression level of the *Kremen1* protein was significantly higher than in control groups. Quantification of three independent experiments revealed that miR-431 reduced *Kremen1* protein levels by 50% when compared with the mimic negative control group. On the other hand, inhibition of endogenous miR-431 resulted in a significant increase of *Kremen1* expression by 45% (**Figures 3D,E**).

Kremen1 EXPRESSION IN DRG *IN VIVO*

After establishing a physical interaction between miR-431 and *Kremen1*, we next investigated the expression patterns of *Kremen1* during axon regeneration. From gene expression array data, *Kremen1* expression in DRG decreased at 4 days after SNI, suggesting its expression was down-regulated as the peripheral nerve regenerated. To further reveal physiological role miR-431 *Kremen1* interaction, we analyzed expression of *Kremen1* at RNA and protein levels from control and regenerating DRGs. RT-qPCR revealed that *Kremen1* RNA expression decreased fourfold at 4 days after sciatic nerve crush, when axons exhibit

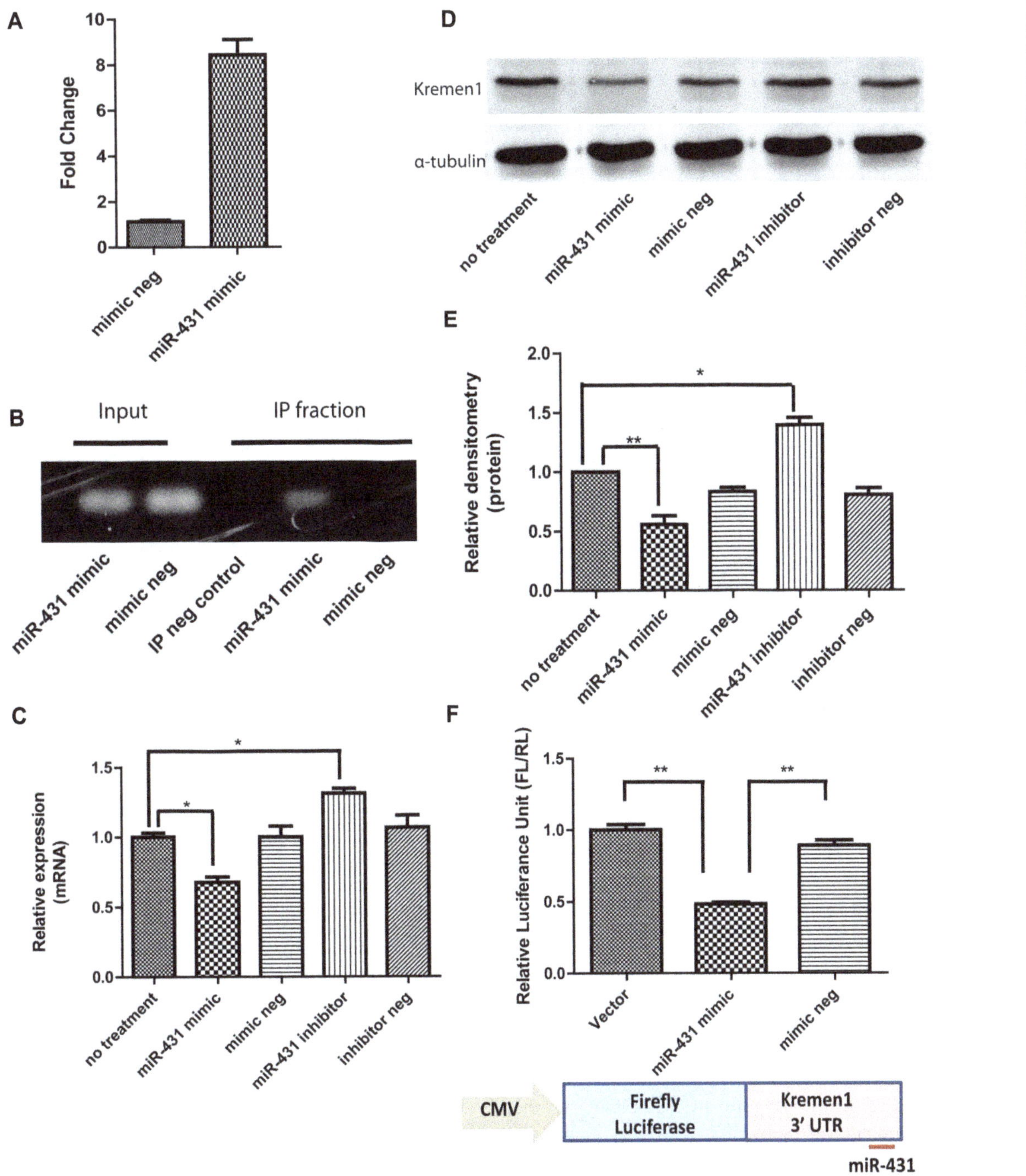

FIGURE 3 | miR-431 regulates *Kremen1* expression. (A) RT-qPCR confirmed the increase of miR-431 level in DRG neuron after the transfection of miR-431 mimic. **(B)** Although overexpression of miR-431 decreased *Kremen1* mRNA in total cell lysates (input), it enhanced the binding between *Kremen1* mRNA and Ago-2 complex. In the Ago immunoprecipitated fractions, there was an increased amount of *Kremen1* mRNA. The lack of signal in the non-specific serum IP sample (IP neg. control) confirmed the specificity of the IP. **(C)** miR-431 negatively regulated *Kremen1* expression at mRNA level. Treatment of miR-431 mimics in DRG neuronal cultures significantly inhibited *Kremen1* expression as compared with that of control groups. On the contrary, suppression of miR-431 activity significantly enhanced the expression of *Kremen1* mRNA. **(D)** Western blot analysis of *Kremen1* expression exhibited similar negative correlation of miR-431 and *Kremen1* expression. Cells transfected with miR-431 mimics had decreased protein level of *Kremen1*, whiles cells transfected with miR-431 inhibitors had an increased expression of *Kremen1*. α-tubulin was used as the loading control and was used to normalize densitometry values. **(E)** The quantification of densitometric levels of *Kremen1*. **(F)** PC12 cells were transfected with *Kremen1* 3'UTR-firefly Luciferase constructs for luciferase assays. Co-transfection with miR-431 mimics significantly reduced the luciferase activity ($^*p < 0.05$, $^{**}p < 0.01$), whereas co-transfection with mimic negative controls did not affect the expression of firefly luciferase gene. A one-way ANOVA followed by Bonferroni's multiple comparison tests was utilized.

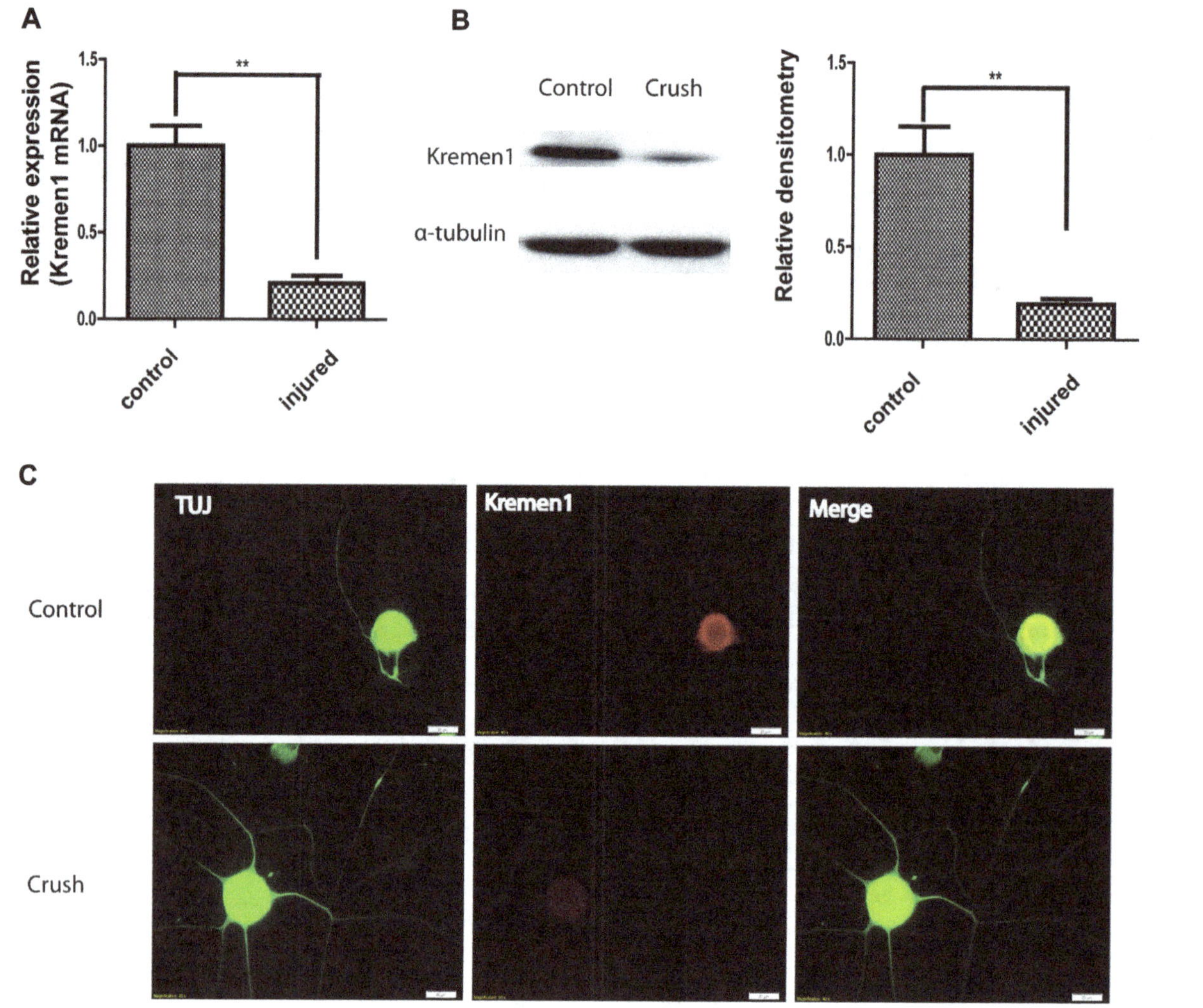

FIGURE 4 | Nerve crush injury reduces *Kremen1* expression. (A) Total RNA was isolated from control or crush-injured mouse DRG, and relative expression of *Kremen1* was determined using RT-qPCR. GAPDH and S12 were used to normalize for RNA loading. **(B)** Western blot analysis of total DRG lysates at 4 days post-crush injury. α-Tubulin was shown as a loading control. As shown in the quantified densitometry data, there was a significant decrease of *Kremen1* expression during nerve regeneration. **(C)** Immunofluorescent staining in dissociated DRG neurons demonstrated the expression of *Kremen1* within neurons. *Kremen1* as a transmembrane receptor was shown to be located in cell bodies, but not axons. TUJ staining was used to visualize neuronal cells. Preconditioning of sciatic nerve clearly promotes regenerative axon growth in DRG neurons, and this phenomenon is accompanied by a decrease in *Kremen1* expression. Scale bar: 20 μm. (**$p < 0.01$) For the analysis of two independent groups, Student's *t*-test was used.

robust regenerative growth (**Figure 4A**). Similarly, we found that *Kremen1* protein was reduced in DRGs at 4 days post-injury. The Western blot data showed a significant 80% decrease in *Kremen1* expression after SNI when compared to control (**Figure 4B**).

The expression of *Kremen1* in DRG neuron was further examined using indirect immunofluorescence (IIF). IIF with antibodies against *Kremen1* revealed the localization of *Kremen1* in dissociated DRG neurons. In both pre-conditioned and control groups, the immunoreactivity of *Kremen1* was detected mainly in neuronal cell bodies, however, there was less *Kremen1* immunostaining in the group with sciatic nerve crush (**Figure 4C**). These data further support a functional relationship between miR-431 and *Kremen1* in regenerating DRG neurons and suggest a role of *Kremen1* in peripheral nerve regeneration.

FUNCTIONAL ANALYSIS OF *Kremen1* ROLE IN AXON REGENERATION

Given the effects of miR-431 on *Kremen1* expression and the role of miR-431 in neurite outgrowth, we investigated the effect of *Kremen1* knockdown on regenerative axon growth. Two groups of DRG neurons were transfected with either siRNA specifically targeting *Kremen1* mRNA, or scrambled siRNA (negative control). The differences in the regenerative growth between *Kremen1* siRNA group and control scrambled siRNA group were quantified based on axon elongation and branching. The experiments revealed that knockdown of *Kremen1* significantly increased axon length in dissociated DRG cultures (**Figure 5**). The axon length in the *Kremen1* knockdown group increased ~30% in comparison to the scrambled siRNA control group. This effect on axon outgrowth is similar to the effect of miR-431 overexpression on axon outgrowth reported earlier (**Figure 2B**). Taken together, these results

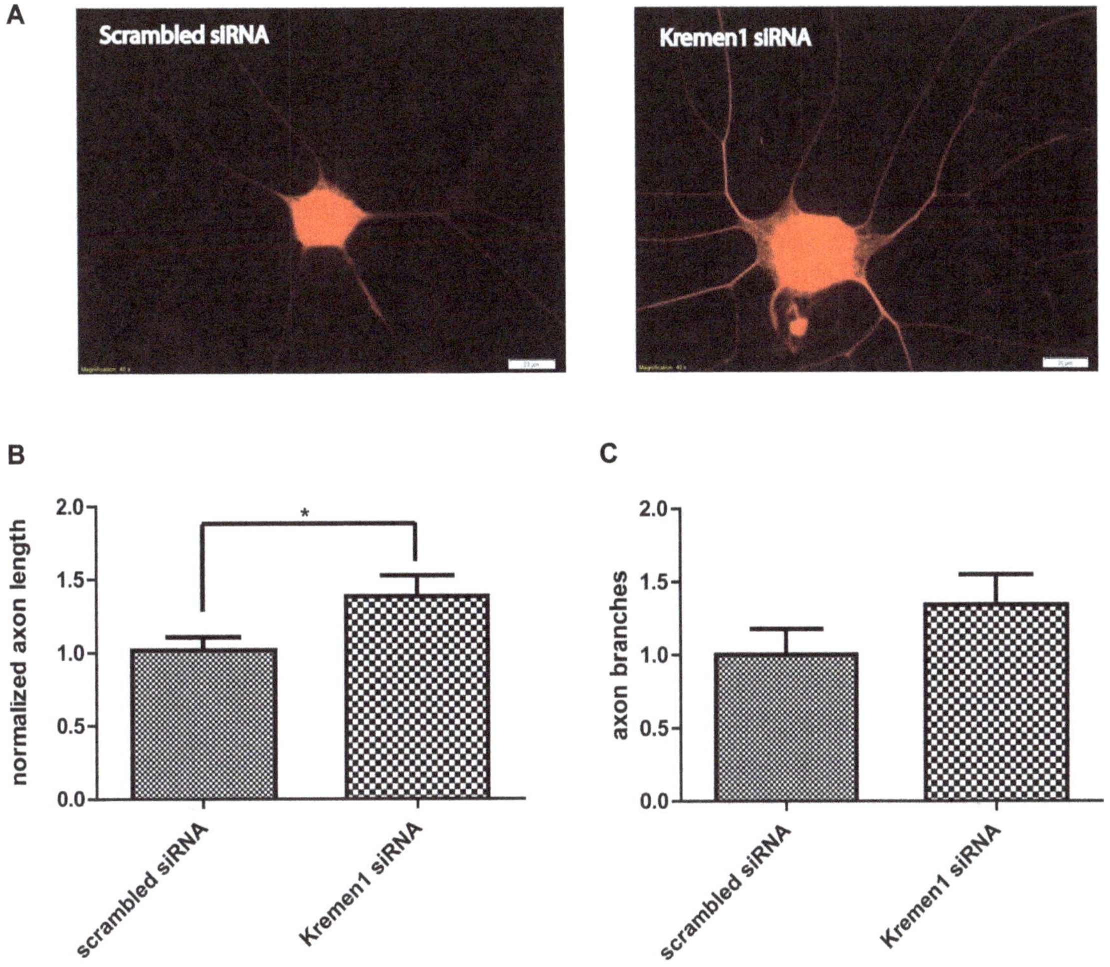

FIGURE 5 | Knockdown of *Kremen1* increases neurite outgrowth. **(A)** Neurite outgrowth in *Kremen1* siRNA and scrambled siRNA treated DRG neurons was detected by TUJ immunostaining. Representative images show that *Kremen1* siRNA significantly decreased *Kremen1* expression level, which was accompanied by an increase of axon outgrowth. Scale bar: 20 μm. As the quantification performed in miR-431 functional analysis, we measured the length of the longest axon for each neuron **(B)** and counted the number of branches for each neuron **(C)**. Inhibition of *Kremen1* significantly increased the length of axon, however, its effect on neurite branching was not significant. $*-p < 0.05$. For the analysis of two independent groups, Student's *t*-test was used. Scale bar: 20 μm.

indicate that miR-431 mediates increase of axon growth through *Kremen1* repression.

DISCUSSION

ALTERED miRNA EXPRESSION FOLLOWING NERVE INJURY

Our microarray experiments identified a group of injury-regulated miRNAs in DRG neurons after conditioning sciatic nerve lesion. Alterations in miRNAs have been recently shown in several studies profiling miRNA expression after nerve injuries in the central nervous system. Microarray based analysis of miRNA in the rat cerebral cortex after traumatic brain injury revealed that a set of miRNAs were differentially expressed at 6, 24, 48, and 72 h after injury. At all-time points post-injury, miR-21 was consistently highly expressed in the cerebral cortex (Lei et al., 2009). Changes in miRNA expression have also been studied by microarray analysis in hippocampus after traumatic brain injury. At three and 24 h after controlled cortical impact injury, 35 miRNA exhibited increased expression levels and 50 miRNA exhibited decreased expression level (Redell et al., 2009). Following a contusive SCI in adult rats (Liu et al., 2009), 60 miRNAs showed significant changes in their expression level in the injured spinal cord at 4 h, 1, and 7 days. Among those 60 miRNAs, 30 were upregulated, 16 were down-regulated, and 14 showed early upregulation at 4 h followed by down-regulation at 1 and 7 days post-SCI (Liu et al., 2009). Recently, observations on miRNA expression have been extended to the PNS. miRNA expression has been profiled following SNI in proximal stumps of injured sciatic nerve and DRG by microarray and deep sequencing in several studies (Strickland et al., 2011; Yu et al., 2011b; Zhou et al., 2012). Following sciatic nerve transection, 20 miRNA transcripts displayed a significant change in expression levels at 7-day post-axotomy in rat DRG (Strickland et al., 2011). Both miR-21 and miR-431 showed significant upregulation in DRG after SNI, comparably to our current data. Taken together, Strickland's and our study, demonstrate that miR-21 and miR-431 are implicated in peripheral nerve regeneration across species. Strickland's study further revealed that miR-21 promoted the regenerative growth of the injured neuron by targeting the Sprouty2 protein (SPRY2; Strickland et al., 2011).

In our studies, we focused on miR-431, which was the most upregulated miRNA in DRG microarray after nerve injury in our experiments. miR-431 was initially identified as central nervous system specific miRNA as it was cloned from brain tissue of mouse embryos (Wheeler et al., 2006). Whole mount *in situ* hybridization revealed miR-431 localization to the developing spinal cord and brain with particularly strong expression in the pons. The pons is particularly rich in synapses because ninety percent of the descending axons passing through the midbrain synapse on neurons in the pons (Wheeler et al., 2006). However, to date, limited information is available about miR-431 physiological function. Recent observation has linked expression of miR-431 to regulation of cell viability (Tanaka et al., 2012). miR-431 was upregulated by the addition of human fibroblast interferon (HuIFN-β) in a non-cancer HuIFN-β sensitive cell line RSa, with concomitant suppression of IGF1R signaling and reduction of cell viability (Tanaka et al., 2012). However, at this time, the function of miR-431 in the nervous system remains uncertain.

THE FUNCTION OF miR-431 IN REGENERATIVE AXON GROWTH

To determine the role of miR-431 in axon regeneration, miR-431 gain- and loss-of-function were investigated in DRG neuronal cultures. Application of miR-431 mimics markedly increased the intracellular miR-431 level and promoted regenerative axon outgrowth. miR-431 gain-of-function correlated with longer axons, more branches, and higher GAP-43 expression, a marker of regeneration. In contrast, transfection of miR-431 inhibitors impaired the regenerative axon growth, as significantly shorter axons and fewer branches were observed in DRG cultures. Analyses of 24 putative targets of miR-431, showed that only six were suppressed in PC12 cells and even less genes were suppressed in DRG primary neurons. This could be related to the specificity of miR-431 to these genes, and to the fact that down-regulation of less specific targets is more easily detected in PC12 cells. The difference may be also related to the fact that the cells were from different species; PC12 were from rat and DRG culture was from mouse.

We have further identified *Kremen1* as the target that mediates the effects of miR-431 on neuronal cells. miR-431 expression inversely relates to *Kremen1*. The direct interaction between miR-431 and *Kremen1* mRNA was confirmed by CLIP, and 3′-UTR luciferase reporter assay. *Kremen1* expression was down-regulated by miR-431 at the mRNA and protein levels. This may mean that miR-431 cleaves the mRNA of this gene rather than repressing its translation. To the best of our knowledge this is the first observation of direct mRNA target cleavage by miR-431. At the same time, our data do not exclude possibility that there is another miRNA or transcription factor that may regulate *Kremen1* too.

Kremen1 was originally discovered as a transmembrane protein containing the kringle domain. Later reports confirmed that both *Kremen1* and its relative Kremen2 were high-affinity receptors for Dickkopf1 (Dkk1), the inhibitor of Wnt/β-catenin signaling (Mao et al., 2002). The canonical Wnt/β-catenin signaling is mediated by two receptor families, Frizzle protein and lipoprotein-receptor-related protein 5 and 6 (LRP5/6). *Kremen1* functionally cooperates with Dkk1 to form a ternary complex composed of *Kremen1*, Dkk1, and LRP5/6, and induces rapid endocytosis and removal of the Wnt receptor LRP5/6 from the cell membrane, which inhibits the transduction of Wnt/β-catenin signaling. Wnt/β-catenin signaling plays a vital role in diverse developmental and physiological processes, including cell-fate determination, tissue patterning, and stem cell regulation (Diep et al., 2004). Wnt/β-catenin signaling pathway also contributes to adult neurogenesis. Blocking Wnt signaling abolishes neurogenesis in adult hippocampal progenitor cells *in vitro* and suppresses neurogenesis *in vivo* (Lie et al., 2005). With ectopic expression of Dkk1, canonical Wnt/β-catenin signaling is markedly reduced in both the hippocampus and cortex (Solberg et al., 2008).

Studies have also established a role for Wnt signaling in regulating synaptic plasticity and axonal growth (Hall et al., 2000; Wang et al., 2006; Budnik and Salinas, 2011). Wnt signaling regulates axon terminal remodeling (Budnik and Salinas, 2011), formation of growth cones and lamellipodia (Hall et al., 2000), microtubules organization (Purro et al., 2008), and synaptic assembly (Ahmad-Annuar et al., 2006). Loss- and gain-of-function studies in animal models demonstrated that loss of Wnt7a results in a strong deficit in the accumulation of synaptic markers at the cell synapses (Ahmad-Annuar et al., 2006). In contrast, in cultured mouse cerebellar granule cells, Wnt7a increased neurite elongation and branching as well as the expression of synaptic markers (Lucas and Salinas, 1997). Likewise, targeted disruption of Wnt receptor genes in mice produced severe defects in axon growth and guidance, resulting in a loss of thalamocortical, nigrostriatal tracts, and the anterior commissure (Wang et al., 2002, 2006). Moreover, SCI induced a time-dependent increase in Wnt expression, phosphorylation of Wnt receptors, and activity of β-catenin protein. Thus, the activation of the Wnt pathway after SCI suggests the involvement of Wnt pathway in nerve regeneration (Fernandez-Martos et al., 2011).

These abundant evidences from studies in animal models, cell and organ culture firmly established an important role of Wnt signaling in neurite outgrowth and axonal guidance. The function of Wnt signaling could potentially link our observation on increased miR-431 and decreased *Kremen1* expression to the enhanced axonal outgrowth. In our study, *Kremen1* loss-of-function produced an increase in axon outgrowth mimicking the effect of miR-431 gain-of-function but did not increase branching. The axon elongation is a critical factor for axon regeneration. The excessive branching can be detrimental to axon regeneration, especially in the PNS. Evidence suggests that axonal elongation and branching are differentially regulated in hippocampal neurons (Pujol et al., 2005).

Taken together, our studies identified miR-431 as an endogenous, injury-regulated inhibitor of *Kremen1*, which promotes regenerative axon growth in adult sensory neurons. Further studies are necessary to fully define the role of miR-431 in axonal regeneration. These findings may not only contribute to our understanding of fundamental biological process, but also could have important implication for improving the therapeutic strategies for nerve injury.

ACKNOWLEDGMENTS

We would like to express sincere gratitude to Dr. Rukiyah T. Van Dross for help with luciferase assay. This research was supported in part by The Wooten Laboratory grant (Alexander K. Murashov), and National Institute of Environmental Health Sciences (NIEHS) Award #A11-0093-001 (Alexander K. Murashov).

REFERENCES

Ahmad-Annuar, A., Ciani, L., Simeonidis, I., Herreros, J., Fredj, N. B., Rosso, S. B., et al. (2006). Signaling across the synapse: a role for Wnt and dishevelled in presynaptic assembly and neurotransmitter release. *J. Cell Biol.* 174, 127–139. doi: 10.1083/jcb.200511054

Bartel, D. P. (2004). MicroRNAs: genomics, biogenesis, mechanism, and function. *Cell* 116, 281–297. doi: 10.1016/S0092-8674(04)00045-5

Benowitz, L. I., and Routtenberg, A. (1997). GAP-43: an intrinsic determinant of neuronal development and plasticity. *Trends Neurosci.* 20, 84–91. doi: 10.1016/S0166-2236(96)10072-2

Budnik, V., and Salinas, P. C. (2011). Wnt signaling during synaptic development and plasticity. *Curr. Opin. Neurobiol.* 21, 151–159. doi: 10.1016/j.conb.2010.12.002

Cheng, L. C., Pastrana, E., Tavazoie, M., and Doetsch, F. (2009). miR-124 regulates adult neurogenesis in the subventricular zone stem cell niche. *Nat. Neurosci.* 12, 399–408. doi: 10.1038/nn.2294

Coleman, M. P., and Perry, V. H. (2002). Axon pathology in neurological disease: a neglected therapeutic target. *Trends Neurosci.* 25, 532–537. doi: 10.1016/S0166-2236(02)02255-5

Dajas-Bailador, F., Bonev, B., Garcez, P., Stanley, P., Guillemot, F., and Papalopulu, N. (2012). MicroRNA-9 regulates axon extension and branching by targeting Map1b in mouse cortical neurons. *Nat. Neurosci.* 15, 697–699. doi: 10.1038/nn.3082

Davare, M. A., Fortin, D. A., Saneyoshi, T., Nygaard, S., Kaech, S., Banker, G., et al. (2009). Transient receptor potential canonical 5 channels activate Ca^{2+}/calmodulin kinase Igamma to promote axon formation in hippocampal neurons. *J. Neurosci.* 29, 9794–9808. doi: 10.1523/JNEUROSCI.1544-09.2009

Diep, D. B., Hoen, N., Backman, M., Machon, O., and Krauss, S. (2004). Characterisation of the Wnt antagonists and their response to conditionally activated Wnt signalling in the developing mouse forebrain. *Brain Res. Dev. Brain Res.* 153, 261–270. doi: 10.1016/j.devbrainres.2004.09.008

Fang, Y., and Bonini, N. M. (2012). Axon degeneration and regeneration: insights from *Drosophila* models of nerve injury. *Annu. Rev. Cell Dev. Biol.* 28, 575–597. doi: 10.1146/annurev-cellbio-101011-155836

Fernandez-Martos, C. M., Gonzalez-Fernandez, C., Gonzalez, P., Maqueda, A., Arenas, E., and Rodriguez, F. J. (2011). Differential expression of Wnts after spinal cord contusion injury in adult rats. *PLoS ONE* 6:e27000. doi: 10.1371/journal.pone.0027000

Filipowicz, W., Bhattacharyya, S. N., and Sonenberg, N. (2008). Mechanisms of post-transcriptional regulation by microRNAs: are the answers in sight? *Nat. Rev. Genet.* 9, 102–114. doi: 10.1038/nrg2290

Forman, D. S., Mcquarrie, I. G., Labore, F. W., Wood, D. K., Stone, L. S., Braddock, C. H., et al. (1980). Time course of the conditioning lesion effect on axonal regeneration. *Brain Res.* 182, 180–185. doi: 10.1016/0006-8993(80)90842-2

Hall, A. C., Lucas, F. R., and Salinas, P. C. (2000). Axonal remodeling and synaptic differentiation in the cerebellum is regulated by WNT-7a signaling. *Cell* 100, 525–535. doi: 10.1016/S0092-8674(00)80689-3

Hebert, S. S., and De Strooper, B. (2007). Molecular biology. miRNAs in neurodegeneration. *Science* 317, 1179–1180. doi: 10.1126/science.1148530

Islamov, R. R., Chintalgattu, V., Pak, E. S., Katwa, L. C., and Murashov, A. K. (2004). Induction of VEGF and its Flt-1 receptor after sciatic nerve crush injury. *Neuroreport* 15, 2117–2121. doi: 10.1097/00001756-200409150-00024

Jaskiewicz, L., Bilen, B., Hausser, J., and Zavolan, M. (2012). Argonaute CLIP – a method to identify in vivo targets of miRNAs. *Methods* 58, 106–112. doi: 10.1016/j.ymeth.2012.09.006

Kim, J., Inoue, K., Ishii, J., Vanti, W. B., Voronov, S. V., Murchison, E., et al. (2007). A microRNA feedback circuit in midbrain dopamine neurons. *Science* 317, 1220–1224. doi: 10.1126/science.1140481

Lei, P., Li, Y., Chen, X., Yang, S., and Zhang, J. (2009). Microarray based analysis of microRNA expression in rat cerebral cortex after traumatic brain injury. *Brain Res.* 1284, 191–201. doi: 10.1016/j.brainres.2009.05.074

Lie, D. C., Colamarino, S. A., Song, H. J., Desire, L., Mira, H., Consiglio, A., et al. (2005). Wnt signalling regulates adult hippocampal neurogenesis. *Nature* 437, 1370–1375. doi: 10.1038/nature04108

Liu, N. K., Wang, X. F., Lu, Q. B., and Xu, X. M. (2009). Altered microRNA expression following traumatic spinal cord injury. *Exp. Neurol.* 219, 424–429. doi: 10.1016/j.expneurol.2009.06.015

Liu, R.-Y., Schmid, R.-S., Snider, W. D., and Maness, P. F. (2002). NGF enhances sensory axon growth induced by laminin but not by the L1 cell adhesion molecule. *Mol. Cell. Neurosci.* 20, 2–12. doi: 10.1006/mcne.2002.1107

Lucas, F. R., and Salinas, P. C. (1997). WNT-7a induces axonal remodeling and increases synapsin I levels in cerebellar neurons. *Dev. Biol.* 192, 31–44. doi: 10.1006/dbio.1997.8734

Mao, B., Wu, W., Davidson, G., Marhold, J., Li, M., Mechler, B. M., et al. (2002). Kremen proteins are Dickkopf receptors that regulate Wnt/beta-catenin signalling. *Nature* 417, 664–667. doi: 10.1038/nature756

McNeill, E., and Van Vactor, D. (2012). MicroRNAs shape the neuronal landscape. *Neuron* 75, 363–379. doi: 10.1016/j.neuron.2012.07.005

Murashov, A. K., Pak, E. S., Hendricks, W. A., Owensby, J. P., Sierpinski, P. L., Tatko, L. M., et al. (2005). Directed differentiation of embryonic stem cells into dorsal interneurons. *FASEB J.* 19, 252–254.

Nakamura, T., and Matsumoto, K. (2008). The functions and possible significance of Kremen as the gatekeeper of Wnt signalling in development and pathology. *J. Cell. Mol. Med.* 12, 391–408. doi: 10.1111/j.1582-4934.2007.00201.x

Natera-Naranjo, O., Aschrafi, A., Gioio, A. E., and Kaplan, B. B. (2010). Identification and quantitative analyses of microRNAs located in the distal axons of sympathetic neurons. *RNA* 16, 1516–1529. doi: 10.1261/rna.1833310

Pujol, F., Kitabgi, P., and Boudin, H. (2005). The chemokine SDF-1 differentially regulates axonal elongation and branching in hippocampal neurons. *J. Cell Sci.* 118, 1071–1080. doi: 10.1242/jcs.01694

Purro, S. A., Ciani, L., Hoyos-Flight, M., Stamatakou, E., Siomou, E., and Salinas, P. C. (2008). Wnt regulates axon behavior through changes in microtubule growth directionality: a new role for adenomatous polyposis coli. *J. Neurosci.* 28, 8644–8654. doi: 10.1523/JNEUROSCI.2320-08.2008

Redell, J. B., Liu, Y., and Dash, P. K. (2009). Traumatic brain injury alters expression of hippocampal microRNAs: potential regulators of multiple pathophysiological processes. *J. Neurosci. Res.* 87, 1435–1448. doi: 10.1002/jnr.21945

Shi, Y., Zhao, X., Hsieh, J., Wichterle, H., Impey, S., Banerjee, S., et al. (2010). MicroRNA regulation of neural stem cells and neurogenesis. *J. Neurosci.* 30, 14931–14936. doi: 10.1523/JNEUROSCI.4280-10.2010

Shioya, M., Obayashi, S., Tabunoki, H., Arima, K., Saito, Y., Ishida, T., et al. (2010). Aberrant microRNA expression in the brains of neurodegenerative diseases: miR-29a decreased in Alzheimer disease brains targets neurone navigator 3. *Neuropathol. Appl. Neurobiol.* 36, 320–330. doi: 10.1111/j.1365-2990.2010.01076.x

Smith, B., Treadwell, J., Zhang, D., Ly, D., Mckinnell, I., Walker, P. R., et al. (2010). Large-scale expression analysis reveals distinct microRNA profiles at different stages of human neurodevelopment. *PLoS ONE* 5:e11109. doi: 10.1371/journal.pone.0011109

Solberg, N., Machon, O., and Krauss, S. (2008). Effect of canonical Wnt inhibition in the neurogenic cortex, hippocampus, and premigratory dentate gyrus progenitor pool. *Dev. Dyn.* 237, 1799–1811. doi: 10.1002/dvdy.21586

Strickland, I. T., Richards, L., Holmes, F. E., Wynick, D., Uney, J. B., and Wong, L. F. (2011). Axotomy-induced miR-21 promotes axon growth in adult dorsal root ganglion neurons. *PLoS ONE* 6:e23423. doi: 10.1371/journal.pone.0023423

Tanaka, T., Sugaya, S., Kita, K., Arai, M., Kanda, T., Fujii, K., et al. (2012). Inhibition of cell viability by human IFN-beta is mediated by microRNA-431. *Int. J. Oncol.* 40, 1470–1476.

Wang, Y., Thekdi, N., Smallwood, P. M., Macke, J. P., and Nathans, J. (2002). Frizzled-3 is required for the development of major fiber tracts in the rostral CNS. *J. Neurosci.* 22, 8563–8573.

Wang, Y., Zhang, J., Mori, S., and Nathans, J. (2006). Axonal growth and guidance defects in

Frizzled3 knock-out mice: a comparison of diffusion tensor magnetic resonance imaging, neurofilament staining, and genetically directed cell labeling. *J. Neurosci.* 26, 355–364. doi: 10.1523/JNEUROSCI.3221-05.2006

Wheeler, G., Ntounia-Fousara, S., Granda, B., Rathjen, T., and Dalmay, T. (2006). Identification of new central nervous system specific mouse microRNAs. *FEBS Lett.* 580, 2195–2200. doi: 10.1016/j.febslet.2006.03.019

Wu, D., Raafat, A., Pak, E., Clemens, S., and Murashov, A. K. (2012). Dicer-microRNA pathway is critical for peripheral nerve regeneration and functional recovery in vivo and regenerative axonogenesis in vitro. *Exp. Neurol.* 233, 555–565. doi: 10.1016/j.expneurol.2011.11.041

Wu, D., Raafat, M., Pak, E., Hammond, S., and Murashov, A. K. (2011). MicroRNA machinery responds to peripheral nerve lesion in an injury-regulated pattern. *Neuroscience* 190, 386–397. doi: 10.1016/j.neuroscience.2011.06.017

Yu, B., Zhou, S., Qian, T., Wang, Y., Ding, F., and Gu, X. (2011a). Altered microRNA expression following sciatic nerve resection in dorsal root ganglia of rats. *Acta Biochim. Biophys. Sin.* 43, 909–915. doi: 10.1093/abbs/gmr083

Yu, B., Zhou, S., Wang, Y., Ding, G., Ding, F., and Gu, X. (2011b). Profile of microRNAs following rat sciatic nerve injury by deep sequencing: implication for mechanisms of nerve regeneration. *PLoS ONE* 6:e24612. doi: 10.1371/journal.pone.0024612

Zhang, H. Y., Zheng, S. J., Zhao, J. H., Zhao, W., Zheng, L. F., Zhao, D., et al. (2011). MicroRNAs 144, 145, and 214 are down-regulated in primary neurons responding to sciatic nerve transection. *Brain Res.* 1383, 62–70. doi: 10.1016/j.brainres.2011.01.067

Zhou, S., Shen, D., Wang, Y., Gong, L., Tang, X., Yu, B., et al. (2012). microRNA-222 targeting PTEN promotes neurite outgrowth from adult dorsal root ganglion neurons following sciatic nerve transection. *PLoS ONE* 7:e44768. doi: 10.1371/journal.pone.0044768

Permissions

All chapters in this book were first published by Frontiers; hereby published with permission under the Creative Commons Attribution License or equivalent. Every chapter published in this book has been scrutinized by our experts. Their significance has been extensively debated. The topics covered herein carry significant findings which will fuel the growth of the discipline. They may even be implemented as practical applications or may be referred to as a beginning point for another development.

The contributors of this book come from diverse backgrounds, making this book a truly international effort. This book will bring forth new frontiers with its revolutionizing research information and detailed analysis of the nascent developments around the world.

We would like to thank all the contributing authors for lending their expertise to make the book truly unique. They have played a crucial role in the development of this book. Without their invaluable contributions this book wouldn't have been possible. They have made vital efforts to compile up to date information on the varied aspects of this subject to make this book a valuable addition to the collection of many professionals and students.

This book was conceptualized with the vision of imparting up-to-date information and advanced data in this field. To ensure the same, a matchless editorial board was set up. Every individual on the board went through rigorous rounds of assessment to prove their worth. After which they invested a large part of their time researching and compiling the most relevant data for our readers.

The editorial board has been involved in producing this book since its inception. They have spent rigorous hours researching and exploring the diverse topics which have resulted in the successful publishing of this book. They have passed on their knowledge of decades through this book. To expedite this challenging task, the publisher supported the team at every step. A small team of assistant editors was also appointed to further simplify the editing procedure and attain best results for the readers.

Apart from the editorial board, the designing team has also invested a significant amount of their time in understanding the subject and creating the most relevant covers. They scrutinized every image to scout for the most suitable representation of the subject and create an appropriate cover for the book.

The publishing team has been an ardent support to the editorial, designing and production team. Their endless efforts to recruit the best for this project, has resulted in the accomplishment of this book. They are a veteran in the field of academics and their pool of knowledge is as vast as their experience in printing. Their expertise and guidance has proved useful at every step. Their uncompromising quality standards have made this book an exceptional effort. Their encouragement from time to time has been an inspiration for everyone.

The publisher and the editorial board hope that this book will prove to be a valuable piece of knowledge for researchers, students, practitioners and scholars across the globe.

List of Contributors

Kira S. Sheinerman and Samuil R. Umansky
DiamiR, LLC, Princeton, NJ, USA

Alexandra Benchoua
Centre d'Etude des Cellules Souches, Institut des cellules Souches pour leTraitement et l'Étude des Maladies monogéniques, Association Française contre les Myopathies, Evry, France

Marc Peschanski
INSERM/UEVE UMR 86, Institut des cellules Souches pour le Traitement et l'Étude des Maladies monogéniques, Association Française contre les Myopathies, Evry, France

Danyella B. Dogini, Simoni H. Avansini, Andre S.Vieira and Iscia Lopes-Cendes
Department of Medical Genetics, School of Medical Sciences, University of Campinas, Campinas, São Paulo, Brazil

Shan Bian
Department of Cell and Developmental Biology, Weill Medical College of Cornell University, NewYork, USA

Jennifer L. Knauss and Tao Sun
Department of Cell and Developmental Biology, Weill Medical College of Cornell University, NewYork, USA
Weill Cornell Graduate School of Medical Sciences, NewYork, USA

Barry B. Kaplan, Amar N. Kar and Anthony E. Gioio
Laboratory of Molecular Biology, National Institute of Mental Health, National Institutes of Health, Bethesda, MD, USA

Armaz Aschrafi
Department of Cognitive Neuroscience, Donders Institute for Brain, Cognition, and Behavior, Radboud University Nijmegen Medical Center, Nijmegen, Netherlands

Emily F. Goodall, Paul R. Heath, Oliver Bandmann, Janine Kirby and Pamela J. Shaw
Department of Neuroscience, Sheffield Institute for Translational Neuroscience, University of Sheffield, Sheffield, UK

Federico Cremisi
Scuola Normale Superiore, Pisa, Italy

Yi-Wen Hsieh and Chiou-Fen Chuang
Division of Developmental Biology, Cincinnati Children's Hospital Research Foundation, Cincinnati, OH, USA

Amel Alqadah
Division of Developmental Biology, Cincinnati Children's Hospital Research Foundation, Cincinnati, OH, USA
Molecular and Developmental Biology Graduate Program, University of Cincinnati, Cincinnati, OH, USA

Amrita D. Chaudhuri, Sowmya V. Yelamanchili and Howard S. Fox
Department of Pharmacology and Experimental Neuroscience, University of Nebraska Medical Center, Omaha, NE, USA

Stephen M. Eacker
Neuroregeneration and Stem Cell Programs, Institute for Cell Engineering, Johns Hopkins University School of Medicine, Baltimore, MD, USA
Department of Neurology, Johns Hopkins University School of Medicine, Baltimore, MD, USA

Ted M. Dawson
Neuroregeneration and Stem Cell Programs, Institute for Cell Engineering, Johns Hopkins University School of Medicine, Baltimore, MD, USA
Department of Neurology, Johns Hopkins University School of Medicine, Baltimore, MD, USA
Solomon H. Snyder Department of Neuroscience, Johns Hopkins University School of Medicine, Baltimore, MD, USA

Valina L. Dawson
Neuroregeneration and Stem Cell Programs, Institute for Cell Engineering, Johns Hopkins University School of Medicine, Baltimore, MD, USA
Department of Neurology, Johns Hopkins University School of Medicine, Baltimore, MD, USA
Solomon H. Snyder Department of Neuroscience, Johns Hopkins University School of Medicine, Baltimore, MD, USA
Department of Physiology, Johns Hopkins University School of Medicine, Baltimore, MD, USA

Janet Hong, Haijun Zhang and Tao Sun
Department of Cell and Developmental Biology, Cornell UniversityWeill Medical College, NewYork, USA

Yoko Kawase-Koga
Department of Cell and Developmental Biology, Cornell UniversityWeill Medical College, NewYork, USA
Department of Oral and Maxillofacial Surgery, The University of Tokyo Hospital, Tokyo, Japan

Carl Ernst
Douglas Hospital Research Institute, Montreal, QC, Canada
Department of Psychiatry, McGill University, Montreal, QC, Canada

Cynthia C. Morton
Department of Obstetrics, Gynecology and Reproductive Biology, Brigham andWomen's Hospital, Harvard Medical School, Boston, MA, USA
Department of Pathology, Brigham andWomen's Hospital, Harvard Medical School, Boston, MA, USA
Medical and Population Genetics Program, The Broad Institute of M.I.T. and Harvard, Cambridge, MA, USA

Gabriele Baj and Enrico Tongiorgi
Department of Life Sciences, Brain Centre for Neurosciences, University of Trieste, Trieste, Italy

Lucia Gardossi
Dipartimento di Scienze Chimiche e Farmaceutiche, University of Trieste, Trieste, Italy

Véronique Dorval and Sébastien S. Hébert
Axe Neurosciences, Centre de Recherche du Centre Hospitalier Universitaire de Québec (Centre Hospitalier de l'Université Laval), Québec, QC, Canada
Département de Psychiatrie et de Neurosciences, Université Laval, Québec, QC, Canada

PeterT. Nelson
Sanders-Brown Center on Aging, University of Kentucky, Lexington, KY, USA

Eduardo Gascon and Fen-Biao Gao
Department of Neurology, University of Massachusetts Medical School,Worcester, MA, USA

Katannya Kapeli
Department of Cellular and Molecular Medicine, University of California San Diego, La Jolla, CA, USA

GeneW.Yeo
Department of Cellular and Molecular Medicine, University of California San Diego, La Jolla, CA, USA
UCSD Stem Cell Program, University of California San Diego, La Jolla, CA, USA
UCSD Moores Cancer Center, University of California San Diego, La Jolla, CA, USA
Institute for Genomic Medicine, University of California San Diego, La Jolla, CA, USA

Davide Carlino
Brain Centre for Neurosciences, University of Trieste, Trieste, Italy
Department of Medical, Surgical, and Health Sciences, Psychiatric Clinic, University of Trieste, Trieste, Italy

Michaela Kress
Department of Physiology and Medical Physics, Division of Physiology, Medical University Innsbruck, Innsbruck, Austria

Alexander Hüttenhofer, Florian Kronenberg and ZlatkoTrajanoski
Medical University Innsbruck, Innsbruck, Austria

Marc Landry and Alexandre Favereaux
UMR 5297, Interdisciplinary Institute for Neuroscience, Centre National de la Recherche Scientifique, University of Bordeaux, Bordeaux, France

Rohini Kuner
University Heidelberg, Heidelberg, Germany

David Greenberg
Hebrew University of Jerusalem, Jerusalem, Israel

Josef Bednarik
University Hospital Brno, Brno, Czech Republic

Paul Heppenstall
The European Molecular Biology Laboratory, Monterotondo, Italy

Marzia Malcangio
King's College London, London, UK

Heike Rittner, Nurcan Üçeyler and Claudia Sommer
University Hospital Würzburg, Würzburg, Germany

Peter Mouritzen
Exiqon A/S, Vedbaek, Denmark

Frank Birklein
University Hospital Mainz, Mainz, Germany

Hermona Soreq
Laboratory of Molecular Neuroscience, Department of Biological chemistry, Hebrew University of Jerusalem, Jerusalem, Israel
Department of Biological Chemistry, The Life Sciences Institute, The Hebrew University of Jerusalem, Jerusalem, Israel
The Edmond and Lili Safra Center for Brain Sciences, The Hebrew University of Jerusalem, Jerusalem, Israel

Adam D. Norris and John A. Calarco
FAS Center for Systems Biology, Harvard University, Cambridge, MA, USA

Stephan Persengiev, Ivanela Kondova and Ronald Bontrop
Biomedical Primate Research Center, Rijswijk, Netherlands

Robert I. Richards, Saumya E. Samaraweera, Clare L. van Eyk and LouiseV. O'Keefe
Molecular Pathology, School of Molecular and Biomedical Science, The University of Adelaide, Adelaide, SA, Australia

Catherine M. Suter
Victor Chang Cardiac Research Institute, Darlinghurst, NSW, Australia

Lilach Soreq
Department of Medical Neurobiology, Hadassah Faculty of Medicine, The Hebrew University of Jerusalem, Jerusalem, Israel

Nathan Salomonis
Division of Genomic Medicine, The Gladstone Institute of Cardiovascular Disease, San Francisco, CA, USA

Michal Bronstein
The Center for Genomic Technologies, The Institute of Life Sciences, The Hebrew University of Jerusalem, Jerusalem, Israel

David S. Greenberg
Department of Biological Chemistry, The Life Sciences Institute, The Hebrew University of Jerusalem, Jerusalem, Israel

Zvi Israel
Department of Neurosurgery, The Centre for Functional and Restorative Neurosurgery, Hadassah University Hospital, Jerusalem

Hagai Bergman
Department of Medical Neurobiology, Hadassah Faculty of Medicine, The Hebrew University of Jerusalem, Jerusalem, Israel
The Edmond and Lili Safra Center for Brain Sciences, The Hebrew University of Jerusalem, Jerusalem, Israel

Melanie D. White, RuthV. J. Milne and Matthew F. Nolan
Centre for Integrative Physiology, University of Edinburgh, Edinburgh, Scotland, UK

Di Wu
Department of Neurobiology and Anatomy, Drexel University College of Medicine, Philadelphia, PA, USA

Alexander K. Murashov
The Harriet and JohnWooten Laboratory for Alzheimer's and Neurodegenerative Diseases Research, Department of Physiology, School of Medicine, East Carolina University, Greenville, NC, USA

Index

N

O

P

R

S

T

V

W

X

www.ingramcontent.com/pod-product-compliance
Lightning Source LLC
LaVergne TN
LVHW060522210926
848485LV00005B/1125